T0178587

Lecture Notes in Computer Science

Lecture Notes in Artificial Intelligence   **14276**

Founding Editor

Jörg Siekmann

Series Editors

Randy Goebel, *University of Alberta, Edmonton, Canada*
Wolfgang Wahlster, *DFKI, Berlin, Germany*
Zhi-Hua Zhou, *Nanjing University, Nanjing, China*

The series Lecture Notes in Artificial Intelligence (LNAI) was established in 1988 as a topical subseries of LNCS devoted to artificial intelligence.

The series publishes state-of-the-art research results at a high level. As with the LNCS mother series, the mission of the series is to serve the international R & D community by providing an invaluable service, mainly focused on the publication of conference and workshop proceedings and postproceedings.

Albert Bifet · Ana Carolina Lorena ·
Rita P. Ribeiro · João Gama · Pedro H. Abreu
Editors

Discovery Science

26th International Conference, DS 2023
Porto, Portugal, October 9–11, 2023
Proceedings

 Springer

Editors
Albert Bifet ⓘ
Waikato University
Hamilton, New Zealand

Ana Carolina Lorena ⓘ
Aeronautics Institute of Technology
São José dos Campos, Brazil

Rita P. Ribeiro ⓘ
University of Porto
Porto, Portugal

João Gama ⓘ
University of Porto
Porto, Portugal

Pedro H. Abreu ⓘ
University of Coimbra
Coimbra, Portugal

ISSN 0302-9743   ISSN 1611-3349 (electronic)
Lecture Notes in Artificial Intelligence
ISBN 978-3-031-45274-1   ISBN 978-3-031-45275-8 (eBook)
https://doi.org/10.1007/978-3-031-45275-8

LNCS Sublibrary: SL7 – Artificial Intelligence

Preface

The Discovery Science Conference is an open forum for in-depth discussions of innovative ideas related to the development and analysis of artificial intelligence methods for discovering scientific knowledge. The conference's scope includes methods from machine learning, data mining, intelligent data analysis, big data analytics, and their application in various domains.

This volume contains the papers selected for presentation at the 26th International Conference on Discovery Science (DS 2023), held in Porto, Portugal, during October 9–11, 2023. DS 2023 received 133 international submissions. Each submission was reviewed by at least three Program Committee (PC) members in a single-blind manner. The PC decided to accept 37 regular papers and 10 short papers. This resulted in an acceptance rate of 35% for regular papers. The conference also included three keynote talks. Mihaela van der Schaar (University of Cambridge) contributed a talk titled "Time: The next frontier in discovery science"; Amílcar Cardoso (University of Coimbra) contributed a talk titled "Computational Creativity: from autonomous generation to co-creation"; Nathalie Japkowicz (American University) contributed a talk titled "Lifelong Anomaly Detection". The invited talks' abstracts are included in these proceedings' frontmatter.

We are grateful to Springer for their continued long-term support. Springer publishes the conference proceedings, as well as a regular special issue of the journal Machine Learning on Discovery Science. The latter offers authors a chance to publish significantly extended and reworked versions of their DS conference papers in this prestigious journal, while being open to all submissions on DS conference topics. This year, Springer also supported a best student paper award.

On the program side, we would like to thank all the authors of the submitted papers and the PC members for their efforts in evaluating the submitted papers, as well as the keynote speakers. A word of appreciation to the Publicity Chairs, Carlos Abreu Ferreira, Ricardo Cerri, and Wenbin Zhang who helped in the dissemination and contributed to the high level of the submissions. On the organization side, we would like to thank all the members of the Organizing Committee, in particular José Pedro Amorim, Ricardo Cardoso Pereira, Joana Cristo Santos, Miriam Seoane Santos, and Bruno Veloso, for their help in all conference-associated activities. We are also grateful to the people behind Microsoft CMT for developing the conference organization system that proved to be an essential tool in the paper submission and evaluation process.

August 2023

Albert Bifet
Ana Carolina Lorena
Rita P. Ribeiro
João Gama
Pedro H. Abreu

Organization

General Chairs

João Gama   University of Porto and INESC TEC, Portugal
Pedro Henriques Abreu   University of Coimbra and CISUC, Portugal

Program Committee Chairs

Albert Bifet   University of Waikato, New Zealand
Ana Carolina Lorena   Aeronautics Institute of Technology, Brazil
Rita P. Ribeiro   University of Porto and INESC TEC, Portugal

Steering Committee Chair

Michelangelo Ceci   University of Bari, Italy

Publicity Chairs

Carlos Abreu Ferreira   Polytechnic Institute of Porto, INESC TEC,
     Portugal
Ricardo Cerri   Federal University of São Carlos, Brazil
Wenbin Zhang   Michigan Technological University, USA

Local Arrangement Chairs

José Pedro Amorim   University of Coimbra, Portugal
Ricardo Cardoso Pereira   University of Coimbra, Portugal
Joana Cristo Santos   University of Coimbra, Portugal
Miriam Seoane Santos   University of Coimbra, Portugal
Bruno Veloso   University of Porto and INESC TEC, Portugal

Program Committee

Reza Akbarinia	Inria, France
Leman Akoglu	Carnegie Mellon University, USA
Amparo Alonso-Betanzos	Universidade da Coruña, CITIC, Spain
José Pereira Amorim	University of Coimbra, CISUC, Portugal
Thiago Andrade	INESC TEC/University of Porto, Portugal
Giuseppina Andresini	University of Bari Aldo Moro, Italy
Annalisa Appice	University of Bari Aldo Moro, Italy
Martin Atzmueller	Osnabrück University, Germany
Colin Bellinger	NRC and Dalhousie University, Canada
Michael R. Berthold	KNIME, Germany
Concha Bielza	Universidad Politécnica de Madrid, Spain
Hendrik Blockeel	KU Leuven, Belgium
Robert Bossy	INRA MaIAGE, France
Henrik Bostrom	KTH Royal Institute of Technology, Sweden
Paula Branco	University of Ottawa, Canada
Ivan Bratko	University of Ljubljana, Slovenia
Paula Brito	Universidade do Porto, INESC TEC, Portugal
Dariusz Brzezinski	Poznan University of Technology, Poland
Humberto Bustince	Universidad Publica de Navarra, Spain
Gustau Camps-Valls	Universitat de València, Spain
Brais Cancela	Universidade da Coruña, Spain
Alberto Cano	Virginia Commonwealth University, USA
Jaime Cardoso	University of Porto, Portugal
Ricardo Cardoso Pereira	University of Coimbra, CISUC, Portugal
Michelangelo Ceci	University of Bari, Italy
Mattia Cerrato	JGU Mainz, Germany
Ricardo Cerri	Federal University of São Carlos, Brazil
Simon Colton	Queen Mary University of London, UK
Paulo Cortez	University of Minho, Portugal
Vitor Santos Costa	University of Porto, Portugal
Bruno Cremilleux	Université de Caen Normandie, France
Claudia d'Amato	University of Bari, Italy
Narjes Davari	INESCTEC, Portugal
Andre C. P. L. F. de Carvalho	São Paulo State University, Brazil
Marcilio de Souto	LIFO/University of Orleans, France
José del Campo-Ávila	Universidad de Málaga, Spain
Wouter Duivesteijn	TU Eindhoven, The Netherlands
Inês Dutra	University of Porto, Portugal
Saso Dzeroski	Jožef Stefan Institute, Slovenia
Tapio Elomaa	Tampere University, Finland

Hadi Fanaee-T	Halmstad University, Sweden
Elaine Ribeiro Faria	Universidade Federal de Uberlândia, Brazil
Ad Feelders	Universiteit Utrecht, The Netherlands
Carlos Ferreira	INESC TEC, Polytechnic Institute of Porto, Portugal
Cèsar Ferri	Universitat Politècnica València, Spain
Peter Flach	University of Bristol, UK
Johannes Fürnkranz	JKU Linz, Austria
Mohamed Gaber	Birmingham City University, UK
Sabrina Gaito	Università degli Studi di Milano, Italy
Mikel Galar	Universidad Pública de Navarra, Spain
Dragan Gamberger	Ruder Bošković Institute, Croatia
Rui Jorge Gomes	University of Coimbra, Portugal
Rafael Gomes Mantovani	Federal University of Technology - Paraná, Brazil
Lawrence Hall	University of South Florida, USA
Howard J. Hamilton	University of Regina, Canada
Barbara Hammer	CITEC, Bielefeld University, Germany
Rui Henriques	IST/INESC-ID, University of Lisbon, Portugal
Alberto Fernandez Hilario	University of Granada, Spain
Martin Holena	Institute of Computer Science, Czech Republic
Jaakko Hollmén	Aalto University, Finland
Tamas Horvath	University of Bonn and Fraunhofer IAIS, Germany
Eyke Hüllermeier	University of Munich, Germany
Dino Ienco	INRAE, France
Roberto Interdonato	CIRAD, France
Alipio M. G. Jorge	INESC TEC/University of Porto, Portugal
Frank Klawonn	Helmholtz Centre for Infection Research, Germany
Dragi Kocev	Jožef Stefan Institute, Slovenia
Stefan Kramer	Johannes Gutenberg University Mainz, Germany
Tetsuji Kuboyama	Gakushuin University, Japan
Vincent Labatut	Université d'Avignon, France
Vincenzo Lagani	KAUST, Saudi Arabia
Pedro Larranaga	Technical University of Madrid, Spain
Anne Laurent	University of Montpellier, France
Nada Lavrač	Jožef Stefan Institute, Slovenia
Fátima M. Leal	University Portucalense, Portugal
Philippe Lenca	IMT Atlantique, France
Francesca Alessandra Lisi	University of Bari "A. Moro", Italy
Gjorgji Madjarov	Saints Cyril and Methodius University, North Macedonia

Nikola Simidjievski	University of Cambridge, UK
Tomislav Šmuc	Institut Ruder Bošković, Croatia
Carlos Soares	University of Porto, Portugal
Marina Sokolova	University of Ottawa, Canada
Larisa Soldatova	Goldsmiths, University of London, UK
Arnaud Soulet	University of Tours, France
Myra Spiliopoulou	Otto-von-Guericke-University Magdeburg, Germany
Jerzy Stefanowski	Poznan University of Technology, Poland
Mahito Sugiyama	National Institute of Informatics, Japan
Andrea Tagarelli	University of Calabria, Italy
Maguelonne Teisseire	INRAE - UMR Tetis, France
Sónia Teixeira	INESC TEC, Portugal
Ljupco Todorovski	University of Ljubljana, Slovenia
Alicia Troncoso	Pablo de Olavide University, Spain
Antti Ukkonen	University of Helsinki, Finland
Peter van der Putten	Leiden University, The Netherlands
Cor Veenman	Leiden University, The Netherlands
Davide Vega	Uppsala University, Sweden
Bruno Veloso	University of Porto and INESC TEC, Portugal
Szymon Wilk	Poznan University of Technology, Poland
Amelia Zafra	University of Cordoba, Spain
Wenbin Zhang	Michigan Tech, USA
Matteo Zignani	Università degli Studi di Milano, Italy
Albrecht Zimmermann	Université de Caen Normandie, France
Blaz Zupan	University of Ljubljana, Slovenia

Sponsors

The organizing committee thanks the sponsors of DS 2023:

Plenary Talks

DS 2023 had three plenary talks. The organizing committee is thankful for the presentation of the invited speakers and fruitful discussions.

Time: The Next Frontier in Discovery Science

Mihaela van der Schaar

University of Cambridge, UK

Abstract. In this talk, Prof. van der Schaar illuminated an underemphasized yet critical dimension in machine learning: time. Time harbors the potential to revolutionize machine learning methodologies, particularly within healthcare. The presentation underscored the opportunities and challenges that emerge from integrating temporal dynamics into machine learning models, enriching prediction accuracy, inference robustness, and conceptual understanding.

In this talk, Prof. van der Schaar aimed to answer questions such as:

- What new challenges are we encountering as we try to uncover dynamical systems over time and can we overcome them?
- How might the increased precision and accuracy in early disease detection afforded by integrating temporal dynamics into machine learning models reduce healthcare costs and improve the quality of life for patients?
- How can we effectively balance the robustness of Bayesian methods with the necessary frequentist guarantees when predicting and managing uncertainties over time?
- How can learning from informative sampling over time help us counteract biases inherent in non-random data collection methods? Could this method be the key to unraveling the subtle, yet critical temporal patterns in health data?
- How might our approach to causal deep learning need to change as we incorporate temporal data?

Computational Creativity: From Autonomous Generation to Co-creation

Amílcar Cardoso

University of Coimbra, Portugal

Abstract. Computational Creativity (CC) is a field of research in Artificial Intelligence that focuses on the study and exploitation of computers' potential to act as autonomous creators and co-creators. The field is a confluence point for contributions from multiple disciplines, such as Artificial Intelligence, which provides most of its methodological framework, and also Cognitive Science, Psychology, Social Sciences, and Philosophy, as well as creative domains like the Arts, Music, Design, Poetry, etc. In this talk, a historical perspective on the field was presented, along with key concepts and abstract models to characterize some common modes of creativity, providing context for understanding how these concepts are being applied in the development of creative systems, particularly in co-creative contexts, in light of the latest advances in AI.

Lifelong Anomaly Detection

Nathalie Japkowicz

American University, USA

Abstract. This talk presented a task-agnostic unsupervised lifelong learning scheme for anomaly detection. The approach builds a long-term memory through hierarchical growth and uses change-point detection to set thresholds autonomously for new concepts and anomaly detection. A new change-point detection method designed for high-dimensional settings and for performing identification in addition to detection was introduced. Finally, a version of the system outfitted with memory consolidation, memory summarization, and experience replay was presented. The different versions of the system and its components are tested on cybersecurity, energy, weather prediction, and gravitational wave data as well as on the TCPD Benchmark for change-point detection.

Contents

Interpretability and Explainability in AI

Data Analysis and Optimization

Graph Theory and Network Analysis

Time Series and Forecasting

Healthcare and Biological Data Analysis

Anomaly, Outlier and Novelty Detection

Machine Learning Methods and Applications

Ensembles of Classifiers and Quantifiers with Data Fusion for Quantification Learning

Adriane B. S. Serapião[1,2(✉)], Zahra Donyavi[2], and Gustavo Batista[2]

[1] São Paulo State University, Rio Claro, Brazil
adriane.serapiao@unesp.br
[2] University of New South Wales, Sydney, Australia
{z.donyavi,g.batista}@unsw.edu.au

Abstract. Quantification is a supervised Machine Learning task that estimates the class distribution in an unlabeled test set. Quantification has practical applications in various fields, including medical research, environmental monitoring, and quality control. For instance, medical research often estimates the prevalence of a particular disease in a population. Despite being a thriving research area, most existing quantification methods are limited to binary-class problems. Moreover, recent experimental evidence suggests that modern state-of-the-art quantifiers do not perform well for multi-class problems, which are prevalent in quantification. This paper proposes two novel multi-class ensemble quantifiers, FMC-SQ and FMC-MQ, that use data fusion methods at the classifier and quantifier levels. We conducted experiments with 12 state-of-the-art (single and ensemble) quantifiers to evaluate our models on 31 multi-class datasets. Our experimental results indicate that FMC-MQ is the best-performing quantifier outperforming other single and ensemble methods. Also, aggregating quantifier outputs seem to be a more promising research direction than aggregating classification scores for quantification.

Keywords: Quantification · prevalence estimation · class probability estimation · ensembles · multi-class · machine learning

1 Introduction

Quantification is a supervised learning task that proposes methods to predict the class distribution for an unlabeled test set [1]. Quantification learning finds applications in several real-world domains that involve predicting the behaviour of groups. One well-known example is sentiment analysis, in which the main objective is to predict how the collective opinion about a product, person or institution varies across time [2].

A simple approach to the quantification problem is to count a classifier's output by class labels, a method known as *Classify & Count* (CC) [3]. A perfectly accurate classifier results in an equally perfect quantifier. However, Forman [3]

© The Author(s), under exclusive license to Springer Nature Switzerland AG 2023
A. Bifet et al. (Eds.): DS 2023, LNAI 14276, pp. 3–17, 2023.
https://doi.org/10.1007/978-3-031-45275-8_1

shows that CC is a biased quantifier for application domains in which the classification is imperfect. CC's prediction error linearly increases as the actual class prevalence moves away from the distribution it perfectly quantifies. This phenomenon often makes CC inaccurately predict the most extreme distributions. This flaw has motivated a thriving research community to propose new quantification algorithms that can accurately count across the whole spectrum of class prevalences.

Recent experimental evidence has shown that modern state-of-the-art quantifiers do not perform well for multi-class problems [4]. However, multi-class problems are dominant in quantification. For instance, sentiments are often classified into "positive", "negative" and "neutral". A promising solution to this problem is using ensembles of multi-class quantifiers [5,6]. However, no previous work in the literature has made an in-depth study of how the ensemble architecture and design decisions impact the performance of quantifiers.

In this paper, we make three contributions to the use of ensembles in quantification learning: (i) we introduce two novel ensembles of quantifiers, (ii) we assess the fusion of classifier scores only, and (iii) we evaluate combined fusion of classifiers and quantifiers. Our results show that fusing quantification probabilities is a more promising research direction than aggregating classification scores for quantification.

This paper is organized as follows. Section 2 introduces concepts and notation used throughout this paper. Section 3 reviews the related work. Section 4 presents the proposed architecture of ensembles for quantification. Section 5 describes the experimental setup to evaluate the ensembles' performance. Section 6 discusses the experimental results for multi-class settings. Section 7 analyses the influence of the fusion operators. Finally, Sect. 8 presents our conclusions and directions for future work.

2   Background

A *dataset* $\mathcal{D}$ is a collection of samples such that $D = \{(\mathbf{x}_1, y_1), \ldots, (\mathbf{x}_n, y_n)\}$, where $\mathbf{x}_i \in \mathcal{X}$ is an instance in the m-dimensional feature space, $y_i \in \mathcal{Y} = \{c_1, \ldots, c_l\}$ is the corresponding class label of $\mathbf{x}_i$.

We can train a predictive model from a dataset D. The *classification* goal is to accurately predict the class labels of unlabeled instances based on their feature values. Hence, the classifier is a predictive model h_c induced from $\mathcal{D}$ such that:

$$h_c : \mathcal{X} \to \mathcal{Y}$$

In the quantification task, a *quantifier* is a supervised model that learns to estimate from dataset $\mathcal{D}$ the relative frequency of classes in an unlabelled set of instances. Therefore, the quantifier is a function h_q such that:

$$h_q : 2^{\mathcal{X}} \to \Delta^l$$

where $2^{\mathcal{X}}$ is the power set of $\mathcal{X}$, i.e., the set with all possible sets of samples under the representation $\mathcal{X}$, and Δ^l is the l-probability simplex defined as:

$$\Delta^l = \{\{p_i\}_{i=1}^l | p_i \in [0,1], \sum_{i=1}^{l} p_i = 1\}$$

Given an unlabeled set $\mathbf{S} \in 2^{\mathcal{X}}$, h_q outputs a vector $\hat{\mathbf{p}} = [\hat{p}(c_1), \ldots, \hat{p}(c_l)]$, with the estimated class prevalences, subject to the constraints $\hat{p}(c_i) \geq 0$ and $\sum_{i=1}^{l} \hat{p}(c_i) = 1$. The objective is to minimize the difference between the predicted probabilities $\hat{p}(c_1), \ldots, \hat{p}(c_l)$ and the true classes prevalence $p(c_1), \ldots, p(c_l)$ in $\mathbf{S}$.

Comparing h_c and h_q, classification and quantification tasks use the same data representation and a labeled attribute-value dataset $\mathcal{D}$ to train their models. However, their objectives are distinct. A classifier predicts a class label for each input instance, whereas a quantifier predicts the class prevalence for a given *sample* of instances.

The instances are *independent* of each other in both classification and quantification so that the occurrence of one instance does not change the probability of the other instances. However, training and test samples are not *identically distributed* in quantification problems, as we expect that the class distribution will change.

Also, we define a scorer since several quantifiers employ it as an intermediate step in their computation. A *scorer* is a model induced from $\mathcal{D}$ such that:

$$h_s : \mathbf{X} \longrightarrow \mathbb{R}^l$$

For a given input instance, a scorer produces a vector $\mathbf{s} = [s_1, \ldots, s_l]$ of real values called *scores*. Each score s_i has a positive correlation with the posterior probability of the class y_i, i.e., $P(Y = y_i | \mathbf{x})$. Consequently, a higher s_i value means an increased chance for an instance belonging to the class y_i.

3   Related Work

This section reviews the most relevant work in quantification and ensemble learning. Quantification research has a thriving community that has proposed several methods in the last decade. A recent survey by González at al. [7] categorized quantifiers into three groups: *classify, count & correct, adaptations of traditional classification algorithms*, and *distribution matching*. Classify, count & correct methods use a classifier to label each instance. Then these methods count the number of instances predicted in each class and calculate the class ratios. These methods often apply a correction factor to their predictions to improve the quantification accuracy. Adaptations of traditional classification algorithms are approaches that modify the mechanics of traditional classification learning methods so that they become quantifiers. These methods use loss functions adapted to quantification tasks such as SVM-K (KLD loss) [8] and SVM-Q (Q-measure-loss) [9]. Distribution matching algorithms parametrically model the training distribution and later search the parameters that produce the best match against the test distribution. In the following, we provide a brief explanation of the baseline quantifiers and the ones we utilized in our experiment.

CC. The Classify & Count (CC) method is a straightforward approach for quantification. This method trains a classifier using a dataset and a standard learning algorithm. Once the classifier is trained, it can be used to classify the items in a sample set **S**. By counting the fraction of examples in the sample that are predicted to belong to each class, we can estimate the class's prevalence. This estimation process corresponds to the computation of the following equation:

$$\hat{p}_{CC}(y = c_i) = \frac{|\{\mathbf{x} \in S \mid h(\mathbf{x}) = c_i\}|}{|S|} \tag{1}$$

Forman [10] shows that CC contains a systematic bias. This flaw has motivated the community to propose more accurate quantifiers.

ACC. Adjusted Classify & Count (ACC) [3] is a binary-class quantifier that applies a correction factor to the output of CC. The correction factor for the positive class ($\oplus$) is defined as:

$$\hat{p}_{ACC}(y = \oplus) = \frac{\hat{p}_{CC}(y = \oplus) - fpr}{tpr - fpr} \tag{2}$$

where tpr and fpr are the true and false positive rates.

ACC is often implemented using cross-validation to obtain unbiased tpr and fpr estimates from training data. Forman [3] shows that ACC is a perfect quantifier if the true tpr and fpr are known, but inaccuracies introduced by the estimation process can make ACC often less accurate than the state-of-the-art quantifiers.

PCC and PACC. Probabilistic Classify & Count (PCC) and Probabilistic Adjusted Classify & Count (PACC) [11] are variations of CC and ACC, respectively, using a probabilistic classifier. PCC averages the probabilities to estimate the class prevalence, and PACC uses Eq. 2 to correct PCC's estimate. PCC and PACC suffer from chicken-and-egg problem as getting calibrated probability estimates requires knowing the class distribution in the test sample [10].

GACC and GPACC. The Generalized Adjusted Classify & Count (GACC) and Generalized Probabilistic Adjusted Classify & Count (GPACC) are multi-class extensions of ACC and PACC, respectively [12]. The methods utilize a system of equations and apply constrained least-squares regression to solve them.

FM. Friedman's method (FM) [13] constructs a system of equations similar to GPACC, but FM focuses only on a subset of test instances that have probabilities greater than the training class prevalences.

EMQ. The Expectation-Maximization Quantifier (EMQ) [14] uses the classic EM algorithm to adjust the outputs of a probabilistic classifier in the presence of distribution shift between the training and test data. While the primary objective of this method is classification, it can also obtain the target class prevalence as a by-product.

3.1   Ensembles

An ensemble is a widely used strategy to improve classification performance. Ensembles are a machine learning approach to combine a set of models, each of which solves the same original task, to obtain a solution that outperforms that obtained from using a single model [15]. The models are referred to as base estimators or base learners. Ensemble methods assume that each model performs well in certain domains while being sub-optimal in others [16]. The main reason to use an ensemble method over a single model is to make better predictions by reducing the variance component of the prediction error and by adding bias to the model [17].

A key point of an ensemble learning method is combining the predictions from multiple models. It is a data fusion perspective and depends heavily on the contributing models to the ensemble and the prediction problem. In the classification, voting is the most popular combination method for crisp class labels [16]. In the case of predicted class scores or probabilities [17], the independently predicted probabilities can be combined directly by an algebraic function as the median. However, if labels are needed, the combined probabilities can be converted to a class label using a softmax function. In regression or quantification tasks, combining numerical predictions often involves using simple statistical methods, and the average is the most common combination method for numeric outputs.

An essential property of an accurate ensemble is the diversity of the predictions made by multiple contributing models. One of the ways to promote diversity is to consider models of different natures. An intuitive explanation for why an ensemble produces accurate models is that when models are combined, uncorrelated errors of individual models can be eliminated. Thus, we propose the usage of ensembles of multiple base models in two different levels (classifiers and quantifiers) to expand diversity for the quantification problem.

Although ensembles are a well-known technique used in many Machine Learning tasks, their application in quantification learning is relatively recent. The current research involving ensembles of quantifiers is limited but has shown promising results. To our knowledge, only two ensemble methods have been proposed for quantification [5,6]. Next, we review these two approaches.

Figure 1a illustrates the ensemble of quantifiers proposed by Pérez-Gállego *et al.* [6]. The idea is to train a set of classifiers using different class distributions, thus promoting diversity. The same classifier C is replicated several times, with each replication training in a data sample D_i with a distinct class distribution to model the class prevalence shift. Each training set D_i is generated using sub-sampling. A single base quantifier Q transforms the output of each classifier into a quantification prediction, producing a prediction matrix $\mathbf{M}$:

$$\mathbf{M} = \begin{bmatrix} \hat{p}_{1,1} & \hat{p}_{1,2} & \cdots & \hat{p}_{1,l} \\ \hat{p}_{2,1} & \hat{p}_{2,2} & \cdots & \hat{p}_{2,l} \\ \vdots & \vdots & \ddots & \vdots \\ \hat{p}_{b,1} & \hat{p}_{b,2} & \cdots & \hat{p}_{b,l} \end{bmatrix} = \begin{bmatrix} \hat{\mathbf{p}}_1 \\ \hat{\mathbf{p}}_2 \\ \vdots \\ \hat{\mathbf{p}}_b \end{bmatrix}$$

where b is the number of base classifiers trained with different class distributions. The final quantification is obtained by applying an aggregation function $\mathcal{F}$, such as median or average, over the columns of matrix $\mathbf{M}$ to produce a single prediction, which is normalized with the function $\mathcal{N}$ for the class prevalences sum to 1. We refer to it as the class-prevalence ensemble (CPE). The authors showed that using an ensemble outperforms the results of using a single classifier for quantification learning.

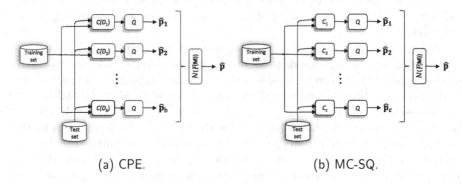

(a) CPE.   (b) MC-SQ.

Fig. 1. Existing ensembles architectures for quantification.

Donyavi *et al.* [5] conducted a study on the performance of multi-class quantifiers and proposed an ensemble approach to enhance their performance. The architecture, as shown in Fig. 1b, consists of c base classifiers, each paired with a base quantifier Q of a single type. To reduce the number of parameters, Donyavi *et al.* [5] suggests $c = 7$ and the following base classifiers: Random Forest (RF), Naïve Bayes (NB), Gradient Boosting (GB), Support Vector Machines (SVM), Linear Discriminant Analysis (LDA), Light Gradient Boosting Machines (LGBM), and Logistic Regression (LR). The final output combines the outcomes of all classifier-quantifier pairs using the median and then normalizes the results. Unlike the CPE method, this model trains all classifiers using training data without changing the class distribution. This ensemble approach is referred to as MC-SQ, representing Multiple Classifiers with Single Quantifier.

Inspired by MC-SQ, our novel ensemble approach uses data fusion to merge the classifiers' outputs and base quantifiers of different types to improve the final quantification.

4 Proposed Approaches

We propose two new ensembles for quantification. Both approaches are extensions of our previous work, MC-SQ [5]. The proposals have in common an aggregation of the classification scores[1] into a single score tensor $\mathcal{S}$. The difference between the two proposals is that the first uses a single base quantifier Q, while the second uses q quantifiers $Q_1, \ldots, Q_q$. Figure 2 illustrates both approaches.

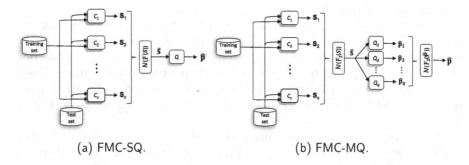

(a) FMC-SQ.   (b) FMC-MQ.

Fig. 2. Proposed ensemble architectures.

Our simplest proposal is named *Fusioned Multiple Classifiers with Single Quantifier* (**FMC-SQ**). It comprises a collection $\mathcal{C} = \{C_1, C_2, \ldots, C_c\}$ of c different classifiers. Each classifier C_i provides a score matrix $\mathbf{S}^{n \times l}$ for the n examples in the test sample:

$$\mathbf{S} = \begin{bmatrix} s_{1,1} & s_{1,2} & \cdots & s_{1,l} \\ s_{2,1} & s_{2,2} & \cdots & s_{2,l} \\ \vdots & \vdots & \ddots & \vdots \\ s_{n,1} & s_{n,2} & \cdots & s_{n,l} \end{bmatrix}$$

where $s_{i,j}$ is the score assigned to the i-th test example and j-th class label.

The output of the c classifiers are combined in a tensor $\mathcal{S}^{c \times n \times l}$ in the following form:

$$\mathcal{S} = \begin{bmatrix} \mathbf{S}_1 \\ \mathbf{S}_2 \\ \vdots \\ \mathbf{S}_c \end{bmatrix}$$

where $\mathbf{S}_i$ is the score matrix provided by the classifier C_i.

Our approach uses a fusion operator $\mathcal{F}$ to convert the c matrices $\mathbf{S}_i$ into a single score matrix which is provided to a quantifier to estimate the class distribution. As we work with normalized scores, the normalization function $\mathcal{N}$

---

[1] Although we use the term *classification score*, in our experiments we use classification probabilities, *i.e.* normalized scores in the range $[0, 1]$ such that the scores for all classes sum to 1.

is also applied to guarantee the scores for all class labels of a single instance sum
to 1. Therefore:

$$\bar{\mathbf{S}} = \mathcal{N}(\mathcal{F}(\mathcal{S})) \tag{3}$$

A single quantifier is applied to the fused classification scores in order to
compute the final quantification.

Figure 2b shows our second ensemble architecture. Like FMC-SQ, it has
a collection of classifiers $\mathcal{C}$, whose classification scores are merged into a single output matrix $\bar{\mathbf{S}}$. However, such a matrix is evaluated by a collection
$\mathcal{Q} = \{Q_1, Q_2, ..., Q_q\}$ of q quantifiers, generating multiple quantifications. These
outputs are then aggregated again to deliver the final quantification. We denomi-
nate this model *Fusioned Multiple Classifiers with Multiple Quantifiers* (**FMC-
MQ**).

The aggregation operation of multiple outputs may be required at one or two
processing steps, depending on the architecture of the ensemble. Thus, there are
two possibilities: *i)* aggregation of the classifiers' outcomes, and *ii)* aggregation
of the quantifiers' outcomes. Each aggregation step is composed by two func-
tions: a data fusion operation ($\mathcal{F}$) and a normalization function ($\mathcal{N}$). The fusion
operation combines continuous-value outcomes delivered by all classifiers or by
all quantifiers for each class, producing a unique output. A typical combiner is
average. The normalization function scales the output values to sum to 1.

This work evaluates the influence of the fusion operators at classification
and quantification levels. We assess FMC-SQ and FMC-MQ architectures with
different fusion operators and compare their results with the state-of-the-art
ensembles.

5   Experimental Setup

We use the Artificial Prevalence Protocol (APP) [18,19] to evaluate quantification methods on classification datasets. APP modifies the class distribution in
the training and test samples through random sampling without replacement.
We extend the experimental setup used in [4] by considering different train-
ing/test ratios $\{(0.1, 0.9), (0.3, 0.7), (0.5, 0.5), (0.7, 0.3)\}$. Table 1 shows all used
distributions. We repeat each run ten times and report the average results.

Table 2 presents the 31 multi-class datasets used in the experiments. These
datasets were collected from various repositories, including UCI[2], Kaggle[3], UEA
& UCR[4] and USP[5] We employ 17 multi-class datasets (3 and 4 classes) from [4,5]
and 14 new datasets (5, 6, 7, and 10 classes) to evaluate quantifier performance
across a wider range of classes. Also, we apply pre-processing steps from [4],
including feature encoding, attribute rescaling, and missing value removal for
the new datasets.

---

[2] https://archive.ics.uci.edu/ml/datasets.php.
[3] https://www.kaggle.com/datasets.
[4] https://timeseriesclassification.com/dataset.php.
[5] https://sites.google.com/view/uspdsrepository.

Table 1. List of training distributions $P_{Train}(Y)$ and test distributions $P_{Test}(Y)$.

Classes	$P_{Train}(Y)$	$P_{Test}(Y)$
3	(0.2, 0.5, 0.3), (0.05, 0.8, 0.15), (0.35, 0.3, 0.35)	(0.1, 0.7, 0.2), (0.55, 0.1, 0.35), (0.35, 0.55, 0.1), (0.4, 0.25, 0.35), (0, 0.05, 0.95)
4	(0.5, 0.3, 0.1, 0.1), (0.7, 0.2, 0.1, 0.1), (0.25, 0.25, 0.25, 0.25)	(0.65, 0.25, 0.05, 0.05), (0.2, 0.25, 0.3, 0.25), (0.45, 0.15, 0.2, 0.2), (0.2, 0, 0, 0.8), (0.3, 0.25, 0.35, 0.1)
5	(0.2, 0.15, 0.35, 0.1, 0.2), (0.35, 0.25, 0.15, 0.05, 0.1), (0.2, 0.2, 0.2, 0.2, 0.2)	(0.15, 0.1, 0.65, 0.1, 0), (0.45, 0.1, 0.3, 0.05, 0.1), (0.2, 0.25, 0.25, 0.1, 0.2), (0.35, 0.05, 0.05, 0.05, 0.5), (0.05, 0.25, 0.15, 0.15, 0.4)
6	(0.1, 0.2, 0.1, 0.1, 0.25, 0,25), (0.05, 0.1, 0.3, 0.4, 0.1, 0.05), (017, 0.16, 0.16, 0.17, 0.16, 0.16)	(0.15, 0.1, 0.55, 0.1, 0, 0.1), (0.4, 0.1, 0.25, 0.05, 0.1, 0.1), (0.2, 0.2, 0.2, 0.1, 0.2, 0.1), (0.35, 0.05, 0.05, 0.05, 0.05, 0.45), (0.05, 0.25, 0.15, 0.15, 0.1, 0.3)
7	(0.2, 0.3, 0.2, 0.15, 0.05, 0.05, 0,05), (0.05, 0.1 0.05, 0.05, 0.25, 0.3, 0.2), (015, 0.14, 0.14, 0.15, 0.14, 0.14, 0.14)	(0.1, 0.1, 0.1, 0.5, 0.1, 0, 0.1), (0.4, 0.1, 0.2, 0.05, 0.1, 0.1, 0.05), (0.15, 0.2, 0.15, 0.1, 0.2, 0.1, 0.1), (0.3, 0.05, 0.05, 0.05, 0.05, 0.05, 0.45), (0.05, 0.25, 0.1, 0.15, 0.1, 0.3, 0.05)
10	(0.05, 0.2, 0.05, 0.1, 0.05, 0.25, 0.05, 0,05, 0.1, 0.1), (0.15, 0.05, 0.2, 0.05, 0.1, 0.05, 0.2, 0.1, 0.05, 0.05), (0.1, 0.1, 0.1, 0.1, 0.1, 0.1, 0.1, 0.1, 0.1, 0.1)	(0.1, 0.2, 0.1, 0.1, 0.2, 0.1, 0, 0.1, 0.05, 0.05), (0.2, 0.05, 0.15, 0.05, 0.1, 0.15, 0.05, 0.05, 0.1, 0.1), (0, 0.1, 0.05, 0.1, 0.05, 0.1, 0.1, 0.15, 0.15, 0.2), (0.05, 0.05, 0.05, 0.35, 0.15, 0.05, 0, 0.1, 0.1, 0.1), (0.05, 0.1, 0.1, 0.15, 0.1, 0.15, 0.05, 0.1, 0.1, 0.1)

Our experiments utilize the same base classifiers as in [5]: Logistic Regression (LR), Linear Discriminant Analysis (LDA), Random Forest (RF), Support Vector Machine (SVM), Light Gradient Boosting Machine (LGBM), Gradient Boosting (GB), and Naive Bayes (NB). We use these classifiers with their default parameter values provided by the Scikit-learn library [20] to simplify the experimental setup. Also, we employ the base quantifiers EM, FM, GACC, and GPACC, since these were the best performing multi-class quantifiers in [4]. These methods require computing scores from training examples, so we use 10-fold cross-validation on the training set to obtain unbiased scores. Additionally, our paper website [21] stores code, figures, tables, and detailed results, including for the binary datasets not included in this paper.

Our proposals, FMC-SQ and FMC-MQ, use all seven mentioned base classifiers. FMC-MQ also incorporates all four base quantifiers (EM, FM, GACC, GPACC). On the other hand, CPE and MC-SQ employ a single base quantifier, so we execute these methods four times, each time with a different quantifier, to report the corresponding quantification errors. CPE uses LR as the base classifier, following the implementation in [6]. LR is trained with 50 samples having diverse class distributions, as recommended by the CPE authors. Our experiments utilize the CPE implementation available in QuaPy [22]. Additionally, for a comprehensive comparison, we include standalone quantifiers S̲ingle C̲lassifier with S̲ingle Q̲uantifier (SC-SQ), where LR is employed as the base classifier following [4].

We use absolute error (AE) as the evaluation measure to assess the results. AE is easily interpreted and restrained in the interval $[0, 2]$ independently of the number of classes [23]. AE is defined according to Eq. 4.

Table 2. Description of the multi-class datasets.

Dataset (Abbreviation)	# Features	# Instances	# Labels	Source
Spoken Arabic Digit (arab)	27	8800	10	UCI
Dry Bean (beans)	16	13611	7	UCI
Bike Sharing Dataset (bike)	59	17379	4	UCI
BlogFeedback (blog)	280	52397	4	UCI
Concrete Compressive Strength (conc)	8	1030	3	UCI
Superconductivity Data (cond)	89	21263	4	UCI
Contraceptive Method Choice (contra)	13	1473	3	UCI
SkillCraft1 Master Table (craft)	18	3338	3	UCI
Electric Devices (device)	96	24348	7	UEA & UCR
Diamonds (diam)	22	53940	3	Kaggle
Drug Consumption (drugs)	136	1885	3	UCI
Appliances Energy Prediction (ener)	25	19735	3	UCI
Epileptic Seizure Recognition (epil)	178	11500	5	Kaggle
FIFA 19 Complete Player Dataset (fifa)	117	14751	4	Kaggle
Gas Sensor Array Drift (gasd)	128	13910	6	UCI
Gesture Phase Segmentation (gest)	32	10356	5	UCI
Human Activity Recognition with Smartphones (hars)	562	10166	6	Kaggle
Insects (insec)	49	5325	5	USP
Insect Sound (insecs)	600	25000	10	UEA & UCR
Microbes (micro)	24	18176	7	Kaggle
News Popularity in Multiple Social Media Platforms (news)	60	39644	4	UCI
Nursery (nurse)	27	12960	3	UCI
Optical Recognition of Handwritten Digits (optd)	63	5620	5	UCI
Pen-Based Recognition of Handwritten Digits (pend)	16	10992	10	UCI
Rice MSC (rice)	106	5000	5	Kaggle
Statlog (Landsat Satellite) (satel)	36	6435	6	UCI
First-order Theorem Proving (thrm)	51	6117	3	UCI
Turkiye Student Evaluation (turk)	31	5820	3	UCI
Video Game Sales (vgame)	133	6825	4	Kaggle
Wine Quality (wine)	14	6497	4	UCI
Yeast (yeast)	9	1484	4	UCI

$$AE(\mathbf{p}, \hat{\mathbf{p}}) = \frac{1}{l} \sum_{i=1}^{l} |\hat{\mathbf{p}}[i] - \mathbf{p}[i]| \tag{4}$$

where $\mathbf{p}$ and $\hat{\mathbf{p}}$ are the vectors with true and predicted prevalence, respectively.

6   Results and Discussion

This section compares the performance of the proposed FMC-SQ and FMC-MQ ensemble methods to the competing CPE, MC-SQ ensembles, and base quantifiers. We execute all ensembles with fusion operators to eliminate parameters. FMC-SQ and FMC-MQ use the median for the fusion of classification scores. MC-SQ and FMC-MQ use the median fusion operator to estimate quantification. CPE uses the average as a fusion operator following the recommendations of [6]. Section 7 investigates the impact of fusion operators on the performance of the proposed methods.

Due to a lack of space, we only include numerical results for multi-class datasets. As pointed out by [4], multi-class quantification is a much more complex problem than binary quantification. Table 3 presents the results for the multi-class datasets. The results show that the ensemble methods often outperform the individual base quantifiers. CPE has, on average, smaller MAE values than the associated single quantifier (SC-SQ). One exception is CPE with EM, which

Table 3. Experimental results for multi-class datasets for all architectures.

Dataset	SC-SQ EM	GACC	GPACC	FM	CPE (avg) EM	GACC	GPACC	FM	MC-SQ (median) EM	GACC	GPACC	FM	FMC-SQ (median) EM	GACC	GPACC	FM	FMC-MQ (median, median)
arab	0.205	0.113	0.088	0.215	0.381	0.075	0.088	0.166	0.061	0.066	0.062	0.077	0.094	0.069	0.069	0.069	**0.058**
beans	0.424	0.179	0.095	0.254	0.512	0.106	0.089	0.121	0.046	0.048	0.047	0.050	0.097	0.052	0.052	0.052	**0.045**
bike	0.082	0.113	0.073	0.102	0.117	0.101	0.096	0.104	0.096	0.068	0.059	0.065	0.208	0.070	0.070	0.070	**0.058**
blog	0.196	0.360	0.236	0.285	0.256	0.238	0.249	0.264	0.167	0.173	**0.115**	0.122	0.277	0.125	0.125	0.125	0.173
conc	0.498	0.486	0.473	0.510	0.410	0.407	0.381	0.389	0.256	0.275	0.266	0.245	0.407	**0.239**	**0.239**	**0.239**	0.253
cond	0.059	0.155	0.066	0.088	0.085	0.078	0.064	0.074	0.054	0.054	**0.045**	0.047	0.199	0.062	0.062	0.062	0.047
contra	0.396	0.600	0.515	0.512	0.409	0.468	0.411	0.402	**0.391**	0.470	0.424	0.419	0.568	0.485	0.485	0.485	0.419
craft	0.191	0.296	0.190	0.190	0.271	0.264	0.206	0.218	0.225	0.186	0.168	**0.156**	0.413	0.179	0.179	0.179	0.172
device	0.228	0.384	0.242	0.340	0.345	0.271	0.291	0.343	0.156	0.129	0.115	0.127	0.362	0.117	0.117	0.117	**0.114**
diam	0.214	0.197	0.098	0.118	0.183	0.196	0.110	0.100	0.042	0.029	**0.027**	**0.027**	0.282	0.030	0.030	0.030	0.030
drugs	0.218	0.256	0.199	0.181	0.229	0.250	0.252	0.259	0.204	0.206	0.181	**0.163**	0.364	0.194	0.194	0.194	0.168
ener	0.131	0.273	0.115	0.129	0.161	0.225	0.120	0.130	0.158	0.108	**0.084**	**0.084**	0.424	0.094	0.094	0.094	0.092
epil	0.796	0.887	0.858	0.704	0.681	0.501	0.530	0.474	0.384	0.268	0.224	0.247	0.361	**0.213**	**0.213**	**0.213**	0.239
fifa	0.127	0.313	0.181	0.216	0.198	0.182	0.202	0.211	0.117	0.145	0.111	**0.104**	0.228	0.135	0.135	0.135	0.126
gasd	0.212	0.097	0.064	0.086	0.241	0.054	0.056	0.069	0.039	0.027	0.027	0.029	0.060	0.027	0.027	0.027	**0.025**
gest	0.464	0.501	0.422	0.474	0.350	0.394	0.310	0.316	0.211	**0.170**	0.158	0.174	0.418	0.182	0.182	0.182	0.172
hars	**0.019**	0.025	0.022	0.030	0.031	0.026	0.032	0.040	0.021	0.022	0.021	0.023	0.040	0.022	0.022	0.022	0.020
insec	0.076	0.070	0.063	0.075	0.127	0.061	0.062	0.073	0.058	0.052	0.051	0.052	0.119	0.055	0.055	0.055	**0.047**
insecs	0.438	0.702	0.695	0.849	0.517	0.416	0.489	0.647	0.390	0.342	0.299	0.397	0.404	**0.256**	**0.256**	**0.256**	0.316
micro	0.334	0.473	0.306	0.431	0.305	0.250	0.252	0.289	0.169	0.143	0.142	0.149	0.227	**0.108**	**0.108**	**0.108**	0.137
news	**0.221**	0.498	0.335	0.376	0.246	0.288	0.249	0.238	0.260	0.325	0.261	0.268	0.508	0.362	0.362	0.362	0.278
nurse	0.022	0.023	0.019	0.020	0.027	0.016	0.017	0.018	0.015	0.011	0.013	**0.009**	0.032	0.010	0.010	0.010	0.011
optd	0.053	0.050	0.044	0.059	0.097	0.046	0.053	0.065	0.031	0.031	0.032	0.035	0.043	0.034	0.034	0.034	**0.028**
pend	0.110	0.057	0.049	0.113	0.185	0.048	0.051	0.104	0.031	0.026	0.027	0.036	0.031	0.028	0.028	0.028	**0.025**
rice	0.041	0.016	0.015	0.024	0.062	0.012	0.019	0.029	0.007	0.009	0.009	0.010	**0.005**	0.008	0.008	0.008	0.007
satel	0.184	0.139	0.086	0.179	0.380	0.123	0.118	0.098	0.063	0.067	0.064	0.065	0.144	0.069	0.069	0.069	**0.058**
thrm	0.494	0.780	0.629	0.663	0.323	0.409	0.337	0.382	0.330	0.344	0.321	**0.302**	0.486	0.401	0.401	0.401	0.342
turk	**0.277**	0.525	0.342	0.392	0.365	0.402	0.338	0.348	0.432	0.408	0.315	0.339	0.616	0.447	0.447	0.447	0.344
vgame	0.322	0.520	0.460	0.474	0.375	0.358	0.364	0.371	**0.315**	0.397	0.391	0.358	0.337	0.372	0.372	0.372	0.375
wine	0.757	0.656	0.575	0.605	0.414	0.416	0.371	0.388	**0.340**	0.440	0.449	0.431	0.460	0.482	0.482	0.482	0.419
yeast	0.613	0.567	0.408	0.413	0.546	0.448	0.401	0.411	0.353	0.450	0.476	0.482	0.320	0.359	0.359	0.359	**0.310**
Mean	0.271	0.333	0.257	0.294	0.285	0.230	0.213	0.230	0.175	0.177	0.161	0.164	0.275	0.170	0.170	0.170	**0.158**
Rank	11.3	15.2	11.1	13.7	13.0	10.7	9.9	11.3	6.0	6.5	**4.1**	5.2	12.7	6.3	6.3	6.3	**3.5**

presented a slight increase in mean MAE than EM, resulting in a lower position in the rank across all datasets.

MC-SQ outperforms CPE as previously observed in [5]. Conversely, FMC-SQ did not outperform MC-SQ. The main difference between FMC-SQ and MC-SQ is that FMC-SQ applies the fusion operator to the classifier scores while MC-SQ applies the fusion operator to the quantification estimates. From our experience, the multi-class quantification approaches present high variance, which means that minor variations in the classifiers' output can lead to significant changes in the quantifier estimation. Thus, the results indicate that averaging the quantification output is a better strategy to reduce variance.

Finally, FMC-MQ is the best-performing approach on average. This approach uses multiple quantifiers to reduce the variance of the estimations further. Overall, using multiple classifiers and quantifiers is the most promising approach to be explored in more complex ensemble architectures. Figure 3 provides the critical difference (CD) diagram for multi-class datasets according to the post hoc Nemenyi test. FMC-MQ outperforms all other methods for multi-class datasets but MC-SQ(GPACC) with statistically significant differences.

7   Analysis of the Data Fusion Operators

We assess nine fusion operators to aggregate the outcomes at the classifiers level and five operators at the quantifiers level. Therefore, we evaluate 45 possible

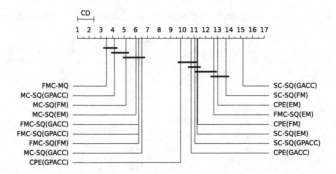

Fig. 3. CD diagrams comparing architectures for multi-class datasets.

combinations for implementing the FMC-MQ (as it uses all quantifiers at once) and nine options for FMC-SQ (for each single quantifier).

Fusion operators for classifiers. We assess well-known algebraic operators, including the arithmetic mean (avg), median, minimum (min), maximum (max), algebraic product (prod), and Cosine similarity (cos). The operators take as input a set of scores (one from each classifier) and output an aggregated score. Also, we employ three other fusion techniques based on a one-hot encoder, where the class with the best similarity is the fused decision, returning 1 for it and 0 for the other classes. These fusion operators are Decision templates (DT), Dempfster-Shafer method (DS), and Maximum Likelihood (ML), as described in [24]. For DT, we use Euclidean distance as a similarity measure.

Fusion operators for quantifiers. The assessed operators are arithmetic mean (avg), median (med), minimum (min), maximum (max) and algebraic product (prod), but using quantification vectors instead of scores vectors. The output vector is normalized for the class prevalence to sum to 1.

Table 4 shows the averaged absolute error for multi-class datasets, considering the use of each classification fusion operator for each quantifier in the FMC-SQ model. DS has the best overall MAE averages. Notice that GACC, GPACC and FM all have the same results for DT, DS and ML operators. This is not a coincidence, as these methods find the same solutions given the one-hot output of these operators and the optimization problem these algorithms solve.

Table 4. Averaged absolute errors of fusion operators for the classification scores and FMC-SQ architecture.

Quantifier	min	max	prod	avg	median	cos	DT	DS	ML
EM	0.477	0.632	0.474	0.398	0.272	0.281	**0.264**	0.275	0.344
GACC	0.302	0.289	0.269	0.214	0.221	0.220	0.174	**0.170**	0.208
GPACC	0.289	0.219	0.257	0.176	0.173	0.176	0.174	**0.170**	0.208
FM	0.291	0.239	0.263	0.200	0.191	0.205	0.174	**0.170**	0.208

Figures 4 and 5 present the results of the combinations of fusion operators for the FMC-MQ architecture, respectively, averaging absolute errors and rank. The critical difference (CD) diagram (Fig. 5) is computed for ranking, employing the non-parametric Friedman test with 95% confidence to determine the presence of significant differences between the operators and the posthoc Nemenyi test to infer which differences are significant.

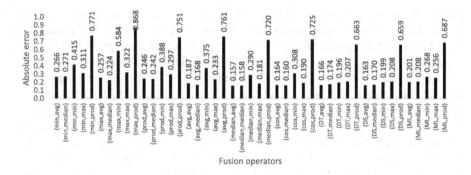

Fig. 4. Averaged absolute errors of fusion operators for FMC-MQ architecture.

The best operators pair with the lowest absolute errors for multi-class datasets with FMC-MQ is (median, avg). However, the best-ranked operators

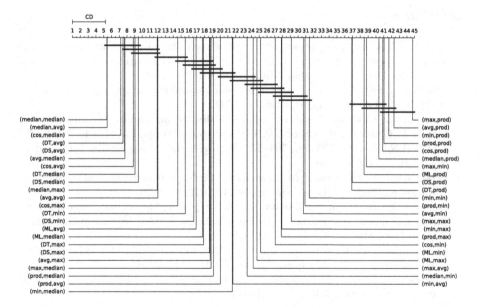

Fig. 5. CD diagrams comparing fusion operators for FMC-MQ with multi-class datasets.

pair for multi-class datasets is (median, median). CD diagrams show no significant differences between (median, avg) and (median, median) for multi-class datasets. Thus, we deduce that (median, median) is the best fusion operators pair for a general case. The median is less liable to be distorted by outliers.

8   Conclusions

This work proposes two ensemble architectures for quantification and assesses several fusion operators to aggregate classification scores and quantification probabilities. Our results show that aggregating quantification probabilities perform better than aggregating classification scores. Thus, FMC-MQ generally outperforms FMC-SQ across all datasets. Similarly, FMC-SQ does not outperform MC-SQ considering each base quantifier, indicating that the fusion of classification scores is not a promising research direction.

In future work, we will investigate other approaches and configurations of ensembles. A promising direction is an ensemble of multiple classifiers and quantifiers that aggregates quantification probabilities only without fusing classification scores. In addition to binary and multi-class quantification, we also intend to tackle the multi-label problem.

Acknowledgment. The authors thank FAPESP for the financial support (Process nº 2021/12278-0). This research was supported by resources supplied by the Center for Scientific Computing (NCC/GridUNESP) of the São Paulo State University (UNESP).

References

1. Forman, G.: Quantifying trends accurately despite classifier error and class imbalance. In: Proceedings of the 12th ACM SIGKDD International Conference on Knowledge Discovery and Data Mining, pp. 157–166 (2006)
2. Moreo, A., Sebastiani, F.: Tweet sentiment quantification: an experimental re-evaluation. PLoS ONE **17**(9), e0263449 (2022)
3. Forman, G.: Counting positives accurately despite inaccurate classification. In: Gama, J., Camacho, R., Brazdil, P.B., Jorge, A.M., Torgo, L. (eds.) ECML 2005. LNCS (LNAI), vol. 3720, pp. 564–575. Springer, Heidelberg (2005). https://doi.org/10.1007/11564096_55
4. Schumacher, T., Strohmaier, M., Lemmerich, F.: A comparative evaluation of quantification methods. arXiv preprint arXiv:2103.03223 (2022)
5. Donyavi, Z., Serapião, A., Batista, G.: MC-SQ: a highly accurate ensemble for multi-class quantification. In: Proceedings of the 2023 SIAM International Conference on Data Mining (SDM), pp. 622–630 (2023)
6. Pérez-Gállego, P., Castano, A., Quevedo, J.R., del Coz, J.J.: Dynamic ensemble selection for quantification tasks. Inf. Fusion **45**, 1–15 (2019)
7. González, P., Castaño, A., Chawla, N.V., del Coz, J.J.: A review on quantification learning. ACM CSUR **50**(5), 1–40 (2017)
8. Esuli, A., Moreo Fernández, A., Sebastiani, F.: A recurrent neural network for sentiment quantification. In: 27th ACM CIKM, pp. 1775–1778 (2018)

9. Barranquero, J., Díez, J., del Coz, J.J.: Quantification-oriented learning based on reliable classifiers. Pattern Recogn. **48**(2), 591–604 (2015)

10. Forman, G.: Quantifying counts and costs via classification. Data Min. Knowl. Disc. **17**(2), 164–206 (2008). https://doi.org/10.1007/s10618-008-0097-y

11. Bella, A., Ferri, C., Hernández-Orallo, J., Ramirez-Quintana, M.J.: Quantification via probability estimators. In: ICDM, pp. 737–742. IEEE (2010)

12. Firat, A.: Unified framework for quantification. arXiv preprint arXiv:1606.00868 (2016)

13. Friedman, J.H.: Class counts in future unlabeled samples. MIT CSAIL Big Data Event (2014). https://jerryfriedman.su.domains/talks/HK.pdf

14. Saerens, M., Latinne, P., Decaestecker, C.: Adjusting the outputs of a classifier to new a *priori* probabilities: a simple procedure. Neural Comput. **14**(1), 21–41 (2002)

15. Zhou, Z.-H.: Ensemble Methods: Foundations and Algorithms, 1st ed. Chapman & Hall/CRC (2012)

16. Rokach, L.: Ensemble Learning: Pattern Classification Using Ensemble Methods, Series in Machine Perception and Artificial Intelligence, 2nd edn. World Scientific Publishing Company (2019)

17. Zhang, C., Ma, Y.: Ensemble Machine Learning: Methods and Applications. Springer, New York (2012). https://doi.org/10.1007/978-1-4419-9326-7

18. Moreo, A., Sebastiani, F.: Tweet sentiment quantification: an experimental re-evaluation. arXiv preprint arXiv:2011.08091 (2020)

19. Hassan, W., Maletzke, A.G., Batista, G.: Pitfalls in quantification assessment. In: CIKM Workshops (2021)

20. Pedregosa, F., et al.: Scikit-learn: machine learning in Python. JMLR **12**, 2825–2830 (2011)

21. Serapião, A., Donyavi, Z., Batista, G.: Ensembles of classifiers and quantifiers with data fusion for quantification learning: paper website (2023). https://sites.google.com/view/fmc-mq

22. Moreo, A., Esuli, A., Sebastiani, F.: QuaPy: a Python-based framework for quantification. In: 30th ACM CIKM, pp. 4534–4543 (2021)

23. Sebastiani, F.: Evaluation measures for quantification: an axiomatic approach. Inf. Retrieval J. **23**(3), 255–288 (2020). https://doi.org/10.1007/s10791-019-09363-y

24. Kuncheva, L.I.: Combining Pattern Classifiers: Methods and Algorithms. Wiley, Hoboken (2004)

Exploring the Intricacies of Neural Network Optimization

Rafael Teixeira[1,2]($\boxtimes$) (iD), Mário Antunes[1,2] (iD), Rúben Sobral[2] (iD),
João Martins[2] (iD), Diogo Gomes[1,2] (iD), and Rui L. Aguiar[1,2] (iD)

[1] Instituto de Telecomunicações, Universidade de Aveiro, Aveiro, Portugal
rafaelgteixeira@av.it.pt
[2] Departamento de Electrónica, Telecomunicações e Informática,
Universidade de Aveiro, Aveiro, Portugal

Abstract. Recent machine learning breakthroughs in computer vision
and natural language processing were possible due to Deep Neural Networks (DNNs) learning capabilities. Even so, applying DNNs is quite challenging, as they usually have more hyperparameters than shallow models.
The higher number of hyperparameters leads to allocating more time for
model optimization and training to achieve optimal results. However, if
there is a better understanding of the impact of each hyperparameter on
the model performance, then one can decide which hyperparameters to
optimize according to the available optimization budget or desired performance. This work analyzes the impact of the different hyperparameters when applying dense DNNs to tabular datasets. This is achieved by
optimizing each hyperparameter individually and comparing their influence on the model performance. The results show that the batch size usually only affects training time, reducing it by up to 80% or increasing it
by 200%. In contrast, the hidden layer size does not consistently affect
the considered performance metrics. The optimizer can significantly affect
the model's overall performance while also varying the training time, with
Adam being the generally the better optimizer. Overall, we show that the
hyperparameters do not equally affect the DNN and that some can be discarded if there is a constrained search budget.

Keywords: Neural Networks · Hyperparameter Optimization · Neural
Architecture Search · Hyperparameter Importance · HPO · NAS

1 Introduction

Machine Learning (ML) has recently witnessed various breakthroughs in complex tasks such as Computer Vision and Natural Language Processing (NLP).
The catalyst for these breakthroughs was the use of Deep Neural Networks
(DNNs) [1]. These DNNs differed significantly on various hyperparameters, such
as the optimizer, learning rate, number of layers, nodes per layer, or the node
type in each layer. Since DNNs have more hyperparameters than typical shallow

A. Bifet et al. (Eds.): DS 2023, LNAI 14276, pp. 18–32, 2023.
https://doi.org/10.1007/978-3-031-45275-8_2

models, achieving a suitable configuration is harder and can even directly limit scientific progress [2]. With the trend of using ever deeper DNNs, these difficulties become more noticeable, and manually optimizing a neural network becomes less suitable. So researchers created automated approaches for hyperparameter and architecture tuning to help them find efficient models [1].

Although results show that some automated approaches outperformed human experts [1–3], they are computationally demanding [2,4]. This disadvantage is the result of two factors. First, every time a configuration is tested, the entire DNNs must be trained, which is, on its own, computationally demanding, especially when the DNNs have millions of parameters. Second, given the number of hyperparameters and their possible values, the search space is enormous. Consequently, some other models can be more suitable when the search budget is small. For example, [5] shows how Random Forest (RF) models outperform DNNs in tabular data using the same search budget.

To mitigate this considerable disadvantage, one must either reduce the training time or the search space. Since some hyperparameters have been shown to have a suboptimal impact on other models' performance [2,6], finding which hyperparameters are meaningful to optimize in DNNs and removing the others can significantly reduce the search space.

This paper provides a step in that direction, analyzing the impact of different dense DNNs' hyperparameters on six tabular datasets. We only consider tabular datasets because the typical DNN used is dense, and the paper's objective is to understand the impacts of each hyperparameter in a dense DNN. We considered six datasets to obtain the average impact of each hyperparameter. After analyzing the individual impact of seven hyperparameters on the performance, training time, and prediction time, only one didn't show a meaningful impact on at least one of the metrics. The activation function presents the most significant influence on performance and the batch size on the training and prediction times.

The main contribution of this work is a list of the expected training and prediction time and accuracy variance for each hyperparameter in a tabular dataset when using a dense DNN, which can be used to guide the training of dense DNN, considering the available search budget and the desired final model performance.

The remainder of this paper is organized as follows. Section 2 presents an overview of the various automated hyperparameter optimization approaches and reviews previous work on hyperparameter relevance. Section 3 defines the experiments, describing the datasets used, the base DNN, what hyperparameters were considered, and the metrics used for evaluation. The results are presented in Sect. 4, and their discussion is elaborated in Sect. 5. The conclusions are drawn in Sect. 6.

2   Background

Automated hyperparameter tunning is vital for non-expert users to develop quality models [6]. Given its importance, several approaches have been proposed to improve its results. The difference between these approaches is how they analyze the hyperparameter space. In DNNs, besides the typical hyperparameters like learning rate or optimizer, that can be changed independently, there are interdependent architecture hyperparameters, which invalidate the use of some approaches. To answer these constraints, the Neural Architecture Search (NAS) subfield was proposed with dedicated algorithms to optimize DNNs architecture [1].

Since the various algorithms are only concerned with searching a given space, the space is usually left for the user to decide. To help aid in the search space decision, several analyses have been performed on various models regarding their hyperparameter importance.

Given that the focus of this paper is on the hyperparameter relevance and not the algorithms used, this section introduces the most common techniques for automated hyperparameter tuning, skipping the dedicated algorithms for NAS, and presents previous studies on the importance of some ML model hyperparameters.

2.1   Automated Hyperparameter Optimization

Hyperparameter optimization aims to find a set of hyperparameters that return the best performance of a model given a validation set [2]. This process's automation consists of creating a searchable space, usually defined by what hyperparameters are optimized and the range of values they assume, and a way to traverse the space to find the hyperparameters' combination that presents the best results.

Besides allowing non-experts to achieve good models, automated hyperparameter optimization provides better reproducibility, as there is an algorithm behind the decisions made, and in some cases, even obtains better performance by experimenting with new combinations. The optimization algorithms can be divided according to how they search the space.

Model-Free Algorithms. Model-free algorithms do not leverage any model to choose the next point in the search space to be tested, requiring brute force to find a reasonable solution. Given their simplicity, they are the most common approaches when developing a hyperparameter search. The two most common methods are grid search [6] and random search [4].

Gradient-Based Algorithms. One approach to consider the previous values tried is to use gradient descent. This traditional optimization technique uses the gradients of variables to find a promising direction. This approach always finds a local minimum given enough epochs, and when the optimization function is convex, the local minimum is the global minimum. The main disadvantage of this approach is that it only supports continuous hyperparameters, as other hyperparameters, like categorical ones, do not have a gradient direction [4].

Bayesian Optimization Algorithms. Similarly to gradient-based algorithms, the Bayesian optimization ones consider previously tried configurations when deciding which values to try next. The decision on which configuration to try next relies on two key components: the surrogate model, which fits the currently-observed points into the objective function, and the acquisition function, which determines the next point to evaluate by balancing the trade-off between exploration and exploitation [7]. Depending on the objective function, different surrogate models can be used. The three main approaches are Bayesian Optimization - Gaussian Process, Bayesian Optimization - Random Forest, and Bayesian Optimization - Tree-structured Parzen estimator [4].

Multi-fidelity Optimization Algorithms. As we mentioned in the introduction, the main disadvantage of automated hyperparameter optimization is that it is computationally demanding. Multi-fidelity optimization techniques tackle this problem by leveraging low-fidelity and high-fidelity evaluations [8]. In low-fidelity evaluations, only a subset of the search space and dataset are evaluated. This provides results at a low cost, although with low generalization. In high-fidelity evaluations, a more extensive set of the dataset is used for evaluation, providing a better generalization, although at a higher cost. The optimization algorithm aims to perform multiple low-fidelity optimizations in the complete search space and a few high-fidelity ones on the best-performing low-fidelity results. Two typical algorithms are successive halving and Hyperband [4].

Metaheuristics Algorithms. Based on biological theories, metaheuristics algorithms are widely used for optimization, with their main advantage being the capability to solve a wide range of problems [9]. One major category of these algorithms is the Population-based optimization algorithms, where the algorithms start by creating a population and then update it consecutively based on the best-performing individuals until it reaches the global optimum. In hyperparameter optimization, two popular approaches are genetic algorithms and particle swarm optimization [4].

2.2 Hyperparameter Importance

Although approaches such as Bayesian optimization algorithms provide good results for most models, blind reliance on these methods can deprive users of valuable information on the importance of each hyperparameter [2]. In turn, to ensure good results, every hyperparameter must be optimized, and their ranges must be considerably broad. When considering a Support Vector Machine (SVM) implemented using Scikit-learn, which has seven hyperparameters (when using the 'rbf' kernel), and the hyperparameters have a relatively small range of values, the search space is small enough that the impact of this approach is not significant. On the other hand, when considering DNNs, which has more than seven hyperparameters and an optimizable architecture, the search space can reach sizes of up to 10^{20} [1].

Although, to the best of the authors' knowledge, there are no studies performed on DNNs hyperparameter importance, several studies provide insights on the hyperparameter importance for other models such as RFs [10] and SVMs [3]. The insights extracted from these studies are outside this paper's scope. However, the tools used to obtain them are still relevant. For instance, one can borrow from the field of feature selection and adapt the sensitivity analysis metrics used to decide which features are relevant to find which hyperparameters present the highest variability [6]. The problem with these approaches is that they only consider the impact of individual hyperparameters. The Functional Analysis of Variance (fANOVA) algorithm [10] aims to tackle this problem using the prediction marginals of an RF trained on the output from the hyperparameter search. Another approach based on the feature selection algorithm Relief that can evaluate the impact of hyperparameters interaction is the N-RReliefF algorithm [3]. This algorithm uses a similar approach to the fANOVA to obtain the initial dataset composed of hyperparameter configurations and obtained performance. However, instead of relying on the RF to obtain the hyperparameter importance, it adapts the Relief algorithm to infer importance based on a continuous value instead of a class and infer the importance of interaction between hyperparameters instead of only assessing it individually.

3   Experiments

To understand if the importance of the hyperparameters is generalizable, one must use various datasets. Furthermore, besides the model's accuracy, other performance metrics are relevant. This section describes the experiments performed, stating which datasets were used, the base models, the hyperparameters evaluated, and finally, the performance metrics considered.

The results presented were obtained on a virtual machine with 24 VCPUs, 32 GBs of RAM, and one GPU (NVIDIA RTX 2080), and the neural networks were implemented with the Keras API from Tensorflow in Python. The source code is publicly available on GitHub[1].

3.1   Dataset Description

As mentioned in the introduction, we considered six tabular datasets in the experiments, three for classification tasks and three for regression tasks. These datasets are part of a more extensive benchmark presented in [5], which selected the datasets based on various criteria, such as difficulty to solve and amount of data. Please refer to the original paper for further discussion on how the benchmark was created. Since the initial criteria to select the datasets provided a varied benchmark. The subset selected was based on the amount of data available (small, medium, and large number of examples) and the features available, where the priority was datasets with categorical and numeric features. Table 1 presents

---

[1] https://github.com/rgtzths/mlp_hpp_analysis..

a summary of the characteristics of the datasets. If the dataset has more than 100 000 examples in the training set, then similarly to [5], the training set is truncated to 100 000 examples.

Table 1. Datasets' specifications. ('-' in № classes means that it is not applicable)

Regression			
Dataset Name	Total examples	№ Features	№ Classes
Bike Sharing	17379	11	-
Abalone	4177	8	-
Delays Zurich Transport	5465575	8	-
Classification			
Dataset Name	Total examples	№ Features	№ Classes
Compass	4966	11	2
Covertype	423680	54	2
Higgs	940160	24	2

3.2   Baseline Models

Since we needed a base model for each dataset with good performance, the baseline models were obtained through a simple random search that only varied architecture hyperparameters. Figure 1 presents the search space considered for the networks. The random search performed 100 trials, and the network was trained for 200 epochs in each trial. The data used for training was 80% of the complete dataset, where 20% was used as validation data. Given that the focus of the paper is on the hyperparameters' influence, not the models, the final baseline model for each dataset will not be discussed and is presented in Appendix B. The performance of said models and the comparison with the best and worst performing ones is available on the project's GitHub (See Footnote 1).

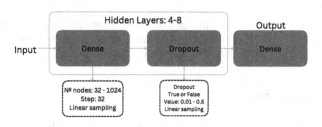

Fig. 1. DNN random search space definition to obtain the baseline models.

3.3 Hyperparameter Search Space

Since the paper's objective is to obtain the importance of each hyperparameter, the search space must be as broad as possible to avoid misleading results because the search space did not include the optimal solution. The complete list of hyperparameters analyzed and their ranges are presented in Table 2. All hyperparameters were analyzed individually except for two cases: the number of layers and layer sizes, as these are usually related, and their relationship is not linear (a network with double the layers with half the size each is not equivalent to the original network) [11], and the loss function and activation layer for the classification datasets, as otherwise the neural network would not work. Furthermore, the values available for each hyperparameter were chosen based on availability. If it was a possible option, then we used it, even in cases that might not make sense. Although it might seem an unfair approach to evaluate the importance, if the hyperparameter search is done by someone with no background on DNNs, the search would follow a similar approach.

Table 2. Hyperparameter values considered.

Hyperparameter	Values
General	
Activation function (hidden layers)	relu, sigmoid, softmax, softplus, softsign, tanh, selu, elu
Batch Size	64, 128, 256, 512, 1024, 2048, 4096
Learning rate	0.00001, 0.0001, 0.001, 0.01, 0.1
Optimizer	Adam, SGD, RMSprop, Adadelta, Adagrad, Adamax, Nadam, Ftrl
Hidden layer size	64, 128, 256, 512, 1024
Nº hidden layers	4, 5, 6, 7, 8
Classification	
Loss function	Binary crossentropy, Sparse categorical crossentropy, Categorical crossentropy
Activation function (last layer)	Softmax, Sigmoid
Regression	
Loss function	Mean squared error, Mean absolute error, Mean absolute percentage error, Mean squared logarithmic error, Cosine similarity

3.4 Performance Metrics

Considering that only some DNNs will be run on a high-end machine and that only some users have a powerful cluster to train the DNNs, taking into account only the model's accuracy as the metric on which we base the importance of the hyperparameter leaves out important information. To that extent, we considered

Mathews Correlation Coefficient (MCC) and Mean Squared Error (MSE) as model accuracy metrics for the classification and regression tasks and the model training and inference time as metrics to regard when aiming to develop models on constrained scenarios.

3.5 Individual Hyperparameter Testing

To evaluate the individual hyperparameter importance, the baseline model was changed one hyperparameter at a time and trained in a cross-validation setting of five over the complete training dataset during 200 epochs with early stopping with ten epochs of patience. The final result was the average score of the performance metrics over the five runs.

4 Results

This section presents the results of the individual hyperparameter importance testing averaged over all the datasets and by dataset type. The fANOVA [10] metric was considered to evaluate the hyperparameter importance. This metric performs a functional analysis of variance by dividing the observed variance of the algorithm's performance (accuracy, training time, or prediction time) into various components (the hyperparameters or their combination). The analysis of the hyperparameter importance for each dataset and the model performance will not be addressed due to page limitations, although the results obtained can be consulted in the Appendix A or in the project GitHub (See Footnote 1).

4.1 General Importance

Aggregating the importance of each hyperparameter for every dataset allows us to find the most impactful hyperparameters that should be prioritized when improving the model. The results from the aggregation are presented in Table 3, where the activation functions were the most critical hyperparameter for model accuracy and batch size was the most important for training and prediction time. Besides these, the loss functions impacted the accuracy. Similarly, the optimizer was also relevant in the accuracy, although it also affected the training time. The learning rate presented significant results in accuracy and the number of hidden layers in the prediction time.

4.2 Importance by Dataset Type

Ideally, the results are generalizable to every dense DNN. However, depending on the dataset and its characteristics, different hyperparameters might be more important than others. One characteristic that might change the importance of the hyperparameters is the type of task it solves. Table 4 presents the results of the hyperparameter importance grouped by task. From this table, it is clear that, indeed, the relevance of most hyperparameters varies. However, batch size

Table 3. Hyperparameter general importance. (Bold indicates the best result)

Hyperparameter	Importance (%)		
	Accuracy	Training time	Prediction time
Activation function (hidden layers)	**18.42**	3.20	6.99
Batch Size	0.95	**55.94**	**37.67**
Loss function	12.23	0.33	2.1
Optimizer	14.88	5.17	2.16
Learning rate	17.65	3.38	1.34
Nº hidden layers	3.94	3.85	16.62
Hidden layer size	3.94	3.61	6.29

continues to be the hyperparameter with the most impact on the training and prediction time, and the importance of the activation function in accuracy is very similar. The hyperparameter whose relevance varied the most was the loss function in accuracy, where the results changed from 19.51% to 9.16%, followed by the optimizer, whose impact on the accuracy in the classification tasks is much more significant.

5 Discussion

Although some results were expected, such as the relevance of the optimizer or learning rate in the accuracy or the batch size in the training time, others, like the poor significance of the Nº of hidden layers and the hidden layer sizes in the accuracy and training time, were surprising. To understand why some results were against expectations, this section discusses the results presented in the previous section.

The most striking result was, as just mentioned, the importance of the hidden layer hyperparameters for accuracy and training time. By intuition, one would think that the model size would significantly affect performance. However, adding more layers will not be very useful if a smaller model can already solve the problem presented. That said, factors like the learning rate or optimizer that can lead an excellent model to perform extremely badly end up having a more significant impact. The same goes for the activation functions in the hidden layers, as a bad combination can make the model perform poorly. For example, using the exact same model, but instead of the 'relu' activation function, the 'linear' one is used, then the model losses all its non-linear capability, and in a non-linear problem, it will perform poorly. Regarding the training time, the results are most likely a consequence of the early stopping, as the deeper DNNs likely use fewer epochs to train, making a trade-off between the number of epochs and time per epoch. When the models are compared directly in the prediction time, the number of layers starts to matter, as each model analyzes the same amount of data, and fewer layers are faster to compute.

Another interesting result was the difference between the accuracy's importance of the loss function in the regression and classification. This can be justified

Table 4. Hyperparameter importance by task. (Bold indicates the best result)

Regression

Hyperparameter	Importance (%)		
	Accuracy	Training time	Prediction time
Activation function (hidden layers)	**23.66**	2.87	15.98
Batch Size	4.49	**64.87**	**37.22**
Loss function	19.51	0.12	0.01
Optimizer	7.4	8.33	0.12
Learning rate	18.09	1.38	3.26
Nº hidden layers	2.1	2.2	12.22
Hidden layer size	3.32	1.48	4.18

Classification

Hyperparameter	Importance (%)		
	Accuracy	Training time	Prediction time
Activation function (hidden layers)	17.59	2.91	2.28
Batch Size	1.31	**57.3**	**37.43**
Loss function	9.16	0.01	3.76
Optimizer	17.11	3.78	4.53
Learning rate	**21.4**	4.69	0.01
Nº hidden layers	6.13	0.67	19.37
Hidden layer size	3.04	5.2	8.54

by looking in detail at the individual results. In classification, the three loss function options all perform similarly, while in regression, one of the loss functions performs poorly in all the datasets, creating a stronger relationship between the performance and the loss function used.

With the unexpected results analyzed, one question remains: "What hyperparameters to optimize?". The answer to this question is highly constrained by the budget available for optimization. The key idea is to optimize as many hyperparameters as the budget allows. However, some decisions must be made when on a limited budget. First, one must define the optimization objectives. If the aim is to improve the model results, optimizing the learning rate and optimizer are two promising approaches. If the objective is to improve training time, then experimenting with increasing batch size is the best approach. Similarly, for improvements in the prediction time, having a bigger batch size can be helpful. However, that is not always possible, so experimenting with a reduced number of hidden layers or layers with smaller sizes can help.

Nevertheless, although the results do not show a meaningful reason to experiment with the model architecture, if the baseline model architecture used is not as solid as the ones presented, optimizing it can significantly improve the performance and training times.

6   Conclusion

Hyperparameter optimization is crucial to any ML pipeline and is usually the most time-consuming part. If the optimized model is a DNN, hyperparameter optimization can take even longer. This paper analyzed individual hyperparameter importance for DNNs in tabular datasets to understand which hyperparameters could be discarded from optimization to reduce the time spent on it.

Six tabular datasets with different characteristics and baseline models were considered during the experiments, and seven hyperparameters were analyzed. From them, the activation function presented the most impact on the model accuracy, followed by the loss function, optimizer, and learning rate. Although the regression datasets highly influence the results of the loss function. The batch size was the most significative hyperparameter for the training and prediction time, followed by the optimizer in the training time and the number of hidden layers in the prediction time.

These results show that for accuracy improvement, one should focus on analyzing the optimizer, learning rate, and activation functions. On the other hand, if the training and prediction times need refinement, then increasing the batch size, experimenting with other optimizers, or reducing the number of layers can significantly improve results.

Nonetheless, there are still more hyperparameters to evaluate, especially related to the model architecture. In future work, we intend to increase the number of hyperparameters considered and analyze their relationship on a set of datasets with different characteristics (images, time series, etc.).

Acknowledgements. This work is supported by the European Union/Next Generation EU, through Programa de Recuperação e Resiliência (PRR) [Project Nr. 29: Route 25].

A   Hyperparameter Importance per Dataset

See Table 5.

Table 5. Hyperparameter importance by dataset. (Bold indicates the best result)

Hyperparameter	Importance (%)		
	Accuracy	Training time	Prediction time
Abalone			
Activation function (hidden layers)	14.77	1.39	4.39
Batch Size	0.55	**56.72**	**21.61**
Loss function	0.0	1.62	0.0
Optimizer	2.96	7.99	3.5
Learning rate	**30.02**	6.9	0.07

(continued)

Table 5. (*continued*)

Hyperparameter	Importance (%)		
	Accuracy	Training time	Prediction time
Nº hidden layers	7.16	0.12	15.69
Hidden layer size	11.55	4.35	11.04
Bike Sharing			
Activation function (hidden layers)	**51.26**	0.59	24.54
Batch Size	0.74	**72.21**	**29.71**
Loss function	0.06	0.0	0.0
Optimizer	17.86	6.28	0.02
Learning rate	11.6	5.17	7.14
Nº hidden layers	0.0	1.98	14.41
Hidden layer size	2.62	1.16	0.82
Compass			
Activation function (hidden layers)	3.4	0.4	0.08
Batch Size	1.16	**43.0**	6.23
Loss function	**33.98**	0.19	0.0
Optimizer	21.68	4.02	4.16
Learning rate	9.59	6.06	0.02
Nº hidden layers	0.76	2.92	**49.31**
Hidden layer size	3.61	7.49	20.06
Covertype			
Activation function (hidden layers)	**29.22**	12.77	4.01
Batch Size	0.77	**56.92**	**41.6**
Loss function	0.06	0.0	10.34
Optimizer	8.29	1.65	4.67
Learning rate	23.64	0.32	0.17
Nº hidden layers	13.27	0.2	3.32
Hidden layer size	1.84	4.79	0.62
Delays Zurich			
Activation function (hidden layers)	0.37	3.57	5.2
Batch Size	0.0	**58.2**	**57.82**
Loss function	**39.27**	0.0	0.01
Optimizer	14.39	2.42	0.0
Learning rate	0.18	0.58	0.5
Nº hidden layers	2.37	10.18	12.22
Hidden layer size	3.81	0.48	3.92
Higgs			
Activation function (hidden layers)	11.51	0.49	3.73
Batch Size	2.46	**48.6**	**69.07**
Loss function	0.01	0.14	2.25
Optimizer	24.08	8.67	0.63
Learning rate	**30.84**	1.22	0.15
Nº hidden layers	0.09	7.68	4.75
Hidden layer size	0.18	3.39	1.25

B   Baseline Models Architecture

See Figs. 2, 3, 4, 5, 6 and 7.

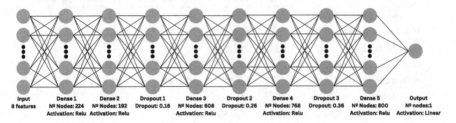

Fig. 2. Baseline model used for the Abalone dataset.

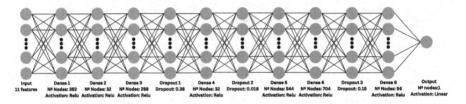

Fig. 3. Baseline model used for the Bike Sharing dataset.

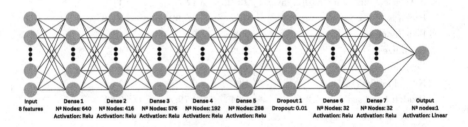

Fig. 4. Baseline model used for the Delays Zurich dataset.

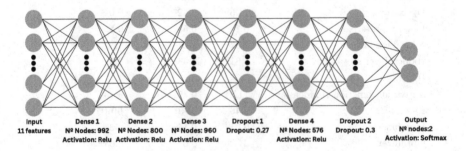

Fig. 5. Baseline model used for the Compass dataset.

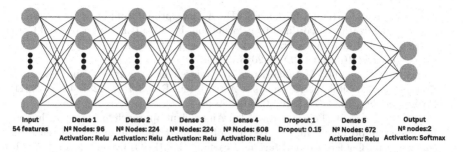

Fig. 6. Baseline model used for the Covertype dataset.

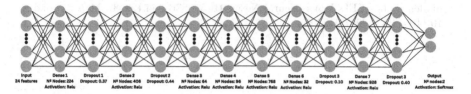

Fig. 7. Baseline model used for the Higgs dataset.

References

1. White, C., et al.: Neural architecture search: insights from 1000 papers. ArXiv arXiv:2301.08727 (2023)
2. Bergstra, J., Bardenet, R., Bengio, Y., Kégl, B.: Algorithms for hyper-parameter optimization. In: Proceedings of the 24th International Conference on Neural Information Processing Systems, NIPS 2011, pp. 2546–2554. Curran Associates Inc., Red Hook (2011)
3. Sun, Y., Gong, H., Li, Y., Zhang, D.: Hyperparameter importance analysis based on N-RReliefF algorithm. Int. J. Comput. Commun. Control **14**, 557–573 (2019). https://doi.org/10.15837/ijccc.2019.4.3593
4. Yang, L., Shami, A.: On hyperparameter optimization of machine learning algorithms: theory and practice. Neurocomputing **415**, 295–316 (2020). https://doi.org/10.1016/j.neucom.2020.07.061. https://www.sciencedirect.com/science/article/pii/S0925231220311693
5. Grinsztajn, L., Oyallon, E., Varoquaux, G.: Why do tree-based models still outperform deep learning on typical tabular data? In: Koyejo, S., Mohamed, S., Agarwal, A., Belgrave, D., Cho, K., Oh, A. (eds.) Advances in Neural Information Processing Systems, vol. 35, pp. 507–520. Curran Associates, Inc. (2022). https://proceedings.neurips.cc/paper_files/paper/2022/file/0378c7692da36807bdec87ab043cdadc-Paper-Datasets_and_Benchmarks.pdf
6. Andonie, R.: Hyperparameter optimization in learning systems. J. Membr. Comput. **1**(4), 279–291 (2019). https://doi.org/10.1007/s41965-019-00023-0
7. Injadat, M., Salo, F., Nassif, A.B., Essex, A., Shami, A.: Bayesian optimization with machine learning algorithms towards anomaly detection. In: 2018 IEEE Global Communications Conference (GLOBECOM), pp. 1–6 (2018). https://doi.org/10.1109/GLOCOM.2018.8647714

8. Zhang, S., Xu, J., Huang, E., Chen, C.H.: A new optimal sampling rule for multi-fidelity optimization via ordinal transformation. In: 2016 IEEE International Conference on Automation Science and Engineering (CASE), pp. 670–674 (2016). https://doi.org/10.1109/COASE.2016.7743467
9. Ezugwu, A.E., et al.: Metaheuristics: a comprehensive overview and classification along with bibliometric analysis. Artif. Intell. Rev. **54**(6), 4237–4316 (2021). https://doi.org/10.1007/s10462-020-09952-0
10. Hutter, F., Hoos, H., Leyton-Brown, K.: An efficient approach for assessing hyperparameter importance. In: Proceedings of the 31st International Conference on International Conference on Machine Learning, ICML 2014, vol. 32, pp. I-754–I-762. JMLR.org (2014)
11. Trenn, S.: Multilayer perceptrons: approximation order and necessary number of hidden units. IEEE Trans. Neural Netw. **19**(5), 836–844 (2008). https://doi.org/10.1109/TNN.2007.912306

Exploring the Reduction of Configuration Spaces of Workflows

Fernando Freitas[1]([⊠])(iD), Pavel Brazdil[2,3](iD), and Carlos Soares[1,4,5](iD)

[1] Faculdade de Engenharia, Universidade do Porto, Porto, Portugal
fmlfreitas@gmail.com, csoares@fe.up.pt
[2] INESCTEC, Porto, Portugal
pbrazdil@inesctec.pt
[3] Faculdade de Economia, Universidade do Porto, Porto, Portugal
[4] Fraunhofer AICOS Portugal, Porto, Portugal
[5] Laboratory for Artificial Intelligence and Computer Science (LIACC),
Porto, Portugal

Abstract. Many current AutoML platforms include a very large space of alternatives (the *configuration space*) that make it difficult to identify the best alternative for a given dataset. In this paper we explore a method that can reduce a large configuration space to a significantly smaller one and so help to reduce the search time for the potentially best workflow. We empirically validate the method on a set of workflows that include four ML algorithms (SVM, RF, LogR and LD) with different sets of hyperparameters. Our results show that it is possible to reduce the given space by more than one order of magnitude, from a few thousands to tens of workflows, while the risk that the best workflow is eliminated is nearly zero. The system after reduction is about one order of magnitude faster than the original one, but still maintains the same predictive accuracy and loss.

Keywords: Configuration spaces · Portfolios of workflows · Reduction of complexity

1 Introduction

One of the common machine learning problems, and more specifically in the areas of metalearning and AutoML, is elaborating a workflow for a specific task. The aim is to come up with a workflow that has the potentially best performance (measured, for instance, by predictive accuracy). Many platforms exist nowadays that facilitate the task of constructing workflows. The AutoML systems can explore the space of alternatives, often referred to as a *configuration space* to come up with good workflows [8,9,15]. One problem with this approach is that the configuration space may include alternatives that are not useful for any task. This has the consequence that the system may spend a long time searching for the right alternative. The methods that can reduce a large configuration space to a significantly smaller one help to reduce the search time for the potentially best workflow. It is important, however, that the potentially best workflows are not eliminated in this process. If this happened, this would affect the performance on new tasks. In the past, various authors have examined the usefulness of different algorithms (workflows) in a given portfolio, while taking into account a given set of tasks [2]. The previous methods had, however, various shortcomings, which were corrected in this paper. Also, in this paper the configuration space explored is much larger than in the previous work.

The rest of this paper is organized as follows. Section 2 describes some related work in this area. Section 3 starts by discussing different reductions methods for the given portfolio of workflows. This is followed by a description of the system used to search through the given set of workflow for the potentially best workflow. The final subsection describes the experiments carried out and the results. Section 4 presents a discussion, future work and conclusions.

2   Relation to Other Work

The problem of determining which hyperparameters or algorithms are important has been studied by various researchers in the past. In this section we reviews some of these approaches.

Establishing Hyperparameter Importance. Ablation analysis [7], for instance, requires that the best possible hyperparameter setting for a given algorithm is determined first. Then all hyperparameters are considered, one at a time, and for each one, the optimal setting is substituted by the default value and the effect on performance is recorded. Functional ANOVA [12,14] determines how much each hyperparameter (or their combination) contributes to the variance of the performance. Many of studies were carried out in a post-hoc manner, i.e., determine which settings led to the best performance. This knowledge cannot, however, be used directly by the recommendation system to determine the most promising workflows.

Marginal Contribution of Algorithms/Workflows. The aim of so-called *marginal contribution* of algorithms/workflows is to determine how much the performance of an existing portfolio of algorithms/workflows can be improved by adding a new algorithm/workflow to it [17]. A more general notion is the notion of a *Shapley value* that determines a marginal contribution of a algorithm/workflow with respect to a given portfolio or any of its subsets [10].

Learning Multiple Defaults for ML Algorithms. Pfisterer et al. [13] have investigated the problem of defining a set of defaults for some ML algorithms. These represent discrete choices that can be reused in future problems. The authors have shown that this has advantages over the random search method applied to the full space of alternatives. The work assumes the existence of pre-defined number of alternative defaults (the maximum is 6). Our work has shown that for some algorithms we may need more settings than others. Also, this method does not deal with situations when several ML algorithms are used at the same time.

Reduction of Portfolio of Algorithms. The method proposed by Abdulrahman et al. [2] (see also [5]) involves two phases, similarly as the method proposed in this paper. In the first one, the aim is to identify the most competitive algorithms for each dataset used in the past.

The second phase used a so-called *covering method* whose aim was to associate one workflows with each dataset. The assumption was that one workflow would be sufficient. Our analysis showed that this assumption was wrong, as this strategy could omit the potentially best workflow, and consequently end up with a rather significant loss. We use a similar approach here, but the second phase is different. The aim is to admit more workflows for each dataset, but at the same time try to reduce their number by eliminating redundant variants. Our experimental results show that the final loss of the new method is either zero or rather negligible. The work of [11] was concerned with the task of reducing the given workflows that included outlier elimination method (OEM) step in preprocessing. The first step was similar to the method presented here, but the second step was different. The aim was to eliminate all workflows which include rather infrequent OEM variants (i.e., those that appear in less than P% workflows). This strategy enabled the authors to identify three most important OEM methods out of the initial set of twelve methods.

3   Reducing the Configuration Space of Workflows

3.1   Variants of the Reduction Method Considered

The configuration space, represented by the initial portfolio of workflows, can be very large. Pruning the search space can be useful to accelerate such search. Our method is somewhat similar to [2] but it overcomes some of its shortcomings. Pruning eliminates the workflows with low performance (step PL) and in addition also the workflows potentially redundant (step PR). Both steps are described in more detail below.

Eliminating Workflows with Low Performance. The aim of the pruning step is to eliminate (prune out) workflows with low performance. This is done by considering the set of existing set of workflows and identifying, for each dataset, the top P% (e.g., 5%) of workflows based on the chosen performance measure. These workflows can be regarded as *specialists* for the particular dataset in question. All other workflows are pruned out.

Regarding the performance measure used in this process, one possibility is to use the normalized accuracy discussed further on. However, if the aim is to include also fast, but not necessarily the most accurate workflows, it is possible to also include the top P% of workflows using A3R measure that combines accuracy and time [1,4]. Our preliminary experiments provided evidence that this is a good option and hence we have decided to use it in all the experiments reported further on in the paper.

After the *PruneLow (PL)* step has been completed, the system returns a list of datasets, in which each element is accompanied by top performing workflows. So each element in this list is of the form $(D_i, \overrightarrow{W_j})$, where D_i represents a dataset and $\overrightarrow{W_j}$ the list of top performing workflows.

Identifying the Generally Useful Workflows. In general, the most useful workflows are those that are are top performers in many datasets. So the list discussed above is rearranged by identifying, for each workflow, the top performers. Each element in this rearranged list is of the form $(W_i, \overrightarrow{D_j})$, where W_i represents a workflow and $\overrightarrow{D_j}$ the vector of datasets in which this workflow achieved the top performance.

In the following we will sometimes use the phrase *"W_i covers datasets $\overrightarrow{D_j}$"* as a shorthand for *"W_i is a top performer in datasets $\overrightarrow{D_j}$"*. In order to give preference to generally useful workflows, the list of pairs is reordered according to the size of the set of datasets covered.

Eliminating Redundant Workflows Using a Cover Test. The goal of *eliminating redundant workflows*, is achieved by constructing a list of non-redundant workflows *WS* in a gradual fashion starting with an empty list. This method is referred to as *PL.PR.C*.

Let us see the method in more detail. The pairs $(W_i, \overrightarrow{D_j})$ are processed sequentially, one by one. Each workflow *Wi* is included in *WS* only if it extends the coverage of *WS*. If this situation is verified, it is taken as an indication that the new workflow *Wi* is non-redundant and so it is added to *WS*, and *DS* is updated to include all the datasets covered. If the workflow *Wi* did not extend the coverage, it is assumed that it is potentially redundant, and no updates are made. This test is of course not totally reliable, and so we will discuss another alternative further on.

Let us consider an example. Suppose at some point of processing we have *WS* = {*W1*} and *DS* = {*d1, d2, d3*} and workflow *W2* is encountered whose coverage is *Cov(W2)* = {*d2, d3*}. As the datasets associated with *W2* do not extend the coverage of *DS*, this workflow is ignored. Suppose that, in the next step, we encounter *W3* and *Cov(W3)* = {*d2, d4*}. As this workflow would introduce a new element to *DS*, the following updates are made: *WS* = {*W1, W3*} and *DS* = {*d1, d2, d3, d4*}. The relationship between *WS* and *DS* can be captured by *Cov(WS)* = *DS*.

Using an Additional Accuracy Test. As we have mentioned before, the cover method is not entirely reliable. The problem can occur when the new workflow does not extend the current coverage of *WS*. The cover method would not add this workflow to *WS* under the assumption that it is potentially redundant. However, the new workflow may not, in fact, be redundant. This can occur, when the new workflow and one of the existing workflows exhibit different performance on some dataset. This problem can be avoided by an additional step which tests whether the average performance values on all datasets are virtually the same within a given tolerance limit ϵ. This method is referred to as *PL.PR.A*.

Scheduling Reduction Using Batch/Incremental Mode. In practice, the given workflows may be separated out into different subsets. For instance, one subset may include workflows that include SVM, another Logistic Regression (LogR) etc. There are basically two ways we can proceed when dealing with a set of workflows that includes various subsets. One can be regarded as a *batch mode*, where all the possible workflows are joined into one large set and the reduction is carried out on this set. This can be captured by the function *PL.PR.Cb* (see Algorithm 1). The symbol *WS* represents the initial set of workflows, and *WR* the reduced set.

Algorithm 1. Function PL.PR.Cb

1: $WS' \leftarrow \cup_{i=1}^{n} WS$
2: $WR \leftarrow PL.PR.C(WS')$
3: end

The other possibility is to do this in an incremental way. In each iteration, the additional (not yet reduced) workflows are added to the already reduced ones, and the combined set is reduced further. This can be captured by function PL.PR.Ci (see Algorithm 2).

Algorithm 2. Function PL.PR.Ci

1: $WR \leftarrow \{\}$
2: **for all** $Wi \in WS$ **do**
3:     $WR' \leftarrow WR \cup PL.PR.C(Wi)$
4:     $WR \leftarrow PL.PR.C(WR')$
5: **end for**
6: end

The incremental mode is particularly useful when we are dealing with workflows that include different classification algorithms with their hyperparameters and settings. The incremental scheme permits to reduce each subsets of workflows to a smaller one before joining it to the other workflows. All three methods discussed earlier can be applied either in batch or incremental mode.

Concluding Remarks. Eliminating workflows is potentially beneficial, as we end up with a simpler configuration space which facilitates search for the truly best workflows. Eliminating workflows has however a disadvantage that the truly best workflows can be wrongly eliminated affecting loss on new tasks. So, a question arises which of these variants should be used in practical settings. The experiments presented further on shed light on this issue.

3.2   Using a Given Configuration of Workflows for Recommendation

In this section we discuss how the recommendation are obtained, and also describe the evaluation methodology.

Using the Average Ranking Method for Recommendation. The given set of workflows (sometimes called a portfolio of workflows) is normally used by a given metalearning system to generate recommendations for the target dataset. Here we use a system based on the average ranking method (AR) to provide recommendations [1,4]. We have opted to use this method because of its simplicity. It uses metadata in the form of test results of a given set of algorithms on a given set of datasets.

The method calculates an average rank for each algorithm based on a given performance measure (accuracy, A3R etc.). The ranks are used to construct the average ranking. The average ranking would normally be followed on the new target dataset: first, the algorithm with rank 1 is evaluated and used to initialize the so-called current best algorithm (sometimes called *the incumbent*). Then the algorithm in rank 2 is evaluated and if its performance is better than that of the current best algorithm, it is used as the new current best algorithm. This process is repeated for all algorithms in the ranking, or until a given termination condition has been achieved. The average ranking can thus be regarded to as the recommended ranking for the target dataset.

Normalization of Performance Values. As accuracy values are normally not comparable across datasets (e.g., 90% accuracy can be a good result in one dataset but bad in other), it is common to adopt measures to avoid this problem. Here we re-scale all values obtained into an interval spanning between 0 to 1. The lowest possible accuracy for a given dataset is the default accuracy equivalent to accuracy obtained by selecting the majority class.

$$Acc_{norm} = (Acc - Acc_{def})/(1 - Acc_{def}) \qquad (1)$$

where Acc represents the measured accuracy, Acc_{def} the default accuracy and Acc_{norm} the normalized accuracy. If the default accuracy was, for instance, 0.5%, the measured accuracy 0.9%, then the normalized value would be 0.8%.

Evaluation Methodology. We are interested to evaluate the effects of reducing portfolios of workflows on a target dataset. This can be done by comparing the loss at corresponding time points for two recommendations, where one is based on a full (non-reduced) portfolio and the other one on the reduced one. We can of course do this for different pairs of time points and this way obtain two loss curves. Ideally, the loss curve relative to the reduced portfolio should exhibit a lower loss or be within ϵ distance of the reference curve, relative to the original (non-reduced) portfolio.

To obtain a non-biased result, we take care that the target dataset is different from the datasets used to obtain the metadata used by the AR method. So, to do this, we use a leave-one-out (LOO) approach in conjunction with N datasets. In each LOO cycle one dataset is selected as the target dataset and the remaining N − 1 datasets are used to obtain the test results (metadata) that is used both by the reduction method and the recommendation system. So, this process results in N pairs of loss curves. As it is difficult to analyze all these curves before carrying out the analysis, we follow the usual strategy and elaborate a curve representing a mean at each time point.

The curve can be characterized by various measures. One is *mean interval loss (MIL)* representing effectively the area under the curve. Another measure is the loss at some time point that is considered as the extreme time point (e.g., at 10^6 s). The aim is that this loss is as low as possible, ideally zero, indicating that the potentially best workflow was identified.

3.3   Experiments

Experimental Setup. The experimental setup included workflows that consist of various ML algorithms shown in Table 1. All algorithms were retrieved from scikit-learn python library. All hyperparameter settings followed a given grid determined before running all experiments. So, for instance, there were 300 workflows that included SVM algorithm, each one relative to particular hyperparameter setting. More details about the settings used for each algorithm are given in several tables listed in the last column of Table 1. As for the hyperparameter settings of SVM, for instance, we see that we should consult Table 2. All experiments were carried on 41 datasets, representing a subset of 72 datasets of benchmarking suite OpenML-CC18 [16]. The list of these datasets is shown in the Appendix.

Table 1. Metadatabase of workflows

Classif. Alg.	Description	#Workflows	Details
SVM	SVM	300	Table 2
LogR	Logistic Regression	220	Table 3
LD	Linear Discriminant	90	Table 4
RF	Random Forest	1080	Table 5

Table 2. Metadatabase of 300 workflows that include SVM

#WFs	kernel	c	gamma	degree
168	rbf	$[0.01–1000]$ 12 settings	$[1x10^6 − 1]$ 14 settings	-
120	poly	- idem -	-	$[0–9]$ 10 settings
12	linear	- idem -	-	-

Table 3. Metadatabase of 220 workflows that include LogR

#WFs	(solver, penalty)	c
40	[(lbfgs, l2), (lbfgs, none)]	$[0.0001–10000]$ 20 settings logspace
40	[(liblinear, l1), (liblinear, l2)]	- idem -
40	[(newton-cg, l2), (newton-cg, none)]	- idem -
40	[(sag, l2), (sag, none)]	- idem -
60	[(saga, l1), (saga, l2), (saga, none)]	- idem -

Experiments. In this section we report on the experiments carried out with all variants of the reduction methods discussed earlier in Sect. 3. The methods are listed in the first column of Table 6. The first line in this table refers to the initial situation before applying any of the reduction methods. For instance, the number of initial workflows that includes SVM is 300.

Various subsequent columns show the sizes of the subsets of workflows that include a particular ML algorithm before and after reduction. The methods are ordered starting with the least aggressive reduction methods. The most aggressive reduction method is *PL.PR.Ci* that is shown in the last line. We note that from the initial 300 workflows for SVM only 24 were retained after reduction with this method.

The total number of workflows for all four ML algorithms is shown in column *"Size total"*. We can see that the initial number of 1690 workflows was reduced to 45 by the most aggressive method, representing only 2.7% of the original workflows (see column *"Size %"*).

Column *"MIL %"* characterizes the corresponding loss curve, as it shows the so-called *mean interval loss* in the interval between 1 and 10^6 s. We note that all reduction methods reduce this value. Low MIL value indicates that good workflows were identified early. The method that reduces the number of workflows most is *PL.PR.Ci*. However, the methods *PL.PR.Cb* and *PL.PR.Ai* are not far off. The benefit of using these three methods can be seen also from Fig. 1 that shows the mean loss curves discussed before (see Evaluation Methodology). The loss curves of the three methods are far below the loss curve of the initial portfolio with 1690 workflows.

Column *"Loss 10^6 %"* shows the mean loss at that time point. Ideally, this loss should be 0, as this would indicate that the potentially best workflow was identified. A non-zero value indicates that the reduction method was *too aggres-*

Table 4. Metadatabase of 90 workflows that include LD

#WFs	solver	tol	shrinkage
8	svd	0.0001, 0.0005, 0.001, 0.005 0.01, 0.05, 0.1, 0.5	-
41	lsqr	-	[0.01–0.975] step 0.025
41	eigen	-	- idem -

Table 5. Metadatabase of 1080 workflows that include Random Forest (RF)

#WFs	bootstrap	max depth	max features	min sample	min sample split	num estimators
1080	[True, False]	[10, 20, 50, 70, 100, None]	[auto, sqrt]	[1, 2, 4]	[2, 5, 10]	[200, 400, 800 1000, 2000]

sive and eliminated incorrectly some workflow that would be useful if kept. We note that the final loss is 0 for *PL*, *PL.PR.Ab* and *PL.PR.Cb* and very small (a fraction of %) for the other the reduction methods (*PL.PR.Ai* and (*PL.PR.Ci*).

Applying Reduction Iteratively. It is interesting to observe how the iterative reduction methods proceed, as more workflows specific to a given ML algorithms are added to the current reduced set. Figure 2 relative to the method *PL.PR.Ci*. The figure shows mean values across all folds of LOO procedure; the numbers differ somewhat from one fold to another of the LOO cycles. We note that, for instance, the 300 initial workflows of SVM with different hyperparameters and their settings were reduced to 59 workflows, when using just SVM alone. In this step of iterative reduction the SVM workflows were competing against other SVM workflows. This competition is not as stringent as the one when other ML algorithms are considered. When LogR is added, this number was reduced further. The process of reduction continued when the other ML algorithms were added. The final number of SVM workflows was 24. This means that the more stringent competition with other ML algorithms eliminated further 35 workflows (59–24). Similar observations can be made about all the other ML algorithms used in this work. Similarly, Fig. 3 shows the progress relative to the method *PL.PR.Ai*. This method shows a similar trend as the previous method.

Qualitative Analysis of Reduced Workflows. We have examined the workflows that remained after the reduction to obtain better understanding of the composition of the reduced sets and the choices made by the reduction methods. We observed that, in general, the reduced sets include different values of the discrete space of the each hyperparameter spread sparsely over the original grid.

As an example, we present a brief analysis of one of the subsets of the 45 workflows obtained by the PL.PR.Ci reduction method (see Table 6) namely the SVM subset of 24 workflows, representing a significant proportion of the total. These workflows consist of 14 RBF kernel configurations with 6 different values

Table 6. Comparison of different reduction methods

Reduction method	Size SVM	Size LogR	Size LD	Size RF	Size total	Size %	MIL %	Loss 10^6 %
No reduction	300	220	90	1080	1690	100.0	6.57	0.00
PL	247	202	70	729	1248	73.8	6.30	0.00
PL.PR.Ab	216	81	37	137	471	27.9	3.26	0.00
PL.PR.Ai	124	41	17	88	270	16.0	2.57	0.02
PL.PR.Cb	25	1	3	25	54	3.1	2.99	0.00
PL.PR.Ci	24	2	4	15	45	2.7	2.40	0.01

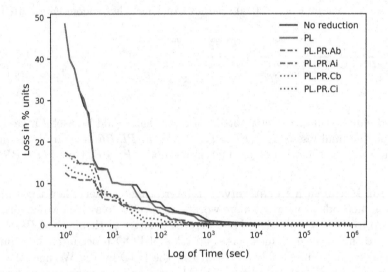

Fig. 1. Loss curves for different reduction methods

of hyperparameter C. As the original grid includes 12 settings, we see that half of these was included in reduced subset. Similar observations can be made about the hyperparameter gamma and other kernels types, namely poly and linear kernel, with 8 and 2 configurations respectively. These results confirm that SVM is a rather difficult algorithm to use, as it requires that different settings are considered on a new task. Still, the original number of 300 was reduced to a much smaller number (24) by the proposed approach.

Ablation Analysis. The given workflows can be regarded as a composition of several subgroups, each corresponding to the use of a particular classification algorithm (e.g., SVM). So we are interested to examine how the particular group contributes towards the overall performance. Ablation analysis enables us to do just that [7]. The aim is to determine how much the performance degrades when this group is eliminated. This is done by first eliminating a subset of workflows from the initial set and then running the chosen reduction method. Here we consider two measures that characterize the overall performance. The first one is the *mean interval loss, MIL*, and the second one, *loss at* 10^6.

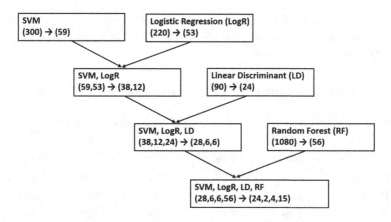

Fig. 2. Applying the iterative reduction method *PL.PR.Ci*

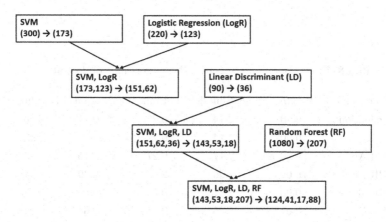

Fig. 3. Applying the iterative reduction method *PL.PR.Ai*

Table 7 shows the effects of eliminating the subgroups that include the chosen classification algorithms on MIL. The value of MIL for the full set of workflows is shown in the first column (column *"Elim.None"*). The second column (column *"Elim.SVM"*) shows how much the value of MIL increases when the corresponding workflows are dropped. For instance, for the reduction method *PL.PR.Ci* the MIL increases to 8.46% when all workflows with SVM are dropped. This is the largest increase when considering different ML algorithms. The ML algorithms can be ordered according to the contribution they have on MIL. The ordering is *(SVM, RF, LogR, LD)*.

Table 8 shows the effects of eliminating certain classification algorithms on the loss at 10^6. This table is organized in a similar way as the previous one but instead of MIL is shows the final loss. The values in this table confirm the results on the relative importance of workflows pertaining to given ML algorithms. The workflows that include SVM appear to be most important, as they affect the final loss most. It is possible to order again the ML algorithms according to the effects on the final loss. We note that the ordering is the same as the one based on MIL that was shown before.

Table 7. Effects of eliminating given ML algorithms on mean interval loss (MIL)

Reduction method	Elim. None	Elim. SVM	Elim. LogR	Elim. LD	Elim. RF
No reduction	6.57	7.99	6.72	6.65	3.83
PL	6.30	7.78	6.50	6.36	3.83
PL.PR.Ab	3.26	7.37	3.24	3.35	3.80
PL.PR.Ai	2.57	8.68	4.87	4.90	5.54
PL.PR.Cb	2.99	7.17	3.15	3.08	3.88
PL.PR.Ci	2.40	8.46	5.14	5.13	5.86

Table 8. Effects of eliminating given ML algorithms on loss at 10^6 s

Reduction method	Elim. None	Elim. SVM	Elim. LogR	Elim. LD	Elim. RF
No reduction	0.00	2.14	0.26	0.11	1.07
PL	0.00	2.14	0.26	0.12	1.07
PL.PR.Ab	0.00	2.17	0.27	0.15	1.07
PL.PR.Ai	0.02	4.96	2.53	2.39	3.50
PL.PR.Cb	0.00	2.41	0.58	0.50	1.75
PL.PR.Ci	0.01	5.25	2.78	2.64	3.86

4   Discussion, Future Work and Conclusions

4.1   Discussion and Future Work

In this work we have used four ML algorithms hyperparameters whose settings followed a fixed grid. Some people argue that the usage of fixed grid has disadvantages, as it can miss good settings that may exist between the settings on the grid [3]. This is true particularly when the grid is rather sparse. A fine grid has disadvantages too, as it may require many tests to be carried out.

Various testing strategies could be employed for this task, including random testing, or more intelligent ones based on the work in AutoML and multiarmed bandits. These strategies could suggest tests of configurations that do not coincide with the grid used by the reduction approach. This problem could be resolved by training a surrogate model [6], and using this model to estimate the performance values on our fixed grid. We have carried a preliminary study that shows the potential of this approach.

The work presented could be extended to include not only other ML algorithms with the corresponding hyperparameters and their setting, but also by including certain preprocessing operations, such as, feature selection. We have initiated this work and will report the results in future publications.

4.2   Conclusions

We have presented a method that can reduce a large configuration space to a significantly smaller one and consequently help to reduce the search time for the potentially best workflow. Unlike many previous approaches, the method deals with different types of workflows at the same time. The workflows used include four different ML algorithms and the corresponding hyperparameters with their settings. Our results show that it is possible to reduce the given space by more than one order of magnitude, from a few thousands to tens of workflows, while the risk that the best workflow is eliminated is nearly zero.

We have presented several variants of the basic approach. Some carry out the reduction in a kind-of batch mode, others in an incremental mode. The methods that employ a simple test (cover test) to detect redundant workflows lead to fewer workflows that the methods that use more elaborate test (both the cover and accuracy test) for this task.

Our results show that if we consider each ML algorithm alone with its possible hyperparameters and settings, the proposed method can identify a subset that can resolve the given set of tasks as well as the full set. The size of the final subset differs from algorithm to algorithm. For instance, Fig. 2 shows that the set of final SVM workflows is larger (24) than the final set of workflows with LogR (2). This means that if the number of workflows associated with a particular algorithm were fixed to some specific relatively small value (e.g. 5 as in [13]), the potentially best workflow could be omitted and consequently, we could end up with rather high loss.

It is interesting to note that if we use more than one ML algorithm in conjunction with their hyperparameters in a given portfolio, reduce each one and then join the reduced subsets, it is possible to apply reduction again. The workflows in different subsets compete with one another for the place in the final set and only the most competitive workflows "survive" so to speak.

The results of ablation analysis can be used to determine the relative importance of different ML algorithms in a given portfolio. As we have shown, the algorithms ordered by their importance are SVM, RF, LogR and LD.

The proposed method could also be used to reorder the given hyperparameters according to their importance. As we have mentioned before (in Sect. 2) various studies exist whose aim was to determine the importance of hyperparameters of a given ML algorithm [14]. This work does not show, however, how to convert these results into a useful set of workflows. The proposed approach discussed in this paper has the advantage that it does just that.

Acknowledgements. This work is financed by National Funds through the Portuguese funding agency, FCT - Fundação para a Ciência e a Tecnologia, within project UIDB/50014/2020. The authors of this paper wish to thank the anonymous referees for their useful comments that helped us to improve the paper.

Appendix

List of 41 datasets, represented by *OpenML-Dataset-Name (OpenML-Dataset-ID)*, used in the experiments. This set is subset of 72 datasets of the benchmarking suite OpenML-CC18 (https://docs.openml.org/benchmark/#openml-cc18):

kr-vs-kp (3), letter (6), balance-scale (11), mfeat-factors (12),
mfeat-fourier (14), mfeat-karhunen (16), cmc (23), optdigits (28),
pendigits (32), diabetes (37), splice (46), tic-tac-toe (50),
vehicle (54), electricity (151), satimage (182), vowel (307),
isolet (300), analcatdata_authorship (458), analcatdata_dmft (469),
Bioresponse (4134), wdbc (1510), phoneme (1489), qsar-biodeg (1494),
wall-robot-navigation (1497), semeion (1501), ilpd (1480), madelon (1485),
ozone-level-8hr (1487), cnae-9 (1468), PhishingWebsites (4534),
GesturePhaseSegmentationProcessed (4538), har (1478), texture (40499),
climate-model-simulation-crashes (40994), wilt (40983), car (40975),
segment (40984), mfeat-pixel (40979), Internet-Advertisements (40978),
dna (40670), churn (40701).

References

1. Abdulrahman, S.M., Brazdil, P., van Rijn, J.N., Vanschoren, J.: Speeding up algorithm selection using average ranking and active testing by introducing runtime. Mach. Learn. **107**, 79–108 (2017). https://doi.org/10.1007/s10994-017-5687-8. Special Issue on Metalearning and Algorithm Selection
2. Abdulrahman, S.M., Brazdil, P., Zinon, M., Adamu, A.: Simplifying the algorithm selection using reduction of rankings of classification algorithms. In: ICSCA 2019 Proceedings of the 8th International Conference on Software and Computer Applications, Malaysia, pp. 140–148. ACM, New York (2019)
3. Bergstra, J., Bengio, Y.: Random search for hyper-parameter optimization. J. Mach. Learn. Res. **13**, 281–305 (2012)
4. Brazdil, P., van Rijn, J., Soares, C., Vanschoren, J.: Metalearning approaches for algorithm selection I (exploiting rankings). In: Brazdil, P., van Rijn, J., Soares, C., Vanschoren, J. (eds.) Metalearning: Applications to Automated Machine Learning and Data Mining, pp. 19–37. Springer, Cham (2022). https://doi.org/10.1007/978-3-030-67024-5_2
5. Brazdil, P., van Rijn, J., Soares, C., Vanschoren, J.: Setting-up configuration spaces and experiments. In: Brazdil, P., van Rijn, J., Soares, C., Vanschoren, J. (eds.) Metalearning: Applications to Automated Machine Learning and Data Mining, pp. 143–168. Springer, Cham (2022). https://doi.org/10.1007/978-3-030-67024-5_8

6. Eggensperger, K., et al.: Towards an empirical foundation for assessing Bayesian optimization of hyperparameters. In: NIPS Workshop on Bayesian Optimization in Theory and Practice, pp. 1–5 (2013)
7. Fawcett, C., Hoos, H.: Analysing differences between algorithm configurations through ablation. J. Heuristics **22**(4), 431–458 (2016). https://doi.org/10.1007/s10732-014-9275-9
8. Feurer, M., Klein, A., Eggensperger, K., Springenberg, J., Blum, M., Hutter, F.: Efficient and robust automated machine learning. In: Advances in Neural Information Processing Systems, pp. 2962–2970 (2015)
9. Feurer, M., Klein, A., Eggensperger, K., Springenberg, J.T., Blum, M., Hutter, F.: Auto-sklearn: efficient and robust automated machine learning. In: Hutter, F., Kotthoff, L., Vanschoren, J. (eds.) Automated Machine Learning. TSSCML, pp. 113–134. Springer, Cham (2019). https://doi.org/10.1007/978-3-030-05318-5_6
10. Fréchette, A., Kotthoff, L., Rahwan, T., Hoos, H., Leyton-Brown, K., Michalak, T.: Using the Shapley value to analyze algorithm portfolios. In: 30th AAAI Conference on Artificial Intelligence (2016)
11. Hetlerovič, D., Popelínský, L., Brazdil, P., Soares, C., Freitas, F.: On usefulness of outlier elimination in classification tasks. In: International Symposium on Intelligent Data Analysis, pp. 143–156 (2022)
12. Hutter, F., Hoos, H., Leyton-Brown, K.: An efficient approach for assessing hyperparameter importance. In: Proceedings of the 31st International Conference on Machine Learning, ICML 2014, pp. 754–762 (2014)
13. Pfisterer, F., van Rijn, J., Probst, P., Müller, A., Bischl, B.: Learning multiple defaults for machine learning algorithms. In: Proceedings of the Genetic and Evolutionary Computation Conference Companion, pp. 241–242 (2021)
14. van Rijn, J.N., Hutter, F.: Hyperparameter importance across datasets. In: KDD 2018: The 24th ACM SIGKDD International Conference on Knowledge Discovery & Data Mining. ACM (2018)
15. Thornton, C., Hutter, F., Hoos, H.H., Leyton-Brown, K.: Auto-WEKA: combined selection and hyperparameter optimization of classification algorithms. In: Proceedings of the 19th ACM SIGKDD International Conference on Knowledge Discovery and Data Mining, pp. 847–855. ACM (2013)
16. Vanschoren, J., van Rijn, J.N., Bischl, B., Torgo, L.: OpenML: networked science in machine learning. ACM SIGKDD Explor. Newsl. **15**(2), 49–60 (2013)
17. Xu, L., Hutter, F., Hoos, H., Leyton-Brown, K.: Evaluating component solver contributions to portfolio-based algorithm selectors. In: Cimatti, A., Sebastiani, R. (eds.) SAT 2012. LNCS, vol. 7317, pp. 228–241. Springer, Heidelberg (2012). https://doi.org/10.1007/978-3-642-31612-8_18

iSOUP-SymRF: Symbolic Feature Ranking with Random Forests in Online Multi-target Regression

Aljaž Osojnik[1]([✉]), Panče Panov[1,2], and Sašo Džeroski[1,2]

[1] Jožef Stefan Institute, Ljubljana, Slovenia
{aljaz.osojnik,pance.panov,saso.dzeroski}@ijs.si
[2] Department of Knowledge Technologies, Jožef Stefan International Postgraduate School, Jamova 39, 1000 Ljubljana, Slovenia

Abstract. The task of feature ranking has received considerable attention across various prediction tasks in the batch learning scenario, but not in the online learning setting. Available methods that estimate feature importances on data streams have thus far focused on ranking the features for the tasks of classification and occasionally multi-label classification. We propose a novel online feature ranking method for online multi-target regression, iSOUP-SymRF, which estimates feature importance scores based on the positions at which a feature appears in the trees of a random forest of iSOUP-Trees. By utilizing iSOUP-Trees, which can address multiple structured output prediction tasks on data streams, iSOUP-SymRF promises feature ranking across a variety of online structured output prediction tasks. We examine the robustness of iSOUP-SymRF and the feature rankings it produces in terms of the methods' parameters: the size of the ensemble and the number of selected features. Furthermore, to show the utility of iSOUP-SymRF and its rankings we use them in conjunction with two state-of-the-art online multi-target regression methods, iSOUP-Tree and AMRules, and analyze the impact of adding features according to the rankings.

Keywords: online learning · feature ranking · multi-target regression

1 Introduction

Predictive modelling tasks are often addressed in both the batch and online learning settings, as predictive models immediately provide a highly desirable ability to predict some values on new unseen examples, potentially bypassing the need to perform timely and/or costly measurements. Less often considered, particularly in the online learning setting, are the related feature ranking tasks. In feature ranking for a predictive modelling task, such as classification, regression or more complex tasks of structured output prediction, we wish to determine which of the descriptive variables, i.e., *features*, are most important to the predictive modelling task at hand. By including only the most informative features,

© The Author(s) 2023
A. Bifet et al. (Eds.): DS 2023, LNAI 14276, pp. 48–63, 2023.
https://doi.org/10.1007/978-3-031-45275-8_4

we can reduce the need for computational resources as well as eliminate the need and cost of measuring less informative features.

A feature ranking method is thus closely related to the underlying predictive task. In predictive modelling, a model is learned using incoming examples to best predict one or more target values, i.e., to generalize the dependence between descriptive and target values. In feature ranking, however, the model that is learned is tasked with ranking the descriptive features in terms of their importance for accurate prediction in the underlying predictive modelling task. Ideally, features that have a higher impact on the prediction should be ranked higher than those with lesser impact. Thus a feature ranking method is a learning procedure that produces a ranking based on the available data examples; in the online learning setting, this process is continuous and the ranking can change with time as potential drift occurs.

Formally, a feature ranking is a list of all features ordered according to their informativeness (importance for predictive modelling), i.e., starting with the most informative feature and ending with the least informative. However, methods for feature ranking often produce a more informative result, where each feature is assigned a numeric score estimating its importance. A ranking can be trivially obtained by sorting the features according to their scores.

In the batch learning setting, a plethora of methods for feature ranking are available across a variety of predictive modelling task, both simple, like classification and regression, as well as structured prediction tasks, such as multi-label classification, multi-target regression, etc. In the online learning setting, however, fewer methods for feature ranking exist, all of which exclusively focus on simple predictive modelling tasks.

A common approach in structured output prediction tasks is to decompose the problem into multiple simple (single-target) sub-problems, e.g., multiple binary classification sub-problems in multi-label classification or multiple single-target regression sub-problems in multi-target regression. Each sub-problem is then addressed using a simple single-target predictor and the predictions of all of these models are the used to solve the original structured problem.

Methods that address the structured problem in its entirety have been shown to have various advantages over this *local* decomposition approach, but in feature ranking they provide an additional benefit. Applying the local approach to feature ranking would yield a ranking per each of the targets. While these could be combined using various aggregation approaches, e.g., averaging, this introduces a non-trivial facet into the feature ranking procedure. Thus, we focus on approaches that consider the complex task as a monolith, i.e., without attempting to decompose it into smaller sub-problems.

In this paper we introduce the *symbolic random forests with iSOUP-Trees* (iSOUP-SymRF) feature ranking method, which utilizes the structure of a random forest of trees to produce the ranks of the observed features. While initially targeted at feature ranking for online multi-target regression, due to the versatility of the base iSOUP-Tree [15] method, the method we propose can be easily extended to other online structured output prediction tasks, such as online multi-label classification [14] or hierarchical multi-target regression [16], as well as to other learning contexts, such as semi-supervised learning [17]. To the best of our knowledge, this paper is the first effort towards online feature ranking for any structured output prediction task.

The rest of this paper is structured as follows. Section 2 presents relevant related work and Sect. 3 introduces the symbolic random forest approach for online feature ranking. Section 4 continues by describing the experimental setup that we use to evaluate the proposed method, while Sect. 5 presents the experiments' results. Finally, Sect. 6 concludes the paper with a summary of the findings and presents avenues for further work.

2   Related Work

In the batch learning setting, there is a variety of feature ranking methods for the classification and regression tasks [26]. Methods for structured output prediction tasks, such as multi-target regression [21] are rarer. However, for the related task of feature selection, where a set of features needs to be selected, not necessarily ranked, several methods for multi-label classification are available [19].

Feature ranking is not addressed often as a standalone task in the online setting. It commonly encompassed under the name of *feature weighing* as part of a method for classification that weighs the input features [3,4,8,23,28]. Perkins et al. [20] introduced the grafting method that combines multiple types of regularization to estimate the importances of features and uses a logistic function of the binomial negative loss function to calculate the probabilities of the class presence. More recently, Razmjoo et al. [22] rank features based on a sensitivity analysis of the performance of a classifier under a potential feature removal.

Several methods that do address online feature ranking have been proposed. Katakis et al. [11,12] introduce a feature-based classifier that uses a system for incremental feature selection (IFS) and explore how IFS impacts the predictive performance of simple online classification methods, such as, e.g., naïve Bayes. Another method that specifically addresses online feature ranking is I-RELIEF [27], which stands for iterative RELIEF, and is an adaptation of the Relief [13] method for batch feature ranking to the online learning setting. Both of these methods operate in the online predictive modeling scenario.

On the other hand, Yoon et al. [29] introduced a method for online feature selection that is unsupervised, i.e., it is not directly tied to a predictive modelling scenario. Their method utilizes the CLeVer method for principal component analysis. Recently, Duarte et al. [5] introduced methods for online feature ranking, designed specifically for methods that use the Hoeffding inequality and used them with AMRules [6], while Karax et al. [10] address the feature ranking for online classification by exploiting heuristic information of decision trees.

Other examples of online feature ranking come from related fields, such computer vision [2] and online image retrieval [9].

3   Symbolic Feature Ranking with Random Forests

In this paper, we adapt the symbolic approach to feature ranking with random forests to the online learning setting. This method was first introduced in the batch learning setting [21]. Petković et al. introduce several feature ranking methods for multi-target regression based on tree ensembles in the batch learning setting. In addition to the symbolic ranking with random forests, the authors introduce the Genie3 ranking method, which calculates the feature importance score based on the heuristic scores produced by the split nodes in the ensemble members, as well as the random forest score feature ranking method, which calculates the scores of the features by looking at the out-of-bag errors and feature value permutations. Note that the Genie3 method employs a method of scoring as similar to that of Karax et al. [10].

Of these three approaches, only the symbolic[1] random forest ranking method is directly applicable to the online learning scenario. To calculate the Genie3 feature importance scores, we need access to the splitting heuristic scores, which are easily accessed in the batch scenario. In online learning, the heuristic score of a split is calculated only on a small sample of the data, and is only partially indicative of the feature importance scores on the entire dataset. The random forest scoring method permutes the values of out-of-bag examples for each tree and observes how the error changes from the original, unpermuted example. This requires the permutation of many example values, after which many predictions must be calculated to estimate the error. While this approach could technically be applied to online learning, it would incur high consumption of computational resources, particularly in terms of processing time.

Symbolic random forest feature ranking with iSOUP-Trees (**iSOUP-SymRF**), however, calculates the feature importance scores using only the structure of trees which are the members of the ensemble. As we are targeting the task of feature ranking for online multi-target regression, we use iSOUP-Trees [15] as a base ensemble model. iSOUP-Tree is a state-of-the-art online learning method that has been applied to a variety of online structured output prediction tasks in addition to multi-target regression, such as multi-label classification [14] and hierarchical multi-target regression [16], thus extending the possible coverage of iSOUP-SymRF to feature ranking for these predictive modelling tasks as well.

Ensemble Construction. As in random forests utilized in batch learning, the main idea is to induce an ensemble of diverse randomized trees. Tree randomization is achieved in two ways, the first of which is example sampling as commonly used in online bagging [18], where each member of the ensemble is updated using a given example for a random number of times drawn from the Poisson distribution. The second way in which the trees are randomized is the selection of feature subspaces in each node when growing the tree. In particular, whenever a

---

[1] *Symbolic* refers to the fact that the method relies only on the qualitative structure of the trees in the random forest and not on other quantitative measures that the remaining proposed methods are based on.

new leaf node is constructed, i.e., at the beginning of the learning procedure or when a leaf node is split into two new leaf nodes, a subset of the input features is randomly selected. The new leaf node only considers those input features for ranking split candidates; thus, statistics are recorded only for the selected input features.

Feature Score Calculation. iSOUP-SymRF is based on the following observation: if a feature's values are important for accurate prediction, the feature will get selected in splits often. More accurately, it will get selected often *when it can be*, as it is not always considered for candidate splits due to the random forest learning process. If the base ensemble was constructed using a regular online bagging approach, the variety in the models would be considerably lesser, as the trees would always be able to select the best feature(s). This would over-focus the scores to only the top features. In random forest tree construction, the best features are sometimes left out of the candidate feature pool, thus allowing the estimation of the importance scores of the remaining (less important) features.

To estimate the importance of a feature we make two observations: (a) the more often a feature appears in the split nodes in the random forest, the more important it is, and (b) the closer to the root of the tree the feature appears, the higher its importance. The first observation follows the reasoning, that, despite the random selection, a feature appearing more often means that splits along this feature increase the predictive performance of the tree, according to standard tree-learning methodology. The second observation is based on the fact that split nodes closer to the tree root will affect more examples than those positioned further from the root. For example, a split at the root node will affect all incoming examples, while a split in one of its children will (on average) only affect half of the examples, i.e., the ones for which the root split node directed them toward this particular child and not the other.

Quantitatively, to calculate a feature importance score of a feature A, we first calculate the feature importance score of A for a given ensemble member T, which is defined as

$$I(A, T) = \sum_{\mathcal{N} \in T(A)} w^{\text{depth}(\mathcal{N})},$$

where w is a predefined weight and $T(A)$ is the set of all split nodes of tree T which have splits on feature A. The total feature importance of feature A is then

$$I(A, E) = \frac{1}{|E|} \sum_{T \in E} I(A, T) = \frac{1}{|E|} \sum_{T \in E} \sum_{\mathcal{N} \in T(A)} w^{\text{depth}(\mathcal{N})},$$

where E is the ensemble of trees. We adapt this method to online learning, as the calculation of the scores is quick, since it requires only the traversal of each tree in the ensemble.

As the scores iSOUP-SymRF calculates are exclusively dependent on the structure of the trees of the random forest, no predictions are needed, which significantly reduces the operational time of the method. Notably, the process of

Fig. 1. Sample trees motivating the selection of the weight parameter w.

calculating the feature ranking is fairly quick and can be executed at any time during the learning process, so the ranking is always available.

What remains is the choice of the weight factor w. When considering its possible values, we note that $w < 1$ gives higher scores to features which appear closer to the root, and, consequently, affect the larger parts of the input space. To settle on a particular value of w, we observe the following example. Consider a leaf in which two best features A_1 and A_2 have the exact same heuristic score. In the first case, we split on the first feature and likewise in the second case, we split on the second feature. Afterwards, in the first case we split both leaves on A_2, and in the second case on A_1 (see Figs. 1a and 1b, respectively).

In both cases, all example traversal paths include splits on A_1 and on A_2. This implies that A_1 and A_2 should have equal importance scores, as they affect the same sets of examples. Under this assumption it follows that

$$I(A_1, T_1) = I(A_1, T_2)$$
$$w^d = w^{d+1} + w^{d+1}$$
$$1 = 2w$$
$$0.5 = w,$$

where d is the depth of the initial twice-split leaf. Hence, we choose $w = 0.5$.

In choosing $w = 0.5$, we note that the total contribution of any level in the tree will equal to 1 and, consequently, the total of all scores of a tree will be about equal to its average depth. Thus, the total scores of a tree (and the random forest) will increase over time as the trees grows.

Parameters. As is standard practice with random forest methods, we define two method parameters for iSOUP-SymRF, ensemble size and subspace size. The first determines the number of base models included in the ensemble, while the second determines how many features are considered as split candidates in each leaf. Ensemble size commonly ranges between 10 and 100, while we use two selection methods for subspace size, either randomly selecting $1 + \lceil \log N \rceil$ or $1 + \lceil \sqrt{N} \rceil$ features in each leaf, where N is the number of all features.

Learning Context. iSOUP-SymRF notably does not have an explicit change detection and adaptation mechanism. While trees have a small innate change adaptation ability, by just growing additional nodes that adhere to the new

Table 1. Datasets used for online feature ranking for multi-target regression.

Dataset	No. of examples	No. and type of features	Targets
Bicycles	17379	12 numeric	3
SCM1d	9803	280 numeric	16
SCM20d	8966	61 numeric	16

concept, this is likely not enough to capture the drift in a reasonable time frame. Thus, in the context of this paper we consider learning (and experiments) in the static context, i.e., we assume no concept drift in the data stream.

4   Experimental Setup

4.1   Datasets

In the interest of brevity, we have selected three multi-target regression datasets, based on their size, primarily looking for diversity in the number of input features. A summary of the datasets and their properties is shown in Table 1, while brief descriptions of the datasets are provided below.

The *Bicycles* dataset is concerned with the prediction of demand for rental bicycles on an hour-by-hour basis [7]. The three targets represent the number of casual (non-registered) users, the number of registered users and the total number of users for a given hour, respectively.

The *SCM1d* and *SCM20d* are datasets derived form the Trading Agent Competition in Supply Chain Management (TAC SCM) conducted in July 2010 [25]. The data examples correspond to daily updates in a tournament – there are 220 days in each game and 18 games per tournament. The 16 targets are the predictions of the next day and the 20 day mean price for each of the 16 products in the simulation, for the SCM1d and SCM20d datasets, respectively.

The Bicycles dataset is available at the UCI Machine Learning Repository[2], the SCM1d and SCM20d datasets are available at the Mulan multi-target regression dataset repository[3].

Even though iSOUP-SymRF is not equipped with explicit change detection and adaptation mechanisms, these datasets do contain drift. As this provides an additional challenge to properly estimating the feature importances, the obtained results are pessimistic estimates of the method's performance in the static context we presuppose in our experiments. Performance on static data streams (or more realistically, in periods without drift) would thus possibly be improved over what is shown in the results of our experiments.

---

[2] URL: https://archive.ics.uci.edu/ml/datasets/Bike+Sharing+Dataset   (accessed 2018/01/22).

[3] URL: http://mulan.sourceforge.net/datasets-mtr.html, (accessed 2018/01/22).

4.2   Experiment: Parameter Stability

In this experiment, we explore which parameter settings produce good rank-ings, in particular, we are interested in parameter configurations that produce stable rankings that "converge" fairly quickly. Notably, this is not a concern in the batch learning scenario, where the learned ranking is static. We perform a small-scale grid-search on the two parameters, ensemble and subspace size. We consider ensembles of sizes 10, 20, 50 and 100, as well as subspaces of logarithmic and square root size. In terms of resource consumption, smaller ensemble and subspace sizes are naturally preferable, thus, the smallest parameter configura-tion (10 models with logarithmic subspaces) will take the baseline role in this experiment.

To evaluate the rate of convergence, we define the *time to final ranking* (TTF), i.e., how many examples it takes for the ranking to reach the same order as the final ranking. Thus, lower values of TTF are more desirable. Notably, TTF considers the ranking only in terms of the feature ranks, ignoring the finer detail of the scores themselves. Furthermore, TTF is not particularly bad at evaluating rankings on data streams that exhibit drift, as the ranking is likely to fluctuate any time a drift occurs. As TTF considers all features equally (according to their rank), it can have large values due to changes in the tail end of the feature ranking. As we are generally more interested in the top ranked features, we also define TTF_n which only considers the top n features. We will then observe TTF_5 to determine how quickly the top five features settle.

Notably, these measures only make sense in a static context. Were drift to occur, the actual importances would possibly get rearranged and the estimated importances would again take time to converge. In the drifting scenario, obser-vation of these two measures would only make sense between drift points.

4.3   Experiment: Ranking Utility

To estimate the utility of the rankings obtained by iSOUP-SymRF, we use *for-ward feature addition* (FFA) [24]. Forward feature addition is performed by observing the performance of a predictive model, while adding features from best to worst. In particular, we first observe the performance of a model learned only using the best ranked feature, then using the two best features, then the three best features, etc.

In the batch learning scenario, observing the performances of the models in these scenarios is fairly simple, as the performance of a model can be easily expressed as a single number, e.g., root mean squared error over a test set. In the online learning setting, this kind of generalization is less informative. Thus, in this paper we examine the progression of the performance (in terms of error) of the observed models.

Furthermore, even though iSOUP-SymRF can produce feature rankings at any point in the learning process (rankings can change throughout the process), for FFA, we need to consider only one static ranking[4]. While this is not ideal,

---

[4] In the batch scenario, this is not an issue as only one ranking is ever obtained.

demonstrating the utility of the rankings is notoriously difficult and using an imperfect method to directly show the impact on the learning process still provides considerable insight into the applicability of the proposed method. To this end, we take the final feature ranking obtained by iSOUP-SymRF, i.e., the ranking after all of the examples have been processed. This biases the results towards optimistic, as in a practical use scenario this information would not be available during learning.

In our case, we select two methods for online multi-target regression to evaluate the rankings obtained by iSOUP-SymRF. In particular, we use a single iSOUP-Tree [15] and AMRules [6], with default learning parameters. As iSOUP-Tree is the method that iSOUP-SymRF is based on, it is natural to expect that this combination will yield better results than when combining iSOUP-SymRF with AMRules. To estimate the error of these models we use the *average relative mean absolute error* ($\overline{\mathrm{RMAE}}$ or aRMAE) [15]:

$$\overline{\mathrm{RMAE}} = \frac{1}{M} \sum_{j=1}^{M} \mathrm{RMAE}_j$$

where M is the number of targets and RMAE_j is the *relative mean absolute error* of the j-th target, defined as

$$\mathrm{RMAE}_j = \frac{\sum_{i=1}^{n} |y_i^j - \hat{y}_i^j|}{\sum_{i=1}^{n} |y_i^j - \bar{y}^j(i)|}$$

where y_i^j and $\hat{y}_i^j$ are the values of target j for data example i, real and predicted by the evaluated model, respectively, while $\bar{y}^j(i)$ is the average of the seen values of the j-th target so far.

As some datasets have many features, we limit ourselves to reporting FFA plots only for the 5 top and 5 bottom features. Using a good ranking, adding the top features in FFA should considerably increase model performance, while adding bottom features should barely affect performance (or even worsen it).

5   Results

5.1   Parameter Stability

The results of the parameter stability experiment are presented in Table 2. The winning result in terms of ensemble size with regards to each of TTF and TTF$_5$ are presented in bold text, while the winning results in terms of subspace size are underlined (a dotted underline indicates a tie).

In terms of TTF, a square root size of the random space is generally preferred, while in terms TTF$_5$ no particular generalization can be made, as logarithmic and square root subspace size win 6 of the contests each. For the Bicycles dataset, we observe that the TTF$_5$ values are generally considerably lower than the TTF values, indicating that the top features converge faster, while the ranks of the

Table 2. The results in terms of TTF and TTF_5.

Size	TTF		TTF_5	
	Log	Sqrt	Log	Sqrt
Bicycles				
10	16835	<u>12504</u>	<u>9809</u>	12504
20	14437	<u>12495</u>	12173	<u>7620</u>
50	16951	**11109**	<u>2600</u>	**3019**
100	**12703**	15866	**2167**	4133
SCM1d				
10	**9670**	<u>9460</u>	<u>8770</u>	9151
20	9799	**9550**	**8171**	<u>7684</u>
50	9787	<u>9773</u>	9741	<u>8342</u>
100	9803	<u>9788</u>	9655	<u>9355</u>
SCM20d				
10	**8911**	**8846**	8676	**7731**
20	8962	<u>8951</u>	<u>7989</u>	8025
50	8964	8964	8771	<u>8735</u>
100	<u>8934</u>	8956	**7970**	8373

remaining features continue to get perturbed. Even though the Bicycles dataset exhibits strong seasonal effects (i.e., drift), the top ranked features appear not to change between seasons, as they stabilize early through the data.

On the other hand, for the SCM1d and SCM20d datasets the TTF_5 are much closer to the TTF values. This indicates that the rankings are turbulent through the entirety of these datasets. Any obtained final rankings are thus possibly not close to converging to the actual underlying feature importances.

The results are even less clear regarding the preferred ensemble size. Larger sizes of the ensemble produce better results on the Bicycles dataset, while smaller ensembles perform better on the SCM1d and SCM20d datasets, with the exception of TTF_5 for 100 trees with a logarithmic subspace size.

A key factor that may influence these results is the total number of features. Regarding the feature set size, in the Bicycles dataset, which is smaller with a total of 12 features, any feature is likelier to get selected as a feature candidate even though the subspace size is relatively large compared to the number of all features (4 vs 12 features for logarithmic, and 5 vs 12 for square root size). This makes it easier to identify key features in this case, resulting in low TTF_5 values. On the other end, SCM1d has 280 features, and the subspaces are of sizes 7 and 18 for logarithmic and square root, respectively. While such low sizes are desirable in the context of random forests to reduce resource consumption, they make it more difficult to identify top features using a random forest based feature ranking method such as iSOUP-SymRF.

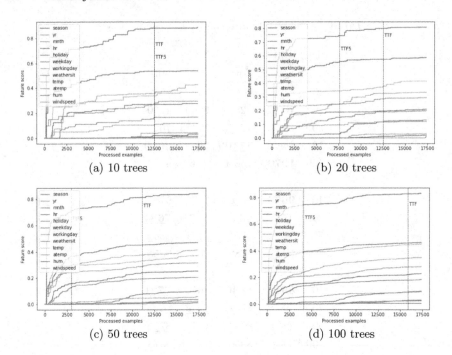

(a) 10 trees   (b) 20 trees

(c) 50 trees   (d) 100 trees

Fig. 2. Feature score progression on the Bicycles dataset in terms of ensemble size. All plots show square root subspace size.

In this context, we can also examine the actual scores and their progression through the learning process. With only 12 features, we can observe the feature scores of the Bicycles dataset directly, as seen in Fig. 2. In addition to the score progression, the plots also show the TTF and TTF$_5$ points. Over all ensemble sizes iSOUP-SymRF identifies 'hr', 'hum', 'workingday', 'temp', 'atemp', 'weekday' and 'weathersit' as the top features, though there is some disagreement about their final ranks. Here, we can see the decreased disambiguation power of smaller ensemble sizes, particularly, in the case of 10 and 20 trees. In these cases, the scores indicate the best features, but their ordering takes longer to establish. Larger ensembles, on the other hand, fairly quickly establish the order, which (for the most informative features) then remains stable. Notably, the choice of the top observed features, i.e., the 5 in TTF$_5$, also impacts the results. Clearly, choosing a lower number lowers the TTF value, but in some cases we could also have observed a larger number of top features and lost little confidence, i.e., TTF$_7$ would be the same as TTF$_5$ in the case of 100 trees.

Ultimately, this experiment does not unilaterally indicate which parameter choices to make. While square root, i.e., larger, subspace sizes seem to be preferred, no clear statement can be made about ensemble size. Thus, motivated by the Bicycles dataset example above, we choose square root subspace size with ensemble size of 100 for our further experiments.

5.2 Ranking Utility

The results using the FFA methodology are presented in Fig. 3. The plots are interpreted in the following way: lines labeled with positive numbers $n \in \{1, 2, 3, 4, 5\}$ depict the $\overline{\text{RMAE}}$ of models trained on the top n features; lines labeled with negative numbers $n \in \{-5, -4, -3, -2, -1\}$ depict the $\overline{\text{RMAE}}$ of models trained with all but the last $|n|$ features.

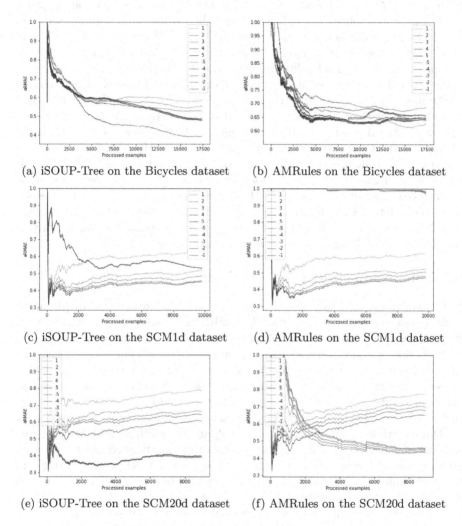

(a) iSOUP-Tree on the Bicycles dataset   (b) AMRules on the Bicycles dataset

(c) iSOUP-Tree on the SCM1d dataset   (d) AMRules on the SCM1d dataset

(e) iSOUP-Tree on the SCM20d dataset   (f) AMRules on the SCM20d dataset

Fig. 3. $\overline{\text{RMAE}}$ of iSOUP-Tree and AMRules using FFA.

In terms of the top features, we wish to see the largest increase when adding higher ranked features, e.g., the increase in performance should be larger when we add the second best feature than the third best. All iSOUP-Tree models exhibit this behaviour, as do the AMRules models, with the exception of the Bicycles dataset. This exception is most likely to AMRules change detection and adaptation mechanism which significantly modifies the model during learning.

Conversely, regarding bottom ranked features, the desired effect is either trivial improvement or even decrease in performance. This is shown in all datasets and methods. In the case of the iSOUP-Tree models, for example, we can see that, both on the Bicycles and SCM1d datasets, adding features toward the bottom actually hurts the overall performance of the model, i.e., all negative labeled lines are above (have higher errors) the highest positively labeled line. On the SCM20d dataset, the addition of the bottom features only has marginal effect. These results are mirrored in quality for the AMRules models, though the effect sizes are considerably different. On the Bicycles dataset, adding the lowest ranked features is either detrimental or has little to no effect on the error of the model, while on the SCM1d and SCM20d datasets we observe the same behaviour as with iSOUP-Tree, except that the decrease in the performance on the SCM1d dataset is significantly higher, where as slight increases on the SCM20d dataset are observed as compared to those in the iSOUP-Tree models.

Naturally, using a tree-based feature ranking method such as iSOUP-SymRF provides good results when used alongside another tree-based method such as iSOUP-Tree. However, iSOUP-SymRF also shows encouraging (if worse) results using the different learning framework in AMRules. Notably, these experiments also indicate that some of the features included in these datasets can be actively detrimental when learning with these two learning methods.

6   Conclusions and Further Work

In this paper, we have introduced a novel method for feature ranking for online multi-target regression called iSOUP-SymRF. It utilizes a random forest of iSOUP-Trees to determine the feature importance scores (and consequently the feature ranking), based on the features' appearance in the split nodes of the trees in the of the forest. We have conducted experiments on a collection of multi-target regression datasets, aiming to (a) determine the methods stability against the values of its parameters and (b) show that the obtained rankings have some utility for increasing the predictive performance of online-multi target regressors.

Our experiments first focused on determining the methods stability over various parameter values. While the experiments were not fully conclusive, we suggest the use of larger subspace sizes (such as square root), while the random forest ensemble size should be further analyzed. The experiments that seek to show the utility in using feature rankings obtained with iSOUP-SymRF show promising results using two different learning methods for online multi-target regression, iSOUP-Tree and AMRules. As expected, the results were better for the related tree-based iSOUP-Tree method, though the feature ranking obtained by iSOUP-SymRF still provided ample utility in terms of predictive performance improvement when used in combination with AMRules.

We identify three key avenues for further work: the main effort will be to equip the iSOUP-SymRF with a change detection and adaptation mechanism, e.g., ADWIN [1]. This will allow us to more accurately capture the evolution of the feature rankings, especially in the presence of concept drift. Another avenue is the adaptation of iSOUP-SymRF to feature ranking for other online structured output prediction tasks, such as online multi-label classification and/or hierarchical multi-target regression, by utilizing the broad coverage of the base iSOUP-Tree method [14,16]. This would also allow a comparison of a wider variety of learners, as methods for online multi-label classification are quite plentiful. Finally, we wish to explore and improve the experimental setup, focusing on better and more concise approaches for the evaluation of feature rankings in the online learning setting.

Acknowledgements. This work was supported by grants funded by the Slovenian Research Agency (P2-0103, J2-2505) and by the European Commission (H2020 TAILOR 952215).

References

1. Bifet, A., Gavaldà, R.: Adaptive learning from evolving data streams. In: Adams, N.M., Robardet, C., Siebes, A., Boulicaut, J.-F. (eds.) IDA 2009. LNCS, vol. 5772, pp. 249–260. Springer, Heidelberg (2009). https://doi.org/10.1007/978-3-642-03915-7_22

2. Collins, R.T., Liu, Y., Leordeanu, M.: Online selection of discriminative tracking features. IEEE Trans. Pattern Anal. Mach. Intell. **27**(10), 1631–1643 (2005)

3. Crammer, K., Dredze, M., Pereira, F.: Confidence-weighted linear classification for text categorization. J. Mach. Learn. Res. **13**, 1891–1926 (2012)

4. Dekel, O., Shamir, O., Xiao, L.: Learning to classify with missing and corrupted features. Mach. Learn. **81**(2), 149–178 (2010)

5. Duarte, J., Gama, J.: Feature ranking in Hoeffding algorithms for regression. In: Proceedings of the Symposium on Applied Computing (SAC 2017), pp. 836–841. ACM (2017)

6. Duarte, J., Gama, J., Bifet, A.: Adaptive model rules from high-speed data streams. ACM Trans. Knowl. Discov. Data (TKDD) **10**(3), 30 (2016)

7. Fanaee-T, H., Gama, J.: Event labeling combining ensemble detectors and background knowledge. Prog. Artif. Intell. **2**(2–3), 113–127 (2013). https://doi.org/10.1007/s13748-013-0040-3

8. Goodman, J., Yih, S.W.: Online discriminative spam filter training. In: Proceedings of the 3rd Conference on Email and Anti-Spam (CAES 2006). CAES (2006)

9. Jiang, W., Er, G., Dai, Q., Gu, J.: Similarity-based online feature selection in content-based image retrieval. IEEE Trans. Image Process. **15**(3), 702–712 (2006)

10. Karax, J.A.P., Malucelli, A., Barddal, J.P.: Decision tree-based feature ranking in concept drifting data streams. In: Proceedings of the 34th ACM/SIGAPP Symposium on Applied Computing, pp. 590–592 (2019)

11. Katakis, I., Tsoumakas, G., Vlahavas, I.: On the utility of incremental feature selection for the classification of textual data streams. In: Bozanis, P., Houstis, E.N. (eds.) PCI 2005. LNCS, vol. 3746, pp. 338–348. Springer, Heidelberg (2005). https://doi.org/10.1007/11573036_32

12. Katakis, I., Tsoumakas, G., Vlahavas, I.: Dynamic feature space and incremental feature selection for the classification of textual data streams. In: Proceedings of

the Fourth International Workshop on Knowledge Discovery from Data Streams (IWKDDS 2006), pp. 107–116. Springer (2006)

13. Kira, K., Rendell, L.A.: A practical approach to feature selection. In: Proceedings of the Ninth International Workshop on Machine learning (ML 1992), pp. 249–256. Morgan Kaufmann (1992)

14. Osojnik, A., Panov, P., Džeroski, S.: Multi-label classification via multi-target regression on data streams. Mach. Learn. **106**(6), 745–770 (2017). https://doi.org/10.1007/s10994-016-5613-5

15. Osojnik, A., Panov, P., Džeroski, S.: Tree-based methods for online multi-target regression. J. Intell. Inf. Syst. **50**, 315–339 (2018). https://doi.org/10.1007/s10844-017-0462-7

16. Osojnik, A., Panov, P., Džeroski, S.: Utilizing hierarchies in tree-based online structured output prediction. In: Kralj Novak, P., Šmuc, T., Džeroski, S. (eds.) DS 2019. LNCS (LNAI), vol. 11828, pp. 87–95. Springer, Cham (2019). https://doi.org/10.1007/978-3-030-33778-0_8

17. Osojnik, A., Panov, P., Džeroski, S.: Incremental predictive clustering trees for online semi-supervised multi-target regression. Mach. Learn. **109**(11), 2121–2139 (2020). https://doi.org/10.1007/s10994-020-05918-z

18. Oza, N.C., Russel, S.J.: Experimental comparisons of online and batch versions of bagging and boosting. In: Proceedings of the Seventh ACM SIGKDD International Conference on Knowledge Discovery and Data Mining (KDD 2001), pp. 359–364. ACM (2001)

19. Pereira, R.B., Plastino, A., Zadrozny, B., Merschmann, L.H.: Categorizing feature selection methods for multi-label classification. Artif. Intell. Rev. **49**, 57–78 (2018). https://doi.org/10.1007/s10462-016-9516-4

20. Perkins, S., Lacker, K., Theiler, J.: Grafting: fast, incremental feature selection by gradient descent in function space. J. Mach. Learn. Res. **3**, 1333–1356 (2003)

21. Petković, M., Kocev, D., Džeroski, S.: Feature ranking for multi-target regression. Mach. Learn. **109**, 1179–1204 (2020). https://doi.org/10.1007/s10994-019-05829-8

22. Razmjoo, A., Xanthopoulos, P., Zheng, Q.P.: Online feature importance ranking based on sensitivity analysis. Expert Syst. Appl. **85**, 397–406 (2017)

23. Salzberg, S.: A nearest hyperrectangle learning method. Mach. Learn. **6**(3), 251–276 (1991). https://doi.org/10.1007/BF00114779

24. Slavkov, I., Karcheska, J., Kocev, D., Džeroski, S.: HMC-ReliefF: feature ranking for hierarchical multi-label classification. Comput. Sci. Inf. Syst. **15**(1), 187–209 (2018)

25. Spyromitros-Xioufis, E., Groves, W., Tsoumakas, G., Vlahavas, I.: Multi-label classification methods for multi-target regression (2012)

26. Stańczyk, U., Jain, L.C.: Feature selection for data and pattern recognition: an introduction. In: Stańczyk, U., Jain, L.C. (eds.) Feature Selection for Data and Pattern Recognition. SCI, vol. 584, pp. 1–7. Springer, Heidelberg (2015). https://doi.org/10.1007/978-3-662-45620-0_1

27. Sun, Y.: Iterative RELIEF for feature weighting: algorithms, theories, and applications. IEEE Trans. Pattern Anal. Mach. Intell. **29**(6), 1035–1051 (2007)

28. Teo, C.H., Globerson, A., Roweis, S.T., Smola, A.J.: Convex learning with invariances. In: Advances in Neural Information Processing Systems (NIPS 2007), vol. 20, pp. 1489–1496. NIPS Foundation (2008)

29. Yoon, H., Yang, K., Shahabi, C.: Feature subset selection and feature ranking for multivariate time series. IEEE Trans. Knowl. Data Eng. **17**(9), 1186–1198 (2005)

Knowledge-Guided Additive Modeling for Supervised Regression

Yann Claes[(✉)] , Vân Anh Huynh-Thu , and Pierre Geurts

University of Liège, 4000 Liège, Belgium
{y.claes,vahuynh,p.geurts}@uliege.be

Abstract. Learning processes by exploiting restricted domain knowledge is an important task across a plethora of scientific areas, with more and more hybrid methods combining data-driven and model-based approaches. However, while such hybrid methods have been tested in various scientific applications, they have been mostly tested on dynamical systems, with only limited study about the influence of each model component on global performance and parameter identification. In this work, we assess the performance of hybrid modeling against traditional machine learning methods on standard regression problems. We compare, on both synthetic and real regression problems, several approaches for training such hybrid models. We focus on hybrid methods that additively combine a parametric physical term with a machine learning term and investigate model-agnostic training procedures. We also introduce a new hybrid approach based on partial dependence functions. Experiments are carried out with different types of machine learning models, including tree-based models and artificial neural networks. Our Python implementations of the hybrid methods are available at https://github.com/yannclaes/kg-regression.

Keywords: Knowledge-guided machine learning · Physics-guided machine learning · Supervised regression · Tree-based methods · Neural networks · Partial dependence plot · Hybrid modeling

1 Introduction

For the past decades, machine learning (ML) models have been developed to tackle a variety of real-life problems, complementing/replacing model-based (MB) approaches, which mostly remain approximations that make stringent assumptions about the system under study. Traditional ML approaches are said to be *data-driven*, i.e. their prediction model is solely built from some learning

This work was supported by Service Public de Wallonie Recherche under Grant No. 2010235 - ARIAC by DIGITALWALLONIA4.AI. Computational resources have been provided by the Consortium des Equipements de Calcul Intensif (CECI), funded by the Fonds de la Recherche Scientifique de Belgique (F.R.S.-FNRS) under Grant No. 2.5020.11 and by the Walloon Region.

dataset, let it be (deep) neural networks or regression trees. While their design comes with great expressiveness, they are likely to be subject to over-fitting without enough training examples and to show a lack of robustness on unseen samples, with predictions that can be inconsistent w.r.t. domain knowledge [5,6,31]. To overcome this generalization issue, hybrid approaches have been introduced to incorporate *a priori* domain knowledge within statistical models, which can be leveraged in a multitude of ways (see [16,25,27] for reviews). The success of these hybrid methods have been shown empirically on a range of synthetic and real-world problems [1,7,18,31]. However, while these models have been mostly applied to dynamical systems, they have not been thoroughly studied in the context of standard regression problems. Furthermore, the majority of ML models that have been considered in these approaches are neural networks and variants of the latter, leaving aside other methods. Our contributions are the following:

- We investigate empirically the performance and benefits of hybrid methods against data-driven methods on *static* regression problems (in opposition to dynamical problems). The static context removes a layer of complexity related to the temporal correlation between observed states, which makes it easier to assess the impact and interaction between the MB and ML components. Specifically, we focus on hybrid models that combine in an additive way a parametric physical term with an ML term.
- We compare different approaches for training such hybrid additive models. We highlight specific assumptions under which these approaches are expected to work well and relate the differences in terms of prediction and parameter recovery performance. We focus on model-agnostic approaches, where the ML term can be of any type, and we compare tree-based methods against neural networks. Tree-based methods have several advantages over neural networks, which motivate their use on static regression problems: they have much less hyperparameters to tune, appear robust to the presence of irrelevant features and have been shown to outperform neural networks on tabular data [12].
- We introduce a new hybrid approach based on partial dependence functions, which makes it easier to find the right balance between the MB and ML components, makes less assumptions than other approaches, and is shown to be competitive in our experiments.

2 Problem Statement

Let us define a regression problem, with $y \in \mathbb{R}$ and $\mathbf{x} \in \mathbb{R}^d$, with $d \in \mathbb{N}_+$, drawn from a distribution $p(\mathbf{x}, y)$ such that $y = f(\mathbf{x}) + \varepsilon$ with $f : \mathbb{R}^d \mapsto \mathbb{R}$ the partially known generating function and $\varepsilon \sim \mathcal{N}(0, \sigma^2)$ the noise term. We focus on problems such that $f(\mathbf{x})$ can be decomposed as:

Assumption 1 (A1, Additivity)

$$y = f_k(\mathbf{x}_k) + f_a(\mathbf{x}) + \varepsilon,$$

where $\mathbf{x}_k$ is a subset of $K \leq d$ input variables. We assume partial knowledge of the generating function through some known algebraic function $h_k^{\theta_k}(\mathbf{x}_k) \in \mathcal{H}_k$ with tunable parameters θ_k, such that for the optimal parameters θ_k^* we have $h_k^{\theta_k^*} = f_k$. The residual term $f_a(\mathbf{x})$ is unknown and is approximated in this work through an ML component $h_a^{\theta_a} \in \mathcal{H}_a$, with parameters θ_a[1]. The final model $h \in \mathcal{H}$ is denoted $h(\mathbf{x}) = h_k^{\theta_k}(\mathbf{x}_k) + h_a^{\theta_a}(\mathbf{x})$, with the function space $\mathcal{H}$ defined as $\mathcal{H}_k + \mathcal{H}_a$. A1 is common when MB methods and ML models are combined [7,22,26,30].

Given a learning sample of N input-output pairs $LS = \{(\mathbf{x}_i, y_i)\}_{i=1}^N$, drawn from $p(\mathbf{x}, y)$, we seek to identify a function $h = h_k^{\theta_k} + h_a^{\theta_a}$, i.e. parameters θ_k and θ_a, that minimizes the following two distances:

$$d(h, y) = \mathbb{E}_{(\mathbf{x},y) \sim p(\mathbf{x},y)} \{(h(\mathbf{x}) - y)^2\}, \tag{1}$$

$$d_k(h_k^{\theta_k}, f_k) = \mathbb{E}_{\mathbf{x}_k \sim p(\mathbf{x}_k)} \{(h_k^{\theta_k}(\mathbf{x}_k) - f_k(\mathbf{x}_k))^2\}. \tag{2}$$

The first distance measures the standard generalization error of the global model h. The hope is that taking h_k into account will help learning a better global model than fitting directly a pure data-driven model on y, especially in the small sample size regime. The second distance d_k measures how well the tuned h_k approximates f_k. The main motivation for this second objective is interpretability: one expects that the algebraic form of h_k will be derived from first principles by domain experts, who will be interested in estimating the parameters of this term from data. An alternative to d_k is a loss that would compare the estimated and optimal parameters $\hat{\theta}_k$ and θ_k^* (e.g., $||\hat{\theta}_k - \theta_k^*||^2$). d_k however has the advantage not to require θ_k^* to be fully identifiable, i.e. there can exist several sets of parameters θ_k^* such that $h_k^{\theta_k^*} = f_k$. In our experiment, we will report both d_k and the relative mean absolute error on the estimated parameters.

The following approximation of (1) can be used as training objective:

$$\hat{d}(h, y; LS) = \frac{1}{N} \sum_{i=1}^N (h(\mathbf{x}_i) - y_i)^2. \tag{3}$$

Minimizing the distance in (2) is expected to be challenging and sometimes even ill-posed. Indeed, if h_a is too powerful, it could capture f entirely and leave little room for the estimation of f_k. Finding the right balance between h_k and h_a is thus very challenging, if not impossible, using only guidance of the learning sample LS. Unlike (1), (2) cannot be estimated from a sample of input-output pairs and hence cannot be explicitly used to guide model training. There are however several scenarios that will make the problem easier. In the following, we will discuss the optimality of the hybrid methods under two additional assumptions:

Assumption 2 (A2, Disjoint features). *Let $\mathbf{x}_a$ be a subset of features disjoint from $\mathbf{x}_k$ ($\mathbf{x}_k \cap \mathbf{x}_a = \emptyset$). There exists a function $f_a^r(\mathbf{x}_a)$ such that $f_a(\mathbf{x}) = f_a^r(\mathbf{x}_a)$ for all $\mathbf{x}$.*

---

[1] In the following, $h_k^{\theta_k}$ and $h_a^{\theta_a}$ will sometimes be denoted simply as h_k and h_a to lighten the notations.

Assumption 3 (A3, Independence). *Features in* $\mathbf{x}_k$ *are independent from features in* $\mathbf{x}_a$ *($\mathbf{x}_k \perp\!\!\!\perp \mathbf{x}_a$).*

A2 makes the problem easier as f_k captures all the dependence of y on $\mathbf{x}_k$. In the absence of A3, it might be hard to distinguish real contributions from $\mathbf{x}_k$ to f from those due to correlations with features not in $\mathbf{x}_k$.

3 Related Work

Hybrid additive modeling methods emerged several decades ago, combining first-principles models with different ML models. Already in the 1990's, approaches in [14,17,24] complemented physics-based models with neural networks, weighting contributions of both components (e.g. through radial basis function networks), to achieve enhanced physical consistency with better generalization properties. More recently, other works applied the same principles to model dynamical systems in various domains, still massively relying on neural networks [21,22,26,30]. In a more standard regression setting, [3,32,33] combined a linear parametric term with a tree-based ML term.

Previous works have introduced regularization of the ML term to reduce parameter identification issues in the decomposition [13,15,28]. Further works on this matter introduced physically-motivated constraints in the learning objective to better control contributions of the MB/ML components [7,31]. Elements of discussion about the well-posedness of this additive decomposition have been introduced in previous works: [31] showed the existence and uniqueness of an optimal pair (h_k, h_a) when the contributions of h_a are constrained to be minimal, and [7] demonstrated the convergence of an algorithm alternating between the optimization of h_k and the optimization of h_a, without however any guarantee about convergence points.

4 Methods

We focus on model-agnostic approaches, i.e. that can be applied with any algebraic function h_k and any type of ML model h_a. For both terms, we only assume access to training functions, respectively denoted fit^{h_k}, $\text{fit}^{h_k+\gamma}$, and fit^{h_a}, that can estimate each model parameters, respectively θ_k, (θ_k, γ) and θ_a, so as to minimize the mean squared error (MSE) over LS (see below for the meaning of γ), where parametric methods rely on gradient descent. Pseudo-codes of methods in Sects. 4.1 and 4.2 and additional illustrations are given on the paper's GitHub[2].

4.1 Sequential Training of h_k and h_a

This baseline approach first fits h_k on the observed output y, then fits h_a on the resulting residuals, as done in [33]. More precisely, we first train $h_k^{\theta_k}$ on y by introducing a constant term $\gamma \in \mathbb{R}$, such that

$$(\hat{\theta}_k, \hat{\gamma}) = \text{fit}^{h_k+\gamma}(LS). \tag{4}$$

---

[2] https://github.com/yannclaes/kg-regression.

Our motivation for introducing the term γ will be explained below. Afterwards, we fit h_a on the output residuals: $\hat{\theta}_a = \text{fit}^{h_a}\{(\mathbf{x}_i, y_i - h_k^{\hat{\theta}_k}(\mathbf{x}_i) - \hat{\gamma})\}_{i=1}^N$.

Let $\hat{\mathcal{F}}_k$ be the set of all functions $\hat{f}_k$ mapping $\mathbf{x}_k \in \mathcal{X}_k$ to some value $y \in \mathbb{R}$, i.e. $\hat{\mathcal{F}}_k = \{\hat{f}_k : \mathcal{X}_k \mapsto \mathbb{R}\}$. Under A2 and A3, it can be shown that $\hat{f}_k^* = \arg\min_{\hat{f}_k \in \hat{\mathcal{F}}_k} d(\hat{f}_k, y)$ is such that $\hat{f}_k^*(\mathbf{x}_k) = f_k(\mathbf{x}_k) + C$, for every $\mathbf{x}_k \in \mathcal{X}_k$, with $C = \mathbb{E}_{\mathbf{x}_a}\{f_a^r(\mathbf{x}_a)\}$ (see Appendix A). Hence, this approach is sound at least asymptotically and justifies the introduction of γ. Note however that even under A2 and A3, we have no guarantee that this approach produces the best estimator for a finite sample size, as $f_a^r(\mathbf{x}_a) + \epsilon$ acts as a pure additive noise term that needs to be averaged out during training. The approaches described in Sects. 4.2 and 4.3 try to overcome this issue by fitting $h_k^{\theta_k}$ on corrected outputs that are expected to be closer to $f_k(\mathbf{x}_k)$. Without A2 and A3, the quality of the estimation of f_k by $h_k^{\hat{\theta}_k}$, according to (2), is not guaranteed as there are regression problems satisfying A1 such that:

$$\nexists \gamma \in \mathbb{R} : \arg\min_{\hat{f}_k \in \hat{\mathcal{F}}_k} d(\hat{f}_k, y) = f_k + \gamma. \tag{5}$$

An example will be given in Sect. 5.2.

4.2   Alternate Training of h_k and h_a

A hybrid additive approach was proposed in [7] that alternates between updating h_k and updating h_a, using neural networks for h_a. Such alternate training was also proposed in [3, 32] with a single decision tree as h_a and a linear h_k. We include this approach in our comparison, but also investigate it with random forests [2] and tree gradient boosting [9]. $\hat{\theta}_k$ is initialized by (fully) fitting $h_k^{\theta_k} + \gamma$ on y. Then, we alternate between: (1) a single epoch of gradient descent on $h_k^{\theta_k} + \gamma$ and (2) either a single epoch for h_a (in the case of neural networks, as in [7]) or a complete fit of h_a (in the case of tree-based models).

While some theoretical results are provided in [7], convergence of the alternate method towards the optimal solution is not guaranteed in general. Despite an initialization favoring h_k, it is unclear whether a too expressive h_a will not dominate h_k and finding the right balance between these two terms, e.g. by regularizing further h_a, is challenging. Under A2 and A3 however, the population version[3] of the algorithm produces an optimal solution. Indeed, h_k will be initialized as the true f_k, as shown previously, making the residuals $y - h_k$ at the first iteration, as well as h_a, independent of $\mathbf{x}_k$. h_k will thus remain unchanged (and optimal) at subsequent iterations.

---

[3] i.e., assuming an infinite training sample size and consistent estimators.

4.3   Partial Dependence-Based Training of h_k and h_a

We propose a novel approach relying on partial dependence (PD) functions [8] to produce a proxy dataset depending only on $\mathbf{x}_k$ to fit h_k. PD measures how a given subset of features impact the prediction of a model, on average. Let $\mathbf{x}_k$ be the subset of interest and $\mathbf{x}_{-k}$ its complement, with $\mathbf{x}_k \cup \mathbf{x}_{-k} = \mathbf{x}$, then the PD of a function $f(\mathbf{x})$ on $\mathbf{x}_k$ is:

$$PD(f, \mathbf{x}_k) = \mathbb{E}_{\mathbf{x}_{-k}}\left[f(\mathbf{x}_k, \mathbf{x}_{-k})\right] = \int f(\mathbf{x}_k, \mathbf{x}_{-k}) p(\mathbf{x}_{-k}) d\mathbf{x}_{-k}, \qquad (6)$$

where $p(\mathbf{x}_{-k})$ is the marginal distribution of $\mathbf{x}_{-k}$. Under A1 and A2, the PD of $f(\mathbf{x}) = f_k(\mathbf{x}_k) + f_a^r(\mathbf{x}_a)$ is [8]:

$$PD(f, \mathbf{x}_k) = f_k(\mathbf{x}_k) + C, \text{ with } C = E_{\mathbf{x}_a}\{f_a^r(\mathbf{x}_a)\}. \qquad (7)$$

The idea of our method is to first fit any sufficiently expressive ML model $h_a(\mathbf{x})$ on LS and to compute its PD w.r.t. $\mathbf{x}_k$ to obtain a first approximation of $f_k(\mathbf{x}_k)$ (up to a constant). Although computing the actual PD of a function using (6) requires in principle access to the input distribution, an approximation can be estimated from LS as follows:

$$\widehat{PD}(h_a, \mathbf{x}_k; LS) = \frac{1}{N} \sum_{i=1}^{N} h_a(\mathbf{x}_k, \mathbf{x}_{i,-k}), \qquad (8)$$

where $\mathbf{x}_{i,-k}$ denotes the values of $\mathbf{x}_{-k}$ in the i-th sample of LS. A new dataset of pairs $(\mathbf{x}_k, \widehat{PD}(h_a, \mathbf{x}_k; LS))$ can then be built to fit h_k. In our experiments, we consider only the $\mathbf{x}_k$ values observed in the learning sample but $\widehat{PD}(h_a, \mathbf{x}_k; LS)$ could also be estimated at other points $\mathbf{x}_k$ to artificially increase the size of the proxy dataset.

In practice, optimizing θ_k only once on the PD of h_a could leave residual dependence of $\mathbf{x}_k$ on the resulting $y - h_k^{\hat{\theta}_k}(\mathbf{x}_k) - \hat{\gamma}$. We thus repeat the sequence of fitting h_a on the latter residuals, then fitting h_k on the obtained $\widehat{PD}(h_a^{\hat{\theta}_a}, \mathbf{x}_k; LS) + h_k^{\hat{\theta}_k}(\mathbf{x}_k) + \hat{\gamma}$, with $\hat{\theta}_k$ and $\hat{\theta}_a$ the current optimized parameter vectors (see Algorithm 1).

The main advantage of this approach over the alternate one is to avoid domination of h_a over h_k. Unlike the two previous approaches, this one is also sound even if A3 is not satisfied as it is not a requirement for (7) to hold. One drawback is that it requires h_a to capture well the dependence of f on $\mathbf{x}_k$ so that its PD is a good approximation of f_k. The hope is that even if it is not the case at the first iteration, fitting h_k, that contains the right inductive bias, will make the estimates better and better over the iterations.

5   Experiments

We compare the different methods on several regression datasets, both simulated and real. Performance is measured through estimates of (1) and (2) (the latter

---

Algorithm 1. Partial Dependence Optimization

Input: $LS = (\mathbf{x}_i, y_i)_{i=1}^N$

$\hat{\theta}_a \leftarrow \text{fit}^{h_a}(LS)$

$(\hat{\theta}_k, \hat{\gamma}) \leftarrow \text{fit}^{h_k + \gamma}(\{(\mathbf{x}_{k,i}, \widehat{PD}(h_a^{\hat{\theta}_a}, \mathbf{x}_{k,i}; LS))\}_{i=1}^N)$

for $n = 1$ **to** $N_{repeats}$ **do**

$\quad \hat{\theta}_a \leftarrow \text{fit}^{h_a}(\{(\mathbf{x}_i, y_i - h_k^{\hat{\theta}_k}(\mathbf{x}_{k,i}) - \hat{\gamma})\}_{i=1}^N)$

$\quad (\hat{\theta}_k, \hat{\gamma}) \leftarrow \text{fit}^{h_k + \gamma}(\{(\mathbf{x}_{k,i}, h_k^{\hat{\theta}_k}(\mathbf{x}_{k,i}) + \hat{\gamma} + \widehat{PD}(h_a^{\hat{\theta}_a}, \mathbf{x}_{k,i}; LS))\}_{i=1}^N)$

end for

$\hat{\theta}_a \leftarrow \text{fit}^{h_a}(\{(\mathbf{x}_i, y_i - h_k^{\hat{\theta}_k}(\mathbf{x}_{k,i}) - \hat{\gamma})\}_{i=1}^N)$

---

only on simulated problems) on a test set TS, respectively denoted $\hat{d}(h, y; TS)$ and $\hat{d}_k(h_k^{\hat{\theta}_k}, f_k; TS)$. In some cases, we also report $\text{rMAE}(\theta_k^*, \theta_k)$, the relative mean absolute error between θ_k^* and θ_k (lower is better for all measures). For the hybrid approaches, we use as h_a either a multilayer perceptron (MLP), gradient boosting with decision trees (GB) or random forests (RF). We compare these hybrid models to a standard data-driven model that uses only h_a. We also compare fitting h_a with and without input filtering, i.e. respectively removing or keeping $\mathbf{x}_k$ from its inputs, to verify convergence claims about h_k in Sect. 4.2. Architectures (e.g. for MLP, the number of layers and neurons) are kept fixed across training methods to allow a fair comparison between them, and are given in Appendix B. We use early stopping of gradient descent training by monitoring the loss on a validation set (except for pure tree-based models, which are trained in the standard way, hence not using gradient descent).

5.1   Friedman Problem (A2 and A3 Satisfied)

We consider the following synthetic regression problem:

$$y = \theta_0 \sin(\theta_1 x_0 x_1) + \theta_2 (x_2 - \theta_3)^2 + \theta_4 x_3 + \theta_5 x_4 + \varepsilon,$$

where $x_j \sim \mathcal{U}(0, 1), j = 0, \ldots 9$, and $\varepsilon \sim \mathcal{N}(0, 1)$ [10]. We generate 10 different datasets using 10 different sets of values for $\theta_0, \ldots, \theta_5$, each with 300, 300 and 600 samples for respectively the training, validation and test sets. For the hybrid approaches, we use the first term as prior knowledge, i.e. $f_k(\mathbf{x}_k) = \theta_0 \sin(\theta_1 x_0 x_1)$.

We see in Table 1 that all hybrid training schemes outperform their data-driven counterpart. They come very close to the ideal $f_k \to h_a$ method, and sometimes even slightly better, probably due to chance. Sequential fitting of h_k and h_a performs as well as the alternate or PD-based approaches, as A2 and A3 are satisfied for this problem (see Sect. 4.1). Filtering generally improves the performance of hybrid schemes as A2 is verified. PD-based optimization yields good approximations of f_k (as shown by a low $\hat{d}_k$). The alternate approach follows closely whereas the sequential one ends up last, which can be expected as fitting h_k only on y induces a higher noise level centered around $\mathbb{E}_{\mathbf{x}_a}\{f_a(\mathbf{x}_a)\}$, while the other approaches benefit from reduced perturbations through h_a estimation,

Table 1. Results on the Friedman problem. We report the mean and standard deviation of $\hat{d}$ and $\hat{d}_k$ over the 10 test sets (TS). "$f_k \to h_a$" fits h_a on $y - f_k(\mathbf{x}_k)$. "Unfiltered" indicates that all the features are used as inputs of h_a, while "Filtered" indicates that the features $\mathbf{x}_k$ are removed from the inputs of h_a.

Method		$\hat{d}(h, y; TS)$		$\hat{d}_k(h_k^{\theta_k}, f_k; TS)$	
		Unfiltered	Filtered	Unfiltered	Filtered
MLP	$f_k \to h_a$	1.58 ± 0.33	1.23 ± 0.10	–	
	Sequential	1.54 ± 0.31	1.43 ± 0.13	0.18 ± 0.16	
	Alternate	1.43 ± 0.09	1.32 ± 0.09	0.10 ± 0.09	0.02 ± 0.02
	PD-based	1.54 ± 0.12	1.38 ± 0.09	0.06 ± 0.07	
	h_a only	2.62 ± 0.75		–	
GB	$f_k \to h_a$	1.73 ± 0.09	1.75 ± 0.12	–	
	Sequential	1.74 ± 0.11	1.81 ± 0.14	0.18 ± 0.16	
	Alternate	1.79 ± 0.11	1.78 ± 0.15	0.91 ± 1.45	0.06 ± 0.06
	PD-based	1.77 ± 0.13	1.78 ± 0.12	0.03 ± 0.02	
	h_a only	3.43 ± 0.94		–	
RF	$f_k \to h_a$	2.03 ± 0.18	1.96 ± 0.17	–	
	Sequential	2.11 ± 0.23	2.05 ± 0.24	0.18 ± 0.16	
	Alternate	2.03 ± 0.19	1.98 ± 0.17	0.04 ± 0.03	0.04 ± 0.04
	PD-based	2.16 ± 0.27	2.09 ± 0.26	0.16 ± 0.15	
	h_a only	5.58 ± 1.91		–	

as explained in Sect. 4.1. Filtering vastly decreases $\hat{d}_k$ for alternate approaches, supporting claims introduced in Sect. 4.2, while this measure remains unimpaired for sequential and PD-based training by construction.

5.2 Correlated Input Features (A3 Not Satisfied)

Correlated Linear Model. Let $y = \beta_0 x_0 + \beta_1 x_1 + \varepsilon$, with $\beta_0 = -0.5, \beta_1 = 1, \mathbf{x} \sim \mathcal{N}(\mathbf{0}, \Sigma)$, and $\varepsilon \sim \mathcal{N}(0.5^2, 1)$. We generate 50, 50 and 600 samples respectively for the training, validation and test sets. We use as known term $f_k(\mathbf{x}_k) = \beta_0 x_0$. Regressing y on x_0 yields the least-squares solution [11]:

$$\mathbb{E}\left[\hat{\beta}_0\right] = \beta_0 + \frac{\text{cov}(x_0, x_1)}{\text{var}(x_0)}\beta_1. \tag{9}$$

We set $\text{cov}(x_0, x_1) = 2.25$ and $\text{var}(x_0) = 2$ so that (9) reverses the sign of β_0 and (5) is satisfied. The sequential approach should hence yield parameter estimates of β_0 close to (9) while we expect the others to correct for this bias.

From Table 2, we observe that, contrary to the PD-based approach, the sequential and alternate methods return very bad estimations of β_0, as A3 is no longer verified. Filtering corrects the bias for the alternate approach but

Table 2. Results for the correlated linear problem. We report $\hat{d}$ and $\text{rMAE}(\beta_0^*, \hat{\beta}_0)$, over 10 different datasets.

Method		$\hat{d}(h, y; TS)$		$\text{rMAE}(\beta_0^*, \hat{\beta}_0)$	
		Unfiltered	Filtered	Unfiltered	Filtered
MLP	Sequential	0.30 ± 0.03	0.74 ± 0.09	224.14 ± 13.48	
	Alternate	0.30 ± 0.02	0.31 ± 0.04	186.65 ± 21.31	15.53 ± 13.57
	PD-based	0.30 ± 0.03	0.29 ± 0.02	26.47 ± 17.32	
GB	Sequential	0.59 ± 0.06	1.38 ± 0.11	224.14 ± 13.48	
	Alternate	0.57 ± 0.06	0.60 ± 0.09	148.75 ± 67.35	24.58 ± 12.20
	PD-based	0.56 ± 0.05	0.64 ± 0.13	36.05 ± 17.50	
RF	Sequential	0.53 ± 0.05	0.90 ± 0.07	224.14 ± 13.48	
	Alternate	0.43 ± 0.04	0.42 ± 0.04	111.04 ± 52.78	45.38 ± 22.39
	PD-based	0.41 ± 0.03	0.43 ± 0.04	57.47 ± 15.55	

degrades the MSE performance for the sequential method as it removes the ability to compensate for the h_k misfit.

Correlated Friedman Problem. The structure is identical to the one in Sect. 5.1 but with correlated inputs drawn from a multivariate normal distribution where $\mu_i = 0.5$ and $\text{var}(x_i) = 0.75, \forall i$, and $\text{cov}(x_i, x_j) = \pm 0.3, \forall i \neq j$ (the covariance sign being chosen randomly). Sizes of the training, validation and test sets are identical to those of Sect. 5.1. Inputs are then scaled to be roughly in $[-1, 1]$. Here again, we use $f_k(\mathbf{x}_k) = \theta_0 \sin(\theta_1 x_0 x_1)$.

As in Sect. 5.1, Table 3 shows that hybrid models outperform their data-driven equivalents. PD-based methods usually yield more robust h_k estimations in the general unfiltered case, but struggle to line up with the alternate scheme in terms of predictive performance, except for GB-related models. For RF, this can be explained by a worse h_k estimation while for MLP we assume that it is due to h_a overfitting: in the alternate approach, it is optimized one epoch at a time, interleaved with one step on h_k, whereas that of PD-based methods is fully optimized (with identical complexities). Sequential and alternate approaches undergo stronger h_k misparameterization without filtering since A3 is not met, but the latter mitigates this w.r.t. the former, as was already observed in Sect. 5.1. Input filtering degrades predictive performance for the sequential methods as they cannot counterbalance a poor h_k.

5.3   Overlapping Additive Structure (A2 and A3 Not Satisfied)

Let $y = \beta x_0^2 + \sin(\gamma x_0) + \delta x_1 + \varepsilon$ with $\varepsilon \sim \mathcal{N}(0, 0.5^2)$, $\beta = 0.2, \gamma = 1.5, \delta = 1$ and $\mathbf{x}$ sampled as in the correlated linear problem. We generate 50, 50 and 600 samples respectively for the training, validation and test sets. We define $f_k(\mathbf{x}_k) = \beta x_0^2$ and $f_a(\mathbf{x}) = \sin(\gamma x_0) + \delta x_1 + \varepsilon$. Hence, A2 and A3 do not hold. Even with

Table 3. Results for the correlated Friedman problem.

Method		$\hat{d}(h, y; TS)$		$\hat{d}_k(h_k^{\theta_k}, f_k; TS)$	
		Unfiltered	Filtered	Unfiltered	Filtered
MLP	$f_k \to h_a$	1.64 ± 0.23	1.51 ± 0.17	–	
	Sequential	2.07 ± 0.40	2.68 ± 1.38	1.35 ± 1.42	
	Alternate	1.95 ± 0.33	1.62 ± 0.24	0.49 ± 0.44	0.14 ± 0.19
	PD-based	2.24 ± 0.31	1.78 ± 0.30	0.17 ± 0.23	
	h_a only	2.77 ± 0.73		–	
GB	$f_k \to h_a$	2.58 ± 0.45	2.53 ± 0.44	–	
	Sequential	2.90 ± 0.39	3.91 ± 1.49	1.35 ± 1.42	
	Alternating	2.67 ± 0.38	2.62 ± 0.43	0.51 ± 0.53	0.22 ± 0.25
	PD-based	2.54 ± 0.35	2.47 ± 0.36	0.03 ± 0.02	
	h_a only	4.49 ± 0.66		–	
RF	$f_k \to h_a$	3.02 ± 0.45	2.93 ± 0.45	–	
	Sequential	3.78 ± 0.78	4.04 ± 1.30	1.35 ± 1.42	
	Alternating	3.06 ± 0.39	2.99 ± 0.38	0.14 ± 0.16	0.15 ± 0.18
	PD-based	3.24 ± 0.38	3.16 ± 0.37	0.27 ± 0.20	
	h_a only	6.70 ± 1.47		–	

$\hat{\beta} = \beta^*$, h_a still needs to compensate for $\sin(\gamma x_0)$. Filtering is thus expected to degrade performance for all hybrid approaches as $h_a(x_1)$ will never compensate this gap, which is observed in Table 4. Results for RF are not shown for the sake of space, but are similar to GB.

Table 4. Results for the overlapping problem.

Method	$\hat{d}(h, y; TS)$		$\hat{d}(h, y; TS)$	
	Unfiltered	Filtered	Unfiltered	Filtered
	MLP		GB	
$f_k \to h_a$	0.35 ± 0.02	0.54 ± 0.04	0.51 ± 0.04	1.00 ± 0.12
Sequential	0.35 ± 0.01	0.59 ± 0.05	0.55 ± 0.07	1.07 ± 0.11
Alternate	0.35 ± 0.02	0.56 ± 0.05	0.54 ± 0.09	1.01 ± 0.11
PD-based	0.34 ± 0.02	0.56 ± 0.05	0.53 ± 0.05	0.99 ± 0.12
h_a only	0.37 ± 0.02		0.55 ± 0.07	

5.4 Real Regression Problems

We now apply all methods on two real-world static datasets. As algebraic term h_k, we chose to use a linear prior on x_k, where x_k is the feature with the highest importance score in a RF model trained on the full dataset (assumed to be inaccessible at training time). As there is no guarantee that any of our assumptions are met (and in particular A1), we do not measure the distance d_k in (2) as it is deemed irrelevant.

We consider two settings for each dataset, inspired by [33]. In the first setting (INT), the training and test sets are sampled from the same distribution $p(\mathbf{x}, y)$ whereas the second one (EXT) evaluates extrapolation performance for samples with unseen target values. If the linear prior is a reasonable assumption or, at the very least, the target increases or decreases monotonically with x_k, then we can expect hybrid methods to yield better results in the latter setting. In the INT setting for each dataset, we randomly select 100 samples for the learning set, 100 samples for the validation set and keep the rest as test set. For the EXT setting, we select the samples (one fourth of the dataset) with the lowest output values as test set. From the remaining samples, we randomly select 100 samples for the learning set and 100 samples for the validation set. For both INT and EXT settings, performance metrics are averaged over 10 different splits. We standardize both input and output variables.

The features for both datasets are described in Table 5. The *Combined Cycle Power Plant* (CCPP) dataset [23] collects 9,568 measurements of net hourly electrical energy output for a combined cycle power plant, along with four hourly average input variables. The *Concrete Compressive Strength (CCS) dataset* [29] is composed of 1,030 samples relating amounts of concrete components with the resulting compressive strength. As done in [33], we introduce a new feature corresponding to the cement-to-water ratio.

Table 5. Variables used for the real-world datasets. For each dataset, the variable indicated in bold type is the one used in the linear prior (x_k).

Dataset	Name	Description
CCPP	**T**	**Ambient temperature [°C]**
	AP	Ambient pressure [mbar]
	RH	Relative humidity [-]
	V	Exhaust vacuum [cmHg]
CCS	Cement	Amount of cement in the mixture [kg/m³]
	Blast Furnace Slag	Amount of blast furnace slag in the mixture [kg/m³]
	Fly Ash	Amount of fly ash in the mixture [kg/m³]
	Water	Amount of water in the mixture [kg/m³]
	Superplasticizer	Amount of superplasticizer in the mixture [kg/m³]
	Coarse Aggregate	Amount of coarse aggregate in the mixture [kg/m³]
	Fine Aggregate	Amount of fine aggregate in the mixture [kg/m³]
	Age	Day (1–365)
	Cement/Water	**Cement to water ratio [-]**

Table 6. Results for the real-world datasets. We report the mean and standard deviation of $\hat{d}$ over the test sets.

	Method	CCPP		CCS	
		INT	EXT	INT	EXT
MLP	Sequential	0.07 ± 0.01	0.23 ± 0.13	0.25 ± 0.04	0.74 ± 0.24
	Alternating	0.07 ± 0.01	0.24 ± 0.10	0.24 ± 0.03	0.89 ± 0.36
	PD-based	0.07 ± 0.01	0.07 ± 0.01	0.25 ± 0.03	0.38 ± 0.04
	h_a only	0.07 ± 0.01	0.36 ± 0.21	0.25 ± 0.05	0.83 ± 0.30
GB	Sequential	0.08 ± 0.01	0.35 ± 0.07	0.21 ± 0.02	0.91 ± 0.24
	Alternating	0.08 ± 0.01	0.41 ± 0.17	0.20 ± 0.03	0.85 ± 0.28
	PD-based	0.08 ± 0.02	0.10 ± 0.01	0.22 ± 0.02	0.31 ± 0.06
	h_a only	0.10 ± 0.01	0.95 ± 0.18	0.25 ± 0.02	1.54 ± 0.28
RF	Sequential	0.07 ± 0.01	0.28 ± 0.07	0.20 ± 0.02	1.06 ± 0.31
	Alternating	0.07 ± 0.01	0.32 ± 0.12	0.20 ± 0.02	1.00 ± 0.32
	PD-based	0.07 ± 0.01	0.08 ± 0.01	0.20 ± 0.02	0.30 ± 0.08
	h_a only	0.08 ± 0.01	0.99 ± 0.19	0.25 ± 0.02	1.89 ± 0.39

From Table 6, it seems that introducing a linear prior does not yield any benefit in the interpolation setting, as all models perform equally well, which suggests that either the prior is not adequate or that A1 is not verified. In the extrapolation scenario, we can however observe that the linear prior allows to mitigate the impact of moving out of the distribution compared to data-driven models. Indeed, compared to the INT setting, the performance of all purely data-driven methods degrades in the EXT scenario, especially for GB and RF as their output predictions are bounded by the minimum target value observed in the training set. PD-based hybrid methods consistently outperform other hybrid approaches and are only slightly impacted, while sequential and alternating methods attain similar results.

6   Conclusion

We study several hybrid methods on supervised regression problems modeled in an additive way, using neural network models and tree-based approaches. We empirically show that trends observed for neural networks also apply for the non-parametric tree-based approaches, in terms of predictive performance as well as in the estimation of the algebraic known function. We introduce claims related to the convergence of these hybrid approaches, under mild assumptions, and verify their soundness on illustrative experiments. We present a new hybrid approach leveraging partial dependence and show its competitiveness against sequential and alternate optimization schemes on both synthetic and real-world problems. We highlight its benefits in estimating the parametric prior and show that it alleviates both the risk of the ML term to dominate the known term and the need for assuming independent input features sets.

As a more general conclusion, hybrid methods are shown in our experiments to improve predictive performance with respect to ML-only models, although not always very significantly. The main benefit of the alternate and PD-based methods over the simple sequential approach is that they provide better estimators of the physical term, especially when filtering can be applied. Why this advantage does not always translate into better overall predictive performance remains to be analyzed as future work. We also plan to investigate further the theoretical properties of the PD-based approach and extend it to dynamical problems.

A   Optimal Model Under A2 and A3

Let us recall the regression problem, where $y \in \mathbb{R}$ can be decomposed into the addition of two independent terms:

$$y = f_k(\mathbf{x}_k) + f_a^r(\mathbf{x}_a) + \varepsilon, \quad \varepsilon \sim \mathcal{N}(0, \sigma^2), \quad \mathbf{x}_k \cup \mathbf{x}_a = \mathbf{x}, \mathbf{x}_k \cap \mathbf{x}_a = \emptyset, \mathbf{x}_k \perp\!\!\!\perp \mathbf{x}_a.$$

For clarity, let us denote by $\mathbb{E}_\mathbf{x}$ the subsequent expectations over the input space $\mathbb{E}_{\mathbf{x} \sim p(\mathbf{x})}\{\cdot\}$. We have:

$$\hat{f}_k^* = \arg \min_{\hat{f}_k \in \hat{\mathcal{F}}_k} d(\hat{f}_k, y) = \arg \min_{\hat{f}_k \in \hat{\mathcal{F}}_k} \mathbb{E}_{\mathbf{x}, \varepsilon}\left\{\left(\hat{f}_k(\mathbf{x}_k) - f_k(\mathbf{x}_k) - f_a^r(\mathbf{x}_a) - \varepsilon\right)^2\right\}$$

$$= \arg \min_{\hat{f}_k \in \hat{\mathcal{F}}_k} \mathbb{E}_{\mathbf{x}_k}\left\{(\hat{f}_k(\mathbf{x}_k) - f_k(\mathbf{x}_k))^2\right\} + \mathbb{E}_{\mathbf{x}_a, \varepsilon}\left\{(f_a^r(\mathbf{x}_a) + \varepsilon)^2\right\}$$

$$- \mathbb{E}_{\mathbf{x}, \varepsilon}\left\{2(\hat{f}_k(\mathbf{x}_k) - f_k(\mathbf{x}_k))(f_a^r(\mathbf{x}_a) + \varepsilon)\right\}.$$

The second term is independent w.r.t. $\hat{f}_k$ and thus has no impact on the minimization. Moreover, since $\mathbf{x}_k \perp\!\!\!\perp \mathbf{x}_a$, the last term writes as the product of two expectations, one of which is constant w.r.t. $\hat{f}_k$. We thus have:

$$\hat{f}_k^* = \arg \min_{\hat{f}_k \in \hat{\mathcal{F}}_k} \mathbb{E}_{\mathbf{x}_k}\left\{(\hat{f}_k(\mathbf{x}_k) - f_k(\mathbf{x}_k))^2\right\} - 2C\mathbb{E}_{\mathbf{x}_k}\left\{(\hat{f}_k(\mathbf{x}_k) - f_k(\mathbf{x}_k))\right\}, \quad (10)$$

with $C = \mathbb{E}_{\mathbf{x}_a, \varepsilon}\{(f_a^r(\mathbf{x}_a) + \varepsilon)\} = \mathbb{E}_{\mathbf{x}_a}\{f_a^r(\mathbf{x}_a)\}$. Cancelling the derivative of (10) w.r.t. $\hat{f}_k$, we obtain the optimal model $\hat{f}_k^*(\mathbf{x}_k) = f_k(\mathbf{x}_k) + C$, for every $\mathbf{x}_k \in \mathcal{X}_k$.

B   Model Architectures

Model hyperparameters are reported in Table 7. We used `PyTorch` [19] for MLP, `scikit-learn` [20] for RF and `xgboost` [4] for GB. Unspecified parameters keep their default values. Learning rates for training h_k and MLP are set to 0.005. H is the number of hidden layers in MLP and W the number of neurons per hidden layer. T is the number of trees in GB and RF, d the maximum tree depth and mss the minimum number of samples required to split an internal tree node.

Table 7. Model hyperparameters, for each experiment.

Problem	MLP	GB	RF
(Correlated) Friedman	$H = 2, W = 15$	$T = 700, d = 2$	$T = 500, mss = 5$
Linear & Overlap	$H = 2, W = 10$	$T = 400, d = 2$	$T = 500, mss = 5$
Real-world problems	$H = 2, W = 30$	$T = 300, d = 2$	$T = 200, mss = 5$

References

1. Ayed, I., de Bézenac, E., Pajot, A., Brajard, J., Gallinari, P.: Learning dynamical systems from partial observations. In: Machine Learning and the Physical Sciences: Workshop at the 33rd Conference on Neural Information Processing Systems (NeurIPS) (2019)
2. Breiman, L.: Random forests. Mach. Learn. **45**(1), 5–32 (2001). https://doi.org/10.1023/A:1010933404324
3. Chen, J., Yu, K., Hsing, A., Therneau, T.M.: A partially linear tree-based regression model for assessing complex joint gene-gene and gene-environment effects. Genet. Epidemiol. Off. Publ. Int. Genet. Epidemiol. Soc. **31**(3), 238–251 (2007)
4. Chen, T., Guestrin, C.: XGBoost: a scalable tree boosting system. In: Proceedings of the 22nd ACM SIGKDD International Conference on Knowledge Discovery and Data Mining, KDD 2016, pp. 785–794. ACM, New York (2016)
5. Daw, A., Karpatne, A., Watkins, W., Read, J., Kumar, V.: Physics-guided neural networks (PGNN): an application in lake temperature modeling. arXiv preprint arXiv:1710.11431 (2017)
6. De Bézenac, E., Pajot, A., Gallinari, P.: Deep learning for physical processes: incorporating prior scientific knowledge. J. Stat. Mech. Theory Exp. **2019**(12), 124009 (2019)
7. Donà, J., Déchelle, M., Levy, M., Gallinari, P.: Constrained physical-statistics models for dynamical system identification and prediction. In: ICLR 2022-The Tenth International Conference on Learning Representations (2022)
8. Friedman, J.H.: Greedy function approximation: a gradient boosting machine. Ann. Stat. **29**(5), 1189–1232 (2001)
9. Friedman, J.H.: Stochastic gradient boosting. Comput. Stat. Data Anal. **38**(4), 367–378 (2002)
10. Friedman, J.H., Grosse, E., Stuetzle, W.: Multidimensional additive spline approximation. SIAM J. Sci. Stat. Comput. **4**(2), 291–301 (1983)
11. Greene, W.H.: Econometric Analysis. Pearson Education (2003)
12. Grinsztajn, L., Oyallon, E., Varoquaux, G.: Why do tree-based models still outperform deep learning on typical tabular data? In: Thirty-Sixth Conference on Neural Information Processing Systems Datasets and Benchmarks Track (2022)
13. Hu, G., Mao, Z., He, D., Yang, F.: Hybrid modeling for the prediction of leaching rate in leaching process based on negative correlation learning bagging ensemble algorithm. Comput. Chem. Eng. **35**(12), 2611–2617 (2011)
14. Johansen, T.A., Foss, B.A.: Representing and learning unmodeled dynamics with neural network memories. In: 1992 American Control Conference, pp. 3037–3043. IEEE (1992)
15. Kahrs, O., Marquardt, W.: Incremental identification of hybrid process models. Comput. Chem. Eng. **32**(4–5), 694–705 (2008)

16. Karniadakis, G.E., Kevrekidis, I.G., Lu, L., Perdikaris, P., Wang, S., Yang, L.: Physics-informed machine learning. Nat. Rev. Phys. **3**(6), 422–440 (2021)
17. Kramer, M.A., Thompson, M.L., Bhagat, P.M.: Embedding theoretical models in neural networks. In: 1992 American Control Conference, pp. 475–479. IEEE (1992)
18. Mehta, V., et al.: Neural dynamical systems: balancing structure and flexibility in physical prediction. In: 2021 60th IEEE Conference on Decision and Control (CDC), pp. 3735–3742. IEEE (2021)
19. Paszke, A., et al.: PyTorch: an imperative style, high-performance deep learning library. In: Advances in Neural Information Processing Systems, vol. 32, pp. 8024–8035. Curran Associates, Inc. (2019)
20. Pedregosa, F., et al.: Scikit-learn: machine learning in Python. J. Mach. Learn. Res. **12**, 2825–2830 (2011)
21. Qian, Z., Zame, W., Fleuren, L., Elbers, P., van der Schaar, M.: Integrating expert ODEs into neural ODEs: pharmacology and disease progression. In: Advances in Neural Information Processing Systems, vol. 34, pp. 11364–11383 (2021)
22. Takeishi, N., Kalousis, A.: Physics-integrated variational autoencoders for robust and interpretable generative modeling. In: Advances in Neural Information Processing Systems, vol. 34, pp. 14809–14821 (2021)
23. Tfekci, P., Kaya, H.: Combined Cycle Power Plant. UCI Machine Learning Repository (2014). https://doi.org/10.24432/C5002N
24. Thompson, M.L., Kramer, M.A.: Modeling chemical processes using prior knowledge and neural networks. AIChE J. **40**(8), 1328–1340 (1994)
25. Von Rueden, L., et al.: Informed machine learning-a taxonomy and survey of integrating prior knowledge into learning systems. IEEE Trans. Knowl. Data Eng. **35**(1), 614–633 (2021)
26. Wehenkel, A., Behrmann, J., Hsu, H., Sapiro, G., Louppe, G., Jacobsen, J.H.: Improving generalization with physical equations. In: Machine Learning and the Physical Sciences: Workshop at the 36th Conference on Neural Information Processing Systems (NeurIPS) (2022)
27. Willard, J., Jia, X., Xu, S., Steinbach, M., Kumar, V.: Integrating scientific knowledge with machine learning for engineering and environmental systems. ACM Comput. Surv. **55**(4), 1–37 (2022)
28. Wouwer, A.V., Renotte, C., Bogaerts, P.: Biological reaction modeling using radial basis function networks. Comput. Chem. Eng. **28**(11), 2157–2164 (2004)
29. Yeh, I.C.: Concrete Compressive Strength. UCI Machine Learning Repository (2007). https://doi.org/10.24432/C5PK67
30. Yin, Y., Ayed, I., de Bézenac, E., Baskiotis, N., Gallinari, P.: LEADS: learning dynamical systems that generalize across environments. In: Advances in Neural Information Processing Systems, vol. 34, pp. 7561–7573 (2021)
31. Yin, Y., et al.: Augmenting physical models with deep networks for complex dynamics forecasting. J. Stat. Mech. Theory Exp. **2021**(12), 124012 (2021)
32. Yu, K., et al.: A partially linear tree-based regression model for multivariate outcomes. Biometrics **66**(1), 89–96 (2010)
33. Zhang, H., Nettleton, D., Zhu, Z.: Regression-enhanced random forests. arXiv preprint arXiv:1904.10416 (2019)

Natural Language Processing and Social Media Analysis

Audience Prediction for Game Streaming Channels Based on Vectorization of User Comments

Takayasu Fushimi$^{(\boxtimes)}$ ⓘ and Kazuki Sakamoto

School of Computer Science, Tokyo University of Technology, 1404-1 Katakuramachi, Hachioji-city, Tokyo 192–0982, Japan
takayasu.fushimi@gmail.com, c0119133b8@edu.teu.ac.jp

Abstract. As live video streaming has become an everyday activity, influential users on video streaming platforms are taking on the role of billboards for product services. Predicting channel growth would be useful for advertising marketing. In this study, we propose a method for predicting future bipartite links between users and channels by treating the relationship between sending and receiving comments as a dynamic bipartite graph. The text information posted by a user is expected to express the user's interests. Since user interests are also a factor in the growth process of a channel, it is an important research topic to verify whether it is possible to predict the growth of a channel using textual information. Specifically, the comments sent by users to live streaming are assumed to be the user's feature vector, and comments received on videos uploaded after live streamings are assumed to be the channel's feature vector, and the relationship between sending and receiving comments is assumed to be a bipartite graph link. The comments are converted into feature vectors by topic extraction using Latent Dirichlet Allocation. Then, a matrix that transforms the feature vectors is learned so that the inner product between the latent vectors of users and channels in the link relationship becomes large. The learned transformation matrix is then used to predict the presence or absence of future links. In our experiments, we evaluate 1. the validity of the bipartite link prediction method for predicting channel growth, 2. the validity of the proposed method for constructing feature vectors from comment data, and 3. the accuracy of link prediction using real data collected from YouTube.

Keywords: Bipartite Graph · Link Prediction · Latent Dirichlet Allocation · Video Streaming

1 Introduction

In recent years, there has been a rapid increase in the number of users utilizing services that allow individuals to livestream video content, such as YouTube, Nico Live, and Twitch. Many people participate in these platforms as viewers or

A. Bifet et al. (Eds.): DS 2023, LNAI 14276, pp. 81–95, 2023.
https://doi.org/10.1007/978-3-031-45275-8_6

streamers. The scale of these video streaming platforms can reach over 100,000 viewers for popular streamers. These streamers have a significant impact on the internet and society, possessing substantial influence as influencers. As a result, companies often leverage influencers in their advertising strategies. Therefore, the effectiveness of advertising in video streaming, the extraction of suitable influencers, and the study of video streaming properties and influencer extraction are important research topics that contribute to a company's profits. Extensive research has been conducted in these areas.

Additionally, there are numerous tasks related to graph prediction, including link prediction and prediction of new nodes. Studies on link prediction have proposed methods that calculate generation probabilities based on the similarity between nodes [4], methods specialized in link prediction for specific nodes [1], and methods that utilize the Holt-Winters method to learn trends in latent vector groups for link prediction [6]. However, these approaches do not consider data in which the presence or absence of nodes may change. Furthermore, studies have addressed this issue by predicting the probabilities of future node appearance, disappearance, and reappearance for use in prediction [7], and by learning distributed representations of unknown node sets [8]. Nevertheless, there has been no research utilizing text information associated with video streaming. In this study, we focus on the problem of link prediction using text information from user posts, based on the assumption that the interests of users are reflected in their posted texts. There are three key reasons highlighting the importance of handling text information posted by users. Firstly, user profiles and demographic attributes are often unavailable or difficult to obtain, making it challenging to reflect the evolving interests of users. Secondly, compared to similarity calculations among all users commonly used in information recommendation, the calculation cost of user interest vectors is relatively low. Thirdly, even in cases where past network structures are not available, the network structure can be predicted solely based on user interest vectors.

This study aims to predict channel growth to enable companies to identify influencers early on. In YouTube, parameters such as the number of channel subscribers and views exist as indicators of growth. The former is unobtainable through third-party sources, and the latter is difficult to predict due to various factors, such as the influence of content types on YouTube. In this study, we attempt to predict channel growth by predicting the number of messages directed from viewers who only appear during live broadcasts to streamers. Specifically, as shown in Fig. 1, we formulate the problem as a link prediction task in a dynamic bipartite graph, determining which users watch which channels. User comments during live channel broadcasts serve as user feature vectors, while comments on uploaded video content after the broadcast serve as channel feature vectors. By treating the sender-receiver relationship between users and channels as links in a bipartite graph, we aim to learn a transformation matrix from feature vectors to latent vectors, ensuring that the dot product between the vectors of linked users and channels is large, while the dot product between vectors of unlinked user-channel pairs is small. Using the learned transformation matrix, we predict whether users will link to the next live broadcast, i.e., send comments.

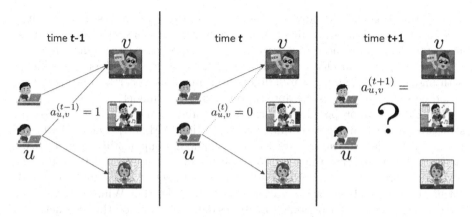

Fig. 1. Modeling as a Dynamic Bipartite Graph Based on Comment Sender-Receiver Relationships

2   Related Work

2.1   Video Streaming

Nonaka et al. proposed a method that uses the page views, edit counts, and inbound link counts of Wikipedia as indicators of content popularity, creates similarity vectors between contents based on individual users' editing histories, and predicts popularity using a multilayer neural network (MLP) [2]. While this method predicts popularity based on content similarity, it differs from this research in terms of link prediction in bipartite graph structures. Additionally, this research focuses on estimating content similarity rather than predicting the users' time-series data.

Okada et al. proposed a system that uses Juman and BERT-CRF to infer and label the meaning of viewer chats during live streams [3]. Although their proposal includes real-time processing and evaluation experiments, the relevance to this research lies in the labeling method. The evaluation experiments combining Juman and BERT-CRF showed lower estimation accuracy for seven out of eight labels compared to the combination of MeCab and CRF. While overfitting was mentioned as a possible cause, this research utilizes MeCab for morphological analysis. Additionally, since the labeling does not aim to characterize each user, CRF labeling is not suitable for this research.

2.2   Link Prediction

Shimura et al. experimentally demonstrated that combining the Generalized Linear Preference (GLP) model, an effective growth model verified by Wang et al.'s experiment [5], with a link prediction index based on network structure improves prediction accuracy [4]. In the GLP model, with a probability of p, m links are generated between existing nodes, and with a probability of $1 -$

p, m links are generated between new nodes and existing nodes. By adopting this growth model, it is possible to generate networks with scale-free properties similar to those of real networks. Additionally, the proposed link prediction index reflects the structural characteristics of the network and predicts future links with high accuracy. This research differs from previous research in terms of the proposed index, as well as the application of the GLP model.

Nakajima et al. proposed a method for link prediction in time-evolving graph structures using Non-negative Matrix Factorization (NMF) [1]. The method represents each adjacency matrix at each time step in a group of adjacency matrices (tensor) that represents the temporal changes in the graph structure. It uses NMF to represent nodes with low-dimensional latent vectors (matrices) and learns the trends of the latent vector group using the Holt-Winters method, a forecasting technique for periodic time series data. The method then predicts the adjacency matrices based on the obtained future latent vector group. The validation experiments have been conducted on data where the number of appearing nodes is fixed, meaning the size of the adjacency matrix is fixed, and it has shown good performance and ease of handling. However, experiments have not been conducted on data where the presence or absence of nodes changes.

Yamaguchi et al. proposed a method that focuses on link prediction for specific nodes, such as corporate accounts in SNS, which is different from predicting links for the entire network [6]. In this method, links are sampled through random walks starting from the target node, and a loss function is designed using weights based on the distance from the target node. The model is then trained using this loss function. The objective of this method aligns with the goal of this study, which attempts to predict the number of viewers for the target channel.

3   Proposed Method

In this study, we model the relationship between sending and receiving comments on a user's live streaming channel as a bipartite graph, and attempt to approximate the number of viewers by predicting the future graph structure based on the past link states and posted comments.

3.1   Bipartite Graph and Feature Vector

We model the relationship between the user set $\mathcal{U}$ and the channel set $\mathcal{V}$ in video streaming sites using a dynamic bipartite graph $B^{(t)} = (\mathcal{U}^{(t)}, \mathcal{V}^{(t)}, \mathcal{E}^{(t)})$. At time step t, $B^{(t)}$ is a bipartite graph, where the link set is defined as $\mathcal{E}^{(t)} \subset \mathcal{U}^{(t)} \times \mathcal{V}^{(t)}$. Specifically, we represent the relationship where a user $u \in \mathcal{U}^{(t)}$ sends a comment during the live streaming of channel $v \in \mathcal{V}^{(t)}$ as a link $e = (u, v) \in \mathcal{E}^{(t)}$. We refer to these comments during the live streaming as dynamic comments. The set of words contained in all dynamic comments posted by user u at time step t is denoted as $\mathcal{X}_u^{(t)}$.

On the other hand, we define comments on recorded videos uploaded after the live streaming of channel v as static comments. The set of words contained

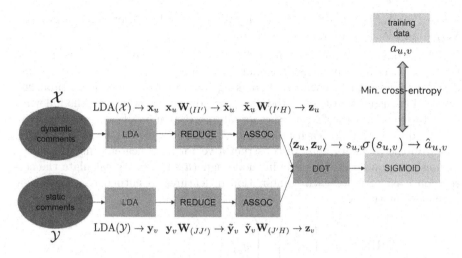

Fig. 2. Prediction model

in the static comments on the videos streamed by channel v at time step t is denoted as $\mathcal{Y}_v^{(t)}$. Based on the word frequencies in the comments, we extract I topics for users and J topics for channels using Latent Dirichlet Allocation, representing users and channels with vectors that consist of topic probabilities as their features:

- $\mathbf{x}_u^{(t)} \leftarrow \text{LDA}(\mathcal{X}_u^{(t)}; \mathcal{X}, I)$,
- $\mathbf{y}_v^{(t)} \leftarrow \text{LDA}(\mathcal{Y}_v^{(t)}; \mathcal{Y}, J)$,

where we use the corpus $\mathcal{X} = \bigcup_{t=1}^T \bigcup_{u \in \mathcal{U}^{(t)}} \mathcal{X}_u^{(t)}$ and $\mathcal{Y} = \bigcup_{t=1}^T \bigcup_{v \in \mathcal{V}^{(t)}} \mathcal{Y}_v^{(t)}$ to train the LDA. $\mathcal{X}$ consists of all dynamic comments from all users across all time steps, while $\mathcal{Y}$ consists of all static comments from all channels across all time steps. We arrange the I-dimensional and J-dimensional row vectors for each user and channel, respectively, into feature matrices $\mathbf{X}^{(t)} = [\mathbf{x}_u^{(t)}]_{u \in \mathcal{U}^{(t)}}$ and $\mathbf{Y}^{(t)} = [\mathbf{y}_v^{(t)}]_{v \in \mathcal{V}^{(t)}}$. From these matrices, we predict the link $a_{u,v}^{(t)}$ indicating whether user u commented on channel v at time step t.

3.2   Prediction Model

Figure 2 depicts an overview of our algorithm. In the proposed prediction model, we pass the feature vectors of users and channels through separate input layers. We perform dimensionality reduction on these feature vectors using weight matrices $\mathbf{W}_{(II')}$ and $\mathbf{W}_{(JJ')}$, respectively.

- $\tilde{\mathbf{x}}_u^{(t)} \leftarrow \text{REDUCE}(\mathbf{x}_u^{(t)}; \mathbf{W}_{(II')}) = \mathbf{x}_u^{(t)} \mathbf{W}_{(II')}$,
- $\tilde{\mathbf{y}}_v^{(t)} \leftarrow \text{REDUCE}(\mathbf{y}_v^{(t)}; \mathbf{W}_{(JJ')}) = \mathbf{y}_v^{(t)} \mathbf{W}_{(JJ')}$,

where I' and J' are the number of dimensions after dimension compression and $\mathbf{W}_{(II')}$ and $\mathbf{W}_{(JJ')}$ are matrices updated by learning.

The obtained $\tilde{\mathbf{x}}_u^{(t)}$ and $\tilde{\mathbf{y}}_v^{(t)}$ are vectors consisting of separately extracted features for users and channels. Therefore, the dimensions of the vectors do not directly correspond to each other, making it impossible to perform similarity calculations using simple dot products. To address this, we follow the following steps to transform the feature vectors into corresponding feature vectors with aligned dimensions. For each link $e \in \mathcal{E}^{(t-1)}$ in the bipartite graph $B^{(t-1)}$ predicted in the previous step, we construct a vector by concatenating the feature dimensions i and j of the endpoint node pair $(u, v) = e$. We calculate the correlation coefficient for each combination of features, resulting in a correlation coefficient matrix $\mathbf{R}^{(t)}$:

$$- \; r_{ij}^{(t)} \leftarrow \text{CORRCOEF}(\begin{pmatrix} \vdots \\ \tilde{x}_{ui}^{(t)} \\ \vdots \end{pmatrix}, \begin{pmatrix} \vdots \\ \tilde{y}_{vj}^{(t)} \\ \vdots \end{pmatrix})$$

$$- \; \mathbf{R}^{(t)} = [r_{ij}^{(t)}]_{\substack{1 \leq i \leq I' \\ 1 \leq j \leq J'}}$$

The number of concatenated node pairs is denoted as $L^{(t-1)} = |\mathcal{E}^{(t-1)}|$. Additionally, when the correlation coefficient $r_{ij}^{(t)}$ is high, it indicates a strong relationship between the reduced-dimensional topics corresponding to dimensions i and j.

We transform the feature vectors of users and channels, which are obtained separately as I'-dimensional user feature vectors and J'-dimensional channel feature vectors, into a common H-dimensional vector. To achieve this, we decompose the matrix $\mathbf{R}^{(t)}$, which consists of the correlations between I' features and J' features, into $\mathbf{W}_{(I'H)}$ and $\mathbf{W}_{(J'H)}$ as follows:

$$- \; \mathbf{R}^{(t)} \rightarrow \mathbf{W}_{(I'H)} \mathbf{W}_{(J'H)}^T$$

Then, we transform the user feature vector $\tilde{\mathbf{x}}_u^{(t)}$ and the channel feature vector $\tilde{\mathbf{y}}_v^{(t)}$ as follows:

$$- \; \mathbf{z}_u^{(t)} \leftarrow \text{ASSOCIATE}(\tilde{\mathbf{x}}_u^{(t)}; \mathbf{W}_{(I'H)}) = \tilde{\mathbf{x}}_u^{(t)} \mathbf{W}_{(I'H)},$$
$$- \; \mathbf{z}_v^{(t)} \leftarrow \text{ASSOCIATE}(\tilde{\mathbf{y}}_v^{(t)}; \mathbf{W}_{(J'H)}) = \tilde{\mathbf{y}}_v^{(t)} \mathbf{W}_{(J'H)}.$$

The resulting vectors $\mathbf{z}_u^{(t)}$ and $\mathbf{z}_v^{(t)}$ have dimensions that correspond to a common meaning.

Finally, we calculate the similarity between user u and channel v by taking the dot product of the obtained vectors:

$$- \; s_{u,v}^{(t)} \leftarrow \text{DOT}(\mathbf{z}_u^{(t)}, \mathbf{z}_v^{(t)}).$$

We use the sigmoid function to compute the link existence probability between user u and channel v based on the obtained dot product:

$$- \; \hat{a}_{u,v}^{(t)} \leftarrow \sigma(s_{u,v}^{(t)}) = \frac{1}{1+\exp(-s_{u,v}^{(t)})}.$$

3.3   Learning Algorithm

We aim to learn various parameters by minimizing the cross-entropy between the actual link existence (comment exchange relationship) $a_{u,v}^{(t)}$ and the predicted probability $\hat{a}_{u,v}^{(t)}$. The objective function is defined as:

$$F(\mathbf{X}^{(t)}, \mathbf{Y}^{(t)}; \mathbf{A}^{(t)}) = \sum_{u \in \mathcal{U}^{(t)}} \sum_{v \in \mathcal{V}^{(t)}} \{-a_{u,v}^{(t)} \log(\hat{a}_{u,v}^{(t)}) - (1 - a_{u,v}^{(t)}) \log(1 - \hat{a}_{u,v}^{(t)})\}. \quad (1)$$

By extracting the term related to user u and channel v, we have:

$$F_{u,v}^{(t)} = -a_{u,v}^{(t)} \log(\hat{a}_{u,v}^{(t)}) - (1 - a_{u,v}^{(t)}) \log(1 - \hat{a}_{u,v}^{(t)}). \quad (2)$$

Differentiating the error function and the activation function (sigmoid function) with respect to the input $s_{u,v}^{(t)}$, we obtain:

$$\begin{aligned}
\frac{\partial F_{u,v}^{(t)}}{\partial s_{u,v}^{(t)}} &= \frac{\partial F_{u,v}^{(t)}}{\partial \hat{a}_{u,v}^{(t)}} \cdot \frac{\partial \hat{a}_{u,v}^{(t)}}{\partial s_{u,v}^{(t)}} \\
&= -\frac{a_{u,v}^{(t)}}{\hat{a}_{u,v}^{(t)}} + \frac{1 - a_{u,v}^{(t)}}{1 - \hat{a}_{u,v}^{(t)}} \cdot \hat{a}_{u,v}^{(t)}(1 - \hat{a}_{u,v}^{(t)}) \\
&= \hat{a}_{u,v}^{(t)} - a_{u,v}^{(t)}.
\end{aligned}$$

Furthermore, by differentiating the DOT layer's output $s_{u,v}^{(t)}$ with respect to $\mathbf{x}_u^{(t)}$ and $\mathbf{y}_v^{(t)}$, respectively, we obtain:

$$-\frac{\partial s_{u,v}^{(t)}}{\partial \mathbf{x}_u^{(t)}} = \mathbf{W}_{(I'H)}^T \mathbf{W}_{(II')}^T,$$
$$-\frac{\partial s_{u,v}^{(t)}}{\partial \mathbf{y}_v^{(t)}} = \mathbf{W}_{(J'H)}^T \mathbf{W}_{(JJ')}^T.$$

Using these gradients, we update various parameters through backpropagation. Then, using the learned parameters at time step t, we predict $\hat{a}_{u,v}^{(t+1)} \leftarrow \sigma(s_{u,v}^{(t)})$ for time step $t + 1$.

4   Experimental Settings

4.1   Dataset

In this study, we evaluate the effectiveness of the proposed method using real data collected from YouTube through the API. The target channels in this study consist of 40 individual channels that engage in more video streaming than video posting.

The data collection period spans six months from October 1, 2021, to March 31, 2022. For this experiment, we obtained chat data by crawling the chat of archived videos streamed on YouTube. Additionally, since viewers can post comments on archived videos after the broadcast, we used the YouTube Data API to retrieve this data. In this study, we distinguish between these two types of text data: the former is referred to as dynamic comments, and the latter as static comments. These data include user IDs uniquely assigned to YouTube users, allowing us to obtain the following statistics: During the target period, there were 6,327,958 commenters and a total of 354,126,994 comments for dynamic comments. For static comments, there were 858,130 commenters and a total of 2,002,441 comments. It should be noted that if either dynamic or static comments cannot be retrieved due to channel settings or other reasons, we do not use the corresponding data as it would be inappropriate.

Next, to treat dynamic comments as user feature vectors and static comments as channel feature vectors, we processed the data using the following steps:

1. Removal of emojis, URLs, symbols, and conversion of full-width characters to half-width, as well as conversion of uppercase to lowercase.
2. Word segmentation using MeCab with the mecab-ipadic-NEologd dictionary.
3. Removal of unnecessary words using the stopwords listed below.
4. Removal of words with a frequency of less than 50 from the obtained set of words.
5. Creation of a dictionary from the obtained set of words.
6. Creation of a corpus using doc2bow from the obtained set of words.
7. Construction of an LDA model using the created dictionary and corpus.

The input to the LDA model obtained above is processed as follows based on the set of words obtained in step 4:

- For dynamic comments: Concatenate the text of the chat a user had on that day and input it.
- For static comments: Concatenate all the texts commented on an archive for a specific date and input it.

In this study, as described above, we represent channel features using text data. However, since it is expected that the viewers differ for each channel, it is necessary to evaluate whether the channel features can express the same or different characteristics among channels with the same or different viewer demographics. Therefore, as a preliminary experiment, we prepare for a comparison to determine if there is a correlation between the channel feature vectors generated by LDA and the game titles played by the channels. The number of times games were played during the experimental period is as follows for the top 10 games: Apex Legends: 8,743, Minecraft: 2,218, VALORANT: 1,095, Ark: Survival Evolved: 930, PUBG: 649, Mario Kart 8 Deluxe: 427, Mahjong Soul: 410, Monster Hunter Rise: 297, Pokémon Brilliant Diamond/Shining Pearl: 266, Shadowverse: 264.

4.2   Comparison Method

For a bipartite graph structure $B^{(t)} = (\mathcal{U}^{(t)}, \mathcal{V}^{(t)}, \mathcal{E}^{(t)})$ at time step t, the latent vectors of users and channels are computed using Singular Value Decomposition

(SVD), Non-negative Matrix Factorization (NMF), and Matrix Factorization. The link existence at time step $t + 1$ is predicted using the inner product of these vectors. Let $M^{(t)} = |\mathcal{U}^{(t)}|$ be the number of users and $N^{(t)} = |\mathcal{V}^{(t)}|$ be the number of channels. The adjacency matrix of users versus channels at time step t is represented by a matrix $\mathbf{A}^{(t)}$ of size $M^{(t)} \times N^{(t)}$.

Singular Value Decomposition:
   In SVD, the adjacency matrix $\mathbf{A}^{(t)}$ is decomposed into a matrix $\mathbf{P}$ containing left singular vectors, a matrix $\mathbf{\Sigma}$ with K singular values in decreasing order on the diagonal, and a matrix $\mathbf{Q}$ containing right singular vectors. Here, K is at most the rank of matrix $\mathbf{A}$, but in this paper, we experiment with $K = 10$ to match the dimensionality of the proposed method. Thus, each user is represented by a 10-dimensional left singular vector $\mathbf{p}_u$, each channel is represented by a 10-dimensional right singular vector $\mathbf{q}_v$, and the prediction is made using $\hat{a}_{u,v}^{(t+1)} \leftarrow \sum_{k=1}^{K} \sigma_k \cdot p_{u,k} \cdot q_{v,k}$.

Non-negative Matrix Factorization:
   In NMF, two non-negative matrices $\mathbf{P}$ and $\mathbf{Q}$ of size $M^{(t)} \times K$ and $N^{(t)} \times K$, respectively, are computed such that the Frobenius norm of the product of the adjacency matrix $\mathbf{A}^{(t)}$ and these matrices is minimized, i.e.,

$$\min_{\mathbf{P} \in \mathbb{R}_{\geq 0}^{M^{(t)} \times K}, \mathbf{Q} \in \mathbb{R}_{\geq 0}^{N^{(t)} \times K}} \sum_{u \in \mathcal{U}^{(t)}} \sum_{v \in \mathcal{V}^{(t)}} \left(a_{u,v}^{(t)} - \langle \mathbf{p}_u, \mathbf{q}_v \rangle \right)^2$$

Matrix Factorization:
   In MF, K-dimensional latent vectors $\mathbf{p}_u$ and $\mathbf{q}_v$ are computed for each link $(u, v) \in \mathcal{E}^{(t)}$, minimizing the error

$$\sum_{(u,v) \in \mathcal{E}^{(t)}} \left(a_{u,v} - \sum_{k=1}^{K} p_{u,k} \cdot q_{v,k} \right)^2 + \lambda \left(\|\mathbf{P}\|_F^2 + \|\mathbf{Q}\|_F^2 \right)$$

where $\lambda \left(\|\mathbf{P}\|_F^2 + \|\mathbf{Q}\|_F^2 \right)$ is the regularization term to prevent overfitting.

5   Experimental Results

5.1   Evaluation of Channel Feature Vectors

Figure 4 is a heatmap representation of dissimilarity between channels. The average feature vector $\bar{\mathbf{y}}_v \leftarrow \frac{1}{T} \sum_{t=1}^{T} \mathbf{y}_v^{(t)}$ for channel v is obtained by averaging the feature vectors $\mathbf{y}_v^{(t)}$ at each time step t. The dissimilarity between channel pairs $(u, v) \in \mathcal{V} \times \mathcal{V}$ is calculated using the L1 distance between their average feature vectors, given by $d_{L1}(\bar{\mathbf{y}}_u, \bar{\mathbf{y}}_v) = \sum_{j=1}^{J} |\bar{y}_{uj} - \bar{y}_{vj}|$. In the heatmap, values closer to white indicate smaller L1 distances and hence closer dissimilarities between channels. This means that the average feature vectors are similar, indicating a higher likelihood of being similar channels. Based on this figure, several examples

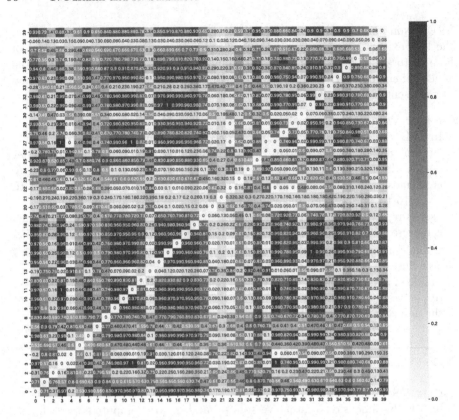

Fig. 3. Inter-channel game title similarity (Cosine similarity)

were identified through the investigation of the actual channel content, revealing similarities or differences between channels.

Channels 4, 5, 7, 13, 18, 20, 21, 22, 24, 25, 26, 30, 33, 37, 38, and 39 appear relatively darker, indicating that they have different content compared to channels 8 to 12. Among these channels, excluding 7, 33, 37, and 38, the remaining 12 channels belong to the same agency A, and it can be observed that the L1 distances between them are low. Since there are no other channels belonging to agency A in the dataset, this distinct feature can be attributed to their affiliation with the agency. On the other hand, channels 7, 33, 37, and 38, which show different characteristics from agency A's content, were confirmed to have distinct features through the actual investigation.

Particularly, channel 6, which includes pairs with low values, has interesting findings. The pair between channel 6 and channel 19 belongs to agency B and collaborates frequently. The pair between channel 6 and channel 34 has different agencies and forms, but they often share similarities in terms of closeness and collaboration, indicating similar content.

Next, a comparison was made to determine whether there is a correlation between the feature vectors obtained from channel titles and the feature vectors generated by LDA. Figure 5 plots the cosine similarity $cos(u, v)$ from Fig. 3

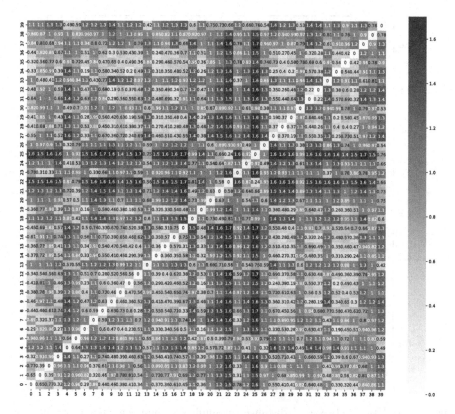

Fig. 4. L1 distance between channel vectors

on the x-axis and the L1 distance $d_{L1}(u, v)$ from Fig. 4 on the y-axis, showing the correlation coefficient. Figure 4 represents the L1 distances (dissimilarity) between channels, while Fig. 3 represents the cosine similarity between channels. Therefore, as the correlation coefficient of the numerical pairs approaches -1, both measures indicate similar properties. The obtained correlation coefficient in this experiment is -0.5, indicating a slight correlation. Thus, it can be concluded that the feature vectors obtained using LDA from static comments are sufficiently suitable for representing the similarity based on the game titles that channels played.

Figure 6 depicts the transition of similarity among sets of users who posted comments during consecutive broadcasts for each channel. The similarity of user sets is calculated using the Jaccard coefficient $\frac{|\Gamma(v)^{(t)} \cap \Gamma(v)^{(t+1)}|}{|\Gamma(v)^{(t)} \cup \Gamma(v)^{(t+1)}|}$, where $\Gamma(v)^{(t)}$ represents the set of commenting users for channel v during time step t. Three randomly selected channels are displayed in the figure, showing that all Jaccard coefficients have small values. It can be observed that the user sets vary considerably between consecutive broadcasts, making it difficult to predict the user set

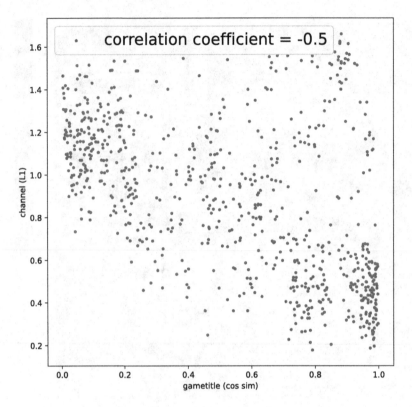

Fig. 5. Correlation between L1 distance (dissimilarity) between channel vectors and cosine similarity of game titles

for the next time step using information from the previous time step. Although not shown in the figure, similar patterns were observed for other channels as well.

5.2 Prediction Accuracy Comparison

Figure 7 illustrates the prediction accuracies of the proposed method and various comparison methods. The x-axis represents each time step, which, in this case, is set to daily intervals. Thus, the prediction results cover a period of 181 days from October 2, 2021, to March 31, 2022. Looking at the figure, we can observe that the proposed method, represented by the red line, consistently achieves high accuracy. On the other hand, the SVD method (green line) and the NMF method (blue line) occasionally exhibit higher prediction accuracy than the proposed method, but on average, they yield lower results. The MF method (yellow line) shows very low prediction accuracy in the given dataset. Finally, the LDA method (pink line) represents the accuracy when using the node's feature vector directly for prediction, and as expected, it yields low values.

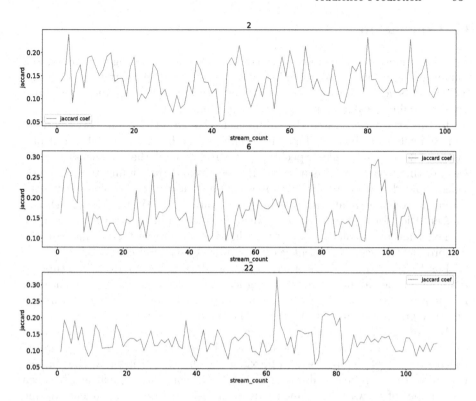

Fig. 6. Transition of similarity of comment posting user group

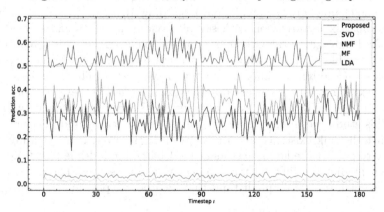

Fig. 7. Prediction accuracy (Color figure online)

Based on these results, it can be concluded that methods that generate topic vectors from text information for prediction are more effective than approaches using adjacency matrices. Furthermore, it suggests that the text information posted by users reflects their interests.

6 Conclusion

In recent times, live streaming of videos has become increasingly common, and influential users on video streaming platforms are taking on the role of brand ambassadors for various products and services. Therefore, predicting channel growth would be valuable in advertising and marketing.

In this study, we treated the comment exchange relationship between users and channels as a dynamic bipartite graph and addressed the problem of predicting future graph links. Specifically, we represented the comments sent by users during live streams as user feature vectors and the comments received on videos uploaded after the live stream as channel feature vectors, considering the comment exchange relationship as bipartite graph links. We used Latent Dirichlet Allocation for topic extraction to vectorize the comments' features. Then, we learned transformation matrices to modify the feature vectors to increase the dot product between the latent vectors of linked users and channels.

For the experiments, we used real-world data collected from YouTube to evaluate: 1) the validity of constructing feature vectors based on comment data in our proposed method, 2) the validity of channel growth prediction through bipartite graph link prediction, and 3) the accuracy of link prediction.

Regarding 1), the correlation coefficient between cosine similarity based on the frequency of mentioned game titles and L1 distance between the generated feature vectors of channels was approximately -0.5, indicating that the channel feature vectors were reasonable. Regarding 2), the correlation coefficient between the number of views on streamed videos and the number of comments during the stream was approximately 0.71, demonstrating that link prediction serves as a pseudo channel growth prediction. Regarding 3), we compared our method based on matrix factorization with several existing methods and showed that our method consistently achieved higher prediction accuracy. This demonstrates that it is possible to predict whether a user will view a channel in the next time step based on the textual information they have posted. Thus, the hypothesis that the user's interests are reflected in the text information they post holds true, even when treating the user's feature vectors in the dynamic bipartite graph.

In this study, we directly used the latent vectors calculated at time step t to predict the link existence at time step $t + 1$. However, it is generally more accurate to predict the latent vectors at $t + 1$ based on multiple past time steps and then predict the link existence based on those vectors. Therefore, in the future, we plan to explore predictions that incorporate past information.

Acknowledgements. This work was supported by JSPS KAKENHI Grant Number JP22K12279.

References

1. Mutinda, F., Nakashima, A., Takeuchi, K., Sasaki, Y., Onizuka, M.: Time series link prediction using NMF. J. Inf. Process. **27**, 752–761 (2019)

2. Nonaka, N., Nakayama, K., Matsuo, Y.: Contents popularity prediction by vector representation learned from user action history. In: Proceedings of the 6th International Conference on Data Analytics (Data Analytics 2017) (2017)
3. Okada, M., Kajioka, S., Yamamoto, D., Takahashi, N.: Filtering audience chat with BERT-CRF for Youtube broadcasters. In: Proceedings of the 14th Data Engineering and Information Management Forum (DEIM2022), pp. C24–1 (2022). (in Japanese)
4. Shimura, K., Ohara, K., Toyoda, T.: Proposal of link prediction method using probabilistic model. In: Proceedings of the 111th Knowledge Base System Study Group (SIG-KBS111), pp. B507–04 (2017). (in Japanese)
5. Wang, W.Q., Zhang, Q.M., Zhou, T.: Evaluating network models: a likelihood analysis. EPL (Europhys. Lett.) **98**(2), 28004 (2012). https://doi.org/10.1209/0295-5075/98/28004
6. Yamaguchi, H., Ogawa, Y., Maekawa, S., Sasaki, Y., Onizuka, M.: Controlling internal structure of communities on graph generator. In: IEEE/ACM International Conference on Advances in Social Networks Analysis and Mining (ASONAM), pp. 937–940 (2020)
7. Yamasaki, S., Harada, K., Sasaki, Y., Onizuka, M.: Prediction of time series graph considering appearance of unknown nodes using deep learning. In: Proceedings of the 11th Data Engineering and Information Management Forum (DEIM2019), pp. D7–6 (2019). (in Japanese)
8. Yamasaki, S., Harada, K., Sasaki, Y., Onizuka, M.: Distributed representation learning of nodes based on bipartite graph maximum weight matching for graph prediction. In: Proceedings of the 12th Data Engineering and Information Management Forum (DEIM2020), pp. F3–3 (2020). (in Japanese)

From Tweets to Stance: An Unsupervised Framework for User Stance Detection on Twitter

Margherita Gambini[ID], Caterina Senette[(✉)][ID], Tiziano Fagni[ID], and Maurizio Tesconi[ID]

Institute of Informatics and Telematics (IIT) - CNR,
Via Giuseppe Moruzzi, 1, 56124 Pisa, Italy
{m.gambini,c.senette,f.fagni,m.tesconi}@iit.cnr.it

Abstract. Current stance inference methods use topic-aligned training data, leaving many unexplored topics due to the lack of training data. Zero-shot approaches utilizing advanced pre-trained Natural Language Inference (NLI) models offer a viable solution when training data is unavailable. This work introduces the *Tweets2Stance - T2S* framework, an unsupervised stance detection framework based on Zero-Shot Learning. It detects a five-valued user's stance on social-political statements by analyzing their Twitter timeline. The ground-of-truth user's stance is obtained from Voting Advice Applications (VAAs), online tools that compare political preferences with party political stances. The *T2S* framework's generalization potential is demonstrated by measuring its performance (F1 and MAE scores) across nine datasets. These datasets were built by collecting tweets from competing parties' Twitter accounts in nine political elections held in different countries from 2019 to 2021. Through comprehensive experiments, an optimal setting was identified for each election. The results, in terms of F1 and MAE scores, outperformed all baselines and approached the best scores for each election. This showcases the ability of T2S to generalize across different cultural-political contexts.

Keywords: user stance detection · transfer learning · unsupervised · Twitter · text content · elections · vaa

1 Introduction

Stance detection (SD) is a text-mining approach that infers the expression of a user's point of view and perception toward a given statement [3]. Unlike sentiment analysis, which categorizes a text as positive, negative, or neutral regardless of a specific target, stance detection focuses on classifying a text based on the user's attitude toward a predetermined target. It is commonly applied in two areas: inferring user agreement/disagreement in social media debates across various contexts (such as political, ideological, and social), and assessing public opinion on products and services [8,15].

ALDayel et al. [2] proposed a recent taxonomy of stance detection tasks. Firstly, the level at which stance is computed must be determined, whether it

© The Author(s), under exclusive license to Springer Nature Switzerland AG 2023
A. Bifet et al. (Eds.): DS 2023, LNAI 14276, pp. 96–110, 2023.
https://doi.org/10.1007/978-3-031-45275-8_7

involves detecting the stance expressed in a piece of text or inferring the stance of a user towards a specific target based on their posted content and context. Secondly, the targets for detecting the stance need to be identified. These targets can be single (e.g., a specific topic), multi-related (where expressing a stance towards one target implies a stance towards similar targets), or claim-based (determining whether a text or user confirms a claim).

In this study, we focus on the investigation and measurement of public opinion on several issues as *user stance detection on multiple unrelated targets*. The analysis of texts extracted from social media users' posts provides valuable information for making such inferences. However, existing literature proposes mixed approaches that partially exploit text analysis in conjunction with user behaviour analysis, such as likes, retweets, and the network of contacts [1,10]. Additionally, user stance detection on unrelated targets poses computational challenges [2]. Limitations in content-based stance detection approaches include the inherent difficulty of processing natural language, the need for large annotated corpora of tweets and language-specific resources, the lack of unsupervised transfer learning to generalize across unrelated targets, and the requirement of training separate classifiers for each target. State-of-the-art research often focuses on two (support, against) or three levels of stance (including the neutral class[1]), and existing unsupervised methods based on clustering techniques in user networks are not suitable for inferring a user's stance for different unrelated targets.

Therefore, in an attempt to address these issues and focus solely on a content-based approach, we present *Tweets2Stance*(T2S), an unsupervised framework for stance detection. T2S analyzes the content of a user's social media (e.g., Twitter) timeline using Zero-Shot Classification (ZSC) techniques [21] to detect their stance towards specific socio-political statements (targets), considering five levels of agreement (completely disagree, disagree, neither disagree/nor agree, agree, completely agree).

To sum up, this work investigates a completely unsupervised solution to user-stance detection by answering the following research questions:

RQ1 – *What are the performances and insights of a completely unsupervised user-stance detection framework leveraging zero-shot classification capabilities on textual contents only?* Here, we also compare T2S's performance when used to detect either five or three stance classes.

RQ2 – *Is there a general framework that performs well across different political contexts?* Here, we explore the generalizing capabilities of T2S.

Contributions. To the best of our knowledge, we *filled the gap* of investigating an *unsupervised content-based-only* model leveraging an advanced Natural Language Processing technique (that is the *Zero-Shot Classification*) to detect a *five-level stance* of a user on *multiple and diverse targets* (the socio-political statements on different political contexts). Furthermore, the framework can be adapted for various scenarios, extending beyond the specific political context

---

[1] The *neutral* level indicates that the user or text did not express a stance on that target or does not take a stance at all.

addressed in this study. Additionally, we offer a set of labeled datasets that can assist other researchers in their endeavors involving unsupervised stance-detection techniques at the user level.

The remainder of this paper is organized as follows: Sect. 2 discusses related work. In Sect. 3, we define the user stance-detection task and dataset collection. Section 4 details the Tweets2Stance framework and experiment settings. Section 5 summarizes and discusses the results, highlighting limitations. Finally, Sect. 6 concludes and suggests future work.

2   Related Work

In the classical definition [3], user-level stance detection involves detecting a user's stance on a given topic based on their authored text. In the following paragraphs, we summarize the literature on *user-based* stance detection in social media, considering the features used and the learning approach.

Content and Behavioural Features. Rashed et al. (2021) [17] focused on user-based stance detection using content features alone. They employed Google's Multilingual Universal Sentence Encoder (MUSE) and a pre-trained CNN to extract tweet embeddings. User representation was obtained by averaging these embeddings and projected onto a two-dimensional plane using the Uniform Manifold Approximation and Projection (UMAP) technique. The authors utilized hierarchical density-based clustering (HDBScan) to classify users into pro and anti stances, achieving an F1 score of 0.86 on a dataset of 168k users. Moreover, interaction patterns and historical behaviour on social media, in addition to content features, can be used as well: Darwish et al. (2020) [4] successfully clustered users based on feature similarities such as retweets, common hashtags, and retweeted accounts; Aldayel et al. (2019) [1] achieved an F1 score of 0.72 by leveraging users' online behaviour cues; Thonet et al. (2017) [18] considered both text and social interactions to uncover topics, user viewpoints, and discourse; Magdy et al. (2016) [14] focused on elements such as retweets, replies, mentions, URLs, and hashtags to predict unexpressed stances (a stance that may or may not have transpired *yet*), not to detect them (an existing stance in past data) (See footnote 1). Lastly, Fraiser et al. (2018) [6] used content-based and social-based proximities in a multi-layer graph, achieving an F1 score of 0.95.

Supervised and Unsupervised Learning. Stance detection techniques using supervised learning rely on large annotated datasets [16]. User-based stance detection has received less attention in these competitions, but notable studies include Aldayel et al. [1] and Magdy et al. (2016) [14]. Aldayel et al. (2019) trained a stance detector for each topic using the SemEval2016 dataset with 3,000 users. Magdy et al. (2016) collected timelines of 44,000 users discussing the Paris terrorist attack, while Fraiser et al. (2018) [6] applied a proximity-based two-level stance detector to different datasets related to political events and gun control. More recently [9,11], the trend in language processing for stance-detection tasks relies on language representation models (e.g., BERT [5]) pre-trained on large un-annotated corpora and *fine-tuned* on labelled and domain-specific datasets [5,21]. The work of Devlin et al. (2018) [5] demonstrated how BERT led to considerable performance improvements for NLP tasks such as sentiment analysis. Ghosh et al. (2019) [9] reported BERT's successful use in stance detection compared to other techniques. Here, the BERT model takes the text as input to generate representations of the words through multiple transformer layers, and then the system is fine-tuned on the task-specific data. Lately, Zang et al. (2023) [22] leveraged ChatGPT for text-based stance detection, achieving state-of-the-art or similar performance on SemEval-2016 [16] and PStance [13] datasets.

To the best of our knowledge, no unsupervised technique for user-based stance detection has yet utilized advanced Transformer-based language models. Existing unsupervised methods, such as Darwish et al. (2019) [4], Trabelsi et al. (2018) [19], and Fraiser et al. (2018) [6], rely on standard linguistic features like n-grams and keyword counts. Recognizing this gap and the increasing use of pre-trained models in stance detection, we propose Tweets2Stance, the first stance detector to work on a five-level stance. We evaluated T2S across diverse political contexts, achieving satisfactory results despite the challenging task.

Comparing the T2S framework to state-of-the-art user-based stance detection methods presents several challenges. Firstly, existing methods (e.g., Rashed et al., 2021 [17]) filtered tweets by removing mentions of specific targets, which is incompatible with our work as our topic lacks a well-defined person or organization. Additionally, these methods rely on timelines of users connected through specific keywords, while T2S aims to infer stance for any random user on any topic without leveraging shared features like retweets or common mentions. Unlike existing methods, updating context for new users in Tweets2Stance does not require recomputing networks and clusters. Moreover, the unavailability of public datasets used by state-of-the-art methods prevents us from evaluating T2S on those datasets. Furthermore, the lack of publicly available labelled datasets for five-level stance further limits the comparison.

3   Task Definition

The task is to detect the stance A_s^u of a Social Media User u with respect to a socio-political statement (or sentence) s making use of the User's textual content timeline (sequence of textual posts) on the considered social media (e.g., the Twitter timeline). The stance A_s^u represents a five-level categorical label: *completely agree* (5), *agree* (4), *neither disagree nor agree* (3), *disagree* (2), *completely disagree* (1). The integer mappings used by the Tweets2Stance framework are shown in parentheses. The label *neither disagree nor agree* encompasses both a not expressed and neutral stance. We refer to the *agreement/disagreement level (or label)* as the stance level (or label). The desired ground-of-truth (GoT) is the label G_s^u, which represents the known agreement/disagreement level of User u regarding sentence s. The GoT is solely used for evaluating our proposed framework and optimizing its parameters; no training step is involved. In this

Table 1. Details of the nine elections under study with the total number of tweets. D_i contains i months of tweets. Values between round brackets are the average number of tweets per Party.

Election	no. of parties	no. of statements	D3	D4	D5	D7
Alberta Provincia Election (AB19)	5	18	5119 (1024)	5701 (1140)	6755 (1351)	8502 (1700)
Australian Federal Election (AU19)	3	17	2538 (846)	3130 (1043)	3368 (1123)	4582 (1527)
Canadian Federal Election (CA19)	6	16	7460 (1243)	9284 (1547)	10750 (1792)	12903 (2151)
Great Britain Election v	5	20	9135 (1827)	10783 (2157)	12074 (2415)	15145 (3029)
British Columbia (BC20)	3	20	3560 (1187)	3751 (1250)	3969 (1323)	4448 (1483)
Saskatchewan Provincial Election (SK20)	2	17	1070 (535)	1245 (623)	1557 (779)	1982 (991)
New Foundland and Labrador Provincial Election (NFL21)	3	12	930 (310)	986 (322)	1070 (357)	1293 (431)
New Scotia Provincial Election (NS21)	3	17	859 (286)	1027 (342)	1454 (485)	1727 (579)
Canadian Federal Election (CA21)	6	16	6752 (1125)	7756 (1293)	8734 (1456)	10931 (1822)

study, users are assumed to be Twitter accounts of various political parties from different countries, as described in the subsequent section.

3.1   Data Collection

A Voting Advice Application (VAA) is an established online tool that helps citizens determine their political leaning by comparing their stance on socio-political statements (e.g., "Brexit was an error") with the positions of political parties. To analyze the Parties' stances, we collected data from eight political elections held between 2019 and 2021 *VoteCompass*[2], including the 2019 Great Britain Election *WhoGetsMyVoteUK*[3]. The statements and corresponding Ground-Of-Truths (GoTs) for each election and Party can be found in the provided repository[4].

For our analysis, we collected the Twitter timelines of the competing Parties using the Full-archive search Twitter API. Since some Parties had significantly fewer tweets compared to others, we removed certain Parties from the analysis and focused on those listed in Table 1. D_i represents the collection of tweets posted within i months before the election day (further details in the Methodology section).

---

[2] https://www.votecompass.com/.
[3] https://www.whogetsmyvoteuk.com/#!/.
[4] https://github.com/marghe943/Tweets2Stance_generalization.

4 Methodology

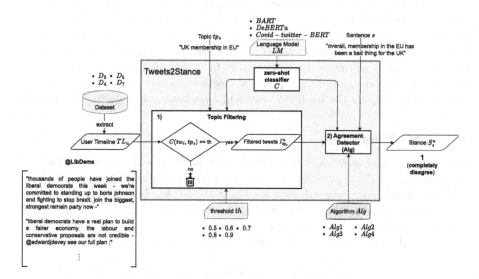

Fig. 1. Our Tweets2Stance framework to compute the agreement/disagreement level A_s^u of User u in regard to sentence s. The inputs are the Twitter timeline TL_u extracted from a certain time-period dataset D_i, the sentence s, the topic tp associated with s, a language model LM, a threshold th and an algorithm Alg. The highlighted components are the parameters that we'll vary during our experiments, as explained in Sect. 4.3.

This section presents the proposed Tweets2Stance (T2S) framework (Fig. 1) to detect the stance A_s^u of a Twitter User u in regard to a sentence s, exploiting its Twitter timeline $TL_u = [tw_1, ..., tw_n]$.

A User might either not talk about a specific political argument (here expressed with sentence s), or debate on an issue not risen by our pre-defined set of statements. For these reasons, our framework executes a preliminary *Topic Filtering* step, exploiting a Zero-Shot Classifier (ZSC) to get only those tweets talking about the topic tp of the sentence s. A ZSC is a language-model-based method that, given a text and a set of labels (e.g., topics), assigns a classification probability score to each label [21]. The higher the score assigned to a label, the higher the likelihood that the input text pertains to that specific label. ZSC does not require further fine-tuning on the target dataset. After obtaining the in-topic tweets $I_{tp_s}^u$ through Topic Filtering, the Agreement Detector module employs the same ZSC to detect the user's agreement/disagreement level.

Figure 1 colour-codes the four parameters of the *T2S* framework to be tuned:

1. the language model (LM) used for Zero-Shot Classification (ZSC) in the *Topic Filtering* and *Agreement Detector* modules to gauge topic agreement and sentence relevance, respectively,
2. the dataset D_i from which extracting the timeline TL_u,

3. the algorithm *Alg* to use in the *Agreement Detector* module,
4. the threshold *th* to get the in-topic tweets $I^u_{tp_s}$ in the *Topic Filtering* module.

The next subsections provide detailed descriptions of the *Topic Filtering* and *Agreement Detector* modules. We will focus on a specific political scenario where the Twitter accounts of interest are those of the political Parties mentioned in Sect. 3.1, and the User u corresponds to the Party p. The choice of the dataset's time period (D_i) as one of the parameters to tune is motivated by the use of T2S for stance detection during political elections, where the proximity to the elections may impact the likelihood of users discussing socio-political topics.

4.1   Topic Filtering

The *Topic Filtering* module extracts the in-topic tweets $I^p_{tp_s}$ from the Twitter Timeline TL_p of Party p, using the topic tp_s associated with sentence s (e.g., the topic for the sentence "*overall, membership in the EU has been a bad thing for the UK*" can be "*UK membership in EU*"). The topic definitions for all considered sentences can be found in the linked repository. The module utilizes the ZSC C to retrieve the in-topic tweets $I^p_{tp_s}$ and their corresponding topic scores $T^p_{tp_s}$.

$$I^p_{tp_s} = \{tw_1, ..., tw_m | C(tw_i, tp_s) >= th\} \tag{1}$$
$$T^p_{tp_s} = \{C(tw_i, tp_s) | tw_i \in I^p_{tp_s}\} \tag{2}$$

$C(tw_i, tp_s) \in [0, 1]$ indicates the degree to which tweet tw_i is associated with topic tp_s. The filtering threshold value th was varied to determine the best and optimal parameter set.

4.2   Agreement Detector

The *Agreement Detector* module (Fig. 1 - Module 2) computes the final five-valued label A^p_s through an algorithm $Alg(T^p_{tp_s}, S^p_s)$, defining

$$S^p_s = \{C(tw_i, s) | tw_i \in I^p_{tp_s}\} \tag{3}$$

as the C scores of tweets $I^p_{tp_s}$ with respect to sentence s, each one indicating the relevance and agreement of tweet tw_i with sentence s.

Each employed algorithm *Alg* exploits one of the following mapping functions:

$$M1(s) = \begin{cases} 1 & \text{if } s \in [0, 0.2) \\ 2 & \text{if } s \in [0.2, 0.4) \\ 3 & \text{if } s \in [0.4, 0.6) \\ 4 & \text{if } s \in [0.6, 0.8) \\ 5 & \text{if } s \in [0.8, 1] \end{cases} \quad (4) \quad M2(s) = \begin{cases} 1 & \text{if } s \in [0, 0.25) \\ 2 & \text{if } s \in [0.25, 0.5) \\ 3 & \text{if } s \in [0.5, 0.75) \\ 4 & \text{if } s \in [0.75, 1] \end{cases} \quad (5)$$

where $M1(s)$ ranges from 1 to 5, corresponding to the five agreement/disagreement labels defined in Sect. 3. Similarly, $M2(s)$ ranges from 1 to 4, representing an intermediate agreement/disagreement scale. Specifically, $M2(s) = \{1, 2\}$ has the same meaning as in Sect. 3, while $M2(s) = 3$ indicates agreement and $M2(s) = 4$ represents complete agreement. The rationale behind this intermediate mapping is explained in Algorithm 4 (Subsect. 4.2).

The proposed algorithms ordered by complexity are the followings:

Algorithm 1 [Alg1] The label A_s^p is computed as

$$A_s^p = \begin{cases} M1(\frac{\sum_{i=1}^{|I_{tp_s}^p|} s_i \cdot t_i}{\sum_{i=1}^{|I_{tp_s}^p|} s_i}) & \text{if } |I_{tp_s}^p| \neq 0 \\ 3 & \text{otherwise} \end{cases} \tag{6}$$

where $s_i \in S_{tp_s}^p$ and $t_i \in T_{tp_s}^p$.

Algorithm 2 [Alg2] First, it maps each tweet $tw_i \in I_{tp_s}^p$ into the label $l_i \in \{1, 2, 3, 4, 5\}$ using its sentence score $s_i \in S_s^p$

$$l_i = M1(s_i) \tag{7}$$

then, A_s^p is

$$A_s^p = \begin{cases} \lfloor \frac{\sum_{i=1}^{|I_{tp_s}^p|} l_i}{|I_{tp_s}^p|} \rceil & \text{if } |I_{tp_s}^p| \neq 0 \\ 3 & \text{otherwise} \end{cases} \tag{8}$$

The step of assigning l_i to each tweet $tw_i \in I_{tp_s}^p$ (Eq. 7) aims to achieve a fairer A_s^p. Tweet normalization aids in aggregating the contribution of each tweet (l_i) through standard mean, employing macro aggregation. Macro-metric aggregation is preferred in multi-class classification setups when class imbalance is suspected. In the current context, the values of l_i are unbalanced with respect to sentence s. Typically, if Party p agrees with a sentence, there will be numerous tweets in agreement (many $l_i = 4$ or $l_i = 5$), and few or no tweets in disagreement (few labels $l_i = 1$, or $l_i = 2$, or $l_i = 3$), and vice-versa.

Algorithm 3 [Alg3] Like $Alg2$, but A_s^p is computed with a slight modification. Introducing V_l as the number of voters for the integer label $l \in \{1, 2, 3, 4, 5\}$

$$V_l = |\{l_i : l_i = l\}_{i=1}^{|I_{tp_s}^p|}| \tag{9}$$

where l_i are the labels computed from Eq. 7. Let's define $v = max(V_l)$, then

$$A_s^p = \begin{cases} l & \text{if } |\{l : V_l = v\}| = 1 & \text{(10a)} \\ \lfloor \frac{\sum_{i=1}^{|I_{tp_s}^p|} l_i}{|I_{tp_s}^p|} \rceil & \text{if } |\{l : V_l = v\}| > 1 & \text{(10b)} \\ 3 & \text{otherwise} & \text{(10c)} \end{cases}$$

where $\lfloor ... \rceil$ is the rounding function. Majority voting (case 10a) potentially contributes more to assigning correct labels than the plain standard mean (case 10b taken from Eq. 8 of $Alg2$) as it effectively accounts for class imbalance.

Algorithm 4 [Alg4] The previous algorithms consider the neutral label $nl = 3$ (*neither disagree, nor agree*) even when $\mid I^p_{tp_s} \mid \neq 0$. However, we explored the scenario where nl is *only* considered when $\mid I^p_{tp_s} \mid = 0$. In such cases, the user might not have taken a position on the sentence s yet, and determining A^p_s based on a single tweet may lack significance. Hence, *Alg4* extends *Alg3* with the following modifications:

$$l_i = M2(s_i) \tag{11}$$

where $l_i \in \{1, 2, 3, 4\}$. Then, we define

$$a^p_s = \begin{cases} 3 & \text{if } \mid I^p_{tp_s} \mid < m \\ \text{majority voting (case 10a)} \\ \text{rounded standard mean (case 10b)} \end{cases} \tag{12}$$

Here, m is the minimum number of tweets required to activate either the majority voting algorithm or the standard mean. The output labels $\{3, 4\}$ from $M2(s)$ correspond to the final labels *agree* and *completely agree*, and they are mapped to the integer labels 4 and 5 as defined in Sect. 3.

$$A^p_s = \begin{cases} a^p_s & \text{if } a^p_s = 1 \vee a^p_s = 2 \\ a^p_s + 1 & \text{if } a^p_s = 3 \vee a^p_s = 4 \end{cases} \tag{13}$$

4.3   Experiment Settings

To validate the T2S's performance we had to choose i) the set of values for each of the four parameters to tune (the dataset size D_i, the language model LM for ZSC, the algorithm Alg, and the topic-filtering threshold th - Fig. 1), ii) the baselines to which compare T2S, and iii) the evaluation metrics.

T2S Parameters. We chose three to seven months of tweets (D_i), a filtering threshold from 0.5 to 0.9, four algorithms for the *Agreement Detector* module (Sect. 4.2), and three language models for the ZSC. The chosen filtering threshold range was set higher than 0.5 to ensure better agreement between a text and a topic. The language models we adopted are[5]: a) BART-large [12] fine-tuned on the MultiNLI dataset [20], b) DeBERTa-v3-base-mnli-fever (*DeBERTa*), and c) covid-twitter-bert-b1-fever-anli (*Covid-twitter-BERT*). Since the majority of collected tweets are in English, we used English language models. Non-English tweets were translated using Google Translate[6]. Our attempts to employ Multi-Language Models resulted in worse performances [7]. BART and DeBERTa were adapted to handle tweets by removing mentions, hashtags, and emojis, while Covid-twitter-BERT, which is already trained on tweets, was evaluated with and without those structures.

---

[5] From huggingface.co: a) facebook/bart-large-mnli, b) MoritzLaurer/DeBERTa-v3-base-mnli-fever-anli, c) digitalepidemiologylab/covid-twitter-bert-v2-mnli.

[6] https://github.com/lushan88a/google_trans_new.

Baselines. To validate T2S's abilities, we compared its performance with two bare baselines: (i) **Random**: the final agreement/disagreement label A_s^p is set to a random integer picked from a discrete uniform distribution of $int \in [1, 5]$; (ii) **Assign-highest-value**: A_s^p is always assigned the highest label (*completely agree*) since our datasets are skewed towards the *agree* and *completely agree* values.

Evaluation Metrics. In assessing the performance of the detection model for this stance detection task, traditional error metrics such as MSE, MAE, R2 Score, Residual Plots, and Macro Averaged Mean Absolute Error are commonly used. However, a custom error metric is needed to account for the varying importance of errors among the stance classes. For example, misclassifying as *agree* instead of *completely disagree* is considered a more acceptable error than misclassifying as *neither disagree, nor agree* instead of *agree*, even though both errors have a magnitude of one. In the absence of such a metric, MAE is the most appropriate choice. Additionally, the F1 weighted score is employed due to the integer nature of the detected labels and the imbalanced distribution of the Ground-of-Truth values among the agreement/disagreement labels.

5   Results and Discussion

Figure 2 shows the F1 and MAE scores over all nine elections respectively. Table 2 indicates the four general optimal settings across the elections by varying the number of labels and the metric considered.

Table 2. The four optimal settings over *no. of labels* and *metric*.

no. of labels	metric	D_i	model	alg	th	avg F1	avg MAE
5	F1	D_3	DeBERTa	alg_4 min no. of tweets: 3	0.9	0.29	1.56
5	MAE	D_4	Covid-twitter-BERT with # and emojis	alg_3	0.9	0.20	1.43
3	F1	D_3	DeBERTa	alg_4 min no. of tweets: 3	0.6	0.53	0.85
3	MAE	D_5	DeBERTa	alg_3	0.9	0.49	0.82

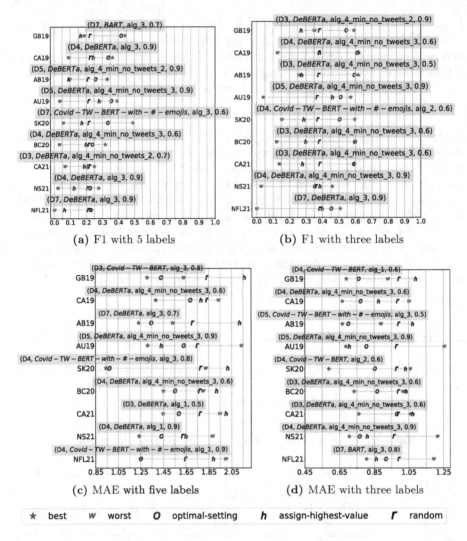

Fig. 2. F1 and MAE scores for all nine elections across baselines (assign-highest-value and random), best and worst setting for each election, and general optimal setting. The green boxes display the best setting for each election. (Color figure online)

5.1   RQ1: What are the Performances and Insights of T2S?

The best setting for each of the nine elections was chosen in two steps: firstly, by varying the algorithm Alg and the threshold th according to Fig. 1, we selected the D_i and LM with the minimum (maximum) MAE (F1), giving priority to MAE. Then we proceeded to choose the filtering threshold (th) and the algorithm (Alg) in a similar manner, while keeping the dataset size and language model fixed. The performance results in Fig. 2 demonstrate that T2S is a strong user stance detection model, surpassing random and assign-highest-value baselines. The best setting for T2S varies across the nine elections, with F1 scores ranging from 0.23 to 0.49 and MAE scores (for five-labelled stance) ranging from 0.94 to 1.45. The selected algorithm alternates between *Alg3* and *Alg4*, indicating that aggregating the tweet contributions (l_i) yields higher detection precision than directly averaging the sentence scores (s_i). However, the chosen filtering threshold, dataset time period, and language model for ZSC differ significantly across the nine datasets.

These differences can be attributed to two intertwined factors: i) the *diverse topic knowledge* of different language models and ii) the *manner* and *timing* of a user's (political party's) *expression on social media*, which influences T2S stance detections. The choice of the language model is crucial, as models not trained or fine-tuned on the topics in the dataset struggle to assign accurate scores to texts containing those topics. This issue could potentially be addressed by using more advanced models like GPT3 or ChatGPT, which have demonstrated state-of-the-art performance on text stance detection [22]. As for how a user expresses themselves on social media, there are three issues: first, if T2S attempts to detect a user's stance on a socio-political statement they haven't tweeted about but have discussed in a conference, T2S may incorrectly assign the *neither agree, nor disagree* label. Second, if a user tweets about a statement using expressions (e.g., acronyms) that T2S's language model hasn't been trained on, T2S is likely to detect an incorrect stance value. Conversely, if another user tweets about the same statement using more common words, T2S is more likely to detect the correct stance. Lastly, the significant variation in the dataset time periods (D_i) suggests that a user may discuss a certain topic either close to or far from the election date. Therefore, obtaining the user's entire timeline, rather than limiting data collection to specific time periods, could be beneficial. In a previous study, we extensively discussed how the writing style of Italian political parties influences T2S's performance [7]. Similar considerations can be made for the results of the three-labelled stance detection. Noticeably, the F1 scores vary less and are closer (around 0.6) to the best F1 score (0.95) of supervised and semi-supervised text-based techniques in the literature [9,16].

To sum up, although T2S's performance is still distant from state-of-the-art user-based stance detection, we believe it represents a valuable starting point for *addressing* the research *gap* in *unsupervised content-based models* leveraging an advanced Natural Language Processing technique (ZSC) to detect a *five-level stance* of the user on *multiple and diverse targets* (the socio-political statements on different political contexts).

5.2   RQ2: Can T2S Generalize over Diverse Political Contexts?

Figure 2 demonstrates that T2S effectively captures the complex five-level stance across diverse political contexts. However, the optimal settings vary for each election. To identify a potential optimal setting, we calculated the average F1 or MAE performance across all nine election datasets. We selected the four best settings based on the metric (F1 or MAE) and the number of stance values (five or three). Analyzing the selected optimal settings (Table 2), we observed that the dataset's time period (D_i) and the filtering threshold value (th) have less influence. Effective algorithms involve majority voting and assign the neutral label based on the presence of a minimum number of in-topic tweets. The best-performing language models for ZSC are either fine-tuned on a larger number of hypothesis-premise pairs or pre-trained on tweets. The inclusion or exclusion of leading mentions, hashtags, and emojis does not significantly affect the results.

Overall, the four optimal settings closely approach the best setting for each election, surpassing the performance of baselines and worst settings, with few exceptions. Despite a fixed setting, T2S exhibits considerable performance variation among the nine election datasets, with a maximum variance of approximately 0.2 points for F1 and 0.8 points for MAE. This variability is attributed to how a user (in our case, a political party) expresses its election program on social media platforms such as Twitter.

In summary, although sacrificing some performance, a general framework setting can achieve satisfactory results across different political contexts, consistently outperforming random and assign-highest-value baselines.

5.3   Potential and Limitations

The Tweets2Stance framework was tested on political parties during election campaigns to detect a user's political orientation. It has potential applications in identifying radicalization and extremism, particularly on topics like vaccines or immigration. The framework can also be applied to social media platforms other than Twitter, such as Facebook. However, T2S has limitations when used in unknown scenarios and different topics, such as the need for domain adaptation, as pre-trained models may not perform well when applied to different domains. Data bias is another issue, as pre-trained models may be biased toward certain topics or demographics, leading to inaccurate stance detections and reinforcing biases. Limited vocabulary is a challenge, as pre-trained models may not understand or classify texts with domain-specific words or phrases. Overfitting can occur when fine-tuning on small datasets, resulting in poor performance on new data. Multilingualism is also a limitation, as pre-trained models trained on one language may not work well for another, requiring multilingual training or alternative methods like automatic translation. Finally, T2S faces a major limitation in transfer-learning as it cannot detect stances when users are not discussing a specific socio-political topic. In these cases, T2S detects a middle stance, which may indicate either neutrality or insufficient data for accurate analysis.

6   Conclusions

The main purpose of this work was to devise and probe the specific and generalizing capabilities of *Tweets2Stance*, an *unsupervised stance detection* framework based on Zero-Shot Learning that detects a *five-labelled* user's *stance* about specific social-political statements by analyzing *content-based analysis* of its Twitter timeline *only*. T2S outperformed the baselines (random and assign-highest-stance-value) on all nine election datasets and demonstrated its ability to generalize across diverse political contexts with a minimum MAE of 0.95 and a maximum F1 of 0.6. However, the scarcity of relevant posts to socio-political statements and the language model's limitations (domain adaptation, data bias, and limited vocabulary) pose constraints on the T2S framework's capabilities.

T2S fills the SOTA gap of unsupervised stance detection models of multiple unrelated targets using content features and innovative language models. While SOTA user-based methods achieve higher F1 scores, they focus on simpler targets (e.g., pro or anti-Trump) with limited stance levels (from two to three); besides, they use a straightforward filtering approach (e.g., excluding tweets mentioning a specific person or organization) or focus on interconnected users through keywords, URLs, and hashtags. In contrast, the T2S framework detects the five-labelled stance of a user on multiple and diverse targets in various contexts, leveraging the unfiltered social media timeline (filtering applied automatically). Lastly, future research could overcome T2S's limitations by employing an advanced language model like GPT-4 or conversational AI like ChatGPT as the ZSC for Topic Filtering and Stance Detector steps, since they showed robust *text* stance detection capabilities.

References

1. Aldayel, A., Magdy, W.: Your stance is exposed! analysing possible factors for stance detection on social media. Proc. ACM Hum.-Comput. Interact. **3**(CSCW), 1–20 (2019)
2. Aldayel, A., Magdy, W.: Stance detection on social media: state of the art and trends. Inf. Process. Manag. **58**(4), 102597 (2021)
3. Biber, D., Finegan, E.: Adverbial stance types in English. Discourse Process. **11**(1), 1–34 (1988)
4. Darwish, K., Stefanov, P., Aupetit, M., Nakov, P.: Unsupervised user stance detection on twitter. In: Proceedings of the International AAAI Conference on Web and Social Media, vol. 14, pp. 141–152 (2020)
5. Devlin, J., Chang, M.W., Lee, K., Toutanova, K.: BERT: pre-training of deep bidirectional transformers for language understanding. In: Proceedings of the 2019 Conference of the North American Chapter of the ACL, Minneapolis, Minnesota, vol. 1, pp. 4171–4186 (2019). https://doi.org/10.18653/v1/N19-1423
6. Fraisier, O., Cabanac, G., Pitarch, Y., Besançon, R., Boughanem, M.: Stance classification through proximity-based community detection. In: Proceedings of the 29th on Hypertext and Social Media, HT 2018, pp. 220–228. ACM, New York (2018). https://doi.org/10.1145/3209542.3209549
7. Gambini, M., Fagni, T., Senette, C., Tesconi, M.: Tweets2Stance: users stance detection exploiting zero-shot learning algorithms on tweets. arXiv preprint arXiv:2204.10710 (2022)
8. Garimella, K., De Francisci Morales, G., Gionis, A., Mathioudakis, M.: Reducing controversy by connecting opposing views. In: Proceedings of the Tenth ACM International Conference on Web Search and Data Mining, pp. 81–90 (2017)

9. Ghosh, S., Singhania, P., Singh, S., Rudra, K., Ghosh, S.: Stance detection in web and social media: a comparative study. In: Crestani, F., et al. (eds.) CLEF 2019. LNCS, vol. 11696, pp. 75–87. Springer, Cham (2019). https://doi.org/10.1007/978-3-030-28577-7_4
10. Gottipati, S., Qiu, M., Yang, L., Zhu, F., Jiang, J.: Predicting user's political party using ideological stances. In: Jatowt, A., et al. (eds.) SocInfo 2013. LNCS, vol. 8238, pp. 177–191. Springer, Cham (2013). https://doi.org/10.1007/978-3-319-03260-3_16
11. Küçük, D., Can, F.: Stance detection: a survey. ACM Comput. Surv. (CSUR) **53**(1), 1–37 (2020)
12. Lewis, M., et al.: BART: denoising sequence-to-sequence pre-training for natural language generation, translation, and comprehension. In: Proceedings of the 58th Annual Meeting of the Association for Computational Linguistics, pp. 7871–7880. ACL, Online (2020). https://doi.org/10.18653/v1/2020.acl-main.703
13. Li, Y., Sosea, T., Sawant, A., Nair, A.J., Inkpen, D., Caragea, C.: P-stance: a large dataset for stance detection in political domain. In: Findings of the Association for Computational Linguistics: ACL-IJCNLP 2021, pp. 2355–2365 (2021)
14. Magdy, W., Darwish, K., Abokhodair, N., Rahimi, A., Baldwin, T.: # isisisnotislam or# deportallmuslims? Predicting unspoken views. In: Proceedings of the 8th ACM Conference on Web Science, pp. 95–106 (2016)
15. Moghaddam, S., Ester, M.: Aspect-based opinion mining from product reviews. In: Proceedings of the 35th International ACM SIGIR Conference on Research and Development in Information Retrieval, p. 1184 (2012)
16. Mohammad, S., Kiritchenko, S., Sobhani, P., Zhu, X., Cherry, C.: SemEval-2016 task 6: detecting stance in tweets. In: Proceedings of the 10th International Workshop on Semantic Evaluation (SemEval-2016), pp. 31–41 (2016)
17. Rashed, A., Kutlu, M., Darwish, K., Elsayed, T., Bayrak, C.: Embeddings-based clustering for target specific stances: the case of a polarized turkey. In: Proceedings of the International AAAI Conference on Web and Social Media, vol. 15, pp. 537–548 (2021)
18. Thonet, T., Cabanac, G., Boughanem, M., Pinel-Sauvagnat, K.: Users are known by the company they keep: topic models for viewpoint discovery in social networks. In: Proceedings of the 2017 ACM on Conference on Information and Knowledge Management, pp. 87–96 (2017)
19. Trabelsi, A., Zaiane, O.: Unsupervised model for topic viewpoint discovery in online debates leveraging author interactions. In: Proceedings of the International AAAI Conference on Web and Social Media, vol. 12 (2018)
20. Williams, A., Nangia, N., Bowman, S.: A broad-coverage challenge corpus for sentence understanding through inference. In: Proceedings of the 2018 Conference of the North American Chapter of the ACL, vol. 1, pp. 1112–1122. ACL (2018). http://aclweb.org/anthology/N18-1101
21. Yin, W., Hay, J., Roth, D.: Benchmarking zero-shot text classification: datasets, evaluation and entailment approach. In: Proceedings of the 2019 Conference on Empirical Methods in Natural Language Processing (EMNLP-IJCNLP), Hong Kong, China, pp. 3914–3923. ACL (2019). https://doi.org/10.18653/v1/D19-1404
22. Zhang, B., Ding, D., Jing, L.: How would stance detection techniques evolve after the launch of ChatGPT? arXiv preprint arXiv:2212.14548 (2022)

GLORIA: A Graph Convolutional Network-Based Approach for Review Spam Detection

Giuseppina Andresini[1,2](✉) [ID], Annalisa Appice[1,2] [ID], Roberto Gasbarro[1], and Donato Malerba[1,2] [ID]

[1] Department of Computer Science, University of Bari "Aldo Moro", Bari, Italy
{giuseppina.andresini,annalisa.appice,donato.malerba}@uniba.it,
r.gasbarro1@studenti.uniba.it
[2] CINI - Consorzio Interuniversitario Nazionale per l'Informatica, Bari, Italy

Abstract. Spam reviews contain untruthful content created with malevolent intent, to affect the overall reputation of a product, service or company. This content is commonly made by malicious users or automated programs (i.e., bots) that mimic human behaviour. With the recent boom of online review systems, performing accurate review spam detection has become of primary importance for a review platform, to mitigate the effect of malicious users responsible for untruthful content. In this work, we propose a review spam classification approach, named GLORIA, that adopts a graph representation of review data and trains a graph convolutional neural network for edge classification as a review spam detection model. In particular, GLORIA represents both users (i.e., authors of reviews) and products (i.e., reviewed items) as nodes of a heterogeneous graph, while it represents reviews as graph edges that connect each author of a review to the reviewed item. Features of users, products and reviews are associated with nodes and edges, respectively.

Experiments performed on publicly available review datasets prove the effectiveness of the proposed approach compared with some state-of-the-art approaches.

Keywords: Review Spam Detection · Graph Convolutional Networks · Heterogeneous Graph Learning · Edge Classification

1 Introduction

With the continuous development of technology and the ubiquitous presence of network-based services in our everyday life, it has become very common to make online purchases of products and services. With the rapid spread of e-commerce services, user reviews have become one of the most influential factors in purchase decisions of customers [14]. Consequently, e-commerce marketplaces are nowadays the most important target of spammers, which have the malicious goal of manipulating the reputation of products and brands, to either promote or criticize products and services. Positive and negative opinions can greatly influence a company's business. For this reason, review spam detection is a crucial problem to address for guaranteeing the reliability of products and services.

Spam reviews are described as untruthful or deceptive opinions that are posted on online commerce platforms in an attempt to manipulate the public perception (in a positive or negative manner) of specific products or services presented on the affected platforms [11]. In the last years, automatic review spam detection has attracted the attention of machine learning, natural language processing and deep learning researchers due to the difficulty of recognizing fake reviews by manually reading their content [9]. On the other hand, over a few years, professional spammers have greatly increased and improved their writing techniques, to evade detection tools that base review spam detection on the analysis of text content only. Most of the existing review spam detection approaches focus on extracting robust, engineered features from both review contents and reviewer behaviours [2,7,17,23], but in the past decade, several approaches have been developed, to leverage the social interaction between users and enhance the feature space of review spam detection problems with contextual information.

In this paper, we perform a step forward in this "social" research direction. In fact, we investigate the use of a heterogeneous graph representation of review data, to capture the relationships between products, users and reviews and gain accuracy in problems of review spam detection. For this purpose, we propose a Graph Convolutional Network (GCN) approach, named **GLORIA** (**G**raph con-vo**L**uti**O**nal Network for **R**ev**I**ew sp**A**m), that learns spatial convolutions on the graph representation of review data, to take advantage of the expressive structural information enclosed in graphs. In particular, the proposed approach implements a heterogeneous bipartite graph used as input to a Crystal GCN [28]. This architecture has been proven effective in the context of chemical material property prediction [8]. Traditional GNN algorithms perform convolutions using a shared weight matrix for all neighbours of a node by neglecting the difference of interaction between neighbours. Instead the Crystal GNN first aggregates neighbour vectors and then performs convolutions on the aggregated neighbour vectors. To the best of our knowledge, this is the first study that explores the use of Crystal GCN in review spam detection problems by showing how the proposed approach can gain accuracy compared to shallow and deep neural models trained neglecting the graph structure of data. An issue of review spam detection problems is that spam data are highly skewed. The imbalance of malicious data is a common condition in several cybersecurity problems (e.g., malware [4], fraud [22] or intrusion [3] detection), as well as in remote sensing problems (e.g., [6]). In this study, we handle the imbalanced condition of review spam data by training the Crystal GCN model with the sigmoid focal loss. This choice bases on [20] that shows how the sigmoid focal loss can help a neural model in focusing on rare samples. In this study, we show that the sigmoid focal loss is better suited than the traditional cross-entropy loss, to handle the imbalance condition of review spam data. Finally, we analyse the topological structure of graphs by showing how the exploration of the centrality of products and users in the graph representation of review data may disclose useful knowledge to explain characteristics of reviews and possible spam patterns.

The paper is organized as follows. Section 2 overviews the related work. Section 3 describes the proposed approach. Section 4 describes the benchmark data collections adopted in the experimental study, describes the experimental setting and discusses the relevant results. Finally, Sect. 5 draws conclusions and outlines the future directions of this work.

2 Related Work

The research in the field of review spam detection has received great attention in the last years. Several machine learning approaches have been recently designed to disentangle spam reviews from non-spam reviews [9,16]. In particular, the seminal study of [17] started the investigation of the task of review spam detection in the context of product reviews.

Recent research trends have started exploring deep learning approaches in problems of review spam detection [2,5,7,24,31]. In [2,5] a multi-view, deep learning approach is described for review spam detection. The proposed approach combines embeddings of textual features, extracted with Word2Vec and BERT models, and behavioural reviewer features to improve the accuracy of a review classifier trained through a multi-input, deep neural network. [7] describes a combination of Word2Vec and Convolutional Neural Networks (CNNs), to learn a document-level representation of reviews. Finally, a Bi-directional LSTM is used for review classification. The work in [24] adopts word embeddings trained on an Amazon review dataset using the Continuous Bag-of-Words (CBOW) algorithm. Finally, it trains a review classification model that combines CNNs and Gated Recurrent Neural Networks. A CNN is also trained in [31], to extract semantic information from the text of reviews by exploiting convolution and pooling operations.

Although all the above studies describe feature-based approaches that rely on an effective way to extract and learn features (from both reviews and reviewers), they ignore relationships between users, products and reviews. On the other hand, a few recent studies have started the investigation of the effectiveness of graphs as data modelling approaches of review spam data. The study of [27] first adopts a heterogeneous graph to represent reviewers, reviews and stores, through different categories of nodes. The review graph is used to infer the truthiness of reviews, honesty of reviewers and reliability of stores. [26] explores an unsupervised review spam detection approach that resorts to clustering, to identify communities of users with similar spam behaviours. [25] describes the use of a heterogeneous graph to connect users to reviews and analyses how graph meta-paths may help in recognizing review spam.

The recent studies that have adopted a graph representation of review data have also paved the way for leveraging GCNs in review spam detection problems. Although GCNs have recently gained great attention in several domains (e.g., recommendation systems [29] and chemical properties predictions [28]), a few studies have explored GCNs in review spam detection problems [1,19,30]. [1] trains a GCN for node classification in spam bot detection problems. This study adopts a social graph representation of relationships between Twitter users (represented as nodes) and leverages both feature nodes and relationships between neighbour nodes for training a GCN that addresses spam bot detection as a problem of graph node classification. A social graph of Twitter user relationships is also adopted in [30] in combination with an Attention-based Graph Neural Network trained for spam bot detection. So both these studies consider a problem of graph node classification, and train a GCN to recognize spam bot communities. They label all messages produced by the member of bot communities in the spam class. Differently, our study accounts that a malicious user does not necessarily only produce review spam. Based upon this consideration, we focus on classifying reviews, instead of classifying reviewers.

Finally, [19] studies the review spam detection problem for Xianyu, that is one of the largest second-hand goods apps in China. In Xianyu, reviews are communication tools for buyers and sellers and the review action usually happens before purchases. As recognised by the authors of [19], this is different from the common use of reviews in other e-commerce systems, also considered in this study, where reviews are usually made by customers who have bought the products. Accounting for the peculiar characteristics of reviews in Xianyu, [19] adopts two graph representations of Xianyu reviews: a heterogeneous graph modelling relationships between users and review items and a homogeneous graph modelling similarities between review items. A review item denotes a review topic (e.g., "iPhone 6s") that is associated in Xianyu with a sequence of review comments produced by (multiple) users on the specific topic. The nodes of the homogeneous graph are associated with the content features extracted from review items. [19] concatenates embeddings extracted through the GCN trained on two graphs to obtain the feature vector for the final classification of the review item. Differently, we consider the traditional e-commerce perspective with reviews written by users on products. So we use a single heterogeneous graph to represent reviews as relationships between users and products, and we associate review features to edges, while characteristics of users and products to nodes. Finally, we train a GCN model for graph edge classification, to classify each single review message.

3   The Proposed Method

In this Section, we describe the GLORIA approach. It adopts a graph representation of review data, where users and products are represented as heterogeneous graph nodes and the reviews as graph edges. Hence, GLORIA implements a GCN for edge classification in heterogeneous graphs.

Let us consider the input graph representation of review data as a heterogeneous bipartite graph defined as $G = (U, P, R)$, where U, P and R correspond to the set of user nodes, product nodes and edges, respectively. Each edge $r = (i, j) \in R$ with $R \subseteq (U \times P) \cup (P \times U)$ defines the undirect relationship between a user node and a product node to express that the user reviewed the product (and the product is reviewed by the reviewer). In addition, let us consider three mapping functions: $\phi_U \colon U \to \mathbf{X^U}$ that associates each node $u \in U$ to a feature vector in $\mathbf{X^U}$, $\phi_P \colon P \to \mathbf{X^P}$ that associates each node $p \in P$ to a feature vector in $\mathbf{X^P}$ and $\phi_R \colon R \to \mathbf{X^R}$ that associates each edge $r \in R$ to a feature vector in $\mathbf{X^R}$. We process this graph representation of review data to train a GCN for edge classification.

The GCN takes both node feature vectors, edge feature vectors and adjacency matrix as input and passes them through a series of L layers. At each layer l, node embeddings are updated according with the Eq. 1 to create an intermediate hidden representation $\mathbf{h}^l$. In particular, at each $\mathbf{h}^l$, the GCN of GLORIA applies a crystal graph convolutional operator [28]. For each node $i \in U \cup P$, for each layer l, this operator learns a function $\mathbf{h}_i^l = \mathbf{h}_i^{l-1} + f(i)$ defined on the previous hidden $l - 1$ layer. $f()$ is formulated as follows:

$$f(i) = AGG_{(j \in \mathcal{N}(i))^k}(\sigma\left(\mathbf{z^l}_{(i,j)^k}\mathbf{W^l}_\sigma + \mathbf{b^l}_\sigma\right) \odot g\left(\mathbf{z^l}_{(i,j)^k}\mathbf{W^l}_g + \mathbf{b^l}_g\right)) \quad (1)$$

where $\mathcal{N}(i)$ is the set of neighbours of node i (i.e., $\mathcal{N}(i) = \{j|(i,j) \in R\}$) and $\mathbf{z^l}_{(i,j)^k} = \mathbf{h}_i^l \oplus \mathbf{h}_j^l \oplus \phi_R(i,j)^k$ is the concatenation of embeddings computed at layer l for the feature vectors associated with node i, neighbour node $j \in \mathcal{N}(i)$ and the feature vector associated with the k-th edge between i and j. If $l = 0$ then embeddings return original feature vectors computed with ϕ_U and ϕ_P for user nodes and product nodes, respectively. $\odot$ denotes the element-wise multiplication, $\mathbf{W}_\sigma$ and $\mathbf{W}_g$ denote the convolution weight matrix at layer l, while b_σ and b_g denote the bias at layer l, for both σ and g functions. In fact, each layer of the Crystal GCN applies both a sigmoid function (σ) [13] and a softplus function (g) [10]. Finally, the operator of aggregation (AGG) denotes the aggregation scheme used for grouping node embeddings generated by different edges relating multiple neighbours j to the same node i. In this study, we use the mean as the aggregation operator.

In particular, GLORIA comprises two graph convolutional layers (i.e., $l = 1, 2$). During the message-passing phase at layer l, the information of each node of the graph is updated based on the aggregation of the messages received from their immediate neighbours achieved in two hops. As such, each message-passing layer increases the receptive field of the GCN by one hop. As we perform two hops in GLORIA, we are able to model relationships between pairs of users, as well as relationships between pairs of products, in addition to the review relationships between users and products.

Figure 1 reports an example of the message-passing realized by GLORIA by considering the user node u_1 as target node. At $l = 0$, all neighbour nodes of u_1 are assigned to initial node feature vectors (by ϕ_U or ϕ_P). At $l = 1$, the information of both node features and edge features are concatenated and aggregated, while $\mathbf{h}_{p_1}^1$ and $\mathbf{h}_{p_2}^1$ are updated based on $f()$ (Eq. 1). In order to get the node embeddings available for target node u_1 at $l = 2$, embeddings of neighbour nodes $\mathbf{h}_{p_1}^1$ and $\mathbf{h}_{p_2}^1$ are concatenated and aggregated to update $\mathbf{h}_{u_2}^2$. Therefore, each node in the graph learns from all the neighbour nodes transitively achieved in two hops.

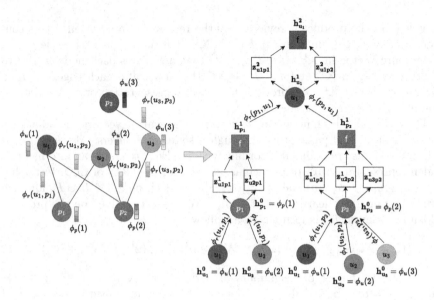

Fig. 1. Example of message-passing with user node u_1 as target in a Crystal GCN with two-hop neighbourhood

Finally, to deal with the expected imbalanced condition of review data in review spam detection problems, we use a sigmoid focal loss [20] for the final prediction:

$$\mathcal{SF} = -\sum_{k=1}^{K} \alpha(1-\hat{y})^{\gamma} \log \hat{y} \tag{2}$$

where K corresponds to the number of classes in the dataset (i.e., spam, non-spam) and parameter γ adjusts the rate and reduces the loss for well-classified samples, to focus learning on hard misclassified samples. α is a weighting factor in range $(0,1)$ to balance spam versus non-spam samples.

4   Experimental Setup

We performed experiments on two benchmark review datasets described in Sect. 4.1. The implementation details of GLORIA architecture, adopted in the experiments, are illustrated in Sect. 4.2. The experimental results are discussed in Sect. 4.3.

4.1   Data

We considered two datasets, namely Hotel and Restaurant, described in [23]. The two datasets contain reviews across 72 hotels and 129 restaurants, respectively, in the Chicago area. Each dataset contains reviews recorded by Yelp.com – a

Table 1. Summary of characteristics of Hotel and Restaurant datasets

Dataset	#spam reviews	#non-spam reviews	#reviewers	#products
Hotel	779	5078	5123	72
Restaurant	8301	58716	16941	129

well-known large-scale online review site. In addition, both datasets were provided with ground-truth labels (spam and non-spam) in [23]. So they can be used for the evaluation of the accuracy of review spam detection approaches. Both datasets include information about products (e.g., category, price range, rating) and reviewers (e.g., number of friends, number of reviews), as well as plain-text reviews. Each review is associated with a reviewer and a product. In this study, we adopted the feature-vector representation of plain-text reviews described in [5]. In this study, each dataset was processed separately, as each domain has specific characteristics to take into account for the review spam analysis. A summary of the characteristics of both datasets is reported in Table 1. We note that the class distribution is imbalanced in both datasets with the "spam" minority class.

To perform the experimental study, we adopted the same split used in [5] with reviews sorted by the post date and the 80% of the oldest reviews selected for the training stage and the 20% of the newest reviews selected for the testing stage.

4.2   Implementation Details

We implemented GLORIA in Python 3[1]. In particular, the GCN architecture was realized using PyTorch Geometric (PyG) 2.3, a geometric deep learning extension library for PyTorch. For each dataset, we conducted an automatic hyper-parameter optimization, using the tree-structured Parzen estimator algorithm, as implemented in the Hyperopt library. In particular, we selected the configuration of the hyper-parameters that achieved the highest F1 computed on the validation set extracted using 20% of the entire training according to the Pareto Principle, by considering spam as the positive class The values of the search space of the hyper-parameters, automatically explored with the tree-structured Parzen estimator, are reported in Table 2.

The neural architecture of GLORIA comprises two Graph Convolutional Layers, a Dense layer and a Sigmoid layer [13] used for the final edge classification. The standard Rectified Linear Unit (ReLU) [12] was selected as the activation function for each hidden layer. A dropout layer was placed before each Graph Convolutional layer, to perform data regularisation and prevent overfitting. The neural network was trained with mini-batches by back-propagation, while the gradient-based optimization was performed using the Adam update rule [18]. The maximum number of epochs was set equal to 300. The early stopping app-

---

[1] https://github.com/robertogasbarro/GLORIA.

Table 2. Hyper-parameter search space for the multi-input neural network

Hyper-parameter	search-space values
Mini-batch size	$\{2^5, 2^6, 2^7, 2^8, 2^9\}$
Learning rate	$[0.0001, 0.001]$
Dropout	$[0, 1]$
γ	$[0, 1]$
α	$[1, 4]$

Table 3. F1 spam, F1 non-spam, Macro-F1 and AUC-ROC of GLORIAby using both BCE loss and SF loss for learning the GCN model. The best results are in bold.

Dataset	Loss	F1 spam	F1 non-spam	Macro-F1	AUC-ROC
Hotel	BCE	0.586	0.898	0.742	0.878
	SF	**0.596**	**0.910**	**0.751**	**0.886**
Restaurant	BCE	0.615	**0.937**	0.776	0.917
	SF	**0.640**	0.931	**0.785**	**0.924**

roach based on the lowest loss on the validation set was used, to obtain the best classification model.

4.3   Results and Discussion

We evaluate the performance of GLORIA to answer the following research questions:

Q1 How does the accuracy of the proposed GCN-based approach change by varying the cost function?

Q2 Does the defined GCN model gain accuracy compared to state-of-the-art review spam detection algorithms that neglect the graph structure of review data?

Q3 Can the graph representation of review data disclose useful knowledge to explain the review domain better?

The accuracy performance of the analysed methods was measured in terms of F1 score computed on both the "spam" class and "non-spam" class, respectively (i.e., F1 spam and F1 non-spam), Macro-F1 (i.e., the average of F1 spam and F1 non-spam) and AUC-ROC. All these metrics were computed on the testing reviews of each dataset.

Sensitivity Analysis. We explored the sensitivity of the accuracy performance of GLORIA to the cost function adopted to learn the GCN model. To this aim, we compare the accuracy results obtained by using the binary cross-entropy (BCE)

Table 4. Competitor analysis: F1 spam, F1 non-spam, Macro-F1 and AUC-ROC of GLORIA, SVM and EUPHORIA. The best results are in bold.

Dataset	Method	F1 spam	F1 non-spam	Macro-F1	AUC-ROC
Hotel	SVM	0.530	0.853	0.692	0.779
	EUPHORIA	0.592	0.887	0.740	0.813
	GLORIA	**0.596**	**0.910**	**0.751**	**0.886**
Restaurant	SVM	0.351	0.751	0.551	0.692
	EUPHORIA	0.372	0.781	0.576	0.706
	GLORIA	**0.640**	**0.931**	**0.785**	**0.924**

(defined as $\mathcal{BCE} = \sum_{k=1}^{K} y \log \hat{y}$) and the sigmoid focal (SF) loss (defined in Eq. 2), to perform the training stage of GLORIA.

Table 3 reports the F1 (spam), F1 (non-spam), Macro-F1 and AUC-ROC measured on the testing data by using both SF and BCE as loss function in both datasets. The results show that the use of the SF loss can help the GCN model to gain accuracy in both datasets. These results confirm the ability of the SF loss to improve the accuracy performance of a deep neural model in the presence of data showing a strong imbalanced condition. In fact, we can observe that the use of SF loss increases almost all the accuracy metrics in both the experimented datasets. The only exception is observed in the F1 (non-spam) calculated in Restaurant dataset, where the BCE loss performs better than SF loss. This is an expected outcome since the BCE loss is a cost function that considers samples of the two classes to have equal weights. Thus, the neural model can be learned with the BCE loss to recognise the majority class better (e.g., non-spam review in this study).

Competitor Analysis. We compare the accuracy performance of GLORIA to that of two competitors: SVM that learns a Support Vector Machine classifier and EUPHORIA that learns a multi-input deep neural model for review spam detection. We consider the SVM as a classification algorithm for this comparison since it has been already adopted in multiple related studies on review spam detection (e.g., [15,21,23]). On the other hand, EUPHORIA is a recent method described in [5] for review spam detection. Both competitors ignore the graph structure of review data.

Table 4 reports the F1 spam, F1 non-spam, Macro-F1 and AUC-ROC, of SVM, EUPHORIA and GLORIA, respectively. The results show that the highest accuracy is achieved by GLORIA, with EUPHORIA as runner-up of this experiment in both datasets. These results contribute to showing the effectiveness of resorting to a graph representation of review data and leveraging the graph structure of data to learn relationships between users and products, to improve the ability of the classification model to predict accurately the review spam.

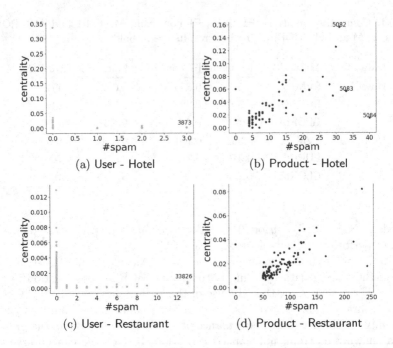

Fig. 2. Betweenness centrality (centrality, axis Y) with respect to the number of edges, which are labeled with the class "spam" on each node of the review graph (#spam axis X). Figures 2a and 2c refer to users, while Figs. 2b and 2d refer to products, in the review graphs of Hotel and Restaurant, respectively.

Qualitative Graph Analysis. Finally, we explore how the graph representation of the review data can disclose useful knowledge to explain the relationships between users, products and reviews in the spam class in the considered datasets. For this purpose, we analyse the betweenness centrality of users and products. The betweenness centrality of a node in a graph measures the amount of influence of the node on the flow of information in the graph. In particular, for a given node within a graph, the betweenness centrality of the node is computed as the number of the shortest paths in the graph, which connect any pair of nodes passing through the node under study, on the total number of the shortest paths which connect any pair of nodes in the graph. A node with a high value of betweenness centrality can be seen as a bridge that, if removed, could disrupt connections between other nodes in the graph.

Due to the main focus of this problem on the class "spam", Fig. 2 shows the betweenness centrality (axis Y) plotted with respect to the number of edges labelled with the class "spam" on each graph node. As GLORIA adopts a heterogeneous graph, we show the betweenness centrality for both users (Figs. 2a and 2c) and products (Figs. 2b and 2d). These results show that the betweenness centrality of a node tends to increase as the number of spam reviews involving the node increases. This trend is more evident in the products than in the users,

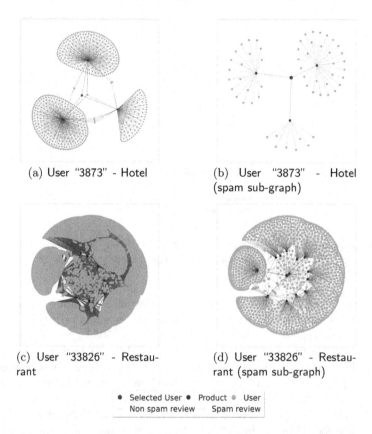

(a) User "3873" - Hotel

(b) User "3873" - Hotel (spam sub-graph)

(c) User "33826" - Restaurant

(d) User "33826" - Restaurant (spam sub-graph)

● Selected User ● Product ● User
Non spam review   Spam review

Fig. 3. Sub-graphs rooted in the users: "3873" of Hotel and "33826" of Restaurant. Both sub-graphs are produced with two hop levels. Figures 3b and 3d report the projection of sub-graphs shown in Figs. 3a and 3c on the edges labeled with the class "spam".

since there is a large number of users who produced zero spam reviews, while a small number of products received zero spam reviews.

Figure 3 shows the sub-graphs rooted in the users: "3873" of Hotel and "33826" of Restaurant. These users are identified according to the plots reported in Figs. 2b for Hotel and 2d for Restaurant as the users who produced the highest number of reviews labeled in the class "spam" in the two datasets. The two sub-graphs are produced with two hop levels. In particular, Figs. 3a and 3c show the entire sub-graphs rooted in the selected users "3873" and "33826", respectively. Figures 3b and 3d show the projection of these sub-graphs on the edges labeled in the class "spam". The sub-graphs rooted in the users "3873" and "33826" show the products for which these two users are spammers into Hotel and Restaurant, respectively. In both cases, the sub-graphs highlight that "multiple" reviewers produced spam reviews on the same target products. This suggests that several spammer profiles co-operated to produce malicious spam

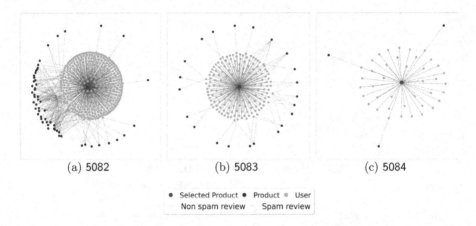

     (a) 5082   (b) 5083   (c) 5084

● Selected Product ● Product ● User
Non spam review   Spam review

Fig. 4. Subgraphs of products "5082", "5083" and "5084" of Hotel dataset. These products have achieved the higher number of spam as reported in Fig. 2b.

on the same target products. Hence, the activity of a spammer on a product may attract the attention of further spammers on the same target product.

Figure 4 shows the sub-graphs rooted in products "5082", "5083" and "5084" of Hotel, which are the three products that received the highest number of spam reviews in Fig. 2b. We note that the product "5082" has the highest betweenness centrality in Fig. 4. Instead, the product "5084" has a low betweenness centrality, while the product "5083" has a medium betweenness centrality in Fig. 4. Consistently with this analysis of betweenness centrality, the node density is higher in the sub-graph rooted in product "5082" than in the sub-graphs rooted in product "5083" and product "5084", respectively. In addition, the sub-graph rooted in product "5084" shows that all reviews produced on this product belong to the class "spam". These malicious reviews were produced by users who created these single reviews (with the exception of two users who created two reviews and both these reviews were spam). This suggests that a possible malicious behaviour is observable in products with low betweenness centrality.

5 Conclusion

In this paper, we illustrate a GCN approach for review spam detection, which takes advantage of relationships between users, products and reviews by resorting to a graph-based representation of review data and training GCN model for edge classification of reviews. The experiments performed on two benchmark datasets prove the accuracy of the proposed approach compared with two baselines that are SVM and a multi-view deep learning-based approach, respectively. In addition, we show that the analysis of the betweeness centrality of products and users allows us to extract useful knowledge to explain review data by disclosing possible review spam patterns. As future work, we plan to continue the investigation of how knowledge explaining the review graph topology can be used during the GCN training stage, to help the learned classification model gain accuracy in detecting review spam. In addition, we plan to investigate the use of the graph-based representation of review data in an online setting, to explore how changes occurring over time in the graph topology may help to keep high accuracy detecting review spam in real-time. Finally, we plan to extend our approach to perform a link prediction task to predict future behaviours of spammers.

Acknowledgments. The work of Giuseppina Andresini and Donato Malerba was supported by the project FAIR - Future AI Research (PE00000013), Spoke 6 - Symbiotic AI, under the NRRP MUR program funded by the NextGenerationEU. The work of Annalisa Appice was partially supported by project SERICS (PE00000014) under the NRRP MUR National Recovery and Resilience Plan funded by the European Union - NextGenerationEU. The authors wish to thank Raffaele Scaringi for the helpful discussion on Graph Neural Networks.

CRediT Authorship Contribution Statement

Giuseppina Andresini. Conceptualization, Methodology, Data curation, Investigation, Validation, Supervision, Visualization, Writing - original draft, Writing - review & editing. **Annalisa Appice:** Conceptualization, Methodology, Investigation, Validation, Supervision, Writing - original draft, Writing - review & editing. **Roberto Gasbarro:** Methodology, Software, Investigation, Data curation, Visualization, Writing - review & editing. **Donato Malerba:** Conceptualization, Writing - review & editing.

References

1. Ali Alhosseini, S., Bin Tareaf, R., Najafi, P., Meinel, C.: Detect me if you can: spam bot detection using inductive representation learning. In: Companion Proceedings of the 2019 World Wide Web Conference, WWW 2019, pp. 148–153. Association for Computing Machinery, New York (2019). https://doi.org/10.1145/3308560.3316504
2. Andresini, G., Iovine, A., Gasbarro, R., Lomolino, M., de Gemmis, M., Appice, A.: Review spam detection using multi-view deep learning combining content and behavioral features. In: CEUR Workshop Proceedings, vol. 3340, pp. 87–98 (2022)
3. Andresini, G., Appice, A., Caforio, F.P., Malerba, D., Vessio, G.: ROULETTE: a neural attention multi-output model for explainable network intrusion detection. Expert Syst. Appl. 117144 (2022). https://doi.org/10.1016/j.eswa.2022.117144
4. Andresini, G., Appice, A., Malerba, D.: Dealing with class imbalance in android malware detection by cascading clustering and classification. In: Appice, A., Ceci, M., Loglisci, C., Manco, G., Masciari, E., Ras, Z.W. (eds.) Complex Pattern Mining. SCI, vol. 880, pp. 173–187. Springer, Cham (2020). https://doi.org/10.1007/978-3-030-36617-9_11
5. Andresini, G., Iovine, A., Gasbarro, R., Lomolino, M., de Gemmis, M., Appice, A.: Euphoria: a neural multi-view approach to combine content and behavioral features in review spam detection. J. Comput. Math. Data Sci. **3**, 100036 (2022). https://doi.org/10.1016/j.jcmds.2022.100036

6. Appice, A., Malerba, D.: Segmentation-aided classification of hyperspectral data using spatial dependency of spectral bands. ISPRS J. Photogramm. Remote. Sens. **147**, 215–231 (2019). https://doi.org/10.1016/j.isprsjprs.2018.11.023

7. Bhuvaneshwari, P., Rao, A., Robinson, H.: Spam review detection using self attention based CNN and bi-directional LSTM. Multimed. Tools Appl. **80**, 1–18 (2021)

8. Cheng, J., Chunkai, Z., Dong, L.: A geometric-information-enhanced crystal graph network for predicting properties of materials. Commun. Mater. **2** (2021). https://doi.org/10.1038/s43246-021-00194-3

9. Crawford, M., Khoshgoftaar, T.M., Prusa, J.D., Richter, A.N., Al Najada, H.: Survey of review spam detection using machine learning techniques. J. Big Data **2**(1), 1–24 (2015). https://doi.org/10.1186/s40537-015-0029-9

10. Dugas, C., Bengio, Y., Bélisle, F., Nadeau, C., Garcia, R.: Incorporating second-order functional knowledge for better option pricing. In: Proceedings of the 13th International Conference on Neural Information Processing Systems, NIPS 2000, pp. 451–457. MIT Press, Cambridge (2000)

11. Ferrara, E.: The history of digital spam. Commun. ACM **62**(8), 82–91 (2019). https://doi.org/10.1145/3299768

12. Glorot, X., Bordes, A., Bengio, Y.: Deep sparse rectifier neural networks. In: AISTATS, pp. 315–323. JMLR.org (2011)

13. Han, J., Moraga, C.: The influence of the sigmoid function parameters on the speed of backpropagation learning. In: Mira, J., Sandoval, F. (eds.) IWANN 1995. LNCS, vol. 930, pp. 195–201. Springer, Heidelberg (1995). https://doi.org/10.1007/3-540-59497-3_175

14. Heydari, A., ali Tavakoli, M., Salim, N., Heydari, Z.: Detection of review spam: a survey. Expert Syst. Appl. **42**(7), 3634–3642 (2015). https://doi.org/10.1016/j.eswa.2014.12.029

15. Hussain, N., Mirza, H., Hussain, I., Iqbal, F., Memon, I.: Spam review detection using the linguistic and spammer behavioral methods. IEEE Access **8**, 53801–53816 (2020). https://doi.org/10.1109/ACCESS.2020.2979226

16. Hussain, N., Turab Mirza, H., Rasool, G., Hussain, I., Kaleem, M.: Spam review detection techniques: a systematic literature review. Appl. Sci. **9**(5) (2019)

17. Jindal, N., Liu, B.: Opinion spam and analysis. In: Proceedings of the 2008 International Conference on Web Search and Data Mining, WSDM 2008, pp. 219–230. Association for Computing Machinery, New York (2008)

18. Kingma, D.P., Ba, J.: Adam: a method for stochastic optimization. In: ICLR (2014)

19. Li, A., Qin, Z., Liu, R., Yang, Y., Li, D.: Spam review detection with graph convolutional networks, pp. 2703–2711 (2019). https://doi.org/10.1145/3357384.3357820

20. Lin, T.Y., Goyal, P., Girshick, R., He, K., Dollár, P.: Focal loss for dense object detection. IEEE Trans. Pattern Anal. Mach. Intell. **42**(2), 318–327 (2020). https://doi.org/10.1109/TPAMI.2018.2858826

21. Lin, Y., Zhu, T., Wu, H., Zhang, J., Wang, X., Zhou, A.: Towards online anti-opinion spam: spotting fake reviews from the review sequence. In: 2014 IEEE/ACM International Conference on Advances in Social Networks Analysis and Mining (ASONAM 2014), pp. 261–264 (2014). https://doi.org/10.1109/ASONAM.2014.6921594

22. Makki, S., Assaghir, Z., Taher, Y., Haque, R., Hacid, M.S., Zeineddine, H.: An experimental study with imbalanced classification approaches for credit card fraud detection. IEEE Access **7**, 93010–93022 (2019). https://doi.org/10.1109/ACCESS.2019.2927266

23. Mukherjee, A., Venkataraman, V., Liu, B., Glance, N.S.: What yelp fake review filter might be doing? In: Kiciman, E., Ellison, N.B., Hogan, B., Resnick, P., Soboroff, I. (eds.) Proceedings of the Seventh International Conference on Weblogs and Social Media, ICWSM 2013, Cambridge, Massachusetts, USA, 8–11 July 2013. The AAAI Press (2013). https://doi.org/10.1609/icwsm.v7i1.14389

24. Ren, Y., Ji, D.: Neural networks for deceptive opinion spam detection: an empirical study. Inf. Sci. **385–386**, 213–224 (2017). https://doi.org/10.1016/j.ins.2017.01.015

25. Shehnepoor, S., Salehi, M., Farahbakhsh, R., Crespi, N.: NetSpam: a network-based spam detection framework for reviews in online social media. IEEE Trans. Inf. Forensics Secur. **12**, 1585–1595 (2017). https://doi.org/10.1109/TIFS.2017.2675361

26. Soliman, A., Girdzijauskas, S.: AdaGraph: adaptive graph-based algorithms for spam detection in social networks. In: El Abbadi, A., Garbinato, B. (eds.) NETYS 2017. LNCS, vol. 10299, pp. 338–354. Springer, Cham (2017). https://doi.org/10.1007/978-3-319-59647-1_25

27. Wang, G., Xie, S., Liu, B., Yu, P.S.: Identify online store review spammers via social review graph. ACM Trans. Intell. Syst. Technol. **3**(4) (2012). https://doi.org/10.1145/2337542.2337546

28. Xie, T., Grossman, J.C.: Crystal graph convolutional neural networks for an accurate and interpretable prediction of material properties. Phys. Rev. Lett. **120**, 145301 (2018). https://doi.org/10.1103/PhysRevLett.120.145301

29. Ying, R., He, R., Chen, K., Eksombatchai, P., Hamilton, W.L., Leskovec, J.: Graph convolutional neural networks for web-scale recommender systems. In: Proceedings of the 24th ACM SIGKDD International Conference on Knowledge Discovery & Data Mining, KDD 2018, pp. 974–983. Association for Computing Machinery, New York (2018). https://doi.org/10.1145/3219819.3219890

30. Zhao, C., Xin, Y., Li, X., Zhu, H., Yang, Y., Chen, Y.: An attention-based graph neural network for spam bot detection in social networks. Appl. Sci. **10**(22) (2020). https://doi.org/10.3390/app10228160

31. Zhao, S., Xu, Z., Liu, L., Guo, M.: Towards accurate deceptive opinion spam detection based on word order-preserving CNN. Math. Probl. Eng. **2018** (2018). https://doi.org/10.1155/2018/2410206

Unmasking COVID-19 False Information on Twitter: A Topic-Based Approach with BERT

Riccardo Cantini[1] , Cristian Cosentino[1] , Irene Kilanioti[2] ,
Fabrizio Marozzo[1](✉) , and Domenico Talia[1]

[1] University of Calabria, Rende, Italy
{rcantini,ccosentino,fmarozzo,talia}@dimes.unical.it
[2] National Technical University of Athens, Athens, Greece
eirinikoilanioti@mail.ntua.gr

Abstract. Every day, many people use social media platforms to share information, thoughts, narratives and personal experiences. The vast volume of user-generated content offers valuable insights into the latest news and trends but also poses serious challenges due to the presence of a lot of false information. In this paper we focus on analyzing the online conversation on Twitter to identify and unveil false information related to COVID-19. To address this challenge, we devised a semi-supervised approach that combines false information detection with a neural topic modeling algorithm. By leveraging a small amount of labeled data, a BERT-based classifier is fine-tuned on the false information detection task and then is used to annotate a large amount of COVID-related tweets, organized in a topic-based clustering structure. This approach allows for effectively identifying the degree of false information in each discussion topic related to COVID-19. Specifically, our approach allows for investigating the presence of false information from a topical perspective, enabling us to examine its impact on specific topics underlying the online discussion. Among the topics with the highest incidence of false information, we found allergic reactions, microchips in vaccines, and 5G- and lockdown-related conspiracy theories. Our findings highlight the importance of leveraging social media platforms as valuable sources of information but at the same time how essential it is to identify and mitigate the impact of false information in online communities.

Keywords: False information · Misinformation · Disinformation · Neural Topic Modeling · COVID-19 · Natural Language Processing · BERT

1 Introduction

In today's digital age, social media has become an integral part of our lives, revolutionizing the way we communicate, share information, and interact with the world around us. With the increasing number of active users across different

A. Bifet et al. (Eds.): DS 2023, LNAI 14276, pp. 126–140, 2023.
https://doi.org/10.1007/978-3-031-45275-8_9

platforms such as Facebook, Instagram, and Twitter, social media has emerged as a vast and rich repository of valuable data [4,36]. This data, generated by billions of users worldwide, holds immense significance and potential for different fields, including business, marketing, research, and even governance [3]. The importance of social media data lies in its ability to provide real-time insights into people's thoughts, opinions, preferences, and behaviors, enabling organizations and individuals to make data-driven decisions and gain a deeper understanding of society at large.

As the influence of social media continues to grow, so does the challenge of dealing with misinformation and fake news, or more generally, false information [21]. False information is false or inaccurate information that is disseminated, either intentionally or unintentionally, leading to confusion, mistrust, and even harm. Detecting and combating false information has become a critical concern, and social media platforms play a central role in this process. For this reason, in recent years more and more researchers and companies are increasingly analyzing this phenomenon, trying to provide new solutions for detecting and mitigating the spread of false information. In this field of research, online discussions on COVID-19 represent one of the main case studies for analyzing and proposing solutions aimed at mitigating the dissemination of false information [35]. False information encompasses a wide range of topics, including vaccine efficacy and safety issues, conspiracy theories on 5G network connection, false claims on the origins of the virus, and many others, that have the potential to spread rapidly and undermine public trust in vaccination efforts. The impact of these falsehoods extends beyond the realm of social media, as they can influence individual decision-making regarding vaccine acceptance and ultimately affect public health outcomes.

In this paper we focus on analyzing the online conversation on Twitter to identify and unmask false information related to COVID-19, interpreting it from a topical viewpoint. To address this challenge, we exploited a semi-supervised approach that combines false information detection with a neural topic modeling algorithm. Our approach is divided into three main phases. Firstly, we exploit a small amount of labeled data to fine-tune a BERT-based false information detection model. Therefore, transfer learning is used to tailor the model to recognize false information in social media tweets about COVID, by adapting its pre-trained features. Subsequently, a neural topic modeling algorithm, namely BERTopic, is used to extract the main topic underlying Twitter discourse related to COVID-19, starting from a large set of unlabeled data. Lastly, we utilize the fine-tuned BERT-based classifier to determine the presence of false information for the different unlabeled posts, organized in a topic-based clustering structure through BERTopic. This process provides detailed information about the nature and extent of false information in the analyzed data, allowing us to quantitatively assess the presence of false information in the main topics discussed by users.

In contrast to state-of-the-art approaches that handle the false information problem within the large and comprehensive scope of COVID-19 discussions as a single entity [10,25,33], our approach allows for more fine-grained analysis, by taking a topical perspective. Specifically, our approach enables us to examine the impact of false information on specific topics generated during discussions. Consequently, we gain a deeper understanding of how false information influences and shapes discussions surrounding particular topics within the broader context of COVID-19. Our findings highlight the importance of leveraging social media platforms as valuable sources of information while addressing the challenges posed by false information. Furthermore, by employing a combination of false information detection and topic modeling, our work can contribute to mitigating the impact of false information in online communities.

The structure of the paper is as follows. Section 2 discusses related work in the fields of false information detection and topic modeling. Section 3 describes the devised approach. Section 4 discusses the achieved results. Finally, Sect. 5 concludes the paper.

2   Related Work

Social media plays a crucial role in information extraction and staying updated on current trends and discussions. However, the reliability of news circulating on social platforms is often questionable and susceptible to various biases. Consequently, we adopt an approach that focuses on effectively identifying topics of discussion while assessing the impact of false information on them, thus characterizing the presence of misleading and false user-generated content from a topical perspective. Therefore, our approach resides at the intersection of false information identification and topic detection. We analyze the main techniques present in the state for both research lines.

2.1   False Information Detection

With the huge amount of user-generated content on social media, assessing the reliability of online published content has become increasingly difficult in recent years. This issue derives from the presence of false information, which can come in different forms. In particular, *misinformation* refers to false information shared unintentionally, while *disinformation* implies the intentional dissemination of false or misleading information, usually for a specific purpose. Furthermore, the term *fake news* is also often used, which is a form of disinformation consisting of fabricated news aimed at deceiving public opinion.

Among the main works in the literature, addressing the detection of either misinformation or fake news, several deep learning-based approaches were proposed, leveraging convolutional neural networks (CNNs) and recurrent neural networks (RNNs) [28,39]. Additionally, natural language processing (NLP) techniques have been increasingly used to detect false information, through the analysis of the linguistic characteristics of news articles or social media posts [15,30]. In this context, the most recent works in the literature leverage transformer-based language representation models such as BERT (Bidirectional Encoder Representation from Transformers) [7]. Such models have proven successful in a wide range of downstream tasks, by demonstrating superior performance in natural language processing and understanding. Among the main examples in the literature, FakeBERT [17] combines deep convolutional neural networks with BERT, while in [19] authors propose a combined approach that jointly leverages BERT and RNNs. In [16], a BERT-based model for fake news detection is presented, which relies on the contextual relationship between the headline and the body text of news. Furthermore, besides assessing the fake content of online news, Transformer-based architectures were also employed for fact-checking and for providing explanations. In particular, in [38] authors proposed a two-stage fake news detection system, that can both estimate the reliability of COVID-19-related claims and provide users with pertinent information about them, in the form of a textual explanation.

2.2 Topic Detection

In recent years topic modeling has emerged as a powerful technique for uncovering latent trends and topics and extracting valuable insights from large text corpora. A wide range of topic modeling techniques have been developed, effectively applicable to a wide range of domains, such as information retrieval, document clustering, and trend detection. Among the first introduced techniques Latent Semantic Analysis (LSA) [6] uses the Singular Value Decomposition (SVD) to compute a low-rank approximation of a document term matrix (DTM) representing the corpus. LSA is simple and efficient, but it assumes a probabilistic generative model where words and documents are Gaussian distributed, which may not align with reality. To address this issue Probabilistic LSA (pLSA) was introduced, which relies on a multinomial generative model [14]. Another method is non-negative matrix factorization (NMF), it is similar to SVD but the decomposition must lead to non-negative values [23]. Latent Dirichlet Allocation (LDA) relies on the concept of mixtures of distributions to model documents as a mixture of latent topics, each of which constitutes a mixture of terms from the corpus vocabulary [18]. Among the main variants of LDA, in [1] a fuzzy version is proposed that relies on the concept of fuzzy Bag-of-Words. This fuzzy representation maps each document to a vector of keywords, where each keyword is assigned to every document with a certain membership degree. This allows for a more nuanced representation of the connections between terms and topics, accommodating the inherent ambiguity of the analyzed corpus. Another variant of LDA, which follows a deep learning approach, is LDA2Vec [27]. It mixes LDA with Word2Vec [26] by learning topic representations and latent vector representations of words simultaneously. This is achieved by modifying the standard Skip-gram model, integrating into the pivot word learnable topical information. Most recent topic modeling techniques, falling into the neural-based category, harness the power of pre-trained transformer-based Large Language

Models (pLLMs) to achieve meaningful semantically-rich sentence representations. Among them, Top2Vec [2] relies on Doc2Vec [22], while BERTopic [12] uses Sentence-BERT [31], based on siamese network architecture. Both approaches rely on the clustering of sentence-level representations projected into a low-dimensional space. Dimensionality reduction is performed using Uniform Manifold Approximation Projection (UMAP), while the HDBSCAN algorithm is used for clustering. Finally, topic representations are extracted from the topic-based clustering structure by selecting the nearest neighbors of the cluster centroid, in the case of Top2VEc, and by applying a class-based tf-idf, in the case of BERTopic.

3   Proposed Approach

This work focuses on the analysis of user-generated content on Twitter to identify and investigate false information related to COVID-19. For this purpose, we devised a semi-supervised approach that leverages a combination of false information detection and topic modeling, to achieve a topic-oriented representation of false information. Specifically, a BERT classifier is fine-tuned on a small set of annotated data, to make it able to identify false information present in a given post. Then, unlabeled data are used to unveil the main COVID-related topics of discussion underlying social media conversation. Specifically, this step relies on BERTopic, one of the most used neural topic modeling methods in the literature, which leverages semantically-rich sentence representations achieved through pre-trained LLMs. Finally, a false information score is computed for each topic identified by BERTopic, through the use of the fine-tuned false information detection model. This process allows for a topic-oriented quantification of the impact of false information on Twitter conversations about COVID-19. Hence, by following this approach, we can highlight the main discussion topics that are most affected by false information, from a quantitative viewpoint, while also finding concrete examples of misinformed user-generated content related to these topics. In the following, we provide a detailed description of the main steps of our approach, whose execution flow is depicted in Fig. 1.

3.1   Fine-Tuning of the False Information Detection Model

In this step, a BERT model is fine-tuned for the false information detection task. Specifically, starting from a small set of labeled posts, we train a binary classifier to detect false information using a transfer learning approach. Indeed, Large Language Models (LLMs) like BERT have proven successful in a wide range of downstream tasks, through the adaptation of pre-trained features to specific purposes. Besides BERT, other optimized variants exist, each introducing improvements in both the architecture and the pre-training phase. Due to this, we tested several BERT-like models, including BERT, ALBERT, BERTWEET, DISTILBERT, and ROBERTA, to find out the best trade-off between classification accuracy and training/inference times. The BERT model as well as the

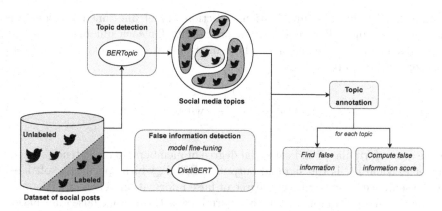

Fig. 1. Execution flow of the proposed approach.

evaluated variants was fine-tuned for detecting false information in social media posts. Specifically, during this step, pre-trained weights are slightly adapted to the binary downstream task under consideration, by using a binary cross-entropy loss, the ADAM optimization algorithm, and a small learning rate, which is crucial to correctly transfer knowledge from BERT by avoiding pre-trained weights to be distorted by large weight updates.

3.2 Topic Detection

In this step, starting from a large set of unlabeled posts, we use topic modeling to extract the main discussion topics underlying social media conversation. To this purpose, we leveraged BERTopic, a neural topic modeling technique that relies on Transformer-based pLLMs to generate semantically-rich vector representations of the sentences in a corpus. As recently demonstrated in the literature, the use of neural approaches like BERTopic for topic modeling leads to superior performance in terms of coherence and diversity [9,11,12]. In particular, in BERTopic Sentence-BERT is utilized for sentence embedding, which uses siamese neural network structures to generate semantically meaningful and comparable sentence representations. Then, such representations are projected in a low-dimensional space using UMAP (Uniform Manifold Approximation and Projection), and clustered into semantically-related groups via HDBSCAN. Following this approach, BERTopic can identify a topic-based clustering structure from which topic representations are computed, one for each cluster, using a class-based version of tf-idf.

3.3 Topic Annotation

Our approach adopts a topic-oriented perspective to thoroughly analyze the impact of false information within social media conversations. Therefore, in this step, we identify the discussion topics that are most affected by false information and quantify the extent of false information prevalent within them.

Specifically, the false information detection model fine-tuned previously is used to determine a false information probability for each unlabeled sentence. Thus, given a cluster, i.e., a topic, a false information score $\mathcal{S}(c)$ associated with that topic is computed as follows.

$$\mathcal{S}(c) = \frac{\sum\limits_{s \in c} p_s^c \cdot p_s^{fi}}{\sum\limits_{s \in c} p_s^c}, \quad \text{where } c \in \mathcal{C} \tag{1}$$

In the above formula, p_s^c indicates the degree of membership of sentence s to the cluster c, while p_s^{fi} is the sigmoid output of BERT, which specifies a soft-label for the sentence s measuring its degree of false information. Therefore, the false information score $\mathcal{S}(c)$ for cluster c is determined as the average false information of the sentence contained in that cluster, weighted on the probability of those sentences.

4   Experimental Results

The COVID-19 pandemic has not only had a profound impact on society but has also led to the widespread dissemination of false information on social media, resulting in increased vaccine hesitancy and the proliferation of conspiracy theories. Therefore, as stated in Sect. 1, the goal of this work is to detect the main false information present in COVID-related discussions, characterizing it from a topical perspective. To this purpose, we applied our approach to the *ANTi-Vax* dataset [13], composed of tweets from December 1, 2020, until July 31, 2021 related to the COVID-19 (SARS-CoV-2) pandemic. The dataset consists of a small portion of labeled data (about 15K) and a large set of unlabeled data (about 15M). Labeled data was manually annotated and validated by health medical experts, into two classes: *false information* for all those tweets that contain common myths and misinformation (e.g., the vaccine contains tracking device), or *reliable content*. It must be noted that all sarcastic and humorous tweets have not been included as false information. Among all unlabeled data, we focused on posts generated in the month of January 2021, encompassing 303,541 tweets. In the following, the experimental results we achieved will be comprehensively discussed, focusing on: (*i*) the choice of the best-suited transformer-based model to be fine-tuned for the false information detection task; (*ii*) the main identified topics that drove COVID-related discourse on Twitter; (*iii*) the analysis of COVID-related false information from a topical perspective.

4.1   Model Selection for False Information Detection

Among the main alternative models that can be effectively used for the binary task of false information detection, describe in detail in Sect. 2.1, we choose to follow a transfer-learning approach, by fine-tuning a BERT-based classifier on our downstream task. To select the most suitable model for our purposes, we conducted a comparative analysis of the following models.

- **BERT:** it is a pre-trained language representation model based on the transformer architecture [7].
- **ALBERT:** it is a lightweight variant of BERT. It introduces some improvements such as factorized embedding parameterization, and inter-sentence coherence loss, by replacing the Next Sentence Prediction with the Sentence order prediction task during pre-training [20].
- **BERTWEET:** it is a Twitter-specific variant of BERT, trained on Twitter text data. It manages unique Twitter features such as hashtags, mentions, URLs, and emojis [29].
- **DISTILBERT:** it is a distilled version of BERT, with about 40% fewer parameters. This reduction in size, achieved through a knowledge distillation approach, allows for faster training, making the model less resource-intensive [34].
- **ROBERTA:** it is an improved version of BERT, which removes the Next Sentence Prediction (NSP) task from the pre-training phase and introduces dynamic masking to vary the masked tokens during language modeling [24].

The performance evaluation of the different BERT-like models (i.e., BERT, ALBERT, BERTWEET, DISTILBERT, ROBERTA) was conducted on a held-out test set, encompassing 3000 samples, considering four metrics: score loss, AUC (Area Under the Curve), binary accuracy, and training time (measured in seconds per epoch). Figure 2 shows the scores obtained from different models for each considered metric.

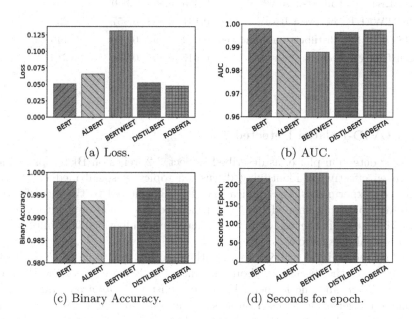

(a) Loss. (b) AUC.

(c) Binary Accuracy. (d) Seconds for epoch.

Fig. 2. BERT-based model comparison for false information detection.

Figure 2(a) shows the loss values achieved by each model, computed with a binary cross-entropy. Models such as BERT, DISTILBERT, and ROBERTA demonstrate lower loss values, indicating a better fit to the test data. Figure 2(b) and Fig. 2(d) display the Area Under the Curve (AUC) values and the binary accuracy for each model, which measure the model's ability to correctly distinguish between negative and positive classes. Also in this case, we observe that BERT, DISTILBERT, and ROBERTA exhibit the highest values. Finally, Fig. 2(c) shows the time required by each model to complete an epoch during training. The DISTILBERT model stands out as the fastest among the compared models, showcasing its capability for fast and efficient training. All achieved results are summarized in Table 1, showing the average values obtained from multiple experiments, which exhibit a negligible variance. Summing up, what emerges from our evaluation is that the DISTILBERT model achieves the best trade-off between accuracy and training time. Consequently, we utilized the DISTILBERT model as the reference model for false information classification throughout all the subsequent experiments.

Table 1. BERT-based model comparison for false information detection.

Model	Version	Loss	AUC	Binary accuracy	Seconds per epoch
BERT	bert-base-uncased	0.050	0,998	0.982	215
ALBERT	albert-base-v2	0.066	0.994	0.980	195
BERTWEET	bertweet-base	0.131	0.988	0.957	230
DISTILBERT	distilbert-base-uncased	0.052	0.997	0.986	146
ROBERTA	roberta-base	0.047	0.997	0.983	210

4.2 COVID-Related Detected Topics

The topic detection phase, as described in Sect. 3.2, relies on BERTopic, which has proven effective in identifying discussion topics in social media data [11, 12]. In our experiments, the application of BERTopic led to the identification of several topics that shaped the COVID-related discussion on Twitter. These topics will be used in the final step, to characterize the identified false information from a topical perspective.

Among the main identified topics, within the broader topic of *COVID vaccines* we found the discussion about the efforts and strategies of the US president *Joe Biden* and the former UK prime minister *B. Johnson*. Additionally, the online conversation focused on specific vaccines such as the *Pfizer vaccine* and *Johnson & Johnson*, discussing their effectiveness and side effects, such as *allergic reactions* and the risks related to *pregnancy and breastfeeding*. The Twitter discourse was also centered on the *European Union*'s approach to managing the pandemic and anti-contagion rules such as *mask wearing* and *lockdown*, also debating the effects on major sporting events like the *Olympic Games and NBA*. Furthermore, users discussed the long-term effects of COVID, especially on *older individuals*, and other conspiracy theories about the presence of *microchip* inside vaccines. Other identified topics include *Dr. A. Fauci*, *vaccine passports*, the impact of *COVID-19 in Florida*, *school-related issues*, and the challenges faced by *workers and employers*.

To evaluate the identified topics we used Topic coherence and diversity. Coherence measures how closely related and meaningful are the words within a topic, thus giving an estimate of how well they express a specific theme or concept. Among the main coherence metrics, we used CV [32] and Normalized Pointwise Mutual Information (NPMI) [5], achieving a value of 0.51 and 0.09 respectively. Differently, topic diversity assesses how different and unrelated the topics are from each other, which is necessary to comprehensively represent the corpus. We used the Percentage of Unique Words (PUW) [8] and the average pairwise Jaccard Distance (JD) [37], achieving a value of 0.97 and 0.99 respectively. Similarly to the experimental evaluation present in [12], we computed each metric by averaging across 10 different runs. In addition, for each run, metrics are averaged by varying the number of topics from 10 to 50, with steps of 10.

4.3 Topic-Oriented False Information Detected in COVID Discussions

In contrast to state-of-the-art approaches that treat the false information problem within the large and comprehensive scope of COVID-19 discussions as a single entity, our approach allows for more granular analysis. Specifically, our approach enables us to examine the impact of false information on the Twitter discourse from a topical perspective. Figure 3 shows the discussion topics ordered according to the level of false information present in them. Specifically, for each topic, we computed a false information score, i.e., $S(c)$ as defined in Sect. 3.3, which quantifies the extent of false information present in it.

From Fig. 3 it can be observed that there is a varying range of false information levels across the different topics. In the following we report three of these topics, characterized by the minimum, maximum, and median value of false information, according to the distribution of the $S(c)$ score.

- *Olympic games and NBA*: this topic refers to the Tokyo Olympics, held in Japan, and specifically to Olympic athletes and NBA basketball players.
- *Dr. A. Fauci*: this topic refers to Dr. Anthony Fauci, a renowned infectious disease expert in the United States, addressing conspiracy theories and spreading scientific information about vaccinations.
- *Allergic reactions*: this topic refers to severe side effects and physical symptoms that may occur after receiving a vaccine injection.

For each of the highlighted topics, we also show the distribution of the output achieved by the fine-tuned DistilBERT classifier. This model, as described

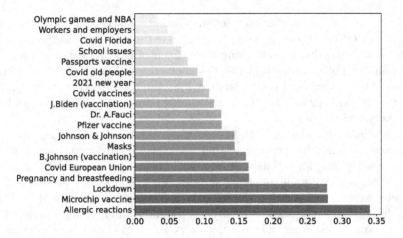

Fig. 3. False information score for each identified topic.

in Sect. 3, computes a probability value, i.e. p_s^{fi}, indicating how likely it is that a given content is false information. Therefore, given a sentence s a value of p_s^{fi} close to 0 indicates a low probability that this sentence contains false information, while values close to 1 represent the opposite case. The achieved results, shown in Fig. 4, are in line with the false information scores computed previously. In particular, Fig. 4(a) referred to the Olympic Games and NBA, shows a distribution whose values are predominantly concentrated toward 0, indicating a higher prevalence of non-false information predictions. A similar unimodal distribution is achieved in Fig. 4(b), related to the topic of Dr. A. Fauci. Differently, by observing Fig. 4(c), related to the topic with the highest false information score (i.e., allergic reactions), a bimodal distribution clearly emerges, indicating a non-negligible presence of predictions very close to 1. This translates into a greater presence of user-generated content identified by DistilBERT as false information, mainly related to untested hypotheses about serious health side effects caused by vaccines.

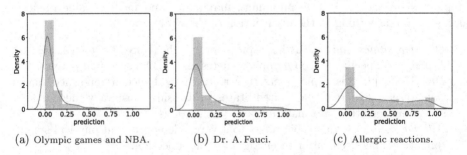

(a) Olympic games and NBA.   (b) Dr. A. Fauci.   (c) Allergic reactions.

Fig. 4. Examples of distribution p_s^{fi} of topics with low, medium and high levels of false information score.

As a final analysis, we focused on the top-3 topics with the highest misinformation score:

- *Allergic reactions*, already described above. False information score: 0.35.
- *Microchip vaccine*: the topic refers to conspiracy theories regarding the presence of microchips in vaccines. False information score: 0.25.
- *Lockdown*: the topic refers to lockdown measures imposed during the COVID-19 pandemic. False information score: 0.24.

For enriching the description of these topics, Fig. 5 presents the word clouds referring to them, which allows for highlighting their main keywords and concepts in a graphical manner. Among the most significant words we found: (*i*) *allergic reaction, vaccine, severe allergic, modern*, and *anaphylaxis* for the allergic reaction topic; (*ii*) *microchip, vaccine, 5g*, and *track* for the microchip topic; (*iii*) *lockdown, vaccine, government*, and *restriction* for the lockdown topic.

(a) Lockdown.   (b) Microchip vaccine.   (c) Allergic reactions.

Fig. 5. Word cloud representations of the top-3 topics per false information.

For the sake of completeness, in Table 2 we report an example of detected false information, for each of the three topics. For each example, we indicate the related topic, the text of the tweet, the false information probability given by DistilBERT (p_s^{fi}), and the degree of membership to the assigned cluster (p_s^c).

Table 2. Example tweets, identified as false information, for each of the top-3 topics per false information score.

Topic	Example of tweets	p_s^{fi}	p_s^c
Lockdown	Gee it's almost like lockdowns are not so much about a virus but are part of a deliberate global financial and social destruction reorganization strategy	0.96	0.99
Microchip vaccine	The new vaccine is going to be a tracking device that emits 5g into your brain!!!	0.98	1.00
Allergic reactions	The vaccine leads to serious allergic crises maybe it's better not to get vaccinated you risk your life less	0.98	0.98

The reported tweets express (*i*) skepticism and concerns about lockdown measures and COVID-19 vaccines; (*ii*) conspiracy theories, including the belief that vaccines emit harmful 5g waves and contain surveillance microchips; (*iii*) serious side effects of vaccines and no-vax instigations.

5   Conclusion

Social media has revolutionized the way we communicate and share information, providing valuable data with high potential for many fields of application. However, alongside these benefits, there has been an alarming surge in the proliferation of false information and fake news, necessitating urgent measures to mitigate their impact.

This paper focuses on the analysis of Twitter conversations to uncover and address false information pertaining to COVID-19. Employing a semi-supervised strategy and harnessing the capabilities of a BERT-based classifier, the study effectively identifies and annotates different topics present in online conversations, while evaluating the extent of false information associated with each topic. These encompass allergic reactions, microchips in vaccines, 5G conspiracy theories, and the impact of lockdown measures.

In contrast to state-of-the-art approaches that treat the false information problem within the large and comprehensive scope of COVID-19 discussions as a single entity, our approach allows for a finer-grained analysis, enabling us to examine the impact of false information on specific topics generated during discussions. Through the employment of transfer learning for false information detection and neural topic modeling, our work not only aids in identifying specific instances of false information but also provides insights into the underlying factors and dynamics contributing to its spread. This understanding is crucial for developing targeted interventions and strategies that effectively combat the dissemination of false information, ultimately strengthening the reliability and trustworthiness of information shared on social media platforms.

Acknowledgements. This work has been partially supported by the "National Centre for HPC, Big Data and Quantum Computing", CN00000013 - CUP H23C22000360005, and by the "FAIR - Future Artificial Intelligence Research" project - CUP H23C22000860006.

References

1. Akhtar, N., Sufyan Beg, M.M., Javed, H.: Topic modelling with fuzzy document representation. In: Singh, M., Gupta, P.K., Tyagi, V., Flusser, J., Ören, T., Kashyap, R. (eds.) ICACDS 2019. CCIS, vol. 1046, pp. 577–587. Springer, Singapore (2019). https://doi.org/10.1007/978-981-13-9942-8_54
2. Angelov, D.: Top2vec: distributed representations of topics. arXiv preprint arXiv:2008.09470 (2020)
3. Belcastro, L., Cantini, R., Marozzo, F.: Knowledge discovery from large amounts of social media data. Appl. Sci. **12**(3) (2022)

4. Belcastro, L., Cantini, R., Marozzo, F., Talia, D., Trunfio, P.: Learning political polarization on social media using neural networks. IEEE Access **8**, 47177–47187 (2020)

5. Bouma, G.: Normalized (pointwise) mutual information in collocation extraction. Proc. GSCL **30**, 31–40 (2009)

6. Deerwester, S., Dumais, S.T., Furnas, G.W., Landauer, T.K., Harshman, R.: Indexing by latent semantic analysis. J. Am. Soc. Inf. Sci. **41**(6), 391–407 (1990)

7. Devlin, J., Chang, M.W., Lee, K., Toutanova, K.: BERT: pre-training of deep bidirectional transformers for language understanding. arXiv preprint arXiv:1810.04805 (2018)

8. Dieng, A.B., Ruiz, F.J., Blei, D.M.: Topic modeling in embedding spaces. Trans. Assoc. Comput. Linguist. **8**, 439–453 (2020)

9. Egger, R., Yu, J.: A topic modeling comparison between LDA, NMF, Top2Vec, and BERTopic to demystify Twitter posts. Front. Sociol. **7** (2022)

10. Enders, A.M., Uscinski, J.E., Klofstad, C., Stoler, J.: The different forms of Covid-19 misinformation and their consequences. Harvard Kennedy School Misinformation Review (2020)

11. Gabarron, E., Dorronzoro, E., Reichenpfader, D., Denecke, K.: What do autistic people discuss on Twitter? An approach using BERTopic modelling (2023)

12. Grootendorst, M.: BERTopic: neural topic modeling with a class-based TF-IDF procedure. arXiv preprint arXiv:2203.05794 (2022)

13. Hayawi, K., Shahriar, S., Serhani, M.A., Taleb, I., Mathew, S.S.: ANTi-Vax: a novel Twitter dataset for Covid-19 vaccine misinformation detection. Public Health **203**, 23–30 (2022)

14. Hofmann, T.: Probabilistic latent semantic indexing. In: Proceedings of the 22nd Annual International ACM SIGIR Conference on Research and Development in Information Retrieval, pp. 50–57 (1999)

15. Jarrahi, A., Safari, L.: Evaluating the effectiveness of publishers' features in fake news detection on social media. Multimed. Tools Appl. **82**(2), 2913–2939 (2023)

16. Jwa, H., Oh, D., Park, K., Kang, J.M., Lim, H.: exBAKE: automatic fake news detection model based on bidirectional encoder representations from transformers (BERT). Appl. Sci. **9**(19), 4062 (2019)

17. Kaliyar, R.K., Goswami, A., Narang, P.: FakeBERT: fake news detection in social media with a BERT-based deep learning approach. Multimed. Tools Appl. **80**(8), 11765–11788 (2021)

18. Korshunova, I., Xiong, H., Fedoryszak, M., Theis, L.: Discriminative topic modeling with logistic LDA. In: Advances in Neural Information Processing Systems, vol. 32 (2019)

19. Kula, S., Choraś, M., Kozik, R.: Application of the BERT-based architecture in fake news detection. In: Herrero, Á., Cambra, C., Urda, D., Sedano, J., Quintián, H., Corchado, E. (eds.) CISIS 2019. AISC, vol. 1267, pp. 239–249. Springer, Cham (2021). https://doi.org/10.1007/978-3-030-57805-3_23

20. Lan, Z., Chen, M., Goodman, S., Gimpel, K., Sharma, P., Soricut, R.: ALBERT: a lite BERT for self-supervised learning of language representations. arXiv preprint arXiv:1909.11942 (2019)

21. Lazer, D.M., et al.: The science of fake news. Science **359**(6380), 1094–1096 (2018)

22. Le, Q., Mikolov, T.: Distributed representations of sentences and documents. In: International Conference on Machine Learning, pp. 1188–1196. PMLR (2014)

23. Lee, D.D., Seung, H.S.: Learning the parts of objects by non-negative matrix factorization. Nature **401**(6755), 788–791 (1999)

24. Liu, Y., et al.: RoBERTa: a robustly optimized BERT pretraining approach. arXiv preprint arXiv:1907.11692 (2019)
25. Loomba, S., de Figueiredo, A., Piatek, S.J., de Graaf, K., Larson, H.J.: Measuring the impact of Covid-19 vaccine misinformation on vaccination intent in the UK and USA. Nat. Hum. Behav. 5(3), 337–348 (2021)
26. Mikolov, T., Chen, K., Corrado, G., Dean, J.: Efficient estimation of word representations in vector space. arXiv preprint arXiv:1301.3781 (2013)
27. Moody, C.E.: Mixing Dirichlet topic models and word embeddings to make lda2vec. arXiv preprint arXiv:1605.02019 (2016)
28. Nasir, J.A., Khan, O.S., Varlamis, I.: Fake news detection: a hybrid CNN-RNN based deep learning approach. Int. J. Inf. Manag. Data Insights 1(1), 100007 (2021)
29. Nguyen, D.Q., Vu, T., Nguyen, A.T.: BERTweet: a pre-trained language model for English tweets. arXiv preprint arXiv:2005.10200 (2020)
30. de Oliveira, N.R., Pisa, P.S., Lopez, M.A., de Medeiros, D.S.V., Mattos, D.M.: Identifying fake news on social networks based on natural language processing: trends and challenges. Information 12(1), 38 (2021)
31. Reimers, N., Gurevych, I.: Sentence-BERT: sentence embeddings using Siamese BERT-networks. arXiv preprint arXiv:1908.10084 (2019)
32. Röder, M., Both, A., Hinneburg, A.: Exploring the space of topic coherence measures. In: Proceedings of the Eighth ACM International Conference on Web Search and Data Mining, pp. 399–408 (2015)
33. Roozenbeek, J., et al.: Susceptibility to misinformation about Covid-19 around the world. R. Soc. Open Sci. 7(10), 201199 (2020)
34. Sanh, V., Debut, L., Chaumond, J., Wolf, T.: DistilBERT, a distilled version of BERT: smaller, faster, cheaper and lighter. arXiv preprint arXiv:1910.01108 (2019)
35. Shu, K., Mahudeswaran, D., Wang, S., Lee, D., Liu, H.: FakeNewsNet: a data repository with news content, social context, and spatiotemporal information for studying fake news on social media. Big Data 8(3), 171–188 (2020)
36. Talia, D., Trunfio, P., Marozzo, F.: Data Analysis in the Cloud: Models, Techniques and Applications. Elsevier (2015). ISBN 978-0-12-802881-0
37. Tran, N.K., Zerr, S., Bischoff, K., Niederée, C., Krestel, R.: Topic cropping: leveraging latent topics for the analysis of small corpora. In: Aalberg, T., Papatheodorou, C., Dobreva, M., Tsakonas, G., Farrugia, C.J. (eds.) TPDL 2013. LNCS, vol. 8092, pp. 297–308. Springer, Heidelberg (2013). https://doi.org/10.1007/978-3-642-40501-3_30
38. Vijjali, R., Potluri, P., Kumar, S., Teki, S.: Two stage transformer model for Covid-19 fake news detection and fact checking. arXiv preprint arXiv:2011.13253 (2020)
39. Yang, Y., Zheng, L., Zhang, J., Cui, Q., Li, Z., Yu, P.S.: TI-CNN: convolutional neural networks for fake news detection. arXiv preprint arXiv:1806.00749 (2018)

Unsupervised Key-Phrase Extraction from Long Texts with Multilingual Sentence Transformers

Hélder Dias[1], Artur Guimarães[1], Bruno Martins[1(✉)], and Mathieu Roche[2]

[1] INESC-ID and IST, University of Lisbon, Lisbon, Portugal
{helder.dias,artur.guimas,bruno.g.martins}@tecnico.ulisboa.pt
[2] CIRAD, TETIS Research Unit, Montpellier, France
mathieu.roche@cirad.fr

Abstract. Key-phrase extraction concerns retrieving a small set of phrases that encapsulate the core concepts of an input textual document. As in other text mining tasks, current methods often rely on pre-trained neural language models. Using these models, the state-of-the-art supervised systems for key-phrase extraction require large amounts of labelled data and generalize poorly outside the training domain, while unsupervised approaches generally present a lower accuracy. This paper presents a multilingual unsupervised approach to key-phrase extraction, improving upon previous methods in several ways (e.g., using representations from pre-trained Transformer models, while supporting the processing of long documents). Experimental results on datasets covering multiple languages and domains attest to the quality of the results.

Keywords: Key-phrase extraction · Multilingual text processing · Transformers

1 Introduction

Key-Phrase Extraction (KPE) can be defined as the task of retrieving a small set of phrases from a given textual document, to best describe its main concepts. The task is useful in the context of discovery science and, similarly to other text mining tasks, recent methods involve the use of text representations produced through neural network models.

Supervised approaches for KPE require quantity and quality of in-domain annotated data, motivating work on transfer learning [18,42], weakly-supervised [40], or unsupervised [5,12,36,38,44] methods, that do not require the usage of expensive annotations nor extensive training procedures to obtain strong results. For instance, EmbedRank [5] was created as a simple, yet very effective, unsupervised KPE method that can be broken down into three main steps: (1) candidate phrase extraction using patterns over parts-of-speech (POS) tags, selecting phrases with zero or more adjectives followed by one or more nouns; (2) using Sent2Vec[1] or Doc2Vec[2] sentence embeddings to represent both

---

[1] https://github.com/epfml/sent2vec.
[2] https://github.com/jhlau/doc2vec.

© The Author(s), under exclusive license to Springer Nature Switzerland AG 2023
A. Bifet et al. (Eds.): DS 2023, LNAI 14276, pp. 141–155, 2023.
https://doi.org/10.1007/978-3-031-45275-8_10

the candidate phrases and the analyzed document; (3) ranking candidate phrases using the cosine similarity measure between representations for each candidate phrase and the document. Results showed that despite its simplicity, EmbedRank could outperform previous unsupervised methods for KPE, mostly based on graph-ranking approaches [13, 23].

SIFRank [38] shares the same methodology of EmbedRank for extracting and ranking candidate phrases, but changes how the candidate phrases and the document are embedded. The ELMo [27] pre-trained language model, based on a deep recurrent neural network, is used to create the embeddings, and instead of directly comparing embeddings SIFRank uses a word weight balancing operation based on contextual information, which compares the domain corpus of the input document and a baseline common corpus, seeking to adapt the model to the specific domain at hand.

Both EmbedRank and SIFRank rely on the assumption that the similarity between a candidate phrase and a document is a good measure of how relevant that candidate phrase is. MDERank [44] subverts this assumption, testing the hypothesis that a relevant candidate phrase maximizes the difference in a document when it is absent. To do so, MDERank ranks candidate phrases by replacing their occurrences in the document with a special [MASK] token, afterwards representing the document by embedding it using a Bidirectional Encoder Representation from Transformers (BERT) pre-trained language model [10]. The cosine distance towards the original document is measured, and candidate phrases having a higher distance are finally ranked as better.

Despite achieving strong empirical results, there are also limitations in previous unsupervised methods. For instance most previous studies focused their evaluation only on the English language, while it would be interesting to see if similar approaches can also generalize across languages. Problems also arise when considering large documents, as pre-trained language models often struggle to process long input sequences (e.g., the base BERT model has a 512 token limit).

This paper explores KPE in an unsupervised multilingual scenario, adapting and re-configuring pre-existing methods (i.e., EmbedRank and MDERank) to work with representations produced with a multilingual Sentence-Transformers [31] model, at the same time also supporting the processing of long documents by converting the Sentence-Transformers model into a Long Document Transformer (Longformer) [4]. The proposed KPE methods were evaluated on different domains (i.e., involving texts of different sizes, types, and languages), and the results show that they offer good generalization and improvements over previous approaches. The source code supporting the experiments is available from a public Github repository[3].

The rest of the paper is organized as follows: Sect. 2 details the proposed approaches, while Sect. 3 presents the experimental evaluation methodology and the obtained results. Section 4 summarizes our contributions and discusses possibilities for future work.

---

[3] https://github.com/araag2/KP_Extraction.

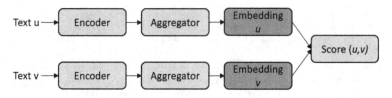

Fig. 1. The bi-encoder architecture for assessing similarity between two textual inputs.

2   The Proposed Approaches

Given a text document d belonging to a dataset $\mathfrak{D}$, we seek to extract a set $\mathfrak{C}$ of candidate phrases c that contains as many relevant key-phrases as possible, for describing the contents of d. After extracting $\mathfrak{C}$, our second goal is to rank the top-k candidates within that set.

The first task is addressed using models from the spaCy[4] library, according to the language of the documents within dataset $\mathfrak{D}$, to tokenize and perform parts-of-speech tagging of each document d. A regular expression over sequences of universal parts-of-speech tags (<PROPN|NOUN|ADJ>*<PROPN|NOUN>+<ADJ>*) is used as a heuristic method to extract candidate phrases, relying only on a simple, although coarse, parts-of-speech tagset that is common to different languages [28]. We also perform lemmatization to join candidates with slight differences into a single representation, through the simplemma[5] library which offers complete multilingual options. We keep a mapping between each possible form and the corresponding lemmatized candidates, so that matches in the text can be aggregated into the lemmatized versions.

On what regards the ranking task, we first need to find suitable representations for the documents and the candidate phrases, adhering to some constraints: computational efficiency to perform multiple comparisons between documents and candidate phrases, support for multilingual textual contents, and adequate handling of potentially large documents.

2.1   Text Representations from a Longformer Model Built from a Multilingual Sentence-Transformer

Transformer encoder models like BERT [10] and RoBERTa [21] can produce effective text representations, but they are also computationally demanding. They can be used as cross-encoders to assess the similarity between a pair of input texts (i.e., processing the concatenation of both texts, and directly outputting a similarity score), but a more efficient approach is to instead consider a bi-encoder setting, in which the texts to be compared are modeled separately, and then a similarity score is computed over aggregates (e.g., token averages) from

---

[4] https://spacy.io/models/.
[5] https://github.com/adbar/simplemma/.

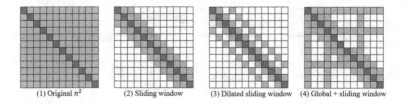

Fig. 2. Regular self-attention versus the attention patterns in a Longformer model.

the resulting representations – see Fig. 1. Moreover, these models also struggle when processing long documents, due to the quadratic complexity associated to the self-attention mechanism computed over all pairs of positions from input sequences. Approaches such as the Longformer [4] or BigBird [43] address this limitation, slightly changing the self-attention operations in order to limit how the different positions interact – see the illustration on Fig. 2.

In our Key-Phrase Extraction (KPE) methods, we use text representations obtained with a multi-lingual model based on RoBERTa, adapted from a model available from the Sentence-Transformers library[6] and pre-trained as a bi-encoder for assessing multilingual sentence similarity [31]. The RoBERTa-based model was adapted into a Longformer without any additional training, extending the input sequence limit to 4096 tokens (i.e., initializing the additional position embeddings by copying and interpolating from the embeddings of the first positions) and changing the implementation of the self-attention operations within the different layers, while keeping the pre-trained model parameters.

In brief, Sentence-Transformers bi-encoders process strings independently through the same Transformer encoder, followed by mean pooling aggregation to create fixed-sized sentence embeddings. These models are trained either to directly predict sentence similarity scores as given in training data corresponding to annotated sentence pairs, or to predict similarity relations between sentences (e.g., given an anchor sentence a, a positive sentence p with high similarity towards a, and a negative sentence n, we can consider a loss function that tunes the network such that the distance between a and p is smaller than the distance between a and n). We specifically started from the Sentence-Transformers model named `paraphrase-multilingual-mpnet-base-v2`, i.e. a pre-existing model built from a multi-lingual RoBERTa and trained to mimic the results of another mono-ligual Sentence-Transformers bi-encoder, through a knowledge distillation objective [32]. This model was then adapted through the procedure described in the Longformer paper to build a Long Document Transformer starting from a RoBERTa checkpoint [4].

As can be seen in Fig. 2, three attention patterns can be combined within the Longformer architecture: sliding window, focusing on the local context and examining a fixed-size window w around each token; dilated sliding window, which adds a gap of size d between each token considered in the sliding window,

---

[6] https://www.sbert.net/.

with d varying across layers and attention heads; and global attention, in which some specific input locations (e.g., the initial [CLS] token) will attend to (and be attended by) all other tokens.

Our Longformer model employs a sliding window attention with window size of 512 tokens, thus involving approximately the same amount of computation as a standard RoBERTa, and also behaving like RoBERTa when the input has fewer than 512 tokens. One additional attention pattern was also considered, in which the specific positions corresponding to tokens associated to the occurrences of the key-phrase candidates were also considered for global attention, when representing candidates or documents.

In the remaining parts of this paper, we refer to the proposed text representation model as the Multilingual Sentence-Longformer (MSL). Using the word representations from MSL, we built different approaches to address the candidate ranking problem for KPE.

2.2   LMEmbedRank

Longformer Multilingual EmbedRank (LMEmbedRank) corresponds to an adaptation of EmbedRank [5] that represents documents through MSL embeddings, and candidate phrases as the average of all MSL token embeddings that form the multiple occurrences of the candidate (first averaging the token representations from each occurrence, and then averaging across occurrences). As seen in Fig. 3, where colored words represent the tokens that are considered, LMEmbendRank averages all token representations in order to create the embedding representation of a document, whilst to embed a candidate phrase (e.g., *core concepts*) it performs an average pooling operation over all tokens that form each occurrence over the document, and then averages over all occurrences. For candidates that only occur after the first 4096 tokens (i.e., the Longformer limit for input sequences), we still manage to generate a representation with a back-off procedure that processes the candidate string alone, without any additional contextual information.

Following the standard EmbedRank procedure, we use the cosine measure to rank candidate phrases according to the similarity of their representations towards the document representation, in descending order of similarity.

2.3   LMMaskRank

Longformer Multilingual MaskRank (LMMaskRank) corresponds to an adaption of MDERank [44] that also represents documents and candidate phrases through MSL embeddings. As can be seen again in Fig. 3, where colored words represent the considered tokens, in order to create the embedding representation of a document LMMaskRank uses the same mechanism as LMEmbedRank, whilst to embed a candidate phrase (e.g., *core concepts*) the method starts by replacing all of the candidate occurrences by the [MASK] token, and then embeds the entirety of the document. As described in the original paper, candidate phrases are then ranked using the cosine distance measure, in descending order of distance (i.e.,

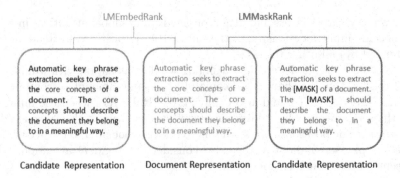

Fig. 3. Overview on how LMEmbedRank and LMMaskRank represent the candidate key-phrase corresponding to *core concepts*.

candidate representations that are further away from the representation of the input document are preferred).

It is interesting to note that the most computationally expensive operation, in both the LMEmbedRank and LMMaskRank methods, corresponds to obtaining the MSL embeddings (i.e., one forward pass over the Longformer model). LMMaskRank is thus much more demanding, given that LMEmbedRank only needs to compute MSL embeddings once for each input document, while LMMaskRank needs a separate computation for each candidate (i.e., replacing the candidate occurrences with [MASK] tokens, before computing the corresponding text representations through a forward pass).

2.4 Combining Both Ranking Approaches

Longformer Multilingual Rank (LMRank) corresponds to a hybrid approach that uses weighted averages of scores obtained by both previous methods, based on the hypothesis that each method would be better suited to handle different types of textual documents, and thus together they could probably perform better. This general approach can be implemented through different combination schemes, and we tested both the arithmetic and harmonic averages of LMEmbedRank and LMMaskRank scores.

3 Experimental Evaluation

This section starts by introducing the datasets that were used in the experiments, together with the considered evaluation metrics. It then follows with an overview on the experimental results across all datasets. We also provide a comparison with previous methods, as well as against ablated versions of the proposed approaches.

Table 1. Statistics for the considered datasets.

Dataset	Language	Average #KPs	Candidate Recall	Absent KPs	Average #Words	#Docs.
DUC	EN	8	87.2%	6.8%	740	308
NUS	EN	11	88.2%	4.3%	5201	209
Inspec	EN	10	58.7%	35.6%	128	2000
SemEval	EN	16	95.0%	3.2%	8332	243
PubMed	EN	15	80.2%	15.8%	3992	1320
PT-KP	PT	24	53.6%	5.2%	304	110
CACIC	ES	5	72.3%	7.3%	3985	888
WICC	ES	5	74.3%	5.9%	1955	1640
FR-WIKI	FR	12	79.1%	4.4%	293	100
TeKET	DE	5	93.5%	0.0%	11524	10

3.1 Metrics and Datasets

To evaluate the performance of our models in different languages and domains, we relied on a wide variety of datasets used in previous unsupervised key-phrase extraction studies: five English datasets, namely NUS [25], DUC-2001 [39], Inspec [15], SemEval [19] and PubMed [3]; an European Portuguese dataset named 110-PT-BN-KP (PT-KP) [22]; two Spanish datasets, namely CACIC and WICC [2]; a French dataset named WikiNews (FR-WIKI) [7]; and a German dataset (TeKET) [29]. Basic statistical information about each dataset is presented in Table 1, and the reader can refer to the original publications, introducing each of the datasets, for additional details.

The candidate-phrase extraction component was initially evaluated in terms of recall, by comparing our extractions with the ground truth key-phrases of each document. Notice that without a high recall it will be impossible to accurately rank the key-phrase candidates so as to recover the ground truth, as the correct key-phrases will not be available to be ranked.

It is interesting to note that there exists an upper bound on the possible recall value, as the candidate extraction method is unable to find correct key-phrases that do not appear within the input text documents (although the lematization operation does help in this regard). The candidate extraction recall, for each dataset, is also shown in Table 1, together with the percentage of ground-truth key-phrases that do not occur in the text.

Overall, we can see that the proposed candidate extraction method is able to correctly find a large percentage of the ground-truth key-phrases in the majority of the datasets, with exceptions for Inspec (where we also have a very large number of absent key-phrases from the textual contents of the documents) and PT-KP. In this latter case, our regular expression pattern was unable to find many of the ground-truth key-phrases, which do not always correspond to a noun phrase in the input text.

Using the set of extracted candidates, we can then evaluate the ranking of the candidates in terms of its ability to place relevant key-phrases in the top positions. The candidate rankings are handled as an ordered list, and a specific cut-off point k can then be defined, comparing the top k ranked candidates with the ground-truth key-phrases. The performance on the ranking task is measured with the F_1-score ($F_{1,k}$) metric, at the cut-off points $k = \{5, 10, 15\}$. Additionally, we also use the Normalized Discounted Cumulative Gain (nDCG) metric over the complete ranked list of candidate key-phrases.

Following most previous studies in the area, both the extracted and the ground-truth key-phrases are processed through a stemming algorithm, prior to performing comparisons for ranking evaluation.

3.2   Experimental Results over the Different Datasets

Table 2 presents experimental results over the multiple datasets, comparing the alternatives discussed in the previous section. The lines named $MRank_{avg_a}$ and $LMRank_{avg_h}$ correspond to using an arithmetic or harmonic average of LMEmbedRank and LMMaskRank scores, as described in Subsect. 2.4.

Although the different evaluation metrics mostly agree on how the methods should be ranked according to result quality, different methods can perform slightly better on some of the datasets:

- LMEmbedRank, which is also the simpler and computationally more effective method, performs clearly better than LMMaskRank on the DUC, Inspec, and CACIC datasets.
- In turn, LMMaskRank clearly performs better than LMEmbedRank on the Pubmed, PT-KP, WICC, FR-WIKI, and TeKET datasets.
- Datasets like NUS, SemEval, PubMed, and particularly TeKET, feature very long documents, going beyond the 4096 token limit in Longformer. In these cases, LMMaskRank tends to perform better, although the relation between result quality and the characteristics of the documents (e.g., size, language, or candidate recall) is not entirely clear. Note that LMMaskRank is biased towards preferring candidates in the first 4096 tokens, since occurrences beyond this limit will not impact the representations (i.e., the representations for these key-phrase candidates are exactly equal to those from the documents, and hence they will be ranked below the other candidates).
- The combination of both approaches, particularly when considering the harmonic mean, is beneficial in most cases. In the NUS, SemEval, PubMed, WICC, and FR-WIKI datasets, the best results are achieved with a combined method. On the other datasets, the combination performs similarly to the best method, improving over the LMEmbedRank or LMMaskRank strategies.

3.3   Results for Ablation Experiments

Table 3 presents results for ablated versions of the LMEmbedRank, LMMaskRank, and $LMRank_{avg_h}$ methods, specifically assessing the impact of the different ideas introduced in our proposal. The following alternatives were tested on 5 of the datasets also seen in Table 2:

- Using the regular Sentence-Transformers model based on a multi-lingual RoBERTa, instead of converting the model into a Longformer. In the case of LMEmbedRank, the candidates that occur after the maximum token limit of the model were also represented with the back-off procedure that processes the candidate string alone;
- Using a standard English Longformer[7], instead of the Sentence-Transformers model pre-trained only for Masked Language Modeling (MLM). This way, we can assess the impact of model pre-training with sentence similarity tasks, noting also that previous studies such as MDERank [44] have only explored the use of regular Transformer encoders pre-trained for MLM.
- Removing the lemmatization procedure that aggregates similar candidates appearing in the text with a slightly different surface form;
- Removing the Longformer attention pattern that considers a global attention for the tokens that correspond to candidate occurrences, instead leaving only the [CLS] token with the global attention over all other tokens.

The results show that all the four previous aspects, and particularly the pre-training over sentence similarity tasks (i.e., using an adapted Sentence-Transformers model) and the conversion of the Sentence-Transformers model into a Longformer, contribute to improved results. Higher differences in the result quality are also seen in the case of the datasets involving longer documents.

Besides the aforementioned ablations, we also considered extensions over the methods in Table 2, leveraging ideas advanced in previous studies. These included (a) post-processing the document/candidate embeddings prior to computing similarities [14,16,33,37], or (b) weighting the individual tokens when computing the representations within LMEmbedRank, e.g. proportionally to attention scores produced by the Longformer model [12]. Still, results were consistently worse, and we decided not to report these scores.

One particular extension that we tested involves weighting the scores of the LMEmbedRank and LMMaskRank methods prior to their combination, in an attempt to further improve results. In a first step towards doing this, we started by analyzing the distribution of the similarity scores between candidate and document representations, for the two different methods (i.e., LMEmbedRank and LMMaskRank) and over the different datsets. Figure 4 illustrates the results of this analysis, specifically for the DUC, Inspec, NUS, and PT-KP datasets (similar patterns could be observed also for the other datasets). The results showed that LMEmbedRank produces similarity scores that are more evenly spread, whereas LMMaskRank mostly produces results in the interval $[0, 0.5]$. Both methods also produce two peaks in terms of the distribution for the similarity values, corresponding to a good distinction between the relevant and the irrelevant candidates (i.e., the top charts in Fig. 4 correspond to the entire sets of candidates, whereas the bottom charts show only the similarity scores for the subset of relevant candidates).

---

[7] https://huggingface.co/allenai/longformer-large-4096.

Table 2. Results on each dataset and for each of the proposed methods.

Model	DUC				NUS				Inspec				SemEval				PubMed			
	nDCG	$F1_{5}$	$F1_{10}$	$F1_{15}$	nDCG	$F1_{5}$	$F1_{10}$	$F1_{15}$	nDCG	$F1_{5}$	$F1_{10}$	$F1_{15}$	nDCG	$F1_{5}$	$F1_{10}$	$F1_{15}$	nDCG	$F1_{5}$	$F1_{10}$	$F1_{15}$
LMEmbedRank	**45.01**	**30.28**	**33.72**	**34.10**	28.55	20.61	22.87	23.12	**45.95**	**31.25**	**36.97**	**38.05**	22.16	14.49	18.74	**20.93**	13.14	7.94	10.22	10.28
LMMaskRank	37.00	24.41	28.31	28.15	30.53	18.58	20.01	17.38	41.19	27.75	33.31	34.78	21.10	15.38	17.48	18.10	27.72	17.36	18.93	18.14
LMRank$_{avg_a}$	39.57	25.58	29.96	31.66	28.16	22.76	24.49	24.86	43.54	29.66	35.34	36.45	19.78	14.59	16.22	16.27	23.65	15.1	16.46	**20.35**
LMRank$_{avg_h}$	40.98	29.68	31.29	30.05	**38.11**	**28.17**	**32.69**	**31.79**	43.43	30.10	34.67	35.14	**22.97**	**16.69**	**19.17**	18.25	**29.48**	**17.94**	**20.15**	18.74

Model	PT-KP				CACIC				WICC				FR-WIKI				TeKET			
	nDCG	$F1_{5}$	$F1_{10}$	$F1_{15}$	nDCG	$F1_{5}$	$F1_{10}$	$F1_{15}$	nDCG	$F1_{5}$	$F1_{10}$	$F1_{15}$	nDCG	$F1_{5}$	$F1_{10}$	$F1_{15}$	nDCG	$F1_{5}$	$F1_{10}$	$F1_{15}$
LMEmbedRank	36.78	23.91	32.91	37.29	40.15	26.49	37.29	37.63	24.88	15.31	16.93	13.20	43.64	24.15	32.92	26.50	6.48	0.00	6.95	6.15
LMMaskRank	39.90	25.96	34.38	37.63	22.52	16.75	26.19	26.50	27.50	17.76	14.72	14.07	49.62	37.06	37.65	34.83	10.44	10.44	**10.95**	**10.75**
LMRank$_{avg_a}$	**41.26**	28.04	**36.62**	**39.59**	39.59	11.09	26.25	27.55	17.99	15.37	13.13	**15.37**	50.84	37.53	**39.97**	36.19	15.87	15.87	10.66	10.15
LMRank$_{avg_h}$	41.09	**28.06**	35.79	38.91	25.64	19.11	27.01	27.37	**31.43**	**20.57**	**15.98**	14.77	**49.95**	**37.93**	38.43	35.86	**17.65**	10.44	**10.95**	**10.75**

Table 3. Results with ablated versions of the proposed key-phrase extraction methods.

Model	DUC				NUS				Inspec				SemEval				PT-KP			
	nDCG	$F1_{5}$	$F1_{10}$	$F1_{15}$	nDCG	$F1_{5}$	$F1_{10}$	$F1_{15}$	nDCG	$F1_{5}$	$F1_{10}$	$F1_{15}$	nDCG	$F1_{5}$	$F1_{10}$	$F1_{15}$	nDCG	$F1_{5}$	$F1_{10}$	$F1_{15}$
LMEmbedRank	45.01	30.28	33.72	34.10	28.55	20.61	22.87	23.12	45.95	31.25	36.97	38.05	22.16	14.49	18.74	20.93	36.78	23.91	32.91	37.29
- Longformer	40.92	27.56	31.26	31.23	18.12	11.14	13.52	12.49	44.51	29.70	35.40	36.46	12.35	1.82	8.69	12.05	34.82	21.41	30.95	35.37
- Sentence-BERT	26.86	15.75	22.10	22.50	17.54	12.23	12.57	11.90	23.68	17.54	19.80	18.27	10.27	2.03	7.07	11.32				
- Lematization	42.10	29.09	31.85	31.93	25.79	19.31	20.72	21.14	43.42	29.88	35.01	36.31	19.21	12.78	16.48	18.88	33.84	22.43	30.89	35.28
- Global Attention	44.15	29.59	33.16	33.37	27.56	19.93	22.22	22.41	45.23	30.88	36.57	37.60	21.18	13.77	18.05	20.34	35.87	23.03	32.32	36.64
LMMaskRank	37.00	24.41	28.31	28.15	30.53	18.58	20.01	17.38	41.19	27.75	33.31	34.78	21.10	15.38	17.48	18.10	36.42	25.96	34.38	37.63
- Longformer	33.89	22.46	22.71	26.50	20.79	9.75	10.72	7.78	39.83	27.09	31.75	33.11	12.21	2.99	7.91	8.69				35.03
- Sentence-BERT	32.92	22.39	27.71	26.50	23.74	12.36	14.09	11.83	27.09	20.45	22.82	23.03	15.64	10.93	12.99	14.21				
- Lemmatization	34.23	23.1	26.19	26.25	27.55	17.49	18.09	15.28	38.29	26.32	31.47	32.87	18.13	13.90	15.22	16.50	33.65	20.56	28.82	30.97
- Global Attention	36.04	23.50	27.37	27.01	29.60	18.00	19.25	16.61	40.58	27.46	32.94	34.27	20.26	14.36	16.96	17.44	39.05	23.71	31.55	35.07
LMRank$_{avg_h}$	40.98	29.68	31.29	30.05	38.11	28.17	32.69	31.79	43.43	30.10	34.67	35.14	22.97	16.69	19.17	18.25	41.09	28.06	35.79	38.91
- Longformer	37.62	27.22	29.14	27.14	29.37	21.40	27.32	25.38	41.25	28.86	32.87	33.21	15.27	8.92	11.28	11.91	37.87	25.15	33.33	36.13
- Sentence-BERT	36.52	26.24	27.55	25.11	29.92	20.71	28.09	26.13	24.91	19.27	21.09	21.11	13.88	5.48	9.23	10.56				
- Lemmatization	38.32	28.14	29.49	28.13	36.09	27.09	30.88	29.95	41.31	28.91	32.78	33.31	20.67	15.04	17.44	16.51	33.39	20.71	29.63	31.58
- Global Attention	39.68	28.63	30.14	28.78	37.12	26.92	31.65	30.47	42.60	28.72	33.90	34.30	22.10	15.66	18.48	17.07	40.13	26.42	34.79	37.64
+ Weighting	40.57	29.06	30.82	29.33	38.81	28.65	33.23	32.38	42.64	29.62	34.14	34.52	22.36	16.32	18.85	17.65	41.83	28.45	36.25	39.53

Table 4. Comparison between our best methods and previously published results.

Model	DUC			NUS			Inspec			SemEval			PT-KP			TeKET		
	$F1_5$	$F1_{10}$	$F1_{15}$	$F1_5$	$F1_{10}$	$F1_{15}$	$F1_5$	$F1_{10}$	$F1_{15}$	$F1_5$	$F1_{10}$	$F1_{15}$	$F1_5$	$F1_{10}$	$F1_{15}$	$F1_5$	$F1_{10}$	$F1_{15}$
TF-IDF [17,26,29,36,38]	9.21	10.63	11.60	11.60	14.20	12.50	24.20	28.00	24.80	16.10	16.70	15.30	—	17.9	—	7.50	8.60	9.60
TopicRank [7,26,29,44]	19.97	21.73	20.97	4.54	7.93	9.37	12.20	17.24	19.33	9.93	12.52	12.26	—	14.80	—	6.30	7.60	8.10
EmbedRank [5,44]	21.75	25.09	24.68	2.13	2.94	3.56	14.51	21.02	23.79	9.63	13.90	14.79	—	—	—	—	—	—
SIFRank [38,44]	24.30	27.60	27.96	3.01	5.34	5.86	29.38	39.12	**39.82**	11.16	16.03	18.42	—	—	—	—	—	—
MDERank [44]	13.05	17.31	19.13	15.24	18.33	17.95	26.17	33.81	36.17	10.16	15.32	17.76	—	—	—	—	—	—
AttentionRank [12]	—	—	—	—	—	—	24.45	32.15	34.49	11.39	15.12	16.66	—	—	—	—	—	—
YAKE! [8,29,44]	11.99	14.18	14.18	7.85	11.05	13.09	8.02	11.47	13.65	6.82	11.01	12.55	—	10.70	—	8.83	**12.30**	**13.80**
KeyGames [34]	24.22	28.28	29.27	—	—	—	**32.12**	40.48	**40.94**	11.93	14.35	14.62	—	—	—	—	—	—
Liang et al. [20]	28.62	35.52	36.29	—	—	—	32.61	40.17	41.09	13.02	19.35	21.72	—	—	—	—	—	—
Multipartite [6,26,29,44]	21.70	24.10	23.62	6.17	8.57	10.82	13.41	18.18	20.52	10.13	12.91	13.24	12.05	15.60	—	7.10	9.10	9.70
CDKGen [11]	—	—	—	41.20	38.10	—	33.10	34.70	—	34.20	35.50	—	—	—	—	—	—	—
SEG-Net [1]	—	—	—	39.60	—	—	21.60	28.30	—	28.30	—	—	—	—	—	—	—	—
SKE-Base-Rank [24]	—	—	—	38.90	36.50	—	28.90	32.10	—	35.40	33.70	—	—	—	—	—	—	—
LMEmbedRank	30.28	33.72	**34.10**	20.61	22.87	23.12	31.25	36.97	38.05	14.49	18.74	20.93	23.91	32.91	37.29	0.0	6.95	6.15
LMRank$_{avg_h}$	29.68	31.29	30.05	**28.17**	**32.69**	**31.79**	30.10	34.67	35.14	**16.69**	**19.17**	18.25	**28.06**	**35.79**	**38.91**	**10.44**	10.95	10.75

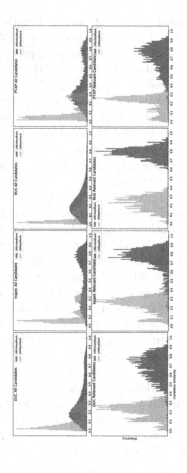

Fig. 4. Distribution of similarity scores between candidates and documents, considering all candidates and documents for two different datasets (on top), or only the subsets of relevant candidates (at the bottom).

With basis on the aforementioned analysis, we then tested a combination method in which a constant of 0.5 is added to the scores from LMMaskRank procedure, prior to the combination with LMEmbedRank. The results are shown in the bottom row from Table 2, although no noticeable improvements were seen. Overall, the proposed methods have also the interesting property of not involving many parameters to tune, which is often the case with unsupervised approaches.

3.4   Comparison to Previous Methods

Table 4 compares the best proposed methods, specifically LMEmbedRank (i.e., the simplest and fastest method) and $LMRank_{avg_h}$, against the results reported in publications presenting and using previous methods (including results for the original EmbedRank [5] and MDERank [44] methods). We present results for the datasets over which more previous methods have been tested (i.e., mostly by re-using results presented on previous comparisons [44]), also including some recent supervised approaches (i.e., the second set of rows in Table 4).

The results in Table 4 show that the proposed approaches are very competitive within the realm of unsupervised KPE, outperforming most previous unsupervised methods in the majority of the considered datasets and often by a very large margin, while simultaneously being simple, multilingual, and thus easy to generalize to different types of applications. Notice that the unsupervised methods considered for the comparison include representatives from different types of approaches, including simple heuristics based on term-frequency statistics (e.g., TF-IDF [17] or YAKE! [8]), approaches based on graph ranking which are also not limited in the processing of long documents (e.g., TopicRank [7] or Multipartite Ranking [6]), and approaches based on neural embeddings [5,38,44].

Notable exceptions correspond to the Inspec and the TeKET datasets. In the specific case of Inspec, SIFRank [38] outperforms the proposed methods in $F1_{10}$ and $F1_{15}$, while KeyGames [34] (i.e., a recently proposed method that tackles key-phrase extraction though a game-theoretic framework) performs even better for all the metrics. This is likely due to the small size of the documents (128 words on average) which offset the advantages of using a Longformer approach. On TeKET, YAKE! [8] outperforms the proposed approaches also in $F1_{10}$ and $F1_{15}$, but in this case it is difficult to draw many conclusions because the dataset only features 10 very long documents (11524 words on average), and hence the results can be very noisy.

It is also important to notice that the differences towards recent supervised methods are still very significant. Previous methods such as CDKGen [11], SEG-NET [1], or SKE-Base-Rank [24] are, usually, still significantly better than the best unsupervised approaches, although this also varies depending on characteristics of the datasets (e.g., on Inspec, the best results in terms of $F1_{10}$ and $F1_{15}$ are obtained with unsupervised methods).

4   Conclusions and Future Work

We proposed new unsupervised methods for key-phrase extraction, extending the previous EmbedRank [5] and MDERank [44] approaches in different directions. We tested the proposed approaches over multiple datasets, with results showing a very competitive performance against state-of-the-art unsupervised methods, while also generalizing across different languages and domains.

For future work, we can consider other text embedding models[8], and other methods for handling long inputs besides the Longformer (e.g., memory efficient attention implementations [9, 30], or other sparse attention patterns such as those in the Hypercube Transformer [41]). We would also like to perform experiments on scenarios that involve multi-document key-phrase extraction [35], this way further stressing the length of the textual inputs that need to be analyzed.

Acknowledgement. This research was supported by the European Union's H2020 research and innovation programme, under grant agreement No. 874850 (MOOD), as well as by Fundação para a Ciência e Tecnologia (FCT), namely through the INESC-ID multi-annual funding with reference UIDB/50021/2020, and through the project grants with references PTDC/CCI-CIF/32607/2017 (MIMU), DSAIPA/DS/0102/2019 (DEBAQI), and POCI/01/0145/FEDER/031460 (DARGMINTS).

References

1. Ahmad, W., Bai, X., Lee, S., Chang, K.W.: Select, extract and generate: neural keyphrase generation with layer-wise coverage attention. In: Proceedings of the Annual Meeting of the Association for Computational Linguistics (2021)
2. Aquino, G.O., Lanzarini, L.C.: Keyword identification in Spanish documents using neural networks. J. Comput. Sci. Technol. **15** (2015)
3. Aronson, A.R., et al.: The NLM indexing initiative. In: Proceedings of the American Medical Informatics Association Symposium (2000)
4. Beltagy, I., Peters, M.E., Cohan, A.: LongFormer: the long-document transformer. arXiv preprint arXiv:2004.05150 (2020)
5. Bennani-Smires, K., Musat, C., Hossmann, A., Baeriswyl, M., Jaggi, M.: Simple unsupervised keyphrase extraction using sentence embeddings. In: Proceedings of the Conference on Computational Natural Language Learning (2018)
6. Boudin, F.: Unsupervised keyphrase extraction with multipartite graphs. arXiv preprint arXiv:1803.08721 (2018)
7. Bougouin, A., Boudin, F., Daille, B.: TopicRank: graph-based topic ranking for keyphrase extraction. In: Proceedings of the International Joint Conference on Natural Language Processing (2013)
8. Campos, R., Mangaravite, V., Pasquali, A., Jorge, A., Nunes, C., Jatowt, A.: YAKE! keyword extraction from single documents using multiple local features. Inf. Sci. **509**, 257–289 (2020)
9. Dao, T., Fu, D.Y., Ermon, S., Rudra, A., Ré, C.: FlashAttention: fast and memory-efficient exact attention with IO-awareness. arXiv preprint arXiv:2205.14135 (2022)

---

[8] https://huggingface.co/spaces/mteb/leaderboard

10. Devlin, J., Chang, M.W., Lee, K., Toutanova, K.: BERT: pre-training of deep bidirectional transformers for language understanding. arXiv preprint arXiv:1810.04805 (2018)
11. Diao, S., Song, Y., Zhang, T.: Keyphrase generation with cross-document attention. arXiv preprint arXiv:2004.09800 (2020)
12. Ding, H., Luo, X.: AttentionRank: unsupervised keyphrase extraction using self and cross attentions. In: Proceedings of the 2021 Conference on Empirical Methods in Natural Language Processing (2021)
13. Florescu, C., Caragea, C.: PositionRank: an unsupervised approach to keyphrase extraction from scholarly documents. In: Proceedings of the 55th Annual Meeting of the Association for Computational Linguistics (2017)
14. Huang, J., et al.: WhiteningBERT: an easy unsupervised sentence embedding approach. arXiv preprint arXiv:1801.04470 (2021)
15. Hulth, A.: Improved automatic keyword extraction given more linguistic knowledge. In: Proceedings of the Conference on Empirical Methods in Natural Language Processing (2003)
16. Jégou, H., Chum, O.: Negative evidences and co-occurences in image retrieval: the benefit of PCA and whitening. In: Fitzgibbon, A., Lazebnik, S., Perona, P., Sato, Y., Schmid, C. (eds.) ECCV 2012. LNCS, vol. 7573, pp. 774–787. Springer, Heidelberg (2012). https://doi.org/10.1007/978-3-642-33709-3_55
17. Jones, K.S.: A statistical interpretation of term specificity and its application in retrieval. J. Doc. **28**(1), 11–21 (1972)
18. Joshi, R., Balachandran, V., Saldanha, E., Glenski, M., Volkova, S., Tsvetkov, Y.: Unsupervised keyphrase extraction via interpretable neural networks. arXiv preprint arXiv:2203.07640 (2022)
19. Kim, S.N., Medelyan, O., Kan, M.Y., Baldwin, T.: SemEval-2010 task 5: automatic keyphrase extraction from scientific articles. In: Proceedings of the International Workshop on Semantic Evaluation (2010)
20. Liang, X., Wu, S., Li, M., Li, Z.: Unsupervised keyphrase extraction by jointly modeling local and global context. In: Proceedings of the Conference on Empirical Methods in Natural Language Processing (2021)
21. Liu, Y., et al.: RoBERTa: a robustly optimized BERT pretraining approach. arXiv preprint arXiv:1907.11692 (2019)
22. Marujo, L., Viveiros, M., da Silva Neto, J.P.: Keyphrase cloud generation of broadcast news. arXiv preprint arXiv:1306.4606 (2013)
23. Mihalcea, R., Tarau, P.: TextRank: bringing order into text. In: Proceedings of the 2004 Conference on Empirical Methods in Natural Language Processing (2004)
24. Mu, F., et al.: Keyphrase extraction with span-based feature representations. arXiv preprint arXiv:2002.05407 (2020)
25. Nguyen, T.D., Kan, M.-Y.: Keyphrase extraction in scientific publications. In: Goh, D.H.-L., Cao, T.H., Sølvberg, I.T., Rasmussen, E. (eds.) ICADL 2007. LNCS, vol. 4822, pp. 317–326. Springer, Heidelberg (2007). https://doi.org/10.1007/978-3-540-77094-7_41
26. Papagiannopoulou, E., Tsoumakas, G.: A review of keyphrase extraction (2019)
27. Peters, M.E., et al.: Deep contextualized word representations. arXiv preprint arXiv:1802.05365 (2018)
28. Petrov, S., Das, D., McDonald, R.: A universal part-of-speech tagset. arXiv preprint arXiv:1104.2086 (2011)
29. Rabby, G., Azad, S., Mahmud, M., Zamli, K.Z., Rahman, M.M.: TeKET: a tree-based unsupervised keyphrase extraction technique. Cogn. Comput. **12**(4) (2020)

30. Rabe, M.N., Staats, C.: Self-attention does not need $o(n^2)$ memory. arXiv preprint arXiv:2112.05682 (2021)
31. Reimers, N., Gurevych, I.: Sentence-BERT: sentence embeddings using Siamese BERT-networks. arXiv preprint arXiv:1908.10084 (2019)
32. Reimers, N., Gurevych, I.: Making monolingual sentence embeddings multilingual using knowledge distillation. arXiv preprint arXiv:2004.09813 (2020)
33. Sajjad, H., Alam, F., Dalvi, F., Durrani, N.: Effect of post-processing on contextualized word representations. arXiv preprint arXiv:2104.07456 (2021)
34. Saxena, A., Mangal, M., Jain, G.: KeyGames: a game theoretic approach to automatic keyphrase extraction. In: Proceedings of the International Conference on Computational Linguistics (2020)
35. Shapira, O., Pasunuru, R., Dagan, I., Amsterdamer, Y.: Multi-document keyphrase extraction: a literature review and the first dataset. arXiv preprint arXiv:2110.01073 (2021)
36. Shen, X., Wang, Y., Meng, R., Shang, J.: Unsupervised deep keyphrase generation. arXiv preprint arXiv:2104.08729 (2021)
37. Su, J., Cao, J., Liu, W., Ou, Y.: Whitening sentence representations for better semantics and faster retrieval. arXiv preprint arXiv:2103.15316 (2021)
38. Sun, Y., Qiu, H., Zheng, Y., Wang, Z., Zhang, C.: SIFRank: a new baseline for unsupervised keyphrase extraction based on pre-trained language model. IEEE Access **8**, 10896–10906 (2020)
39. Wan, X., Xiao, J.: Single document keyphrase extraction using neighborhood knowledge. In: Proceedings of the Conference of the Association for the Advancement of Artificial Intelligence (2008)
40. Wang, X., Song, X., Li, B., Guan, Y., Han, J.: Comprehensive named entity recognition on CORD-19 with distant or weak supervision. arXiv preprint arXiv:2003.12218 (2020)
41. Wang, Y., Lee, C.T., Qipeng Guo, Z.Y., Zhou, Y., Huang, X., Qiu, X.: What dense graph do you need for self-attention? arXiv preprint arXiv:2205.14014 (2022)
42. Xiong, L., Hu, C., Xiong, C., Campos, D., Overwijk, A.: Open domain web keyphrase extraction beyond language modeling. In: Proceedings of the Conference on Empirical Methods in Natural Language Processing (2019)
43. Zaheer, M., et al.: Big bird: transformers for longer sequences. In: Proceedings of the Annual Meeting on Neural Information Processing Systems (2020)
44. Zhang, L., et al.: MDERank: a masked document embedding rank approach for unsupervised keyphrase extraction. arXiv preprint arXiv:2110.06651 (2021)

Interpretability and Explainability in AI

Counterfactuals Explanations for Outliers via Subspaces Density Contrastive Loss

Fabrizio Angiulli[ID], Fabio Fassetti[ID], Simona Nisticó$^{(\boxtimes)}$[ID],
and Luigi Palopoli[ID]

DIMES Department, University of Calabria, Rende, Italy
{fabrizio.angiulli,fabio.fassetti,simona.nistico,
luigi.palopoli}@dimes.unical.it

Abstract. Explainable AI refers to techniques by which the reasons underlying decisions taken by intelligent artifacts are single out and provided to users. Outlier detection is the task of individuating anomalous objects within a given data population they belong to. In this paper we propose a new technique to explain why a given data object has been singled out as anomalous. The explanation our technique returns also includes counterfactuals, each of which denotes a possible way to "repair" the outlier to make it an inlier.

Thus, given in input a reference data population and an object deemed to be anomalous, the aim is to provide possible explanations for the anomaly of the input object, where an explanation consists of a subset of the features, called *choice*, and an associated set of changes to be applied, called *mask*, in order to make the object "behave normally". The paper presents a deep learning architecture exploiting a *features choice module* and *mask generation module* in order to learn both components of explanations. The learning procedure is guided by an *ad-hoc loss function* that simultaneously maximizes (minimizes, resp.) the isolation of the input outlier before applying the mask (resp., after the application of the mask returned by the mask generation module) within the subspace singled out by the features choice module, all that while also minimizing the number of features involved in the selected choice. We present experiments on both artificial and real data sets and a comparison with competitors validating the effectiveness of the proposed approach.

Keywords: Outlier Explanation · Explainable Artificial Intelligence · Deep Learning

1 Introduction

The anomaly detection problem is a main task in data analysis and has applications in several different real contexts. The huge interest on this problem is witnessed by the number of papers appeared in the last years on the topic (see, e.g., [2,6,7,26,33]). Among the application contexts in which the the outlier

A. Bifet et al. (Eds.): DS 2023, LNAI 14276, pp. 159–173, 2023.
https://doi.org/10.1007/978-3-031-45275-8_11

detection task is relevant we cite environmental monitoring [15,19,29], cyber-security [18,37], fraud detection [1,14], healthcare [10,13] and others as well [25].

On the other hand, anomalies (either known or computationally detected) need to be explained for the user to get full understanding of their nature but this specific aspect is much less explored in the literature. We refer to the problem of finding which characteristics distinguish an anomalous point from a normal one as the *Outlier Explanation* problem. Finding an explanation of an outlier is important for many reasons: for instance, users benefit from explanations since this will enhance their comprehension of the outliers, which is a crucial factor for taking right decisions, domain experts can find new knowledge by the interpretation of the outliers since this can potentially retrieve unknown paths from data, and so on. It is also worth noting that the outlier explanation problem can be looked at from slightly different perspectives in terms of little differences in the formal setting associated with explanations and several names have been correspondingly adopted for the task, such as *subspace selection, outlier explanation, object explanation, outlier interpretation* or *outlying subspaces detection*.

According to [27], there are many types of outlier explanations, and each of them aims to ensure different desiderata and fit various use cases. In this work, we focus on explanations defined as that of looking for feature subset(s) (and related values) on which the given data point is anomalous w.r.t. the population of data objects it belongs to [31].

In order to tackle the problem introduced above, we propose a new approach which exploits a neural network trained by an ad-hoc Subspaces Density Contrastive Loss. Given in input a reference data population and an object deemed to be anomalous, the aim is to provide possible explanations for the anomaly of the input object, where an explanation consists of a subset of the features, called *choice*, and an associated set of modifications to be applied to the outlier, called *mask*, in order to make that object behave normally. In the rest of the paper and whenever no ambiguities arise, we will use the terms "attribute", "feature" and "dimension" basically interchangeably. To summarize the contribution of this work:

- A new perspective for the Outlier Explanation problem where a concept of restoration is embedded inside the explanation is proposed.
- The technique **Masking Models for Outlier Explanation**, shortly M^2OE, is presented. An innovative aspect of this technique, closely related to the new perspective adopted, is that the explanations also includes counterfactuals, which shows possible ways to "repair" the outlier to make it an inlier.
- A study of the effectiveness of our proposal is performed considering synthetic as well as real data sets, comparing its results with the ones achieved by competitors.

The rest of the paper is organized as follows: Sect. 2 gives a brief description of the state of the art, Sect. 3 expounds the M^2OE technique, Sect. 4 comments on the experimental results and, finally, Sect. 5 concludes this work.

2   Related Works

In this section we give an overview of the techniques proposed in the literature aimed at constructing an outlier explanation expressed as that those object feature subsets which characterize the most a given point outlierness, since this is the closest setting to the one considered in this work.

The user-centred perspective, which characterizes the anomaly explanation task, originates from several definitions of the concept of explanation quality [24,27], each of which is relevant to certain classes of users and such a variety translates into different ways of tackling the considered problem.

According to [35], the class of outlier explanation methods that we are considering can be summarized into two macro-categories, which distinguish the techniques based on *feature selection* from that based on a *score-and-search*.

Features selection approaches are widely used for dimensionality reduction in order to improve model performances in classification tasks. Here, the goal is to retrieve the most relevant features among the ones that characterise the samples. In the considered context, this strategy is used with the objective of retrieving the features that best characterize a given point outlierness. Subspace Outlier Degree (SOD) [17] is an outlier detection technique that also returns the set of features associated to the outlier selection. This is done using a reference set, which is seen as a possible subspace cluster. Thus, if a point deviates considerably, as far as the subspace which characterizes it is considered, from the other objects in the reference set then it can be considered an outlier in that subspace, so that subspace represents an explanation. The approach presented in [23] considers a two-class classification problem and uses its outputs as a starting point to obtain an explanation for outlier samples. The basic idea here is that a good explanatory subspace determines a good separability between outliers and inliers and that a good proxy for this concept is the accuracy reached in the classification problem. Local Outliers with Graph Projection (LOGP), proposed in [8], considers both the detection and the explanation problems, by exploiting concepts borrowed from the spectral graph embedding theory and is based on the idea of projecting data points into a space having lower dimensionality. In particular, by using an invertible function, some subspaces of reduced dimensionality are constructed (without changing the neighbourhood structure) in which the outlier is easily recognisable. The paper [4] presents a technique that, given a categorical data set and an oulier q, finds the top k attributes associated with the largest outlier score for q. That approach is extended to numerical data in [3].

Score-and-search methodologies groups all the methods that use a measure to choose the explanatorial features for an outlier so that, in this context, to each retrieved feature a "relevance level" is associated. The High-dimensional Outlying Subspace Miner (HOS-Miner) [39] technique, which exploits an *Outlying Degree distance function* to detect the most outlying subspace. Density is used as subspace score criterion in [9], where a kernel-density estimation is employed to rank the attribute subspaces. In [35] two dimensional-unbiased measures are proposed: the Z-score and the Isolation Path score (iPath), which is inspired by Isolation Forest outlier detection algorithm [20]. Together with them, it proposes a searching procedure which performs an exhaustive search for 1-D and 2-D subspaces and a beam search for subspaces with dimension greater than 2. The method sGrid [36] uses a grid-based density estimator to search for anomalous features, leveraging the results of [32]. Its objective is to speed up the mining algorithm, which is attained at the cost of the estimation space unbiasedness. A notable further form of score is the Simple Isolation score Using Nearest Neighbor Ensamble (SiNNE) [30], whose definition is related to an outlier detection algorithm named Isolation using Nearest Neighbor Ensembles(iNNE) [5]. COIN [21], which stands for Contextual Outlier INterpretation, exploits the neighbourhood of anomalous points dividing it into clusters, then trains on each of them a simple classification model, and finally computes a score for each attribute by combining the learned parameters.

Finally, the technique proposed in [34] represents an hybrid [31], in that it combines the characteristics of both features-selection-based and score-and-search-based approaches. It consists of a two-stage approach where, in the first stage, features are ranked according to their potential to make the query object an outlier and then, in the second stage, which is optional, score-and-search exploration is performed on a smaller subset of the top-ranked m features.

Outside this categorization there are two outlier explanation methods focused on building explanations for group of anomalies. LookOut [12] which proposes pictorical explanations based on 2D plot, and x-PACS [22] which is aimed to cluster anomalies having the same characteristic subspace and to find rules which describe the patterns associated to those clusters of anomalies.

3   The Proposed Technique

The purpose of this section is to describe our technique for solving the outlier explanation problem introduced in Sect. 1.

We start by defining next what we mean by "explaining an outlier" in our context.

Definition 1. *Given a data set DS and an object o known to be an outlier, the goal of* outlier explanation *problem is to find an* explanation *e for o. An explanation e for o is a pair* $\langle c, m \rangle$ *where c denotes a set of attributes and m denotes a set of values, one value for each element in c, such that for o to take the values m for the attributes c would make o an inlier in DS.*

Thus, among the various forms that an explanation could possibly take, in our context the information content of e is represented by an attribute set c, called *choice* and realized as a binary vector implementing c's characteristic function over the set of all attributes, and a value set m, called *mask*, which is realized as a real valued vector of the same size as c specifying how much each attribute value must be changed in order for the outlier o to become an inlier. In detail, the binary values of c indicate if the corresponding attribute (aka, "feature" in the following) of o is anomalous (1) or not (0), while the real values of m indicate how to modify that feature to eliminate the point outlierness. The decision to decouple those two pieces of information has been taken to avoid possible biases related to the magnitudes of the modifications in the outlying features selection process. An example follows.

Example 1. Figure 1 depicts an example of what have been said above. The anomalous data point (the red one) is anomalous on the dimension showed on the ordinate but the magnitude of the modification that needs to be applied to move the anomaly closer to the normal data points is higher on the abscissa dimension compared to the ordinate one.

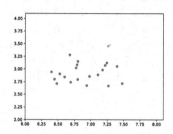

In this paper, in order to search for the possible explanations associated to a given input outlier o (there may exist more than one), we propose the M^2OE approach based on the idea of exploiting an ad-hoc Neural architecture which we have designed to compute them, which is described next.

3.1   Architecture Pipeline Description

Fig. 1. An example in which there is a possible bias related to the magnitude of the modification. Blue points represents normal data, while the red point is the outlier. (Color figure online)

As just anticipated, M^2OE uses a neural architecture in order to construct outlier explanations. Figure 2 depicts the structure of the pipeline underlying our proposal. This consists of a concatenation of an operation and two modules (each of them is in charge of constructing one of the two components required to get an explanation), followed by a postprocessing step.

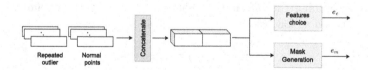

Fig. 2. Model architecture.

Both modules are realized as neural networks. The first module, called *Feature Choice*, has the purpose of finding the outlying features. The latter one, named *Mask Generation module*, is devoted to searching the associated mask values. The system is fed with a tuple containing a subset of inlier points (selected via a suitable sampling over DS) and the outlier to be explained. The concatenation operation (displayed as a green rectangle in Fig. 2) rearranges those two data collections and returns a 2-dimensional data structure as the output in

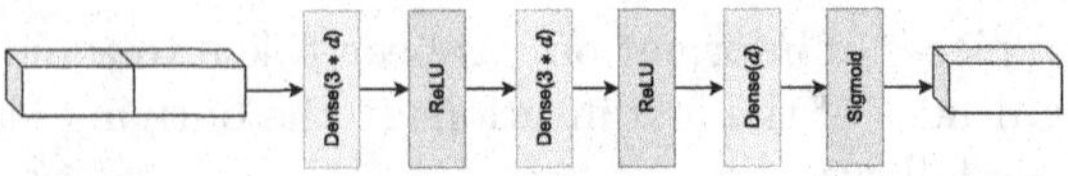

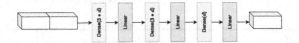

Fig. 3. Feature choice module structure.

which each row contains an outlier-inlier pair. The $2d$ structure is then fed in input to the modules *Feature Choice* and *Mask Generation module*.

As depicted in Fig. 3, the *Feature Choice* module consists in a three-layered neural network in which the number of units for each layer is automatically determined on the basis of the number of features d of the data points. More in detail, for the first two layers, the number of neurons is equal to 3 times d whereas, for the output layer, it is equal to d. We adopt the sigmoid as the activation function (that notoriously returns values ranging from 0 to 1), which are then converted into binary values through a thresholding operation. We will refer to (real valued) output of the sigmoid as $\tilde{c}$, where c is, again, the $\{0,1\}$-valued choice vector.

Fig. 4. Mask generation module structure.

Figure 4 shows the structure of the *Mask Generation* module, which is similar to that of the *Feature Choice*, except for the adopted activation functions: indeed, in this module, each single layer is followed by a linear activation function.

As already stated, the output of the system denotes a transformation that, if applied to the outlier o, would make o an inlier o'. In particular, o' is obtained from o by combining the information content carried by the explanation and "applying" it to o as follows:

$$o' = t(o, c, m) = o + c * m \tag{1}$$

where c is the choice, m is the mask and $*$ denotes the element-wise product of the elements in c and m.

Example 2. Considering again Example 1, the ideal explanation is the one in which only the second feature (the one depicted in the ordinate of Fig. 1) is highlighted as anomalous, so the desired choice vector is $c_e = [1, 0]$. A possible mask could be the following: $m_e = [0.5, -0.3]$. Given the outlier point, which is $o = [7.25, 3.5]$, and the previously defined c_e and m_e, the patched sample o' obtained from this transformation according to Equation (1) is $o' = [7.25 + 0 * 0.5, 3.5 + 1 * (-0.3)] = [7.25, 3.2]$.

In the following section the explanation computation process is illustrated.

3.2 Explanation Computation

Let a data set DS and a point $o \in DS$ known to be an outlier be given. Moreover, let $X_n \subset DS$ be the set of inliers (aka, normal points in the following) in DS. To find an explanation e for o, M^2OE first computes a differences statistics vector s, which contains the mean feature-wise squared differences between normal points. Then, it selects the k-nearest normal points $X_n^k \subset X_n$ to o using the euclidean distance as proximity criteria and replicates the outlier o k times – we will refer to the thus obtained structure as X_o^k. Finally, the tuple (X_o^k, X_n^k) (see Fig. 2) is the data used to train the architecture described in Sect. 3.1.

The training process is lead by the following loss function:

$$\mathcal{L}(o, x, \tilde{c}, m) = \alpha_1 \sum_{i=1}^{d} \frac{s_i \cdot \tilde{c}_i}{(o_i - x_i)^2 \cdot \tilde{c}_i^2 + \epsilon} + \alpha_2 \sqrt{\frac{\sum_{i=1}^{d} (p_i - x_i)^2 \cdot \tilde{c}_i}{d}} + \alpha_3 \cdot ||\tilde{c}|| \quad (2)$$

where $o \in X_o^k$ is the outlier to explain, $x \in X_n^k$ is a normal point, $\tilde{c}$ is the choice vector, m is the mask vector, s is differences statistics vector, ϵ is a small constant to avoid division by 0 and, finally, $p = t(o, \tilde{c}, m)$, where $t(\cdot)$ is the transformation defined by Eq. (1). The first term of the loss aims at finding the group of features that maximises the ratio between normal points distances, codified by the s vector, and the distance between the outlier o and the normal point x, so that this term leads the choice of outlying features. The second term aims at minimizing the distance between the transformed version of the point resulting from $t(o, c, m)$ and x_n, so that it guides the construction of the mask. The last term of the loss has the purpose of reducing the number of selected outlying features, in compliance with the Occam razor principle.

Once the two neural networks have been trained, to obtain the explanation, the architecture is queried on the training data (X_o^k, X_n^k) obtaining a list of choices C and a list of masks M.

Explanation Post-processing. To combine the information gathered after the training procedure, a post-processing procedure is carried out the goal of which is to build the explanations to be returned to user. In particular, first of all, an algorithm for frequent itemsets mining is executed on C to find the set of features that commonly appear together as outlying features. Then, for each set of features c^i detected by this mining algorithm, the clustering algorithm DBSCAN [11] is applied to all the normal points associated to tuples which corresponding explanation contains c^i. For each cluster j found by the algorithm, M^2OE finds its medoid, which is a representative point for the cluster, and queries the model using the tuple containing the outlier and the medoid to produce a mask m_j^i related to that cluster.

The need for this additional clustering step arises from the chance to have explanations characterized by masks with different, in some extreme cases opposite, transformations. Figure 5 shows an example of this kind of situation. Clearly, the outlying feature is the one represented on the abscissa, but there are two possible ways to restore the outlier. The first one, in which the value on the abscissa is lowered, is highlighted by the green arrow and the second one, in which the value on the abscissa is increased, is represented by the red row. Each of these two possible options brings the outlier near to one of the two clusters that are present in the considered normal data (the blue points). Note that, if we were to combine both explanations, it would be unclear how to disguise the point. Another objective of the post-processing step is to summarize the information collected in order to build diverse explanations for each outlier.

The final outcome of the explanation process is a list of tuples which contains a choose vector c^i and a "restored" version of the outlier p_j^i, obtained though c^i and the mask m_j^i, which is a counterfactual that shows an example supposed to be an inlier.

Fig. 5. A toy example in which to the same points can be associated different explanations characterized by the same group of features but by masks with different transformations.

4 Experiments

In this section, an assessment of the quality of our technique is performed through an experimental campaign which focuses both on studying how hyper-parameter values affect M^2OE performances and comparing its results with the one reached by its state-of-art competitors, that are, ATON [38] and COIN [21]. We note that, since none of the techniques presented to date gives information about how to modify the outlier (see Sect. 2), only the choice c is taken into account for the comparison purposes.

For experiments, we consider both synthetic and real data sets. Synthetic data comes from the collection of benchmarking data sets published in [16], which contain subspace clusters and outliers generated in subspaces having dimensions ranging from 2 to 5. These are used as ground truth for explanations. The peculiarity of these data sets is that the outliers have been generated in such a way that they are not observable in any lower dimensional sub-space projection, which results in making building suitable explanations more challenging. This collection contains data sets with 10, 20, 30, 40, 50 and 75 dimensions and, in particular, it contains three data sets for each dimensions size. Each data set consists of 100 data points including from 19 to 111 outliers.

For what real data are concerned, we have considered 5 data sets from the ODDS repository [28], that are, *cardio, breast, ionosphere, musk* and *arrhythmia*, which number of dimensions ranges from 21 to 274. More details about these data sets are provided in Table 1.

The rest of the section is organized as follows: Sect. 4.1 describes the metrics used for the evaluation, Sect. 4.2 shows the results of the parameter study and Sect. 4.3 compare M^2OE with its competitors.

Table 1. Real data sets overview.

Name	Dimensions	Points	Outliers	Brief description
cardio	21	1831	176	Measurements of fetal heart rate (FHR) and uterine contraction (UC) features on cardiotocograms classified by expert obstetricians
breast	30	569	212	Features are computed from a digitized image of a fine needle aspirate (FNA) of a breast mass. They describe characteristics of the cell nuclei present in the image
ionosphere	33	351	126	Radar data collected by a system in Goose Bay, Labrador. This system consists of a phased array of 16 high-frequency antennas
musk	166	3062	97	Set of 92 molecules of which 47 are judged by human experts to be musks or non-musks, its records describe them depend upon the exact shape, or conformation
arrhythmia	274	452	66	Data extracted from ECGs of different patient which the aim of distinguish the one containing arrhytmia episodes

4.1 Employed Metrics

In order to globally evaluates our techniques, it is needed to measure both the quality of the set of features retrieved and the effectiveness of the mask suggested to obtain a normal point, so that the experimental campaign follows this two-fold objective. Indeed, given an object o and an explanation e for o, in order for e to be significant, o should be clearly an outlier in the subspace e_c and the object o', obtained by transforming o through e, should be normal in e_c and, furthermore, e_c should be as small as possible.

In order to evaluate the set of features highlighted as anomalous, it is necessary to distinguish the situation where an annotation of which are the outlying features is available from the case in which there are no available information about it. In the first case, to evaluate the performances of our techniques, we use the precision and recall measures

$$Prec(c_p, c_t) = \frac{|c_p \cap c_t|}{|c_p|} \quad \text{and} \quad Rec(c_p, c_t) = \frac{|c_p \cap c_t|}{|c_t|}.$$

where e_t is the true choice while e_c is the choice predicted by the method.

Counter-wisely, if there are no information available about which are the true choices (as it happens for the real data sets considered in this work) a proxy measure for subspaces outlierness is needed. In this work we employ the iForest score [20] to evaluate the quality of the set of features retrieved in this setting. The idea underlying this method is that anomalies are better isolated from normal data so that, if binary trees are considered, they need fewer splits to be isolated as compared to normal data.

The iForest score is used also to evaluate the quality of the transformation suggested by the model (which gives an indication of mask effectiveness). Particularly, the score s_o of the outlier o and the score $s_{o'}$ of its "patched" version are considered. A transformation is considered successful if the value $s_{o'}$ is lower then s_o: indeed, in this situation, the transformation has lowered the point outlierness of at least the 5% of the original value. For each of the considered data set, we will take the percentage of successful transformation to measure the quality of the mask.

4.2   Parameters Tuning

This section aims at illustrating how the performances of the method changes for different hyper-parameters configuration. In particular, we consider the value of α_1, α_2 and α_3, which weights the terms of the loss function and consider all the synthetic data sets. For each dimensionality value, we computed the result as the arithmetic mean taken on all the data sets which points have that certain number of features.

Figure 6 shows how precision and recall measures change for different values of α_1, α_2 and α_3. The values of the first two hyper-parameters range from 0.8 to 1.2, while the value of the latter one ranges from 0.1 to 0.5. Note that while within the range of values associated to α_1 and α_2, no relevant difference in precision and recall values is induced, when the value of α_3 increases, the precision value significantly increases. Conversely, large values of α_3 produce a decrease in the recall value (since adding too many features enhances too much the third loss term). The trend of the transformation success is depicted in Fig. 7 which shows that α_2 and α_3 affects most the value of this metric.

4.3   Comparison with ATOM and COIN

In this section we compare our techniques with two outlier explanation methods ATOM and COIN referred to in Sect. 2. The evaluation is carried out first on the same synthetic data sets used to study the behaviour of the method for different hyper parameters and then on some real world data sets.

Synthetic Data Sets. We begin by considering again the HiCS data sets where the ground truths for anomalous features are available so that performances are computed in terms of precision and recall. Results are obtained following the same methodology of Sect. 4.2.

The results displayed as Fig. 8 show that our technique outperforms its competitors even when the dimensionality grows up. The quality of the results produced by all the considered methods shows a constant decreasing as the number

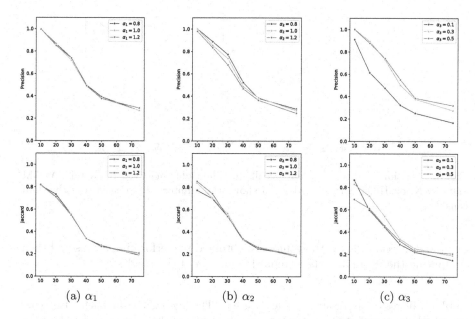

Fig. 6. Trend of precision and recall measures for different values of α_1 (a), α_2 (b) and α_3 (c). Results are obtained by changing one hyper-parameters at a time, where their default values, when fixed, are, respectively: $\alpha_1 = 1.0$, $\alpha_2 = 1.0$ and $\alpha_3 = 0.3$. Abscissa shows the number of dimensions of the data considered.

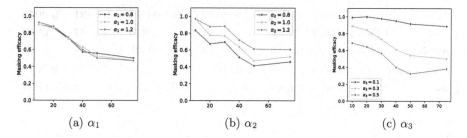

Fig. 7. Trend of transformation success for different values of α_1 (a), α_2 (b) and α_3 (c). Results are obtained by changing one hyper-parameters at a time, where their default values, when fixed, are, respectively: $\alpha_1 = 1.0$, $\alpha_2 = 1.0$ and $\alpha_3 = 0.3$. Abscissa shows the number of dimensions of the data considered.

of the dimensions of the data sets increase. This can be explained by the presence of some alternative subspaces, not listed in the ground truth, which also explain the outlier.

Real Data Sets. Here we considered the real data sets introduced in Sect. 4 and described in Table 1. Since in this setting no information related to the outlying features are available, the iForest score is employed to perform evaluation, as

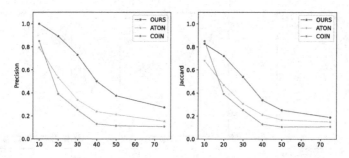

Fig. 8. Precisions (on the left) and recalls (on the right) achieved by M^2OE, ATOM and COIN, on HiCS data set. Abscissa shows the number of dimensions of the data considered.

described in Sect. 4.1. All the values are obtained by performing 5 runs and then computing the mean and the standard deviation of the results.

Table 2. iForest score results for real data sets. The first column highlights the score on the whole feature space, the other ones the score on the subsets retrieved by the considered techniques. For each data set, the maximum value is highlighted in bold while the second maximum is highlighted in italics.

Data set	Global baseline	ATOM	COIN	M^2OE
Cardio	0.5158 ± 0.0007	0.6734 ± 0.0012	0.6402 ± 0.0012	$\mathbf{0.6921 \pm 0.0014}$
Breast	0.4563 ± 0.0027	0.5393 ± 0.0075	0.5354 ± 0.0007	$\mathbf{0.5647 \pm 0.0008}$
Ionosphere	0.5258 ± 0.0015	0.6160 ± 0.0030	$\mathbf{0.6292 \pm 0.0011}$	0.6180 ± 0.0027
Arrhythmia	0.4162 ± 0.0070	0.7077 ± 0.0174	0.7518 ± 0.0013	$\mathbf{0.7577 \pm 0.0018}$
Musk	0.5603 ± 0.0057	0.6864 ± 0.0038	0.7419 ± 0.0014	$\mathbf{0.7674 \pm 0.0027}$

Table 2 shows the quality score achieved by the subspaces retrieved by each considered method. For each data set, the maximum value is highlighted in bold while the second maximum is highlighted in italics, the global baseline value represents the anomaly score computed in the whole feature space. From this, it is possible to notice how M^2OE almost always reaches the maximum score among his competitors and, when this does not happen, it ranks second. We also note that the value of all the techniques is greater than the global baseline.

Table 3. Percentage of outliers correctly fixed by M^2OE for each real data set.

Cardio	Breast	Ionosphere	Arrhythmia	Musk
0.8872 ± 0.0159	0.8566 ± 0.0233	0.9104 ± 0.0150	1.0000 ± 0.0000	0.9913 ± 0.0092

To check if M^2OE is able to patch even real data set outliers, the percentage of successfully-patched outliers is computed and results are reported in Table 3. The table allows us to conclude that M^2OE is able to correctly patch samples even in this context.

5   Conclusions

In this paper a new technique which employs a deep neural-network to explain why a given data object has been singled out as anomalous has been presented. One main novelty of our proposal is that its explanations include counterfactuals, each of which denotes a possible way to "repair" the outlier to make it an inlier, which can provide users with a deeper understanding of data point characteristics. Thanks to the summarization performed by its post-processing step, M^2OE is able to provide diverse explanations which account for the peculiarities of the outlier w.r.t the normal points leveraged for the explanation.

The experimental campaign carried out on synthetic and real data sets provide empirical evidence demonstrating the quality of the proposed technique both for what concerns retrieved subspaces and for what concerns counterfactuals returned to users.

As future development, we consider to extend the technique to other data types in order to enlarge its applicability and, furthermore to extend experiments with user studies which might show how much returned explanations improves data comprehension.

Acknowledgment. We acknowledge the support of the PNRR project FAIR - Future AI Research (PE00000013), Spoke 9 - Green-aware AI, under the NRRP MUR program funded by the NextGenerationEU.

References

1. Abdallah, A., Maarof, M.A., Zainal, A.: Fraud detection system: a survey. J. Netw. Comput. Appl. **68**, 90–113 (2016)
2. Angiulli, F., Fassetti, F., Ferragina, L.: LatentOut: an unsupervised deep anomaly detection approach exploiting latent space distribution. Mach. Learn. 1–27 (2022)
3. Angiulli, F., Fassetti, F., Manco, G., Palopoli, L.: Outlying property detection with numerical attributes. Data Min. Knowl. Disc. **31**(1), 134–163 (2017)
4. Angiulli, F., Fassetti, F., Palopoli, L.: Detecting outlying properties of exceptional objects. ACM Trans. Database Syst. (TODS) **34**(1), 1–62 (2009)
5. Bandaragoda, T.R., Ting, K.M., Albrecht, D., Liu, F.T., Zhu, Y., Wells, J.R.: Isolation-based anomaly detection using nearest-neighbor ensembles. Comput. Intell. **34**(4), 968–998 (2018)
6. Bhuyan, M.H., Bhattacharyya, D.K., Kalita, J.K.: Network anomaly detection: methods, systems and tools. IEEE Commun. Surv. Tutor. **16**(1), 303–336 (2014). https://doi.org/10.1109/SURV.2013.052213.00046
7. Chandola, V., Banerjee, A., Kumar, V.: Anomaly detection for discrete sequences: a survey. IEEE Trans. Knowl. Data Eng. **24**(5), 823–839 (2012). https://doi.org/10.1109/TKDE.2010.235

8. Dang, X.H., Assent, I., Ng, R.T., Zimek, A., Schubert, E.: Discriminative features for identifying and interpreting outliers. In: 2014 IEEE 30th International Conference on Data Engineering, pp. 88–99. IEEE (2014)

9. Duan, L., Tang, G., Pei, J., Bailey, J., Campbell, A., Tang, C.: Mining outlying aspects on numeric data. Data Min. Knowl. Disc. **29**(5), 1116–1151 (2015). https://doi.org/10.1007/s10618-014-0398-2

10. Duraj, A., Chomatek, L.: Supporting breast cancer diagnosis with multi-objective genetic algorithm for outlier detection. In: Kościelny, J.M., Syfert, M., Sztyber, A. (eds.) DPS 2017. AISC, vol. 635, pp. 304–315. Springer, Cham (2018). https://doi.org/10.1007/978-3-319-64474-5_25

11. Ester, M., Kriegel, H.P., Sander, J., Xu, X., et al.: A density-based algorithm for discovering clusters in large spatial databases with noise. In: KDD, vol. 96, pp. 226–231 (1996)

12. Gupta, N., Eswaran, D., Shah, N., Akoglu, L., Faloutsos, C.: Beyond outlier detection: LookOut for pictorial explanation. In: Berlingerio, M., Bonchi, F., Gärtner, T., Hurley, N., Ifrim, G. (eds.) ECML PKDD 2018, Part I. LNCS (LNAI), vol. 11051, pp. 122–138. Springer, Cham (2019). https://doi.org/10.1007/978-3-030-10925-7_8

13. Hauskrecht, M., Batal, I., Valko, M., Visweswaran, S., Cooper, G.F., Clermont, G.: Outlier detection for patient monitoring and alerting. J. Biomed. Inform. **46**(1), 47–55 (2013)

14. Hilal, W., Gadsden, S.A., Yawney, J.: A review of anomaly detection techniques and applications in financial fraud. Expert Syst. Appl. **193**, 116429 (2021)

15. Hill, D.J., Minsker, B.S.: Anomaly detection in streaming environmental sensor data: a data-driven modeling approach. Environ. Modell. Softw. **25**(9), 1014–1022 (2010)

16. Keller, F., Muller, E., Bohm, K.: HiCS: high contrast subspaces for density-based outlier ranking. In: 2012 IEEE 28th International Conference on Data Engineering, pp. 1037–1048. IEEE (2012)

17. Kriegel, H.-P., Kröger, P., Schubert, E., Zimek, A.: Outlier detection in axis-parallel subspaces of high dimensional data. In: Theeramunkong, T., Kijsirikul, B., Cercone, N., Ho, T.-B. (eds.) PAKDD 2009. LNCS (LNAI), vol. 5476, pp. 831–838. Springer, Heidelberg (2009). https://doi.org/10.1007/978-3-642-01307-2_86

18. Kruegel, C., Vigna, G.: Anomaly detection of web-based attacks. In: Proceedings of the 10th ACM Conference on Computer and Communications Security, pp. 251–261 (2003)

19. Leigh, C., et al.: A framework for automated anomaly detection in high frequency water-quality data from in situ sensors. Sci. Total Environ. **664**, 885–898 (2019)

20. Liu, F.T., Ting, K.M., Zhou, Z.H.: Isolation forest. In: 2008 Eighth IEEE International Conference on Data Mining, pp. 413–422. IEEE (2008)

21. Liu, N., Shin, D., Hu, X.: Contextual outlier interpretation. In: Proceedings of the 27th International Joint Conference on Artificial Intelligence, IJCAI 2018, pp. 2461–2467. AAAI Press (2018)

22. Macha, M., Akoglu, L.: Explaining anomalies in groups with characterizing subspace rules. Data Min. Knowl. Disc. **32**(5), 1444–1480 (2018). https://doi.org/10.1007/s10618-018-0585-7

23. Micenková, B., Ng, R.T., Dang, X.H., Assent, I.: Explaining outliers by subspace separability. In: 2013 IEEE 13th International Conference on Data Mining, pp. 518–527. IEEE (2013)

24. Molnar, C.: A guide for making black box models explainable (2018). http://christophm.github.io/interpretable-ml-book

25. Narayanan, V., Bobba, R.B.: Learning based anomaly detection for industrial arm applications. In: Proceedings of the 2018 Workshop on Cyber-Physical Systems Security and PrivaCy, pp. 13–23 (2018)
26. Pang, G., Shen, C., Cao, L., Hengel, A.V.D.: Deep learning for anomaly detection: a review. ACM Comput. Surv. **54**(2) (2021). https://doi.org/10.1145/3439950
27. Panjei, E., Gruenwald, L., Leal, E., Nguyen, C., Silvia, S.: A survey on outlier explanations. VLDB J. **31**(5), 977–1008 (2022)
28. Rayana, S.: ODDS library (2016). http://odds.cs.stonybrook.edu
29. Russo, S., Lürig, M., Hao, W., Matthews, B., Villez, K.: Active learning for anomaly detection in environmental data. Environ. Modell. Softw. **134**, 104869 (2020)
30. Samariya, D., Aryal, S., Ting, K.M., Ma, J.: A new effective and efficient measure for outlying aspect mining. In: Huang, Z., Beek, W., Wang, H., Zhou, R., Zhang, Y. (eds.) WISE 2020. LNCS, vol. 12343, pp. 463–474. Springer, Cham (2020). https://doi.org/10.1007/978-3-030-62008-0_32
31. Samariya, D., Ma, J., Aryal, S.: A comprehensive survey on outlying aspect mining methods. arXiv preprint arXiv:2005.02637 (2020)
32. Silverman, B.W.: Density Estimation for Statistics and Data Analysis. Routledge (2018)
33. Steinwart, I., Hush, D., Scovel, C.: A classification framework for anomaly detection. J. Mach. Learn. Res. **6**(2), 211–232 (2005)
34. Vinh, N.X., Chan, J., Bailey, J., Leckie, C., Ramamohanarao, K., Pei, J.: Scalable outlying-inlying aspects discovery via feature ranking. In: Cao, T., Lim, E.-P., Zhou, Z.-H., Ho, T.-B., Cheung, D., Motoda, H. (eds.) PAKDD 2015. LNCS (LNAI), vol. 9078, pp. 422–434. Springer, Cham (2015). https://doi.org/10.1007/978-3-319-18032-8_33
35. Vinh, N.X., Chan, J., Romano, S., Bailey, J., Leckie, C., Ramamohanarao, K., Pei, J.: Discovering outlying aspects in large datasets. Data Min. Knowl. Disc. **30**(6), 1520–1555 (2016). https://doi.org/10.1007/s10618-016-0453-2
36. Wells, J.R., Ting, K.M.: A new simple and efficient density estimator that enables fast systematic search. Pattern Recogn. Lett. **122**, 92–98 (2019)
37. Xu, H., et al.: Unsupervised anomaly detection via variational auto-encoder for seasonal KPIs in web applications. In: Proceedings of the 2018 World Wide Web Conference, pp. 187–196 (2018)
38. Xu, H., et al.: Beyond outlier detection: interpreting outliers by attention-guided triplet deviation network. In: Proceedings of The Web Conference 2021 (WWW 2021). ACM (2021)
39. Zhang, J., Lou, M., Ling, T.W., Wang, H.: HOS-miner: a system for detecting outlying subspaces of high-dimensional data. In: Proceedings of the 30th International Conference on Very Large Data Bases (VLDB 2004), pp. 1265–1268. Morgan Kaufmann Publishers Inc. (2004)

Explainable Spatio-Temporal Graph Modeling

Massimiliano Altieri[1] , Michelangelo Ceci[1,2] , and Roberto Corizzo[3]([envelope])

[1] Department of Computer Science, University of Bari Aldo Moro, Bari, Italy
{massimiliano.altieri,michelangelo.ceci}@uniba.it
[2] Department of Knowledge Technologies, Jožef Stefan Institute, Ljubljana, Slovenia
[3] Department of Computer Science, American University,
Washington, DC 20016, USA
rcorizzo@american.edu

Abstract. Explainable AI (XAI) focuses on designing inference explanation methods and tools to complement machine learning and black-box deep learning models. Such capabilities are crucially important with the rising adoption of AI models in real-world applications, which require domain experts to understand how model predictions are extracted in order to make informed decisions. Despite the increasing number of XAI approaches for tabular, image, and graph data, their effectiveness in contexts with a spatial and temporal dimension is rather limited. As a result, available methods do not properly explain predictive models' inferences when dealing with spatio-temporal data. In this paper, we fill this gap proposing a XAI method that focuses on spatio-temporal geo-distributed sensor network data, where observations are collected at regular time intervals and at different locations. Our model-agnostic method performs perturbations on the feature space of the data to uncover relevant factors that influence model predictions, and generates explanations for multiple analytical views, such as features, timesteps, and node location. Our qualitative and quantitative experiments with real-world forecasting datasets show the effectiveness of the proposed method in providing valuable explanations of model predictions.

Keywords: Explainable AI · graph data · sensor networks

1 Introduction

The large availability of sensor network data paves the way for its exploitation for decision-making processes in many real-world sectors. Relevant examples of machine learning and deep learning methods for sensor networks data analysis have been adopted in traffic prediction [24], change detection in smart-grids [5], pollution forecasting [11], and energy forecasting [7], to mention a few. A large number of methods have been proposed in the last few years, reaching remarkable prediction performance. However, the black-box nature of many of these methods reduces their applicability in practical decision support processes. End users, practitioners, and domain experts may require an explanation of

A. Bifet et al. (Eds.): DS 2023, LNAI 14276, pp. 174–188, 2023.
https://doi.org/10.1007/978-3-031-45275-8_12

model predictions to increase their confidence in the decision making process. This capability is crucial in domains such as medical healthcare, smart grids, and credit risk assessment, where a wrong decision may result in a strongly negative impact on people's health, result in damages to the underlying infrastructure, or yield economic losses.

In this context, Explainable AI (XAI) approaches are becoming highly popular in recent years [20,26], since they complement model predictions with explanations that make domain experts more confident about their decisions. A large number of XAI approaches have been proposed for tabular, image, and graph data. However, their effectiveness in contexts with spatial and temporal dimensions is rather limited. Approaches for tabular data include instance-based methods [15,19], which present wide applicability in many domains, but provide only local explanations for single observations. Some tabular methods also offer global interpretability [16], providing a holistic understanding of the overall model behavior by analyzing feature importance globally, but present an extremely high computational cost [3]. Moreover, tabular approaches are unable to consider spatial and temporal dimensions of the data. Image-based approaches [12,21,22] are an excellent solution to explain predictions extracted by Convolutional Neural Networks (CNNs) when using image datasets. However, their adoption is cumbersome and not ideal in domains with time series data. Graph-based approaches [10,25] take the spatial dimension into consideration, but they are limited to the analysis of node features without the notion of a temporal dimension. As a result, they are only suitable for static node prediction tasks.

Overall, available methods do not provide explanations at a sequence level and, consequently, do not properly explain predictive model inferences when dealing with spatio-temporal data. In this paper, we fill this gap by proposing an XAI method that focuses on geo-distributed sensor network data, where observations are collected at regular time intervals and at different locations. Our model-agnostic method performs perturbations on the feature space of the data to uncover relevant factors that influence model predictions and generates explanations for multiple analytical views, such as features, timesteps, and node location. Our qualitative and quantitative experiments with real-world forecasting datasets show the effectiveness of the proposed method in providing valuable explanations of model predictions. The paper is structured as follows. In Sect. 2 we review existing relevant works in the literature. In Sect. 3 we present the proposed method. In Sect. 4 we describe the experimental setting and the considered datasets, and we discuss the experimental results. Finally, we draw our conclusions and provide directions for further work in Sect. 5.

2 Background

2.1 XAI Methods

Approaches for tabular data include instance-based methods [15,19], which present wide applicability in many domains, but provide only local explanations for single observations. In this context, a popular approach is LIME [19], which explains the predictions of any classifier by fitting a regression model on a single model's prediction. This approach is general since it supports different models, such as random forests and neural networks. Some approaches for tabular data with global interpretability capabilities [16] provide a holistic understanding of the overall model behavior by analyzing feature importance globally, but typically suffer from an extremely high computational cost [3]. Moreover, such approaches are unable to consider spatial and temporal dimensions of the data. Image-based approaches are used in supporting decisions in several domains [8,12,21,22]. Among them, gradient-based localization [8,21] is a popular approach for image data which exploits the gradients flowing to the final convolutional layer of the discriminator to generate a localization map highlighting the salient regions of the image that are responsible for the prediction. A variant of this approach is proposed in [22], where variants of an image are modified by adding noise and averaging the resulting sensitivity maps. Despite the significant explainability capabilities that these methods provide for image data, they are not appropriate for time series data.

Graph-based approaches [10,25] take the spatial dimension of data into consideration. Such approaches study the importance of node features in the graph typically resorting to the optimization of soft masks. The optimization process usually seeks to maximize the mutual information between the predictions of the original graph and the predictions of the masked graph. These methods treat models as black-boxes, since they do not require access to their model specifications, and they represent a natural fit for graph neural networks. However, they are designed for the node classification task, which lacks the notion of a temporal dimension.

2.2   Forecasting Methods for Sensor Networks

Sensor networks give the opportunity to gather data observations for a set of properties of interest in multiple geographical locations. In sensor networks, forecasting methods can be used to forecast future values for such properties, leading to relevant decision support capabilities. Recent literature suggests that accurate forecasting capabilities for geo-distributed sensor data require the ability for the models to deal with spatio-temporal autocorrelation [1,6], which is typically achieved through multi-node analysis. However, a higher degree of model sophistication to deal with this aspect also corresponds to increased challenges from the explainability viewpoint. Initial attempts leverage attention maps as indicators of relevance for time steps and features [9,13].

An interpretable flood forecasting method using temporal and spatial attention is proposed in [9]. In [13], an attention-based architecture combines multi-horizon forecasting with interpretable insights using recurrent and self-attention layers. A different approach [23] proposes multilevel wavelet decomposition to embed frequency analysis in deep time series forecasting models. In [17] the authors adopt series saliency, i.e. a mask learned using a perturbation strategy, to improve both accuracy and interpretability.

One important limitation of these approaches is that visual artifacts are not particularly easy to understand for end users and domain experts, and rarely lead to actionable insights. Moreover, the quality of the explanations returned by such methods cannot be directly evaluated using quantitative metrics. Consequently, model interpretability in the context of forecasting models learned from sensor networks is still in its infancy.

3   Method

3.1   Spatio-Temporal Forecasting Setting

In this paper, we focus on the multi-node multivariate forecasting scenario, which is characterized by time series consisting of observations generated at regular time intervals by each node in a sensor network, where relationships of different types are possible among nodes. Considering a network of N nodes, the generated data has a 3-D time series structure, where a sequence $\mathbf{x}_k$ at the time window k can be defined as a tensor $\mathbf{x}_k \in \mathbb{R}^{T \times N \times F}$, for a discrete number of time points T determining the sequence length, and a number F of observed independent features. Note that sequences at time point k are constructed from T consecutive observations, arranged in a non-overlapping way. For example, T can represent 24 observations, one for each hour, and k can represent the index of the day.

Therefore, we can define the *sequence of tensors* $\mathcal{X} \in \mathbb{R}^{k \times T \times N \times F}$, containing contiguous and chronologically-ordered sequences:

$$\mathcal{X} = [\mathbf{x}_0, \mathbf{x}_1, \ldots, \mathbf{x}_{k-1}],$$

which are used as training data for the models.

Given a new sequence $\mathbf{x}_k$, the observed forecast data for $\mathbf{x}_k$ is described as the corresponding sequence y_k containing the subsequent T observations for the target property of interest. The forecasting task consists in approximating the function $f : \mathbf{x}_k \mapsto \mathbf{y}_k$ using a model Ψ. The learning task can be formalized as identifying the model that minimizes a given forecasting loss function

$$\Psi^* = \arg\min_{\Psi} \mathcal{L}\big(\Psi(\mathcal{X}), f(\mathcal{X})\big),$$

where $\Psi(\mathcal{X})$ extracts forecasts $\hat{\mathbf{y}}_k$ for the target variable of interest of each input sequence $\mathbf{x}_k$ in $\mathcal{X}$, and $f(\mathcal{X})$ is the ground truth on historical data for the target variable of interest: $\mathbf{y}_0, \mathbf{y}_1, \ldots, \mathbf{y}_{k-1}$.

3.2   Explainability Problem Definition

The goal of our method is to provide human-interpretable explanations that justify the forecasting output of an arbitrary model on a given input sequence. For a model Ψ trained on historical time series $\mathcal{X} = [\mathbf{x}_0, \mathbf{x}_1, \ldots, \mathbf{x}_{k-1}]$, we define the explanation process as a function $\mathcal{E}$:

$$\mathcal{E}(\Psi, \mathbf{x}_k) = \mathcal{M},$$

where $\mathcal{M} \in \mathbb{R}^{T \times N \times F}$ constitutes the explanation for $\Psi(\mathbf{x}_k)$ and it is a binary mask tensor ($\mathcal{M}_{t,n,f} = 1$ or $\mathcal{M}_{t,n,f} = 0$, for $1 \leq \frac{t}{f} \leq \frac{T}{N}$) such that $\mathcal{M} \odot$

$\mathbf{x}_k$ highlights salient components across any axis of tensor $\mathbf{x}_k$, and $\odot$ denotes element-wise multiplication.

This explanation framework allows for the categorization of existing explanation methods by imposing different constraints on the obtained explanation $\mathcal{M}$. Explanation methods that can natively only operate on tabular data, such as LIME [19] position themselves within our framework using the following constraint on the resulting explanation:

$$(1) \quad \mathcal{M}_{t,n,f} = 1 \iff \mathcal{M}_{t',n',f} = 1, \quad t \neq t', n \neq n',$$

which means that if a given feature is deemed relevant, it will be relevant for all timesteps and nodes, and all entries will be selected. This behavior also implies that the user is provided with a single explanation that only considers the feature axis, without direct consideration of the temporal and spatial axes.

At the same time, graph-based and spatially aware explanation methods, such as GNNExplainer [25], can be categorized using the same framework as being generally constrained by:

$$(2) \quad \mathcal{M}_{t,n,f} = 1 \iff \mathcal{M}_{t',n,f'} = 1, \quad t \neq t', f \neq f',$$

which means that a given node is highlighted as relevant considering all time steps and features, and the user is provided with a single explanation that does not analyze time points (i.e. t) and features (i.e. f) in isolation.

Similarly, explanations of methods that are not able to properly consider the temporal dimension are constrained by:

$$(3) \quad \mathcal{M}_{t,n,f} = 1 \iff \mathcal{M}_{t,n',f'} = 1, \quad n \neq n', f \neq f'.$$

In all these methods, one or multiple axes of the explanation $\mathcal{M}$ are not taken into consideration, and if a given dimension is highlighted by the explanation method, it will be highlighted for all values of the other axes that are not considered. On the contrary, our explanation method poses no such constraints on $\mathcal{M}$ and it allows all dimensions of the spatio-temporal output sequence to be highlighted independently, for a complete 3D multi-level explanation of the inference process.

3.3   Multi-level Explanation Process

Our explanation method extracts inference explanations for the three levels (feature, timesteps and nodes) using a multi-stage process, focusing on one perspective at a time. The approach is model-agnostic, in that it supports any forecasting model able to work with multivariate, multi-step, and multi-node sequences $\mathcal{X}$, regardless of the model architecture. A graphical representation of our method and its workflow is shown in Fig. 1. Additionally, our method is able to provide explanations for models that exploit the topology of the sensor network in a spatially-aware way. Such a topology is usually defined using an adjacency matrix $\mathcal{A} \in \mathbb{R}^{N \times N}$ to represent relationships among nodes, such that entries $\mathcal{A}_{ij}$

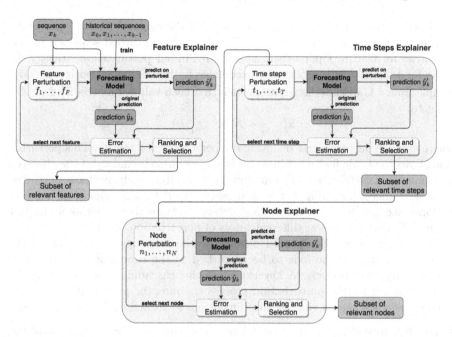

Fig. 1. Graphical representation of our proposed explainable spatio-temporal graph modeling method.

constitute a numerical property such as spatial closeness, correlation, etc., for a given pair of nodes i and j. This feature is common in models such as Graph-Convolutional Long-Short Term Memory (GCN-LSTM) neural networks. We recall that our method always provides explanations for the node contributions towards the final forecast, even for forecasting models that do not capture graph topology or that fully ignore spatial information, e.g. LSTM, as long as they are capable to process multi-node observations in a single inference step. However, in the case of graph-based forecasting models, the node-level explanations generated by our method are potentially more meaningful, since node relationships are explicitly represented and exploited to extract model predictions, allowing our method to provide even greater insights to the end user.

Perturbations. Our explanation process is perturbation-based. We follow the rationale that perturbing a relevant dimension of the input sequence should generate a larger impact on model predictions compared to perturbing a less relevant dimension, and monitoring the difference in model predictions should allow us to uncover the relevance of each feature. This concept was shown to be a feasible strategy in previous works [22]. However, existing approaches either apply this technique on flat 2D tabular data [19], or node features in graph data in the context of node classification [10]. In our method, we apply perturbations to all axes of the input sequence, following a three-stage process.

Starting with the feature axis, we perturb the original input sequence $\mathbf{x}_k$ to generate F new sequences $\tilde{\mathbf{x}}_k^1, \ldots, \tilde{\mathbf{x}}_k^F$, where the perturbed sequence $\tilde{\mathbf{x}}_k^f$ has all entries for feature f perturbed. Specifically, in our perturbation strategy the value for feature f in the perturbed sequence $\tilde{\mathbf{x}}_k^f$ is determined with a random sample from a Gaussian distribution, such that each value of matrix $\tilde{\mathbf{x}}_k^f \in \mathbb{R}^{T \times N}$ is set to:

$$\tilde{\mathbf{x}}_{k;t_i n_i}^f \sim \mathcal{N}(\mathbf{x}_{k;t_i n_i}^f, \sigma),$$

where σ is a hyperparameter denoting the perturbation intensity. This perturbation allows to obtain feature values within a reasonable boundary of the observed phenomenon's distribution. In our approach, we iteratively perform perturbations with increasing values of σ, and assess their average impact on model's performance. By doing so, the method detects features that generate an impact on model's performance at different levels of perturbation intensity.

The same process is repeated during the timesteps and, subsequently, the nodes stage, varying the axis to be perturbed to the time axis and the nodes (spatial) axis, respectively. A key aspect of this cascading process is that the input sequence for the perturbation process contains the perturbations applied in the previous stage for the relevant dimensions identified in the previous stage.

Error Estimation. This process allows us to compare the magnitude of the impact of perturbations on model's prediction values for the entire sequence x_k (error estimation). Specifically, in our work we consider Mean Absolute Error (MAE) since it is a standard metric for evaluating forecasting models. Following the above-mentioned rationale, we measure the impact that perturbing each dimension along the current axis brings to model predictions. We define the impact as $\Delta_d = |\Psi(\mathbf{x}_k) - \Psi(\tilde{\mathbf{x}}_k^d)|$, for $d \leq T, N$ or F depending on the axis under analysis, such that $\arg\max_d \Delta_d$ is the most impactful, and thus relevant, dimension.

Ranking and Selection. From the impact of each perturbation, our method carries out a ranking process to sort dimensions in decreasing order of impact. This ranking can be formalized as a permutation matrix $R \in \mathbb{R}^{D \times D}$, for D number of dimensions, that sorts the dimension vector in the appropriate descending order with $R \cdot [1, 2, \ldots, D]$.

After ranking, the model selects the most relevant dimensions. To this aim, the model evaluates the transition of consecutive values of feature importance, analyzing their ratio of relevance. Let's define with $\Delta = [\Delta_1, \Delta_2, \ldots, \Delta_D]$ the vector of dimension impacts for each dimension along the current axis. Then, the method computes the incremental relative decrements $\frac{(R\Delta)_j}{(R\Delta)_i}$ for each pair of consecutive dimensions in the ranking $(i, j) : j - i = 1$, and selects the cutoff dimension using:

$$d_c = \arg\max_{(i,j)} \frac{(R\Delta)_j}{(R\Delta)_i}.$$

This behavior is conceptually similar to the adoption of an elbow rule for the determination of the optimal number of clusters in previous research works [14], representing an effective heuristic in such scenarios. At this point, the set of relevant dimensions is defined as $S = \{d : (R\mathbf{\Delta})_{d_c} < (R\mathbf{\Delta})_d\}$.

The whole process described in this subsection is repeated for the three considered levels: Feature, Time Steps, and Nodes, in a sequential manner. Specifically, as we move from Feature to Time Steps explanations, the subset of relevant features extracted by the Feature Explainer represents the input to the Time Steps Explainer. By doing so, we are able to focus Time Steps perturbations and the subsequent operations to the most relevant features, which are fixed. Similarly, the subset of relevant Time Steps returned by the Time Steps Explainer, represents the input to the Node Explainer, which restricts Node perturbations to the subset of the most relevant Features and Time Steps, which are fixed by the previous levels. This choice allows us to reduce the computational complexity of our method, and to reduce the end user's overhead, providing summarized and actionable explanations. Overall, our method generates, for each evaluation sequence, three visual explanations in the form of bar plots, which separately highlight Feature, Time Steps, and Node relevances.

4 Experiments

In this section, we describe our experimental evaluation. Specifically, in the next subsections, we first describe the considered datasets and the experimental setting, including the considered competitor systems. Then, we report the obtained results and discuss them.

4.1 Datasets

The datasets considered in our experiment consist of weather variables (such as temperature, humidity, etc.) monitored at hourly granularity by sensors placed on renewable energy plants, located in different geographical areas. In particular, we consider the following datasets:

- *Lightsource.* Data observations related to energy production of 7 solar power plants located in the United Kingdom. Values are aggregated hourly. [7]
- *PV Italy.* Data collected from 17 photovoltaic power plants. Values are aggregated hourly (from 2:00 AM to 8:00 PM) and are collected during the period from January 1st, 2012 to May 4th, 2014. More details about data preprocessing steps can be found in [4].
- *Wind NREL.* Dataset generated using an environmental model by the Weather Research & Forecasting (WRF). It consists of five power plants obtaining the time series of wind speed and production. Values are aggregated hourly, and are observed in the time period Jan 1st, 2005 to Dec 31st, 2006).

For the three datasets, we consider the following features: latitude and longitude of the different power plant; day and time of the observation; altitude and azimuth; several weather conditions, including ambient temperature, irradiance, pressure, wind speed, and several others. The weather conditions that complement the energy production are extracted from Forecast.io, except for the expected altitude and azimuth, that are extracted from SunPosition, and the expected irradiance (PV Italy dataset only), that is extracted from PVGIS.

4.2 Experimental Setup

In our experiments, we train a pool of base models (LSTM, GRU, SVD-LSTM, CNN-LSTM, GCN-LSTM) using all available historical data for the three considered sensor network datasets. Taking into account the chronological order of the data, predictions are extracted for the last sequence, and are fed to the explainer to extract the results. We remark that our explainer is model-agnostic and therefore does not favor the adoption of any specific model.

For all models, we use the following hyperparameter configuration: Lightsource: $\{$batch_size $= 16,$ learning_rate $= 10^{-2}\}$, PV Italy and Wind NREL: $\{$batch_size $= 16,$ learning_rate $= 10^{-2}\}$. An early stopping criterion with a patience of 20 epochs is adopted to prevent overfitting. We assess the effectiveness of the proposed method using popular metrics XAI metrics [26]. Following the idea that the explanations should identify input features that are relevant for the model, the $Fidelity^+$ [18] metric highlights the observed change in model predictions when a subset of the features is removed. On the contrary, the $Fidelity^-$ metric assesses the change in model predictions when relevant input features are kept and features deemed irrelevant are removed. Both $Fidelity^+$ and $Fidelity^-$ can be applied at the Model level (without any access to the ground truth) and at the Phenomenon level (comparing model's predictions with ground truth). In the following, we use the M and the P subscripts to denote the two variants applied at the Model and at the Phenomenon levels, respectively. Additionally, the $Sparsity$ metric indicates the ratio of features selected as important by explanation methods over the entire number of available features. Conceptually, good explanations should be sparse, since they are able to capture the most relevant features while ignoring the irrelevant ones.

These metrics allow for the assessment and comparisons of different explainability methods. Specifically, a XAI method that is able to discover relevant features should present high values of $Fidelity+$ and low values of $Fidelity-$.

$$Fidelity_M^+ = \frac{1}{S} \sum_{i=1}^{S} \left| \Psi\left(\mathbf{x}_i\right) - \Psi\left((1 - \mathcal{M}_i) \odot \mathbf{x}_i\right) \right|$$

$$Fidelity_M^- = \frac{1}{S} \sum_{i=1}^{S} \left| \Psi\left(\mathbf{x}_i\right) - \Psi\left(\mathcal{M}_i \odot \mathbf{x}_i\right) \right|$$

$$Fidelity_P^+ = \frac{1}{S} \sum_{i=1}^{S} \left| \left| \Psi\left(\mathbf{x}_i\right) - \mathbf{y}_i \right| - \left| \Psi\left((1 - \mathcal{M}_i) \odot \mathbf{x}_i\right) - \mathbf{y}_i \right| \right|$$

$$Fidelity_P^- = \frac{1}{S} \sum_{i=1}^{S} \left| \left| \Psi\left(\mathbf{x}_i\right) - \mathbf{y}_i \right| - \left| \Psi\left(\mathcal{M}_i \odot \mathbf{x}_i\right) - \mathbf{y}_i \right| \right|$$

$$Sparsity = \frac{1}{S} \sum_{i=1}^{S} \left(1 - \frac{\sum \mathcal{M}_{i_{t,n,f}}}{T \cdot N \cdot F} \right),$$

where $\mathbf{x}_i$ denotes a test sequence for evaluation (made of features, nodes, and timesteps), $\mathcal{M}_{i_{t,n,f}} \odot \mathbf{x}_i$ denotes the non-zero entries corresponding to the subset of relevant dimensions selected by the explanation method for the i-th sequence, and S denotes the size of the batch of sequences under evaluation.

4.3   Results and Discussion

In Tables 1 and 2 we report all the results obtained in our experiments, comparing the results obtained with our proposed method with two core baselines. The first is a Random Forest (RF) model with permutation-based feature importance as an explanation technique for the feature perspective. Feature importance is defined as the decrease in the model's classification performance when a single feature value is randomly shuffled, and a reduction of performance is a signal that such a feature is important for the model [2]. For evaluation, we leverage the same feature importance cutoff approach described in Sect. 3.3 to select the subset of the most relevant features. The second baseline is a LIME explainer [19] paired with Linear Regression and Gradient Boosting regression models. For the latter, we extract one explanation for each timestep and node in the evaluation sequence, and average the obtained feature rankings. For evaluation, we choose the 5 features with the highest ranking returned by the explainer.

Results in Table 1 show that our method coupled with either the GCN-LSTM model or the GRU model achieves a significantly better performance than RF in terms of Fidelity+ with all three datasets. LIME outperforms all other approaches in terms of Fidelity+ (Model) with two out of three datasets (Lightsource, PV Italy). Our proposed method (GRU) outperforms LIME with Wind NREL, while achieving the same results in terms for Fidelity+ (Phenomenon) with PV Italy[1]. Such results should be seen as a preliminary investigation that confirms the correctness and the calibration of the explanations extracted by our method with respect to classical explainers. While RF and LIME explicitly optimize explanations for the Feature perspective, our method focuses on the holistic extraction of multi-perspective explanations also involving Time Steps, and Nodes, which are neglected by other methods.

Results for the Time Steps and Nodes perspectives are shown in Table 2. Considering Fidelity+ (Model), the best values for the Features perspective are observed for LSTM (LightSource - 0.095), CNN-LSTM (0.135 - PV Italy), GRU (0.242 - Wind NREL). However, considering these results in relationship with the Sparsity metric, highlights very low values for the first two cases, which is undesirable in practice. This behavior shows that low Sparsity may facilitate achieving high results in terms of Fidelity+ (Model), and both metrics should be considered in combination to properly assess different models. Considering

---

[1] For brevity, we restrict our analysis of the Features comparison to Fidelity+ metrics.

Table 1. Experimental results and comparison with methods restricted to the analysis of the Features perspective. High values of F+ (Model/Phenomenon) and low values of F− (Model/Phenomenon) indicate a better performance. High values of Sparsity are preferred in combination with satisfactory values of F+ and F−. The best performance for each metric and dataset is marked in bold.

Model	Dataset	Model F+	Model F−	Phenom F+	Phenom F−	Sparsity
Random Forest (Feature Importance)	Lightsource	0.023	0.043	0.017	0.043	0.545
	PV Italy	0.109	0.107	0.067	0.106	**0.583**
	Wind NREL	0.213	0.224	0.209	0.215	0.250
LIME (Linear Regression)	Lightsource	0.055	0.066	0.044	0.053	0.545
	PV Italy	**0.230**	**0.026**	0.093	0.022	**0.583**
	Wind NREL	0.082	0.063	0.079	0.062	0.375
LIME (Gradient Boosting)	Lightsource	**0.082**	**0.009**	**0.046**	**0.008**	0.545
	PV Italy	0.178	0.018	**0.094**	**0.017**	**0.583**
	Wind NREL	0.161	**0.036**	0.159	**0.035**	0.375
Proposed (GCN-LSTM)	Lightsource	0.033	0.081	0.029	0.070	**0.909**
	PV Italy	0.047	0.076	0.044	0.066	0.417
	Wind NREL	0.089	0.230	0.083	0.208	0.375
Proposed (GRU)	Lightsource	0.025	0.042	0.019	0.034	0.364
	PV Italy	0.116	**0.026**	**0.094**	0.024	**0.583**
	Wind NREL	**0.242**	0.277	**0.235**	0.248	**0.625**

the Time Steps perspective, a satisfactory trade-off between Fidelity+ (Model) and Sparsity is obtained by GRU (0.143 and 0.737 for PV Italy, respectively). Considering the Nodes perspective, examples of satisfactory trade-offs between Fidelity+ (Model) and Sparsity are obtained by SVD-LSTM (0.075 and 0.800 for Wind NREL and PV Italy, respectively), as well as CNN-LSTM (0.218 and 0.800 for Wind NREL and PV Italy, respectively).

Moving our focus to Fidelity+ (Phenomenon) the best values for the Features perspective are LSTM (LightSource - 0.059), CNN-LSTM (0.111 - PV Italy), GRU (0.235 - Wind NREL). These results confirm what we observed in terms of Fidelity+ (Model). Considering the Time Steps perspective, a satisfactory trade-off between Fidelity+ (Phenomenon) and Sparsity is achieved by LSTM (0.220 and 0.417 for Wind NREL, respectively) and GRU (0.099 and 0.737 for PV Italy, respectively). From the Nodes perspective, a notable result is GCN-LSTM (0.134 and 0.600 for Wind NREL), as well as LSTM (0.089 and 0.471 for PV Italy, respectively).

One general consideration is that, in the presence of large values of Sparsity, similar values for Fidelity+ and Fidelity− should be regarded as a positive, even if they deviate from the canonical case of high Fidelity+ and small Fidelity−. The features identified by the explainer are in this case particularly relevant since,

Table 2. Experimental results (proposed method - all perspectives). High values of F+ (Model/Phenomenon) and low values of F− (Model/Phenomenon) indicate a better performance. High values of Sparsity are preferred.

Model	Dataset	Level	Model F+	Model F-	Phenom F+	Phenom F-	Sparsity
LSTM	Lightsource	Features	0.095	0.037	0.059	0.033	0.091
		Timesteps	0.090	0.021	0.038	0.020	0.263
		Nodes	0.090	0.028	0.053	0.026	0.143
	PV Italy	Features	0.052	0.185	0.049	0.124	0.667
		Timesteps	0.161	0.021	0.093	0.020	0.316
		Nodes	0.117	0.129	0.089	0.109	0.471
	Wind NREL	Features	0.059	0.236	0.056	0.222	0.625
		Timesteps	0.236	0.035	0.220	0.035	0.417
		Nodes	0.332	0.135	0.311	0.130	0.400
GRU	Lightsource	Features	0.025	0.042	0.019	0.034	0.364
		Timesteps	0.053	0.011	0.041	0.009	0.632
		Nodes	0.081	0.048	0.067	0.042	0.571
	PV Italy	Features	0.116	0.026	0.094	0.024	0.583
		Timesteps	0.143	0.063	0.099	0.056	0.737
		Nodes	0.042	0.103	0.036	0.094	0.706
	Wind NREL	Features	0.242	0.277	0.235	0.248	0.625
		Timesteps	0.052	0.000	0.052	0.000	0.125
		Nodes	0.062	0.154	0.059	0.149	0.600
SVD-LSTM	Lightsource	Features	0.032	0.059	0.027	0.045	0.455
		Timesteps	0.030	0.046	0.026	0.034	0.895
		Nodes	0.017	0.058	0.015	0.041	0.857
	PV Italy	Features	0.066	0.113	0.063	0.101	0.333
		Timesteps	0.152	0.002	0.120	0.002	0.158
		Nodes	0.060	0.099	0.058	0.088	0.706
	Wind NREL	Features	0.140	0.059	0.135	0.048	0.250
		Timesteps	0.100	0.017	0.094	0.017	0.542
		Nodes	0.075	0.136	0.074	0.129	0.800
CNN-LSTM	Lightsource	Features	0.071	0.046	0.047	0.039	0.455
		Timesteps	0.049	0.067	0.039	0.048	0.842
		Nodes	0.036	0.101	0.029	0.074	0.571
	PV Italy	Features	0.135	0.042	0.111	0.041	0.167
		Timesteps	0.149	0.000	0.093	0.000	0.105
		Nodes	0.028	0.136	0.027	0.114	0.824
	Wind NREL	Features	0.080	0.158	0.074	0.154	0.375
		Timesteps	0.061	0.014	0.056	0.014	0.375
		Nodes	0.218	0.110	0.212	0.106	0.800
GCN-LSTM	Lightsource	Features	0.033	0.081	0.029	0.070	0.909
		Timesteps	0.081	0.008	0.061	0.007	0.263
		Nodes	0.011	0.050	0.010	0.041	0.714
	PV Italy	Features	0.047	0.076	0.044	0.066	0.417
		Timesteps	0.072	0.059	0.064	0.055	0.737
		Nodes	0.021	0.070	0.020	0.064	0.706
	Wind NREL	Features	0.089	0.230	0.083	0.208	0.375
		Timesteps	0.167	0.008	0.162	0.008	0.292
		Nodes	0.137	0.067	0.134	0.064	0.600

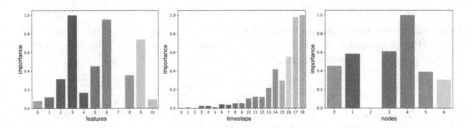

Fig. 2. Features, timesteps, and nodes relevance (Lightsource dataset - GCN-LSTM).

when removed (Fidelity+), they generate a numerically similar error to that obtained when removing all the others (Fidelity−). For example, considering the Timesteps perspective, an interesting example is GCN-LSTM, where the value of Fidelity+ (Model) is close to that of Fidelity− (Model) (0.072 and 0.059, respectively) with PV Italy, with a high Sparsity value of 0.737. Similarly, the value of Fidelity+ (Phenomenon) is close to that of Fidelity− (Phenomenon) (0.064 and 0.055, respectively). Another case is provided by SVD-LSTM for the Timesteps perspective, where the value of Fidelity+ (Model) is close to that of Fidelity− (Model) (0.030 and 0.046, respectively) with Lightsource, with a high Sparsity value of 0.895. Correspondingly, the value observed for Fidelity+ (Phenomenon) is close to that of Fidelity− (Phenomenon) (0.026 and 0.034, respectively).

Figure 2 shows the visualizations of the output generated by our explainer using a GCN-LSTM model[2]. Notably, our method returns visual explanations for all three analytical views (Features, Timesteps, Nodes) instead of a single one as in other approaches, providing an information-rich summary of the relevance of different aspects for the currently predicted sequence. The visualization is easier to interpret thanks to the normalization post-processing scheme adopted in our method. It is also interesting to observe that, for the Timesteps perspective, the most recent timesteps are highlighted as the most relevant for the prediction. This behavior is in line with the GCN-LSTM model adopted for this result, which consists of an encoder-decoder architecture. It is in fact in line with our expectations that the last encoder steps have more influence for the decoding phase, while least recent encoder steps present an increasingly reducing influence. On the other hand, the Nodes perspective provide useful additional highlights, supporting domain experts in the understanding of which nodes provided the relevant information for predicting the current time series. The average execution time of the proposed method was 110 s for the extraction of results for all perspectives with a single dataset. We note that using different predictive models does not significantly impact the execution time of our method, as they are pre-trained and only used for inference.

---

[2] Additional visual artifacts generated by our method (histograms, perturbation plots, and rank plots) are available at: https://www.rcorizzo.com/graph-xai/.

5   Conclusion

In this paper we tackled the problem of Explainable AI (XAI) in the context of spatio-temporal graph data. Motivated by the shortcomings of available tabular, image, and graph XAI approaches in this context, we proposed a novel explainability method specialized for sensor network data observations collected at regular time intervals and at different locations. Our experiments have shown that our perturbation-based model-agnostic method is effective in uncovering relevant factors influencing model predictions, generating useful explanations for multiple analytical views, such as features, timesteps, and nodes. Limitations of our work include the adoption of metrics that are cumbersome to interpret when used in isolation and the lack of direct exploitation of fidelity in the explanation algorithm. In future work, we will address the issue of jointly evaluating XAI methods from multiple perspectives, through the design of appropriate metrics. Moreover, we will investigate optimization approaches based on model fidelity.

Acknowledgement. This work was partially supported by the project FAIR - Future AI Research (PE00000013), Spoke 6 - Symbiotic AI, under the NRRP MUR program funded by the NextGenerationEU and by NVIDIA with the donation of a Titan V GPU.

References

1. Altieri, M., Corizzo, R., Ceci, M.: Scalable forecasting in sensor networks with graph convolutional LSTM models. In: 2022 IEEE International Conference on Big Data (Big Data), pp. 4595–4600. IEEE (2022)
2. Breiman, L.: Random forests. Mach. Learn. **45**, 5–32 (2001)
3. Van den Broeck, G., Lykov, A., Schleich, M., Suciu, D.: On the tractability of SHAP explanations. J. Artif. Intell. Res. **74**, 851–886 (2022)
4. Ceci, M., Corizzo, R., Fumarola, F., Malerba, D., Rashkovska, A.: Predictive modeling of PV energy production: how to set up the learning task for a better prediction? IEEE Trans. Industr. Inf. **13**(3), 956–966 (2016)
5. Ceci, M., Corizzo, R., Japkowicz, N., Mignone, P., Pio, G.: ECHAD: embedding-based change detection from multivariate time series in smart grids. IEEE Access **8**, 156053–156066 (2020)
6. Ceci, M., Corizzo, R., Malerba, D., Rashkovska, A.: Spatial autocorrelation and entropy for renewable energy forecasting. Data Min. Knowl. Disc. **33**(3), 698–729 (2019). https://doi.org/10.1007/s10618-018-0605-7
7. Corizzo, R., Ceci, M., Fanaee-T, H., Gama, J.: Multi-aspect renewable energy forecasting. Inf. Sci. **546**, 701–722 (2021)
8. Corizzo, R., Dauphin, Y., Bellinger, C., Zdravevski, E., Japkowicz, N.: Explainable image analysis for decision support in medical healthcare. In: 2021 IEEE International Conference on Big Data (Big Data), pp. 4667–4674. IEEE (2021)
9. Ding, Y., Zhu, Y., Feng, J., Zhang, P., Cheng, Z.: Interpretable spatio-temporal attention LSTM model for flood forecasting. Neurocomputing **403**, 348–359 (2020)
10. Huang, Q., Yamada, M., Tian, Y., Singh, D., Chang, Y.: GraphLIME: local interpretable model explanations for graph neural networks. IEEE Trans. Knowl. Data Eng. **35**(7), 6968–6972 (2023). https://ieeexplore.ieee.org/abstract/document/9811416

11. Kalajdjieski, J., Zdravevski, E., Corizzo, R., et al.: Air pollution prediction with multi-modal data and deep neural networks. Remote Sens. **12**(24), 4142 (2020)
12. Lapuschkin, S., Wäldchen, S., Binder, A., Montavon, G., Samek, W., Müller, K.R.: Unmasking clever Hans predictors and assessing what machines really learn. Nat. Commun. **10**(1), 1096 (2019)
13. Lim, B., Arık, S.Ö., Loeff, N., Pfister, T.: Temporal fusion transformers for interpretable multi-horizon time series forecasting. Int. J. Forecast. **37**(4), 1748–1764 (2021)
14. Liu, F., Deng, Y.: Determine the number of unknown targets in open world based on elbow method. IEEE Trans. Fuzzy Syst. **29**(5), 986–995 (2020)
15. Lundberg, S.M., et al.: From local explanations to global understanding with explainable AI for trees. Nat. Mach. Intell. **2**(1), 56–67 (2020)
16. Lundberg, S.M., Lee, S.I.: A unified approach to interpreting model predictions. In: Advances in Neural Information Processing Systems, vol. 30 (2017)
17. Pan, Q., Hu, W., Chen, N.: Two birds with one stone: series saliency for accurate and interpretable multivariate time series forecasting. In: Proceedings of the Thirtieth International Joint Conference on Artificial Intelligence, IJCAI-21, pp. 2884–2891 (2021)
18. Pope, P.E., Kolouri, S., Rostami, M., Martin, C.E., Hoffmann, H.: Explainability methods for graph convolutional neural networks. In: Proceedings of the IEEE/CVF Conference on Computer Vision and Pattern Recognition, pp. 10772–10781 (2019)
19. Ribeiro, M.T., Singh, S., Guestrin, C.: "Why should i trust you?" Explaining the predictions of any classifier. In: Proceedings of the 22nd ACM SIGKDD International Conference on Knowledge Discovery and Data Mining, pp. 1135–1144 (2016)
20. Saeed, W., Omlin, C.: Explainable AI (XAI): a systematic meta-survey of current challenges and future opportunities. Knowl.-Based Syst. **263**, 110273 (2023)
21. Selvaraju, R.R., Cogswell, M., Das, A., Vedantam, R., Parikh, D., Batra, D.: Grad-CAM: visual explanations from deep networks via gradient-based localization. In: Proceedings of the IEEE International Conference on Computer Vision, pp. 618–626 (2017)
22. Smilkov, D., Thorat, N., Kim, B., Viégas, F., Wattenberg, M.: SmoothGrad: removing noise by adding noise. arXiv preprint arXiv:1706.03825 (2017)
23. Wang, J., Wang, Z., Li, J., Wu, J.: Multilevel wavelet decomposition network for interpretable time series analysis. In: Proceedings of the 24th ACM SIGKDD International Conference on Knowledge Discovery & Data Mining, pp. 2437–2446 (2018)
24. Yang, H.F., Dillon, T.S., Chen, Y.P.P.: Optimized structure of the traffic flow forecasting model with a deep learning approach. IEEE Trans. Neural Netw. Learn. Syst. **28**(10), 2371–2381 (2016)
25. Ying, Z., Bourgeois, D., You, J., Zitnik, M., Leskovec, J.: GNNExplainer: generating explanations for graph neural networks. In: Advances in Neural Information Processing Systems, vol. 32 (2019)
26. Yuan, H., Yu, H., Gui, S., Ji, S.: Explainability in graph neural networks: a taxonomic survey. IEEE Trans. Pattern Anal. Mach. Intell. **45**, 5782–5799 (2022)

Probabilistic Scoring Lists
for Interpretable Machine Learning

Jonas Hanselle[1](✉), Johannes Fürnkranz[2], and Eyke Hüllermeier[3,4]

[1] Paderborn University, Paderborn, Germany
jonas.hanselle@uni-paderborn.de
[2] Johannes-Kepler-University Linz, Linz, Austria
juffi@faw.jku.at
[3] LMU Munich, Munich, Germany
eyke@lmu.de
[4] Munich Center for Machine Learning (MCML), Munich, Germany

Abstract. A scoring system is a simple decision model that checks a set of features, adds a certain number of points to a total score for each feature that is satisfied, and finally makes a decision by comparing the total score to a threshold. Scoring systems have a long history of active use in safety-critical domains such as healthcare and justice, where they provide guidance for making objective and accurate decisions. Given their genuine interpretability, the idea of learning scoring systems from data is obviously appealing from the perspective of explainable AI. In this paper, we propose a practically motivated extension of scoring systems called probabilistic scoring lists (PSL), as well as a method for learning PSLs from data. Instead of making a deterministic decision, a PSL represents uncertainty in the form of probability distributions. Moreover, in the spirit of decision lists, a PSL evaluates features one by one and stops as soon as a decision can be made with enough confidence. To evaluate our approach, we conduct a case study in the medical domain.

1 Introduction

Predictive models generated by modern machine learning algorithms, such as deep neural networks, tend to be complex and difficult to comprehend, and may not be appropriate in applications where a certain degree of transparency of a model and explainability of decisions are desirable. Besides, depending on the situation and application context, time and computational resources for applying decision models might be limited. For example, a human's resources to collect, validate, and enter data might be scarce, or decisions must be taken quickly, in the extreme case even by the human herself without any technical device.

Therefore, being interested in simple, genuinely interpretable model classes, we focus on so-called *scoring systems* in this paper. In a nutshell, a scoring system is a decision model that checks a set of features, adds (or subtracts) a certain number of points to a total score for each feature that is satisfied, and finally makes a decision by comparing the total score to a threshold. Scoring systems

© The Author(s), under exclusive license to Springer Nature Switzerland AG 2023
A. Bifet et al. (Eds.): DS 2023, LNAI 14276, pp. 189–203, 2023.
https://doi.org/10.1007/978-3-031-45275-8_13

have a long history of active use in safety-critical domains such as healthcare (Six et al., 2008) and justice (Wang et al., 2022), where they provide guidance for making objective and accurate decisions. Given their genuine interpretability, scoring systems are appealing from the perspective of explainable AI, which is why the idea of learning such systems from data has recently attracted attention in machine learning.

In this paper, we propose a practically motivated extension of scoring systems called *probabilistic scoring lists* (PSL), as well as a method for learning PSLs from data. First, to increase uncertainty-awareness, a PSL produces predictions in the form of probability distributions (instead of making deterministic decisions). Second, to increase cost-efficiency, a PSL is conceptualised as a *decision list*: It evaluates features one by one and stops as soon as a decision can be made with enough confidence.

Following a brief overview of related work in the next section, we introduce PSLs in Sect. 3 and address the problem of learning such models from data in Sect. 4. To evaluate our approach, we conduct as case study in the realm of medical decision making, which is presented in Sect. 5, prior to concluding the paper with an outlook on extensions and future work in Sect. 6.

2   Related Work

In a series of papers, Ustun and Rudin developed the so-called Supersparse Linear Integer Model (SLIM) for inducing scoring systems from data, as well as an extension called RiskSLIM (Ustun and Rudin, 2016; 2017; 2019). Their methods are based on formalising the learning task as an integer linear programming problem, with the objective to find a meaningful compromise between sparsity (number of variables included) and predictive accuracy. The problem can then essentially be tackled by means of standard ILP solvers.

In several applied fields, one also finds methods of a more heuristic nature. Typically, standard machine learning methods, such as support vector machines or logistic regression, are used to train a (sparse) linear model, and the real-valued coefficients of that model are then turned into integers, e.g., through rounding or by taking the sign. Obviously, approaches of that kind are rather ad-hoc, and indeed, can be shown to yield suboptimal performance in practice (Subramanian et al., 2021). From a theoretical perspective, certain guarantees for the rounded solutions can nevertheless be given (Chevaleyre et al., 2013).

A related research direction is the learning of simple decision heuristics that are considered plausible from the perspective of cognitive psychology. Again, however, this is a relatively unexplored field, in which only a few publications can be found so far—Simsek and Buckmann (2017) collect and empirically compare some of these heuristics.

Decision lists have been primarily used in inductive rule learning (Fürnkranz et al., 2012), where each term consists of a conjunction of conditions, which are sufficient to make a prediction in case the conditions are satisfied, or else continue with the next rule. They have been shown to generalize both, k-term CNF and DNF expressions, as well as decision trees with a fixed depth k (Rivest, 1987). Practically, they represent a simple way for tie-breaking in situations where multiple rules cover the same example: in that case, the first rule in the list is given priority. They can be easily learned, as their structure mirrors the commonly covering or separate-and-conquer strategy (Fürnkranz, 1999), which learns one rule at a time, typically by appending rules to the list, assuming that most important rules are tried first, but prepending has also been tried (Webb, 1994). While rules are typically used for classification, they may also be viewed as simple probability estimators, using the class distribution among the covered examples as the basis for various estimation techniques (Sulzmann and Fürnkranz, 2009). However, these are known to be overly optimistic, because the way the conditions are selected results in a bias towards the positive examples during learning (Možina et al., 2018). Also, in decision lists in rule learning, the probability estimates are derived from the last rule in isolation, practically ignoring all previous rules, whereas, as will be seen later, the probability distributions in PSLs are successively refined.

3   From Scoring Systems to Probabilistic Scoring Lists

Consider a scenario where decisions need to be made in different contexts, which are characterised in terms of a set of variables or features $\mathcal{F} = \{f_1, \ldots, f_K\}$. A concrete situation is specified by a vector $\boldsymbol{x} = (x_1, \ldots, x_K)$, where x_i is the value observed for the feature f_i, and the set of all conceivable vectors of that kind forms the instance space $\mathcal{X}$. Features can be of various kind, i.e., binary, (ordered) categorical, or numeric. Decisions are taken from a decision space $\mathcal{Y}$, which is normally finite, typically comprising a small to moderate number of alternatives to choose from.

A decision model is a mapping $h : \mathcal{X} \longrightarrow \mathcal{Y}$, i.e., $y = h(\boldsymbol{x})$ is the decision suggested by h in the context $\boldsymbol{x}$. Note that such models can be represented in different ways. For the reasons already explained, we shall focus on scoring systems in this paper. In a nutshell, scoring systems consist of a set of simple criteria (presence or absence of certain characteristics or features) that are checked, and if satisfied, contribute a certain number of points to a total score. The final decision is then based on comparing this score to one or more thresholds. Formally, scoring systems can be seen as a specific type of generalised additive models (Hastie, 2017) defined over a set of features.

Definition 1 (Scoring system). *A scoring system over a set of (binary) candidate features $\mathcal{F}$ and score set $\mathcal{S} \subset \mathbb{Z}$ is a triple $h = \langle F, S, t \rangle$, where $F = \{f_1, \ldots, f_K\} \subset \mathcal{F}$ is a subset of the candidate features, $S = (s_1, \ldots, s_K) \in \mathcal{S}^K$ are scores assigned to the corresponding features, and $t \in \mathbb{Z}$ is a decision threshold. For a given decision context $\boldsymbol{x} = (x_1, \ldots, x_K)$, i.e., the projection of an instance to the feature set F, the decision prescribed by h is given by*

$$h(\boldsymbol{x}) = \left[\!\!\left[\sum_{i=1}^{K} s_i x_i \geq t \right]\!\!\right], \tag{1}$$

where $[\![\cdot]\!]$ is the indicator function[1].

In the following, we generalise scoring systems in two ways: from deterministic to probabilistic, and from a single decision model to a decision list.

As for the first extension, the idea is to return a probability distribution over $\mathcal{Y}$ instead of a binary decision (1), i.e., to assign a probability $p(y)$ to each decision $y \in \mathcal{Y}$. The latter can be interpreted as the probability that y is the best or correct decision, which (implicitly) presupposes the existence of a kind of ground truth. Without loss of generality, we can assume that the ground truth distinguishes between a class of positive cases and a class of negative cases, and that the decision is a prediction of the correct class. Therefore, we shall use the terms "decision" and "class" interchangeably.

We contextualise the distribution p, not directly with $\boldsymbol{x}$, but rather with the total score $T(\boldsymbol{x})$ assigned to $\boldsymbol{x}$. In other words, we consider conditional probabilities $p(\cdot \,|\, T(\boldsymbol{x}))$ on $\mathcal{Y}$. This appears meaningful and is in line with the assumption that the total score is indicative of the class—in fact, standard scoring systems can be seen as a special case, returning probability 1 for the positive class when exceeding the threshold and probability 0 otherwise.

Definition 2 (Probabilistic scoring system, PSS). *A probabilistic scoring system (PSS) over candidate features $\mathcal{F}$ and score set $\mathcal{S} \subset \mathbb{Z}$ is a triple $h = \langle F, S, q \rangle$, where $F = \{f_1, \ldots, f_K\} \subset \mathcal{F}$, $S = (s_1, \ldots, s_K) \in \mathcal{S}^K$, and q is a mapping $\Sigma \longrightarrow [0,1]$, where*

$$\Sigma := \left\{ T = \sum_{i=1}^{K} s_i\, x_i \;\middle|\; s_1, \ldots, s_K \in \mathcal{S},\, x_1, \ldots, x_K \in \{0,1\} \right\}$$

is the set of possible values for the total score that can be obtained by any instance $\boldsymbol{x} \in \mathcal{X}$, and $q(T) = p(y = 1 \mid T)$ is the (estimated) probability for the positive class ($y = 1$) given that the total score is T (and hence $1 - q(T)$ the probability for the negative class).

Note that an increase in the total score should only increase but not decrease the probability of the positive decision, so that probabilistic scoring systems should satisfy the following monotonocity constraint:

$$\forall T, T' \in \Sigma : (T < T') \Rightarrow q(T) \leq q(T'). \tag{2}$$

This property is again in line with standard scoring systems and appears to be important from an interpretability perspective: A violation of (2) would be considered as an inconsistency and compromise the acceptance of the decision model. Therefore, in the remainder of the paper, we consider only monotonic probabilistic scoring systems.

Our second extension combines probabilistic scoring systems with the notion of decision lists. The underlying idea is as follows: Instead of determining all K

---

[1] $[\![P]\!] = 1$ if predicate P is true (positive decision) and $[\![P]\!] = 0$ if P is false (negative decision).

Stage	0	1	2	3	4
Feature	–	f_3	f_1	f_2	f_4
Score	–	+1	−2	+1	+2
$T = -2$	–	–	0.1	0.1	0.1
$T = -1$	–	–	0.2	0.2	0.1
$T = 0$	0.3	0.2	0.5	0.6	0.2
$T = +1$	–	0.4	0.6	0.7	0.6
$T = +2$	–	–	–	0.9	0.7
$T = +3$	–	–	–	–	0.9
$T = +4$	–	–	–	–	0.9

Fig. 1. Example of a PSL with feature set $\mathcal{F} = \{x_1, x_2, x_3, x_4\}$ and score set $\mathcal{S} = \{0, \pm 1, \pm 2\}$.

feature values x_i right away, these values are determined successively, one after the other, in a predefined order. Each time a new feature is added, the total score T is updated, and the probability $q(T)$ of the positive class is determined. Depending on the latter, the process is then continued or stopped: If the probability is sufficiently high or sufficiently low, the process is stopped, because a decision can be made with enough confidence; otherwise, the process is continued by adding the next feature.

Example 1. Figure 1 depicts a PSL with four features $F = \{f_1, f_2, f_3, f_4\}$. As can be seen from the assigned scores, all features except f_1 are indicative of the positive class, i.e., the presence of f_2, f_3 or f_4 increases the probability of the positive class, wheres the presence of f_1 decreases the probability.

The decision process starts with an empty feature set and a prior probability of 0.3 for the positive class. After seeing the first feature f_3 with a weight $s_3 = +1$, the possible scores are $T = 0$ if the feature does not hold (the value of the feature is $x_3 = 0$), or $T = +1$, if $x_3 = 1$. In the former case, the probability for the positive decision decreases to 0.2, in the latter case it increases to 0.4. The next feature is f_1 with a weight of $s_1 = -2$, resulting in a total of four possible scores, ranging from $T = -2$ (if $x_3 = 0$ and $x_1 = 1$) to $T = +1$ (if $x_3 = 1$ and $x_1 = 0$). Note that the absence of f_3, in this case, may increase the probability of a positive score to 0.6. Adding the remaining features continues this process, until we get a diverse set of seven probability estimates (five of which are different) corresponding to the seven different score values we can obtain for the $2^4 = 16$ possible instances. For example, the instance $\boldsymbol{x} = (1, 1, 1, 1)$ would be assigned a probability of $q(2) = 0.7$, based on its total score of $T(\boldsymbol{x}) = +2$.

Note the monotonicity (2) in the scores in each column (higher score values result in higher probabilities for the positive decision). Also note that if the final maximal probability of 0.9 is considered to be sufficiently high for making a positive decision, the process could already have been stopped after seeing the first three features for any instance $\boldsymbol{x} = (0, 1, 1, *)$, irrespective of its value x_4 for the fourth feature f_4.

Formally, we can define a probabilistic scoring list as follows:

Definition 3 (Probabilistic scoring list, PSL). *A probabilistic scoring list over candidate features $\mathcal{F}$ and score set $\mathcal{S} \subset \mathbb{Z}$ is a triple $h = \langle F, S, p \rangle$, where $F = (f_1, \ldots, f_K)$ is a list of (distinctive) features from $\mathcal{F}$, $S = (s_1, \ldots, s_K) \in \mathcal{S}^K$, and q is a mapping*

$$q : \bigcup_{k=0}^{K} (k, \Sigma_k) \longrightarrow [0, 1] \tag{3}$$

such that

$$\forall \, k \in \{0, 1, \ldots, K\}, T, T' \in \Sigma_k : (T < T') \Rightarrow q(k, T) \leq q(k, T'). \tag{4}$$

Here, Σ_k is the set of possible values for the total score at stage k, i.e.,

$$\Sigma_k = \left\{ T = \sum_{i=1}^{k} s_i \, x_i \;\middle|\; s_1, \ldots, s_k \in \mathcal{S}, \, x_1, \ldots, x_k \in \{0, 1\} \right\}.$$

A value $q(k, T)$ is interpreted as the probability of the positive decision if the total score at stage k is given by T.

Note that $k = 0$ is included in (3). This case corresponds to the empty list, where no feature has been determined at all. The corresponding value $q(0, 0)$ can be considered as a default probability of the positive class.

4   Learning Probabilistic Scoring Lists

While standard scoring systems have often been handcrafted by domain experts in the past, more recent methods for the data-driven construction of scoring systems aim to achieve a good trade-off between the complexity of models and the quality of their recommendations (Ustun and Rudin, 2016). This is crucial for the successful adoption of decision models in practice, as overly complex models are difficult to analyse by domain experts and impede the manual application by human practitioners.

Instead of learning standard scoring systems, we are interested in the task of learning probabilistic scoring lists, i.e., in constructing a PSL h from training data

$$\mathcal{D} = \left\{ (\boldsymbol{x}_i, y_i) \right\}_{i=1}^{N} \subset \mathcal{X} \times \mathcal{Y}. \tag{5}$$

This essentially means determining the following components:

- the subset of features to be included and the order of these features;
- the score assigned to each individual feature;
- the probabilities for the resulting combinations of stage and total score.

A first question in this regard concerns the quality of a model h: What do we actually mean by a "good" probabilistic scoring list? Intuitively, a good PSL allows for making decisions that are quick and confident at the same time. Thus, we would like to optimise two criteria simultaneously, namely, to minimise the number of features that need to be determined before a decision is made, and to maximise the confidence of the resulting decision. This compromise could be formalised in different ways, but regardless of how an overall performance measure is defined, the problem of optimising that measure over the space of possible PSLs will be computationally hard (Chevaleyre et al., 2013).

4.1   A Greedy Learning Algorithm

As a first attempt, we therefore propose a heuristic learning procedure that is somewhat inspired by decision tree learning. Starting with the empty list, the next feature/score combination (x_k, s_k) is added in a greedy way, i.e., so as to improve performance the most[2], and this is continued until no improvement is obtained anymore. To this end, each (remaining) feature/score combination is tried and evaluated in terms of the *expected entropy*: Suppose that, after adding (x_k, s_k) in stage k, the set of possible values for the total score is Σ_k. The expected entropy is then defined as

$$E = \sum_{T \in \Sigma_k} \frac{N_T}{N} \cdot H\big(\hat{q}(k, T)\big), \tag{6}$$

where $N = |\mathcal{D}|$ is the total number of training examples, N_T is the number of training examples with total score T, the $\hat{q}(k, T)$ are the estimated probabilities, and H is the Shannon entropy

$$H(q) = -q \cdot \log(q) - (1 - q) \log(1 - q).$$

4.2   Probability Estimation

As for the estimation of the probabilities $q(k, T)$, the most obvious idea would be a standard frequentist approach, i.e., to estimate them in terms relative frequencies P_T / N_T, where N_T is again the number of training examples with total score T, and P_T is the number of examples with total score T and class $y = 1$. However, as these estimates are obtained independently for each score T, they may violate the monotonicity condition (2). A better idea, therefore, is to estimate them jointly using a probability calibration method (Silva Filho et al., 2021). To this end, the original data $\mathcal{D}$, or a subset $\mathcal{D}_{cal}$ specifically reserved for calibration (and not used for training), is first mapped to the data

$$\mathcal{C} := \big\{(T(\boldsymbol{x}), y) \mid (\boldsymbol{x}, y) \in \mathcal{D}_{cal}\big\} \subset \Sigma_k \times \mathcal{Y},$$

to which any calibration method can then be applied. In our approach, we make use of isotonic regression (Niculescu-Mizil and Caruana, 2005) for that purpose,

---

[2] As the importance of a feature x_k, and hence the score s_k, can only be decided relative to other features, the choice of the score for the first feature is ambiguous; assuming this feature to be important, we given it the largest score possible.

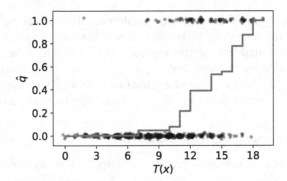

Fig. 2. Example of calibration with isotonic regression, using the medical data set introduced in Sect. 5. The values on the x-axis correspond to the total scores. As class labels are either 0 or 1, the data points $\mathcal{C}$ are plotted with jittering for better visualisation.

which amounts to finding values $\hat{q}(k, T)$ solving the following constrained optimisation problem:

$$\text{minimise} \quad \sum_{(T,y)\in\mathcal{C}} \big(\hat{q}(k,T) - y\big)^2$$

$$\text{s. t.} \quad \forall T, T' \in \Sigma_k : (T < T') \Rightarrow (\hat{q}(k,T) \leq \hat{q}(k,T'))$$

An illustration is shown in Fig. 2.

Note that, from a probability estimation point of view, the estimation of one distribution per total score $T \in \Sigma_k$ is a meaningful compromise between a global probability estimate (not taking any context features into account) and a *per-instance* estimation, i.e., the prediction of an individual distribution $p(\cdot \,|\, \boldsymbol{x})$ tailored to any specific instance $\boldsymbol{x}$. Obviously, the former is not informative enough, while the latter is very difficult to obtain, due to a lack of statistical information related to a single point (Foygel Barber et al., 2021). According to our assumption, all instances with the same total score T share the same probability. Therefore, those instances in the training data with the same score form a homogeneous statistical subgroup

$$\mathcal{D}_T := \big\{(\boldsymbol{x}_i, y_i) \in \mathcal{D} \mid T(\boldsymbol{x}) = T\big\},$$

to which statistical estimation methods can be applied. While this is in line with other local prediction methods, such as probability estimation trees (PETs) (Provost and Domingos, 2003), the distinguishing feature here is the way in which the instance space $\mathcal{X}$ is partitioned. For example, compared to PETs, PSLs appear to have a more rigid structure, because the succession of tests (features) is fixed and can not vary depending on the value of the features (like in trees). Moreover, the size of the partition, $|\Sigma_k|$, will normally be smaller than the (up to) 2^k different leaf nodes in a tree (leaves with same scores are merged). Both factors contribute to the increased interpretability of a single sequence of k feature tests, as opposed to up to the 2^k different paths through a PET.

4.3   Beyond Probabilities: Capturing Epistemic Uncertainty

Going beyond standard probabilistic prediction, various methods have recently been proposed in machine learning that seek to distinguish between so-called aleatoric and epistemic uncertainty (Senge et al., 2014; Hüllermeier and Waegeman, 2021). Broadly speaking, aleatoric uncertainty refers to inherent randomness and stochasticity of the underlying data-generating process. This type of uncertainty is relevant in our case, because the dependence between total score T and decision/class assignment y is presumably non-deterministic. Aleatoric uncertainty is properly captured in terms of probabilities, i.e., by the approach introduced above.

Epistemic uncertainty, on the other side, refers to uncertainty caused by a lack of knowledge, e.g., the learner's uncertainty about the true distribution $p = p(\cdot \mid T)$. In a machine learning context, this uncertainty could be caused by insufficient or low-quality training data. Obviously, it is relevant in our case, too: Proceeding further in the decision list, the training data will be more and more fragmented, because the number of possible values for the total score increases. Consequently, the estimation $\hat{q}$ of a conditional probability $p(y = 1 \mid T)$ will be based on fewer and fewer data points, so that the epistemic uncertainty increases (even if the joint estimation of these probabilities for all scores T alleviates this effect to some extent).

Representing this uncertainty is arguably important from a decision making point of view. For example, proceeding in the list and adding another variable may imply that the (predicted) distribution becomes better in the sense of having lower entropy, but at the same time, the prediction itself may become more uncertain. In that case, it is not clear whether the current stage should be preferred or maybe the next one—the answer to this question will depend on the attitude of the decision maker (toward risk), and probably also on the application. Interestingly, a method for probability calibration has recently been proposed that properly represents (epistemic) uncertainty by producing interval-predictions, and these predictions come with a guarantee of correctness that can be pre-specified by the user. This method, called Venn-ABERS predictors (Vovk and Petej, 2014), generalises isotonic regression and can be seen as a special case of the more general Venn predictors (Vovk et al., 2004).

5   A Case Study in Medical Decision Making

In this section, we present a case study in medical decision making meant as a first evaluation of our approach. This case study is aimed at the diagnosis of coronary heart disease.

5.1   Coronary Heart Disease Data

The data set for this case study has originally been used to evaluate the diagnostic accuracy of symptoms and signs for coronary heart disease (CHD) in patients presenting with chest pain in primary care. Chest pain is a common complaint in primary care, with CHD being the most concerning of many potential causes. Based on the medical history and physical examination, general practitioners (GPs) have to classify patients into two classes: patients in whom an underlying CHD can be safely ruled out (the negative class) and patients in whom chest pain is probably caused by CHD (the positive class).

Briefly, 74 general practitioners (GP) recruited consecutively patients aged ≥ 35 who presented with chest pain as primary or secondary complaint. GPs took a standardised history and performed a physical examination. Patients and GPs were contacted six weeks and six months after the consultation. All relevant information about course of chest pain, diagnostic procedures and treatments had been gathered during six months. An independent expert panel of one cardiologist, one GP and one research staff member reviewed each patient's data and established the reference diagnosis by deciding whether or not CHD was the underlying reason of chest pain. For details about the design and conduct of the study, we refer to Bösner et al. (2010).

Overall, the data set is comprised of 1199 (135 CHD and 1064 non-CHD) patients described by ten binary attributes: (f_1) patient assumes pain is of cardiac origin, (f_2) muscle tension, (f_3) age gender compound, (f_4) pain is sharp, (f_5) pain depends on exercise, (f_6) known clinical vascular disease, (f_7) diabetes, (f_8) heart failure, (f_9) pain is not reproducible by palpation, (f_{10}) patient has cough. Note that, by way of domain knowledge, all these features can be encoded in such a way that the presence of a feature does always increase the likelihood of the positive class. Therefore, scoring systems can be restricted to positive scores.

5.2 Expected Entropy Minimisation

Figure 3 visualizes an exemplary run of the greedy learning algorithm for PSLs with a score set of $\mathcal{S} = \{1, 2, 3\}$. The algorithm iteratively selects the feature-score-pair that minimizes the expected entropy, as defined in Eq. 6, in the next stage. As can be seen, the improvements diminish stage by stage, and almost vanish after the fifth stage. Interestingly, this result is very much in agreement with previous studies on this data, and the top-5 features in Fig. 3 exactly correspond to those features that have eventually been included in the the "Marburg Heart Score" (MHS), a decision rule that is now in practical use.[3]

As our algorithm minimises expected entropy on the training data in a greedy way, one may wonder to what extent expected entropy is also minimised globally, i.e., across all stages. To get an idea, we compared the expected entropy curve produced by the greedy algorithm with the curves produced by all other PSLs on the five top-features and with score set $\mathcal{S} = \{1, 2\}$—a complete enumeration of the resulting set of PSLs is still feasible. As can be seen from Fig. 4, the greedy approach (shown in black) performs quite well, at least on this example, and indeed leads to a globally optimal solution. Needless to say, this result cannot

---

[3] https://www.mdcalc.com/calc/4022/marburg-heart-score-mhs.

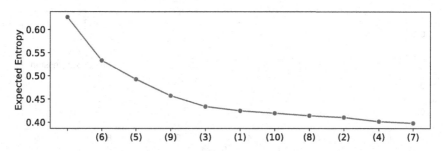

Fig. 3. Example run of the greedy learning algorithm with the score set $\mathcal{S} = \{1, 2, 3\}$, showing the decrease in expected entropy for the feature added in each stage (numbers on the x-axis). Recall that no feature is used at stage zero.

be generalised, but at least suggests that scoring systems can be extended in a greedy manner without losing too much in performance.

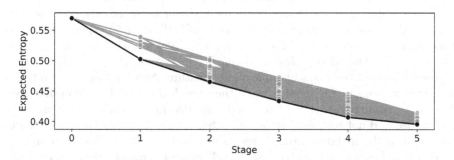

Fig. 4. Evaluation of the greedy learning algorithm (black line) on the top five top-features from the coronary heart disease dataset. Expected entropy curves are shown for all PSLs possible on these features and score set $\mathcal{S} = \{1, 2\}$.

5.3 Expected Loss Minimisation

In medical diagnosis, the consequences of a false negative prediction, i.e., not treating an ill patient, are typically far more severe than of a false positive. This asymmetry can be captured by a loss function that assigns a loss of 1 to a false positive and a loss of $M \gg 1$ to a false negative. In the medical domain, this also goes under the notion of "diagnostic regret", and various empirical methods for eliciting preferences in decision-making (i.e., the cost factor M) have been proposed in the literature (Tsalatsanis et al., 2010; Moreira et al., 2009).

Given M and a prediction $\hat{p}$ for the positive class (and hence $1 - \hat{p}$ for negative), the risk-minimising decision is given by

$$\hat{y} = \begin{cases} 1 & \text{if } 1 - \hat{p} < M \cdot \hat{p} \\ 0 & \text{otherwise} \end{cases},$$

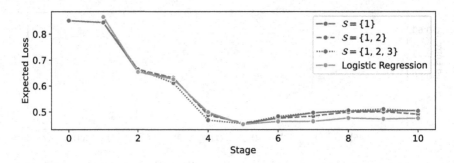

Fig. 5. Average loss of the PSLs predictions for each stage of the PSL averaged over 50 MCCV repetitions.

and the (estimated) expected loss itself by $E(\hat{y}) = \min\{1-\hat{p}, M\cdot\hat{p}\}$. This decision strategy nicely emphasises the importance of (accurate) probabilistic predictions and, more generally, uncertainty-awareness, in safety-critical domains.

To evaluate our learning algorithm for PSLs in terms of expected loss, we conducted a Monte Carlo cross-validation (MCCV) with 50 repetitions, each time using a fraction of $\frac{2}{3}$ of the coronary heart disease data as training data and $\frac{1}{3}$ as test data. Missing feature values have been imputed using the mode, representing the most frequent value of each feature. As a baseline, we also train a logistic regression (LR) model, using the same features as PSL in each stage. Note that, compared to PSL, LR is more flexible in the sense that scores are real-valued and not restricted to (small) integers. On the other hand, it is more restricted in the mapping from scores to probabilities: In LR, this mapping is accomplished by a logit transformation, and hence of parametric nature, whereas the isotonic regression in PSL is non-parametric.

Figure 5 shows the loss for $M = 10$, averaged over all 50 MCCV repetitions. The PSL has been configured with three different score sets, $\mathcal{S} = \{1\}$, $\mathcal{S} = \{1,2\}$ and $\mathcal{S} = \{1,2,3\}$. We can see that all three PSL variants perform quite similarly, with small improvements for larger score sets. Moreover, they are all on a par with LR, sometimes even a bit better, which is quite remarkable. For all variants, we observe a monotonic decrease in loss, up until the fifth feature is added. Again, in the large majority of cases, the five top-features correspond to the features also included in the Marburg heart score. Adding further features leads to a slight deterioration for both PSL and LR. As more features increase the capacity of the learner, this might be due to a standard overfitting effect. Note that the deterioration is a bit more pronounced for PSL than for LR, which can be explained as follows: While LR can modulate the influence of any additional feature in a very flexible way, by appropriately tuning the weight coefficient, PSL does not have this ability. Instead, it can only weight all features in (more or less) the same way. Therefore, in cases where adding another feature might be useful, but with a weight much smaller than the others, it might be better to omit it completely instead of giving it the same influence as the more important features.

As another interesting observation, note that the steepest decrease of expected loss is observed at the second stage, not at the first stage. Assuming that the first feature is the most important one, this would actually be expected—more generally, one may expect the loss curve to be convex. Here, however, the first feature does actually not yield a significant improvement. This can be explained by the fact that our learning algorithm is not specifically tailored to the (cost-sensitive) loss used for evaluation. Instead, for the reason of generality, we deliberately use expected entropy as an optimisation criterion for constructing a PSL—very much like in decision tree learning. However, as suggested by the experiments, tailoring the learner to a specific loss might be useful in cases where such a loss is known beforehand and can be given as additional input to the learner.

6   Summary and Conclusion

In this paper, we introduced probabilistic scoring lists, a probabilistic extension of scoring systems. Their main advantage and intended use is that they not only allow one to obtain probability estimates that correspond to the scores of the underlying scoring system, but that these estimates can be gradually refined by adding more features. This may, e.g., be important if features are expensive or time-consuming to obtain, so that rough estimates can be obtained cheaply and quickly, and be further refined once additional evidence comes in. In particular, it also allows one to end the decision making process once a certain probability threshold has been surpassed, thereby allowing a dynamic adjustment of the number of features needed for a positive or negative decision.

Building on the approach presented in this paper, we plan to address the following extensions in future work:

- Although the greedy learning algorithm proposed in this paper seems to perform quite well, more sophisticated algorithms for learning PSLs should be developed, including algorithms tailored to specific loss functions.
- We would also like to try other calibration techniques, especially beta calibration (Kull et al., 2017), which is parametric and hence less prone to overfitting than isotonic regression, but at the same time more flexible than a logit transform. Likewise meaningful is an extension toward more uncertainty-aware ("epistemic") calibration (cf. Sect. 4.3).
- Scoring systems check conditions in the form of binary features, which necessitates a binarisation of numerical or categorical features; this binarisation should not be done independently as a preprocessing step, but rather be integrated with the learning of scoring systems.
- So far, we only considered the case of binary decisions, which is common for scoring systems; yet, an extension to decision spaces of higher cardinality (polychotomous classification) is practically relevant.

Acknowledgment. We gratefully acknowledge funding by the German Research Foundation (Deutsche Forschungsgemeinschaft, DFG): TRR 318/1 2021 – 438445824 and the German Research Foundation (DFG) within the Collaborative Research Center "On-The-Fly Computing" (SFB 901/3 project no. 160364472).

References

Bösner, S., et al.: Accuracy of symptoms and signs for coronary heart disease assessed in primary care. Br. J. Gener. Pract. **60**(575), e246–e257 (2010)

Chevaleyre, Y., Koriche, F., Zucker, J.D.: Rounding methods for discrete linear classification. In: Proceedings of ICML, International Conference on Machine Learning, pp. 651–659 (2013)

Foygel Barber, R., Candes, J., Emmanuel, J., Ramdas, A., Tibshirani, R.J.: The limits of distribution-free conditional predictive inference. Inf. Inference **10**(2), 455–482 (2021). https://doi.org/10.1093/imaiai/iaaa017

Fürnkranz, J.: Separate-and-conquer rule learning. Artif. Intell. Rev. **13**(1), 3–54 (1999)

Fürnkranz, J., Gamberger, D., Lavrač, N.: Foundations of Rule Learning. Springer, Heidelberg (2012). https://doi.org/10.1007/978-3-540-75197-7. ISBN 978-3-540-75196-0

Hastie, T.J.: Generalized Additive Models. Routledge (2017)

Hüllermeier, E., Waegeman, W.: Aleatoric and epistemic uncertainty in machine learning: an introduction to concepts and methods. Mach. Learn. **110**(3), 457–506 (2021). https://doi.org/10.1007/s10994-021-05946-3

Kull, M., Silva Filho, T., Flach, P.: Beta calibration: a well-founded and easily implemented improvement on logistic calibration for binary classifiers. In: Proceedings of AISTATS, 20th International Conference on Artificial Intelligence and Statistics, vol. 54, pp. 623–631. PMLR (2017)

Moreira, J., Bisig, B., Muwawenimana, P., Basinga, P., Bisoffi, Z., Haegeman, F.: Weighing harm in therapeutic decisions of smear-negative pulmonary tuberculosis. Med. Decis. Making **3**, 380–390 (2009)

Možina, M., Demšar, J., Bratko, I., Žabkar, J.: Extreme value correction: a method for correcting optimistic estimations in rule learning. Mach. Learn. **108**(2), 297–329 (2018). https://doi.org/10.1007/s10994-018-5731-3

Niculescu-Mizil, A., Caruana, R.: Predicting good probabilities with supervised learning. In: Proceedings of ICML, 22nd International Conference on Machine Learning, New York, USA, pp. 625–632 (2005)

Provost, F.J., Domingos, P.: Tree induction for probability-based ranking. Mach. Learn. **52**(3), 199–215 (2003)

Rivest, R.L.: Learning decision lists. Mach. Learn. **2**, 229–246 (1987)

Senge, R., et al.: Reliable classification: learning classifiers that distinguish aleatoric and epistemic uncertainty. Inf. Sci. **255**, 16–29 (2014)

Silva Filho, T., Song, H., Perelló-Nieto, M., Santos-Rodríguez, R., Kull, M., Flach, P.A.: Classifier calibration: how to assess and improve predicted class probabilities: a survey. CoRR, abs/2112.10327 (2021). https://arxiv.org/abs/2112.10327

Simsek, O., Buckmann, M.: On learning decision heuristics. In: Imperfect Decision Makers: Admitting Real-World Rationality, pp. 75–85 (2017)

Six, A., Backus, B., Kelder, J.: Chest pain in the emergency room: value of the heart score. Neth. Hear. J. **16**(6), 191–196 (2008)

Subramanian, V., Mascha, E.J., Kattan, M.W.: Developing a clinical prediction score: comparing prediction accuracy of integer scores to statistical regression models. Anesth. Analg. **132**(6), 1603–1613 (2021)

Sulzmann, J.-N., Fürnkranz, J.: An empirical comparison of probability estimation techniques for probabilistic rules. In: Gama, J., Costa, V.S., Jorge, A.M., Brazdil, P.B. (eds.) DS 2009. LNCS, vol. 5808, pp. 317–331. Springer, Heidelberg (2009). https://doi.org/10.1007/978-3-642-04747-3_25

Tsalatsanis, A., Hozo, I., Vickers, A., Djulbegovic, B.: A regret theory approach to decision curve analysis: a novel method for eliciting decision makers' preferences and decision-making. BMC Med. Inform. Decis. Mak. 10, 51 (2010). https://doi.org/10.1186/1472-6947-10-51

Ustun, B., Rudin, C.: Supersparse linear integer models for optimized medical scoring systems. Mach. Learn. 102(3), 349–391 (2016)

Ustun, B., Rudin, C.: Optimized risk scores. In: Proceedings of 23rd ACM SIGKDD International Conference on Knowledge Discovery and Data Mining, pp. 1125–1134 (2017)

Ustun, B., Rudin, C.: Learning optimized risk scores. J. Mach. Learn. Res. 20(150), 1–75 (2019)

Vovk, V., Petej, I.: Venn-Abers predictors. In: Proceedings of UAI, 30th Conference on Uncertainty in Artificial Intelligence (2014)

Vovk, V., Shafer, G., Nouretdinov, I.: Self-calibrating probability forecasting. In: Proceedings of NIPS, Advances in Neural Information Processing Systems, vol. 16, pp. 1133–1140 (2004)

Wang, C., Han, B., Patel, B., Rudin, C.: In pursuit of interpretable, fair and accurate machine learning for criminal recidivism prediction. J. Quant. Criminol. 39, 1–63 (2022)

Webb, G.I.: Recent progress in learning decision lists by prepending inferred rules. In: Proceedings of the 2nd Singapore International Conference on Intelligent Systems, pp. B280–B285 (1994)

Refining Temporal Visualizations Using the Directional Coherence Loss

Pavlin G. Poličar[1](✉) ⓘ and Blaž Zupan[1,2] ⓘ

[1] Faculty of Computer and Information Science, University of Ljubljana,
Ljubljana, Slovenia
`pavlin.policar@fri.uni-lj.si`
[2] Department Education, Innovation and Technology, Baylor College of Medicine,
Houston, TX, USA

Abstract. Many real-world data sets contain a temporal component or include transitions from state to state. For exploratory data analysis, we can present these high-dimensional data sets in two-dimensional maps, using embeddings of data objects under exploration and representing their temporal relations with directed edges. Most existing dimensionality reduction techniques, such as t-SNE and UMAP, disregard the temporal or relational nature of the data during embedding construction, leading to cluttered visualizations obscuring potentially interesting temporal patterns. To address this issue, we introduce Directional Coherence Loss (DCL), a differentiable loss function that we can incorporate into existing dimensionality reduction techniques. We have designed DCL to highlight the temporal aspects of the data, revealing temporal patterns that might otherwise remain unnoticed. By encouraging local directional coherence of the directed edges, the DCL produces more temporally-meaningful and less-cluttered visualizations. We demonstrate the effectiveness of our approach on a real-world multivariate time-series data set tracking the progression of the COVID-19 pandemic in Slovenia. We show that incorporating the DCL into the t-SNE algorithm elucidates the time progression of the pandemic in the embedding and reveals interesting cyclical patterns otherwise hidden in standard embeddings.

Keywords: Temporal-data visualization · Dimensionality reduction · Data visualization

1 Introduction

A common method for analyzing the structure of high-dimensional data involves representing it in two-dimensional, point-based visualizations. We can use dimensionality reduction approaches such as principal component analysis, multi-dimensional scaling, or t-SNE to obtain such data maps. Additionally, we can overlay these data maps with arrows indicating temporal dependence between data points to present temporal relations between data points. This approach has been used extensively for the visualization of dynamic graphs [4], multi-variate time-series [1], and gene-expression data [5].

ⓒ The Author(s) 2023
A. Bifet et al. (Eds.): DS 2023, LNAI 14276, pp. 204–215, 2023.
https://doi.org/10.1007/978-3-031-45275-8_14

Existing approaches for visualizing temporal data via two-dimensional embeddings rely primarily on off-the-shelf embedding techniques, which do not incorporate the temporal aspects of the data. Commonly-used dimensionality reduction approaches, however, may be semantically constrained. Principal component analysis (PCA) [10], for example, relies on the linear transformation of attribute space and may fail to reveal complex patterns with non-linear interactions of input features. Non-linear data embedding techniques, such as multi-dimensional scaling [3], t-SNE [6], and UMAP [7], may overcome this limitation and introduce distortions into the embedding. None of these techniques, however, explicitly incorporates the available temporal information into the embedding construction process, resulting in embeddings that fail to reflect or even obscure the temporal patterns in the underlying data.

This report introduces the *directional coherence loss* (DCL), which integrates available temporal information into the embedding construction process. The result of DCL is embeddings designed to facilitate the discovery of temporal patterns in the two-dimensional embedding space. The DCL is differentiable, and we can incorporate it into existing dimensionality reduction techniques. Adding the DCL to the existing data embedding approach reveals temporal patterns in the resulting embeddings, aiding in discovering temporal patterns in the data.

2 Related Work

There are a plethora of approaches that we can use for the visualization of high-dimensional, temporal data. Rauber et al. [11] developed Dynamic t-SNE, which constructs a series of t-SNE embeddings and stacks them stacked along a third dimension corresponding to time. A similar approach has been proposed for UMAP, termed AlignedUMAP [7].

Alternatively, van den Elzen et al. [4] portray the progression of time in two-dimensional embeddings by connecting data points with arrows. Their approach focuses on visualizing dynamic graphs. At each point in time, the graph adjacency matrix is treated as a high-dimensional data point. This high-dimensional collection of graph snapshots is subsequently embedded into a two-dimensional visualization using an off-the-shelf embedding technique. Ali et al. [1] apply a similar approach to multivariate time-series data, where each sliding time window is treated as a single high-dimensional data point. In this way, they embed temporal sequences into two dimensions, where arrows connect consecutive time points. Unlike dynamic t-SNE and AlignedUMAP, which construct a three-dimensional embedding by stacking multiple two-dimensional embeddings along a time dimension, these approaches illustrate the entire temporal progression into two dimensions and indicate dependence using arrows.

In bioinformatics, single-cell RNA velocity [5] may accompany more standard gene expression data and requires a different visualization approach. Each data point corresponds to the gene expression of a single cell, characterized by tens of thousands of genes. Then, for each cell, single-cell RNA velocity estimates the likely transitions between different cell states, for instance, during differentiation.

The resulting visualization typically consists of a two-dimensional embedding constructed using t-SNE or UMAP overlaid with arrows to indicate likely cell-to-cell transitions. This approach is conceptually similar to van den Elzen et al. [4] and Ali et al. [1], where we deal only with a single time-step for every cell.

Another notable approach, Time Curves [2], offers general guidelines for visualizing the temporal progression of a single entity. The framework may be viewed as a generalization of the work by Ali et al. [1], allowing for arbitrary time steps between snapshots.

3   Methods

Consider a high-dimensional data set $\mathbf{X} \in \mathbb{R}^{N \times d}$, where N is the number of data points and d is the dimensionality of each data point. Let G be a directed graph $G = (V, E)$, where V denotes the set of vertices v_i corresponding to individual data points $\mathbf{x}_i$. E is the set of edges e_{ij} representing the temporal connections between data points i and j. When visualizing high-dimensional data sets, our primary objective is to find a low-dimensional embedding $\mathbf{Y} \in \mathbb{R}^{N \times 2}$ that accurately reflects the topological features of $\mathbf{X}$. In two-dimensional visualizations, we represent the connections e_{ij} as directed line segments $\mathbf{p}_{ij}$ (depicted as arrows) linking two related data points i and j in the embedding space such that $\mathbf{p}_{ij} = [\mathbf{y}_i, \mathbf{y}_j]$.

3.1   t-SNE

t-distributed stochastic neighbor embedding (t-SNE) is a non-linear dimensionality reduction technique commonly used to visualize high-dimensional data [6]. t-SNE aims to find a low-dimensional representation $\mathbf{Y}$ such that if two data points are close in the high-dimensional space $\mathbf{X}$, then they are also close in the low-dimensional space $\mathbf{Y}$.

Formally, the t-SNE algorithm aims to find a low-dimensional representation $\mathbf{Y}^*$, such that the Kullback-Leibler (KL) divergence between similarities $\mathbf{P}$ between data points in the high-dimensional space $\mathbf{X}$ and the similarities $\mathbf{Q}$ between data points in the low-dimensional space $\mathbf{Y}$ is minimized, such that

$$\mathbf{Y}^* = \arg\min_{\mathbf{Y}} \mathrm{KL}(\mathbf{P} \parallel \mathbf{Q}). \tag{1}$$

The similarities $\mathbf{P} = [p_{ij}]$ between data points in $\mathbf{X}$ are obtained using the Gaussian kernel,

$$p_{ij} = \frac{p_{j|i} + p_{i|j}}{2N}, \quad p_{j|i} = \frac{\exp\left(-\mathcal{D}(\mathbf{x}_i, \mathbf{x}_j)/2\sigma_i^2\right)}{\sum_{k \neq i} \exp\left(-\mathcal{D}(\mathbf{x}_i, \mathbf{x}_k)/2\sigma_i^2\right)}, \quad p_{i|i} = 0, \tag{2}$$

where $\mathcal{D}$ is some distance measure and the bandwidth of each Gaussian kernel σ_i is selected such that the perplexity u of each conditional distribution matches a user-specified parameter value,

$$\log(u) = -\sum_j p_{j|i} \log\left(p_{j|i}\right) \tag{3}$$

In the low-dimensional representation $\mathbf{Y}$, the similarities $\mathbf{Q} = [q_{ij}]$ are characterized by the t-distribution,

$$q_{ij} = \frac{\left(1 + ||\mathbf{y}_i - \mathbf{y}_j||^2\right)^{-1}}{\sum_{k \neq l} \left(1 + ||\mathbf{y}_k - \mathbf{y}_l||^2\right)^{-1}}, \quad q_{ii} = 0. \tag{4}$$

3.2 Directional Coherence Loss (DCL)

The key idea behind the *directional coherence loss* (DCL) is that arrows close to one another in the embedding space should point in approximately the same direction. Since each arrow is defined as a line segment parameterized by points $\mathbf{y}_i$ and $\mathbf{y}_j$, we can achieve this directional coherence by adjusting the positions of points $\mathbf{y}_i$ and $\mathbf{y}_j$ accordingly.

Let $\mathbf{u}_{ij}$ be the unit vector corresponding to the line segment $\mathbf{p}_{ij} = [\mathbf{y}_i, \mathbf{y}_j]$,

$$\mathbf{u}_{ij} = \tilde{\mathbf{u}}_{ij}/||\tilde{\mathbf{u}}_{ij}||, \quad \tilde{\mathbf{u}}_{ij} = \mathbf{y}_j - \mathbf{y}_i \tag{5}$$

Then, for each pair of edges e_{ij} and e_{kl} in E, we can determine the directional coherence of their corresponding arrows in the embedding by computing the dot product $\mathbf{u}_{ij} \cdot \mathbf{u}_{kl} = ||\mathbf{u}_{ij}|| \, ||\mathbf{u}_{kl}|| \cos \theta$, where θ denotes the angle between the two vectors. In our case $||\mathbf{u}_{ij}|| = ||\mathbf{u}_{kl}|| = 1$, so their dot product simplifies to $\mathbf{u}_{ij} \cdot \mathbf{u}_{kl} = \cos \theta$. When $\mathbf{u}_{ij}$ and $\mathbf{u}_{kl}$ point in the same direction, their dot product is 1. Conversely, when $\mathbf{u}_{ij}$ and $\mathbf{u}_{kl}$ point in opposite directions, their dot product is -1. Therefore, to achieve good directional coherence for any pair of arrows in E, we must maximize the dot product of their corresponding directional vectors.

To make directional coherence compatible with existing dimensionality reduction loss functions, we convert the directional coherence into a strictly positive minimization loss. To convert the maximization into a minimization objective, we multiply the equation with -1. To enforce strict-positivity and avoid negative penalties, we add a $+1$ term to the above formulation and shift the domain from $[-1, 1]$ to $[0, 2]$. Additionally, we have found it beneficial to square the resulting equation, leading to faster convergence and more visually appealing visualizations. The directional coherence loss between edges pair of edges e_{ij} and e_{kl} then becomes

$$\text{DCL}(\mathbf{p}_{ij}, \mathbf{p}_{kl}) = \left(-\left(\mathbf{u}_{ij} \cdot \mathbf{u}_{kl}\right) + 1\right)^2 \tag{6}$$

We penalize only nearby arrow pairs in order to enforce the local penalization of the DLC. The distance between two line segments $\mathbf{p}_{ij} = [\mathbf{y}_i, \mathbf{y}_j]$ and $\mathbf{p}_{kl} = [\mathbf{y}_k, \mathbf{y}_l]$ is defined as

$$d(\mathbf{p}_{ij}, \mathbf{p}_{kl}) = \arg\min_{s,t} || [s \cdot \mathbf{y}_i + (1 - s) \cdot \mathbf{y}_j] - [t \cdot \mathbf{y}_k + (1 - t) \cdot \mathbf{y}_l] ||, \tag{7}$$

where $s, t \in [0, 1]$. Intuitively, their distance corresponds to the distance between the two closest points on these line segments. If the line segments intersect, then their distance is 0.

We penalize nearby arrow pairs using a Gaussian kernel on the obtained pairwise line-segment distances,

$$w(\mathbf{p}_{ij}, \mathbf{p}_{kl}) = \frac{1}{\sqrt{2\pi\sigma^2}} \exp\left(-d(\mathbf{p}_{ij}, \mathbf{p}_{kl})/2\sigma^2\right), \tag{8}$$

where σ^2 is the variance of the Gaussian distribution. The variance σ^2 determines the region around each arrow where we wish the arrows to point in the same direction. This parameter can greatly affect the final embedding, as a large value of σ^2 will enforce the DCL across the entire embedding. In contrast, small values of σ^2 will have a limited effect on the point positions. It is also worth noting that this parameter should depend on the scale of the embeddings, which can change during optimization and vary across different dimensionality reduction algorithms. In our experiments, we use $\sigma^2 = 1$.

Combining the directionality penalty from Eq. 6 and the weights from Eq. 8, we obtain the final directional coherence loss,

$$\mathcal{L}_{DCL} = \frac{1}{\binom{|E|}{2}} \sum_{e_{ij}\in E} \sum_{e_{kl}\in E} w(\mathbf{p}_{ij}, \mathbf{p}_{kl})\left(-(\mathbf{u}_{ij}\cdot\mathbf{u}_{kl})+1\right)^2, \quad (i,j)\neq(k,l). \tag{9}$$

We can incorporate the DCL loss into various dimensionality reduction methods. In our case, we augment the t-SNE algorithm with the DCL loss,

$$\mathcal{L} = \mathcal{L}_{t\text{-}SNE} + \lambda\mathcal{L}_{DCL} \tag{10}$$

where λ is the trade-off parameter between the two loss functions. In our experiments, we used $\lambda = 10$.

4   Results and Discussion

Below, we demonstrate the conceptual idea and expected results of our approach using a toy example. Additionally, we include a real-world case study on the progression of the COVID-19 pandemic in Slovenia. We conclude this section with a discussion of the potential shortcomings and limitations of the proposed approach.

4.1   Toy Example

We first consider a toy example to demonstrate that adding the DCL to the t-SNE dimensionality reduction algorithm elucidates trajectories or transitions between different clusters. This synthetic data set consists of seven distinct, non-overlapping clusters at equal distances from one another, each containing 50 points sampled from unit-Gaussian distributions. To simulate transitions between clusters, we connect each point from a given cluster c to a randomly chosen point from the subsequent cluster $c+1$. The data points in the last cluster from a sequence of connected clusters are connected to the data points from the first cluster. This toy example can be thought of as a cyclic process containing seven distinct states where transitions are only possible between adjacent states. This may correspond to, for instance, single-cell data containing gene expression profiles corresponding to four different cell-cycle states in cell division.

We optimize the embedding using batch gradient descent as implemented in `pytorch` [9] for 10,000 iterations using a learning rate of 10. We use `ReduceLROnPlateau` to reduce the learning rate once the loss has not improved for 3,000 iterations. We use a perplexity value of 30 in the t-SNE loss function.

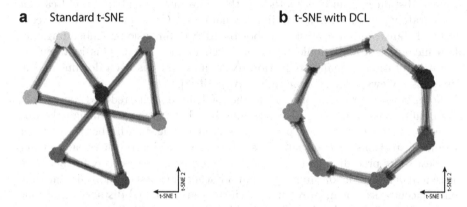

a   Standard t-SNE   **b** t-SNE with DCL

Fig. 1. The toy example demonstrates that incorporating the directional coherence loss (DCL) can help highlight the temporal transitions between data points. We construct a standard t-SNE embedding in (**a**), which can recover the seven distinct clusters. However, the arrows between clusters cross over one another, making it challenging to observe the underlying cyclic pattern. Incorporating the DCL in (**b**) helps untangle the crossing arrows and highlights the cyclic pattern in the underlying data set while still recovering the seven clusters.

Figure 1a shows that while t-SNE can recover the seven distinct clusters from the high-dimensional space, overlaying the embedding with arrows clutters the visualization, concealing the cyclic pattern in the underlying data set. On the other hand, augmenting the standard t-SNE loss function with the DCL untangles the arrows and highlights the cyclic pattern as shown in Fig. 1b. Combining the t-SNE dimensionality reduction algorithm, which can identify the distinct clusters, with the DCL, which positions the clusters so that the transitions between the clusters are most apparent, considerably enhances the interpretability of the embedding and the underlying temporal pattern.

The t-SNE algorithm aims to preserve distances to a user-specified number of neighbors. However, accurately preserving distances obtained from high-dimensional data sets in a two-dimensional embedding is only possible in some of the most straightforward data sets. Using a perplexity value of 30, t-SNE does its best to preserve distances to each point's 30 nearest neighbors in the high-dimensional space. However, t-SNE also attempts to preserve distances to other data points, albeit to a much lesser extent. In our synthetic data set, each cluster comprises 50 data points, meaning that, in addition to the points in the same cluster, t-SNE also attempts to preserve at least some distances from the other clusters. In Fig. 1a, the purple cluster is positioned centrally to other clusters, roughly at equal distances from the remaining clusters. Here, t-SNE can preserve the distances reasonably well. On the other hand, the top-left yellow cluster appears close to the central purple cluster and the light-green cluster below it, suggesting that these clusters are closer to one another than to, for instance, the right-most green cluster. However, by design, all seven clusters are at equal distances from one another in the high-dimensional space and cannot be accurately embedded in a two-dimensional plane. Consequently, the between-cluster distances in all nearest-neighbor-based two-dimensional embeddings are often meaningless and should never be taken at face value. This is a general limitation of dimensionality reduction techniques and has been documented in numerous reviews, e.g., by Nonato and Aupetit [8].

Note, however, that incorporating the DCL necessarily reduces the embedding quality regarding the t-SNE loss function. For instance, although the distances between clusters were poorly preserved in Fig. 1a, the between-cluster distance distortions were arguably less severe than in Fig. 1b, where each cluster is closest to its preceding and subsequent cluster, and progressively further from the remainder. This layout indicates that adjacent clusters are more similar than non-adjacent ones when, in reality, all clusters are at equal distances from one another. Nonetheless, despite this embedding being quantitatively worse at preserving distances between clusters, we argue that it provides a more informative visualization. When constructing embeddings for high-dimensional data sets, distances between clusters in the embedding should never be taken at face value, regardless of the dimensionality reduction technique. While the spatial relationships between clusters can aid in hypothesis generation, they should always be validated using alternative techniques.

Given that the spatial relationships between clusters lack informative value and can even mislead, it would be more sensible to position clusters in a temporally coherent manner. In this way, at least, the temporal relationships are more clearly highlighted, and the user is more directly aware of the limitations of interpreting spatial relationships, an often overlooked limitation of non-linear dimensionality methods. This way, the embedding algorithm can still recover well-defined clusters of data points in the high-dimensional space. Still, we explicitly decide that the spatial positions will reflect the temporal component of the embedding and not the spatial relationships between clusters.

4.2   COVID-19 Pandemic in Slovenia

We obtain Slovenian national data on the COVID-19 pandemic spanning from the beginning of March 2020 up until the end of March 2022[1]. Although the data

---

[1] National Slovenian data on the COVID-19 pandemic is available at https://covid-19.sledilnik.org/en/stats.

includes many variables, we limit our analysis to three time-series variables: the daily number of tests performed, the daily number of confirmed cases, and the daily number of hospital patients. We plot the individual time series in Fig. 2a. The line plots indicate the progression of the COVID-19 pandemic in Slovenia, with visible distinct phases of the pandemic.

To construct a two-dimensional visualization of the pandemic progression through time, we follow the approach from Ali et al. [1]. We first convert this multi-variate time series into a high-dimensional data set by constructing vectors from a sliding window with window size 7. Thus, the 160 21-dimensional data points represent one week of the pandemic. We connect data points corresponding to subsequent weeks with arrows.

We construct a t-SNE visualization of the high-dimensional data set in Fig. 2b. While the plot indicates a clear progression through time, the plot fails to reveal any underlying patterns in the data. Figure 2c depicts the results of our approach. While the embedding has not changed much structurally, the visualization reveals two clear cyclic patterns in the upper-right and lower regions of the embedding space.

We investigate the top-right cyclic pattern in Fig. 2c. Inspecting the two corresponding time spans highlighted in the original time series in Fig. 2a, it appears that this cyclic pattern coincides with high hospitalization rates, moderate levels of testing, and a moderate number of positive tests. Interestingly, both periods occurred during the spring season, one in 2021 and one in 2022. The first of these periods was substantially longer, lasting to the end of May, while the second lasted only a month and a half. It is also interesting to inspect which COVID-19 variants were prevalent in the country at that time[2]. During the first period in 2021, we were dealing with the initial 20A strain. The second period coincides with the transition from the Delta strain to the Omicron strain. The highlighted region in Fig. 2a corresponds to the final weeks of the Delta variant, which had higher mortality rates than the Omicron variant [12]. These strain prevalence and dynamics may explain the subsequent peak in the positive test cases and lower hospitalization rate following the highlighted region.

Finally, adding the DCL to the t-SNE algorithm elucidates the time progression of the time series. For instance, in Figs. 2e and 2e, we focus on a particular region of the embedding space, where it first appears as though the standard t-SNE embedding better highlights the temporal progression than with the addition of the DCL. Upon closer inspection, however, it is challenging to trace the arrows denoting the temporal progression of the pandemic as the arrow seems to veer off to the right, then cycle back, only to make another cycle back to the originating point. It is unclear which of these cycles occurred first and which second. With the addition of the DCL, it becomes easy to trace the temporal progression, as indicated by the red arrow drawn on top of the arrows to facilitate reading the embedding.

---

[2] The prevalence of the different COVID-19 variants in different countries is available at https://covariants.org/per-country.

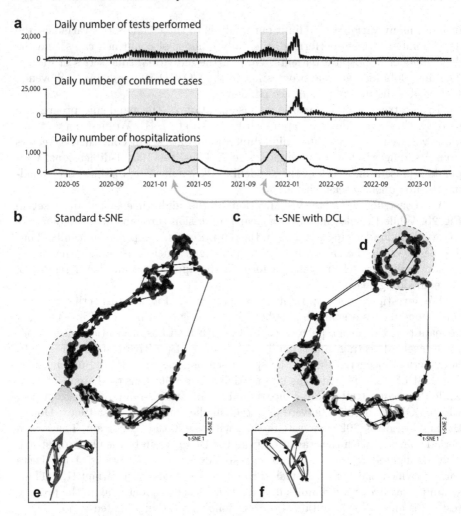

Fig. 2. We plot the progression of the COVID-19 pandemic in Slovenia from March 2020 to March 2022. **(a)** depicts individual line plots of the three variables under consideration. We construct a t-SNE embedding of the multivariate time series in **(b)** and augment the t-SNE loss function with our directional coherence loss in **(c)**. Individual points correspond to one week of the time series. We indicate the chronological progression by point colors where dark, purple colors correspond to the start of the pandemic, while lighter, yellow colors coincide with later stages of the pandemic. We connect consecutive weeks by arrows. Incorporating the directional coherence loss uncovers interesting temporal patterns in the visualization. We highlight one such cyclic region in **(d)** and mark the corresponding time spans in the original line plots. Panels **(e)** and **(f)** provide close-up views of regions of the original and augmented t-SNE embedding. We clarify the time progression by superimposing a red arrow onto the plot.

4.3   Hyperparameter and Evaluation Considerations

Incorporating the DCL into existing algorithms introduces two additional hyper-parameters to the visualization procedure. The kernel bandwidth σ determines the radius in which the DCL is enforced. A larger bandwidth emphasizes global coherence, while a lower value results in more locally consistent arrows. Additionally, the parameter λ determines the trade-off between the visualization loss and the DCL. Placing a greater emphasis on temporal coherence highlights temporal progression, enabling a clearer visualization of temporal dependencies. However, this may obscure the underlying structure in the resulting visualization. Therefore, finding optimal parameter settings for the DCL is crucial to achieving a well-balanced visualization and likely varies from dataset to dataset.

We evaluate our approach using a toy dataset designed to illustrate the conceptual motivation behind our method. While this example supports the validity of our approach, real-world data may not display such straightforward temporal patterns. For a more comprehensive evaluation, we could create other synthetic datasets to test various scenarios and temporal patterns. While we could also apply our approach to real-world, multivariate time-series data, the interpretation of such study outcomes might be subjective. An ideal evaluation would involve an objective measure of visualization quality. However, devising such a quantitative metric is challenging even for non-temporal, two-dimensional embeddings. Adding temporal coherence to this metric introduces additional complexities and challenges.

5   Conclusion

The work presented here was motivated by the difficulties of identifying temporal patterns in presentations of multi-variate data in a low-dimensional, non-linear embedding. There, we may expose the temporal relations using arrows to indicate the transitions. These visualization elements often clutter the data presentations and obscure the underlying temporal patterns. Existing dimensionality reduction techniques do not account for the temporal nature of the data. To this end, we propose the *directional coherence loss* (DCL), which can be incorporated into existing dimensionality reduction techniques. Uniquely, the DCL explicitly integrates the temporal information into the embedding construction process and produces embeddings highlighting the temporal patterns in the underlying data more clearly.

This presented work opens up several avenues for future research. First, the DCL enforces directional coherence by affecting the positions of the data points in the two-dimensional embedding. While this approach is viable for simpler data sets, such an arrangement may be difficult to achieve in the presence of more complex patterns. Secondly, the DCL is applicable when the arrows begin at one data point and end at another. This is not the case in data such as those from bioinformatics that include RNA velocity, where arrows originate from data points but end in an average position of multiple data points. The DCL must be extended to make it applicable to this case. Thirdly, in its current form, the DCL exhibits quadratic scaling in the number of connections between data points, making it unsuitable for visualizing large data sets. Due to the local nature of the DCL, approximation schemes could be developed which would only compute the interaction between nearby line segments. Lastly, we could find better optimization schemes leading to faster convergence.

Acknowledgements. This work was supported by the Slovenian Research Agency Program Grant P2-0209 and Project Grant V2-2272.

References

1. Ali, M., Jones, M., Xie, X., Williams, M.: Towards visual exploration of large temporal datasets. In: 2018 International Symposium on Big Data Visual and Immersive Analytics (BDVA), pp. 1–9 (2018). https://doi.org/10.1109/BDVA. 2018.8534025
2. Bach, B., Shi, C., Heulot, N., Madhyastha, T., Grabowski, T., Dragicevic, P.: Time curves: folding time to visualize patterns of temporal evolution in data. IEEE Trans. Visual Comput. Graphics **22**(1), 559–568 (2016). https://doi.org/10.1109/ TVCG.2015.2467851
3. Borg, I., Groenen, P.J.: Modern Multidimensional Scaling: Theory and Applications. Springer, New York (2005). https://doi.org/10.1007/0-387-28981-X
4. Van den Elzen, S., Holten, D., Blaas, J., van Wijk, J.J.: Reducing snapshots to points: a visual analytics approach to dynamic network exploration. IEEE Trans. Visual Comput. Graphics **22**(1), 1–10 (2016). https://doi.org/10.1109/TVCG. 2015.2468078
5. La Manno, G., et al.: RNA velocity of single cells. Nature **560**(7719), 494–498 (2018). https://doi.org/10.1038/s41586-018-0414-6
6. Van der Maaten, L., Hinton, G.: Visualizing data using t-SNE. J. Mach. Learn. Res. **9**(Nov), 2579–2605 (2008)
7. McInnes, L., Healy, J., Melville, J.: UMAP: uniform manifold approximation and projection for dimension reduction. ArXiv e-prints (2018)
8. Nonato, L.G., Aupetit, M.: Multidimensional projection for visual analytics: linking techniques with distortions, tasks, and layout enrichment. IEEE Trans. Visual Comput. Graphics **25**(8), 2650–2673 (2019)
9. Paszke, A., et al.: PyTorch: an imperative style, high-performance deep learning library. In: Wallach, H., Larochelle, H., Beygelzimer, A., d' Alché-Buc, F., Fox, E., Garnett, R. (eds.) Advances in Neural Information Processing Systems, vol. 32. Curran Associates, Inc. (2019)
10. Pearson, K.: On lines and planes of closest fit to systems of points in space. Phil. Mag. **2**(11), 559–572 (1901). https://doi.org/10.1080/14786440109462720
11. Rauber, P.E., Falcão, A.X., Telea, A.C.: Visualizing time-dependent data using dynamic t-SNE. In: EuroVis 2016 - Short Papers, pp. 73–77. The Eurographics Association (2016). https://doi.org/10.2312/eurovisshort.20161164
12. Wrenn, J.O., et al.: COVID-19 severity from Omicron and Delta SARS-CoV-2 variants. Influenza Other Respir. Viruses **16**(5), 832–836 (2022). https://doi.org/ 10.1111/irv.12982

Semantic Enrichment of Explanations of AI Models for Healthcare

Luca Corbucci[1]([⊠])[iD], Anna Monreale[1][iD], Cecilia Panigutti[1][iD],
Michela Natilli[2][iD], Simona Smiraglio[1], and Dino Pedreschi[1][iD]

[1] Department of Computer Science, University of Pisa, Pisa, Italy
`luca.corbucci@phd.unipi.it`
[2] ISTI-CNR Pisa, Pisa, Italy

Abstract. Explaining AI-based clinical decision support systems is crucial to enhancing clinician trust in those powerful systems. Unfortunately, current explanations provided by eXplainable Artificial Intelligence techniques are not easily understandable by experts outside of AI. As a consequence, the enrichment of explanations with relevant clinical information concerning the health status of a patient is fundamental to increasing human experts' ability to assess the reliability of AI decisions. Therefore, in this paper, we propose a methodology to enable clinical reasoning by semantically enriching AI explanations. Starting with a medical AI explanation based only on the input features provided to the algorithm, our methodology leverages medical ontologies and NLP embedding techniques to link relevant information present in the patient's clinical notes to the original explanation. Our experiments, involving a human expert, highlight promising performance in correctly identifying relevant information about the diseases of the patients.

1 Introduction

Recent efforts in Artificial Intelligence (AI) have shown great potential in helping physicians in several of their daily clinical practices, for example, the interpretation of medical scans [30] and the accurate assessment of prognosis [9] and treatment recommendation [5]. While some worries have been raised about AI systems replacing the role of doctors, human reasoning and oversight remain indispensable for the proper functioning of such systems [10]. Indeed, current AI applications focus on narrow tasks and have been shown to be sensitive to adversarial attacks [23] and biased datasets and algorithms [26]. These shortcomings raised several concerns about the trustworthiness of such systems, especially because most state-of-the-art AI-based solutions are hardly interpretable by humans. The transparency of AI systems in high-stakes domains such as healthcare has been subject to many recent European regulatory efforts like the GDPR and the recent proposal to regulate AI (AI Act).

For example, the European General Data Protection Regulation (GDPR), which came into full effect in May of 2018, prescribes providing the data subject of any automated decision-making process with "meaningful information about

A. Bifet et al. (Eds.): DS 2023, LNAI 14276, pp. 216–229, 2023.
https://doi.org/10.1007/978-3-031-45275-8_15

the logic involved, as well as the significance and the envisaged consequences of such processing for the data subject" [14]. Furthermore, the recent proposal for a European regulation of AI (AI Act) prescribes high-risk AI systems to be developed in such a way that they enable users to interpret their output correctly and use them appropriately [15]. In response to these ethical and legal issues, in the past years, the research community has been very active in developing several techniques to explain the reasoning of *black box* AI models, i.e., models whose internal decision-making process is obscure. The research field that studies the interpretability of AI systems is that of eXplainable Artificial Intelligence (XAI) [4]. Most XAI techniques offer interpretations of the black box behaviour by providing *explanations*, i.e., interfaces between humans and algorithms that allow the user to understand the AI decision-making process. Developing AI systems able to support medical decision-making requires creating appropriate human-computer interfaces to enable clinical reasoning. However, most XAI explanations are designed to provide insights on model behaviour to AI developers [3].

In this paper, we present a novel methodology that exploits access to the patient's clinical notes and the domain knowledge encoded in medical ontologies to semantically enrich the explanations provided by a state-of-the-art XAI technique for clinical decision support systems (DSS). While the original explanation considers only patient features that the AI algorithm received as input, our methodology exploits medical ontologies to link such features to an external source of knowledge on the patient. The result is an augmented explanation that allows the physician to reason over the clinical context. Our experiments, involving a human expert, show promising performance in correctly identifying relevant information about the diseases of the patients.

The paper is structured as follows. In Sect. 2 we briefly present the field of XAI, its applications in the healthcare context and the related uses of ontologies. In Sect. 3 we formalize the problem we address in the paper, while in Sect. 4 we describe the details of our methodology. Section 5 presents the experiments used to validate our methodology. Finally, in Sect. 6 we discuss our results and we present our ideas for future developments of our methodology.

2 Related Work

In this section, we overview some research work linked to our methodology.

XAI in Healthcare. XAI research studies how to provide explanations for AI systems behaviour in *human-understandable* terms [4]. The need for XAI techniques stems from the fact that many AI systems have an opaque internal reasoning process, i.e., they are considered black boxes. In the literature, the transparency of AI systems is achieved mainly in two ways: by building *transparent-by-design* models and by extracting explanations from black box models [16]. Some examples of transparent-by-design models employed in healthcare are models that allow the visualization of the relationships between input features and model output [6] and case-based reasoning models where the decision-making process is entirely interpretable [2]. However, it is not always possible to build transparent-by-design models for the task at hand. Therefore it might be necessary to extract explanations from black box models. The two most known examples of such XAI techniques are LIME [27] and SHAP [22]. While LIME trains a local linear model on a feature space neighbourhood of the data point to be explained and uses its weights as a local explanation for the model classification, SHAP assigns to each feature an importance value using a game theory approach. In the healthcare field, an example of an explainer is MARLENA [25], a model-agnostic solution to explain classifiers that perform multi-label tasks such as multi-morbidity classification or unknown genes functional expressions. Another example is Doctor XAI [24], the XAI algorithm employed in our experiments which we detail in the next paragraph. However, none of these works really takes into consideration end-user needs and domain expertise in the design of their explanations.

Some examples of transparent-by-design models employed in healthcare are the ones presented in [6] and [2]. In [6], the authors use Generalized Additive Models (GAM) with pairwise interactions to predict the probability of 30-days readmission to the hospital and the probability of death from pneumonia. GAM allows the visualization of the relationships between single and pairs of input features with the output, enabling the user to inspect what the model has learned. In [2], the authors develop a case-based interpretable Deep Learning model to classify mass lesions in mammographies. The case-based reasoning, highlighting the classification-relevant parts of the image used to make the decision, makes the model interpretable. However, it is not always possible to build transparent-by-design models for the task at hand. Therefore it might be necessary to extract explanations from black box models. The type of XAI technique that we employ in this paper is post-hoc and model-agnostic. Post-hoc XAI techniques extract explanations from trained models, and model-agnostic ones can extract such explanations from any type of black box model because they do not use any of its internal parameters in the explanation extraction process.

This kind of XAI techniques are agnostic w.r.t. the black box model, however, they are not agnostic w.r.t. the type of input and output data processed by the model. Therefore, they are considered specific for healthcare when they are able to deal with the peculiarities of healthcare data.

Doctor AI and Doctor XAI. In this paper we semantically enrich the explanations of Doctor XAI, which is a post-hoc model-agnostic XAI technique able to deal with multi-label classification tasks and ontology-linked sequential data. Doctor XAI exploits medical ontologies in its explanation extraction process. Once the user selects one data point whose outcome needs an explanation, Doctor XAI first finds a set of semantically close neighbours of that data point from a set of available instances by employing an ontological distance metric. Then, it augments such neighbourhood by *ontologically perturbing* the neighbour's data points, i.e. it masks ontologically similar features and queries the black box on such perturbed data points. Finally, it learns a multi-label decision tree on such an augmented neighbourhood and extracts an explanation from it in the form of a decision rule matching the decision path for that data point on the tree.

Doctor XAI is used to explain the outcomes of Doctor AI [12], a Recurrent Neural Network (RNN) trained on the sequential representation of patients' clinical histories encoded using International Classification of Diseases (ICD) codes. Doctor AI predicts patients' next clinical events, i.e. the set of diseases (represented as ICD codes) that each patient will have in future visits to the hospital.

Therefore, we use it in our experiments as clinical DSS and study how to improve Doctor XAI explanations of Doctor AI predictions to enable clinical reasoning.

Ontologies Use in XAI. Some XAI works already explored how to use ontologies (or knowledge graphs) to improve the explanation process or to tailor explanations to specific user needs or characteristics. Besides Doctor XAI, also the authors of *Trepan Reloaded* [13] use the ontology in the explanation extraction process. In particular, they use ontology to constrain the training of the decision tree acting as a local interpretable model. Closer to our research, other works use ontologies to tailor the explanation to user-specific needs [7, 21]. The authors of [8] use an ontology that encodes all types of explanations to find the most appropriate one for user questions.

In [28] the authors use an ontology to customize the explanation to user needs. However, to the best of our knowledge, ours is the first attempt to enrich explanations of clinical DSS to enable clinical reasoning and the first method that extracts sentences from clinical notes guided by ontology and ICD-9 codes (the ninth revision of ICD). We are aware of the existence of semantic annotation tools like [1], and [19]. However, our method is different, and it does not tag each sentence in the clinical note with a corresponding entity. Our method highlights only the relevant sentences of the note based on the associated ICD-9 codes and the relations extracted from the ontology. This difference did not allow us to compare our method with the already existing tools.

3   Problem Statement

Our aim is to use medical ontologies and external sources of medical knowledge to semantically enrich the explanation provided by state-of-the-art explainability techniques for clinical DSS. In particular, we are interested in augmenting explanations, that consider only the features given as input to the model, with external sources of knowledge in order to present to the end-user the complete clinical picture relevant for a particular algorithmic decision and enabling clinical reasoning. We focus our effort on the post-hoc explanations provided by Doctor XAI [24] and use clinical notes representing the patient's discharge summary as an external source of knowledge. We have already presented Doctor XAI in Sect. 2 and now we provide more details on its explanations. In Fig. 1, we

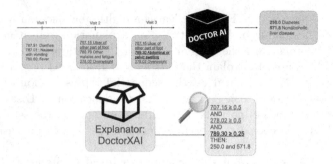

Fig. 1. An example of Doctor XAI Explanation.

show an example of an explanation of a Doctor AI outcome for a patient having three visits. Each visit is represented by a set of ICD-9 codes and the explanation for the multi-label classification provided by Doctor AI is the decision rule depicted in the bottom right. Each conjunction of the rule premise follows the following pattern: ICD_code $\geq$ *threshold_value*. Here, the *threshold_value* is a split value assigned by a decision tree to that ICD code. The internal encoding of Doctor XAI allows giving a temporal interpretation of such value, e.g., *threshold_value* = 0.5 means that the ICD code was present in the last visit. At the top of the image, we have a more readable representation of the explanation. The ICD-9 codes of the patient's clinical history identified as meaningful by Doctor XAI have been coloured to enhance the readability. However, the final user who wants to exploit this explanation has to analyse the description associated with each highlighted code and derive the possible relationships and their meaning. Furthermore, the explanation does not provide any information on the clinical context of the patient. The method we are proposing aims to enrich the Doctor XAI explanation with information derivable from clinical notes associated with each visit and written by nurses and physicians.

Our methodology enriches such an explanation by highlighting the parts of the patient's clinical notes mostly correlated with the ICD-9 codes and uses medical ontologies to identify if, in that clinical note, there are references to clinically relevant information such as the ICD-9 description, the parts of the body affected by the disease, its causes and its effects.

4   Methodology

Our methodology exploits the SNOMED-CT medical ontology [29] to semantically enrich XAI explanations. The SNOMED-CT ontology contains a comprehensive representation of clinical healthcare terminology including diseases, symptoms, signs, diagnoses, medications and procedures. Our methodology first finds all the SNOMED-CT concepts related to each ICD-9 code in the explanation, then it selects some clinically relevant ontological relationships associated with these concepts (more details in Sect. 4.1), and finally uses clinical embeddings to find the parts of the patient's clinical note most related to these relationships and highlights them on the clinical note itself (more details in Sect. 4.2). A bird-view of our methodology is provided in Fig. 2b. In Fig. 2a we show an example of some of the concepts and relations contained in the SNOMED-CT Ontology. In particular, for the concept "Bacterial Pneumonia" we have two different relations, the *Finding Site*, which is "Lung Structure", and the *Due to* which is "Bacteria". Note that all the diseases are also involved in a Parent-Child relation where the parent node represents a more general disease than the child e.g. "Pneumonia" is more general than "Infective Pneumonia".

4.1 SNOMED-CT Relationships Extraction

Each ICD-9 of the explanation has a one-to-many mapping to the concepts in the SNOMED-CT ontology. For example, consider the ICD-9 code 707.15, which stands for "Ulcer of other part of foot". This code is mapped to a set of SNOMED-CT concepts such as "Ulcer of foot", "Ulcer of big toe" and "Diabetic foot". For providing the clinician with the most accurate clinical context related to the decision, we first consider all of these possibilities and for each of them, we extract all the relevant clinical information. We focus on three SNOMED-CT ontological relationships: (a) *Finding Site*, i.e., the body site affected by a condition; (b) *Associated Morphology*, i.e., the morphological changes seen at the tissue or cellular level that are characteristics of a disease; and (c) *Due to*, i.e., the cause of the clinical finding, might be another clinical finding or a procedure.

More formally, we define a function g that given an ICD-9 code cd and the SNOMED-CT ontology O returns the corresponding set of the SNOMED-CT concepts SC, i.e., $g(cd, O) = SC$. Then, starting from the set of concepts SC, our method navigates the SNOMED-CT ontology and derives:

- A set of descriptions $D = \cup_{s \in SC} d_s$, where each d_s is the description associated with the SNOMED-CT concept s;
- A set of finding sites $F = \cup_{s \in SC} F_s$, where $F_s = f_s^1, f_s^2, \ldots, f_s^n$ is the set of finding sites associated with the SNOMED-CT concept s;

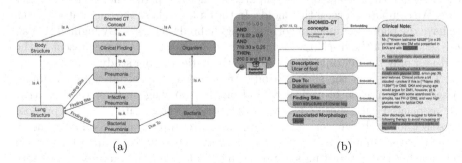

(a)   (b)

Fig. 2. (a) SNOMED-CT Ontology relationships and (b) Bird-view of our methodology.

- A set of associated morphology caused by the disease $M = \cup_{s \in SC} M_s$, where $M_s = m_s^1, m_s^2, \ldots, m_s^k$ is the set of associated morphology associated with the SNOMED-CT concept s;
- A set of causes of the disease $C = \cup_{s \in SC} C_s$, where $C_s = c_s^1, c_s^2, \ldots, c_s^h$ is the set of causes associated to the SNOMED-CT concept s.

We denote by $f \in F$, $m \in M$ and $c \in C$ any of the finding sites, associated morphology and causes extracted from the ontology.

4.2   Information Extraction from Clinical Notes

We exploit biomedical word embeddings to encode the description of each clinically relevant piece of information found in the previous step and find the most similar piece of text in the clinical note associated with the patient. Given an ICD-9 code and a clinical note N, our methodology, by using the function g (defined above), first extracts from the ontology O the set of descriptions D related to the concepts SC in SNOMED-CT, or the corresponding sets of finding sites F, associated morphology M, causes C. Then, for each $d_s \in D$, $f \in F$, $m \in M$ or $c \in C$ To this end, we use a sliding window of length r that generates a set of word sequences W composed of r contiguous words that can be used to represent the note N. We then embed each element of W obtaining the corresponding set of pairs $\langle embedding, sentence \rangle$ denoted by E. We also compute the embedding for each d_s, $f \in F$, $m \in M$, or $c \in C$ and for each of them we identify the most similar embedded sentence E_w corresponding to the pair $\langle E\_w, w \rangle \in E$.

   We use the cosine similarity metric to compute a similarity score between these embeddings and those generated using the sliding window. Given two embeddings A and B, the similarity is computed as follows: $Similarity = \frac{A \cdot B}{||A|| ||B||}$. Thus, we obtain that each element of D, F, M and C is associated with the most similar sentence of the note and a similarity score i.e., we have four score vectors D_{score}, F_{score}, M_{score} and C_{score}.

   To identify the descriptions in D, the finding sites in F, the associated morphology in M and the causes in C referred to in the note N, we select from these sets only the elements with a similarity score higher than a threshold τ. In our experiments, the threshold τ is computed as the 90th percentile of the score vectors. We compared several thresholds. In the end, we chose the one that allows us to have the highest number of correctly highlighted sentences. By highlighting all the sentences of the discharge summary having $T\_scores \geq \tau$, we present to the end-user only the information relevant to the patient under study.

   The length r of the sliding window has a clear impact on the embedding-based representation of each note and on the resulting parts of the text that are associated with specific concept descriptions, finding sites, etc. We propose to select for each type of relationship the more appropriate r value by using a data-driven approach. In particular, it finds the suitable r value for a given relationship type by testing several sliding window length values on a separate set of clinical notes and by selecting the value leading on average to the highest similarity score.

5   Experiments and Results

In this section, we experimentally show the ability of our approach to identify the correct sentences of the patient's clinical notes for explanation enrichment.[1]

We carried out two types of experiments with the help of a human expert. In the first experiment, the human expert manually annotated a set of clinical notes with the ICD-9 descriptions and their relevant ontological relationships. This allowed us to build a ground truth for the automatic extraction of our methodology. In the second experiment, we used our methodology to extract the sentences from another set of clinical notes and then, we asked the human expert to validate whether the identified sentences were correct.

5.1   Dataset

We tested our methodology on the Medical Information Mart for Intensive Care database (Mimic-III) [18]. This dataset contains de-identified data of approximately 40.000 patients collected between 2001 and 2012 in the Beth Israel Deaconess Medical Center data in Boston. Data is stored in 26 different tables; in particular, we used the NoteEvents table which contains all the clinical notes written by nursing and clinicians during a patient's stay in the hospital.

Note Cleaning. We applied a pre-processing to the clinical notes to clean them and reduce noise: we have lower-cased the text; we have removed numbers; we have substituted odd characters with space; we have removed stopwords; we have removed the punctuation; and we have replaced the contractions in the text with an extended form using a dictionary of possible contractions.

5.2   Implementation Details

We trained Doctor AI for 50 epochs, splitting MIMIC-III using 70% of its patients as a training set, 15% as a validation set, and 15% as a test set. We then used Doctor XAI as detailed in the original paper. To navigate the ontology, we used a Python Library called PyMedTermino [20]. For the embedding of the clinical notes' sentences, we used three different methods:

- *BioWordVec* [31], a pre-trained word embedding for biomedical natural language processing trained on PubMed and Mimic-III;
- *ClinicalBert* [17], a Bert based embedding trained on Mimic-III;
- and *BioSentVec* [11], a biomedical sentence embedding with sent2vec trained on Mimic-III and PubMed.

---

[1] Code available at: https://github.com/lucacorbucci/Semantic-Enrichment. Hardware used: NVIDIA Quadro RTX 6000 GPU, Intel(R) Core(TM) i9-10980XE CPU @ 3.00 GHz, 256 gb of RAM.

BioSentVec and BioWordVec are based on Word2Vec, the embeddings are context-independent and we can use them without the model that generated them because we just have $<key, value>$ pairs where the keys are the words and the values are the embeddings. ClinicalBert is based on Bert, the embeddings are generated considering the context of a word, this means that we have to give a sentence as input to the model and it will return the embedding. This is computationally more expensive than the Word2Vec model.

Before applying the embedding we identified the suitable length value r of the sliding window for each type of relationship: $r = 7$ for the *Finding Site* and *Associates Morphology* relationships, $r = 9$ for *Due to* relationship and $r = 10$ for the *Description*. To this end, we tested values in the range between 3 and 30 using a subset of 500 clinical notes contained in the original dataset.

5.3 Human Validated Experiment

Clinical Notes Manual Annotation. The domain expert took into account each ICD-9 code associated with the clinical notes and highlighted by Doctor XAI. The notes were manually annotated highlighting the most similar sentences to the following information: *i)* Code description; *ii)* Cause of the disease associated with the code; *iii)* Finding site of the disease associated with the code; and *iv)* Associated morphology of the disease associated with the code. In particular, we considered the clinical histories of nine different patients, involving a total of 32 clinical notes. These patients have been diagnosed with ICD-9 code 250.00 i.e. diabetes, 584.9 i.e. Acute kidney failure, 428.0 i.e. Congestive heart failure and 401.9 i.e. Unspecified essential hypertension, among other diseases. Once the domain expert annotated the notes, we tested our method to compare the extracted sentences with the manually annotated ones. In Table 1, we report *Accuracy, F1-Score, Precision* and *Recall* for each ontological relationship and the corresponding confidence interval at $1 - \alpha = 0.95$ confidence level. The results are divided according to the type of relationship and in bold we highlight the best performance. The confidence intervals for all the metric values are tight meaning that the performances of the methods are reliable. To evaluate these metrics, we have defined:

- *True Positive* as the number of sentences that were manually annotated in the clinical notes and are correctly annotated by our method;
- *False Negative* as the number of sentences in the clinical notes that our method does not annotate because they have a similarity score lower than the input threshold and that were manually annotated by the domain expert.
- *False Positive* as the number of sentences in the clinical notes that our method annotates and that do not have a corresponding manual annotation.
- *True Negative* as the number of sentences in the clinical notes that our method does not annotate because they have a similarity score lower than the input threshold and that do not have a corresponding manual annotation.

Table 1 shows that BioWordVec presents the best performance across all relationships, for this reason, we chose to employ it in our methodology. Furthermore,

Table 1. Validation on 32 manually annotated clinical notes of 9 patients. Confidence of Accuracy, Precision, Recall and F1-score at $1 - \alpha = 0.95$ of confidence level.

Relationship	Embedding			Accuracy	F1-score	Precision	Recall
Description	BioWordVec	Value		**0.718**	**0.707**	**0.819**	**0.622**
		Confidence		0.715–0.719	0.704–0.708	0.815–0.819	0.619–0.624
Description	BioSentVec	Value		0.662	0.664	0.804	0.566
		Confidence		0.659–0.663	0.661–0.665	0.800–0.804	0.563–0.567
Description	ClinicalBert	Value		0.640	0.602	0.654	0.557
		Confidence		0.637–0.641	0.599–0.603	0.651–0.655	0.555–0.559
Finding site	BioWordVec	Value		**0.743**	0.274	0.170	**0.708**
		Confidence		0.740–0.744	0.273–0.277	0.169–0.173	0.705–0.709
Finding site	BioSentVec	Value		0.726	**0.294**	**0.200**	0.555
		Confidence		0.723–0.727	0.293–0.297	0.199–0.203	0.553–0.557
Finding site	ClinicalBert	Value		0.686	0.214	0.150	0.375
		Confidence		0.683–0.687	0.213–0.217	0.149–0.153	0.373–0.377
Due to	BioWordVec	Value		**0.666**	**0.451**	**0.350**	**0.636**
		Confidence		0.647–0.673	0.440–0.466	0.342–0.368	0.618–0.644
Due to	BioSentVec	Value		0.600	0.091	0.050	0.500
		Confidence		0.582–0.609	0.091–0.119	0.050–0.080	0.486–0.513
Due to	ClinicalBert	Value		0.568	0.214	0.150	0.375
		Confidence		0.552–0.579	0.211–0.238	0.149–0.176	0.366–0.392
Associated morphology	BioWordVec	Value		**0.856**	**0.577**	**0.464**	**0.764**
		Confidence		0.845–0.856	0.571–0.581	0.459–0.470	0.755–0.766
Associated morphology	BioSentVec	Value		0.803	0.409	0.321	0.562
		Confidence		0.793–0.803	0.405–0.415	0.318–0.329	0.556–0.566
Associated morphology	ClinicalBert	Value		0.734	0.339	0.321	0.360
		Confidence		0.726–0.736	0.336–0.347	0.318–0.329	0.356–0.367

BioWordVec computational runtime was an order of minutes shorter if compared with ClinicalBert, as previously observed in [19]. Our experiment pointed out that the *Description* is the easiest relation to search for and in most of the cases, our methodology is able to extract the same sentence highlighted during the manual annotation phase. On the contrary, it is not easy to deal with *Finding Site* and *Due to*. As explained by our domain expert usually this information is often underlined by the clinicians and is not explicitly written in the notes. This sometimes led to the extraction of wrong sentences.

A Kruskal-Wallis test was used to determine whether or not there are statistically significant differences between the medians of accuracy, precision, recall and F1-score of the different embedding methods reported in Table 1 (BioWordVec, BioSentVec, and ClinicalBert). The Kruskal-Wallis test is the non-parametric test considered equivalent to the One-Way ANOVA and, given the low number of observations that we are comparing, it is the best fitting for our setting. The Kruskal-Wallis test uses the following null and alternative hypotheses: H_0: "The median is equal across all embedding methods", H_1: "The median is not equal across all embedding methods". For the *accuracy* we obtained that the Kruskal-Wallis Statistics is 2.192 with a p-value of 0.334 (> 0.05) meaning no statistically significant difference among the accuracy medians, so the H_0 hypothesis cannot be rejected. For the *recall* we obtained that the Kruskal-Wallis statistic is 8.800

Table 2. Kruskal-Wallis test for the recall: results.

	BioWordVec	BioSentVec	ClinicalBert
BioWordVec	stat: 1.000 p: 0.000	stat: 3.036 p: 0.0814	stat: 5.398 p: **0.0202**
BioSentVec		stat: 1.000 p: 0.000	stat: 5.333 p: **0.0209**
ClinicalBert			stat: 1.000 p: 0.000

Table 3. Human validation of the extracted sentences using a 90th percentile threshold.

Relationship Embedding	Description		Finding site		Associated Morphology		Due to	
	valid	non-valid	valid	non-valid	valid	non-valid	valid	non-valid
BioWordVec	75 (75%)	25 (25%)	**65 (65%)**	35 (35%)	**21 (75%)**	7 (25%)	**13 (65%)**	7 (35%)
BioSentVec	**77 (77%)**	23 (23%)	56 (56%)	44 (44%)	**21 (75%)**	7 (25%)	**13 (65%)**	7 (35%)
ClinicalBert	67 (67%)	33 (33%)	32 (32%)	68 (68%)	18 (64%)	10 (36%)	9 (45%)	11 (55%)

with a p-value of 0.012 meaning a statistically significant difference among the recall medians, so the H_0 hypothesis has been rejected. We performed both a Kruskal-Wallis test and a Mann Whitney test on the pairs to verify which are the pairs with a significant difference. In Table 2 we report the results of the pairwise comparisons of the Kruskal-Wallis test (equal results were obtained with the Mann Whitney test).

Looking at Table 2, it is interesting to note how the pairwise comparisons between BioWordVec *vs* ClinicalBert and BioSentVec *vs* ClinicalBert give statistically significant differences between the pairs (always lower for ClinicalBert). For the *F1-score* (Kruskal-Wallis Statistics 1.505 with a p-value of 0.471) and the *precision* (Kruskal-Wallis Statistics 1.462 with a p-value of 0.481) we found no significant difference among at least one of the medians (both for F1-score and precision), so the H_0 hypotheses have to be accepted in both cases. Thus, to summarize the three embedding methods present statistically significant differences only for the recall: BioWordVec and BioSentVec perform statistically better than ClinicalBert, reinforcing the choice of one of these two embeddings.

Classification of Extracted Sentences. We made a second experiment exploiting the knowledge of the domain expert. We selected almost 100 notes classified with the ICD-9 250.00 (diabetes), 584.9 (Acute kidney failure), 428.0 (Congestive heart failure) and 401.9 (Unspecified essential hypertension).

Then, we ran our method on each note to highlight the most similar sentences to the *Description, Finding Site, Associated Morphology* and *Due To* relation associated with all the ICD-9 with which the note is associated. We used the previously mentioned method and the three different embeddings to compute the similarity. After extracting the sentences with our method, the domain expert analysed each sentence evaluating the correlation with the relation with which similarity was calculated and if the highlighted sentence provided helpful information about the patient's clinical history. Each of the extracted sentences was classified as *"Valid Sentence"* or *"Non-Valid Sentence"*. In Table 3 we report the result of this experiment with a similarity threshold of the 90th percentile. The results show that the performances using embeddings BioWordVec and BioSentVec are very similar while those using ClinicalBert are slightly worse.

6 Conclusion and Future Work

We presented a methodology to semantically enrich the explanation of an XAI technique in the healthcare context by exploiting SNOMED-CT ontology and clinical notes. In particular, it highlights the relevant clinical information related to one algorithmic decision directly on the patient's clinical note. Thanks to the domain expert, we were able to annotate a small part of the dataset and to have a preliminary "human validation" of our methodology. The presence of a "human validator" was crucial in our methodology. Unfortunately, we have not found any pre-annotated dataset that could fit our needs and that could be used as a ground truth. The "human-validated" experiment showed promising results concerning the identification of sentences related to the description of the disease and the associated morphology while selecting the correct finding site and cause of the disease is more challenging. We studied many different approaches to extract the information, and we compared different embeddings to have a better representation of our notes. In terms of embeddings, we compared the performances achieved with BioWordVec, BioSentVec and ClinicalBert, and we concluded that, for the same performance, BioWordVec performs slightly better in general and it is faster in computing embeddings. A limitation of an approach that involves the use of pre-trained embeddings is that we would not be able to generalise this task with the same performances when using a completely different medical dataset. In that context, an embedding like ClinicalBert would probably perform better. However, it would have a high computational cost to the embedding computation.

In the future, we would like to validate our method on a larger quantity of clinical notes and exploit our methodology to generate explanations expressed by natural language.

In addition, we would like to test the methodology to understand if the semantically enriched explanation could improve the interpretability of the explanation. Lastly, we plan to investigate the opportunity to exploit our methodology to generate explanations expressed by natural language.

Acknowledgments. This work is supported by: the EU NextGenerationEU programme under the funding schemes PNRR-PE-AI FAIR (Future Artificial Intelligence Research); the EU - Horizon 2020 Program under the scheme "INFRAIA-01-2018-2019 - Integrating Activities for Advanced Communities" (G.A. n.871042) "SoBig-Data++: European Integrated Infrastructure for Social Mining and Big Data Analytics" (http://www.sobigdata.eu); PNRR-"SoBigData.it - Strengthening the Italian RI for Social Mining and Big Data Analytics" - Prot. IR0000013; TAILOR (G.A. 952215), HumanE-AI-Net (G.A. 952026) and EU H2020 project XAI (Grant Id 834756).

References

1. Aronson, A.R.: Effective mapping of biomedical text to the UMLS metathesaurus: the MetaMap program. In: Proceedings AMIA Symposium, pp. 17–21 (2001). ISSN 1531-605X. eprint 11825149. https://pubmed.ncbi.nlm.nih.gov/11825149
2. Barnett, A.J., et al.: IAIA-BL: a case-based interpretable deep learning model for classification of mass lesions in digital mammography. arXiv preprint arXiv:2103.12308 (2021)
3. Bhatt, U., et al.: Explainable machine learning in deployment. In: Proceedings of the 2020 Conference on Fairness, Accountability, and Transparency, pp. 648–657 (2020)
4. Bodria, F., et al.: Benchmarking and survey of explanation methods for black box models. CoRR abs/2102.13076 (2021)
5. Boominathan, S., et al.: Treatment policy learning in multiobjective settings with fully observed outcomes. In: ACM SIGKDD 2020, pp. 1937–1947 (2020)
6. Caruana, R., et al.: Intelligible models for healthcare: Predicting pneumonia risk and hospital 30-day readmission. In: Proceedings of the 21th ACM SIGKDD International Conference on Knowledge Discovery and Data Mining, pp. 1721–1730 (2015)
7. Celino, I.: Who is this explanation for? Human intelligence and knowledge graphs for eXplainable AI. arXiv preprint arXiv:2005.13275 (2020)
8. Chari, S., Seneviratne, O., Gruen, D.M., Foreman, M.A., Das, A.K., McGuinness, D.L.: Explanation ontology: a model of explanations for user-centered AI. In: Pan, J.Z., et al. (eds.) ISWC 2020. LNCS, vol. 12507, pp. 228–243. Springer, Cham (2020). https://doi.org/10.1007/978-3-030-62466-8_15
9. Cheerla, A., et al.: Deep learning with multimodal representation for pancancer prognosis prediction. Bioinformatics $35(14)$, i446–i454 (2019)
10. Chekroud, A., et al.: The perilous path from publication to practice. Mol. Psychiatry $23(1)$, 24–25 (2018)
11. Chen, Q., et al.: BioSentVec: creating sentence embeddings for biomedical texts. In: 2019 IEEE IICHI (2019)
12. Choi, E., et al.: Doctor AI: predicting clinical events via recurrent neural networks. In: Machine learning for healthcare conference. PMLR (2016)
13. Confalonieri, R., et al.: Trepan reloaded: a knowledge-driven approach to explaining artificial neural networks (2019)
14. EU General Data Protection Regulation. European Commission (2018). https://ec.europa.eu/commission/sites/beta-political/files/data-protection-factsheet-changes_en.pdf. Accessed 17 June 2019
15. European Parliament. Proposal for a Regulation of the European Parliament and of the Council laying down harmonised rules on Artificial Intelligence (Artificial Antelligence Act) and amending certain union legislative acts (2021). https://eur-lex.europa.eu/legal-content/EN/TXT/?qid=1623335154975&uri=CELEX%3A52021PC0206. 11 June 2021
16. Guidotti, R., et al.: A survey of methods for explaining black box models. ACM Comput. Surv. (CSUR) $51(5)$, 1–42 (2018)
17. Huang, K., et al.: ClinicalBERT: modeling clinical notes and predicting hospital readmission. arXiv:1904.05342 (2019)

18. Johnson, A.E., et al.: MIMIC-III, a freely accessible critical care database. Sci. Data **3**, 160035 (2016)
19. Kraljevic, Z., et al.: Multi-domain clinical natural language processing with Med-CAT: the medical concept annotation toolkit (2020)
20. Lamy, J.-B., et al.: PyMedTermino: an open-source generic API for advanced terminology services. Stud. Health Technol. Inform. **210** (2015)
21. Longo, L., Goebel, R., Lecue, F., Kieseberg, P., Holzinger, A.: Explainable artificial intelligence: concepts, applications, research challenges and visions. In: Holzinger, A., Kieseberg, P., Tjoa, A.M., Weippl, E. (eds.) CD-MAKE 2020. LNCS, vol. 12279, pp. 1–16. Springer, Cham (2020). https://doi.org/10.1007/978-3-030-57321-8_1
22. Lundberg, S.M., et al.: A unified approach to interpreting model predictions. In: Proceedings of the 31st International Conference on Neural Information Processing Systems, pp. 4768–4777 (2017)
23. Ma, X., et al.: Understanding adversarial attacks on deep learning based medical image analysis systems. Pattern Recognit. **110**, 107332 (2021)
24. Panigutti, C., et al.: Doctor XAI: an ontology-based approach to black-box sequential data classification explanations. In: ACM FAccT (2020)
25. Panigutti, C., Guidotti, R., Monreale, A., Pedreschi, D.: Explaining multi-label black-box classifiers for health applications. In: Shaban-Nejad, A., Michalowski, M. (eds.) W3PHAI 2019. SCI, vol. 843, pp. 97–110. Springer, Cham (2020). https://doi.org/10.1007/978-3-030-24409-5_9
26. Panigutti, C., et al.: FairLens: auditing black-box clinical decision support systems. Inf. Process. Manage. **58**(5), 102657 (2021)
27. Ribeiro, M.T., et al.: "Why should i trust you?" Explaining the predictions of any classifier. In: Proceedings of the 22nd ACM SIGKDD International Conference on Knowledge Discovery and Data Mining, pp. 1135–1144 (2016)
28. Rožanec, J.M., et al.: Semantic XAI for contextualized demand forecasting explanations. arXiv preprint arXiv:2104.00452 (2021)
29. U. T. Services. SNOMED CT International Edition. https://www.nlm.nih.gov/healthit/snomedct/international.html
30. Signoroni, A., et al.: BS-net: learning COVID-19 pneumonia severity on a large chest X-ray dataset. Med. Image Anal. **71**, 102046 (2021)
31. Zhang, Y., et al.: BioWordVec: improving biomedical word embeddings with subword information and MeSH. Sci. Data **6**, 52 (2019)

Text to Time Series Representations: Towards Interpretable Predictive Models

Mattia Poggioli[1], Francesco Spinnato[2,3]($\boxtimes$) (iD), and Riccardo Guidotti[1,3] (iD)

[1] University of Pisa, Pisa, Italy
{mattia.poggioli,riccardo.guidotti}@unipi.it
[2] Scuola Normale Superiore, Pisa, Italy
francesco.spinnato@sns.it
[3] ISTI-CNR, Pisa, Italy
francesco.spinnato@isti.cnr.it

Abstract. Time Series Analysis (TSA) and Natural Language Processing (NLP) are two domains of research that have seen a surge of interest in recent years. NLP focuses mainly on enabling computers to manipulate and generate human language, whereas TSA identifies patterns or components in time-dependent data. Given their different purposes, there has been limited exploration of combining them. In this study, we present an approach to convert text into time series to exploit TSA for exploring text properties and to make NLP approaches interpretable for humans. We formalize our Text to Time Series framework as a feature extraction and aggregation process, proposing a set of different conversion alternatives for each step. We experiment with our approach on several textual datasets, showing the conversion approach's performance and applying it to the field of interpretable time series classification.

Keywords: Time Series Classification · Interpretable Machine Learning · Natural Language Processing · Explainable AI

1 Introduction

In recent years, both Time Series Analysis (TSA) and Natural Language Processing (NLP) have seen a surge in popularity [2,8,14]. NLP has found numerous applications, including machine translation, email spam detection, information extraction and summarization, and question-answering [14]. Meanwhile, the development of time series classifiers [2] and the increasing availability of time-dependent data such as electrocardiogram records, motion sensor data, climate measurements, and stock indices [8] have fueled interest in TSA. Despite the individual growth of NLP and TSA, there has been limited exploration into combining these two fields, which usually have different goals. NLP focuses mainly on enabling machines to manipulate and generate human language, whereas TSA identifies local patterns or components in time-dependent data. However, they also share similarities since the text, from a human perspective, remains a

A. Bifet et al. (Eds.): DS 2023, LNAI 14276, pp. 230–245, 2023.
https://doi.org/10.1007/978-3-031-45275-8_16

sequence of spoken or written expressions rather than a comprehensive machine-readable representation. This work explores the benefits of integrating TSA into NLP to make it more interpretable for humans.

NLP relies on text representation techniques to convert text into machine-readable input for classification, clustering, and sentiment analysis [31]. Typically, text encoding involves transforming text into vectors representing its content. Traditional approaches build an explicit representation of the distributional properties of the text, using raw frequencies like bag-of-words [9], or exploiting tf-idf [12], and n-grams [28]. State-of-the-art techniques are based on transformers [26], which can better capture semantic and syntactic relationships between words and sentences. The vectorial representations generated by these models, known as embeddings, are denser and lower-dimensional than previous models, even if not interpretable [22]. On the other hand, TSA techniques can capture the temporal evolution of sequences of data points measured at regular intervals [6]. By considering temporal dependencies, TSA can be used for various purposes, such as descriptive analysis, clustering, classification, and forecasting [16].

In this paper, we investigate the impact of converting text observations into time series observations to solve interpretable text classification through time series representations. In particular, we exploit interpretable models originally developed for time series [25] as interpretable text classifiers. In the literature, state-of-the-art time series classifiers are mainly black-box models [2], not interpretable from a human standpoint. We instead focus on time series classification through shapelets [30], i.e., subsequences that allow for interpretable predictions based on local similarities in shape. Hence, we propose using shapelets in NLP by turning texts into time series. To perform this transformation, we design and implement TOTS, a framework to turn Text tO Time Series. TOTS exploits a range of alternatives for converting a text into multivariate time series, including sentence embeddings [17,20], sentiment scores [11], and linguistic features [5,13,19]. In this way, we can leverage implicit representations of language and concrete linguistics variables to represent language vectorially. Then, TOTS adopts aggregation techniques to compress multivariate time series into univariate ones. By compressing time series, we can identify shapelets on 1-d signals that are easier to analyze and interpret than multivariate ones, shedding light into many domains, such as sentiment analysis over time, event detection, social media trends, etc.

Overall, this work contributes to the fields of TSA and NLP by *(i)* proposing a formalization of text to time series conversion, *(ii)* exploring dimensionality reduction and aggregation techniques that can effectively convert multivariate time series into univariate, *(iii)* testing different text to time series conversions through a novel evaluation metric, and *(iv)*, showing the effectiveness of the approach in the field of interpretable text/time series classification. The rest of this work is structured as follows. Section 2 discusses related works at the intersection between TSA and NLP. Section 3 introduces notions useful for describing the proposed transformations, detailed in Sect. 4. Section 5 presents the experimental results, and Sect. 6 concludes the paper.

2   Related Works

A proper intersection between TSA and NLP lies in analyzing texts produced within a time window. Works in this domain build time series by extracting features from each text and condensing them within each timestep to represent a time-dependent phenomenon. In [10], the authors constructed a sentiment scoring rule from the count of positive and negative words in multiple social media texts, resulting in an event-driven, irregularly spaced time series. In [1], the authors combined text mining and time series to analyze sequences of dated documents, such as news articles, and extracted correlations and patterns among frequently used words. Differently from the described approaches, we analyze documents individually and introduce time by splitting each text content.

Other works are focused on the relationship between features (such as market sentiment) extracted from Twitter data and financial trends. In [21], the authors analyzed correlations between stock-market events and features extracted from micro-blogging messages, relying on overall activity measures (e.g., number of posts, re-posts) and graph-related indices (e.g., number of connected components, degree distribution). In [27], the authors used market sentiment and text mining techniques for financial time series, proposing a hybrid model that combines the conventional ARIMA model with a support vector regressor method to extract valuable insights from the market sentiment. Similarly, in [18], it was proposed ST-GAN, combining financial news texts and numerical data to predict stock trends. In [4], the authors used a flexible multiple-output Gaussian process to analyze multimodal statistical causality between cryptocurrency market sentiment and price processes, proposing an NLP framework for interpretable sentiment indices as inputs for time-series models. Differently from these approaches, we convert each text into individual time series representations, moving away from the financial domain and focusing on classification rather than forecasting.

To the best of our knowledge, the only approach for mapping text to time series is T^3 [29]. T^3 uses combinations of granularity, n-grams, and different space-filling curves to assign appropriate numeric values to each character. When applied to the "record linkage" problem, T^3 achieved good accuracy with considerable speed-ups. Our study goes beyond this work by focusing on mapping text at the sentence level, allowing for incorporating multiple types of features, including advanced models like sentence embeddings.

3   Setting the Stage

In order to keep our paper self-contained, we report in this section a brief overview of concepts necessary to comprehend our proposal. We define a text corpus and each of its components, i.e., documents and sentences, as follows:

Definition 1. A *corpus* is a structured set of textual documents, represented as a collection $\mathcal{T} = \{T_1, T_2, ..., T_n\}$ where $\mathcal{T}$ is the corpus and T_i an individual document within the corpus. A *document* T is a sequence of sentences, where each sentence is denoted by S_j, and the entire text is represented as $T = \{S_1, S_2, ..., S_m\}$, where m is the total number of sentences in the text. Finally, a *sentence* S is an ordered sequence of words.

For example, a corpus might be a set of film or theater scripts, a document might be a scene or an opera, and a sentence might be an actor's line. Also, we can consider a set of songs as a corpus, a song as a document, and a verse of the song as a sentence. We can now define the Text Classification problem as follows:

Definition 2. Given a corpus T with a vector of finite integer labels (or classes) assigned to each document $\mathbf{y} \in \mathbb{N}^n$, the *Text Classification Problem* is the task of training a function f from the space of possible inputs T to a probability distribution over the class values in $\mathbf{y}$.

In the following, we establish a connection between text and time series, defining them in similar and coherent ways.

Definition 3. A *time series dataset* $\mathcal{X} = \{X_1, \ldots, X_n\} \in \mathbb{R}^{n \times d \times m}$ is a collection of n time series. A *time series* $X = \{\mathbf{x}_1, \ldots, \mathbf{x}_d\} \in \mathbb{R}^{d \times m}$ is a set of d signals. A *signal*, or dimension, $\mathbf{x} = \{x_1, \ldots, x_m\} \in \mathbb{R}^m$ is a sequence of m real-valued observations sampled at equal time intervals. When $d = 1$, a time series is *univariate*, while if $d > 1$, the time series is *multivariate*.

Consequently, we define the Time Series Classification problem as follows:

Definition 4. Given a time series dataset $\mathcal{X}$ with a vector of finite integer labels $\mathbf{y} \in \mathbb{N}^n$, *Time Series Classification* is the task of training a function f from the space of possible inputs $\mathcal{X}$ to a probability distribution over the class values in $\mathbf{y}$.

Given the formulations above, a parallel can be drawn between a text corpus T and a time series dataset $\mathcal{X}$, and a document T and a time series X. Consequently, the only difference between the two problems is in the type of dataset used, i.e., T vs $\mathcal{X}$. By exploiting the parallelism between time series and text, our intuition is that we can solve the Text Classification Problem through TSA approaches. Our idea is to exploit interpretable machine learning methods on time series [25] to build algorithms able to identify the most discriminative subsequences of a time series and project them back into the original text. This would allow us to perform text classification in a human-understandable way.

We focus on interpretable classification through shapelets, i.e., time series subsequences representing a particular class within a dataset [30]. A subsequence is an ordered and contiguous subpart of a signal, formally:

Definition 5. Given a signal $\mathbf{x}$, a *subsequence* $\mathbf{s} = \{x_j, \ldots, x_{j+l-1}\}$ of length l is an ordered sequence of values such that $1 \leq j \leq m - l + 1$.

To extract shapelets from a dataset, candidate shapelets are generated, and their distances to the time series in the dataset are calculated. Then, their quality is assessed based on how well they separate different classes, and the best shapelets are selected based on their quality scores. After that, each time series is represented as a feature vector, where each feature corresponds to the distance between the time series and one of the shapelets [7]. Formally:

Definition 6. Given a time series dataset $\mathcal{X} \in \mathbb{R}^{n \times d \times m}$, a shapelet discovery function, $shp\_discovery$, extracts a set Q of q discriminative shapelets, i.e., $shp\_discovery(\mathcal{X}) = Q \in \mathbb{R}^{q \times l}$. Then, a transform function, $shp\_transform$, converts $\mathcal{X}$ into a real-valued tabular dataset, D, obtained by taking the minimum Euclidean distance between each time series in $\mathcal{X}$, and each shapelet in Q, via a sliding-window, i.e., $shp\_transform(\mathcal{X}, Q) = D \in \mathbb{R}^{n \times q}$.

Once the time series dataset is converted through the shapelet transform, an interpretable classifier such as a Decision Tree [3] can be used, having the advantage of an interpretable feature representation. Given these notions, we can now easily link the concept of a time series shapelet to that of a *subdocument* by defining it as an ordered and contiguous subpart of a document, formally:

Definition 7. Given a document T, a *subdocument* $P = \{S_j, \ldots, S_{j+l-1}\}$ is an ordered sequence of l sentences, such that $1 \leq j \leq m - l + 1$.

Therefore, by finding important subsequences in a time series, i.e., shapelets, we can find the most discriminative parts in a corresponding text. Consequently, the real challenge we face in this paper consists in converting a text into a time series. Our proposal to accomplish this task, and to allow solving interpretable text classification through time series classification, is illustrated in the next section.

4   Text to Time Series Conversion

In this section, we describe TOTS, a framework to turn Text tO Time Series. The TOTS framework is a text to time series conversion workflow formed by three core steps: tokenization, feature extraction, and aggregation. We regard TOTS as a framework because every step can be implemented differently. In this work, we defined its main steps and realized only some possible variants. However, the TOTS structure leaves space to integrate various alternatives easily.

A summary of TOTS is illustrated in Algorithm 1. Given a text corpus $\mathcal{T}$, TOTS returns a time series dataset $\mathcal{X}$ where the ith time series $X \in \mathcal{X}$ is the time series representation of the corresponding i^{th} document $T \in \mathcal{T}$. First, TOTS initializes the empty time series dataset $\mathcal{X}$ (line 1). Then, for each document $T \in \mathcal{T}$, it runs the conversion of T into X and adds it to $\mathcal{X}$ (lines 2–10). The first step of TOTS on T is tokenization, in which the document is split into sentences and readied for further analysis (line 4). After that, for each sentence S in the tokenized document T' (lines 5–7), TOTS extracts characteristic features describing S through the features extraction function $feat\_extr$ and places them into a vector $\mathbf{v} \in \mathbb{R}^d$ where d is the number of features. Supposing a document T' formed by m sentences, the sequence of m vectors $\mathbf{v}$ is concatenated into the matrix $X \in \mathbb{R}^{m \times d}$. This matrix $X \in \mathbb{R}^{m \times d}$ can be viewed as a multivariate time series. Thus, we can transpose X in order to have a proper multivariate time series $X' \in \mathbb{R}^{d \times m}$ where different rows model different signals, i.e., features in this case, and different columns capture different timesteps (line 8). The various signals of the multivariate time series are aggregated into a univariate one

---

Algorithm 1: TOTS($\mathcal{T}$, *tokenize*, *feat_extr*, *aggregate*)

Input : $\mathcal{T}$ - text corpus, *tokenize* - splitting function,
  feat_extr - feature extraction function, *aggregate* - aggregation function
Output: $\mathcal{X}$ - time series dataset

1	$\mathcal{X} \leftarrow \emptyset$;	// init. time series dataset
2	**for** $T \in \mathcal{T}$ **do**	// for each document
3	$\quad X \leftarrow \emptyset$;	// init. time series
4	$\quad T' \leftarrow tokenize(T)$;	// tokenize document
5	$\quad$ **for** $S \in T'$ **do**	// for each sentence
6	$\quad\quad \mathbf{v} \leftarrow feat\_extr(S)$;	// extract feature vector
7	$\quad\quad X \leftarrow X \cup \{\mathbf{v}\}$	// store feature vector
8	$\quad X' \leftarrow X^\mathsf{T}$;	// transpose feature vector matrix
9	$\quad X' \leftarrow aggregate(X')$;	// aggregate multivariate time series
10	$\quad \mathcal{X} \leftarrow \mathcal{X} \cup \{X'\}$;	// store time series
11	**return** $\mathcal{X}$	

---

through the aggregation function *aggregate* (line 9). The function *aggregate* has no effect on X' when the time series is already univariate, i.e., $d = 1$.

Once a given text corpus $\mathcal{T}$ is converted into a time series dataset $\mathcal{X}$ through TOTS, we can run any TSA approach exploiting the advance of a clear correspondence between texts and time series. In particular, we can use an interpretable shapelet-based time series classifier. In the remainder of this section, we illustrate some alternatives to implement the three functions used by TOTS.

Tokenization. The first step in our approach consists in defining the granularity of the final time series by splitting the original text. Tokenization involves breaking up a given text into units, called *tokens*, that can be individual words, phrases, or whole sentences [13]. In TOTS, we tokenize at the sentence level.

Definition 8 (Sentence Tokenization). Given a text document, T, a tokenization function, *tokenize*, splits the document into tokens, creating a set of m sentences $T' = \{S_1, S_2, ..., S_m\} = tokenize(T)$.

Here, we use the term "sentence" loosely, i.e., not as a sequence of words ending with a punctuation mark but as a grammatically complete sequence expressing a full thought. Text splitting is a crucial step that may vary depending on the nature of the text and the specific problem. For example, in a dialogue, a timestep may correspond to a speaker's turn, while in a book, it may correspond to a whole paragraph. If the focus is on song lyrics, line splitting is instead the most sensible option. We use a real 66-line-long rap lyric from the Song Lyrics dataset as a running example to illustrate the various step of the proposed framework (see Sect. 5 for further details). The newline character (/) is adopted as the splitting criterion for this example. The following are the first six lines, i.e., $T_{0:5}$:

$T_{0:5}$ Say brah / In this game called life / It's charces , decisions, and consequences / I decided to change my life, for the better / So anybody that's out there seeking conviction / because of profanity in my music /

Feature Extraction. The second step consists in extracting features from each token, i.e., each sentence in our setting. We present here alternatives to extract features from a document and to implement the *feat_extr* function defined as:

Definition 9 (Feature Extraction). The feature extraction function *feat_extr* takes as input a sentence S, and returns a vector **v** containing d characteristics of S, i.e. $feat\_extr(S) = \{v_1, \ldots, v_d\} = \mathbf{v} \in \mathbb{R}^d$.

There exist many different feature extraction approaches in NLP. We design and implement three alternatives: one based on linguistic features, one based on sentence embeddings, and an approach relying on sentiment/emotions.

Linguistic Features. Computational linguistics offers several methods for extracting meaningful *linguistic features* within a text [5,13,19]. Features such as type-token ratio can be extracted through tokenization [13], measuring the lexical diversity of a text by calculating the ratio of unique words (types) to the total number of words (tokens) in a text. Readability scores, such as the Flesch-Kincaid and Dale-Chall formulas [5], can also be viewed as features assessing the complexity of a text. Part-of-speech tagging, such as the Universal POS tagsets [19], can instead be used to identify the grammatical category of each word in a text. These linguistic features provide valuable information about the structure and complexity of a text and can be used in conjunction with other features to improve NLP tasks. In TOTS, we define the function *feat_extr* to extract the following features **v** from a given sentece S: sentence_length (`snl`), monosyl_words_count (`mwc`), polysyl_words_count (`pwc`), avg_token_length (`atl`), readability_score (`rs`), normalized_sentence_freq (`nsf`), sentence_ttr (`st`), avg_token_freq (`atf`), alliteration_score (`as`), verb_count (`vc`), noun_count (`nc`), adj_count (`adj`), adv_count (`adv`), intj_count (`ic`). With linguistic feature extraction, a dynamic representation of text characteristics emerges as a multivariate time series. This provides insights into the changing grammatical and phonological qualities from sentence to sentence. Figure 1 (left) shows the linguistic feature-based conversion for the rap text above (all features are normalized). We notice that `nsf` and `atf` can recognize repeating patterns in the text, identifying the three changes between chorus and verse (lines 12:15, 34:36, 56:59). Further, monosyllabic words (`mwc`) have a generally low frequency in the text, except for the first few sentences. The only constant feature throughout the text is the number of interjections (`ic`).

Sentence Embeddings. Sentence embeddings are high-dimensional vectors that encode the semantic meaning of a sentence into a space where similar sentences are spatially closer [20]. Several NLP models have been developed to output embeddings. In TOTS, we implement the function *feat_extr* through Sentence-BERT (SBERT) [20] and Doc2Vec [17], in order to extract the embedding **v** from a given sentence S. Doc2Vec takes a document as input and outputs embeddings capturing context, while Sentence-BERT uses Siamese and Triplet networks to derive semantically meaningful sentence embeddings that can be compared using cosine-similarity. The sentence embeddings S of these models are "static" vectorial representations, **v**, of sentences. However, considering the sequence of embeddings X', we can capture the relationship between subsequent sentences. Figure 1 (center) shows an example of a 100-dimensional embedding vector of Doc2Vec for each input sentence. Sentence embeddings are not directly

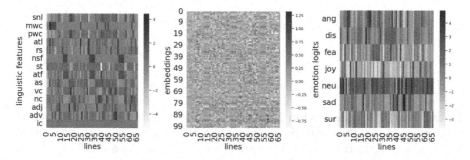

Fig. 1. From left to right: multivariate time series obtained through feature extraction via *(i)* linguistic features, *(ii)* text embeddings, *(iii)* sentiment analysis.

interpretable by humans, but they can capture complex semantic information, which is extremely useful for machine learning predictors.

Sentiment/Emotion Features. The logit layer of a sentiment/emotion analysis model produces a vector of scores or activations for each possible output class, indicating the model's confidence or belief that the input sentence corresponds to each possible sentiment/emotion. Examples of such models are VADER (Valence Aware Dictionary and sEntiment Reasoner) [11], a lexicon and rule-based sentiment analysis tool attuned explicitly to sentiments expressed in social media, and RoBERTa (Robustly Optimized BERT Pretraining Approach) [15], a variant of BERT that has been shown to achieve state-of-the-art performance on several NLP tasks, including sentiment analysis. While originating from a transformer model, logits are more interpretable than embeddings as they provide a sort of expectation of an input sentence for a certain sentiment/emotion.

In TOTS, we implement the function *feat_extr* to extract the list of sentiments/emotions **v** from a given sentence S through both VADER and RoBERTa, which, in a time series context, can track the fluctuation of sentiment and emotions within the text, providing dynamic information instead of static analysis. VADER provides a single sentiment score, while RoBERTa outputs logits for the following emotions: Anger (`ang`), Disgust (`dis`), Fear (`fea`), Joy (`joy`), Neutral (`neu`), Sadness (`sad`), and Surprise (`sur`). Figure 1 (right), shows an example of a multivariate time series obtained with RoBERTa. This series depicts a mostly neutral document, with a high peak of sadness on line 42 (*"Rest in peace and then deceased but we still strugglin while you sleep"*).

Aggregation. In order to use a shapelet-based interpretable machine learning model, we need to reduce multivariate time series into univariate ones. We accomplish this task by defining an aggregation function, *aggregate*, that takes a multivariate time series as input and "compresses" it into a univariate time series without changing the number of observations m. Formally:

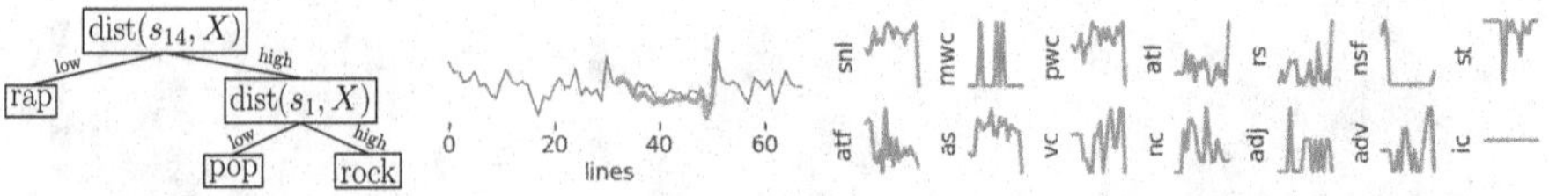

Fig. 2. Shapelet analysis approach on a linguistic time series aggregated with PCA.

Definition 10 (Aggregation Function). An aggregation function *aggregate* takes as input a multivariate time series $X \in \mathbb{R}^{d \times m}$, with $d > 1$, and compresses it into a univariate time series $X' = aggregate(X)$, where $X' \in \mathbb{R}^{1 \times m}$.

In this work, we experiment with two naive approaches such as *average* and *max* aggregation, and with a complex dimensionality reduction method such as Principal Component Analysis (PCA) [24]. Aggregation by taking the average may be sufficient when multivariate dimensions represent the same phenomenon detected by different models, such as the sentiment or the emotion computed by two different transformers, which is averaged for a more robust prediction. On the other hand, aggregation by taking the maximum could be enough when the different signals in a time series represent logits of different sentiments, highlighting the intensity of the prevalent emotion at a specific timestep. More sophisticated approaches like PCA may be required for more complex signals, like those resulting from embeddings. PCA dynamically detects the significant time series signals that include characteristic patterns of the original data because the significance of each signal is represented in each component of the transformation [24].

Figure 2 (left) displays in blue the univariate time series resultant from the PCA aggregation from the linguistic features. The signal is hardly interpretable at first look, but, as illustrated in the following, the contribution of each signal toward the final component can be retrieved, providing insights into the most relevant signals at specific timesteps, i.e., for specific sentences in our setting.

Time Classification. Once a given text corpus, $\mathcal{T}$, has been converted into the corresponding time series dataset, $\mathcal{X}$, by TOTS, i.e., $\mathcal{X} = \text{TOTS}(\mathcal{T})$, we can extract a set of q shapelets, Q, from $\mathcal{X}$ with Learning Shapelets (LS) [7]. LS learns shapelets through gradient descent optimization and is regarded as a state-of-the-art approach. In the example in Fig. 2, we use the extracted shapelets with a decision tree classifier to distinguish between rap and rock lyrics transparently. The resulting tree is extremely simple and, using only two of the extracted subsequences ($\mathbf{s}_{14}$ and $\mathbf{s}_1$), can discern between the two genres, by looking at the distance between the shapelets and the text conversion. Here, to aid interpretability, we present distances as "high" or "low" instead of specific values. Hence, there are only three rules to classify songs: *if* dist($\mathbf{s}_{14}, X$) is low *then* the class = rap, else, *if* dist($\mathbf{s}_1, X$) is low *then* class = pop, else class = rap.

Figure 2 (center) displays our running example for shapelet $\mathbf{s}_{14}$ (in orange). We notice that the best alignment of the shapelet with the time series begins at index 32 and ends at index 55 included. With this information, the shapelet can be mapped back to its multivariate components, i.e., the subsequences between 32 and 55 of each signal depicted in Fig. 2 (right). Furthermore, the same indexes can be mapped back to the original text by unveiling the lines between 32 and 55

Table 1. Dataset information.

dataset	id	classes	records	labels
SongLyrics	pprc	2	24000	pop, rock
SongLyrics	rcrp	2	24000	rock, rap
SongLyrics	lyr3	3	36000	pop, rap, rock
SongLyrics	rsub	5	25000	pop-rock, metal, indie, hard-rock, punk
WikipediaMoviePlots	mplt	4	7512	drama, comedy, thriller, horror
20Newsgroups	20ng	5	1775	talk, religion, sci, rec, comp

in the original lyric, which the model uses to make a prediction. In the following, we show the document T from sentence 28 to sentence 59 to better appreciate the text highlighted by the shapelet.

$T_{28:59}$ Sometime we do bad, but we all in it / You gotta learn to dream, cause there's No Limit, ya heard me? / - singing / Y'all don't know what we goin through / Y'all don't know what they put us through / Y'all don't know what we goin through / Y'all don't know what they put us through / Don't treat me like a disease, cause my skin darker than yers / And my environment is hostile, nuttin like your suburbs / I'm from the ghetto, home of poverty - drugs and guns / Where hustlers night life for funds but, makin crumbs / in the slums in the street, in the cold in the heat / Rest in peace and then deceased but we still strugglin while you sleep / And the game never change it's still the same since you passed / We get beat and harassed, whenever them blue lights flash / To the little homies in the hood, claimin wards and wearin rags / Tryin to feel a part of a family he never had / And it's sad, I feel his pain, I feel his wants / To avoid bein locked up, there's do's and don'ts / Use your head little soldier, keep the coke out your system / that ? out your veins, that won't do away with the pain / Only prayers will get you through, ain't no use to bein foolish / Ain't got one life to live, so be careful how you use it / - singing / Y'all don't know what we goin through / Y'all don't know what they put us through / Y'all don't know what we goin through / Y'all don't know what they put us through /

From the comparison between the shapelet and the text, we can observe how the text evolves. For instance, at the beginning of the shapelet, the normalized sentence frequency drops (**nsf**), indicating the end of the chorus and the beginning of the verse. A slight increase at the end highlights the beginning of a new chorus. Further, the alliteration score seems to grow in the verse, with the more rhythmic repetition of sounds (*"To the little **h**omies in the **h**ood, claimin **w**ards and **w**earin rags"*). **snl** and **pwc** represent the higher length of sentences in the verse w.r.t. the chorus. Other subsequences are harder to interpret in this instance, such as the number of adjectives (**adj**) and verbs (**vc**).

5   Experiments

We experiment with TOTS[1] on three datasets to assess the correctness and effectiveness of the proposed transformation.

Datasets. The first dataset is Song Lyrics, containing lyrics associated with the artist's genres. We created four different balanced subsets of this dataset, **pprc**, **rcrp**, **lyr3**, and **rsub**, containing different labels, as described in Table 1. We split Song Lyrics line-by-line with *tokenize*, removing duplicates and non-English text (e.g., *Chorus 2x*). For **20ng**, we used sentences as tokens, removing hyperlinks, HTML tags, email addresses, symbol repetitions, and expanding contractions. For **mplt**, we merged coherent genre labels, tokenizing at the sentence

---

[1] Code available at: https://github.com/mattiapggioli/lyrics2ts.

level. We discarded sentences with less than 20 lines for all datasets to avoid generating very short time series and performed an 80/20% train/test split.

Experimental Setting. We detail here the alternative implementations adopted to realize the function *feat_extr* and *aggregate*. *Linguistic features* are derived using the textstats and NLTK packages. Regarding *sentence embedding* methods, for Sentence-BERT [20] (SBE) we used the *all-MiniLM-L12-v2* model provided by SentenceTransformers, while for Doc2Vec [17] (D2V) we used Gensim after using its tokenizer with lowercasing[2]. For *sentiment* features (SEN), we used VADER [11] through the NLTK library, which outputs a compound score, ranging from -1 (extremely negative) to $+1$ (extremely positive). Thus, the resulting time series are univariate and require no aggregation. Finally, for emotion features (EMO), we used *emotion-english-distilroberta-base*, extracting the emotion logits of the last layer. As *aggregate* functions, we tested naive *avg* and *max* by simply applying the respective numpy functions column-wise and PCA by adopting the scikit-learn implementation. We experimented with PCA by *(i)* fitting and transforming each time series separately (*pca*), and *(ii)*, by fitting a global PCA model on the entire multivariate time series dataset and using it to transform each time series into a univariate one (*gpca*). In the latter, the idea is to consider timesteps as individual observations in a vector space that we want to reduce in one dimension and time series as movements within it.

Assessing Conversions Correctness. In this experiment, we assess the correctness of the different conversion workflows that can be realized through the TOTS framework. We measured the correctness by checking if similar texts are mapped to similar time series after the conversion in a controlled experiment on the lyr3 dataset. Formally, given a document T from the corpus $\mathcal{T}$, a document $T' \neq T$ that by construction is similar to T, i.e., is obtained by alterating T, and a document $T'' \neq T$ randomly selected from $\mathcal{T}$, our desiderata is that the distance between TOTS(T) and TOTS(T') is smaller than the distance between TOTS(T) and TOTS(T''). Thus, similar documents should be converted in similar time series. Since we are comparing time series, we adopt the Dynamic Time Warping (DTW) distance [23]. Hence, given a corpus $\mathcal{T}$, a corpus of similar documents $\mathcal{T}'$, and a randomly shuffled corpus $\mathcal{T}''$, we define the correctness score CS as:

$$CS = \frac{1}{n} \sum_{i}^{n} \mathbb{1}\left[dtw(\text{TOTS}(T_i), \text{TOTS}(T_i')) < dtw(\text{TOTS}(T_i), \text{TOTS}(T_i''))\right]$$

where CS is the percentage of times the desiderata holds. In practice, we sampled 50 song lyrics per genre from lyr3, i.e., $\mathcal{T}$ and, for each of them, we created a similar lyric by applying text augmentation line by line, i.e., $\mathcal{T}'$. For this purpose, we used the *ContextualWordEmbsAugmenter* of the nlpaug library, which replaces words in a text with their contextually similar counterparts using a pre-

---

[2] We set the following parameters: dm = 1, vector_size = 100, min_count = 2, epochs = 20, window = 5.

trained contextual word embedding model. Then, we associated each original text in $\mathcal{T}$ with a randomly selected one, i.e., $\mathcal{T}''$. Finally, we computed CS.

Table 2. CS metric for `lyr3`. The best *aggregate* for each *feat_extr* method are in bold.

	feat_extr	D2V	SBE	LIN	EMO	SEN
aggregate	*avg*	**0.740**	0.667	0.840	**0.813**	
	max	**0.740**	0.693	0.713	0.693	**0.800**
	pca	0.593	0.540	0.633	0.713	
	gpca	0.720	**0.767**	**0.997**	0.786	

Table 2 shows the results of this experiment w.r.t. different types of feature extraction and aggregation functions (the higher, the better). Excluding *gpca*, D2V performs better than SBE, with an average difference of about 0.05. However, with *gpca* applied, SBE demonstrates the highest performance among sentence embedding approaches, outperforming D2V. The *gpca* method demonstrates significant superiority among those based on linguistic features. Traditional *pca* demonstrates poor results not only against *gpca* but also to *max* and *avg*. In summary, the best aggregation approaches seem to be *avg* and *gpca*. However, the single sentiment signal SEN, without any aggregation, scores surprisingly high. As for runtime performance, the fastest method is SEN, with a runtime of 5.3 ms per sentence, followed by LIN and D2V with an average execution time of 44–48 ms. SBE takes longer, with an average execution time of 1.120 s, and the slowest model is EMO taking on average 2.83 s[3].

Given these results, we chose one instance for each feature extraction method to experiment with the classification task. In particular, we selected SBE with *gpca*, which produced the best results among the embeddings, despite being less efficient than D2V. For LIN, we also picked the *gpca* method, which proved extremely accurate during validation. Finally, for the sentiment/emotion method, we selected SEN, given that it performed well, with extremely fast runtimes.

Classification Benchmark. This section evaluates the performance of interpretable ML models applied to solve the text classification problem. Regarding our proposal, after having selected the most promising functions *feat_extr* and *aggregate* as described in the previous section, we applied TOTS on the text corpus obtaining the corresponding time series datasets, i.e., $\mathcal{X} = \text{TOTS}(\mathcal{T})$. To achieve our goal of interpretable text classification with explanations based on the dynamical properties of text, we extracted the shapelets from $\mathcal{X}$ through a *shp_discovery* function, and we turned $\mathcal{X}$ into D with a *shp_transform* function. In particular, we obtained *shp_discovery* and D with the *LearningShapelets* function of tslearn[4]. Then, we trained the following ML models selected for their

---

[3] Experiments were run on a ThinkPad E595. AMD Ryzen 5 3500U CPU, 8 gb RAM.

[4] We set the number of shapelets to extract q using the provided heuristic, and Adam as optimizer training for 2000 epochs per dataset.

Table 3. Classification accuracy (higher is better). The best results by column, i.e., by TOTS conversion, are bolded, best results by dataset are underlined.

		pprc			rcrp			lyr3			rsub			mplt			20ng		
		LIN	SEN	SBE	LIN	SEN	SBE	LIN	SEN	SBE	LIN	SEN	SBE	LIN	SEN	SBE	LIN	SEN	SBE
shp	DT	.53	.54	.57	.79	.73	.74	.52	.50	.48	.22	.23	.24	.28	.33	.31	.27	.31	.32
	RF	.59	**.60**	**.64**	**.86**	.81	**.82**	**.61**	.60	.56	.24	.27	**.28**	.33	**.39**	**.39**	**.32**	.34	**.41**
	LG	.60	**.60**	**.64**	**.86**	.81	**.82**	**.61**	.60	**.57**	.24	**.28**	**.28**	.31	**.39**	.37	.29	.35	.38
feat	DT	.55	.54	.55	.80	.76	.71	.53	.52	.47	.21	.24	.23	.29	.33	.31	.25	.29	.34
	RF	.60	.59	.62	**.86**	**.83**	.78	.60	.60	.54	**.25**	.27	.26	**.34**	**.39**	.36	.26	**.39**	**.41**
	LG	**.61**	**.60**	**.64**	**.86**	**.83**	.79	**.61**	**.61**	.56	**.25**	.27	**.28**	.33	.38	.35	.27	.34	.40
knn	EUC	.52	.55	.56	.56	.51	.55	.38	.37	.38	.22	.22	.24	.32	.35	.33	.27	.24	.33
	DTW	.54	.54	.60	.75	.68	.67	.51	.47	.44	.22	.23	.26	.29	.33	.34	.28	.30	.32

interpretability properties on the shapelet-transformed dataset D (*shp*), i.e., a Decision Tree (DT), a Random Forest (RF), and LightGBM (LG). As a competitor, we extracted global time series statistics (*feat*) such as the minimum, maximum, mean, variance, skewness, and kurtosis on $\mathcal{X}$, and then we train the tree-based models DT, RF, and LG. In this setting, classifiers are only statically interpretable because all the temporal references given by the time series are completely lost. Finally, in line with instance-based explanation approaches [8], we experimented also with k-Nearest-Neighbors (*knn*) trained directly on $\mathcal{X}$. In particular, we experiment with *knn* with $k = 5$ using the Euclidean distance (EUC) and DTW with a 3-window Sakoe Chiba band [23], adopting the `pyts` library[5].

Table 3 presents the accuracy of the various classifiers. The column header represents the different dataset conversions of TOTS, i.e., we convert each of the six datasets using the three best approaches from the previous section for a total of 18 dataset representations. The rows represent different classifiers, i.e., based on shapelets (*shp*), static global features (*feat*), and distances (*knn*). The best results in each column are in bold, highlighting the best feature extraction and aggregation. The best approach overall for each dataset is underlined.

At first glance, the best-performing classifiers are RF and LG, with DT, EUC, and DTW always having subpar performance. In general, shapelets and global features perform similarly, with their respective best models tieing in all of the six datasets. However, as shown in the example rap lyric, the advantage of using shapelets is to look at the importance of specific paragraphs in the text, which is impossible with global features. Regarding TOTS conversion alternatives, SBE wins in 20ng and pprc, while LIN is the overall best for rcrp. Classifiers trained on SEN have slightly lower performance, likely because they are based on a single

---

[5] Given the computational complexity of DTW on large datasets, we first used Piecewise Aggregate Approximation (PAA) to reduce the length of the time series by 80% and then kept one-third of the records for each class, selected using the ClusterCentroids method of `imblearn`.

sentiment, which may not be sufficient for the classification. The similar performance of embeddings and linguistic features is promising for explainability. It demonstrates that, for specific problems, using domain knowledge to extract interpretable features can achieve similar results to non-interpretable embeddings.

As a final note, we highlight that the purpose of TOTS at this stage is not to beat standard NLP approaches applied to the whole text but to define a way of using TSA approaches for text classification. While our approach may not perform as well as standard NLP classifiers, we offer a unique way to analyze text by taking into account local patterns rather than relying solely on the properties of the entire text. This allows for a more nuanced understanding of the text and its underlying dynamics. Overall, a sentence-based explanation can provide a more fine-grained and interpretable classification. For example, when analyzing a song, a sentence-based explanation can help identify the most relevant lines or sections to the classification result. Finally, in datasets such as 20ng, containing multiple topics, a sentence-based explanation can provide insights into how different parts of the text contribute to the classification result.

6 Conclusion

We have introduced TOTS, a method that represents text as a time series using TSA techniques and NLP approaches. Our formalization enables the conversion between text and time series, enhancing interpretability by capturing local textual patterns. Additionally, TOTS allows for easy transformation back to text, facilitating human interpretation. Through experiments, we showed that our text to time series conversion uncovers new insights and patterns not easily observable with traditional NLP approaches. A potential limitation of TOTS is its reliance on multiple independent steps, where the quality of underlying models can influence the overall performance. For example, we acknowledge that aggregating time series into univariate ones is a strong simplification, and directly analyzing multivariate text-time series could be more effective. Moreover, instead of exclusively relying on shapelets, alternative patterns could be tested for the classification task. Combining features and patterns offers a promising approach to extracting local characteristics and global dynamic trends, capturing the entire document's semantic context. After further improvements, we plan to compare TOTS against state-of-the-art NLP models and study possible avenues of integration with Large Language Models. Finally, we plan on extending text shapelets' interpretability to unsupervised analyses like clustering or topic modeling, where sequentiality can be incorporated by localizing the analysis on extracted sequential patterns.

Acknowledgment. This work is partially supported by the EU NextGenerationEU programme under the funding schemes PNRR-PE-AI FAIR (Future Artificial Intelligence Research), PNRR-SoBigData.it - Strengthening the Italian RI for Social Mining and Big Data Analytics - Prot. IR0000013, H2020-INFRAIA-2019-1: Res. Infr. G.A.

871042 *SoBigData++*, G.A. 761758 *Humane AI*, G.A. 952215 *TAILOR*, ERC-2018-ADG G.A. 834756 *XAI*, and CHIST-ERA-19-XAI-010 SAI, and by the Green.Dat.AI Horizon Europe research and innovation programme, G.A. 101070416.

References

1. Badea, I., Trausan-Matu, S.: Text analysis based on time series. In: ICSTCC 2013, pp. 37–41. IEEE (2013)
2. Bagnall, A., et al.: The great time series classification bake off: a review and experimental evaluation of recent algorithmic advances. DAMI **31**, 606–660 (2017)
3. Breiman, L., Friedman, J.H., Olshen, R.A., Stone, C.J.: Classification and Regression Trees. Routledge (2017)
4. Chalkiadakis, I., Zaremba, A., Peters, G.W., Chantler, M.J.: On-chain analytics for sentiment-driven statistical causality in cryptocurrencies. Blockchain: Res. Appl. **3**(2), 100063 (2022)
5. Dale, E., Chall, J.S.: A formula for predicting readability: instructions. Educ. Res. Bull. **27**, 37–54 (1948)
6. Fu, T.C.: A review on time series data mining. Eng. Appl. Artif. Intell. **24**(1), 164–181 (2011)
7. Grabocka, J., Schilling, N., Wistuba, M., Schmidt-Thieme, L.: Learning time-series shapelets. In: SIGKDD 2014, pp. 392–401. ACM (2014)
8. Guidotti, R., Monreale, A., Spinnato, F., Pedreschi, D., Giannotti, F.: Explaining any time series classifier. In: CogMI 2020, pp. 167–176. IEEE (2020)
9. Harris, Z.S.: Distributional structure. Word **10**(2–3), 146–162 (1954)
10. Hassani, H., Beneki, C., Unger, S., Mazinani, M.T., Yeganegi, M.R.: Text mining in big data analytics. Big Data Cogn. Comput. **4**(1), 1 (2020)
11. Hutto, C., Gilbert, E.: Vader: a parsimonious rule-based model for sentiment analysis of social media text. In: ICWSM 2014, vol. 8, pp. 216–225 (2014)
12. Jing, L.P., Huang, H.K., Shi, H.B.: Improved feature selection approach TFIDF in text mining. In: ICMLC 2002, vol. 2, pp. 944–946. IEEE (2002)
13. Kaplan, R.M.: A method for tokenizing text. Inquiries into words, constraints and contexts 55 (2005)
14. Khurana, D., Koli, A., Khatter, K., Singh, S.: Natural language processing: state of the art, current trends and challenges. Multim. Tools Appl. **82**(3), 3713 (2023)
15. Liu, Y., et al.: Roberta: a robustly optimized BERT pretraining approach. arXiv preprint arXiv:1907.11692 (2019)
16. Makridakis, S., Wheelwright, S.C., Hyndman, R.J.: Forecasting Methods and Applications. Wiley, Hoboken (2008)
17. Mikolov, T., Sutskever, I., Chen, K., Corrado, G.S., Dean, J.: Distributed representations of words and phrases and their compositionality. ICML **2013**, 26 (2013)
18. Muthukumar, P., Zhong, J.: A stochastic time series model for predicting financial trends using NLP. arXiv preprint arXiv:2102.01290 (2021)
19. Petrov, S., Das, D., McDonald, R.: A universal part-of-speech tagset. In: LREC'12, pp. 2089–2096 (2012)
20. Reimers, N., Gurevych, I.: Sentence-bert: Sentence embeddings using siamese bert-networks. arXiv preprint arXiv:1908.10084 (2019)
21. Ruiz, E.J., Hristidis, V., Castillo, C., Gionis, A., Jaimes, A.: Correlating financial time series with micro-blogging activity. In: WSDM 2012, pp. 513–522 (2012)
22. Şenel, L.K., Utlu, I., Yücesoy, V., Koc, A., Cukur, T.: Semantic structure and interpretability of word embeddings. IEEE/ACM TASLP **26**(10), 1769–1779 (2018)

23. Senin, P.: Dynamic time warping algorithm review. Information and Computer Science Dept. University of Hawaii at Manoa Honolulu, USA 855(1–23), 40 (2008)
24. Tanaka, Y., Iwamoto, K., Uehara, K.: Discovery of time-series motif from multi-dimensional data based on mdl principle. Mach. Learn. **58**, 269–300 (2005)
25. Theissler, A., Spinnato, F., Schlegel, U., Guidotti, R.: Explainable AI for time series classification: a review, taxonomy and research directions. IEEE Access **10**, 100700–100724 (2022)
26. Vaswani, A., Shazeer, N., Parmar, N., Uszkoreit, J., Jones, L., Gomez, A.N., Kaiser, Ł, Polosukhin, I.: Attention is all you need. NIPS **2017**, 30 (2017)
27. Wang, B., Huang, H., Wang, X.: A novel text mining approach to financial time series forecasting. Neurocomputing **83**, 136–145 (2012)
28. Wang, X., McCallum, A., Wei, X.: Topical n-grams: phrase and topic discovery, with an application to information retrieval. In: ICDM, pp. 697–702. IEEE (2007)
29. Yang, T., Lee, D.: T3: on mapping text to time series. In: AMW (2009)
30. Ye, L., Keogh, E.: Time series shapelets: a new primitive for data mining. In: SIGKDD 2009, pp. 947–956 (2009)
31. Zhang, W., Yoshida, T., Tang, X.: A comparative study of TF* IDF, LSI and multi-words for text classification. Expert Syst. Appl. **38**(3), 2758–2765 (2011)

Data Analysis and Optimization

Enhancing Intra-modal Similarity in a Cross-Modal Triplet Loss

Mario Mallea$^{(\boxtimes)}$ 🆔, Ricardo Nanculef 🆔, and Mauricio Araya 🆔

Universidad Federico Santa María, Valparaíso, Chile
mario.mallea@sansano.usm.cl, jnancu@inf.utfsm.cl, mauricio.araya@usm.cl

Abstract. Cross-modal retrieval requires building a common latent space that captures and correlates information from different data modalities, usually images and texts. Cross-modal training based on the triplet loss with hard negative mining is a state-of-the-art technique to address this problem. This paper shows that such approach is not always effective in handling intra-modal similarities. Specifically, we found that this method can lead to inconsistent similarity orderings in the latent space, where intra-modal pairs with unknown ground-truth similarity are ranked higher than cross-modal pairs representing the same concept. To address this problem, we propose two novel loss functions that leverage intra-modal similarity constraints available in a training triplet but not used by the original formulation. Additionally, this paper explores the application of this framework to unsupervised image retrieval problems, where cross-modal training can provide the supervisory signals that are otherwise missing in the absence of category labels. Up to our knowledge, we are the first to evaluate cross-modal training for intra-modal retrieval without labels.

We present comprehensive experiments on MS-COCO and Flickr30K, demonstrating the advantages and limitations of the proposed methods in cross-modal and intra-modal retrieval tasks in terms of performance and novelty measures. Our code is publicly available on GitHub https://github.com/MariodotR/FullHN.git.

Keywords: cross-modal retrieval · triplet loss · hard negative mining · unsupervised image retrieval

1 Introduction

Content-Based Image Retrieval (CBIR) is an approach to searching an image database where users can input an image as a query and retrieve related images based on their visual content [11]. Cross-modal retrieval (CMR) is a related approach that enables users to query the system with data in one modality (e.g.,

This research was partially funded by National Agency for Research and Development (ANID, Chile), grant numbers FONDEF IT21I0019, ANID PIA/APOYO AFB180002 and ANID-Basal Project FB0008.

a textual description) and retrieve data in another modality (e.g., an image) [5]. With the rapid growth of image and multimedia data, CBIR and CMR have become essential technologies for building effective information systems in various domains, including social networks [23], online retail [14], remote sensing [1], and medicine [17].

In the last decade, deep learning has significantly enhanced hand-crafted feature extraction algorithms such as SIFT and BoVW [4]. However, deep learning-based methods require training with appropriate loss functions to learn the semantics of images and their relationships [9]. One state-of-the-art technique for training CBIR models is triplet-based learning [4,11]. A training triplet typically consists of an anchor image, a positive image that is similar to the anchor, and a negative image that is dissimilar or less similar to the anchor. The triplet loss function is then used to encourage the anchor to be closer to the positive image than the negative image in the latent space. In CMR [5], where different data types are involved, representation learning requires building a common latent space that captures and correlates information from these modalities. The cross-modal triplet loss addresses this challenge by taking the anchor from one modality but the positive and negative samples from another modality . In this way, learning seeks to ensure that the similarity between cross-modal pairs representing the same concept in different modalities is higher than between cross-modal pairs unobserved in the training data. In practice, it is well-known that this approach is governed by cases where the largest error occurs. These elements are called hard negatives [5]. Recent research has been largely focused on methods to select negative samples that improve learning efficiency [3,35], techniques that use more than one negative to constrain data representations [21,28], and adaptive margin formulations [27,30].

In this paper we investigate triplet-based cross-modal training with hard negative mining for CBIR in label-free scenarios. This research is motivated by a real-world medical application in which we must deploy a CBIR engine to support the process of differential diagnosis which helps physicians to distinguish between multiple conditions or diseases that share similar characteristics. The training data includes images and radiology reports but lacks explicit category labels such as diagnoses. In the absence of labels, we can leverage text reports and train image representations through cross-modal learning.

We found that cross-modal triplet-based learning alone does not effectively handle intra-modal similarities and can lead to inconsistent similarity relationships in the latent space. This phenomenon can introduce spurious neighbors in the latent space that hurt retrieval performance. Previous research on this issue has focused on labelled data, studying how intra-class concentration and inter-class dispersion contribute to the engine's overall performance [6,25,28,35]. Without class annotations, intra-class and inter-class similarities cannot be measured and the accuracy of intra-modal searches cannot easily computed. To address these limitations, we propose two novel loss functions that utilize intra-modal similarity constraints within a training triplet. Unlike the classic formulation, the proposed losses explicitly encourages ground-truth cross-modal pairs to be closer in the latent space compared to intra-modal pairs for which no relationship is observed in the data. Additionally, we propose evaluating intra-modal retrieval results using non-binary relevance metrics that take advantage of the additional modalities to measure the semantic relationship between items [15,16]. This approach enables intra-modal evaluation of cross-modal methods without resorting to class labels.

Our main contributions are:

1) We show that the cross-modal triplet loss with hard negative mining can lead to inconsistent similarity orderings in the latent space. Specifically, we show that intra-modal pairs unobserved in the training data can be closer than cross-modal pairs representing the same concept in different modalities.

2) We propose two new triplet losses that enhance the use of hard negatives that exploit the order relationships between inter-modal and intra-modal similarities. Experimental results demonstrate that our proposals can enable the construction of more effective and novel cross-modal and intra-modal retrieval systems.

3) We design a novel way of evaluating unlabeled image-to-image retrieval tasks based on two semantic relevance measures using modern cross-modal retrieval approaches. Furthermore, we use semantic information to measure the novelty of the retrieval lists. Besides, we use self-information to measure general surprise [36] in the retrieved results.

2   Related Work

Triplet-based learning and hard negative mining techniques are widely used techniques in different areas [4]. In deep metric learning [9,21,31] propose a novel loss that exploits all pairwise relationships within a batch. [3] suggests a linearization of the triplet loss by using fixed class centroids that guarantee linear run-time complexity.

Focusing on the role of the margin in selecting hard negatives, [6] proposes a dynamically computed margin using a class tree constructed from the data to estimate intra-class and inter-class separations. Additionally, [27] defines an adaptive margin that is computed as a non-linear (and non-learnable) function of the average distances among positive and adversarial negative pairs. It aims to compact the intra-class and separate the inter-class.

In particular for similarity learning [29,30] proposes an adaptive triplet margin that takes advantage of intra-modal information. Given a text, the margin is defined as a convex combination of the distance among images both in the feature space and in the label space (one-hot representation) assuming a labeled data scenario. In the task of person re-identification [33]. [28] introduces a quadruplet loss that extends the triplet loss by incorporating a second negative of a class different than the classic negative, and promotes that the distance between these two negatives is greater than the distance of the positive pair. This implicitly promotes that the minimum inter-class distance exceeds the maximum intra-class distance. On the other hand, [35] also optimizes the intra/inter-class distance and reduces the cost associated with computing hard negatives by learning the centers of the classes and using them as anchors to calculate the hard negatives.

For the image retrieval task applied to global descriptors, [26] extends the triplet loss with hard-negative mining by incorporating second-order similarity regularization [25]. This regularization enforces positive pairs exhibit similar distances with respect to other points in the embedding space.

Specifically in cross-modal retrieval, VSE++ [5] is a model that demonstrates the effectiveness of triplet loss with hard negative mining. Conceptually, this is our starting point. However, it should be mentioned that the state of the art for this task, particularly in our experimental setup, is given by [15,16]. Nevertheless, these improvements on VSE++ are obtained by designing better neural network architectures, not by modifying the triplet loss with hard negatives. These works propose a semantic $nDCG$ metric that allows evaluation beyond non-binary (exact) retrieval, through two different sentence similarity functions that capture different aspects of the sentence.

Based on our knowledge, we deal with a problem that has been identified previously in other contexts [19,32]. We are the first in several contributions: describing the intra-modal problem for cross-modal triplet loss, proposing novel loss formulations, and evaluating the image-to-image retrieval task with label-free cross-modal training and evaluation [15,16].

3 Preliminaries

The goal of the cross-modal retrieval task is to accept a query in one modality and retrieve relevant data in another modality. Specifically, let us consider N pairs of images and texts $P = \{(i_n, c_n)\}_{n=1}^N$ as the training data, and let p_{data} be the distribution of these data. We refer to these tuples as positive pairs, and call negative pair to any tuple that does not belong to P.

The classical loss for cross-modal retrieval is based on triplet learning for the task of image retrieval from text (t2i) and text retrieval from images (i2t). Triplet learning formalizes the intuition that the similarity between a positive pair (i_n, c_n) should be greater than the similarity between the query text c_n and some other image $\bar{i}_n$ (t2i). Similarly, for the i2t task. The loss (represented in Fig. 4a) combines the two tasks as follows:

$$\mathcal{L}(i_n, \bar{i}_n, c_n, \bar{c}_n) = \mathcal{L}_{i2t} + \mathcal{L}_{t2i}. \tag{1}$$

$$\mathcal{L}_{i2t} := [\alpha + s(i_n, \bar{c}_n) - s(i_n, c_n)]_+, \tag{2}$$

$$\mathcal{L}_{t2i} := [\alpha + s(\bar{i}_n, c_n) - s(i_n, c_n)]_+, \tag{3}$$

where $[x]_+ = max(0, x)$, α is a hyperparameter known as margin, and the similarity s is measured as the dot product $s(i, c) = f_v(i)^t f_t(c)$ between the normalized representations [5] f_v and f_t assigned to images and texts, respectively. The visual encoder is $f_v(i) = E^V g_i$ and the textual encoder is $f_t(c) = E^T t_c$, where $g_i \in \mathbb{R}^{n_V}$ is the feature vector associated with the image and $t_c \in \mathbb{R}^{n_T}$ with the text. If pre-trained and untrainable neural networks are used to obtain these feature vectors, the trainable parameters θ are the projection matrices $E_\phi^V \in \mathbb{R}^{k \times n_V}$ and $E_\phi^T \in \mathbb{R}^{k \times n_T}$, with k the hyperparameter of the latent space dimension. The optimization problem is formulated as follows:

$$\min_{\theta} \mathop{\mathbb{E}}_{\substack{(i_n, c_n) \sim p_{data} \\ (\bar{i}_n, \bar{c}_n) \sim \bar{p}_{data}}} \mathcal{L}(i_n, \bar{i}_n, c_n, \bar{c}_n). \tag{4}$$

In practice, the expectation is approximated by sampling from the training data. In the Random Negative approach (RN), negative pairs are randomly sampled. A well-known improvement to this technique is the Hard Negative (HN) approach that dynamically chooses the negative examples according to the current model's state [5]. That is, $\bar{p}_{data} = p_{data}^{max}$, which means taking as negatives the cases that are the most problematic for each task: $\bar{i}_n = \underset{x \neq i_n}{argmaxs}(x, c_n)$ and $\bar{c}_n = \underset{x \neq c_n}{argmaxs}(i_n, x)$. For performance reasons, the negatives are sampled from the mini-batch and not globally. See Fig. 1 for an illustration of the resulting framework.

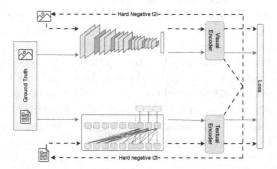

Fig. 1. Cross-modal retrieval training framework. Two models are jointly trained on the image-text dataset in the visual and textual modalities, with the goal of maximizing the similarity between an image and its corresponding text. Based on the positive pair, hard negative elements are generated and processed by models with shared weights in each modality.

4 Proposed Methods

4.1 Motivation

One of the drawbacks of the cross-modal triplet loss is that it only considers inter-modal similarity relationships and neglects the impact of intra-modal similarities. To illustrate the limitations of this formulation, consider a t2i retrieval task. Given a training triplet $(i, c, \bar{i})$, the triplet loss (Eq. 3) enforces the t2i constraint $s(i, c) > s(\bar{i}, c) + \alpha$, which ensures that i will be ranked higher than $\bar{i}$ when c is presented as a query. However, this constraint does not impose conditions on the relationship between i and $\bar{i}$.

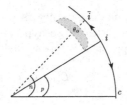

Fig. 2. Example of an inconvenient vector distribution for the t2i task. Although the query text c is more similar to its image i than to the hard negative image $\bar{i}$, in the face of a possible insufficient margin it is possible that $\bar{i}$ is more similar to i than to c.

As depicted in Fig. 2, there may exist multiple $(i, c, \bar{i})$ such that:

$$s(i, \bar{i}) > s(i, c) > s(\bar{i}, c) + \alpha . \tag{5}$$

This situation is problematic for several reasons:

1. While the pair (i, c) is known to represent the same concept, the pair $(i, \bar{i})$ may or may not correspond to semantically similar concepts. Without explicit supervision, such as class labels, it is challenging to differentiate between a spurious relationship and a valid one. Thus, when only (i, c) is observed in the data, we should expect $s(i, c)$ to be higher than $s(i, \bar{i})$.
2. Let $c_{\bar{i}}$ be the instance corresponding to $\bar{i}$ in the other modality. If both $s(i, \bar{i})$ and $s(i, c)$ are high, c can be ranked higher than $c_{\bar{i}}$ when $\bar{i}$ is presented as a query, resulting in an i2t retrieval error. While the triplet loss can handle this scenario through the i2t constraint, it may require additional training steps to sample $c_{\bar{i}}$ to correct the mistake, slowing down the learning process.
3. Without additional forms of supervision, we expect a high value of $s(i, \bar{i})$ (similar image representations) if and only if $s(c, c_{\bar{i}})$ is high (similar text representations). However, if Eq. 5 holds, it is possible that $s(c, c_{\bar{i}})$ is small but $s(i, \bar{i})$ is large. Indeed, if $\bar{i}$ is very close to $c_{\bar{i}}$, it may happen that $s(c, c_{\bar{i}}) \approx s(c, \bar{i})$, resulting in the following inequality: $s(i, \bar{i}) > s(i, c) > s(c, c_{\bar{i}}) + \alpha$, which shows that the relationship in the visual modality $s(i, \bar{i})$ is inconsistent with the relationship in the textual modality $s(c, c_{\bar{i}})$.

Assuming normalized latent representations, we can identify the triplets that satisfy the problem identified above by solving Eq. 5, as shown in Fig. 3 The existence of such triplets is naturally influenced by the magnitude of the margin. A greater margin helps reducing the probability of selecting these triplets (note the smaller volume in Fig. 3b, but it is not sufficient to eliminate the problem. In the classic HN model, these problematic triplets are indirectly handled by the margin. However, a fixed margin is a suboptimal strategy because it may not adapt to the different possible triplets, leading to triplets that satisfy the described problem and therefore do not favour the discriminative capacity or generalization of the model.

We propose two novel extensions of the triplet loss with hard negative mining that address the above-identified problem using the identical training triplets, i.e., without increasing the complexity of the sampling strategy.

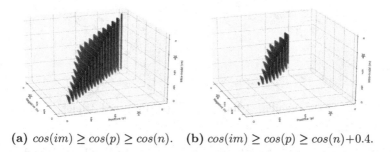

(a) $cos(im) \geq cos(p) \geq cos(n)$.   (b) $cos(im) \geq cos(p) \geq cos(n)+0.4$.

Fig. 3. We define the angles p, n and im, as the angle of a positive pair, a negative pair and between a query and a hard negative in the same modality, respectively. Visualization of cross-sectional cuts of p of the triplets that satisfy the inequalities. We also consider triangular inequality $im \geq n - p$ [20].

4.2   Full Hard Negative Method (F-HN):

We propose to make a better use of the hard negative by extending the triplet loss to consider all the similarity constraints that can be derived from a training triplet. The first loss we propose is defined as:

$$\mathcal{L}(i_n, \bar{i}_n, c_n, \bar{c}_n) = \mathcal{L}_{i2t} + \mathcal{L}_{t2i} + \mathcal{L}_{vc} + \mathcal{L}_{tc} + \mathcal{L}_{sc}. \tag{6}$$

The new components of the loss will be called *visual constraint* (vc), *textual constraint* (tc), and *structural constraint* (sc). They are defined as follows:

$$\mathcal{L}_{vc} := [\alpha + s(i_n, \bar{i}_n) - s(i_n, c_n)]_+, \tag{7}$$

$$\mathcal{L}_{tc} := [\alpha + s(c_n, \bar{c}_n) - s(i_n, c_n)]_+, \tag{8}$$

$$\mathcal{L}_{sc} := (\mathbb{1}_{(\bar{i}_n, \bar{c}_n) \notin P})[\alpha + s(\bar{i}_n, \bar{c}_n) - s(i_n, c_n)]_+. \tag{9}$$

Here $\mathcal{L}_{vc}$ and $\mathcal{L}_{tc}$ aim to ensure that intra-modal similarities $s(i_n, \bar{i}_n)$ and $s(c_n, \bar{c}_n)$ are lower than that of positive pairs. Meanwhile, L_{sc} acts only in cases where the hard negatives for each cross-modal task do not correspond to each other, and therefore should not be more similar than the ground truth pair. We can see the difference with the vanilla triplet model in Fig. 4.

4.3   Intra-modal Margin Hard Negative Control Method (M-HN)

We propose the use of an adaptive margin for each training triplet, this margin allows for the selection of locally informative samples, capturing local similarity structures in the latent space and making the training process more efficient.

For a given triplet $(i_n, c_n, \bar{c}_n)$, we determine the margin values based on the intra-modal similarities. In the case of the i2t loss, we seek to ensure that the cross-modal constraint holds with a margin that is at least the similarity $s(i_n, \bar{i}_n)$. Similarly for the t2i loss. These choices lead to the following definitions:

$$\mathcal{L}(i_n, \bar{i}_n, c_n, \bar{c}_n) = \mathcal{L}^*_{i2t} + \mathcal{L}^*_{t2i}, \tag{10}$$

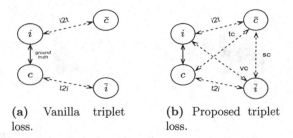

(a) Vanilla triplet loss.

(b) Proposed triplet loss.

Fig. 4. We represent the order relations promoted by the vanilla (HN) and proposed (F-HN) triplet loss. F-HN exploits all possible similarity relations between the same elements of the vanilla formulation.

$$\mathcal{L}_{i2t}^* := [s(i_n, \bar{i}_n) + s(i_n, \bar{c}_n) - s(i_n, c_n)]_+, \tag{11}$$

$$\mathcal{L}_{t2i}^* := [s(c_n, \bar{c}_n) + s(\bar{i}_n, c_n) - s(i_n, c_n)]_+. \tag{12}$$

Compared to the vanilla cross-modal triplet loss, the proposed formulation introduces a looser i2t constraint to triplets $(i_n, c_n, \bar{c}_n)$ with similar image representations and a tighter i2t constraint to triplets with dissimilar image representations. Equivalently for the t2i constraint. This approach promotes a greater consistency between the visual and textual modalities, leading to an increase or decrease in both cross-modal and intra-modal similarity. Such consistency is beneficial for cross-modal retrieval, as well as for intra-modal retrieval scenarios where the only available supervisory signal for learning a representation is obtained from the other modality. Finally, note that this formulation removes the margin as a hyperparameter, which makes it easier to deploy and less prone to overfitting.

5   Experimental Setting

We conduct experiments to evaluate the proposed algorithms in cross-modal and intra-modal retrieval tasks. Below experimental settings adopted to this end.

5.1   Datasets

Although we extend evaluation in several directions, we adopt experimental settings widely used in cross-modal retrieval research [5,7,15,16]. In particular, we use popular datasets in the area: MS-COCO [13] and Flickr30k [34] (see Table 1 for statistics). Each dataset consists of a set of images with 5 human-written captions describing it. We used the splits introduced by [8][1].

---

[1] For MS-COCO, we report results for 5k images.

Table 1. Datasets statistics (#Images)

	Train	Validation	Test	Total
Flickr30k	29,000	1,014	1,000	31,014
MS-COCO	113,287	5,000	5,000	123,287

5.2 Models and Implementation Details.

We compare our methods against the following baselines:

- RN: This method uses the classic cross-modal loss randomly selecting the negative from the current mini-batch [5]. RN is equivalent to VSE0 [5] but, as detailed below, we use more recent architectures to extract image and text representations.
- HN: This is equivalent to VSE++ [5] with updated image and text representations.
- TERAN: Transformer encoder reasoning and alignment network is a state-of-the-art model for cross-modal retrieval [15].

To simplify experiments, we used pre-trained neural nets for feature extraction. We only trained the encoders (see Sect. 3), mapping the visual and textual representations to the cross-modal latent space. We adopted state-of-the-art models for both modalities. For the images, we used the Efficient Net V2L model ($n_V = 1280$) [24], and for the text we used the MPNET model ($n_T = 768$) [22]. Visual features were computed on the entire image without data augmentation or random crop extraction as in [5].

Our methodology starts by selecting the best hyperparameters according to the sum of recall in the cross-modal retrieval tasks [5]. For Flickr30K, we randomly selected half of the training data and evaluated on the validation set. To compare against the best possible version of the HN method, we tuned the margin between 0.2–0.4, and the dimensionality of the latent space in [768, 896, 1024]. The best results were obtained for a margin of 0.4 and a dimensionality of 1024^2. These values were transferred to the other models. We explored modifying the margin in the new components of F-HN considering the grid $[1e - 3, 1e - 3, 0.2, 0.4, 0.6]^3$. The best result was obtained with the same margin of 0.4 selected for the classic triplet loss's components. For MS-COCO, we adopted hyperparameter values commonly used in the literature: a margin of 0.2 and a dimensionality of 1024. We applied the recommendation to keep the margins of the new constraints in F-HN equal and worked with a mini-batch of 512. The models were trained for 30 epochs with the Adam optimizer [10] using a learning rate of 0.0002 for the first 15 epochs and 0.00002 for the remaining 15 epochs [5].

---

[2] With results of $[351.3 - 350.3, 351.0 - 349.3, 350.4 - 351.4]$, respectively.

[3] With results of $[367.0, 368.1, 375.9, 377.5, 370.4]$, respectively.

5.3   Evaluation

Cross-modal retrieval is commonly evaluated using Recall among the first 1, 5, and 10 retrieved results. Besides this classic setting, we extend evaluation using a non-binary relevance metric recently proposed in [15,16], which we applied for the top $K = 25$ recoveries. Below we explain how we extend this metric for evaluating intra-modal retrieval without category labels. We also propose extending the evaluation by considering novelty metrics.

nDCG. For a query q, the nDCG@K is defined as $\text{nDCG}_K = \text{DCG}_K/\text{IDCG}_K$, where $\text{DCG}_K = \sum_{v=1}^{K} \frac{rel(q,v)}{\log_2(v+1)}$, and IDCG_K is a model-agnostic normalization constant that makes $\text{nDCG}_K = 1$ for ideal retrieval lists. Besides, $rel(q,v)$ is a relevance function computed according to the modality of q:

- Image retrieval. In t2i and i2i tasks, we compute $rel(q, v) = \tau\left(\bar{C}_v, C_q\right)$, where τ is defined below, C_q is the query caption and $\bar{C}_v$ is the set of all captions associated to the image I_v. In i2i tasks, the metric is averaged among the different captions.
- Text retrieval. In i2t and t2t tasks, we compute $rel(q, v) = \tau\left(\bar{C}_q, C_v\right)$, where $\bar{C}_q$ is the set of captions associated to the query image I_q. For the t2t task, each textual top is evaluated using the relevance function associated with the image corresponding to the caption query.

Following [15,16] we use ROUGE-L [12] and SPICE [18] as sentence similarity functions τ for computing caption similarities.

Novelty-Biased nDCG. We propose the following adaptation of $\alpha - nDCG@K$ [2]. For a query q, the metric is defined as follows:

$$\frac{1}{IDCG} \sum_{v=1}^{K} \frac{rel(q,v)(1-\alpha)^{r(q,v-1)}}{\log_2(v+1)}, \text{ where } r(q, v-1) = \sum_{z=1}^{v-1} rel(q, z),$$

where we compute the normalization constant $IDCG$ by considering the most relevant results for each query and $\alpha(= 0.5)$ is a constant penalizing long repetitions of relevant results in favor of novelty.

Novelty via Self-information. We adapt a metric used in Recommender Systems [36] for measuring to which extent retrieved results are unexpected or unusual. By defining the *self-information* or *surprise* of a retrieved item as the inverse of its prior selection's probability, novelty can be measured as

$$\frac{1}{K} \sum_{v=1}^{K} \log_2\left(\frac{n_q}{count(v)}\right),$$

where $count(v)$ counts the number of times that v was retrieved among n_q evaluation queries.

Note that self-information rewards results that are globally infrequent among the retrieval lists. Novelty-biased nDCG, in contrast, rewards results that break the monotonicity of an individual retrieval list.

6   Results and Discussion

Tables 2, 3, 4, and 5 present the results[4] for i2t, t2i, i2i and t2t tasks, respectively.

6.1   Cross-Modal Retrieval Performance

For i2t task, it can be observed that HN outperforms the other algorithms in terms of Recall at various cutoffs, except for recall@10 on Flickr30k, where F-HN is the superior method. F-HN ranks as the second-best algorithm overall. It is worth noting that our results for RN and HN surpass the publicly available results for VSE0 (ResNet) and VSE++ (ResNet) [5], respectively. These findings provide evidence of the exceptional performance of the neural architectures utilized to obtain visual and text representations. Nonetheless, it is important to acknowledge that the performance of all methods, including ours, can be further enhanced by fine-tuning the neural networks.

For t2i task, we can see that F-HN achieves the best results on Flickr30K, while, on MS-COCO, RN gets the best retrieval performance followed by F-HN. HN systematically obtains the worst retrieval performance. We attribute this result to the problem identified in Sect. 4.1: HN has a problem with handling intra-modal similarities. For example, as shown in Fig. 5, HN ranks $s(\bar{c}, c)$ higher than $s(i, c)$ during the entire learning process. In contrast, both F-HN and M-HN require a few training epochs to correct this problem. F-HN is clearly more effective. The RN model is often competitive and robust to outliers [5], because it samples with high probability negatives that are harder than 90% of the entire training set [5]. This effect is enhanced by the batch size used in this research.

6.2   Intra-modal Retrieval Performance

For intra-modal retrieval tasks, F-HN achieves the best retrieval performance according to nDCG (except for t2t on MS-COCO where RN dominates). We

Table 2. Results of experiments for image to text retrieval.

Metric		Flickr30K					MS-COCO				
		RN	HN	M-HN	F-HN	TERAN	RN	HN	M-HN	F-HN	TERAN
Recall	@1	40.5	**48.2**	41.8	47.3	<u>75.8</u>	20.34	**23.66**	18.14	18.82	<u>55.6</u>
	@5	67.5	**77.1**	72.3	75.5	<u>93.2</u>	45.62	**48.84**	41.32	42.42	<u>83.9</u>
	@10	80.6	84.7	82.7	**85.0**	<u>96.7</u>	59.5	**61.18**	54.3	55.9	<u>91.6</u>
nDCG@25	Rouge-L	0.566	**0.594**	0.553	0.585	<u>0.687</u>	0.516	**0.531**	0.492	0.497	<u>0.643</u>
	Spice	0.480	**0.509**	0.460	0.503	<u>0.614</u>	0.475	**0.483**	0.433	0.448	<u>0.606</u>
Novelty@25	Rouge-L	0.803	**0.830**	0.803	0.821	<u>0.911</u>	0.769	**0.781**	0.758	0.759	<u>0.875</u>
	Spice	0.665	**0.700**	0.657	0.694	<u>0.811</u>	0.676	**0.685**	0.648	0.658	<u>0.805</u>
	Self-information	7.292	**7.345**	7.291	7.249	7.285	9.224	**9.306**	8.676	8.569	<u>9.667</u>

---

[4] The best value is underlined and the best without considering TERAN is highlighted in bold.

Table 3. Results of experiments for text to image retrieval.

Metric		Flickr30K					MS-COCO				
		RN	HN	M-HN	F-HN	TERAN	RN	HN	M-HN	F-HN	TERAN
Recall	@1	32.9	28.5	30.6	**39.4**	<u>59.5</u>	16.48	9.88	14.31	**19.48**	<u>42.6</u>
	@5	65.1	58.6	59.9	**69.0**	<u>84.9</u>	40.5	26.2	34.76	**42.68**	<u>72.5</u>
	@10	76.8	71.3	71.4	**79.2**	<u>90.6</u>	**54.71**	37.76	46.23	54.7	<u>82.9</u>
nDCG@25	Rouge-L	0.617	0.603	0.599	**0.629**	<u>0.686</u>	**0.618**	0.581	0.595	0.615	<u>0.682</u>
	Spice	0.498	0.482	0.471	**0.517**	<u>0.695</u>	**0.558**	0.503	0.521	0.551	<u>0.610</u>
Novelty@25	Rouge-L	0.816	0.802	0.804	**0.830**	<u>0.880</u>	0.801	0.770	0.786	**0.804**	<u>0.868</u>
	Spice	0.660	0.642	0.634	**0.675**	<u>0.721</u>	**0.720**	0.668	0.689	0.715	<u>0.778</u>
	Self-information	**5.187**	4.959	5.186	5.145	<u>5.258</u>	**7.357**	6.734	7.133	7.334	<u>7.576</u>

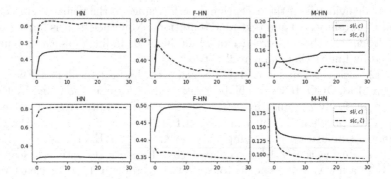

Fig. 5. Average similarity by training epochs in Flickr30K (top) and MS-COCO (bottom). The solid line corresponds to the cross-modal similarity between ground-truth pairs. The dashed line corresponds to the intra-modal similarity between unobserved pairs.

Table 4. Results of experiments for image to image retrieval.

Metric		Flickr30K					MS-COCO				
		RN	HN	M-HN	F-HN	TERAN	RN	HN	M-HN	F-HN	TERAN
nDCG@25	Rouge-L	0.701	0.699	0.695	**0.703**	<u>0.718</u>	0.705	0.700	0.699	**0.708**	<u>0.719</u>
	Spice	0.574	0.574	0.563	**0.578**	<u>0.582</u>	0.618	0.609	0.609	**0.623**	<u>**0.623**</u>
Novelty@25	Rouge-L	0.925	0.924	0.923	**0.925**	<u>0.930</u>	0.931	0.930	0.930	**0.932**	<u>0.936</u>
	Spice	0.713	0.642	0.704	**0.716**	<u>0.722</u>	0.779	0.773	0.772	**0.783**	<u>0.788</u>
	Self-information	5.185	5.097	**5.219**	5.104	5.197	**7.393**	7.166	7.271	7.214	<u>7.452</u>

must note also that performance results are always higher for intra-modal tasks compared to cross-modal tasks. In particular, our intra-modal retrieval metrics outperform the SOTA results in terms of cross-modal nDCG [15]. Due to fine-grained reasoning, TERAN achieves superior results in almost all cases. F-HN clearly reduces the difference with TERAN especially in the intra-modal tasks. As future work we will use our full hard negative loss with an attentional model.

Table 5. Results of experiments for text to text retrieval.

Metric		Flickr30K					MS-COCO				
		RN	HN	M-HN	F-HN	TERAN	RN	HN	M-HN	F-HN	TERAN
nDCG@25	Rouge-L	0.716	0.726	0.646	**0.733**	0.758	**0.682**	0.671	0.586	0.662	0.714
	Spice	0.635	**0.645**	0.511	**0.645**	0.677	**0.650**	0.633	0.424	0.587	0.667
Novelty@25	Rouge-L	0.941	0.945	0.918	**0.946**	0.953	**0.933**	0.928	0.904	0.927	0.940
	Spice	0.766	0.642	0.652	**0.767**	0.795	**0.837**	0.826	0.688	0.805	0.848
	Self-information	7.455	7.342	7.510	**7.528**	7.292	9.613	9.401	9.771	**9.795**	9.670

6.3 Novelty Results

For t2i task, the F-HN model obtains the best novelty-biased nDCG scores in the t2i task, except for MS-COCO with Spice, where it takes the second place. Using the self-information-based novelty measure and ignoring the RN model, F-HN and M-HN achieve the best novelty score on MS-COCO and Flickr30K, respectively. It is worth noting that HN obtains the worst novelty score across all metrics on both datasets (except with SPICE on Flickr30K). However, for i2t task, HN outperforms the other models. F-HN achieves the second-best results regarding the novelty-biased nDCG metrics on Flickr30K. RN also produces competitive results regarding the three novelty metrics introduced in this work, often obtaining better novelty than models with better nDCG scores. This result resembles the classic exploration-exploitation (novelty-accuracy) trade-off.

When considering the intra-modal retrieval tasks, F-HN obtains the highest novelty scores most of the time. In the i2i task, it achieves the best novelty-biased nDCG scores in both datasets. In the t2t task, F-HN outperforms the other models regarding the self-information-based novelty score in both datasets and achieves the best novelty-biased nDCG scores on Flickr30K. One more time, it is worth noting that for both intra-modal retrieval tasks, HN obtains the worst novelty scores concerning the self-information metric.

Once again, the performance of intra-modal surpasses that observed in the cross-modal scenario. These observations show that cross-modal learning approaches can lead to accurate, novelty-aware intra-modal retrieval systems. Our F-HN proposal enhances this property. We provide a visual example in our github repository.

7 Conclusions and Future Work

We have introduced two methods to leverage intra-modal relationships in cross-modal training for IR systems. Our evaluation of these techniques went beyond binary relevance, considering both the content of retrieved items and the novelty of the retrieved lists. Furthermore, we explored the benefits of cross-modal training in intra-modal retrieval tasks. Importantly, our methodology does not rely on category labels for either training or evaluation. Moreover, our methods are encoder-agnostic and can be applied in various tasks.

Experimental results indicate that the proposed F-HN method, which extends the triplet loss to enforce accurate intra-modal similarity orderings, yields significant improvements in text-to-image retrieval compared to the conventional cross-modal training. Besides, this method often ranks second for image-to-text retrieval, where the conventional approach is a more challenging baseline. The second proposed method, which uses intra-modal similarities to replace the margin hyper-parameter required by the conventional approach, produced more varied outcomes. Nevertheless, it holds promise for time-constrained applications where hyper-parameter tuning is problematic. Beyond classic cross-modal evaluation, our experiments on image-to-image and text-to-text retrieval tasks revealed that our F-HN method is particularly suitable for intra-modal tasks, often providing more accurate and novel retrieval lists than classic cross-modal training. Overall, these findings highlight the importance of considering intra-modal similarities in cross-modal learning, especially when the task involves retrieving items within the same modality, but relevance must be determined using another modality (e.g., because explicit feedback is not given). As future work, we plan to investigate the effect of including intra-modality similarity relationships for other negative sampling techniques beyond hard-negative. Also, we propose to investigate the impact of the proposed loss functions with more advanced visual and textual encoders (such as transformers [15]).

References

1. Chaudhuri, U., Banerjee, B., Bhattacharya, A., Datcu, M.: CMIR-NET: a deep learning based model for cross-modal retrieval in remote sensing. Pattern Recogn. Lett. **131**, 456–462 (2020)
2. Clarke, C.L., et al.: Novelty and diversity in information retrieval evaluation. In: SIGIR 2008 ,p p. 659–666. ACM, New York (2008)
3. Do, T.T., Tran, T., Ian, R., et al.: A theoretically sound upper bound on the triplet loss for improving the efficiency of deep distance metric learning. In: IEEE CVPR, pp. 10404–10413 (2019)
4. Dubey, S.R.: A decade survey of content based image retrieval using deep learning. IEEE Trans. Circ. Syst. Video Technol. **32**, 2687–2704 (2020)
5. Faghri, F., Fleet, D.J., Kiros, J.R., Fidler, S.: VSE++: improving visual-semantic embeddings with hard negatives. In: Proceedings of BMVC (2017)
6. Ge, W., Huang, W., Dong, D., Scott, M.R.: Deep metric learning with hierarchical triplet loss. In: Ferrari, V., Hebert, M., Sminchisescu, C., Weiss, Y. (eds.) ECCV 2018. LNCS, vol. 11210, pp. 272–288. Springer, Cham (2018). https://doi.org/10.1007/978-3-030-01231-1_17
7. Gong, Y., Cosma, G.: Improving visual-semantic embeddings by learning semantically-enhanced hard negatives for cross-modal information retrieval. Pattern Recogn. **137**, 109272 (2023)
8. Karpathy, A., Fei-Fei, L.: Deep visual-semantic alignments for generating image descriptions. IEEE Trans. Pattern Anal. Mach. Intell. **39**(4), 664–676 (2017)
9. Mahmut, K., Şakir, H.: Deep metric learning: a survey. Symmetry **11**(9), 1066 (2019)
10. Kingma, D.P., Ba, J.: Adam: a method for stochastic optimization (2017)

11. Li, X., Yang, J., Ma, J.: Recent developments of content-based image retrieval (CBIR). Neurocomputing **452**, 675–689 (2021)
12. Lin, C.Y.: Rouge: a package for automatic evaluation of summaries. In: Text summarization branches out, pp. 74–81 (2004)
13. Lin, T.-Y., et al.: Microsoft COCO: common objects in context. In: Fleet, D., Pajdla, T., Schiele, B., Tuytelaars, T. (eds.) ECCV 2014. LNCS, vol. 8693, pp. 740–755. Springer, Cham (2014). https://doi.org/10.1007/978-3-319-10602-1_48
14. Ma, H., et al.: Ei-clip: entity-aware interventional contrastive learning for e-commerce cross-modal retrieval. In: CVPR, pp. 18051–18061 (2022)
15. Messina, N., et al.: Fine-grained visual textual alignment for cross-modal retrieval using transformer encoders. ACM Trans. Multimedia Comput. Commun. Appl. (TOMM), **17**(4), 1–23 (2021)
16. Messina, N., Falchi, F., Esuli, A., Amato, G.: Transformer reasoning network for image-text matching and retrieval. In: 2020 25th International Conference on Pattern Recognition (ICPR), pp. 5222–5229. IEEE (2021)
17. Molina, G., et al.: A new content-based image retrieval system for SARS-CoV-2 computer-aided diagnosis. In: Su, R., Zhang, Y.-D., Liu, H. (eds.) MICAD 2021. LNEE, vol. 784, pp. 316–324. Springer, Singapore (2022). https://doi.org/10.1007/978-981-16-3880-0_33
18. Anderson, P., Fernando, B., Johnson, M., Gould, S.: SPICE: semantic propositional image caption evaluation. In: Leibe, B., Matas, J., Sebe, N., Welling, M. (eds.) ECCV 2016. LNCS, vol. 9909, pp. 382–398. Springer, Cham (2016). https://doi.org/10.1007/978-3-319-46454-1_24
19. Ren, R., et al.: Pair: leveraging passage-centric similarity relation for improving dense passage retrieval, pp. 2173–2183 (2021)
20. Schubert, E.: A triangle inequality for cosine similarity. In: Reyes, N., et al. (eds.) SISAP 2021. LNCS, vol. 13058, pp. 32–44. Springer, Cham (2021). https://doi.org/10.1007/978-3-030-89657-7_3
21. Song, H.O., Xiang, Y., Jegelka, S., Savarese, S.: Deep metric learning via lifted structured feature embedding. In: IEEE CVPR, pp. 4004–4012 (2016)
22. Song, K., Tan, X., Qin, T., Lu, J., Liu, T.Y.: Mpnet: masked and permuted pre-training for language understanding. NIPS **33**, 16857–16867 (2020)
23. Song, Y., Soleymani, M.: Polysemous visual-semantic embedding for cross-modal retrieval. In: CVPR, pp. 1979–1988 (2019)
24. Tan, M., Le, Q.V.: Efficientnetv2: smaller models and faster training. CoRR abs/2104.00298 (2021)
25. Tian, Y., et al.: Sosnet: second order similarity regularization for local descriptor learning, pp. 11008–11017 (2019)
26. Ng, T., Balntas, V., Y, Tian., Mikolajczyk, K.: Solar: Second-order loss and attention for image retrieval. ArXiv (2020)
27. Wang, Z., et al.: Adaptive margin based deep adversarial metric learning. In: IEEE BigDataSecurity/HPSC/IDS 2020, pp. 100–108 (2020)
28. Chen, W., Chen, X., Zhang, J., Huang, K.: Beyond triplet loss: a deep quadruplet network for person re-identification. In: IEEE CVPR, pp. 1320–1329 (2017)
29. Wu, Y., Wang, S., Huang, Q.: Online asymmetric similarity learning for cross-modal retrieval. In: IEEE CVPR, pp. 3984–3993 (2017)
30. Wu, Y., Wang, S., Huang, Q.: Online fast adaptive low-rank similarity learning for cross-modal retrieval. IEEE Trans. Multimedia **22**(5), 1310–1322 (2020)
31. Xuan, H., Stylianou, A., Liu, X., Pless, R.: Hard negative examples are hard, but useful. In: Vedaldi, A., Bischof, H., Brox, T., Frahm, J.-M. (eds.) ECCV 2020.

LNCS, vol. 12359, pp. 126–142. Springer, Cham (2020). https://doi.org/10.1007/978-3-030-58568-6_8

32. Yang, J., et al.: Vision-language pre-training with triple contrastive learning. In: 2022 IEEE/CVF Conference on Computer Vision and Pattern Recognition (CVPR), pp. 15650–15659 (2022)

33. Ye, M., et al.: Deep learning for person re-identification: a survey and outlook. IEEE Trans. Pattern Anal. Mach. Intell. 44(6), 2872–2893 (2021)

34. Young, P., Lai, A., Hodosh, M., Hockenmaier, J.: From image descriptions to visual denotations: new similarity metrics for semantic inference over event descriptions. TACL 2, 67–78 (2014)

35. Zhao, C., et al.: Deep fusion feature representation learning with hard mining center-triplet loss for person re-identification. IEEE Trans. Multimedia 22(12), 3180–3195 (2020)

36. Zhou, T., et al.: Solving the apparent diversity-accuracy dilemma of recommender systems. PNAS 107, 4511–4515 (2010)

Exploring the Potential of Optimal Active Learning via a Non-myopic Oracle Policy

Christoph Sandrock[✉] , Marek Herde , Daniel Kottke ,
and Bernhard Sick

University of Kassel, Wilhelmshöher Allee 73, 34121 Kassel, Germany
{christoph.sandrock,marek.herde,daniel.kottke,bsick}@uni-kassel.de

Abstract. Active learning aims to reduce the amount of labeled data while maximizing machine learning models' performances. Currently, there is sparse research on the potential of an optimal active learning strategy. Therefore, we propose a non-myopic oracle policy that accesses the true labels of the data pool to approximate an optimal active learning strategy. We evaluate how the hyperparameters of this oracle policy influence its performance and empirically demonstrate that it is an upper baseline for common active learning strategies while being faster than a state-of-the-art oracle policy. For the sake of reproducibility, all the code related to our research is publicly available on our GitHub repository at https://github.com/ies-research/non-myopic-oracle-policy.

Keywords: Active Learning · Oracle Policy · Classification

1 Introduction

Active learning (AL) aims to reduce the amount of required labeled data for training well-performing machine learning models [9]. The idea of pool-based AL is that the learner uses a specific *selection strategy* to find the most beneficial instances to be annotated for training. While many articles present such selection strategies, there needs to be more research exploring the potential of AL. Concretely, assessing the gap between the performance of selection strategies and an optimal instance selection is difficult. Moreover, it is unclear whether selection strategies can outperform a random instance selection in certain cases. Finding an optimal selection strategy would help to investigate these issues. Even if the labels of all instances are known in advance, this is computationally intractable in practice due to the high number of possible instance selections [14]. This motivates *oracle policies*, which are hypothetical selection strategies knowing the labels of all unlabeled instances in advance. They leverage this knowledge to approximate an optimal selection strategy and to construct an upper baseline for actual selection strategies without that prior knowledge.

In this article, we propose a novel iterative non-myopic oracle policy, which resolves issues of a myopic (greedy) instance selection by considering future label acquisitions. Further, we ablate the hyperparameters of this oracle policy, compare it to existing oracle policies, and show that it builds an upper baseline for common selection strategies in the context of classification tasks.

© The Author(s), under exclusive license to Springer Nature Switzerland AG 2023
A. Bifet et al. (Eds.): DS 2023, LNAI 14276, pp. 265–276, 2023.
https://doi.org/10.1007/978-3-031-45275-8_18

2   Problem Setting

We consider pool-based AL [9] with $\mathcal{X}$ as the instance space and $\mathcal{Y}$ as the set of class labels. Further, let $p(x, y)$ be the data generating distribution over $\mathcal{X} \times \mathcal{Y}$, and let $\mathcal{D} \sim p(x, y)$ be the observed data pool whose elements, i.e., instance-label pairs, are drawn independently from $p(x, y)$. A budget $B \in \mathbb{N}_{>0}$ specifies the maximum number of annotated instances. A selection strategy or oracle policy defines a *selection order* as an injective function $\sigma : \{1, \ldots, B\} \to \{1, \ldots, |\mathcal{D}|\}$ mapping indices of AL iterations to indices of instances in the data pool $\mathcal{D}$. This order induces a sequence of labeled sets

$$\mathcal{L}_0 := \emptyset, \quad \mathcal{L}_k := \{(x_{\sigma(i)}, y_{\sigma(i)}) : 1 \leq i \leq k\} \tag{1}$$

and unlabeled sets

$$\mathcal{U}_0 := \{x : (x, \cdot) \in \mathcal{D}\}, \quad \mathcal{U}_k := \{x : (x, \cdot) \in \mathcal{D} \backslash \mathcal{L}_k\} \tag{2}$$

for iteration $k \in \{1, \ldots, B\}$. Oracle policies know all labels in $\mathcal{D}$ from the start, while selection strategies only know the labels in $\mathcal{L}_{k-1}$ at iteration k. Given a classifier $f_{\mathcal{L}} : \mathcal{X} \to \mathcal{Y}$, we aim to find a selection order minimizing the widespread area under the learning curve (AULC)

$$\frac{1}{B} \sum_{k=1}^{B} R(f_{\mathcal{L}_k}) \text{ with } R(f_{\mathcal{L}_k}) := \mathbb{E}_{(x,y) \sim p(x,y)} L(f_{\mathcal{L}_k}(x), y) \tag{3}$$

as the expected risk for a loss function L, such as the 0–1 loss. Since the true data generating distribution $p(x, y)$ is unknown, we approximate the expected risk with a test set $\mathcal{T} \sim p(x, y)$, which results in the empirical risk

$$\hat{R}_{\mathcal{T}}(f_{\mathcal{L}_k}) := \frac{1}{|\mathcal{T}|} \sum_{(x,y) \in \mathcal{T}} L(f_{\mathcal{L}_k}(x), y) . \tag{4}$$

3   Related Work

The first oracle policy proposed by Koshorek et el. [6] uses a fixed batch size $D \in \mathbb{N}_{>0}$ and a sample size $S \in \mathbb{N}_{>0}$ to build the labeled set iteratively. The policy randomly draws S candidate sets of D unlabeled instances in each itera-tion. It retrains the classifier for all sets and evaluates its performance on a given validation set. Then, it selects the set with the best performance and starts the next iteration. There are a total of $S \cdot \lceil B/D \rceil$ classifier validations. The authors tested only the sampling sizes $S \leq 5$, for which the policy does not consistently outperform random sampling. Zhou et al. [14] propose an oracle policy for deep AL in a pool-based setting. Unlike the previous one, this policy does not build the labeled set iteratively. Instead, it uses a simulated annealing (SA) search to opti-mize the selection order. The search starts with a random order and iteratively exchanges a batch of instances in this order to optimize the AULC. Therefore, the policy proceeds in two phases: In the first phase ($T_S \in \mathbb{N}_{>0}$ iterations), the policy randomly draws two indices and exchanges them with probability proportional to the performance improvement. In the second phase ($T_G \in \mathbb{N}_{>0}$ iterations), it greedily optimizes the order, i.e., accepts the change if and only if the performance of the order increases. The optimization regarding the complete learning curve means that each step has a linear run time in the number of annotated instances B. This results in $(T_S + T_G) \cdot B$ classifier validations making it computationally very costly for large budgets. The authors' experiments show that their oracle policy clearly outperforms all evaluated selection strategies.

4   Non-myopic Oracle Policy

Even if the labels in the data pool $\mathcal{D}$ are known in advance, the difficulty in finding an optimal selection strategy is the fast-growing number of possible selection orders $\binom{|\mathcal{D}|}{B} \cdot B!$ with an increasing budget B. Thus, an exhaustive search exceeds computing capacities for common data sets and budgets. There are two approaches for approximating an optimal selection order: The first approach [14] is to start with any selection order σ and optimize it by iteratively changing this order, which has the advantage that the AULC can be directly optimized. However, such AULC computations are very expensive (cf. Sect. 3).

Algorithm 1: non_myopic_oracle_policy

---

input : budget B, data pool $\mathcal{D}$, validation set $\mathcal{V}$, sample size S, lookahead M
output: selection order σ
// Initialize labeled and unlabeled set
$\mathcal{L}_0 \leftarrow \emptyset$
$\mathcal{U}_0 \leftarrow \{x : (x,y) \in \mathcal{D}\}$
for $k = 1, \ldots, B$ **do**   // Select until budget is reached
 // Initialize risk table
 $\forall x \in \mathcal{U}_{k-1}, \forall m \in \{1, \ldots, M\} : T_k[x,m] \leftarrow \infty$
 for $m = 1, \ldots, M$ **do**
  for $s = 1, \ldots, S$ **do**
   Draw $\mathcal{D}_s \subseteq \mathcal{D} \backslash \mathcal{L}_{k-1}$ with $|\mathcal{D}_s| = m$ s.t. $\mathcal{D}_s \neq \mathcal{D}_i \, \forall i < s$
   // Extend labeled set by candidate set
   $\mathcal{L}^+ \leftarrow \mathcal{L}_{k-1} \cup \mathcal{D}_s$
   for $(x,y) \in \mathcal{D}_s$ **do**
    // Update risk table
    $T_k[x,m] \leftarrow \min(T_k[x,m], \hat{R}_\mathcal{V}(f_{\mathcal{L}^+}))$

 $(\hat{x}, \hat{y}) \leftarrow$ best_candidate$(T_k, \mathcal{D} \backslash \mathcal{L}_{k-1})$
 // Get index to update oracle policy
 $\sigma(k) \leftarrow$ index_of$(\hat{x}, \mathcal{D})$
 // Update labeled and unlabeled set
 $\mathcal{L}_k \leftarrow \mathcal{L}_{k-1} \cup \{(\hat{x}, \hat{y})\}$
 $\mathcal{U}_k \leftarrow \mathcal{U}_{k-1} \backslash \{\hat{x}\}$

The second approach [6] avoids these expensive computations by building the selection order iteratively. Due to its greedy nature, this approach is vulnerable to local minima, which may result in bad long-term performances. We propose an iterative non-myopic oracle policy (NOP) in Algorithm 1, which avoids greedy behavior through a lookahead $M \in \mathbb{N}_{>0}$. That is, the policy considers candidate sets $\mathcal{D}_s \subset \mathcal{D}$ of size $|\mathcal{D}_s| = m \in \{1, \ldots, M\}$ and evaluates their performance gains. Although our policy evaluates sets of instance-label pairs, the final selection concerns a single instance-label pair differing from the batch AL setting in [6]. Due to a large number of sets, the policy randomly samples S candidate sets of size m from the pool. The candidate sets are not disjoint since it would limit the number of candidate sets. For each candidate set in iteration k, the policy updates a risk table $T_k \in (\mathbb{R} \cup \{\infty\})^{|\mathcal{U}_{k-1}| \times M}$ to track the respective improvement.

Algorithm 2: `best_candidate`

---

input : table T, data pool $\mathcal{D}$
output: instance-label pair $(\hat{x}, \hat{y})$
// Compute minimum risks
$\forall (x, y) \in \mathcal{D} : R_{\min}(x) \leftarrow \min_{m \in \{1, \ldots, M\}} T[x, m]$
// Initialize candidates
$\mathcal{C} \leftarrow \{(\tilde{x}, \tilde{y}) \in \mathcal{D} : R_{\min}(\tilde{x}) = \min_{(x,y) \in \mathcal{D}} R_{\min}(x)\}$
for $m = 1, \ldots, M$ **do**   // Reverse order for backward filtering
  | // Filter candidates
  | $\mathcal{C} \leftarrow \{(x, y) \in \mathcal{C} : T[x, m] = \min_{(x,y) \in \mathcal{C}} T[x, m]\}$
$(\hat{x}, \hat{y}) \leftarrow$ `random_selection`$(\mathcal{C})$

After evaluating all sampled candidate sets, the `best_candidate` function in Algorithm 2 describes how the policy selects an instance-label pair $(\hat{x}, \hat{y})$ to update the selection order, based on the risk table. For this, it iteratively filters the rows (instances) of T_k. Figure 1 illustrates this procedure. In the first step, `best_candidate` calculates the minimum risk $R_{\min}(x) := \min_m T_k[x, m]$ of every instance-label pair $(x, y) \in \mathcal{D} \setminus \mathcal{L}_{k-1}$ (cf. the $R_{\min}$ columns in Fig. 1). This value is the minimum risk of all drawn sets containing the respective instance. The policy then only keeps candidates with minimum $R_{\min}$ among all candidates.

If there are multiple candidates with minimum risk (which always happens if a candidate set with a size greater than one leads to minimum risk), we propose two different filterings to select one candidate, depending on whether we aim to optimize the AULC or only the performance on a fixed budget. If the goal is to optimize the AULC, focusing on the short-term reward is beneficial as it leads to multiple small improvements in the learning curve. Therefore, we iteratively filter the remaining rows of the table from left to right (*forward filtering*, cf. the left side of Fig. 1). In the i-th iteration, we only keep the instances with minimum risk among the remaining candidates in the i-th column. This way, we focus on instances that immediately lead to a lower risk. If the goal is to

Risk Table for Forward Filtering					Risk Table for Backward Filtering					Legend
T_k	R_{min}	$m=1$	$m=2$	$m=3$	$m=4$	T_k	$m=1$	$m=2$	$m=3$	$m=4$	R_{min}	Candidates $\mathcal{C}$ after the
x_1	0.1	0.6	0.4	0.5	0.1	x_1	0.6	0.4	0.5	0.1	0.1	initialization
x_2	0.2	0.7	0.6	0.2	0.3	x_2	0.7	0.6	0.2	0.3	0.2	1st iteration
x_3	0.1	0.6	∞	0.1	0.3	x_3	0.6	∞	0.1	0.3	0.1	2nd iteration
x_4	0.1	0.6	0.4	0.2	0.1	x_4	0.6	0.4	0.2	0.1	0.1	3rd iteration
$\vdots$	$\vdots$	$\vdots$	$\vdots$	$\vdots$	$\vdots$	$\vdots$	$\vdots$	$\vdots$	$\vdots$	$\vdots$	$\vdots$	
x_n	0.3	0.8	∞	0.3	0.7	x_n	0.8	∞	0.3	0.7	0.3	

Fig. 1. Exemplary risk table T_k for $M = 4$ and $|\mathcal{U}_{k-1}| = n$ as basis for the candidate selection. The instances $x \in \{x_1, x_3, x_4\}$ have the minimum risk $R_{min}(x) = 0.1$ and are therefore the only ones we consider in the selection. Colored cells show how long an instance remains in the candidate set $\mathcal{C}$. We query the instance x_4 for both filterings.

optimize performance at a fixed budget, the performance at the final iteration is all that matters. To focus on such a long-term reward, we propose to filter the table from right to left (*backward filtering*, cf. the right side of Fig. 1), thereby ignoring the short-term rewards. In both cases, if two instances have the same entry in every cell, we randomly select one of these instances because we cannot distinguish them through the risk table T_k.

The **lookahead** M describes the maximum size of the candidate sets to be considered. Small values (especially $M = 1$) might lead to a bad approximation because a greedy search with small lookahead can easily run into local minima. To achieve a truly optimal algorithm, we need to set $M = B$, such that all permutations of the candidates are considered and the best one is found. But since the number of permutations grows fast in M, larger values are computationally infeasible without restricting to a sub-sample of all candidate sets.

The **sampling size** S determines the number of randomly drawn candidate sets per lookahead. That is, for every $m = 1, \ldots, M$, the algorithm draws up to S candidate sets of size m and only evaluates the performance of these sets. This may lead to some instances not being considered if they appear in none of the drawn sets. Koshorek et el. [6] only evaluate their greedy oracle policy for $S \leq 5$. We suppose that this is too small to outperform random selections since the largest part of the candidates is not covered.

The **validation set** $\mathcal{V}$ is required to determine the performance of the selection order. It can be chosen in two ways: The first option is to draw a new set $\mathcal{V} \sim p(x, y)$, and the second option is to optimize the performance directly on the test set by setting $\mathcal{V} := \mathcal{T}$. Drawing a new set yields a more realistic baseline for selection strategies since they do not know the test set. However, the optimal strategy for $\mathcal{V}$ may differ from the optimal strategy for $\mathcal{T}$, especially if the sets are small. This may lead to an underestimation of the potential of AL.

The **run time** of our NOP is a function of $M \cdot S \cdot B$ corresponding to the maximum number of classifier validations. In comparison, the SA search requires $(T_S + T_G) \cdot B$ classifier validations. For $M \cdot S = T_S + T_G$, both algorithms have the same number of classifier validations. In this case, the iterative strategy evaluates B-times as many candidate sets, resulting in better candidate space coverage.

5 Experiments

Our experiments are divided into two parts. The first part ablates the NOP's hyperparameters. The second part compares NOPs to the SA search [14] and common selection strategies. Both analyses evaluate the policies' and strategies' learning curves and AULCs. Although we only present a selected set of learning curves here, an exhaustive list of learning curves can be found in our repository.

5.1 Experimental Setup

For our experiments, we use 13 OpenML [13] data sets listed in Table 1, which are preprocessed by standardizing the instances' features. We repeat each experiment 25 times. In each repetition, we randomly split the data set into a data pool $\mathcal{D}$ containing 20% of the instances, a validation set $\mathcal{V}$ with 40% of the instances, and a test set $\mathcal{T}$ containing the remaining 40% of the instances. Following the setup of Zhou et al. [14], we set $B := \lfloor |\mathcal{U}|/2 \rfloor$, and only the oracle policies optimize their selection on the validation set. After each iteration k, we compute the 0–1 loss of the classifier fitted with $\mathcal{L}_k$ both on the test and the validation set. The results for the validation set are only given in the repository. For the SA search, we compute the selection order σ beforehand and select $x_{\sigma(k)}$ in iteration k. For classification, we use a Parzen window classifier [2] with a radial basis function kernel, whose bandwidth hyperparameter is specified according to the mean bandwidth criterion [3]. This classifier allows us to compare the policies to selection strategies such as probabilistic AL and epistemic uncertainty sampling. Moreover, the classifier's fast (re-)training speeds up the AL process, in particular the instance selection of the oracle policies. Still, the oracle policies can be combined with any other classifier.

Table 1. Description of data sets by (reference) name, number of instances, number of features, number of classes, and OpenML identifier (ID).

data set name	reference name	instances	features	classes	OpenML ID
chscase_vine2	VINE	468	2	2	814
kc2	KC2	522	22	2	1063
wdbc	WDBC	569	30	2	1510
balance-scale	BAL	625	5	3	11
blood-transfusion	BLOOD	748	5	2	1464
diabetes	DIAB	768	9	2	37
vehicle	VEH	846	18	4	54
qsar-biodeg	QSAR	1022	42	2	1494
banknote-authentication	BANK	1372	4	2	1462
steel-plates-fault	STEEL	1941	34	2	1504
mfeat-pixel	PIXEL	2000	240	10	20
segment	SEG	2310	19	7	36
satimage	SAT	6430	36	6	182

5.2 Ablation Studies

Our experiments' first part compares different configurations of the hyperparameters introduced in Sect. 4. Note that policies cannot always sample S candidate sets since there are settings where the number of available candidate sets is smaller than S. This is particularly true for the greedy oracle policy (GOP) with $M = 1$ because then the number of candidate sets corresponds to the number of unlabeled instances. We use mean rank tables to evaluate the configurations. Accordingly, we compute the ranking of the configurations regarding their AULCs for each data set, averaged over all repetitions. Further, we compute the mean across all ranks (cf. column mean). Based on these tables, we examine the following four hypotheses.

H1: NOPs strongly outperform a GOP. Table 2 shows the ranks of nine selected hyperparameter configurations with the same maximum number of classifier validations. The GOP ranks worst on most data sets. The only exceptions are WDBC, DIAB, KC2, and BLOOD. The only data set where the GOP performs better than most NOPs is KC2. On this data set, the Parzen window classifier can reach its maximum accuracy with only two labeled instances (cf. Fig. 2). Thus, the greedy selection of the GOP is not disadvantageous in this case. In most other cases, the GOP's rank is much worse than the rank of the NOPs. The learning curves in Fig. 2 support these results. Here, the GOP often performs substantially worse than all NOPs. The remaining learning curves in the repository show similar results. This confirms H1 that a non-myopic selection is crucial for an oracle policy to approximate an optimal instance selection.

H2: The forward filtering achieves a better AULC. We check if traversing the risk table T_k via forward filtering results in a better AULC than traversing it via backward filtering. As expected (cf. Sect. 4), the mean rank with forward filtering is better than the one with backward filtering for every hyperparameter configuration in Table 2, respectively. Only for two data sets (SAT and STEEL), the best configuration uses backward filtering. However, the differences are fairly small in most cases. Overall, the filtering type does not seem to impact the performance majorly. For all the following hypotheses, we only consider forward filtering.

Table 2. Ranks of nine selected hyperparameter configurations for all data sets regarding their AULCs. The bold red numbers highlight the best rank in each column.

M	S	filtering	VINE	WDBC	VEH	BANK	PIXEL	SEG	SAT	BAL	DIAB	KC2	BLOOD	QSAR	STEEL	mean
1	1000	–	6.36	5.86	7.56	8.84	7.28	8.64	8.76	6.60	5.20	3.66	5.68	7.52	8.88	6.99
2	500	forward	5.44	4.46	3.64	3.12	4.50	4.76	5.44	4.44	4.22	3.94	4.70	5.28	3.96	4.45
3	333	forward	3.92	4.22	4.28	4.36	4.64	4.46	3.72	4.04	5.52	4.10	5.28	4.00	3.68	4.32
4	250	forward	4.80	3.44	4.44	4.00	3.96	4.12	4.40	4.32	4.08	4.16	3.88	5.32	4.76	4.28
5	200	forward	4.36	4.74	5.20	4.92	4.36	5.12	4.76	4.76	5.00	3.42	4.08	4.44	5.48	4.66
2	500	backward	4.84	4.76	4.60	3.76	4.74	4.42	5.28	4.96	5.34	4.02	4.42	4.48	3.64	4.56
3	333	backward	4.44	5.64	5.24	5.56	5.76	4.48	3.48	4.96	4.84	6.92	5.32	5.08	5.04	5.14
4	250	backward	5.24	5.92	4.56	4.56	4.28	4.48	3.92	5.36	5.00	6.82	5.72	4.40	4.56	4.99
5	200	backward	5.60	5.96	5.48	5.88	5.48	4.52	5.24	5.56	5.80	7.96	5.92	4.48	5.00	5.61

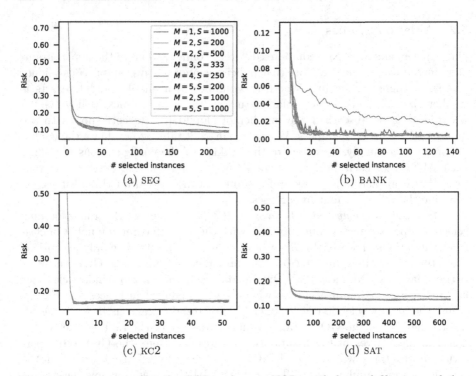

Fig. 2. Learning curves for the GOP and seven NOPs with forward filtering and the shown hyperparameters.

H3: The trade-off between sample size and lookahead influences the performance. As shown in Sect. 4, the NOPs' run times are determined by $B \cdot M \cdot S$ classifier validations. Hence, we expect that a good trade-off between the lookahead M and the sample size S is crucial for good performance and fast run time. Table 2 compares hyperparameter configurations with $M \cdot S = 1000$ candidate sets per iteration. Thereby, all compared NOPs have similar run times. The best values for M and S vary between the data sets. Each NOP with forward filtering performs best on at least one data set such that there is no clear winner across all data sets. The mean ranks of the NOPs with forward filtering differ only by less than 0.4. In contrast, the ranks on the individual data sets differ more strongly. Thus, choosing a suitable trade-off depends mainly on the specific data set. Yet, the learning curves in Fig. 2 and our repository show that the performance improvement of a suitable trade-off for an NOP is fairly small.

H4: Higher sample size and lookahead improve the policy's performance. The experiments with more candidate sets show that increasing M and S improves the performance. As one can see in Table 3, the NOP with the highest number of considered sets (i.e., $M = 5$ and $S = 1000$) performs best on 10 of the 13 data sets. As before, the ranks do not differ much. That is, the higher sample size still does not persistently improve the performance. The learning curves in Fig. 2 show that all curves of the NOPs are similar or nearly identical. Even

Table 3. Ranks of NOPs with varying computational effort for all data sets regarding their AULCs. The bold red numbers highlight the best rank in each column.

M	S	filtering	VINE	WDBC	VEH	BANK	PIXEL	SEG	SAT	BAL	DIAB	KC2	BLOOD	QSAR	STEEL	mean
2	200	forward	2.64	2.64	3.00	2.44	3.12	2.80	2.76	2.92	2.72	2.52	2.48	2.56	2.76	2.72
2	500	forward	2.96	2.62	2.36	2.40	2.52	2.80	2.92	2.64	2.56	2.12	2.44	2.96	2.56	2.60
2	1000	forward	2.20	2.14	2.68	2.44	2.64	2.32	2.32	2.32	2.60	2.84	2.68	2.32	2.44	2.46
5	1000	forward	2.20	2.60	1.96	2.72	1.72	2.08	2.00	2.12	2.12	2.52	2.40	2.16	2.24	2.22

more candidate sets do not change the learning curve notably. This suggests that the tested NOPs are already close to convergence toward an optimal instance selection. All plots for the remaining data sets show similar behavior.

Summary: The most crucial parameter for the performance is to set $M > 1$, i.e., to use an NOP. As expected, the forward filtering outperforms the backward filtering. However, the difference is fairly small. The trade-off between lookahead and sample size depends on the data set. Increasing the lookahead and the sample size improves the performance only slightly. Hence, one can save a lot of computation time by choosing a small lookahead and a relatively small sample size, e.g., $M = 2, S = 200$, without notably losing performance.

5.3 Comparison with SA Search and Selection Strategies

This section compares the GOP/NOPs with a random baseline (Rand), and with the popular selection strategies uncertainty sampling (LC) [10], query by committee (QBC) [12], epistemic uncertainty sampling (Epis) [11], probabilistic active learning (PAL) [8], discriminative active learning (DAL) [4], and querying informative and representative examples (QUIRE) [5]. Since epistemic uncertainty sampling works only for binary classification tasks, we evaluate it only on eight data sets. For all selection strategies, we use the implementations of scikit-activeml [7] with default hyperparameters. Further, we compare the GOP/NOPs with SA search [14] with the proposed annealing factor of $\gamma := 0.1$. To the best of our knowledge, this is the only comparable oracle policy for sequential pool-based AL. We investigate two different configurations: one with the hyperparameters $T_S := 25000, T_G := 5000$ as proposed by Zhou et al. [14] (which takes much more time than our tested GOP/NOPs, as shown on right side of Table 4) and one with $T_S := 2000, T_G := 400$ (SA fast) being more comparable to the run time of the tested GOP/NOPs. Table 4 (left side) compares each GOP/NOP to its competitors by computing the win percentage across all data sets and repetitions, where a win corresponds to a lower AULC. For example, a win percentage larger than 50% indicates that the GOP/NOP has more wins than its respective competitor. Following the idea of Benavoli et al. [1], we test the statistical significance of the pairwise performance differences between each GOP/NOP and each competitor via a two-sided Wilcoxon signed-rank test. For this purpose, we compute the mean AULC per data set and algorithm. We target a global significance level of 5% as the upper bound for the familywise error rate. Therefore, we apply a Bonferroni correction for the comparisons of each GOP/NOP to its

Table 4. Left: Comparison of GOP/NOPs vs. competitors for AULC. The column titles denote M/S. The table entries denote the percentages of GOP/NOP wins over all data sets and repetitions. The symbol denotes whether the performance results provide sufficient evidence that the GOP/NOP is significantly better (+) or worse (−) than its respective competitor. If no symbol is given, the performance results do not provide sufficient evidence regarding a significant performance difference between the GOP/NOP and its respective competitor. **Right:** Time comparison on the data set SAT. Our repository shows the times for the other data sets and selection strategies.

	GOP 1/1000	NOP 2/200	NOP 4/250	NOP 5/1000		time
PAL	88.6% +	98.2% +	97.5% +	97.8% +	GOP 1/1000	32h
QBC	96.6% +	98.2% +	98.8% +	98.2% +	NOP 2/200	13h
LC	89.2% +	98.2% +	99.1% +	99.1% +	NOP 4/250	38h
Rand	91.7% +	98.2% +	97.5% +	97.8% +	NOP 5/1000	151h
QUIRE	92.9% +	97.2% +	97.2% +	97.2% +	LC	52s
DAL	89.2% +	98.2% +	98.8% +	98.5% +	PAL	502s
Epis	86.5%	97.0%	97.0%	95.5%	QUIRE	54h
SA search	24.9%	43.2%	51.7%	53.8%	SA search	402h
SA fast	52.3%	81.5% +	82.2% +	84.9% +	SA fast	51h

nine competitors, which results in a local significance level of about 0.6%. Based on the results of Table 4, we examine the following two hypotheses.

H5: NOPs are an upper baseline for selection strategies. The win percentages in Table 4 (left side) show that the GOP/NOPs outperform all selection strategies. This observation is especially true for the NOPs, which consistently achieve win rates above 95%. These superior performances are significant except for the comparisons to Epis, since this strategy is limited to the eight binary classification tasks. Figure 3 confirms these superior performances by showing that there is a large gap between oracle policies and selection strategies. The curves of the selection strategies are closer to Rand than to the oracle policies.

H6: NOPs achieve similar performance as SA search while being much faster. Table 4 (right side) exemplarily demonstrates for the data set SAT that SA fast's run time is higher than the run times of the GOP, NOP 2/200, and NOP 4/250. Yet, Table 4 (left side) demonstrates that the GOP performs similarly to SA fast (win rate slightly above 50%), while NOP 2/200 and NOP 4/250 significantly outperform SA fast. SA search takes considerably longer than the GOP and all tested NOPs, as exemplary shown for the data set SAT in Table 4 (right side). However, Table 4 (left side) reveals that only the GOP is clearly inferior to the SA search (win rate about 25%), while the NOP 2/200 is only slightly worse (win rate about 43%). In contrast, the NOP 4/250 and NOP 5/1000 achieve comparable performances (win rates slightly above 50%) to SA search. The learning curves in Fig. 3 confirm the comparable performances between NOPs and SA search, while SA fast performs worse.

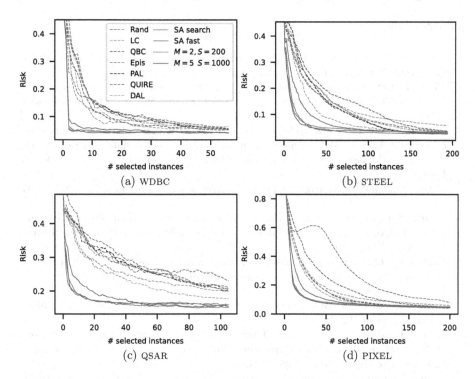

Fig. 3. Learning curves of oracle policies (solid lines) and selection strategies (dashed lines).

Summary: The results in Table 4 support H5 that NOPs are strong upper baselines for selection strategies. Moreover, they verify H6 that the SA search takes much longer to achieve comparable performances.

6   Conclusion

In this article, we proposed a novel non-myopic oracle policy (NOP) to explore the potential of optimal active learning (AL). We showed that NOPs are robust upper baselines for selection strategies while being faster than the simulated annealing search [14] as the only comparable oracle policy. We further demonstrated that, most notably, instance selections with lookaheads benefit AL, which could motivate future research to design non-myopic selection strategies. Exploring NOPs' performances for other classifiers or even regressors would also be an interesting research issue. Finally, combining NOPs with deep neural networks is appealing and requires an extension toward AL with batch instance selections.

References

1. Benavoli, A., Corani, G., Mangili, F.: Should we really use post-hoc tests based on mean-ranks? J. Mach. Learn. Res. **17**(1), 152–161 (2016)
2. Chapelle, O.: Active learning for Parzen window classifier. In: International Conference on Artificial Intelligence and Statistics, pp. 49–56. Bridgetown, Barbados (2005)
3. Chaudhuri, A., Kakde, D., Sadek, C., Gonzalez, L., Kong, S.: The mean and median criteria for kernel bandwidth selection for support vector data description. In: International Conference on Data Mining Workshops, pp. 842–849. New Orleans, LA (2017)
4. Gissin, D., Shalev-Shwartz, S.: Discriminative active learning. arXiv:1907.06347 (2019)
5. Huang, S.J., Jin, R., Zhou, Z.H.: Active learning by querying informative and representative examples. In: Advance in Neural. Information Processing System Vancouver, BC (2010)
6. Koshorek, O., Stanovsky, G., Zhou, Y., Srikumar, V., Berant, J.: On the limits of learning to actively learn semantic representations. In: Conference Comput. Nat. Lang. Learn. Hong Kong (2019)
7. Kottke, D., et al.: scikit-activeml: a library and toolbox for active learning algorithms. Preprints (2021)
8. Kottke, D., Krempl, G., Lang, D., Teschner, J., Spiliopoulou, M.: Multi-class probabilistic active learning. In: European Conference on Artificial Intelligence, pp. 586–594. The Hague, Netherlands (2016)
9. Kumar, P., Gupta, A.: Active learning query strategies for classification, regression, and clustering: a survey. J. Comput. Sci. Technol. **35**(4), 913–945 (2020). https://doi.org/10.1007/s11390-020-9487-4
10. Lewis, D.D., Gale, W.A.: A sequential algorithm for training text classifiers. In: International Conference Research Development in Information Retrieval, pp. 3–12. Dublin, Ireland (1994)
11. Nguyen, V.-L., Destercke, S., Hüllermeier, E.: Epistemic uncertainty sampling. In: Kralj Novak, P., Šmuc, T., Džeroski, S. (eds.) DS 2019. LNCS (LNAI), vol. 11828, pp. 72–86. Springer, Cham (2019). https://doi.org/10.1007/978-3-030-33778-0_7
12. Seung, H.S., Opper, M., Sompolinsky, H.: Query by committee. In: Conference on Learnimg Theory, pp. 287–294. Pittsburgh, PA (1992)
13. Vanschoren, J., van Rijn, J.N., Bischl, B., Torgo, L.: OpenML: networked science in machine learning. SIGKDD Explor. **15**(2), 49–60 (2013)
14. Zhou, Y., Renduchintala, A., Li, X., Wang, S., Mehdad, Y., Ghoshal, A.: Towards Understanding the Behaviors of Optimal Deep Active Learning Algorithms. In: International Conference on Artificial Intelligence and Statistics, pp. 1486–1494. Virtual (2021)

Extrapolation is Not the Same as Interpolation

Yuxuan Wang[1]([⊠]) and Ross D. King[1,2,3] (iD)

[1] Department of Chemical Engineering and Biotechnology, University of Cambridge, Philippa Fawcett Drive, Cambridge CB3 0AS, UK
yw453@cam.ac.uk, rossk@chalmers.se
[2] Department of Computer Science and Engineering, Chalmers University, Gothenburg, Sweden
[3] Alan Turing Institute, London, UK

Abstract. We propose a new machine learning formulation designed specifically for extrapolation. The textbook way to apply machine learning to drug design is to learn a univariate function that when a drug (structure) is input, the function outputs a real number (the activity): $F(\text{drug}) \rightarrow$ activity. The PubMed server lists around twenty thousand papers doing this. However, experience in real-world drug design suggests that this formulation of the drug design problem is not quite correct. Specifically, what one is really interested in is extrapolation: predicting the activity of new drugs with higher activity than any existing ones. Our new formulation for extrapolation is based around learning a bivariate function that predicts the difference in activities of two drugs: $F(\text{drug1}, \text{drug2}) \rightarrow$ signed difference in activity. This formulation is general and potentially suitable for problems to find samples with target values beyond the target value range of the training set. We applied the formulation to work with support vector machines (SVMs), random forests (RFs), and Gradient Boosting Machines (XGBs). We compared the formulation with standard regression on thousands of drug design datasets, and hundreds of gene expression datasets. The test set extrapolation metrics use the concept of classification metrics to count the identification of extraordinary examples (with greater values than the training set), and top-performing examples (within the top 10% of the whole dataset). On these metrics our pairwise formulation vastly outperformed standard regression for SVMs, RFs, and XGBs. We expect this success to extrapolate to other extrapolation problems.

Keywords: machine learning · extrapolation · drug discovery

1 Introduction

The original motivation for this work came from applying machine learning (ML) to drug design, specifically quantitative structure activity relationship (QSAR) learning. The standard way to cast QSAR learning as ML is to learn a univariate function that when a drug (structure) is input, the function outputs a real number (the activity): $F(\text{drug}) \rightarrow$ activity. The PubMed server lists around twenty thousand papers doing this.

Experience in real-world drug discovery suggests that this formulation is not exactly what is really required in practice. Specifically, what one is really interested in is predicting the activity of new drugs with higher activity than any existing ones - extrapolation.

© The Author(s) 2023
A. Bifet et al. (Eds.): DS 2023, LNAI 14276, pp. 277–292, 2023.
https://doi.org/10.1007/978-3-031-45275-8_19

N.B. extrapolation in QSAR learning has two related meanings: one is the ability to make predictions for molecules with descriptor values (x_i) outside the applicability domain defined by the training set of the model (Fig. 1a) [1–3]; the other is the identification of the "extraordinary molecules" with activities (y) beyond the range of activity values in the training data (Fig. 1b) [1, 4]. In drug discovery both types of extrapolation are important. Extrapolating beyond training set descriptor values enables new molecular types (maybe unpatented) to be proposed. Extrapolating beyond the highest observed y values is strongly desired to select more effective drugs.

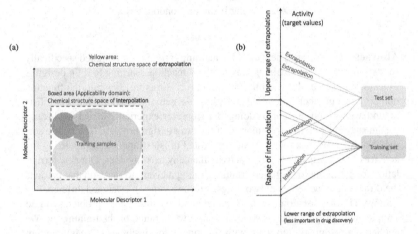

Fig. 1. The illustration of two types of extrapolation in drug discovery. (a) extrapolation outside the applicability domain, (b) extrapolation outside the range of drug activities.

Although many QSAR learning studies have reported advantageous ML methods based on their model prediction accuracy using metrics such as mean squared error, in practice the ability to produce accurate predictions is less valuable than the extrapolation ability in this type of application [4, 5]. In fact, some ML methods can hardly extrapolate beyond the training sets. For example, random forest (RF) is incapable of predicting target values (y) outside the range of the training set because it gives ensembled prediction by averaging over its leaf predictions [4, 6]. Our study is therefore motivated by the purpose to improve ML methods to be better at finding extraordinary samples (Fig. 1b). This will also be a tool that benefits many other applications, such as material sciences, dynamics modelling and system management.

Our extrapolation problem can be defined as following. Consider a training set of N_{tr} samples, its feature vectors of length N_f is $\mathbf{x} \in R^{(N_f \times N_{tr})}$, and its target activity values is $y \in R^{N_{tr}}$. Therefore, the range of the target values for the training set is $\{y_{tr,\min}, y_{tr,\max}\}$. A ML model f is then obtained so that $f(x_i) \approx y_i$. Suppose there exist a test set $\mathbf{x}_{ts}$ of size N_{ts} containing $N_{ex,\text{true}}$ ($N_{ex,\text{true}} <= N_{ts}$) extraordinary samples with target values $y_{ex} > y_{tr,\max}$. The extrapolation problem will be if the test samples with $f(x_{ts}) > y_{tr,\max}$ are truly extraordinary, or if the model f can rank extraordinary test samples above $x_{tr,\max}$ if f is a ranking method. In addition, we also define $N_{\text{top,true}}$ top-performing samples whose rank is within the top 10% of the whole dataset. We would like to know if the model can rank the top-performing test samples as top 10% of the dataset of $(N_{tr} + N_{ts})$ samples, once the model predicts $y_{ts}^{pred} = f(x_{ts})$ and rank the training and test samples by y_{tr} and y_{ts}^{pred} together.

There have been several studies recognising the importance of ranking performance in drug screening. Some have proposed to optimise the ML method directly to achieve higher ranking coefficients [7, 8], while some have instead proposed to boost the ranking performance from non-ML perspectives [9, 10]. Agarwal *et al.* proposed the method, RankSVM, to directly minimise a ranking loss to maximise the number of correctly ordered pairs of molecules for all ranks [7]. Rathke *et al.* reported a new algorithm, StructRank, which also directly solves the ranking problem with better focus and optimisation on the top-*k*-ranked molecules [8]. Al-Dabbagh *et al.* developed a probability ranking approach that employed quantum interference analogy [9]. Liu and Ning improved the ranking performance of SVMrank by leveraging assistance bioassays and compounds [10]. Zhang *et al.* have also deployed "Learning-to-rank" (LTR) from information retrieval successfully to integrate heterogeneous data and to identify compounds by prioritising their relevance to drug targets in a cross-target manner, similar to matching queries and documents in information retrieval applications [11]. Although our new approach also emphasises the importance of ranking to meet the problem specifications, it differs from LTR ranking algorithms. LTR data usually contain a large, fixed number of items matched with different queries. LTR models are trained to rank a fixed set of instances given queries, focusing on if the top-*k* items are correctly placed and extrapolation is not needed. LTR algorithms are therefore designed to incorporate the ranking of the items in the model objective directly, putting emphasis on the relative positions of test samples within the test set, rather than the extrapolation behaviour of a model. Our approach, however, makes common ML methods learn explicitly to distinguish samples' differences, so that it can later rank training and test samples to achieve extrapolation over the training set.

Some recent work has emphasised the importance of extrapolation and proposed new evaluation procedures for extrapolation performance of ML models. Kauwe *et al.* tested the extrapolation ability of several common ML methods by keeping the top 1% of the instances in the test sets for properties calculated from density functional theory [1]. Von Korff and Sander used sorted and shuffled datasets to evaluate extrapolation and interpolation performance, respectively [4]. Xiong *et al.*, Meredig *et al.* and Watson *et al.* have each proposed a new model validation technique to evaluate extrapolation performance of ML methods [6, 12, 13]. However, due to a lack of systematic review for them, it is unclear that these methods are statistically meaningful. Therefore, in this study, we apply standard *k*-fold cross validation [2, 6, 14].

This study proposes a ML configuration approach, the "pairwise approach", to boost the extrapolation ability of a traditional regression learning method. A pairwise model is designed to model the relationship between the differences in the structures of pairs of drugs and the sign of differences in their activity values. The learned pairwise model is a bivariate function, $F(\text{drug1}, \text{drug2}) \rightarrow$ signed difference in activity, whose outputs can give a better ranking of drugs by ranking algorithms. By transforming the learning objective, the pairwise model enables improved performance in extrapolation compared to traditional regression evaluation.

2 Method

2.1 Datasets and Data Pre-processing

ChEMBL is a chemical database of bioactive molecules [15, 16]. It contains a large number of molecules and their measured activities against a variety of targets. Due to their size and scope, these datasets are suitable for benchmarking ML applications in the realm of QSAR [17]. ChEMBL features a number of different activities, in this study we are employing pXC50 as our target values, *i.e.* -log(measured activity). The structure of drug molecules is represented by the commonly employed Morgan fingerprint (1024 bits, r = 2) encoding the molecular substructures by Boolean values [18].

The other large-scale database we used is the human gene expression datasets (accession code GSE70138) from the Library of Integrated Network-based Cellular Signatures data (LINCS) [19]. These datasets were used by Olier *et al.* in transformational ML study [16]. This set of datasets contains the measured gene expression level across different tissue types and drug treatments in cancer cell lines. There are in total of 978 human genes, each of which was measured under 118,050 experimental conditions. Each dataset is the expression levels of a gene, measured and processed as level 5 differential gene expression signatures, under a series of conditions. The conditions are featured into 1,154 Boolean values describing drugs' fingerprints (1024 bits) and experimental settings, which include 83 dosages, 14 cell types and 3 time points.

Before training any ML model, a basic feature selection is performed to reduce the large feature space and accelerate the learning process. For a given dataset, the features were removed if they have the same feature value assigned to every sample in a dataset. The features that repeat to have the same pattern for all the samples were also removed.

2.2 Formulation of Baseline Approach

In this study, when evaluating the performance of the pairwise approach on a specific dataset, in most cases it is compared with that of a baseline ML configuration, addressed as the "standard approach". It refers to the standard way of learning a regression problem. For a dataset, the model is built directly between the feature vector, $\mathbf{x}_i \in R^{(N_f \times N_s)}$, and the target value, $y_i \in R^{N_s}$ of all the samples. With multiple samples of known $(\mathbf{x}, y)$, a ML method can learn the relationship between features, $\mathbf{x}$, and target values, y, establishing a model f which can produce $f(\mathbf{x}_i) \approx y_i$ for the training set. The feature values of the test samples, $\mathbf{x}_{ts}$ are fed into the model f to obtain the predicted target values, $y_{ts}^{pred} = f(\mathbf{x}_{ts})$. The performance of this model is then evaluated using metrics for evaluating the extrapolation performance (see Sect. 2.6).

2.3 Formulation of Pairwise Approach

For a given Boolean dataset, a pair of samples P_{AB} is derived from sample A (S_A) and sample B (S_B). The difference in the ith feature for this pair can be presented in one of the following ways: present in both samples ($x_{A,i} = 1$, $x_{B,i} = 1$), present in S_A but not in S_B ($x_{A,i} = 1$, $x_{B,i} = 0$), present in S_B but not in S_A ($x_{A,i} = 0$, $x_{B,i} = 1$), and absent from both samples ($x_{A,i} = 0$, $x_{B,i} = 0$). To represent each type of difference in a feature, a unique value is assigned to the ith feature of the pair. An example of generating pairwise feature for the ith feature from a ChEMBL dataset is shown in Fig. 2. The unique values used in our experiments are:

$$x_{A,i} = 1, \ x_{B,i} = 1 \rightarrow X_{AB,i} = 2$$
$$x_{A,i} = 1, \ x_{B,i} = 0 \rightarrow X_{AB,i} = 1$$
$$x_{A,i} = 0, \ x_{B,i} = 1 \rightarrow X_{AB,i} = -1$$
$$x_{A,i} = 0, \ x_{B,i} = 0 \rightarrow X_{AB,i} = 0$$

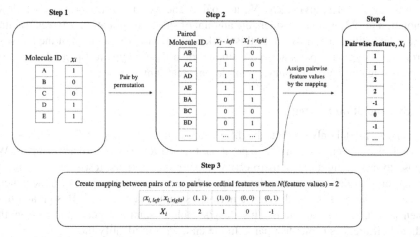

Fig. 2. An example of generating pairwise samples for a ChEMBL dataset.

The way of generating pairwise features is called ordinal encoding. It is often used for categorical features and each category value is assigned an integer value. Another popular way to encode real values for categorical features is one-hot encoding. It assigns Boolean bits to describe the absence or presence of each category. Therefore, it needs to at least double the size of features space. In the pairwise case, one-hot encoding is equivalent to the concatenation of features of two samples to generate the pairwise features. Considering the large expansion of training set by permutation, the further expansion in the feature size can greatly increase training time. Furthermore, our experiments on ChEMBL datasets have shown that one-hot encoding made little difference in the training accuracy. Therefore, we decided to use ordinal encoding for the pairwise features. In ordinal encoding, the choice of the integer value for each category is not restricted [20]. Despite potential doubts regarding the effect of their relative magnitudes under numeric transformations [21], it has been proven not to affect our study through simple tests. We endeavoured to assign each combination listed above with a different value (*e.g.*, $x_{A,i} = 1, x_{B,i} = 1 \rightarrow X_{AB,i} = -1; x_{A,i} = 0, x_{B,i} = 1 \rightarrow X_{AB,i} = 0$). We have also tried a different set of ordinal values, for example, using $\{1, 2, 3, 4\}$ instead of $\{-1, 0, 1, 2\}$. In both tests the results were hardly varied by the choice of ordinal values.

The pairwise target value needs to represent the difference in target values. For a specific pair, P_{AB}, its pairwise target value, Y_{AB}, is equal to $y_A - y_B$. Pairs P_{AB} and

P_{BA} are treated differently as two pairwise samples despite $Y_{AB} = -Y_{BA}$. A ML method can learn to predict the real values of those pairwise differences Y via regression or learn to predict the sign of the pairwise differences, sign(Y) via classification. The latter type of learning was found to be more advantageous to extrapolate the model and find extraordinary samples (see Sect. 2.4).

Suppose a dataset is split into a training set of size N_{tr} and a test set of size N_{ts}. The training samples are paired via permutation, creating N_{tr}^2 pairwise training pairs. This type of pairs is referred to as C1-type training pairs in this study. The test pairs can be obtained in two ways: (1) C3-type test pairs: generate from a permutation of test samples, giving N_{ts}^2 test pairs; (2) C2-type test pairs: generate from pairing test samples with training samples, giving $2N_{tr}N_{ts}$ test pairs. The naming of the pair types follows the notation in [22] which considers the amount of shared information between training and test data within a pair. Because this work is about the extrapolation of the pairwise approach, C2-type test pairs are more studied than the C3-type test pairs due to their ability to compare between training samples and test samples.

2.4 Extrapolation Strategy

The pairwise model only predicts the differences of pairs of samples. Therefore, a conclusive decision needs to be made to point out the predicted extraordinary samples. We propose to use rating algorithms to estimate the ranking of the test samples with the training set. The idea is to treat each predicted difference as the result of "a game match" between two samples. If the difference between sample A and sample B is greater than 0, then sample A wins sample B. This "league table of samples" gets updated from the predicted differences of the test pairs. In the end, we can identify the extraordinary or top-performing test samples.

Most of the generic rating algorithms were developed based on absolute wins or losses to give rating scores to the players, such as Elo's rating algorithm for chess competition and Trueskill for computer games [23, 24]. There has also been some advanced research that enables these methods to take score differences to help the rating [25]. But for their application in the pairwise approach, it was found that the former version can serve our purpose better than the latter advanced version. We have noticed that if the pairwise model is trained on signed differences via classification, the accuracy of the predicted signs (wins, losses and draws) is higher than if it is trained on numerical differences via regression, from which the signs are then extracted. In other words, the accuracy of sign(Y)^{pred} is higher than that of sign(Y^{pred}). This result may come from the fact that there exist pairs with the same differences in features (X) but different differences in target values (Y). Despite some loss of information when taking the signs, the training of the classification model can suffer from less "noise" in pairwise target values than the regression model. For a rating algorithm, correct results of win or loss are more informative in deciding the rank of the samples than the more accurate numerical score differences with potentially wrong signs. Therefore, training the pairwise model via classification and generic version of rating algorithm were used.

We have also experimentally examined several generic ranking algorithms and found that the choice of the generic ranking algorithm can merely affect the ranking accuracy given the same sets of sign(Y)^{pred}, usually by about 1%. It is believed that the main contribution to accurate ranking should come from the accuracy in $\text{sign}(Y)^{\text{pred}}$ rather than the rating algorithm. Therefore, Trueskill is selected and used to rank the samples from the predicted signs. Trueskill is originally designed to rank players in the game "Halo". Because it assumes variances both in players' performance and skill levels, it can deal with potential conflicts in match outcomes, in our case, conflicts in $\text{sign}(Y)^{\text{pred}}$ due to learning errors. For example, when $\text{sign}(Y)^{\text{pred}}_{AB} = -1$ and $\text{sign}(Y)^{\text{pred}}_{BC} = -1$, it implies that sample A $<$ sample B $<$ sample C. But if $\text{sign}(Y)^{\text{pred}}_{AC} = 1$, which implies sample A $>$ sample C, then these predictions are suggesting opposite opinions. This situation is similar to game tournaments, in which a strong player does not necessarily win every time. The python package for Trueskill was already made available [26]. In our experiment, the default Trueskill parameters were used.

In an extrapolation task, the relationship between the test samples and the training samples is important for comparing the training and test data in order to predict the extraordinary samples. So, despite the existence of C3-type test pairs, using them to rank solely can only tell the relative ranks within the test set. On the other hand, C2-type test pairs describe the relative differences between training and test samples. These are better suited for the extrapolation task. Therefore, in the following experiments on extrapolation, the signed differences of C2-type test pairs will be primarily used to rank.

2.5 Machine Learning Methods

The pairwise formulation is potentially ML method agnostic. We utilised the most common ML methods applied to QSAR learning: support vector machines (SVMs), random forests (RFs), Gradient Boosting Machine (XGBs) and K-nearest neighbours (KNN). We did not use deep learning as the datasets were generally too small. The ML methods used in this study are all based on the open-source ML python library, scikit-learn [27]. When a ML method is used to compare the standard and pairwise models, it is used with the default parameter setting from scikit-learn.

The pairwise approach uses classification for the predictions of signed differences, we therefore compared classification version of each ML method versus the standard regression approach. For each evaluation, 10-fold cross validation is used.

2.6 Extrapolation Metrics

To evaluate the extrapolation ability of a ML method, metrics other than the traditional evaluation metrics, such as mean squared error and R2, are required. This is because the common metrics are usually designed to cover predicted results over the whole test set, resulting in an averaged performance evaluation for both interpolation and extrapolation. In a random splitting in cross validation, the test set usually contains more interpolating samples than the extrapolating samples. Therefore, these metrics are good for evaluating the interpolation power of a model, but not very informative in terms of extrapolation power [6]. In this study, we decided to adopt the classification metrics of precision, recall and f1 score to count the identification of extraordinary and top-performing samples [1, 6]. This will give a more direct view of how useful a ML method is in an application where identifying top-performing samples is highly desired.

3 Results and Discussion

3.1 The Pairwise Approach Extrapolates Better

Our extrapolation experiments on 1436 ChEMBL datasets showed a clear advantage of the pairwise approach over the standard approach (Table 1 and Fig. 3). The ChEMBL datasets were sorted by size and experimented in order. When comparing the two approaches, the standard approach uses the regression version of a ML method to predict target values y and rank the test samples with training samples by predicted target values, while the pairwise approach uses the classification version of that ML method to predict $sign(Y_{C2})$ to rank the whole dataset.

Table 1. (a) The percentages of 1436 ChEMBL datasets indicating the pairwise approach had an equal or better performance than the standard approach, *i.e.*, metric(pairwise) > = metric(standard) by each ML method. **(b)** The percentages of ChEMBL datasets indicating the pairwise approach was better than the standard approach, *i.e.*, metric(pairwise) > metric(standard), excluding datasets showing equal performance. All the values have a binomial p-value < 0.05.

Metrics	(a) Percentage of equal or better performance			(b) Percentage of better performance, excluding equally performed datasets		
	RF	SVM	XGB	RF	SVM	XGB
p_{extra}	99.8%	100%	99.4%	99.2%	100%	96.6%
r_{extra}	99.9%	100%	99.5%	99.6%	100%	97.4%
$f1_{extra}$	99.9%	100%	99.4%	99.2%	100%	97.0%
$p_{top10\%}$	78.1%	92.4%	72.4%	66.8%	89.7%	58.8%
$r_{top10\%}$	88.7%	97.2%	86.4%	82.4%	96.3%	78.5%
$f1_{top10\%}$	82.3%	95.4%	76.7%	74.3%	93.9%	66.7%

It was found that the pairwise approach was much better at recognising the extraordinary and top-performing molecules than the standard approach. For all the three ML method (RF, SVM and XGB) tested, the pairwise approach almost always found equally or more extraordinary molecules than the standard approach (Table 1a). It can also identify more test molecules ranked within top 10% of the dataset most of the time, as shown by a high percentage for $r_{top10\%}$. Its outperformance in $p_{top10\%}$ is not as good as that in $r_{top10\%}$, but is still overall better than the standard approach. However, it was noted that this outperformance is less good for XGB or for larger datasets. This means that the ratio of false positives in the top-performing molecules by the pairwise approach can sometimes be similar to that by the standard approach. At the same time the pairwise approach often caused a greater increase in recall, which means it proposed more true positives. Hence, despite an outperformance in $p_{top10\%}$, the pairwise approach could propose slightly more false positives together with more true positives.

As extraordinary molecules do not necessarily exist every time when a train-test split is made, there were many datasets showing $p_{extra} = r_{extra} = f1_{extra} = 0$ or non-existing.

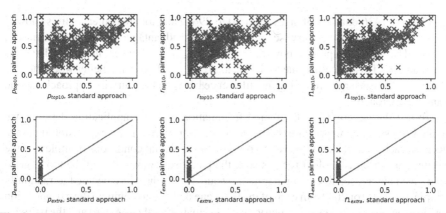

Fig. 3. The six metrics obtained by the pairwise approach versus those metrics obtained by the standard approach over 1436 ChEMBL datasets using SVM.

Therefore, to illustrate outperformance, the datasets showing equal performance were removed. The percentage of datasets suggesting the pairwise approach outperformed the standard approach were re-calculated for the rest of the datasets (Table 1b). Across the three ML methods tested, the pairwise approach did outperform the standard approach in finding both the extraordinary and top-performing molecules. The results also suggested that using RF or XGB had less outperformance than SVM. Through further investigation, we found that the difference among ML methods was due to the variation in extrapolation performance by the standard approach. The standard approach using RF and XGB can evidently produce higher extrapolation metrics than using SVM for the ChEMBL datasets. At the same time, the pairwise approach performed similarly via both ML methods. This gives rise to the higher percentage of datasets showing pairwise approach was better with SVM in Table 1 and Fig. 3.

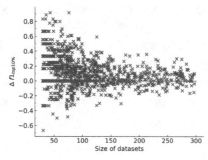

Fig. 4. The increase in f1 score for top-performing molecules versus the size of datasets with RF for 1436 datasets. On y-axis, $\Delta f1_{top10\%} = f1_{top10\%}$ (pairwise) - $f1_{top10\%}$ (standard).

Apart from a statistical overview of the extrapolation power of the pairwise approach, we had a close look at its performance versus the size of the datasets. Figure 4 shows an example of the increase in $f1_{top10\%}$ versus the size of datasets for the experiments with RF for 1436 datasets of sizes from 30 to 298. The plots for other metrics showed a similar trend, that is the pairwise approach is more advantageous on smaller datasets, indicated by more data points above the line of $\Delta f1_{top10\%} = 0$ when the size of dataset is less than 200. This is mainly due to the standard approach learning better when the size of the dataset was larger, reducing the difference between the pairwise approach and the standard approach.

To test the generality of the paired formulation on other application datasets, we applied the same comparison experiment to a set of human gene expression datasets. Because each dataset contains 118050 rows of experimental conditions (samples), if the pairwise approach is applied for this size, the pairwise training set will be too large to train given any reasonable computational resources. We therefore decided to randomly sample a size 100 or 200 from each of the 978 gene datasets to compare the extrapolation performance. The extrapolation metrics were evaluated for the standard and the pairwise approach across four ML methods, random forest (RF), support vector machine (SVM), k-nearest neighbour (KNN) and gradient boosting machine (XGB).

We can see from Table 2 that for the gene expression datasets, the pairwise approach followed the trend seen in the ChEMBL experiments to outperform the standard approach. When the size of the datasets increased from 100 to 200, some of the extrapolation metrics decreased. This is also because the standard approach improved its learning through the additional data at a rate slightly greater than the pairwise approach, resulting in a decrease in the percentage of datasets showing outperformance. This is consistent with observations from Fig. 4.

Table 2. The percentages of gene expression datasets which indicate the pairwise approach had an equal or better performance than the standard approach, *i.e.*, metric(pairwise) > = metric(standard) by each ML method for 978 gene expression datasets (except for KNN which was run on fewer datasets due to computational restriction). All the percentages have a binomial p-value < 0.05.

Metrics	RF-100	RF-200	KNN-100 *313 datasets	KNN-200 *320 datasets	SVM-100	XGB-100
p_{extra}	100%	100%	100%	100%	100%	99.9%
r_{extra}	100%	100%	100%	100%	100%	99.9%
$f1_{extra}$	100%	100%	100%	100%	100%	99.9%
$p_{top10\%}$	87.6%	71.6%	86.3%	77.2%	86.3%	76.6%
$r_{top10\%}$	88.9%	77.3%	85.9%	80.3%	85.7%	78.2%
$f1_{top10\%}$	87.2%	70.6%	85.6%	76.3%	85.3%	75.3%

3.2 The Extrapolation Strategy Improves Extrapolation

As shown in Sect. 2.4, we proposed a strategy to utilise the predictions of the pairwise models to give a ranking of training and test sets combined. This strategy is not exclusive to the pairwise approach. It can be applied to the standard approach to improve its extrapolation performance (Table 3). Once the standard approach has predicted the target values for the test set, the signed differences of C2-type or C3-type test pairs can be calculated from y_{train}^{true} and y_{test}^{pred}. By inputting these signs to the rating algorithm, a ranking of the dataset can be obtained for further extrapolation evaluation. We will abbreviate the results from this procedure as the "standard rank approach". Likewise, we compared the standard approach and the standard rank approach on ChEMBL datasets, which were sorted by size and experimented in order. Each of them was trained and tested via RF with 10-fold cross validation. In this experiment, we also compared the extrapolation results from both rankings obtained from C2-type test pairs and from C2-type test pairs plus C3-type test pairs.

Table 3. (a) The percentages of ChEMBL datasets indicating the standard rank approach had an equal or better performance than the standard approach. (b) The percentages of datasets indicating the standard rank approach was better than the standard approach, among the datasets excluding the ones showing equal performance. The models were obtained by RF from 1456 ChEMBL datasets. Each column represents the type(s) of test pairs used to produce the overall ranking. Bold means a binomial p-value < 0.05.

Metrics	(a) Percentage of equal or better performance		(b) Percentage of better performance, excluding equally performed datasets	
	C2-Type	C2-Type + C3-Type	C2-Type	C2-Type + C3-Type
P_{extra}	**100%**	**100%**	**100%**	**100%**
r_{extra}	**100%**	**100%**	**100%**	**100%**
$f1_{extra}$	**100%**	**100%**	**100%**	**100%**
$P_{top10\%}$	**80.3%**	**85.2%**	36.4%	46.5%
$r_{top10\%}$	**87.2%**	**93.7%**	**54.0%**	**74.7%**
$f1_{top10\%}$	**80.1%**	**86.3%**	41.4%	**55.7%**

The results in Table 3 show that the proposed extrapolation strategy can evidently enable the standard regression to identify more extraordinary samples compared to the direct regression with RF, which in theory is incapable to extrapolate outside the range of training targe values. By taking the signs and re-ranking the samples, despite at a cost of reducing the overall ranking correlation, which might have caused a reduced number of identified top-performing samples, Trueskill had the chance to re-allocate their relative positions by updating the probability distribution for each sample's rating score. Because Trueskill updates the samples' rating scores by numbers of pairwise comparisons, the more comparisons are entered the Trueskill algorithm, the more accurate and confident the rating scores will be. This might account for the increased number of datasets finding more top-performing molecules when C2-type and C3-type pairs are both used to rank. We also tested the case when C1-type training pairs, C2-type and C3-type test pairs are all entered the Trueskill and indeed the extrapolation performance was even better. To

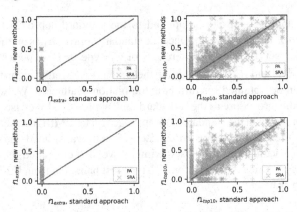

Fig. 5. F1 scores obtained by the pairwise approach (PA) or the standard rank approach (SRA) versus f1 scores obtained by the standard approach over 1456 ChEMBL datasets using RF. The upper row is the results from ranking with C2-type test pairs, whereas the lower row is from ranking with C2-type + C3-type test pairs.

validate properly from a ML methodology standpoint the results are not included due to its use of training pairs.

The main differences that distinguish the standard rank approach from the pairwise approach are that (1) its calculated signed differences are all non-conflicting and consistent with each other, (2) its prediction objective focuses on the accuracy of predicted target values, and (3) the extraordinary samples are more likely to be predicted to draw ($Y = 0$) with the best training samples than to win ($Y = 1$) them. We found that the pairwise approach still can achieve a better extrapolation performance than the standard rank approach (see Fig. 5 and Table 4). This indicates that the pairwise model can produce a set of signed differences that better describes the relative positions of the training and test samples, resulting in the outperformance in extrapolation.

Table 4. *P*-values from Friedman-Nemenyi test for each extrapolation metric among three approaches: standard approach (SA), standard rank approach (SRA) and pairwise approach (PA). Pairs of methods showing a *p*-value < 0.05 are highlighted in green, otherwise in orange.

		*p*_top			*r*_top			*f1*_top			*p*_extra			*r*_extra			*f1*_extra		
		SA	SRA	PA	SA	SRA	PA	SA	SRA	PA	SA	SRA	PA	SA	SRA	PA	SA	SRA	PA
C2-type	SA	1.0000	0.0037	0.0010	1.0000	0.6824	0.0010	1.0000	0.1213	0.0010	1.0000	0.0073	0.0010	1.0000	0.0090	0.0010	1.0000	0.0073	0.0010
	SRA	0.0037	1.0000	0.0010	0.6824	1.0000	0.0010	0.1213	1.0000	0.0010	0.0073	1.0000	0.0010	0.0090	1.0000	0.0010	0.0073	1.0000	0.0010
	PA	0.0010	0.0010	1.0000	0.0010	0.0010	1.0000	0.0010	0.0010	1.0000	0.0010	0.0010	1.0000	0.0010	0.0010	1.0000	0.0010	0.0010	1.0000
C2-type + C3-type	SA	1.0000	0.8891	0.0010	1.0000	0.0010	0.0010	1.0000	0.2568	0.0010	1.0000	0.0010	0.0010	1.0000	0.0011	0.0010	1.0000	0.0010	0.0010
	SRA	0.8891	1.0000	0.0010	0.0010	1.0000	0.0010	0.2568	1.0000	0.0010	0.0010	1.0000	0.0010	0.0011	1.0000	0.0010	0.0010	1.0000	0.0010
	PA	0.0010	0.0010	1.0000	0.0010	0.0010	1.0000	0.0010	0.0010	1.0000	0.0010	0.0010	1.0000	0.0010	0.0010	1.0000	0.0010	0.0010	1.0000

3.3 Discussion

The pairwise formulation is a method of combing model-reconfiguration and feature pre-processing techniques, rather than a new ML algorithm. It can be applied to multiple types of ML. The new formulation transforms the ML learning objective so that the emphasis is placed on the relationship between training and test samples. For a standard approach, when ML algorithms learn from seen examples and try to predict unseen examples from their "experience", it can be difficult to extrapolate out of its "experience" domain. The pairwise approach, on the other hand, learns from the differences in features, which are sometimes more common and generalisable than the original features. It learns to predict the difference between training and test sample, directly aiming to predict if a test sample could win over the training samples. This transformed objective brings about the extrapolation performance of the pairwise formulation.

This study also recommends using classification metrics to evaluate extrapolation performance in a direct way. These metrics suit practical uses when extrapolation is required to identify the extraordinary samples from a test set. For example, they can be used to select ML algorithm for active learning. Active learning (AL) is a learning algorithm that interactively selects unlabelled samples to be labelled to learn the model in a goal-oriented way. In the selection, the exploration and exploitation are usually balanced so that AL can both improve the model's applicability to a larger domain and improve the model's prediction accuracy for the samples with desired properties. Hence, these extrapolation metrics can be used to assess and select ML methods with the desired exploitation property.

We believe that the extrapolation ability of the pairwise approach could be employed directly to fulfil the exploitation duty in an AL task for top-performing samples. Tynes *et al.* have also discovered the advantage of a pairwise approach for uncertainty-driven AL tasks, which encourages the exploration of the wider domain by selecting samples with less confident predictions [28]. We believe that it is possible to develop pairwise-approach-based AL, combining both the exploration and extrapolation traits found by Tynes's study and ours. Despite the difference in how our pairwise approaches generate the pairwise features, it is ultimately the difference induced by data pre-processing techniques, which makes little differences between the two.

The main limitation of the pairwise approach is the additional time and memory requirement to train a pairwise model, as pointed out by Tynes *et al.* [28]. This is because the size of training set needs to be squared for the pairwise approach. Some techniques such as batch training and sub-sampling could certainly mitigate this. More generally, improvements in computer hardware will increasingly remove this limitation. Nevertheless, the pairwise approach can be useful in novel discovery projects with a limited budget or where data is scarce to better explore the surrounding space. In drug design, for example, accurate data points are expensive to generate, so it is important to utilise them efficiently. This study revealed the general applicability of the pairwise approach over thousands of datasets using default ML methods. Our next study will more thoroughly explore the new approach with tuned models on selected problems that demand extrapolation in order to mimic practical applications. To enable reproducibility, the code and datasets used for the experiments have been deposited on: https://anonymous.4open.science/r/pairwise_approach_extrapolation_2023-A188/

4 Conclusion

In this study we proposed a new pairwise configuration by first learning a classification function, F(sample1, sample 2) $\rightarrow$ signed difference in target values, then ranking the samples through Trueskill rating algorithm. We have compared for extrapolation the standard regression approach with our novel pairwise formulation. We found that the pairwise approach can almost always find more extraordinary samples from the test sets than the standard approach, across all the ML methods tested over 2400 ChEMBL and gene expression datasets. The pairwise approach outperformed the standard approach in identifying equally or more top-performing samples on ~ 70% of the datasets. It was also observed that the pairwise approach is more advantageous and effective when applied to smaller datasets. Additionally, we have found that this configuration can be adopted by the standard regression to identify more extraordinary samples. Yet the pairwise approach still outperformed the configured standard approach in all the extrapolation metrics tested.

Acknowledgement. This work was partially supported by the Wallenberg AI, Autonomous Systems and Software Program (WASP) funded by the Alice Wallenberg Foundation. Funding was also provided by the Chalmers AI Research Centre and the UK Engineering and Physical Sciences Research Council (EPSRC) grant nos: EP/R022925/2 and EP/W004801/1. HN is supported by the EPSRC under the program grant EP/S026347/1 and the Alan Turing Institute under the EPSRC grant EP/N510129/1.

References

1. Kauwe, S.K., Graser, J., Murdock, R., Sparks, T.D.: Can machine learning find extraordinary materials? Comput. Mater. Sci. **174**, 109498 (2020). https://doi.org/10.1016/j.commatsci.2019.109498
2. Tong, W., Hong, H., Xie, Q., Shi, L., Fang, H., Perkins, R.: Assessing QSAR Limitations – A Regulatory Perspective
3. Nicolotti, O. ed: Computational Toxicology: Methods and Protocols. Springer New York (2018). https://doi.org/10.1007/978-1-4939-7899-1
4. von Korff, M., Sander, T.: Limits of prediction for machine learning in drug discovery. Front. Pharmacol. **13**, 832120 (2022). https://doi.org/10.3389/fphar.2022.832120
5. Cramer, R.D.: The inevitable QSAR renaissance. J. Comput. Aided Mol. Des. **26**, 35–38 (2012). https://doi.org/10.1007/s10822-011-9495-0
6. Xiong, Z., Cui, Y., Liu, Z., Zhao, Y., Hu, M., Hu, J.: Evaluating explorative prediction power of machine learning algorithms for materials discovery using k-fold forward cross-validation. Comput. Mater. Sci. **171**, 109203 (2020). https://doi.org/10.1016/j.commatsci.2019.109203
7. Agarwal, S., Dugar, D., Sengupta, S.: Ranking chemical structures for drug discovery: a new machine learning approach. J. Chem. Inf. Model. **50**, 716–731 (2010). https://doi.org/10.1021/ci9003865
8. Rathke, F., Hansen, K., Brefeld, U., Müller, K.-R.: StructRank: a new approach for ligand-based virtual screening. J. Chem. Inf. Model. **51**, 83–92 (2011). https://doi.org/10.1021/ci100308f

9. Al-Dabbagh, M.M., Salim, N., Himmat, M., Ahmed, A., Saeed, F.: Quantum probability ranking principle for ligand-based virtual screening. J. Comput. Aided Mol. Des. **31**, 365–378 (2017). https://doi.org/10.1007/s10822-016-0003-4

10. Liu, J., Ning, X.: Multi-assay-based compound prioritization via assistance utilization: a machine learning framework. J. Chem. Inf. Model. **57**, 484–498 (2017). https://doi.org/10.1021/acs.jcim.6b00737

11. Zhang, W., et al.: When drug discovery meets web search: learning to rank for ligand-based virtual screening. J Cheminform. **7**, 5 (2015). https://doi.org/10.1186/s13321-015-0052-z

12. Watson, O.P., Cortes-Ciriano, I., Taylor, A.R., Watson, J.A.: A decision-theoretic approach to the evaluation of machine learning algorithms in computational drug discovery. Bioinformatics **35**, 4656–4663 (2019). https://doi.org/10.1093/bioinformatics/btz293

13. Meredig, B., et al.: Can machine learning identify the next high-temperature superconductor? Examining extrapolation performance for materials discovery. Mol. Syst. Des. Eng. **3**, 819–825 (2018). https://doi.org/10.1039/C8ME00012C

14. King, R.D., Orhobor, O.I., Taylor, C.C.: Cross-validation is safe to use. Nat Mach Intell. **3**, 276 (2021). https://doi.org/10.1038/s42256-021-00332-z

15. Mendez, D., et al.: ChEMBL: towards direct deposition of bioassay data. Nucleic Acids Res. **47**, D930–D940 (2019). https://doi.org/10.1093/nar/gky1075

16. Olier, I., et al.: Transformational machine learning: Learning how to learn from many related scientific problems. Proc. Natl. Acad. Sci. U.S.A. **118**, e2108013118 (2021). https://doi.org/10.1073/pnas.2108013118

17. Mayr, A., et al.: Large-scale comparison of machine learning methods for drug target prediction on ChEMBL. Chem. Sci. **9**, 5441–5451 (2018). https://doi.org/10.1039/C8SC00148K

18. Morgan, H.L.: The generation of a unique machine description for chemical structures-a technique developed at chemical abstracts service. J. Chem. Doc. **5**, 107–113 (1965). https://doi.org/10.1021/c160017a018

19. Koleti, A., et al.: Data portal for the library of integrated network-based cellular signatures (LINCS) program: integrated access to diverse large-scale cellular perturbation response data. Nucleic Acids Res. **46**, D558–D566 (2018). https://doi.org/10.1093/nar/gkx1063

20. Brownlee, J.: Data Preparation for Machine Learning: Data Cleaning, Feature Selection, and Data Transforms in Python. Machine Learning Mastery (2020)

21. Kunanbayev, K., Temirbek, I., Zollanvari, A.: Complex encoding. In: 2021 International Joint Conference on Neural Networks (IJCNN), pp. 1–6. IEEE, Shenzhen, China (2021). https://doi.org/10.1109/IJCNN52387.2021.9534094

22. Park, Y., Marcotte, E.M.: Flaws in evaluation schemes for pair-input computational predictions. Nat. Methods **9**, 1134–1136 (2012). https://doi.org/10.1038/nmeth.2259

23. Herbrich, R., Minka, T., Graepel, T.: TrueSkill(TM): A Bayesian skill rating system. In: Presented at the Advances in Neural Information Processing Systems 20 January 1 (2007)

24. Elo, A.E.: The Rating of Chessplayers, Past and Present. Arco Pub. (1978)

25. Hubáček, O., Šourek, G., železný, F.: Forty years of score-based soccer match outcome prediction: an experimental review. IMA J. Manage. Math. **33**, 1–18 (2022)https://doi.org/10.1093/imaman/dpab029

26. TrueSkill — trueskill 0.4.5 documentation. https://trueskill.org/. Accessed 25 Apr 2023

27. Pedregosa, F., et al.: Scikit-learn: machine learning in python. J. Mach. Learn. Res. **12**, 2825–2830 (2011)

28. Tynes, M., et al.: Pairwise difference regression: a machine learning meta-algorithm for improved prediction and uncertainty quantification in chemical search. J. Chem. Inf. Model. **61**, 3846–3857 (2021). https://doi.org/10.1021/acs.jcim.1c00670

Gene Interactions in Survival Data Analysis: A Data-Driven Approach Using Restricted Mean Survival Time and Literature Mining

Jaka Kokošar$^{(\boxtimes)}$ ⓘ, Martin Špendl ⓘ, and Blaž Zupan ⓘ

Faculty of Computer and Information Science, University of Ljubljana,
Ljubljana, Slovenia
{jaka.kokosar,martin.spendl,blaz.zupan}@fri.uni-lj.si

Abstract. Unveiling gene interactions is crucial for comprehending biological processes, particularly their combined impact on phenotypes. Computational methodologies for gene interaction discovery have been extensively studied, but their application to censored data has yet to be thoroughly explored. Our work introduces a data-driven approach to identifying gene interactions that profoundly influence survival rates through the use of survival analysis. Our approach calculates the restricted mean survival time (RMST) for gene pairs and compares it against their individual expressions. If the interaction's RMST exceeds that of the individual gene expressions, it suggests a potential functional association. We focused on L1000 landmark genes using TCGA na METABRIC data sets. Our findings demonstrate numerous additive and competing interactions and a scarcity of XOR-type interactions. We substantiated our results by cross-referencing with existing interactions in STRING and BioGRID databases and using large language models to summarize complex biological data. Although many potential gene interactions were hypothesized, only a fraction have been experimentally explored. This novel approach enables biologists to initiate a further investigation based on our ranked gene pairs and the generated literature summaries, thus offering a comprehensive, data-driven approach to understanding gene interactions affecting survival rates.

Keywords: survival analysis · censored data · RMST · gene expression · gene interactions · literature mining · large language models

1 Introduction

Survival analysis is a set of statistical methods used to study the time until an event of interest occurs and is commonly used in medical research to estimate life expectancy based on patient-specific data [21]. A pivotal aspect of survival

Supported by the Slovenian Research Agency grants P2-0209 and L2-3170.

A. Bifet et al. (Eds.): DS 2023, LNAI 14276, pp. 293–307, 2023.
https://doi.org/10.1007/978-3-031-45275-8_20

analysis is estimating survival curves and comparing the probability of survival over time between different cohorts [4]. In biomedicine, we can relate the differences in survival to potential markers such as specific genes [2] or groups of genes [22], which can help distinguish patients who respond to treatments from those who do not (see Fig. 1) [27].

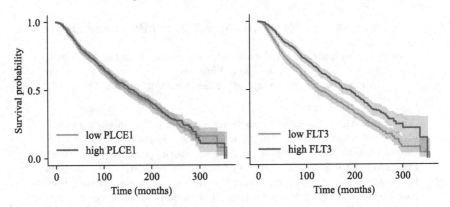

Fig. 1. Example of survival curves representing two conditions dependent on gene expression and associated with patient survival. A group of patients in METABRIC dataset with a highly expressed *FLT3* gene shows a noticeably higher survival function, as depicted by the survival curve on the right. This difference is less prominent and not significant in the case of the *PLCE1* gene, as seen in the survival curve on the left. Part of survival analysis in data with gene expressions is to find markers, that is, genes and sets of genes, whose expression can characterize cohorts of patients with substantially different survival functions.

Rather than a single gene, intricate networks of gene interactions determine the complex nature of diseases such as cancer [26]. Identifying and characterizing these interactions is essential, as they offer critical insights into the onset and progression of a disease, potentially overlooked when analyzing individual genes. Computational discovery of gene interactions is a well-researched area in genome-wide association (GWAS) [17] and gene expression-based phenotype categorization [5]. For the former, a notable approach for handling survival data is the adaption of multifactor dimensionality reduction (MDR) [6,15]. Authors also typically utilize Cox regression analysis to analyze the interaction effects of candidate genes [28,30]. Analyzing survival data is crucial in the clinical domain, highlighting the need for more systematic, data-driven methodologies to unravel intricate gene interactions linked to survival data. Computational methods that specifically address gene interactions from survival data are, at best, scarce, and due to the recent abundance of survival data that includes gene expression, there is a need for their development.

Here we report on a data-driven approach for identifying gene interactions significantly affecting survival rates. In the context of our study, gene interaction refers to the combined effect of two genes on survival, which may be substantially different from their individual effects. Our method aims to measure this interaction effect, quantified as the difference in restricted mean survival time (RMST) when considering the expression of both genes together compared to the expression of individual genes. We then rank gene pairs based on the significance of the difference in RMST. We use top-ranked gene pairs, cross-reference our findings with documented interactions, and synthesize complex literature findings using large language models, thus expanding the exploratory scope of our study.

In Sect. 2, we start with (1) introducing the data, (2) describing how we measured the effect on survival, (3) explaining the measure of interaction and how we define different types of interactions, and (4) describe the utilization of large language models when cross-referencing our findings with existing literature. Section 3 briefly describes our analysis findings, followed by a discussion of limitations and possible future work in Sect. 4.

2   Methods

Our method focuses on two-gene interactions and unfolds through a four-step process. First, we separate samples into evenly sized groups according to the median gene expression value. Subsequently, we estimate survival curves for each group, and for each survival curve, we compute the restricted mean survival time (RMST). We then quantify the difference in RMST between the groups. Lastly, we assess the interaction effect by evaluating how significant is the RMST difference between the interaction term, as discussed in Sect. 2.4 and participating genes. We replicate this procedure for each gene pair in our data set during our discovery-driven analysis and rank them based on their interaction effect. This ranked list paves the way for biologists to initiate their interpretation and investigation. To aid this process, we implement literature mining and harness the utility of large language models to distill complex biological knowledge for assistance and interpretation.

2.1   Data

In this study, we leveraged two sources of survival data:

TCGA. We procured RNA-Seq data, including gene expression matrices and corresponding survival endpoints, for various cancer types from The Cancer Genome Atlas via the GEO portal (GSE62944) [16]. Given the variability in sample size across different datasets, we included only those with more than 100 samples, resulting in 20 TCGA datasets.

METABRIC. We obtained microRNA gene expression matrix and patient survival data from The Molecular Taxonomy of Breast Cancer International Consortium through cBioPortal [3].

Across all datasets in our study, we implemented a log transformation on each gene expression value supplemented with a pseudo count 1. Additionally, z-score normalization was carried out on each gene across samples within a dataset, essentially standardizing the columns of the expression matrix. We utilized clinical metadata for each sample's overall survival (OS) time and event status. OS time refers to the most recent date a patient was confirmed alive. The event is recorded when a patient dies due to the condition under study, in this case, cancer. If a patient's status is unknown or death occurs due to unrelated causes, we classify the event status as censored. Note that sample sizes and event rates vary across datasets Table 1.

Table 1. Statistics about censoring in obtained datasets. The table shows the number of samples and the ratio of censored events. We observe a high rate of censoring across datasets.

Data	HNSC	KIRC	LAML	CESC	BRCA	BLCA	METABRIC
Samples	504	542	178	306	1119	414	1964
Censored	0.67	0.71	0.35	0.81	0.91	0.75	0.42

To limit our exploration scope, we have focused solely on a specific set of genes referred to as L1000 genes [23]. The L1000 gene set contains roughly one thousand landmark genes acting as proxies to infer the expression of other genes. Using this curated set of landmark genes, we significantly reduced the dimensionality of our search space to a set of 1058 genes. Additionally, we removed genes with low expression values to reduce noise before we proceeded with computation. We have disregarded genes with a 75th percentile expression value lower than 10.

2.2   Summary Measure of Survival: Restricted Mean Survival Time

Restricted Mean Survival Time (RMST) is the average survival time up to a pre-specified time point, quantified as the area under the survival curve up to that point (see Fig. 2) [29]. Its primary benefits are that it is interpretable, provides a meaningful summary of survival data, and is considered more robust than measures of median survival time [7].

Building upon its intuitive nature, RMST has gained substantial traction for its versatile utility in comparing differences in survival between cohorts [19]. The difference in RMST is an alternative means to measure gains or losses in the event-free survival between different groups of patients (see Fig. 3). Unlike the log-rank test, which heavily relies on the assumption of proportional hazards and may be sensitive to instances of crossing survival curves, the difference in RMST presents a more flexible and reliable approach [25].

2.3   Interaction Scoring

We have devised a data-driven approach to identify interaction revealing significant RMST differences. This difference implies that the combined influence

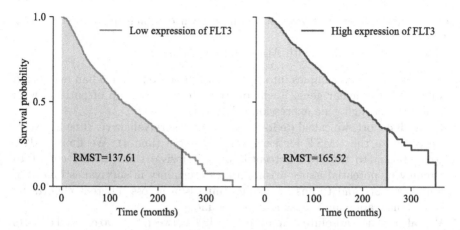

Fig. 2. Illustration of RMST for groups of patients distinguished by varying expression levels of *FLT3*. The group with low expression of *FLT3* has an average survival time of around 137 months compared to 165 months for the other group if we consider the first 250 months of the study.

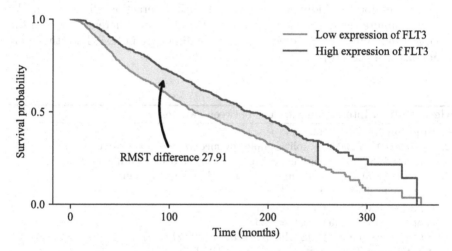

Fig. 3. RMST is a good metric for comparing two survival curves. The absolute difference in RMST represents the area between the curves. Here we illustrate this with survival curves from Fig. 2. The absolute difference in RMST can be considered the measure of time lost/gained between patients that were grouped by the expression values of gene *FLT3*. Quantifiable measure that supports the visual interpretation of the difference in Fig. 1.

of both features on survival differs considerably from the individual influence of each feature. While this technique broadly applies to various types of data, our primary focus here is on gene expression data, which we use to determine the combined influence of gene pairs on survival outcomes compared to their individual effects.

The steps summarized with Algorithm 1 are following:

1. First, we partition samples into two cohorts based on the median expression value of a particular gene. Each cohort represents a group of patients with either low or high gene expression values (line 2).
2. For each cohort, we calculate its Kaplan-Meier survival curve. (line 3). Next, we compute the RMST for each survival curve (line 4). We limit RMST computation to the 75th percentile of all survival times in the cohort to circumvent potential issues arising from uncertainty in survival estimates of long survivors and to ensure a fair comparison across different cohorts by consistently applying the same upper bound.
3. We calculate the absolute difference in RMST between the two created cohorts (line 5). This difference effectively represents the area between the survival curves, providing a measure of the disparity in survival outcomes between the two groups (as shown in Fig. 3).
4. To determine whether an interaction effect exists, we first calculate the RMST differences for the individual genes and their interaction (lines 6- 8). We then compute the interaction measure as the absolute difference between the largest individual RMST difference and the difference in RMST for the inter-action term (line 10).

Algorithm 1. Interaction measure between two genes

1: **function** RMSTDIFFERENCE($gene$)
2:    $cohortA, cohortB \leftarrow$ split samples by median $gene$ expression value
3:    $SurvA, SurvB \leftarrow$ estimated KaplanMeier curves for both cohorts
4:    $rmstA, rmstB \leftarrow$ computed RMST for both survival curves
5:    **return** abs($rmstA$ - $rmstB$)

6: $geneA \leftarrow$ RMSTdifference($geneA$)
7: $geneB \leftarrow$ RMSTdifference($geneB$)
8: $interactionTerm \leftarrow$ RMSTdifference($interactionTerm$)
9: **if** $interactionTerm > \max(geneA, geneB)$ **then**
10:    $interactionMeasure \leftarrow$ abs($\max(geneA, geneB)$ - $interactionTerm$)

2.4 Interaction Types

We define three types of interactions between genes that correspond to different cohort formations (see Fig. 4). Using standardized gene expression values of two genes, we construct a new feature and create cohorts using the approach mentioned earlier. Gene interactions measured with this approach should be interpreted with respect to survival and not as physical interactions.

The first interaction is an **additive** (+) interaction, where standardized gene expression values of both genes are summed together. Such interactions are more common for genes of protein complex subunits.

The second interaction is a **competing** (-) interaction, where standardized gene expression values are subtracted. The cohorts represent which of the two genes was more expressed. Such interactions are more common for activator and inhibitor-type interactions, where both genes regulate the same process.

The last interaction is an **XOR-type** (×) interaction, where we multiply standardized gene expression values. These interactions are more complex and are scarce in nature. They may result from the alternative signaling pathways to the same process influencing survival.

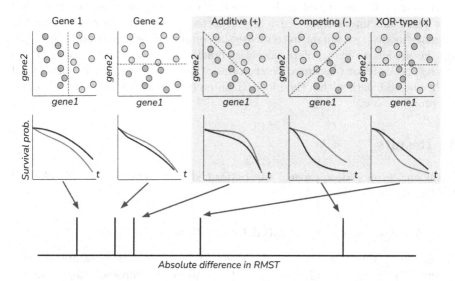

Fig. 4. Interaction definition schema. Cohort formation (top row), RMST difference calculation (middle row), interaction significance according to absolute RMSE difference (bottom row).

2.5 Discovering False Positives with Permutation Test

To identify potential false positive interactions, we performed a permutation test for every data set and interaction type, which involved random shuffling of the survival endpoint and rerunning the experiment 100 times. Given that we conducted 100 such permutation runs per data set and different interaction types, the computation required was extensive due to the sheer volume of potential combinations to examine. Our analysis yielded results that allowed us to isolate the top 0.01% interactions, deemed non-random occurrences. In essence, we consider interactions exceeding the 99.99th percentile as potential interaction hits.

2.6   Literature Mining

We propose to use literature mining to, where possible, explain the interactions and synthesize intricate biological knowledge, leveraging the power of large language models. Specifically, we have used GPT-3.5 and GPT-4 developed by OpenAI. We focused on each data set's top 100 ranked gene interaction pairs and interaction types. These were cross-referenced within STRING [24] and BioGRID [14] databases to ascertain how many gene pairs are in those intricate networks of interactions. We also determined the number of shortest paths and the shortest path length between gene pairs within the BioGRID interaction network. We also incorporated UniProt descriptions of all genes under investigation to supplement our analysis [1].

Having performed initial analyses, we then concentrated on the top 10 ranked gene pairs and interactions previously reported in the literature. Utilizing the language models, we sought to condense the complex biological context, prompting the models to extrapolate potential functional associations between these genes. The UniProt functional descriptions of gene pairs and some genes found in the shortest path within the bioGRID interaction network informed the models' prompts.

3   Results

With our proposed approach we performed the analysis on TCGA and METABRIC datasets.

3.1   Analysis Reveals Potential Interactions

We overlay interaction hits, as described in Sect. 2.5 with permutation test results. The average number of interactions above the threshold for permutations was always 55.9, equivalent to 0.01% of all tested interactions. The tail of the distribution corresponding to the 99.99% of interactions is also visualized (see Fig. 5).

The number of additive and competing interaction hits overwhelmingly exceeded the 56 random interaction threshold for almost all data sets (Table 2). The number of additive interactions is generally lower than the number of competing interactions for the same data set. On the other hand, XOR-type interactions are scarce and found in abundance only in one out of 21 data sets tested. Interestingly, there was no correlation between the number of interaction hits and samples or events in the data set.

3.2   Cross-Referencing with Established Interaction Networks

We have cross-referenced the top 100 ranked gene interactions against known gene interaction networks in STRING and BioGRID. Our findings indicate that many of these interactions have some form of confirmation in these referenced

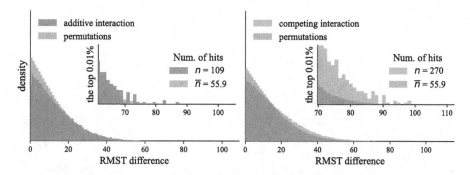

Fig. 5. Permutation test results for TCGA-HNSC dataset. Additive interaction hits against permutations (left), and competing interaction hits against permutations (right).

Table 2. Number of interaction hits for data sets with at least 100 events.

project	additive	competing	XOR-type
BLCA	113	149	58
BRCA	124	196	55
GBM	95	97	37
HNSC	112	272	217
KIRC	527	571	16
LAML	277	339	45
LUAD	113	137	38
LUSC	73	64	23
METABRIC	830	925	17
OV	83	105	69
SKCM	138	122	59

databases. Additionally, we performed these steps using randomly selected pairs of genes instead of our top-ranked list and repeated this random sampling process a thousand times. As illustrated in Fig. 6, competing interactions from HNSC and KIRC emerge as interesting outliers. On average, the top additive and XOR interactions are more scarce in the databases than competing interactions.

Given the surprisingly high number of documented interactions, even among randomly selected gene pairs, we hypothesize that because we are dealing with well-established genes, enhancing the likelihood of their documentation in high-throughput analyses. These analyses are typically characterized by their ability to investigate thousands of genes simultaneously, which are then reported in databases like BioGRID.

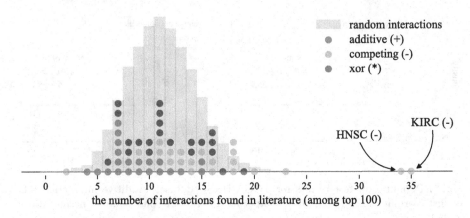

Fig. 6. Number of interactions with conformation in the literature for every dataset used in the analysis. Additive (blue), competing (red), and XOR-type (purple) interactions against 100 randomly selected interactions. (Color figure online)

3.3   Case Study: RHOA-CD44 Competing Interaction

We present one of the top 3 competing-type gene interaction hits from the kidney renal clear cell carcinoma (TCGA-KIRC) data set with confirmed interaction in both STRING and BioGRID database (see Fig. 7a). Competing interaction between RHOA and CD44 genes shows more than five months larger difference between cohorts than any of these genes individually (see Fig. 7b).

CD44 gene produces a cell surface receptor that binds Hyaluronan (HA) and is involved in cell-cell interactions, adhesion, and migration. It serves for signal transduction to different pathways, including cytoskeleton reorganization via RhoA small GTPase [8]. Overexpression of CD44 was related to poor prognosis in glioblastomas [20] and renal cell carcinoma [12] but had no significant effect on breast cancer patient survival [18].

RhoA gene produces small GTPases, which function as molecular switches mainly in cytoskeleton dynamics and cell migration [10]. Increased RhoA-ROCK activities mediate the upregulation of tumor suppressor p53 and induce G1 cell cycle arrest in kidney cell lines [13]. It has been shown that reduced RhoA expression enhances metastasis in breast cancer [9].

Observing Kaplan-Meier plots for both genes' high and low expression cohorts confirms findings from the literature (see Fig. 7c,d). Our method reveals a competing interaction between CD44 and RhoA genes. We interpret this as a competition between CD44 and RhoA-related biology, where the higher expressed gene prevails. Note that we are comparing relative expressions according to the mean expression in the data set (see Fig. 7e). When RhoA is highly expressed, it inhibits the tumor suppression mechanism. Only when CD44 is more expressed than RhoA it sufficiently activates downstream pathways to have a significant effect on survival over the effect of RhoA gene (see Fig. 7f).

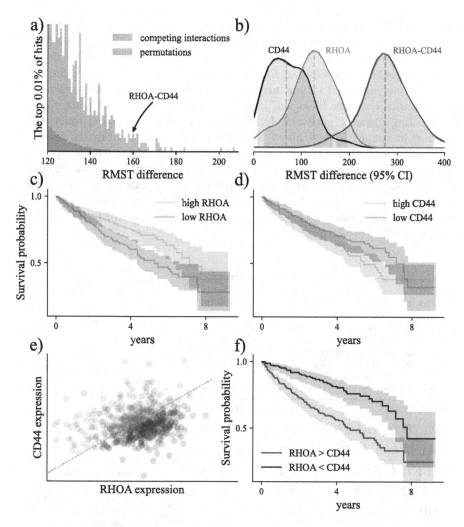

Fig. 7. Case study of the RHOA - CD44 functional association in Kidney renal clear carcinoma (TCGA-KIRC). **a)** permutation distribution tail, **b)** interaction confidence interval, **c, d)** Kaplan-Meier plots for RHOA and CD44 genes, **e)** cohort formation based on gene expression, **f)** Kaplan-Meier plot of the competing interaction.

4   Discussion

Our results suggest a novel ability to identify interactions significantly affecting survival outcomes, thus unveiling insights into the complex landscape of gene interplay and disease prognosis. Even so, our methodology's ranked gene interaction lists should be interpreted cautiously, serving primarily as an exploratory analysis. Due to the vastness of possible gene interactions, we expect some to arise purely by chance. Our preliminary work with permutation tests and literature mining only provides some supportive evidence against these findings. Our analysis identified several potential gene interactions affecting patient survival rates, providing a basis for further in-depth investigations. Particularly noteworthy is the abundance of XOR-type interactions in the HNCS dataset.

Our study also reveals an intriguing potential for large language models to summarize complex biological knowledge when fed with adequate context. By distilling intricate gene pair interactions and their associated functions as informed by resources like UniProt and interaction network databases, the models demonstrated their capacity to reason about known interactions, speculate on potential associations, and guide future exploratory directions (as illustrated with an example in Fig. 8). Although the present analysis should not be regarded as a definitive evaluation of interaction, it establishes an efficient pipeline to facilitate knowledge synthesis and accelerate the pace of scientific discovery, as demonstrated in the case study above.

As there is a shortest path between BCL7B and NR1H2 in the BioGRID protein interaction network, it is possible that these genes functionally interact. BCL7B plays roles in cell cycle progression, nuclear structure maintenance, and stem cell differentiation, while NR1H2 is involved in cholesterol uptake, phospholipid remodeling in the endoplasmic reticulum, and lipid metabolism regulation. However, without further experimental evidence, the specific biological processes in which they may interact cannot be determined.

There is a shortest path of protein-protein interactions between IKBKB and RPN2. IKBKB is a serine kinase involved in the NF-kappa-B signaling pathway, while RPN2 is a subunit of the oligosaccharyl transferase complex. Thus, there may be a functional association between these genes in regulating protein glycosylation and immune response.

Gene pair	Score	Literature
BCL7B & NR1H2	98.5	✗
NR1H2 & SMAD3	98.2	✗
IKBKB & RPN2	93.2	✗
CCNB1 & CDK1	84.2	✓
BIRC5 & NCAPD2	83.2	✓
CEP57 & SNX17	80.5	✗
CDKN2A & IQGAP1	77.4	✓

Possible functional associations: CCNB1 and CDK1 both play essential roles in the control of the cell cycle. CCNB1 is a cyclin subunit that imparts substrate specificity to the complex while CDK1 is a serine/threonine kinase. They interact with each other to form a holoenzyme complex known as maturation promoting factor (MPF) that regulates progression through the G2/M phase of the cell cycle.

Possible functional associations between CDKN2A and IQGAP1 include regulation of cell cycle progression and cytoskeleton organization. CDKN2A acts as a negative regulator of cell proliferation by interacting with CDK4 and CDK6, while IQGAP1 plays a crucial role in regulating the dynamics and assembly of the actin cytoskeleton. There is currently no direct evidence of functional associations between these two genes.

Fig. 8. Example of seven HNSC dataset interactions, ranked by their RMST difference compared to non-interacting terms. The literature column reflects their documentation in public databases. We also display four large language model-generated summaries.

We also recognized noticeable differences in the quality of summaries generated by GPT-3.5 and GPT-4, indicating a trend of improved comprehension and representation of complex biological interactions with newer model iterations. This observation suggests a promising area for future research - the potential of customized language models, fine-tuned on recent, domain-specific literature, which could serve as a more streamlined and context-aware alternative to the vast, generalized models currently accessed via APIs.

While our study presents interesting insights, several limitations present opportunities for future exploration and refinement. The choice of equally-sized cohorts, achieved by splitting at the median, does not account for potential variations in the cohort splits that might optimize the difference in RMST between cohorts. Additionally, we did not consider the potential influence of time limits on RMST calculations, which could significantly impact results and can be very study specific. Lastly, our analysis was constrained by a low number of samples relative to the vast space of possible feature interactions. The enormous space of potential feature interactions may limit the generalizability of our findings. Future work is required to address these limitations and deepen the insights offered by our proposed methodology.

5   Conclusions

The prevalent nature of censored data and molecular fingerprints in clinical environments highlights the need for techniques to illuminate the biological processes regulating disease progression. Unraveling gene interactions is fundamental in understanding these processes, specifically their collective effects on phenotypes.

We report on our work to introduce a data-centric method for detecting gene interactions significantly affecting survival rates, leveraging restricted mean survival times. Using the proposed approach, we can identify possible novel gene interaction candidates on publicly available datasets. We further contextualize the hypothesized gene interactions through literature mining and using large language models to distill complex biological knowledge for assistance and interpretation. In a case study, we show the applicability of such an approach and its potential to uncover and explain potential new interactions.

We have made our method's implementation and the accompanying data and scripts available on GitHub[1] and archived them on Zenodo [11]. These resources include the extended results of permutation tests, summaries produced by the language models, and the prompt used to generate them.

Acknowledgements. This work was supported by the Slovenian Research Agency Program Grant P2-0209 and Project Grants L2-3170 and V2-2272.

References

1. Uniprot: the universal protein knowledgebase in 2023. Nucleic Acids Research 51(D1), D523–D531 (2023)

---

[1] https://github.com/biolab/Discovery-Science-2023.

2. Beer, D.G., et al.: Gene-expression profiles predict survival of patients with lung adenocarcinoma. Nat. Med. **8**(8), 816–824 (2002)

3. Curtis, C., et al.: The genomic and transcriptomic architecture of 2,000 breast tumours reveals novel subgroups. Nature **486**(7403), 346–352 (2012)

4. Dey, T., Mukherjee, A., Chakraborty, S.: A practical overview and reporting strategies for statistical analysis of survival studies. Chest **158**(1), S39–S48 (2020)

5. Evans, L.M., et al.: Transcriptome-wide gene-gene interaction associations elucidate pathways and functional enrichment of complex traits. PLoS Genet. **19**(5), e1010693 (2023)

6. Gui, J., Moore, J.H., Kelsey, K.T., Marsit, C.J., Karagas, M.R., Andrew, A.S.: A novel survival multifactor dimensionality reduction method for detecting gene-gene interactions with application to bladder cancer prognosis. Hum. Genet. **129**, 101–110 (2011)

7. Han, K., Jung, I.: Restricted mean survival time for survival analysis: a quick guide for clinical researchers. Korean J. Radiol. **23**(5), 495 (2022)

8. Hassn Mesrati, M., Syafruddin, S.E., Mohtar, M.A., Syahir, A.: CD44: a multifunctional mediator of cancer progression. Biomolecules **11**(12), 1850 (2021)

9. Kalpana, G., Figy, C., Yeung, M., Yeung, K.C.: Reduced RhoA expression enhances breast cancer metastasis with a concomitant increase in CCR5 and CXCR4 chemokines signaling. Sci. Rep. **9**(1), 16351 (2019)

10. Kim, J.G., et al.: Regulation of RhoA GTPase and various transcription factors in the RhoA pathway. J. Cell. Physiol. **233**(9), 6381–6392 (2018)

11. Kokošar, J., Špendl, M.: biolab/discovery-science-2023: Release 1.0 (2023). https://doi.org/10.5281/zenodo.8023658

12. Li, X.: Prognostic value of CD44 expression in renal cell carcinoma: a systematic review and meta-analysis. Sci. Rep. **5**(1), 13157 (2015)

13. Miyazaki, J., et al.: Progression of human renal cell carcinoma via inhibition of RhoA-rock axis by parg1. Transl. Oncol. **10**(2), 142–152 (2017)

14. Oughtred, R., et al.: The BioGRID database: a comprehensive biomedical resource of curated protein, genetic, and chemical interactions. Protein Sci. **30**(1), 187–200 (2021)

15. Park, M., Lee, J.W., Park, T., Lee, S.: Gene-gene interaction analysis for the survival phenotype based on the kaplan-meier median estimate. BioMed Research International 2020 (2020)

16. Rahman, M., Jackson, L.K., Johnson, W.E., Li, D.Y., Bild, A.H., Piccolo, S.R.: Alternative preprocessing of RNA-sequencing data in the cancer genome atlas leads to improved analysis results. Bioinformatics **31**(22), 3666–3672 (2015)

17. Ritchie, M.D., Van Steen, K.: The search for gene-gene interactions in genome-wide association studies: challenges in abundance of methods, practical considerations, and biological interpretation. Ann. Transl. Med. **6**(8), 157 (2018)

18. Roosta, Y., Sanaat, Z., Nikanfar, A.R., Dolatkhah, R., Fakhrjou, A.: Predictive value of CD44 for prognosis in patients with breast cancer. Asian Pacific J. Cancer Prev. APJCP **21**(9), 2561 (2020)

19. Royston, P., Parmar, M.K.: Restricted mean survival time: an alternative to the hazard ratio for the design and analysis of randomized trials with a time-to-event outcome. BMC Med. Res. Methodol. **13**(1), 1–15 (2013)

20. Si, D., Yin, F., Peng, J., Zhang, G.: High expression of CD44 predicts a poor prognosis in glioblastomas. Cancer Manage. Res. **12**, 769 (2020)

21. Singh, R., Mukhopadhyay, K.: Survival analysis in clinical trials: basics and must know areas. Perspect. Clin. Res. **2**(4), 145 (2011)

22. Špendl, M., Kokošar, J., Praznik, E., Ausec, L., Zupan, B.: Ranking of survival-related gene sets through integration of single-sample gene set enrichment and survival analysis. In: Juarez, J.M., Marcos, M., Stiglic, G., Tucker, A. (eds.) AIME 2023. LNCS, vol. 13897, pp. 328–337. Springer, Cham (2023). https://doi.org/10.1007/978-3-031-34344-5_39
23. Subramanian, A., et al.: A next generation connectivity map: L1000 platform and the first 1,000,000 profiles. Cell **171**(6), 1437–1452 (2017)
24. Szklarczyk, D., et al.: String v11: protein-protein association networks with increased coverage, supporting functional discovery in genome-wide experimental datasets. Nucleic Acids Res. **47**(D1), D607–D613 (2019)
25. Uno, H., et al.: Moving beyond the hazard ratio in quantifying the between-group difference in survival analysis. J. Clin. Oncol. **32**(22), 2380 (2014)
26. Van Steen, K.: Travelling the world of gene-gene interactions. Brief. Bioinform. **13**(1), 1–19 (2012)
27. Vargas, A.J., Harris, C.C.: Biomarker development in the precision medicine era: lung cancer as a case study. Nat. Rev. Cancer **16**(8), 525–537 (2016)
28. Zhang, R., et al.: Independent validation of early-stage non-small cell lung cancer prognostic scores incorporating epigenetic and transcriptional biomarkers with gene-gene interactions and main effects. Chest **158**(2), 808–819 (2020)
29. Zhao, L., et al.: On the restricted mean survival time curve in survival analysis. Biometrics **72**(1), 215–221 (2016)
30. Zhu, J., et al.: A two-phase comprehensive NSCLC prognostic study identifies lncRNAs with significant main effect and interaction. Mol. Genet. Genomics **297**(2), 591–600 (2022)

Joining Imputation and Active Feature Acquisition for Cost Saving on Data Streams with Missing Features

Maik Büttner$^{(\boxtimes)}$, Christian Beyer, and Myra Spiliopoulou

Otto-von-Guericke-University Magdeburg, Magdeburg, Germany
maik.buettner@ovgu.de

Abstract. Replacing missing features in data streams is an important task in order to enable many machine learning algorithms that require feature-complete instances for training and prediction. Two popular methods for dealing with missing features are imputation and active feature acquisition, where in the former missing values are approximated, whereas in the latter, missing features are provided by an expert for a cost and within a limited budget. In this work, we present a hybridized approach, where we employ an active feature acquisition method in the first stage to pick candidate features on which we would require a costly expert and then check in a second stage how well we could impute these candidate features. If the imputation is expected to be of a certain quality, we skip the purchase and impute instead. We provide a framework for such a scenario and used it to run extensive experiments. Our results on 6 data sets show that our proposed method can achieve a similar classification performance while spending 1% to 27% less budget.

Keywords: active feature acquisition · imputation · data streams · stream mining

1 Introduction

Missing features in data are a quite common occurrence in many real-world scenarios and very often pose challenges, when we want to analyze such data and use it for extracting information or making predictions. One of the most common examples is the arrival of a patient at a doctor's office who has a certain issue and has already undergone some tests. The doctor now has to choose, if the tests are sufficient for a diagnosis or whether additional tests are required. It is infeasible in such a scenario to run all the possible tests because of time and budget constraints. Choosing which tests to run under budget constraints, which translates into which features to acquire, is the research topic of Active Feature Acquisition (AFA) [22].

Another established way to address missing values is to estimate the value based on previously available information, which is called imputation [15].

Shared first author position.

A. Bifet et al. (Eds.): DS 2023, LNAI 14276, pp. 308–322, 2023.
https://doi.org/10.1007/978-3-031-45275-8_21

Imputations can be very simple, like replacing a missing feature value with the feature's most common value, or quite complex like training a predictive model, in order to predict missing features from available feature values. All imputation methods have in common that they introduce biases into the data and that we cannot be sure that the imputed value is correct. This can have negative impacts on our data analysis tasks that build on top of the imputed data.

In a data stream setting the issue of the introduced biases becomes even more extreme as concept drift might occur [24] and conducting AFA is also much harder as we have to make acquisition decisions on an instance-to-instance basis and cannot optimize our decisions over the whole data set [4].

In this work, we developed a hybrid approach for data streams that aims to handle missing feature values in a two-stage approach. In the first stage, a set of missing features of an instance is chosen for acquisition, based on the estimated merit of the features. The merit of a feature indicates how much it helps in separating the classes [7,23]. In the second stage, we check for each feature in the acquisition set, and how well it can be predicted by the available features of the respective instance. If the prediction based on available data seems promising, choosing to impute rather than acquiring the missing values might be preferable in order to save budget. The main contributions of this work are a method that saves budget when imputation seems to be an adequate alternative to AFA and secondly, an imputation method that tracks for each feature, how it can be best predicted by another feature.

2   Related Works

Imputation covers methods for the replacement of missing values using estimates based on statistical methods, information-theoretic approaches, or model-based methods [1,15]. These methods can be quite simple, like replacing a missing numeric feature with the mean value of the respective feature or forwarding features of nearest neighbors using the available features of an instance [12], but they can also become complex like the multiple imputation by chained equations (MICE) [1] or deep imputation methods [8]. In the context of data streams, handling missing data becomes challenging due to continuous arrivals and concept drift [24]. Incremental models that can be trained online [17] or windowed approaches [4] are proposed to handle the continuous growth of data and address concept drift. In this work, we propose our own imputation method which is using windowing and is closely related to [17] but instead of using multiple regression, we train a linear regression model for each feature pair and select the best available model for imputation.

In contrast to imputation, Active Feature Acquisition (AFA) tackles missing values by determining them through a costly oracle, such as lab tests or Subject Matter Expert (SME) inquiries [20]. One approach is to purchase features for instances that the current model misclassifies or is uncertain about [16]. Another method estimates the merit of purchasing individual feature values, based on the change in model performance [19]. However, these methods are computationally expensive and not suitable for data streams. A more recent work based on matrix factorization [10] is also not applicable to data streams, as it is trained on the whole data set and is computationally expensive. The method presented in [18] applies an active feature acquisition strategy to relational databases in a static scenario. It keeps track of its decisions through tree-based distances which allows it to explain its decisions to human experts. Recently, an online feature selection metric combined with budgeting for stream active learning has been proposed [2], which was later extended to deal with varying feature costs [3]. A different approach was taken in [21] where the authors employed reinforcement learning (RL) in order to minimize the cost of the feature acquisitions while maximizing classifier performance. Another RL-based method for static and time series data was introduced in [14]. The authors propose the use of a surrogate model in order to provide better rewards to the RL agent and managed to reach state-of-the-art results, albeit not on streams. RL has been used for handling data streams with incomplete information, where P and Q networks are trained together [11]. The P network predicts labels based on acquired information, and the Q network predicts the impact of feature acquisition. However, the deep-learning-based nature of this approach requires an enormous amount of initial data and training time before it can be used on new data. Our work extends the framework introduced in [3] with a hybrid approach that favors imputation over acquisition in certain scenarios.

3   Methodology

In this section, we discuss the underlying method of this paper. Firstly, we present the framework of supervised merit rankings [3], which forms the foundation of our extended work. Secondly, we describe the Incremental Percentile Filter, a crucial component in understanding the decision-making processes of both the regular framework and the feature pair imputer threshold skip (FPITS) which is used to determine acquisition and saving decisions. Thirdly, we introduce a new imputation method called the feature pair imputer (FPI). Algorithm 1 provides an overview of our framework.

3.1   Supervised Merit Ranking Active Feature Acquisition Framework

The supervised merit ranking framework for active feature acquisition has its origins in the publication of [3]. It is a feature importance ranking method by which features are selected and acquired online in a surrounding batch-wise evaluation process. At the heart of its operation is a feature importance method like in [7,23]. This importance metric is used in a merit function that incorporates the costs of individual features to give a cost ratio indicator of a feature's effective usefulness when acquired given its costs (line 7).

By means of a quality function, which evaluates an instance and its known features using merit functions, the framework can make a judgment of how well the current instance helps the classification task given its associated cost.

Algorithm 1. Simplified pseudo-code of the used framework.

Require: Initial data X_{init}, A data stream consisting of batches X, a sliding window of batches W, a classifier C, a budget manager BM, an imputation model I, a feature importance metric afa, the cost of features C, a feature set selection method fss, an initial budget threshold T_{init}, the budget added for each instance B_{gain}

Ensure: $B_{spent} \leftarrow 0$, $B_{given} \leftarrow 0$, $B_{saved} \leftarrow 0$

1: add X_{init} to W
2: train initial model C_0 on W
3: **for** X_i in X, $i \geq 1$ **do**
4:    update I with X_i
5:    adjust budget threshold T
6:    **for** x in X_i **do**
7:       update merits using afa, W and x
8:       $B_{given} \leftarrow B_{given} + B_{gain}$
9:       get acquisition set A of x using fss
10:      calculate quality gain of $x \cup A$
11:      determine acquisition decision of BM according to quality gain of $x \cup A$
12:      determine confidence decision of I given x
13:      **if** (I) BM wants to acquire and I is confident **then**
14:         $B_{spent} \leftarrow B_{spent} + A_C$
15:         $B_{saved} \leftarrow B_{saved} + A_C$
16:      **else if** (IV) BM wants to acquire and I is not confident **then**
17:         $B_{spent} \leftarrow B_{spent} + A_C$
18:         acquire A for x
19:      **end if**
20:   **end for**
21:   impute remaining missing values of X_i using I
22:   evaluate X_i using C_i
23:   add imputed X_i to W
24:   train new model C_{i+1} on W
25: **end for**

A further component selects promising feature sets to consider for acquisition, coined the feature selection strategy returning the acquisition set A (line 9). Assuming one was to acquire A's feature values and include them in the current instance, we can calculate two quality values, before and after acquisition of the set respectively, and calculate the projected gain in quality (line 10). This quality gain value is passed on to the Incremental Percentile Filter (IPF) which acts as a budget manager for the process, making decisions to keep the spent budget within the user-specified limits (line 11). With each batch processed the budget threshold shown to the IPF is dynamically adjusted and the still missing values are imputed and trained on (line 5).

3.2 Incremental Percentile Filter (IPF)

The Incremental Percentile Filter (IPF) [13] is a decision maker that bases its decisions on the recent history of "usefulness" values. Incoming usefulness values are simultaneously stored in two separate windows of size w_{IPF}: the chronologically sorted window W_{IPF_c} and the value sorted window W_{IPF_v}. Only if a newly arriving decision and its associated usefulness value is inserted into W_{IPF_v} at or below a rank based on the threshold θ_{IPF} the IPF makes a positive decision. Since the windows of the IPF keep forgetting their oldest values, they act like a sliding window. Given a constant threshold, enough time, and no constant biases in the incoming decision usefulness values, the fraction of positive to all decisions will approach θ_{IPF}. As a flexible and dynamic way of adjusting decisions based on current circumstances, the IPF is used both as a budget manager and as the decision maker of the imputation mechanism.

3.3   Defining Budget

The framework of [3] distributes budget evenly along the stream to ensure no bias towards time points at which lots of budget is available. To do so budget is distributed instance-wise so that with each arriving instance we gain a bit of budget B_{gain} (line 8). The sum of the B_{gain} at a point in time is B_{given}. Similarly, for each acquisition the expenditure pool B_{spent} is increased. Thus, the budget usage B_{used} is defined as the relative ratio $B_{used} = \frac{B_{given}}{B_{spent}}$. The framework aims to use as much budget as is available or less $B_{used} \leq 1$. Costs are associated with each feature and stored in a cost vector C of size $|F|$. Due to biases and trends in the selection and costs of acquisition sets as well as the budget-agnostic nature of the IPF, the framework dynamically adjusts the budget threshold assigned to the IPF whenever a batch has been processed (line 5). This adjustment is performed using the following formulas:

$$T = \begin{cases} T_{basic} & if\, B_{used} \leq 1 \\ T_{basic}/P & otherwise \end{cases} \tag{1}$$

$$T_{basic} = \frac{B_{gain}}{\hat{A}_C \cdot B_{used}} \tag{2}$$

$$P = \lfloor p_c \cdot (B_{used} - 1) \rfloor + 2 \tag{3}$$

where $p_c = 32$ is a user-specified penalty parameter and $\hat{A}_C$ is the estimated average cost of a presented acquisition set returned from the feature set selection strategy [3].

3.4   Feature Pair Imputer (FPI)

The Feature Pair Imputer (FPI) is an imputation model designed to be usable in fast-acting data streams with the possibility of a quick method for evaluating its current reconstruction performance. At its core it keeps separate sliding windows W_{F_i,F_j} (shortened to $W_{i,j}$ for brevity) with size w_{FPI} for each feature pair (i,j) and each individual feature (i,i) for a total of $\sum_{i=1}^{|F|} i$ windows. Each sliding window of value pairs has two associated imputation models $M_{i,j}$ and $M_{j,i}$ that train on this data. These models can further be evaluated using the same data. The type of model applicable and the method needed to estimate the imputation error depend on the feature types of the input and output features, see Table 1.

Table 1. Methods that are used for the feature pair imputations and how their error is calculated

Feature In	Feature Out	Imputation	Error
numeric	numeric	linear regression model	$\frac{RMSE(M_{i,j})}{max(W_{i,j}-min(W_{i,j}))}$
categorical	numeric	mean of the posterior distribution of output values given the input value	$\frac{RMSE(M_{i,j})}{max(W_{i,j}-min(W_{i,j}))}$
categorical	categorical	mode of the posterior distribution of output values given the input value	$1 - Jaccard(W_{i,j}[:,j], M_{i,j})$
numeric	categorical	1-d nearest neighbour mapping	$1 - Jaccard(W_{i,j}[:,j], M_{i,j})$

The calculated error values are stored in an error matrix $E \in [0,1]^{|F| \times |F|}$ where the rows are mapped to the input feature and the columns to the output feature. Thus, $E_{i,j}$ is the calculated imputation error of model $M_{i,j}$. Given E and an encoding of the known features of x as a vector mask $known(x) = \{1, 0\}^{|F|}$, i.e. $known(x)_i = 1$ iff feature value of feature i of instance x is known, we can estimate the importance of models for imputing the particular missing features of the instance x by means of creating a weight matrix from the errors. The set of known features of x we call K_x.

This weight matrix $weight$ is calculated using the reciprocal of the error values as follows:

$$weight_{i,j} = \begin{cases} (E_{i,j} + \epsilon)^{-1} & if\ \exists a \exists b E_{a,b} = 0 \\ E_{i,j}^{-1} & otherwise \end{cases} \tag{4}$$

If any error value in E is 0, then a small value $\epsilon = 0.001$ is added to all errors to circumvent a division by zero.

Using $weight$, the known feature mask of x $known(x)$, and the predicted values for x using the models we can impute any values of x. We calculate the specific weight matrix Z for instance x:

$$Z_{i,j} = weight_{i,j} \cdot known(x)_j + weight_{i,j} \cdot I_{i,j} \tag{5}$$

where I stands in for the identity matrix of size $|F|$.

Let X' be the matrix of imputed values using the models M given the feature values of x as input. Depending on the importance strategy, the imputed feature value of missing feature j in x $\hat{f}_j$ may be calculated either as:

Weighted Approach

$$\hat{f}_j = \begin{cases} \frac{\sum_{i=1}^{|F|} Z_{i,j} \cdot X'_{i,j}}{\sum_{i=1}^{|F|} Z_{i,j}} & if\ isNumerical(F_j) \\ X'_{m,j}, m = argmax_{1 \le i \le |F|} Z_{i,j} & otherwise \end{cases} \qquad (6)$$

Choose Best Approach

$$\hat{f}_j = X'_{m,j}, m = argmax_{1 \le i \le |F|} Z_{i,j} \qquad (7)$$

We will focus entirely on the latter *choose best approach*.

3.5   Feature Pair Imputer Threshold Skip (FPITS)

The FPI's ability to quickly evaluate its models allows itself to calculate a believed 'confidence' value of imputing a particular instance's missing values. Before any calculation takes place we discern between three cases based on the number of known features of some instance x:

$$FPI_{conf}(x) = \begin{cases} 0 & if\ known(x) = 0 \\ 1 & if\ known(x) = |F| \\ 1 - max(Miss_{min}(x)) & else \end{cases} \qquad (8)$$

For instances with at least one feature value missing and at least one feature being known the following steps are done to calculate the FPI_{conf}: Given an instance x's known features K_x and the current FPI's errors matrix E, we calculate the believed ease of imputing x. The calculation uses the biggest reconstruction error among the missing features, which we would suffer if we chose to impute.

For each unknown feature j in x extract the error vector V_j that contains all errors $E_{i,j}$ with $i \in \{K_x \cup j\}$ that are available as models for the imputation given K_x. According to the *choose best approach* in Eq. 7, we are only interested in the smallest error of each respective unknown feature. Thus, we take the minimum error of each vector and store the results as a new set of minimum errors: $Miss_{min}(x) = \bigcup_{j \notin K_x}^{F} min(V_j)$.

The maximum of these errors is the believed worst error in imputing x, thus its one complement reorders the scale so that values close to one indicate that the FPI believes to be capable of perfectly imputing the missing values of x, $FPI_{conf}(x) = 1 - max(Miss_{min}(x))$.

We exploit these values to create a second IPF-based decision maker alongside the framework's budget manager and further restrict the acquisition of an acquisition set A for instance x to only be acquired if the budget manager's IPF decides to purchase A and the IPF that is given the FPI's goodness estimates for instance x deems x hard to impute (line 12).

We coin this new method Feature Pair Imputer Threshold Skip, or FPITS for short. Ideally, the FPITS' additional constraint will further limit the spending of budget in scenarios in which A is deemed highly informative but the FPI is confident to impute x correctly anyhow.

4   Evaluation

To test the effectiveness of our method we performed multiple experimental runs using the framework[1] as a basis. In the following subsections we will briefly mention the data sets on which we ran our experiments, explain our framework hyperparameters and finally show and discuss the results of these experiments (Table 2).

Table 2. Data sets used in our experiments

Data set	Instances	Labels	Cat. features	Num. features
electricity	45312	2	1	7
adult	32561	2	4	8
magic	19020	2	0	10
cfpdss	13000	2	5	5
nursery	12960	5	8	0
pendigits	10992	10	0	16

The experiments were run on six data sets, four of which are based on static concepts (*adult*, *magic*, *nursery* and *pendigits*) while the other two are stream-based (*cfpdss* and *electricity*). Data sets *magic* and *pendigits* are entirely comprised of numerical features, data set *nursery* contains only categorical features and data sets *adult*, *cfpdss* and *electricity* offer a mix of categorical and numerical features. We chose to test our method on static data sets to validate the usefulness of our imputation method independent of drift.

We highlight these differences since mixed and categorical feature types lead to noticeable performance drops in the FPI's performance and should be considered separately. All but the *cfpdss* data set are available through the UCI site [5].

The *cfpdss* is a synthetic data set generated by us that offers various kinds of feature-to-feature correlations and changes its underlying feature and label generative functions every thousand instances to simulate concept drifts, specifically incremental, gradual, and sudden drift, that may occur in a streaming setting. Since the different feature-to-feature correlations are to be tested these drifts are synced with the various correlation types, i.e. only the features involved in a specific correlation type drift when a concept change occurs in order to evaluate the performance of the FPI and its imputation models. Correlation types present in the data set include linear correlations between two numerical features plus some minor normally distributed noise, a numeric to bi-categorical feature correlation determined by a threshold value and minor noise and vice versa as well as a linear combination of two features determining the value of another feature plus some normally distributed noise. Since the label function depends on three numerical features each linearly correlated with another feature of the

---

[1] https://github.com/Buettner-Maik/caafa-stream.

data set and an additional categorical feature that is similarly correlated with other features, makes their imputation with the linear regressors used by the FPI an easy problem to solve. For more details please refer to the provided code in the github.

4.1 Experiment Setup

Incomplete data streams are generated based on the type of data set. For static data sets the data is shuffled and an initial batch of 50+ one example for each label is created. Stream data sets keep their order and have their first 50 instances serve as the initial batch. The remaining data is split into batches of 50 instances. Based on the missingness parameter m the features of the instances are infused with missing values in a completely at random fashion. This process was performed for 10 iterations and 7 missingness values, giving a total of 70 different incomplete permutations of data streams per data set.

The classification model uses a Support Vector Machine in its default parameterized form from the sklearn library. An initial classifier is trained on the initial batch (line 2) and a new classifier is trained each time after the acquisition and imputation steps have been performed (line 24). We use the prequential evaluation paradigm as it is common for stream evaluation [6]. To do the acquisitions the single window average Euclidean distance function was used as the basis for the merit calculation [23], a budget manager in the form of an incremental percentile filter with a window size $w_{IPF} = 50$ and a 4-Best feature set selection strategy was used. The cost of each feature was homogenous and set to 1.

The FPI follows the *choose best approach* for imputing values and uses a pair window size of $w_{FPI} = 25$. Thresholds for the FPITS method had been chosen to be one of either 0, 0.1, 0.2, ..., 0.9, or 1.0, where the threshold 0.0 is equivalent to a method without the use of the FPITS and a threshold 1.0 is equivalent to a method that does not acquire any feature values at all. The FPI's IPF used a window of size $w_{IPF\_FPITS} = 100$, i.e. its window was completely renewed every two batches. We added a lower bound which shows the minimal performance of a particular experiment when no budget is spent and an upper bound which shows the maximum performance of a particular experiment when every feature value is acquired.

4.2 FPI Performance

In order to evaluate the effectiveness of the FPI and imputation models we compared the performance of both the old imputation model, which is limited to mean and mode imputations coined Simple Imputer (SI), and the FPI for the various data sets under seven different missingness ($m = \{0.125, 0.25, ..., 0.875\}$), four different budget settings ($B_{gain} = \{0, 0.5, 1, 2\}$) and on 10 iterations for a total of 280 comparisons per data set. We chose to perform a Wilcoxon test ($\alpha = 0.05$) to check whether the methods differ in their performance and further evaluate their ranks to determine which of these performs better given a certain data set. Our findings along with the p-values are listed in Table 3.

Table 3. Wilcoxon test performed on each data set with a total of 280 (7 missingness, 4 budgets, and 10 iterations) paired sample points each. Winners are in bold.

Data set	p-value	FPI mean rank	SI mean rank
adult	<0.001	1.896	**1.104**
cfpdss	<0.001	**1.246**	1.754
electricity	0.717	1.529	1.471
magic	0.371	1.504	1.496
nursery	<0.001	1.739	**1.261**
pendigits	<0.001	**1.046**	1.954

The FPI outperforms the simple imputation model on the *pendigits* and *cfpdss* data set. Given that the *cfpdss* data set has multiple linear feature correlations this is not surprising. The FPI performs much better on data set *pendigits* which indicates strong linear dependencies between its features. Additionally, the data set's static nature enables the FPI to outperform the SI even more than on the *cfpdss* data set, as the adaptation of the FPI to new concepts takes longer with increasing missingness.

On the data sets *adult* and *nursery* the SI trumps over the FPI. Data set *nursery* indicates a weakness in the FPI's ability to correctly predict categorical features with only one single other categorical feature. Overall the FPI loses to the SI on data set *adult* the most. Data sets *electricity* and *magic* are undecided.

4.3 FPITS Behavior

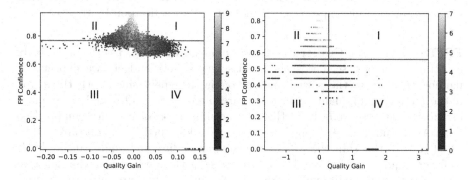

Fig. 1. Scatter plots of the decision variables seen by both decision makers. The left plot shows data set *magic* and the right data set *nursery* each on a single missingness $m = 50\%$ run and the number of known features of an instance encoded in the color. (Color figure online)

To further illustrate the workings of the FPITS method consider Fig. 1 which maps the quality gain values of the budget manager for the selected acquisition set A and the confidence values of the FPI FPI_{conf} for its associated instance x on single runs of the data sets *magic* and *nursery*. Every point in this figure is a single instance and their position in this graph indicates whether they are to be acquired or imputed.

A vertical and horizontal line that represent the mean decision thresholds of the budget manager and FPI respectively split these graphs into four quadrants. The instances in quadrant Q_I and Q_{IV} that pass the threshold on the x-Axis are the candidates for feature acquisition, whereas all instances with missing features in Q_{II} and Q_{III} will only be imputed. Whether an acquisition candidate instance will be imputed (Q_I) or sent for actual acquisition (Q_{IV}) is decided by the FPI and its threshold. The ratio of values falling into Q_I compared to those falling into either Q_I or Q_{IV} determines how much budget is saved. Do note, that during execution the thresholds change depending on the recent history of the stream. The clear steps in the FPI_{conf} values on the right *nursery* figure which entirely consists of categorical features and, given the nature of the imputation models for categorical features, leads to well-defined ratios when calculating their respective FPI_{conf} values. The shape of the combined quality gain and FPI_{conf} value distribution determines how well the user-specified FPI threshold maps onto the resulting budget. Since the FPI's confidence grows stronger the fewer features of an instance it has to impute while also having access to more imputation models, FPI_{conf} positively correlates with the number of features known in an instance. On the other hand, the nature of the averaging function makes changes in the quality smaller and smaller the more features are already known in the instance. Thus, the acquisition sets' quality gain values become more clustered around the mean of the x-axis. We have visualized this correlation through the use of the color bar which encodes the number of known feature values within their respective instance. It is easy to see that most of the instances that have many features known (light color) are to be found in Q_{II}, and therefore have not been selected for acquisition and do not require a decision by the FPITS.

FPITS in Budget-Constrained Scenarios. When faced with increasingly constrained budget scenarios (high missingness, low budget) we expect our method to increase the actual amount of budget spent since both the shape of the FPI_{conf} and quality gain values are tilted towards Q_{IV} and the number of known features similarly furthers an accumulation of instances mapped to Q_{IV}. As the number of skipped acquisitions is described as the ratio of $|Q_I|$ to $(|Q_I \bigcup Q_{IV}|)$, this shift towards more instances ending up in quadrant IV leads to fewer acquisitions being skipped. We consider this desired behavior since we may require more additional information the more values are missing. Figure 2 plots the B_{spent} depending on the chance of missing feature values m and illustrates this behavior on data sets *pendigits* and *nursery*.

However, we can also observe an adverse effect at very high missingness and thresholds. Due to the increase in the likelihood of instances with no known feature values the number of $FPI_{conf} = 0$ values increases as well. Once the FPITS' IPF value-sorted window is sufficiently saturated with these $FPI_{conf} =$

0 values, that is, once more than $w_{IPF\_FPITS} \cdot thr$ fill the window, the IPF will compare the FPI_{conf} of an incoming instance to the current threshold value, which is 0. Since any incoming value will always be greater or equal to 0, the IPF will decide to skip acquisitions. In extreme scenarios, the FPITS will never recover from this state and skip all future acquisitions leading to a sharp drop off in budget expenditure, see high thresholds in Fig. 2. We can calculate the expected average minimum critical threshold at which point the comparison value of the IPF will evaluate to 0 for a missingness chance m as: $thr_{critical} \leq 1 - m^{|F|}$.

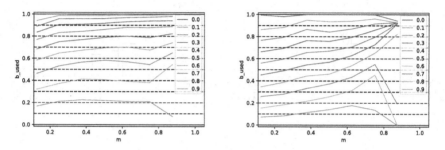

Fig. 2. Budget spent for various FPITS methods with *thr* 0.0 (top most) up to 0.9 (bottom most) on the data set *pendigits* (left) and data set *nursery* (right) at $B_{gain} = 1$

4.4 Budget Comparison at Similar Performance

To validate whether our method can save budget while keeping similar performance we devise a comparison based on the Friedman and post-hoc Nemenyi test. We compared each prequential result of a batch and iteration among 11 methods of the FPITS with thresholds 0.0, 0.1, ..., 1.0 and rank their classification performance, here accuracy.

Using the individual ten iterations and number of batches (i.e. 10 iterations times $(13000-50)/50 = 259$ batches) as comparison points we can calculate whether these methods show different performance characteristics using a Friedman test. If they did, we further evaluated the methods' rankings by means of the post-hoc Nemenyi test which calculates the mean rankings of the respective methods and a corresponding critical distance at confidence level $\alpha = 0.05$.

Intuitively, one expects the methods with the highest budget spending to perform the best, i.e. have the lowest rank, while higher and higher thresholds will result in less and less budget being spent and thus their relative ranking getting worse, i.e. their ranks being higher. If the method with $thr = 0$ performs similarly according to the critical distance of the post-hoc Nemenyi test to methods with higher than $thr = 0$ we can further evaluate their respective budget spending and make claims about which of these methods used less budget and thus was more effective in making acquisition decisions. All Friedman tests performed returned p-values smaller than $2.2 \cdot 10^{-16}$, making the post-hoc Nemenyi test

seen in Fig. 3 applicable. The plots were generated using the autorank package
[9]. For brevity, we also provide our condensed findings in table 4. In our chosen
example we see the old method perform similarly to the $thr = 0.4$ method on the
adult data set. However, the $thr = 0.4$ method only requires 69% of the original
budget per batch to achieve comparable performance. Over the entirety of our
compared scenarios $m = \{0.25, 0.5, 0.75\}$ and $B_{gain} = \{0.5, 1.0, 1.5, 2.0\}$ we saw
budget savings at similar performance in the range of 1% on *nursery* and up to
27% on *adult*.

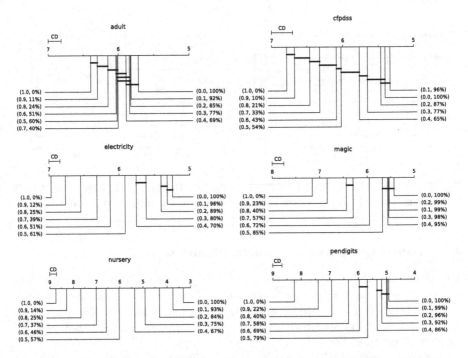

Fig. 3. CD-plots of the six data sets in a $m = 0.5$ and $B_{gain} = 1$ setting. The name
tuples signify the threshold and percentage of budget spent compared to the method
without FPITS.

Table 4. Overview of the relevant information of cd plots in Fig. 3.

	adult	*cfpdss*	*electricity*	*magic*	*nursery*	*pendigits*
cd	0.187	0.297	0.159	0.245	0.297	0.323
$rank(thr = 0)$	5.714	5.408	5.246	5.434	3.305	4.927
thr_{sim}	0.4	0.3	0.1	0.5	0.0	0.3
$rank(thr_{sim})$	5.891	5.655	5.322	5.678	3.305	5.175
B_{saved}	31%	23%	1%	15%	0%	8%

5   Conclusion

The vast majority of past research either solely focused on active feature acquisition (AFA) or solely worked on imputation, often also in conjunction with unclear budget scenarios. In this work, we presented a hybrid method that combines AFA on streams with an intelligent imputation mechanism in order to deal with feature incomplete instances. This mechanism intervenes in costly feature purchase decisions, in cases where there is a good chance that the missing features can be imputed correctly by features that are available. Furthermore, we introduced a novel lightweight imputation model which can easily evaluate itself even on very fast streams and its plug-and-play nature allows it to adapt to various kinds of simple feature-to-feature correlations.

We ran experiments on six different data sets with varying budget scenarios, varying amounts of missing values, and different targeted budget savings. These experiments showed that our new method can save on budget expenditure and adapt these target savings based on the current needs of the stream. Our new method achieved a similar performance to the method which solely relies on AFA while saving between 1% to 27% of our invested budget depending on the data set.

In our experiments, we identified several drawbacks of our simple feature pair imputer which we plan to remedy in the future. This entails using more sophisticated pairwise models, that are not limited to linear concepts, like Gaussian and deep regression models. Furthermore, we plan to replace our static window sizes with adaptive windows and enrich them with feature drift detection in order to keep our imputation models up to date. This also extends to drift-aware feature importance functions similar to what is presented in [7]. In addition, we will investigate the influence of varying feature costs, the initial estimation of good thresholds as well as the possibility to add the imputation confidence to the merit function.

References

1. Adhikari, D., et al.: A comprehensive survey on imputation of missing data in internet of things. ACM Comput. Surv. **55**(7), 77–114 (2022)
2. Beyer, C., Büttner, M., Unnikrishnan, V., Schleicher, M., Ntoutsi, E., Spiliopoulou, M.: Active feature acquisition on data streams under feature drift. Ann. Telecommun. **75**, 597–611 (2020)
3. Büttner, M., Beyer, C., Spiliopoulou, M.: Reducing missingness in a stream through cost-aware active feature acquisition. In: 2022 IEEE 9th International Conference on Data Science and Advanced Analytics (DSAA), pp. 1–10 (2022)
4. Dong, W., Gao, S., Yang, X., Yu, H.: An exploration of online missing value imputation in non-stationary data stream. SN Comput. Sci. **2**, 1–11 (2021)
5. Dua, D., Graff, C.: UCI machine learning repository (2017)
6. Gama, J., Sebastiao, R., Rodrigues, P.P.: On evaluating stream learning algorithms. Mach. Learn. **90**, 317–346 (2013)

7. Gomes, H.M., de Mello, R.F., Pfahringer, B., Bifet, A.: Feature scoring using tree-based ensembles for evolving data streams. In: 2019 IEEE International Conference on Big Data (Big Data), pp. 761–769. IEEE (2019)

8. Hallaji, E., Razavi-Far, R., Saif, M.: DLIN: deep ladder imputation network. IEEE Trans. Cybern. **52**(9), 8629–8641 (2021)

9. Herbold, S.: Autorank: a python package for automated ranking of classifiers. J. Open Source Softw. **5**(48), 2173 (2020)

10. Huang, S.J., Xu, M., Xie, M.K., Sugiyama, M., Niu, G., Chen, S.: Active feature acquisition with supervised matrix completion (2018)

11. Kachuee, M., Goldstein, O., Kärkkäinen, K., Darabi, S., Sarrafzadeh, M.: Opportunistic learning: budgeted cost-sensitive learning from data streams. In: 7th International Conference on Learning Representations, ICLR 2019, New Orleans, LA, USA, 6–9 May 2019. OpenReview.net (2019)

12. Keerin, P., Boongoen, T.: Improved KNN imputation for missing values in gene expression data. Comput. Mater. Contin. **70**(2), 4009–4025 (2021)

13. Kottke, D.: Budget Optimization for Active Learning in Data Streams. Master's thesis, Otto von Guericke University Magdeburg, Germany (10 2014)

14. Li, Y., Oliva, J.: Active feature acquisition with generative surrogate models. In: International Conference on Machine Learning, pp. 6450–6459. PMLR (2021)

15. Lin, W.C., Tsai, C.F.: Missing value imputation: a review and analysis of the literature (2006–2017). Artif. Intell. Rev. **53**, 1487–1509 (2020)

16. Melville, P., Saar-Tsechansky, M., Provost, F., Mooney, R.: Active feature-value acquisition for classifier induction. In: Fourth IEEE International Conference on Data Mining (ICDM'04), pp. 483–486. IEEE (2004)

17. Peng, T., Sellami, S., Boucelma, O.: IoT data imputation with incremental multiple linear regression. Open J. Internet Things (OJIOT) **5**(1), 69–79 (2019)

18. Ramanan, N., Odom, P., Kersting, K., Natarajan, S.: Active feature acquisition via human interaction in relational domains. In: Proceedings of the 6th Joint International Conference on Data Science & Management of Data (10th ACM IKDD CODS and 28th COMAD), pp. 70–78. Association for Computing Machinery, New York, NY, USA (2023)

19. Saar-Tsechansky, M., Melville, P., Provost, F.J.: Active feature-value acquisition. Manag. Sci. **55**(4), 664–684 (2009)

20. Settles, B.: Active learning literature survey. Technical report 1648, University of Wisconsin-Madison Department of Computer Sciences (2009)

21. Shim, H., Hwang, S.J., Yang, E.: Joint active feature acquisition and classification with variable-size set encoding. In: Bengio, S., Wallach, H., Larochelle, H., Grauman, K., Cesa-Bianchi, N., Garnett, R. (eds.) Advances in Neural Information Processing Systems, vol. 31. Curran Associates, Inc. (2018)

22. Tharwat, A., Schenck, W.: A survey on active learning: state-of-the-art, practical challenges and research directions. Mathematics **11**(4) (2023)

23. Yuan, L., Pfahringer, B., Barddal, J.P.: Iterative subset selection for feature drifting data streams. In: Proceedings of the 33rd Annual ACM Symposium on Applied Computing, SAC 2018, Pau, France, 09–13 April 2018, pp. 510–517 (2018)

24. Zhang, P., Zhu, X., Tan, J., Guo, L.: SKIF: a data imputation framework for concept drifting data streams. In: Proceedings of the 19th ACM International Conference on Information and Knowledge Management, pp. 1869–1872 (2010)

Fairness, Privacy and Security in AI

EXPHLOT: EXplainable Privacy Assessment for Human LOcation Trajectories

Francesca Naretto[1]([✉]) [iD], Roberto Pellungrini[2] [iD], Salvatore Rinzivillo[3] [iD],
and Daniele Fadda[3] [iD]

[1] University of Pisa, Pisa, Italy
francesca.naretto@di.unipi.it
[2] Scuola Normale Superiore, Pisa, Italy
roberto.pellungrini@sns.it
[3] ISTI CNR, Pisa, Italy
{salvatore.rinzivillo,daniele.fadda}@isti.cnr.it

Abstract. Human mobility data play a crucial role in understanding mobility patterns and developing analytical services across various domains such as urban planning, transportation, and public health. However, due to the sensitive nature of this data, accurately identifying privacy risks is essential before deciding to release it to the public. Recent work has proposed the use of machine learning models for predicting privacy risk on raw mobility trajectories and the use of SHAP for risk explanation. However, applying SHAP to mobility data results in explanations that are of limited use both for privacy experts and end-users. In this work, we present a novel version of the EXPERT privacy risk prediction and explanation framework specifically tailored for human mobility data. We leverage state-of-the-art algorithms in time series classification, as ROCKET and INCEPTIONTIME, to improve risk prediction while reducing computation time. Additionally, we address two key issues with SHAP explanation on mobility data: first, we devise an entropy-based mask to efficiently compute SHAP values for privacy risk in mobility data; second, we develop a module for interactive analysis and visualization of SHAP values over a map, empowering users with an intuitive understanding of SHAP values and privacy risk.

Keywords: Mobility Data · Privacy · Explainability

1 Introduction

The analysis of human mobility data is very important for the development of analytical services and for supporting decision-making processes in many sectors: urban planning [33], health [16] or tourism [7]. During the COVID-19 pandemic, for example, studying human mobility data helped understand and explain to the public how the infection spreads and propose good practices to stop it. Analyses

© The Author(s) 2023
A. Bifet et al. (Eds.): DS 2023, LNAI 14276, pp. 325–340, 2023.
https://doi.org/10.1007/978-3-031-45275-8_22

in this field are usually conducted on large datasets containing information on the temporal sequences of locations visited by individuals, such as GPS tracks. This type of data, however, is very sensitive, as it can lead to the disclosure of personal information about an individual, such as the home location and place of work. For example, it has been proven that four spatiotemporal points may be sufficient to identify 95% of the individuals within a mobility dataset [20]. To address privacy risks associated with mobility data, various methodologies have been proposed to protect the privacy of the users, but they often involve modifying the data or Machine Learning (ML) models, compromising overall performance. Striking a balance between privacy protection and data quality requires reliable and efficient methods to quantify privacy risk. Pratesi *et al.* [25] proposed a risk assessment framework that computes privacy risk through the definition and simulation of various attack scenarios. While effective, this framework has drawbacks, including high time complexity and the need to re-compute the privacy risk for all data when new samples are added.

To mitigate these problems, Pellungrini et al. [22] proposed a ML approach for the computation of privacy risk based on individual and collective mobility features extracted from the data. Further improvements have been proposed by Naretto *et al.* [21] with the EXPERT framework, which implements a Long Short Term Memory neural network (LSTM) able to predict privacy risk directly from mobility data trajectories. In compliance with the EU General Data Protection Regulation, EXPERT also ensures the "right to explanation", proposing the use of SHAP (SHapley Additive exPlanations) [17], a well-known explainer based on SHAP values, which is commonly used for its stability and robustness. However, EXPERT has several limitations: *L1)* the LSTM training is time demanding and requires deep models to be effective; *L2)* SHAP can be efficiently applied only with specific heuristics tailored on specific ML models, like *DeepExplainer*, whereas general prediction models require a lot of time to be explained, since they rely on the combinatorial evaluation of the SHAP values; *L3)* the explanation provided by SHAP in the context of mobility data is not easy to interpret, given the high number of dimensions, and it gives limited information to non-technical users. Therefore, in this paper, we propose EXPHLOT, a framework tailored towards human mobility data that solves the aforementioned problems. To tackle *L1* we employ state-of-the-art ML models for sequential data (as INCEPTIONTIME [14] and ROCKET [8]) to speed up the training process. For *L2*, we propose a novel optimization heuristic based on entropy masks to execute efficiently SHAP permutation explainer for mobility data. For *L3*, we propose a visualization dashboard specifically tailored for the analysis of human mobility focused on both privacy risk and explanation, thus improving the fruition of the system for non-technical users.

The paper is structured as follows: in Sect. 2, we present the most relevant papers in the related literature; in Sect. 3 are reported the necessary definitions and notation; in Sect. 4 is presented our proposed framework; in Sect. 5 we show an application of our proposed framework to real human mobility data and provide an empirical evaluation.

2   Related Works

Privacy Risk Assessment. In our work, we use the PRUDEnce framework from Pratesi *et al.* in [25], which allows for a systematic computation of privacy risk in a data-driven way. At its core, PRUDEnce is based on the principle of *k-anonymity* [29] as it computes privacy risk based on the size of the *k*-sets for each individual represented in the data. PRUDEnce has been extensively used in privacy risk assessment for a diverse range of data [23,24]. The high computational cost of PRUDEnce lead to the development of ML approaches that try to predict privacy risk instead of computing it. Pellungrini *et al.* [22] developed an approach based on Individual Mobility Profiles extracted from the data. Naretto et al. [21] proposed the EXPERT framework, which improves PRU-DEnce in two ways: first, by developing a ML methodology able to predict risk directly from sequential data, secondly by explaining the privacy risk prediction using a set of methodologies like SHAP [17] and LIME [26]. Our EXPHLOT starts from the EXPERT and adds new improvements by integrating models and solutions that leverage domain-specific characteristics of mobility data. Several works are related to privacy risk assessment, mainly focused on applying classic risk assessment techniques to various privacy problems [32]. One of the most recent and relevant works in the field of privacy risk assessment is the work from Silva *et al.* [30], in which the authors provide an application of CRISP methodology and fuzzy logic to natural language processing tasks. Their work relies on the definition of a sensitivity level for the features possibly extracted from an individual's text and therefore is not entirely data-driven like our approach. For location-based data, Khalfoun *et al.* [15] proposed EDEN, a federated learning approach to location anonymization that is based on the FURIA federated learning framework for re-identification risk assessment. In their setting, they consider three types of attack: AP-Attack, POI-Attack, and PIT-Attack, considering spatial, temporal, and aggregated features. EDEN then selects the best privacy preservation technique with respect to this kind of assessment.

Predictive Models for Human Mobility Data. In this section, we present the latest solution in the context of predictive models for human mobility data. EXPHLOT predicts the privacy risk directly on mobility data. For this task, one of the most applied ML models is the Long Short-Term Memory networks (LSTM) [13], a specific architecture belonging to Recurrent Neural Network (RNN), that are able to overcome some of the shortcomings of RNN, e.g., vanishing gradient in fully connected RNN. LSTM have been applied to human mobility data in many works [7,21,34]. Song *et al.* [31] use a LSTM network to develop a system for simulating and predicting human mobility and transportation model at a citywide level, while Altché *et al.* [1] use a LSTM to model vehicular movement on highways. LSTM have been also applied to predict the privacy risk in human mobility data [21]. However, the application of LSTM requires deep models to be effective and hence also a long training time. Recently, Fawaz *et al.* proposed INCEPTIONTIME [14], an ensemble of deep inception modules. This model achieves comparable performance as the LSTM reducing the learning time. Another recent proposal is ROCKET [8]. It is an ensemble method based on convolutional kernels which transform the time series into features that are then used to train a linear classifier. This approach is very efficient and stable, allowing good generalization capabilities. In EXPHLOT we exploit both INCEPTIONTIME and ROCKET models to overcome the time limitation of EXPERT.

2.1 Explainability

Explainability is one of the most important modern lines of research in AI as it is crucial in achieving Trustworthy Artificial Intelligence. [3] provides a comprehensive overview of existing techniques for interpretability in ML, identifying two main types of explanation models: *global* and *local* explainers. Local explainers focus on explaining the results of predictions on single instance [11,18,27] while global explainers explain the logic of the whole machine learning model [5,6,10]. With EXPHLOT, we aim at explaining to the end user the reasons why he/she has a privacy risk exploiting local explanations for time series. In this context there are many recent methods, however, the majority of them are computationally inefficient and require a long training time [12]. In this work, we provide explanations by using SHAP [17], a well-known explainer based on SHAP values, which is commonly used for its stability and robustness of results.

3 Background

3.1 Privacy Risk Assessment Framework

In this paper, we consider each individual's mobility as a trajectory, i.e., a temporally ordered sequence of pairs, $T_u = (l_1, t_1), (l_2, t_2), \ldots, (l_m, t_m)$, where $l_i = \langle x_i, y_i \rangle$ is the location identified by the latitude x_i and longitude y_i, while t_i $(i = 1, \ldots, m)$ denotes the corresponding timestamp such that $\forall 1 \leq i \leq m$ $t_i < t_{i+1}$. We denote by $\mathcal{D} = T_1, \ldots, T_n$ the *mobility dataset* that describes the movements of n individuals. In this paper, we simulate a privacy attack on human mobility data to acquire the ground truth to train our predictive model. Our attack is simulated using the PRUDEnce framework.

As mentioned in Sect. 2, PRUDEnce is based on *k-anonymity* [29], in which the privacy risk computation relies on the size of k-sets for each individual in the data. PRUDEnce has been utilized for privacy risk assessment in various data domains, such as purchase and mobility data [23,24]. The framework provides an effective approach to quantifying privacy risks and has demonstrated its applicability in diverse contexts. For these reasons, we have chosen the PRUDEnce methodology as the pre-processing step for computing privacy risk on raw mobility data in our work.

Technically, the privacy risk computation procedure of PRUDEnce is general and requires the definition of a privacy attack. The *privacy risk computation* defined in Prudence is the following:

1. Define an attack, based on a specific background knowledge category B;
2. Consider a set of background knowledge configurations $B_1, B_2, \ldots, B_m$;

3. For all the configurations $B_1, B_2, ..., B_m$, compute all the possible instances $b \in B_k$ and its probability of re-identification;
4. For each individual, select the maximum privacy risk, defined as the maximum probability of re-identification across all the instances $b \in B_k$.

Therefore PRUDEnce adopts an *exhaustive* privacy risk evaluation technique, by considering all the possible background knowledge the attacker could have over a given dataset (or dataview of the original dataset). For our purpose, we consider the case where each individual is represented by a single trajectory T_u in $\mathcal{D}$. Formally, given a single individual u, the probability of re-identification is:

$$Pr_{\mathcal{D}}(T_u|b) = \frac{1}{\sum_{T_i \in \mathcal{D}} \{matching(T_i, b)\}} \tag{1}$$

where $\mathcal{D}$ is the dataset under analysis, b the background knowledge instance considered and T_u the trajectory under analysis. In essence, we compute the support for b with respect to each trajectory in the dataset. The *matching* function formalizes how an adversary matches background knowledge b to the data. b is generated systematically, i.e., PRUDEnce performs exhaustive privacy risk assessment, among all possible $b \in B_k$. We simulate an attack where we assume that an adversary has access to some of the points in the trajectory of an individual, knowing a subsequence of the original trajectory with the relative order of the points.

Let h be the number of locations l_j of an individual u known by the adversary and let $L(T_u)$ be the complete sequence of locations $l_j \in T_u$ visited by u (i.e., regardless of time). The location sequence background knowledge is a set of configurations based on h locations, defined as $B_h = L(T_u)^{[h]}$, where $L(T_u)^{[h]}$ denotes the set of all the possible h-subsequences of the elements in the set $L(T_u)$, i.e., each instance $b \in B_h$ is a subsequence of locations of length h. In each b, the order among the elements is preserved and known to the adversary. The *matching* function for this privacy attack is therefore defined as:

$$matching(T_i, b) = \begin{cases} 1, & \text{if } b \subseteq L(T_u) \\ 0, & \text{otherwise} \end{cases} \tag{2}$$

Privacy Risk is the maximum probability of re-identification across all b:

$$Risk(u, \mathcal{D}) = max(Pr_{\mathcal{D}}(d = u|b)) \tag{3}$$

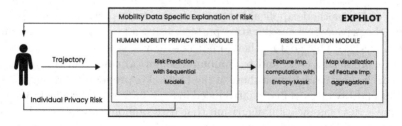

Fig. 1. The general structure of the proposed framework.

3.2 EXPERT

PRUDEnce is not suited for providing personalized recommendations in terms of risks associated with personal mobility: for any new user requiring risk evaluation, the system should re-compute the privacy risk against the whole dataset. In addition, it does not provide any explanation of the privacy risk derived by the system. To overcome these drawbacks, EXPERT [21] predicts the user's privacy risk to increase individual awareness, by also providing an explanation of the derivation of the risk associated with sharing sensitive location information. EXPERT implements a *privacy risk prediction* module which takes as input the user's trajectory and predicts the privacy risk level of that user by means of a ML model. It also uses an *explanation* module to produce the explanation of the predicted risk. The output of the *privacy risk prediction* module is the predicted privacy risk as a binary value (HIGH risk *vs* LOW risk). The output of the *risk explanation* module is an explanation of the ML model for the predicted risk label. EXPERT is modular with respect to the explainer, allowing the use of any explanation method which outputs a local explanation, suitable to the type of data under analysis. The authors use SHAP, and LORE in the original paper [21] (Fig. 1).

4 EXPHLOT

In this paper we propose EXPHLOT, an improved version of EXPERT tailored for Human Mobility Data. Our aim is to provide analysts with an actionable framework to predict and visualize privacy risk with an integrated explanation. The general architecture of EXPHLOT is shown in Fig. 1.

4.1 EXPHLOT Predictive Model

EXPHLOT objective is to predict the privacy risk of a human trajectory while providing the analyst with also an explanation to increase user awareness. Privacy risk is a continuous value in the interval $[0, 1]$. However, we decide to model the problem as a binary classification. Indeed, we are interested in distinguishing between HIGH risk and LOW risk users, in such a way that higher-risk users can be protected. Technically, we discretize the privacy risk obtained from the location-based attack: LOW risk or 0 (privacy risk ≤ 0.5) and HIGH risk or 1 (privacy risk >0.5). The Γ vector generated in this way is then joined to the mobility dataset D and we use $\langle D, \Gamma \rangle$ to train a classification model. To avoid the problem of having to craft and compute features to be used as input data, Naretto *et al.* [21] propose to use methods applicable to raw sequences. In particular, they propose to solve the privacy risk classification problem using a Long-Short Term Memory network (LSTM). Our goal is, therefore, to use novel, state-of-the-art models to solve this prediction task, and to compare the performance and time-efficiency results of the new models with those of the LSTM. We propose two recent models, ROCKET and INCEPTIONTIME, introduced in Sect. 2. ROCKET is a fast and accurate time series classification algorithm that uses random convolutional kernels. It is composed of two parts: a first part in which k randomly generated convolutional kernels are used to calculate a feature map from which, for each kernel, two aggregated features are extracted (*ppv* and *maximum value*); a second part in which the aggregated features are passed to a linear classification algorithm to obtain the actual result. The number k of kernels is the only hyper-parameter of the model. In theory, ROCKET can be used for both variable-length and fixed-length time series. To be applied to variable length time series, the kernels must be shorter than the length of the shortest time series. In the case where the length of the series varies greatly, as in our case, this approach is very inconvenient, as finding the right kernel would be time-consuming. We, therefore, chose a fixed-length approach, using low amplitude or zero padding to keep the result of the convolution operation on those segments close to zero and constant, cutting it off the calculation of the features (*ppv* and *maximum value*). We chose ROCKET over MINIROCKET [9] as the latter eliminates the random component in the choice of kernels' characteristics. Therefore, even though MINIROCKET is generally faster, we believe that a set of varied kernels fits better for our case, to capture the most diverse pattern possible. INCEPTIONTIME is an ensemble time series classification algorithm based on an ensemble of inception architectures. The Inception model is composed of convolutional layers and simultaneously applies several filters of different lengths to the input time series. This structure alleviates the vanishing gradient problem by enabling a direct flow of the gradient. It cannot be used on time series of variable length. To choose the best models, we focused on the recall of both classes, giving priority to class 1, and the precision of both classes. This is because we want to protect high-risk users by preventing them from being classified as LOW-risk, so that their sensitive data would not be threatened. Moreover, we wanted to maximize the possibility of sharing the data of LOW-risk users, thus preventing them from being classified as HIGH-risk.

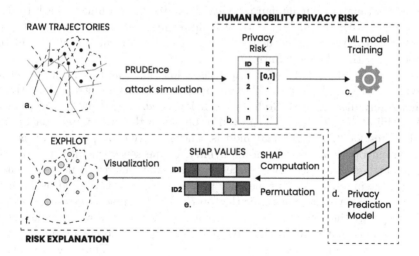

Fig. 2. EXPHLOT analytical pipeline. Starting from the generalized trajectories (a) a privacy prediction model (d) is trained from a set of observations generated by a privacy risk model (b). The prediction is explained by means of SHAP values (e) that are visualized within an analytical dashboard (f)

4.2   Exphlot Risk Explanation Module

For the Explanation Module of EXPHLOT our goal is to provide an explanation that is informative for experts and users in the dominion of Human Mobility data (Fig. 2). We chose to employ SHAP to generate an attribution-based explanation for our models. Our aim is to indicate, for each individual, what parts of his movement lead to higher privacy risks. Given the nature of our specific ML models, we must employ the *Kernel Explainer*, which is the agnostic explainer of the SHAP library. Clearly, depending on the size of the given data, the computation is more accurate but also longer in time. One possible solution, suggested also by the authors of SHAP, is to exploit K-means clustering by selecting a large k and then feeding all the centroids obtained to the *Kernel Explainer*. In this way, we are able to represent all the space under analysis by considering a small number of trajectories. However, this solution for mobility data is not enough: SHAP considers each location of the trajectory as a variable and for computing the SHAP values all the permutations of variables are calculated as well as their relative interactions. This procedure is exponential in time if the number of variables is high, as in our case. Computation of SHAP values becomes therefore unfeasible in a reasonable time. Mitchell et al. [19] propose several sampling strategies that can in theory speed up SHAP values computation. However, many of the proposed strategies work under assumptions of bounds to the possible values or shape of the data. For human mobility, these bounds may not hold. For these reasons, we decide to apply the *PermutationExplainer* with a dynamic mask. This method can take as input a user-defined mask that allows certain features to be hidden, thus decreasing the individual evaluations made on these and the complexity of the calculation. In our setting, each feature corresponds to a location of the geographical map of our human mobility data. We used a binary mask to hide the features with the highest entropy, fully evaluating the locations with the lowest entropy. We formally define location entropy for each location i in the dataset with the Shannon Entropy equation: $E_i = -\sum_{u \in U_i} p_u \log_2 p_u$, where p_u is the probability that individual u visits location i and U_i is the set of all individuals visiting location i. The importance of location entropy for privacy is thoroughly discussed by Rodriguez-Carrion *et al.* [28], while in the work of Pellungrini *et al.* [22] entropy is proven to be one of the most important predictive features/locations also in ML models. The intuitive concept behind it is that location entropy is a measure of anonymity, in the sense that if a user passes through high-entropy locations, where therefore many different other people pass through, the uniqueness of his mobility profile is lost as it is blurred by the general movement. We, therefore, hide the top 70% of the highest entropy locations, evaluating only the 30% with the lowest entropy. In this way, we are focusing on those locations that have fewer individuals visiting in a more sporadic way and thus we are focusing on explaining HIGH-risk predictions. Thus, we are able to speed up the computation of the SHAP values.

4.3   Exphlot Risk and Explanation Visualization Module

The effective visualization of mobility properties can provide a boost to gaining deeper insights into spatial and temporal patterns. To manage the complexity of spatial resolutions, a widely adopted solution leverages spatial aggregation based on spatial partitioning [2,4]. The process organizes close entities into groups and, for each group, a single centroid point is determined. Then the centroid points are used as seeds to partition the territory. In the scope of our work, the data related to geography is linked to multiple dimensions and attributes, like mobility indicators, privacy risk prediction, and feature relevance. Moreover, many of these indicators may have multiple spatial scales, for example ranging from an urban building block resolution to a city district.

Thus, we designed a visual interface where the set of locations of each trajectory is presented within two linked displays: a *dynamic map* with embedded graphics and a *bubble chart* (see Fig. 4). The *dynamic map* shows for each location a visual mark, a circle, whose visual properties are linked to internal indicators of the location it represents. Each circle is driven by two visual variables, the area of the circle and the fill color, which both encode the same quantitative value. Without loss of generality, we can assume that these quantitative values are mapped to the $[0, 1]$ interval, in order to implement a pair of scale functions to determine the area and the color of each circle. The *Bubble Chart* contains the same set of circles of the map (to create conceptual links between the two displays) located accordingly to the respective values on the two axes. The user can decide which attributes are associated with which value. Any selection/filter activated on the Bubble Chart is propagated to the map (and *viceversa*).

The SHAP values are computed for every single individual trajectory. However, the domain expert is interested in the analysis of collective behavior. Thus, we aggregate the individual explanations into a global one using the aggregation procedure available within the SHAP library. This is especially important for all those instances where the data is not public or is under strict confidentiality constraints. From a geographical point of view, we considered for each location l the set of all the trajectories crossing l. For this subset of trajectories, a set of indicators is computed, such as *number of trajectories*, and *risk of re-identification*. For the latter, we compute statistical indicators to have a compact representation of the distribution: min, max, first quartile, third quartile, median, and average.

This design achieves multiple objectives. First, it provides a user-driven exploration of the SHAP values, since the analyst can evaluate and compare the contribution of each location to the risk prediction and let the user visually identify zones containing locations with similar characteristics. Second, the possibility of navigating the map allows for a deeper investigation of local areas and provides a solution to limit cluttering when the number of locations is high. Third, geographic mapping allows a topological exploration of close locations, enabling the identification of general patterns, i.e. urban areas versus rural areas. Fourth, the expert can exploit the linked display to investigate relevant cases that are not directly evident from the map. The possibility of cross-selecting visual elements enables better identification of patterns and rules of the data.

5 Experiments

For validating EXPHLOT we used GPS tracks of private vehicles, provided by Octo Telematics[1], an insurance company. We selected trajectories from the city area of Prato and Pistoia (Italy), with 8651 users observed in a period of one month, from 1st May to 31st May 2011[2]. The dataset considered is composed of a trajectory for each user. Hence, each trajectory contains all the points visited by the user in temporal order. On these trajectories, we applied a transformation, in the following called VORONOI, in which the territory is split in tiles based on a data-driven Voronoi tessellation [2]. This approach considers the traffic density of an area to create the tiles. Then, we used the cells of this tessellation to generalize the original trajectories. The algorithm applies interpolation between non-adjacent points[3]. The outliers were removed using DBScan algorithm obtaining 1473 different locations, with an average length of 240.2 per trajectory. Given the processed dataset D, for an in-depth validation of EXPHLOT, we considered *four background knowledge* configurations B_h using $h = 2, 3, 4, 5$ obtaining four different risk datasets, $\Gamma_{h=2,3,4,5}$ where, we recall, h represents the length of the background knowledge of the simulated attacker. We discretized the risk values in two classes: LOW, when the privacy risk is in $[0, 0.5]$ and HIGH in $]0.5, 1]$. At this point, we merged the privacy risk data with the trajectories to obtain the classification datasets for our supervised learning task, following the methodology explained in Sect. 3.1. Hence, we obtained 4 different datasets for our experiments. We remark that the datasets with the highest and lowest background knowledge are highly imbalanced, having the $D_{h=2}$ with the 71% of users belonging to the LOW class, while for $D_{h=5}$ has the 63% of trajectories in the HIGH class. This is to be expected, as when the knowledge of the attacker is small, such as $h = 2$, the attack is less effective, having fewer people re-identified. In addition, we remark that we compute the privacy risk of the entire dataset D, splitting the data after privacy risk computation. This decision is based on the fact that if we calculate the privacy risk separately for the training and testing sets, the final result will differ from the computation performed on the complete dataset, due to k-*anonymity* (Sect. 3.1). It has been demonstrated that the models still generalize well and possess transfer learning capabilities [22].

5.1 Exphlot Privacy Risk Prediction Module

For all the models we split our datasets into 80% for training and validation (10%) and 20% for testing. The predictive performance of ROCKET, INCEPTIONTIME, and LSTM are reported in Table 1. All the models perform well, achieving good precision and recall for both classes, even in unbalanced settings. For the most unbalanced case, which is the $h = 2$, ROCKET and INCEPTIONTIME

---

[1] https://www.octotelematics.com/it/.

[2] Data are collected by GPS devices that detect the position every 30 s, if the vehicle is not in motion the device automatically stops recording.

[3] Voronoi tessellation obtained using http://geoanalytics.net/V-Analytics.

Table 1. Metrics of ROCKET (R), INCEPTIONTIME (IT) and LSTM (LS) compared for each dataset h. For precision P and recall R we present the values for both classes (*high* and *low* risk. From a privacy perspective R_{high} is the most important value as it represents the fraction of correctly predicted HIGH risk individuals.

	$h = 2$			$h = 3$		
	ROCKET	INCEPTIONTIME	LSTM	ROCKET	INCEPTIONTIME	LSTM
Acc	0.81	**0.84**	0.80	**0.88**	0.87	**0.88**
P_{low}	**0.91**	0.88	0.90	0.89	0.86	**0.90**
P_{high}	0.63	**0.72**	0.62	**0.88**	**0.88**	**0.88**
R_{low}	0.81	**0.89**	0.81	0.84	**0.85**	0.84
R_{high}	**0.80**	0.70	0.76	0.91	0.89	**0.92**
F1	0.78	**0.80**	0.76	**0.88**	0.87	**0.88**

	$h = 4$			$h = 5$		
	ROCKET	INCEPTIONTIME	LSTM	ROCKET	INCEPTIONTIME	LSTM
Acc	**0.90**	0.89	0.89	0.91	0.90	**0.92**
P_{low}	**0.90**	0.87	**0.90**	0.87	0.86	**0.89**
P_{high}	0.90	**0.91**	0.89	0.93	0.93	**0.94**
R_{low}	0.86	**0.89**	0.84	0.88	0.88	**0.89**
R_{high}	**0.93**	0.90	0.92	0.92	0.92	**0.93**
F1	**0.90**	0.89	0.89	0.90	0.90	**0.91**

Table 2. Training and test times for ROCKET and InceptionTime. Overall ROCKET is the fastest model in training.

	INCEPTIONTIME		ROCKET		LSTM	
Dataset	Training	Test Time	Training	Test	Training	Test
$h = 2$	16h49min	6sec	2min32sec	44sec	8h50min	60sec
$h = 3$	20h7min	6sec	3min	40sec	5h30min	60sec
$h = 4$	4h	4sec	7min	16sec	5h50min	60sec
$h = 5$	9h24min	5sec	8min	17sec	6h15min	60sec

perform better than LSTM, showing better generalization capabilities. However, ROCKET achieves the highest recall on class HIGH, which is the most important class for our setting, being the class of the users with HIGH risk of privacy. INCEPTIONTIME, instead, while having generally good metrics, does not perform well on the recall for HIGH class. The real benefit of ROCKET over other models is in training time, as can be seen in Table 2. While training the LSTM can take many hours, the other models are faster. ROCKET is the quickest, with a training time of just a few minutes, allowing us to achieve the *online* interaction with the end user we are aiming at.

Fig. 3. Shap Force Plot visualization of the contributions towards HIGH risk. The standard visualization does not provide significant information to domain experts.

5.2   Mobility Privacy Risk Explanation

Applying SHAP we obtain a local explanation based on feature importance: for each feature we have a value associated to it that represents how important the feature is for the prediction at hand. Local explanations can be summed up to obtain a global explanation as shown in Fig. 3. This plot represents the explanation for all trajectories predicted as HIGH risk. A large number of features makes it very difficult for the analyst to understand which are the most relevant locations that contribute to the HIGH (or LOW) risk. Clearly, this linear layout has two main limitations: first, the high number of features does not allow a clear reading of those locations with smaller contributions; second, the topological and spatial relations among locations are not evident. The visual interface introduced in Sect. 4.3 addresses these two limitations. Figure 4 shows a screenshot of the interface showing the SHAP values associated with the prediction of HIGH risk for each location[4]. This visualization allows an analyst to immediately understand which areas of the map present the highest contribution for the model towards risk classification. Our map allows for a more intuitive understanding of the contributions of each location with respect to the original SHAP visualization. Our visualization can help the analyst understand the dependence of privacy risk on the mobility behaviors of the collectivity. In the figure, there is a cluster of locations along a country road with a high contribution to the HIGH risk, confirming the intuition that low-traffic roads are more prone to privacy exposures. Moreover, the urban surroundings present a lower level of risk, even if it is possible to visually detect different privacy levels in two close municipalities: the south-east town has very low-risk levels; the north-west town has a higher risk level.

---

[4] The interactive maps of the experiments in this paper are available at this link.

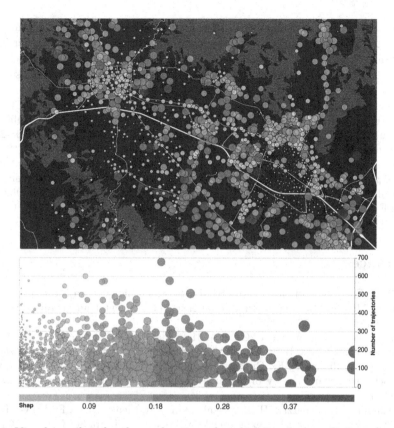

Fig. 4. Visual interface for the exploration of explanation and prediction of privacy risk. Each circle represents the contribution to the prediction of HIGH risk

6   Conclusion

In this paper, we proposed EXPHLOT, a privacy assessment prediction and explanation framework tailored towards human mobility data. We improve on previous privacy risk assessment frameworks by employing specific ML models for sequential data and develop custom heuristic techniques for computing SHAP values in feasible times and a visualization tool tailored for human mobility data analysis. Our framework can accurately predict privacy risk in human mobility data and effectively explain the predictive models with fast SHAP value calculation and an intuitive and interactive visualization tool that maps the essential contribution and information about the problem onto a dynamic map. We validated our framework on real, confidential human mobility data and showed how it is possible to immediately gain new insight into the nature of privacy risk. Our work provides privacy analysts and experts in the field with an interactive and actionable tool to understand the privacy risk of human mobility data in an interactive and fast way. As a future work, we are working on exploiting our visual analytics environment to validate the effeect of different privacy mitigation techniques. This would be a *"what-if"* simulation module to allow analysts to interactively assess privacy risk, providing a new tool in the development of privacy protection measures based on generalization or deletion. Another interesting direction is the integration of additional data quality measures, to allow further experimentation of protection measures on the data before release.

Acknowledgments. This work is supported by: the EU NextGenerationEU programme under the funding schemes PNRR-PE-AI FAIR (Future Artificial Intelligence Research); the EU - Horizon 2020 Program under the scheme "INFRAIA-01-2018-2019 - Integrating Activities for Advanced Communities" (G.A. n.871042) "SoBig-Data++: European Integrated Infrastructure for Social Mining and Big Data Analytics" (http://www.sobigdata.eu); PNRR-"SoBigData.it - Strengthening the Italian RI for Social Mining and Big Data Analytics" - Prot. IR0000013; TAILOR (G.A. 952215), HumanE-AI-Net (G.A. 952026) and EU H2020 project XAI (Grant Id 834756); CREX-DATA (G.A. 101092749).

References

1. Altché, F., de La Fortelle, A.: An LSTM network for highway trajectory prediction. In: 2017 IEEE 20th International Conference on Intelligent Transportation Systems (ITSC), pp. 353–359 (2017)
2. Andrienko, N.V., Andrienko, G.L.: Spatial generalization and aggregation of massive movement data. IEEE Trans. Vis. Comput. Graph. **17**(2), 205–219 (2011)
3. Bodria, F., Giannotti, F., Guidotti, R., Naretto, F., Pedreschi, D., Rinzivillo, S.: Benchmarking and survey of explanation methods for black box models. DAMI (2023)
4. Buchmüller, J., Janetzko, H., Andrienko, G.L., Andrienko, N.V., Fuchs, G., Keim, D.A.: Visual analytics for exploring local impact of air traffic. Comput. Graph. Forum **34**(3), 181–190 (2015). https://doi.org/10.1111/cgf.12630
5. Craven, M., Shavlik, J.W.: Extracting tree-structured representations of trained networks. In: NIPS, pp. 24–30 (1996)
6. Craven, M.W., Shavlik, J.W.: Using sampling and queries to extract rules from trained neural networks. In: JMLR, pp. 37–45. Elsevier (1994)
7. Crivellari, A., Beinat, E.: LSTM-based deep learning model for predicting individual mobility traces of short-term foreign tourists. Sustainability **12**(1) (2020). https://doi.org/10.3390/su12010349
8. Dempster, A., Petitjean, F., Webb, G.I.: ROCKET: exceptionally fast and accurate time series classification using random convolutional kernels. Data Min. Knowl. Disc. **34**(5), 1454–1495 (2020). https://doi.org/10.1007/s10618-020-00701-z
9. Dempster, A., Schmidt, D.F., Webb, G.I.: MiniRocket. In: Proceedings of the 27th ACM SIGKDD Conference. ACM (2021). https://doi.org/10.1145/3447548.3467231
10. Deng, H.: Interpreting tree ensembles with inTrees. Int. J. Data Sci. Anal. **7**(4), 277–287 (2019)
11. Guidotti, R., Monreale, A., Giannotti, F., Pedreschi, D., Ruggieri, S., Turini, F.: Factual and counterfactual explanations for black box decision making. IEEE Intell. Syst. **34**(6), 14–23 (2019)
12. Guidotti, R., Monreale, A., Spinnato, F., Pedreschi, D., Giannotti, F.: Explaining any time series classifier. In: CogMI 2020 (2020)
13. Hochreiter, S., Schmidhuber, J.: Long short-term memory. Neural Comput. **9**(8), 1735–1780 (1997)

14. Ismail Fawaz, H., et al.: InceptionTime: finding alexnet for time series classification. Data Min. Knowl. Discov. **34**, 1936–1962 (2020)
15. Khalfoun, B., Ben Mokhtar, S., Bouchenak, S., Nitu, V.: Eden: Enforcing location privacy through re-identification risk assessment: a federated learning approach (2021). https://doi.org/10.1145/3463502
16. Lucchini, L., et al.: Living in a pandemic: changes in mobility routines, social activity and adherence to COVID-19 protective measures. Sci. Rep. (2021). https://doi.org/10.1038/s41598-021-04139-1
17. Lundberg, S.M., Lee, S.: A unified approach to interpreting model predictions. CoRR abs/1705.07874 (2017). http://arxiv.org/abs/1705.07874
18. Lundberg, S.M., Lee, S.I.: A unified approach to interpreting model predictions. In: NIPS, pp. 4765–4774 (2017)
19. Mitchell, R., Cooper, J., Frank, E., Holmes, G.: Sampling permutations for shapley value estimation. J. Mach. Learn. Res. **23**, 1–46 (2022)
20. Montjoye, Y.A., Hidalgo, C., Verleysen, M., Blondel, V.: Unique in the crowd: the privacy bounds of human mobility. Sci. Rep. (2013). https://doi.org/10.1038/srep01376
21. Naretto, F., Pellungrini, R., Nardini, F.M., Giannotti, F.: Prediction and explanation of privacy risk on mobility data with neural networks. In: ECML PKDD 2020 Workshops (2020)
22. Pappalardo, L., Pellungrini, R., Pratesi, F., Monreale, A.: A data mining approach to assess privacy risk in human mobility data. ACM Trans. Intell. Syst. Technol. (2017). https://doi.org/10.1145/3106774
23. Pellungrini, R., Pappalardo, L., Pratesi, F., Monreale, A.: Analyzing privacy risk in human mobility data (2018)
24. Pellungrini, R., Pratesi, F., Pappalardo, L.: Assessing privacy risk in retail data (2017)
25. Pratesi, F., Monreale, A., Trasarti, R., Giannotti, F., Pedreschi, D., Yanagihara, T.: Prudence: a system for assessing privacy risk vs utility in data sharing ecosystems. Trans. Data Priv. **11**, 139–167 (2018)
26. Ribeiro, M.T., Singh, S., Guestrin, C.: Why should i trust you?: explaining the predictions of any classifier (2016)
27. Ribeiro, M.T., Singh, S., Guestrin, C.: Why should i trust you?: explaining the predictions of any classifier. In: ACM SIGKDD, pp. 1135–1144 (2016)
28. Rodriguez-Carrion, A., et al.: Entropy-based privacy against profiling of user mobility. Entropy **17**(6), 3913–3946 (2015). https://doi.org/10.3390/e17063913
29. Samarati, P.: Protecting respondents identities in microdata release. IEEE Trans. Knowl. Data Eng. (2001). https://doi.org/10.1109/69.971193
30. Silva, P., Gonçalves, C., Antunes, N., Curado, M., Walek, B.: Privacy risk assessment and privacy-preserving data monitoring. Expert Syst. Appl. **200** (2022)
31. Song, X., Kanasugi, H., Shibasaki, R.: Deeptransport: prediction and simulation of human mobility and transportation mode at a citywide level. In: IJCAI'16 (2016)
32. Tang, J., Cui, Y., Li, Q., Ren, K., Liu, J., Buyya, R.: Ensuring security and privacy preservation for cloud data services. ACM Comput. Surv. (CSUR) **49**, 1–39 (2016)
33. Wang, J., Kong, X., Xia, F., Sun, L.: Urban human mobility: data-driven modeling and prediction. SIGKDD Explor. Newsl. **21**, 1–19 (2019)
34. Wu, F., Fu, K., Wang, Y., Xiao, Z., Fu, X.: A spatial-temporal-semantic neural network algorithm for location prediction on moving objects. Algorithms (2017). https://doi.org/10.3390/a10020037

Fairness-Aware Mixture of Experts with Interpretability Budgets

Joe Germino[ID], Nuno Moniz[ID], and Nitesh V. Chawla[(✉)][ID]

Lucy Family Institute for Data & Society, University of Notre Dame,
Notre Dame, IN 46656, USA
{jgermino,nuno.moniz,nchawla}@nd.edu

Abstract. As artificial intelligence becomes more pervasive, explainability and the need to interpret machine learning models' behavior emerge as critical issues. Discussions are usually bounded by those who defend that interpretable models must be the rule or that non-interpretable models' ability to capture more complex patterns warrants their use. In this paper, we argue that interpretability should not be viewed as a binary aspect and that, instead, it should be viewed as a continuous domain-informed notion. With this aim, we leverage the well-known Mixture of Experts architecture with user-defined budgets for the controlled use of non-interpretable models. We extend this idea with a counterfactual fairness module to ensure the selection of consistently *fair* experts: **FairMOE** . We compare our proposal to contemporary approaches in fairness-related data sets and demonstrate that FairMOE is competitive with the state-of-the-art methods when considering the trade-off between predictive performance and fairness while providing competitive scalability and, most importantly, greater interpretability .

Keywords: Interpretability · Mixture of Experts · Counterfactual Fairness · Scalability

1 Introduction

Explainable AI (XAI) has been studied for over three decades [8], with the objective of providing explanations for learning models' outcomes such that it i) guarantees the highest level possible of model accuracy, and ii) that human actors can understand [3]. Efforts within XAI are divided into two groups [3], influenced by the concept of interpretability: for non-interpretable models, post-hoc explanations extract information from their behavior where the inputs-outputs relation is complex; for interpretable models, methods are applied to provide a more transparent view of how model decisions are carried out. Naturally, there are trade-offs in each of these types of models: non-interpretable models deployed in high-risk decision-making environments may incur costly mistakes, but building interpretable models requires extensive time and effort from domain experts and they sometimes fail to uncover "hidden patterns" within the data that black-box (i.e., non-interpretable) models may specialize in finding [29].

© The Author(s), under exclusive license to Springer Nature Switzerland AG 2023
A. Bifet et al. (Eds.): DS 2023, LNAI 14276, pp. 341–355, 2023.
https://doi.org/10.1007/978-3-031-45275-8_23

We posit that defining interpretability as a binary notion is severely limiting. Instead, we define it as a domain-informed and user-defined parameter, allowing for models with varying levels of interpretability, capable of extracting the benefits of complex models but retaining interpretability for higher-risk predictions. However, this objective hinges on accurately anticipating such high-risk cases, e.g. those related to decisions concerning non-privileged groups in protected classes. This basis should allow for models that better balance interpretability, fairness, and performance trade-offs, avoiding focus on a single one.

Contributions. We introduce **FairMOE**, a Mixture of Experts (MOE) architecture using interpretable and non-interpretable experts, where a single expert is chosen per prediction. To the traditional MOE architecture we add i) *Performance meta-learners* to anticipate the probability of a given expert prediction being correct; ii) a *Counterfactual Fairness Module* to identify highest-risk samples and ensure they are handled fairly, and; iii) an *Assignment Module* for expert selection, using results from the previous components within constraints of maximum levels of non-interpretability, i.e., the maximum amount of predictions from non-interpretable experts.

2   Related Work

Our work intersects four topics: *i)* interpretability: definitions and contradictions; *ii)* mixture of experts, the basis for our proposal; *iii)* meta-learning, and how to anticipate predictive performance, and; *iv)* fairness, and how to improve interpretability, fairness, and predictive performance trade-offs.

Interpretability. Despite a significant level of research, there is still no single agreed-upon definition of interpretability. Miller [24] defines interpretability as "the degree to which a human can understand the cause of a decision", while Kim et al. [20] define it as "the degree to which a human can consistently predict the model's result". One consistency is that models are either interpretable or not. In this paper, we use a continuous notion of interpretability, envisioning an architecture capable of minimizing the number of non-interpretable errors.

Mixture of Experts. Proposed over 30 years ago, MOE [16] has been extensively explored within regression and classification tasks [34]. Recently, sparse MOE has been used as layers to large neural networks [30] and as a vision transformer [28] to increase large, deep learning tasks' efficiency. Closer to our work, Ismail et al. [15] applied an interpretable MOE approach to structured and time series data, using an Assignment Module to pick individual expert for predictions and variable percentage of samples assigned to interpretable experts. Our approach leverages meta-learners to predict the accuracy of each expert given a specific sample, inspired by Cerqueira et al. [6] work on time series forecasting.

Meta-learning. Meta-learning has been applied to domains such as transfer learning, neural networks, and few-shot learning [31]. We use meta-learning for error anticipation, towards selecting the best model. Khan et al. [19] detail meta-learners' usage for classifier selection. In an error-anticipation context, meta-learners are trained to predict model performance using a combination of the original feature space, meta-features, and model predictions. Using meta-learners, our proposal creates a fully-interpretable pipeline for selecting individual models and allows us to exploit each model's strengths. However, by optimizing our proposal for predictive performance, this might create additional issues with regard to model fairness.

Fairness. There are two main approaches to analyzing fairness. Group fairness measures disparate treatment in protected groups over predictions, including pre-processing, in-processing, and post-processing methods [14]. Pre-processing includes methods such as relabeling data [17], perturbation, and sampling [7]. Post-processing methods include input correction [1], classifier correction [13], and output correction [18]. In-processing methods attempt to train a model to learn fairness concepts. Agarwal et al. [2] use adversarial learning. Zafar et al. [35] apply constraints to the loss function to ensure fairness. Other approaches include a composition of multiple classification models [27] and adjusted learning [36]. Fairness can also be measured on an individual or sample-wise basis. Kusner et al. [21] proposed the notion of counterfactual fairness, which uses the tools from causal inference to establish a prediction as fair if an individual's prediction remains the same with changing protected attributes. Counterfactual fairness has been adopted in several domains as a viable approach toward fairness. For example, Garg et al. [12] apply counterfactual fairness to text classification by considering perturbations obtained by substituting words within specific identity groups. Our approach uses counterfactual fairness to ensure our selected model predicts samples consistently. That is, selected experts should not discriminate against different protected attribute values. We separately evaluate our results with group fairness.

3   Fair Mixture of Experts

This section describes our fairness-aware MOE-based proposal. FairMOE has four main components: i) individual experts, where each predicts each sample; ii) performance meta-learners, which predict the probability of each expert's prediction accuracy; iii) a counterfactual fairness model, to assess predictive consistency regardless of protected attribute values in each case and, iv) an assignment module, combining the outcome of the previous two components and solving for non-interpretable model usage constraints. This high-level workflow is illustrated in Fig. 1, and components are described below.

3.1   FairMOE Components

1) Experts. FairMOE leverages a set of diverse expert learners trained using half the training data, including interpretable and non-interpretable models.

2) Performance Meta-learner. A performance meta-learner per expert is trained to predict the probability of an accurate prediction. For interpretability, meta-learners use one of the following algorithms: Logistic Regression, Naive

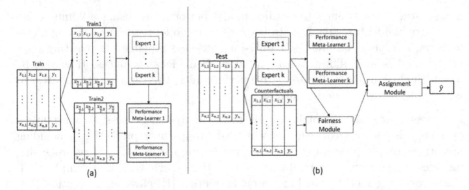

Fig. 1. (a) FairMOE training. Train data is split into two halves: *Train1* and *Train2*. Experts are trained with *Train1* and performance meta-learners on *Train2* using experts' predictions. (b) FairMOE testing. Experts predict the test data, which feeds into the respective performance meta-learners. Counterfactuals are generated around the protected attributes and assessed for consistency regarding expert predictions (*Fairness Module*). Finally, the *Assignment Module* uses the output from the *Fairness Module* and Performance Meta-Learners to select an expert and make the final prediction.

Bayes, Decision Tree, or K-Nearest Neighbors. They are trained using 10-fold cross-validation with grid search. The expert prediction is included as a feature within training, and the ground truth is a binary value indicating whether the expert correctly classified the sample. The learners are fit using the unused half of the training data to ensure they are trained using out-of-sample predictions.

3) Counterfactual Fairness Module. To assess the fairness of individual models in a given sample, FairMOE uses a counterfactual fairness approach inspired by Kusner et al. [21]. Let A, X, Y and (U, V, F) represent protected features, remaining features, the output of interest, and a causal model where U is a set of latent background variables, V a set of observable variables, and F a set of structural equations. We define a predictor $\hat{Y}$ as counterfactually-fair if:

$$P(\hat{Y}_{A\leftarrow\alpha}(U) = y|X = x) = P(\hat{Y}_{A\leftarrow\alpha'}(U) = y|X = x) \qquad (1)$$

where $\mathcal{A}$ is the set of all possible combinations of values within A, α is a given combination, and y is a given label.

The *Counterfactual Fairness Module* creates counterfactuals per sample to assess models with regard to individual (counterfactual) fairness. It generates a counterfactual for all possible permutations of Privileged/Unprivileged across protected classes, minus the original sample combination while holding all non-protected features constant[1]. Each expert then predicts them, being evaluated with a consistency score to determine their level of fairness, defined as:

---

[1] Counterfactuals are created by *i)* binarizing the features as privileged/unprivileged, *ii)* creating the permutations as described, and *iii)* transforming the binary value into a categorical or continuous variable by picking a random value following the distribution from the original training data.

$$CS = \frac{\sum_{\alpha' \in \mathcal{A} \setminus \alpha} I(\hat{Y}_{\mathcal{A} \leftarrow \alpha}(U)|X = x, \hat{Y}_{\mathcal{A} \leftarrow \alpha'}(U)|X = x)}{|\mathcal{A}| - 1}, \qquad (2)$$

where I is an indicator function returning one if the two values match and 0 otherwise. Then, the Module selects the set of models per sample with the maximum consistency score:

$$M_x = \{\forall e \in E : CS(e) = max(CS(E))\} \qquad (3)$$

where E is the set of experts. This set of counterfactually fair experts M_x is then used by the *Assignment Module* to pick the best fair expert per prediction.

4) Assignment Module. Finally, the *Assignment Module* considers the performance meta-learners, the *Fairness Module*, and the *Non-Interpretable Budget* to select an expert for each sample. It has two stages. In the first stage, each sample within the test data is considered individually. The *Fairness Module* returns the fair experts for each sample to be predicted. Using the performance meta-learners, vectors are created with the interpretable and non-interpretable experts with the highest probability of an accurate prediction (HPAP), and the difference between the probabilities is calculated. In the second stage, the test data is considered as a whole. Samples with the highest positive difference, i.e., the probability of an accurate prediction is higher for the non-interpretable model, are assigned to the non-interpretable expert until the budget is exhausted. All remaining samples are assigned to the interpretable expert. Samples with a negative difference are always assigned to the interpretable expert, so the total budget is not always used. The designated expert's prediction is the final FairMOE prediction. The selection procedure is described in Algorithm 1.

Algorithm 1. Assignment Module

Require: L: Meta-Learners, E: Experts, b: Budget, X
1: **for** $x \in X$ **do**
2:     $M_x =$ FairnessModule(x, E, L)
3:     $M^I \leftarrow$ HPAP$(M_x \cap E_I)$, $E_I \in E$: subset of interpretable experts
4:     $M^{NI} \leftarrow$ HPAP$(M_x \cap E_{NI})$, $E_{NI} \in E$: subset of non-interpretable experts
5:     Δprob $\leftarrow L_{M^{NI}}(x) - L_{M^I}(x)$
6: **end for**
7: $PNI =$ SelectPositive$(\Delta$prob, b$)$
8: **return** $x \in X$: if $x \in PNI$: $M^{NI}(x)$; otherwise $M^I(x)$

---

4   Experimental Evaluation

First, we present the data and methods used. Then, we proceed to assess the performance of FairMOE regarding predictive accuracy, interpretable decision-making, and fair behavior. We compare such performance against state-of-the-art baselines, aiming to answer the following research questions:

RQ1 Does the *Non-Interpretable Budget* impact predictive performance?
RQ2 Does FairMOE improve the predictive performance and fairness trade-off?
RQ3 What is the impact of the *Counterfactual Fairness Module*?
RQ4 Does FairMOE scale well with larger datasets?

Table 1. Data sets used in the experimental evaluation

Name	Prediction Task	Cases	Feat.	Protected Attributes	Privileged Classes
Adult [11]	Annual income exceeds $50,000	45222	94	Sex, Race, Age	Male, White, 25–60
German Credit [11]	Bank Account is high credit risk	1000	47	Sex, Age	Male, 25+
Dutch Census [5]	Person's occupation is prestigious	60420	50	Sex	Male
Bank Marketing [25]	Client subscribes with deposit	45211	42	Age, Marital Status	25–60, Married
Credit Card Clients [33]	Client will default in next month	30000	82	Sex, Marital Status	Male, Single
OULAD [22]	Student will pass class	21562	40	Sex	Male
Lawschool [32]	Student will pas bar on first attempt	20798	18	Sex, Race	Male, White

4.1 Data

We use seven fairness-oriented and public data sets [23] (Table 1), following the
pre-processing steps, protected class definitions, and privileged groups described
in Le Quy et al. [23]. The pre-processing steps include removing missing values, dropping non-predictive columns, and when necessary binarizing the target
variable. The majority class was designated as privileged when lacking a defined
privileged group. Categorical variables were one-hot encoded.

4.2 Algorithms

We compare FairMOE against each expert and four fairness-aware algorithms.
To build FairMOE we used seven algorithms as experts, optimized using grid
search with 10-fold cross-validation (Table 2): Logistic Regression, Decision Tree,
Naive Bayes, K-Nearest Neighbors (KNN) are interpretable, and Random Forest,
LightGBM (LGBM), and XGBoost (XGB) are not. Concerning fairness-aware
algorithms, we used the solutions proposed by Hardt et al. [13] (post-processing
optimization of equalized odds), Zafar et al. [35] (builds models using covariance
between a sample's sensitive attributes to measure the decision boundary fairness, which guarantees disparate impact's business necessity clause, by maximizing fairness subject to accuracy constraints), Agarwal et al. [2] (reduces a fairness
classification task to a series of cost-sensitive classification problems, where the
final outcome is a randomized classifier optimized for the most accurate classifier subject to fairness constraints) and xFAIR [26] (aims to mitigate bias and
identify its cause by relabeling protected attributes in test data through extrapolation models designed to predict protected attributes through other independent variables). Hardt et al. and Agarwal et al. methods are implemented using
the Fairlearn python package [4] with an underlying optimized LGBM model.
For xFAIR, we used a Decision Tree as the extrapolation model and a Random
Forest as the classification model suggested in the original paper [26]. The Zafar

et al. baseline was implemented using a Logistic Regression loss function. Of these alternatives, only the method proposed by Zafar et al. is interpretable. We adapted the authors' code to allow for multiple protected classes as necessary. Protected classes were encoded as binary features for the baselines incompatible with categorical or continuous features.

We evaluate six versions of our method. The most basic version (noted as "Mode") considers the experts as an ensemble that predicts the most common prediction from all experts. Next, we consider an ensemble method that prioritizes fairness over performance (noted as "FairMode") by using the *Counterfactual Fairness Module* and predicting the most common prediction from only the counterfactual-fairest models, i.e., with a maximum consistency score. Alternatively, we consider the Mixture of Experts approach using performance meta-learners without the *Counterfactual Fairness Module* to test the interpretability aspect of our proposal, noted as "MOE". Finally, our full proposal "FairMOE", combines performance meta-learners, the *Counterfactual Fairness Module* and the *Assignment Module*. For MOE and FairMOE, we examined non-interpretable budgets of 0% (fully interpretable model) and 100% (no interpretability constraints), noted as $MOE_{0.0}$, $FairMOE_{0.0}$, $MOE_{1.0}$, and $FairMOE_{1.0}$, respectively.

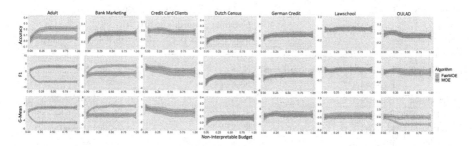

Fig. 2. Predictive performance of FairMOE and MOE at varying budgets. The performance lines represent the average percentage change in Accuracy, F1, and G-Mean scores over ten runs compared to the fully interpretable FairMOE. Higher scores represent better performance. Note that the Y-axes are not on the same scale.

4.3 Evaluation Metrics

For thoroughness, we evaluate our results with Accuracy, F1-score, and G-mean. To measure fairness, we used Statistical Parity (SP) [9] and Equalized Odds (EO) [13]. SP is derived from the legal doctrine of Disparate Impact [10] but disregards ground truth labels, while EO considers them [23].

FairMOE is evaluated by running each dataset 10 times with different 80%/20% train-test splits. For each iteration, the models were ranked by performance across all five metrics. With multiple protected classes, EO and SP are calculated for each protected class. The metrics are grouped by performance (Accuracy, F1, G-mean) and fairness (SP, EO), and assessed as to the model's average ranking across these groups.

Table 2. Overview of the solutions used as benchmarks including their name, underlying model(s), parameters, and whether or not the solution is interpretable.

Model	Underlying Algorithm(s)	Tuning Parameters	Interpretable?
Expert 1	Logistic Regression	N/A	Yes
Expert 2	Decision Tree	Max. Depth: [3, 5, 10, 15], Min. Samples per Leaf: [5, 10, 25]	Yes
Expert 3	Naïve Bayes	N/A	Yes
Expert 4	KNN	Weights: distance, Neighbors: [5, 9, 13, ..., 53]	Yes
Expert 5	Random Forest	Estimators: [10, 50, 100, 250], Min. Samples per Leaf: [5, 10, 25]	No
Expert 6	LGBM	Estimators: [10, 50, 100, 250], Learning Rate: [.001, .01, .1], Min. Samples per Leaf: [5, 10, 25]	No
Expert 7	XGB	Estimators: [10, 50, 100, 250], Learning Rate: [.001, .01, .1], Max. Depth: [3, 5, 10]	No
Agarwal [2]	LGBM	Estimators: [10, 50, 100, 250], Learning Rate: [.001, .01, .1], Min. Samples per Leaf: [5, 10, 25]	No
Hardt [13]	LGBM	Estimators: [10, 50, 100, 250], Learning Rate: [.001, .01, .1], Min. Samples per Leaf: [5, 10, 25]	No
Zafar [35]	Logistic Regression	N/A	Yes
xFAIR [26]	Decision Tree, Random Forest	N/A	No
Mode	Experts 1-7	N/A	No
Fair Mode	Experts 1-7	N/A	No
$MOE_{0.0}$	Experts 1-7	N/A	Yes
$MOE_{1.0}$	Experts 1-7	N/A	Partially
$FairMOE_{0.0}$	Experts 1-7	N/A	Yes
$FairMOE_{1.0}$	Experts 1-7	N/A	Partially

4.4 Results

Levels of Interpretability (RQ1). To measure the impact of interpretability on predictive performance, we test how Accuracy, G-Mean and F1 scores change as the *Non-Interpretable Budget* is increased (0% to 100% in 5pp) within each dataset. Results (Fig. 2) show that increasing the *Non-Interpretable Budget* can lead to predictive performance increases, but the magnitude of the effect is usually small. In many cases, the increase in performance is less than 1%. For example, the largest increase in accuracy from increasing the budget is less than 2% within the German Credit dataset. Additionally, in some cases increasing the use of more complex (non-interpretable) models worsens performance.

Importantly, results show that FairMOE performs well even in contexts where strict transparency is necessary. And, even when allowed to use the *Non-Interpretable Budget*, every metric quickly stabilizes when increasing the budget.

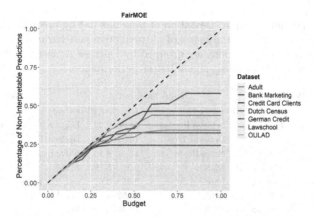

Fig. 3. Average total percentage of non-interpretable predictions for each budget in FairMOE and MOE. The dashed line indicates maximum budget usage.

We illustrate this in Fig. 3, showing that FairMOE does not need to resort to the total allotted non-interpretable predictions: with no interpretability constraints, FairMOE only used an average of 39.7% of the budget. Results with the version MOE are similar (40% of non-interpretable predictions budget). This suggests that, in the majority of instances, fully interpretable models are capable of producing accurate predictions with high confidence. While non-interpretable models offer some performance benefits, these improvements occur on the margins supporting our theory that peak performing models can be achieved while maintaining high interpretability.

Performance and Fairness (RQ2). Next, we compare how well FairMOE balances the predictive performance and fairness trade-off compared to other baselines, studying each baseline's Accuracy, F1-score, G-mean, SP, and EO rankings. The results depicted in Table 3 (grouped by metric type) show that:

1. Adding the *Counterfactual Fairness Module* notably increases group fairness at the cost of predictive performance;
2. Performance meta-learners add interpretability and fairness to our model with only a minor impact on predictive performance;
3. FairMOE is competitive with state-of-the-art baselines in predictive performance and fairness while increasing consistency and adding interpretability;
4. The *Non-Interpretable Budget* increases FairMOE's predictive performance without sacrificing fairness, demonstrating a cumulative advantage.

On predictive performance, XGB and LGBM are the best individual experts. While both are competitive with FairMOE overall, they produce non-interpretable models and poorly balance fairness and predictive performance (see the rightmost column in Table 3), limiting their utility in domains with fairness

Table 3. Average and Standard Deviation of rankings (R) by predictive performance and fairness metrics. "All" is the mean of all predictive and fairness metric rankings and ΔR their difference. Solutions are grouped by Fairness Agnostic, Fairness Aware, and our proposal. Lower rankings signal better performance.

	Predictive Performance			Group Fairness			All		
	solution	$\overline{R}$	$sd(R)$	solution	$\overline{R}$	$sd(R)$	solution	$\overline{R}$	ΔR
Agnostic	Logistic Regression	7.57	4.00	Logistic Regression	10.90	4.32	Logistic Regression	9.24	3.33
	Decision Tree	9.45	4.23	Decision Tree	10.52	4.69	Decision Tree	9.99	1.07
	Naive Bayes	9.75	5.78	Naive Bayes	14.11	4.14	Naive Bayes	11.93	4.36
	KNN	13.73	3.23	KNN	9.56	5.19	KNN	11.65	-4.17
	Random Forest	8.36	4.60	Random Forest	9.35	3.67	Random Forest	8.85	0.99
	LGBM	4.89	3.87	LGBM	10.52	4.11	LGBM	7.70	5.63
	XGB	**4.53**	**3.24**	XGB	12.31	3.49	XGB	8.42	7.78
Aware	Agarwal [2]	9.29	4.68	**Agarwal**	**5.44**	**4.31**	**Agarwal**	**7.36**	**-3.85**
	Hardt [13]	11.49	6.85	Hardt	5.83	5.02	Hardt	8.66	-5.67
	Zafar [35]	11.00	5.59	Zafar	7.66	6.24	Zafar	9.33	-3.35
	xFAIR [26]	9.20	4.93	xFAIR	8.63	4.65	xFAIR	8.91	-0.58
Proposal	Mode	7.53	3.65	Mode	10.17	4.11	Mode	8.85	2.64
	Fair Mode	10.17	4.11	Fair Mode	5.94	3.98	Fair Mode	8.05	-4.23
	$MOE_{0.0}$	8.91	3.83	$MOE_{0.0}$	8.96	3.84	$MOE_{0.0}$	8.93	0.05
	$MOE_{1.0}$	8.31	3.95	$MOE_{1.0}$	9.48	3.82	$MOE_{1.0}$	8.90	1.17
	$FairMOE_{0.0}$	9.46	3.11	$FairMOE_{0.0}$	6.86	3.52	$FairMOE_{0.0}$	8.16	-2.60
	$FairMOE_{1.0}$	9.35	3.59	$FairMOE_{1.0}$	6.78	3.30	$FairMOE_{1.0}$	8.06	-2.57

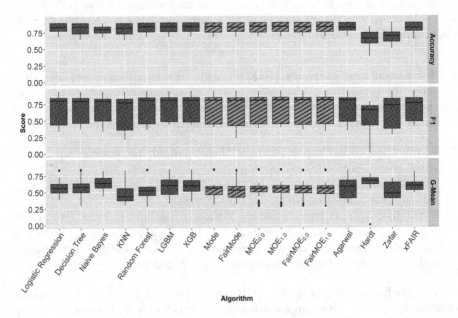

Fig. 4. Accuracy, F1, and G-mean scores per solution across all trials. Higher scores signal better performance.

concerns. As for fairness-aware approaches, Agarwal and Hardt are the top models in group fairness. However, although Hardt is the second worst in predictive performance, Agarwal is competitive with FairMOE. Regardless, Agarwal's performance is less consistent than FairMOE, i.e., higher ΔR, and, importantly, both Agarwal and Hardt produce non-interpretable models.

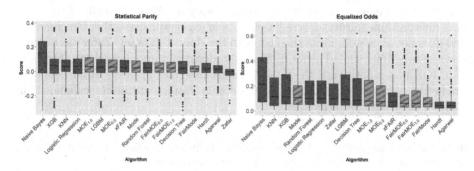

Fig. 5. SP and EO for each solution across all trials. Lower scores represent better performance. Note that the y-axes are not on the same scale.

FairMOE, with and without interpretability constraints, shows competitive performance with regard to predictive and fairness. Figure 4 shows the magnitude of between model disparity with regard to predictive power beyond their rankings. FairMOE and MOE are consistently in the middle or top-half of the accuracy and F1 box plots, suggesting they are competitive with the baselines. This is also observed concerning fairness metrics (Fig. 5).

Ultimately, FairMOE is competitive with the state-of-the-art baselines at striking a balance between fairness and predictive performance and can do so while maintaining interpretability. Even for high-risk domains, results show that a fully interpretable FairMOE (FairMOE$_{0.0}$) is competitive with baselines.

Counterfactual Fairness Module (RQ3). Comparing the results of Mode and FairMode (Table 3), it is evident that the *Counterfactual Fairness Module* improves group fairness. Mode is one of the worst-performing models regarding group fairness, while FairMode is the third-best. On the other hand, FairMode is the fourth-worst in predictive performance and Mode the third-best, demonstrating the significant trade-off between fairness and predictive performance. The differences between MOE and FairMOE further support these findings. However, adding the performance meta-learners mitigates the loss in predictive performance. Additionally, removing the interpretability constraints from MOE, leads to a significant drop in group fairness as predictive performance is prioritized. However, in FairMOE, the model is able to maintain roughly equivalent levels of fairness and predictive performance. Overall, the *Counterfactual Fairness Module* successfully improves fairness while the performance meta learners add predictive performance and interpretability.

Scalability (RQ4). Results show that FairMOE is competitive in predictive performance and fairness with regard to state-of-the-art baselines while producing consistent results. However, scalability is key. Table 4 shows the average total train and predict time per fairness-aware model, by dataset. It shows that, while FairMOE is slower than Hardt and xFAIR, it improves over both alternatives in combined predictive performance and group fairness. Also, FairMOE is faster than Agarwal, the leading fairness-aware algorithm. Finally, Zafar, the only interpretable fairness-aware baseline, is much slower than other benchmarks and does not scale well. From the fastest to the slowest data set, Zafar has a slow-down of over 100x while FairMOE has a slowdown of approximately 8x. Overall, FairMOE is competitive with state-of-the-art baselines in terms of fairness and predictive performance trade-off, interpretable, faster, and more scalable than some of the leading alternatives.

Table 4. Median processing time (seconds) to train and predict each solution in 10 trials per dataset. Asterisks (*) denote fully interpretable solutions.

Dataset	Samples	Features	Median Time (s)							
			Agarwal	Hardt	*Zafar	xFAIR	$*MOE_{0.0}$	$MOE_{1.0}$	$*FairMOE_{0.0}$	$FairMOE_{1.0}$
German Credit	1000	47	22.68	0.70	12.63	1.23	11.52	11.47	12.17	12.18
Lawschool	20798	18	60.36	1.18	91.71	10.94	31.37	31.32	46.43	46.35
OULAD	21562	40	39.16	1.31	140.00	5.63	28.85	29.03	31.34	31.28
Credit Card Clients	30000	82	94.61	3.17	805.51	24.54	53.51	53.37	80.20	80.73
Bank Marketing	45211	42	122.83	2.35	711.34	29.42	68.39	68.22	108.40	107.78
Adult	45222	94	371.84	2.92	1489.74	45.22	93.93	93.91	157.16	156.92
Dutch Census	60420	50	106.23	3.08	1025.92	17.12	90.21	90.35	96.97	97.18

5 Discussion

This work intersects three essential concepts: predictive performance, fairness, and interpretability. The interactions between each of these are complex, and each has its own set of unique challenges. Importantly, FairMOE challenges the paradigm that interpretability is a binary aspect of modeling. Instead, with Fair-MOE, we introduce the idea of interpretability as a continuous domain-informed notion to exploit the typical performance interpretability trade-off best. Fair-MOE is currently only applicable to classification problems, though we intend to expand it to regression in future work.

FairMOE utilizes a *Non-Interpretable Budget* to address the trade-off between predictive performance and fairness. FairMOE balances the predictive performance of complex, non-interpretable models with the user-specified interpretability requirements with this budget. As our results demonstrate, FairMOE is capable of maintaining interpretability on more than 60% of predictions (average) without noticeable drops in performance. More importantly, FairMOE with strict interpretability performs competitively with FairMOE without any interpretability constraints. This finding shows that FairMOE is applicable even in highly-regulated domains with strict transparency requirements. Introducing a user-defined, domain-specific *Non-Interpretable Budget* allows FairMOE to be amendable to different domain requirements.

Next, the *Counterfactual Fairness Module* within FairMOE addresses the trade-off between interpretability and fairness. By limiting our results to our counterfactually fair learners, FairMOE confines itself to making fair predictions even if such a result leads to a non-interpretable prediction. The results illustrate that, by adding the *Counterfactual Fairness Module*, we improve group fairness results. This is an intriguing result that we aim to further explore in future work.

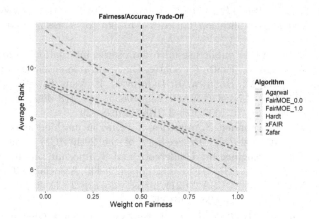

Fig. 6. The average global ranks of each fairness-aware baseline based on the weight given to fairness. Lower average ranks signal better performance.

Finally, we established FairMOE's success at balancing the predictive performance and fairness trade-off: it is the second-best option to Agarwal. To extend our understanding of how FairMOE handles this trade-off, in Fig. 6, we show how each solution performs with varying weights on performance and fairness. Our results show that FairMOE attains its success via consistent performance in both prediction and fairness. We make all the data and code available for reproducibility purposes at https://github.com/joegermino/FairMOE.

6   Conclusion

In this paper, we propose FairMOE, a fairness-aware solution based on the mixture of experts' architecture. By combining three components, predictive meta-learners, the counterfactual fairness module, and the assignment module, we demonstrate how it is possible to obtain a good trade-off between predictive performance and fairness while being scalable. Importantly, FairMOE challenges the paradigm that interpretability is a binary aspect of modeling. Instead, with Fair-MOE, we introduce the idea of interpretability as a continuous domain-informed notion to exploit the typical performance interpretability trade-off best.

References

1. Adler, P., et al.: Auditing black-box models for indirect influence. Knowl. Inf. Syst. **54**(1), 95–122 (2018)
2. Agarwal, A., Beygelzimer, A., Dudík, M., Langford, J., Wallach, H.: A reductions approach to fair classification. In: International Conference on Machine Learning, pp. 60–69. PMLR (2018)
3. Arrieta, A.B., et al.: Explainable artificial intelligence (XAI): concepts, taxonomies, opportunities and challenges toward responsible AI. Inf. Fusion **58**, 82–115 (2020)
4. Bird, S., et al.: Fairlearn: a toolkit for assessing and improving fairness in AI. Technical report MSR-TR-2020-32, Microsoft (2020). https://www.microsoft.com/en-us/research/publication/fairlearn-a-toolkit-for-assessing-and-improving-fairness-in-ai/
5. Center, M.P.: Integrated public use microdata series international (2013)
6. Cerqueira, V., Torgo, L., Pinto, F., Soares, C.: Arbitrated ensemble for time series forecasting. In: Ceci, M., Hollmén, J., Todorovski, L., Vens, C., Džeroski, S. (eds.) ECML PKDD 2017. LNCS (LNAI), vol. 10535, pp. 478–494. Springer, Cham (2017). https://doi.org/10.1007/978-3-319-71246-8_29
7. Chakraborty, J., Majumder, S., Menzies, T.: Bias in machine learning software: why? How? What to do? In: Proceedings of the 29th ACM Joint Meeting on European Software Engineering Conference and Symposium on the Foundations of Software Engineering, ESEC/FSE 2021, pp. 429–440. Association for Computing Machinery, New York, NY, USA (2021). https://doi.org/10.1145/3468264.3468537. https://doi.org/10.1145/3468264.3468537
8. Chandrasekaran, B., Tanner, M., Josephson, J.: Explaining control strategies in problem solving. IEEE Expert **4**(1), 9–15 (1989). https://doi.org/10.1109/64.21896
9. Cynthia, D., Moritz, H., Toniann, P., Omer, R., Richard, Z.: Fairness through awareness. In: Proceedings of the 3rd innovations in theoretical computer science conference, ITCS'12, pp. 214–226. Association for Computing Machinery, New York, NY, USA (2012)
10. Davis, K.R.: Age discrimination and disparate impact-a new look at an age-old problem. Brook. L. Rev. **70**, 361 (2004)
11. Dua, D., Graff, C.: UCI machine learning repository (2017). http://archive.ics.uci.edu/ml
12. Garg, S., Perot, V., Limtiaco, N., Taly, A., Chi, E.H., Beutel, A.: Counterfactual fairness in text classification through robustness. In: Proceedings of the 2019 AAAI/ACM Conference on AI, Ethics, and Society, AIES '19, pp. 219–226. Association for Computing Machinery, New York, NY, USA (2019). https://doi.org/10.1145/3306618.3317950
13. Hardt, M., Price, E., Srebro, N.: Equality of opportunity in supervised learning. Adv. Neural Inf. Process. Syst. **29**, 3315–3323 (2016)
14. Hort, M., Chen, Z., Zhang, J.M., Sarro, F., Harman, M.: Bias mitigation for machine learning classifiers: a comprehensive survey. arXiv preprint arXiv:2207.07068 (2022)
15. Ismail, A.A., Arik, S.Ö., Yoon, J., Taly, A., Feizi, S., Pfister, T.: Interpretable mixture of experts for structured data. arXiv preprint arXiv:2206.02107 (2022)
16. Jacobs, R.A., Jordan, M.I., Nowlan, S.J., Hinton, G.E.: Adaptive mixtures of local experts. Neural Comput. **3**(1), 79–87 (1991)
17. Kamiran, F., Calders, T.: Data preprocessing techniques for classification without discrimination. Knowl. Inf. Syst. **33**(1), 1–33 (2012)

18. Kamiran, F., Karim, A., Zhang, X.: Decision theory for discrimination-aware classification. In: 2012 IEEE 12th International Conference on Data Mining, pp. 924–929 (2012). https://doi.org/10.1109/ICDM.2012.45

19. Khan, I., Zhang, X., Rehman, M., Ali, R.: A literature survey and empirical study of meta-learning for classifier selection. IEEE Access **8**, 10262–10281 (2020). https://doi.org/10.1109/ACCESS.2020.2964726

20. Kim, B., Khanna, R., Koyejo, O.O.: Examples are not enough, learn to criticize! criticism for interpretability. Adv. Neural Inf. Process. Syst. **29**, 2280–2288 (2016)

21. Kusner, M.J., Loftus, J., Russell, C., Silva, R.: Counterfactual fairness. Adv. Neural Inf. Process. Syst. **30**, 4066–4076 (2017)

22. Kuzilek, J., Hlosta, M., Zdrahal, Z.: Open university learning analytics dataset. Sci. Data **4**(1), 1–8 (2017)

23. Le Quy, T., Roy, A., Iosifidis, V., Zhang, W., Ntoutsi, E.: A survey on datasets for fairness-aware machine learning. Wiley Interdiscip. Rev. Data Min. Knowl. Discov. **12**(3), e1452 (2022)

24. Miller, T.: Explanation in artificial intelligence: insights from the social sciences. Artif. Intell. **267**, 1–38 (2019)

25. Moro, S., Cortez, P., Rita, P.: A data-driven approach to predict the success of bank telemarketing. Decis. Support Syst. **62**, 22–31 (2014)

26. Peng, K., Chakraborty, J., Menzies, T.: Fairmask: better fairness via model-based rebalancing of protected attributes. IEEE Trans. Softw. Eng. 1–14 (2022). https://doi.org/10.1109/TSE.2022.3220713

27. Pleiss, G., Raghavan, M., Wu, F., Kleinberg, J., Weinberger, K.Q.: On fairness and calibration. Adv. Neural Inf. Process. Syst. **30**, 5680–5689 (2017)

28. Riquelme, C., et al.: Scaling vision with sparse mixture of experts. Adv. Neural. Inf. Process. Syst. **34**, 8583–8595 (2021)

29. Rudin, C.: Stop explaining black box machine learning models for high stakes decisions and use interpretable models instead. Nat. Mach. Intell. **1**(5), 206–215 (2019)

30. Shazeer, N., et al.: Outrageously large neural networks: the sparsely-gated mixture-of-experts layer. arXiv preprint arXiv:1701.06538 (2017)

31. Vanschoren, J.: Meta-learning: a survey. arXiv preprint arXiv:1810.03548 (2018)

32. Wightman, L.F.: LSAC national longitudinal bar passage study. LSAC research report series (1998)

33. Yeh, I.C., Lien, C.H.: The comparisons of data mining techniques for the predictive accuracy of probability of default of credit card clients. Expert Syst. Appl. **36**(2), 2473–2480 (2009)

34. Yuksel, S.E., Wilson, J.N., Gader, P.D.: Twenty years of mixture of experts. IEEE Trans. Neural Netw. Learn. Syst. **23**(8), 1177–1193 (2012). https://doi.org/10.1109/TNNLS.2012.2200299

35. Zafar, M.B., Valera, I., Rogriguez, M.G., Gummadi, K.P.: Fairness constraints: mechanisms for fair classification. In: Artificial Intelligence and Statistics, pp. 962–970. PMLR (2017)

36. Zhang, W., Bifet, A., Zhang, X., Weiss, J.C., Nejdl, W.: FARF: a fair and adaptive random forests classifier. In: Karlapalem, K., et al. (eds.) PAKDD 2021. LNCS (LNAI), vol. 12713, pp. 245–256. Springer, Cham (2021). https://doi.org/10.1007/978-3-030-75765-6_20

GenFair: A Genetic Fairness-Enhancing Data Generation Framework

Federico Mazzoni(✉) , Marta Marchiori Manerba , Martina Cinquini ,
Riccardo Guidotti , and Salvatore Ruggieri

University of Pisa, Pisa, Italy
{federico.mazzoni,marta.manerba,martina.cinquini}@phd.unipi.it,
{riccardo.guidotti,salvatore.ruggieri}@unipi.it

Abstract. Bias in the training data can be inherited by Machine Learning models and then reproduced in socially-sensitive decision-making tasks leading to potentially discriminatory decisions. The state-of-the-art of pre-processing methods to mitigate unfairness in datasets mainly considers a single binary sensitive attribute. We devise GENFAIR, a fairness-enhancing data pre-processing method that is able to deal with two or more sensitive attributes, possibly multi-valued, at once. The core of the approach is a genetic algorithm for instance generation, which accounts for the plausibility of the synthetic instances w.r.t. the distribution of the original dataset. Results show that GENFAIR is on par or even better than state-of-the-art approaches.

Keywords: Fairness · Pre-processing · Bias mitigation · Genetic Algorithm · Synthetic Data Generation · Supervised Learning

1 Introduction

Machine Learning (ML) models can inherit biases from the training data, leading to discriminatory outcomes in automated decision-making tasks [17]. For instance, historical and selection biases have led recidivism prediction models to discriminate against Afro-American defendants[1]. The fast-growing literature on fairness in ML has considered the issues of assessing and mitigating the bias in training data against social groups, as characterized by a *sensitive attribute* (e.g., *Gender* or *Race*) [14,15]. However, most data pre-processing techniques are affected by two issues: *(i)* they deal only with binary sensitive attributes, and *(ii)* they are unable to account for multiple sensitive attributes at once.

This paper addresses the aforementioned issues by proposing GENFAIR, a data pre-processing method for unfairness mitigation, which leverages a genetic algorithm to generate synthetic data. GENFAIR operates on tabular data, by eliminating discriminatory instances and identifying the most appropriate feature combinations for synthetically generated instances to balance the dataset with respect to multiple sensitive attributes. The synthetic generation process is carried out by a genetic algorithm. Experimental evaluation of various biased

---

[1] https://www.propublica.org/article/how-we-analyzed-the-compas-recidivism-algorithm.

© The Author(s), under exclusive license to Springer Nature Switzerland AG 2023
A. Bifet et al. (Eds.): DS 2023, LNAI 14276, pp. 356–371, 2023.
https://doi.org/10.1007/978-3-031-45275-8_24

datasets demonstrates that our method is either on par or better than competitors, increasing the fairness of a classifier with a low impact on its performance.

The rest of the paper is organized as follows. Section 2 provides an overview on fairness, while Sect. 3 formalized the problem after providing the necessary background. GENFAIR is presented in Sect. 4, and experimental results are resported in Sect. 5. Finally, Sect. 6 summarizes our contributions and discusses open research directions.

2 Related Work

Fairness Definitions. The concept of "fairness" lacks an established framing within the ML community, as it is often not adequately formalized, revealing inconsistency and lack of normativity. Various definitions have been proposed in the literature to quantify models' fairness [18]. According to the *group-based notion of fairness*, each value of a sensitive attribute should receive a similar treatment. Following this definition, the most common statistical metrics are [25]: Statistical Parity Difference (SPD), which checks the difference between the *Positive Rates* of the privileged and discriminated classes [1]; Disparate Impact (DI), which compares the treatment w.r.t. performance received by a privileged and a discriminated value of a sensitive attribute; Equal Opportunity Difference (EOD) [9], which switches its focus to *True Positive Rates* (TPR); Average Odds Difference (AOD), which takes into account *False Positive Rate* (FPR) and TPR by calculating the mean of their difference. Enhancing the fairness of a model should reduce discrimination, as measured by these metrics.

Fairness Algorithms. Bias mitigation strategies can be traced to three categories [15]. *Pre-processing* methods adjust the data on which the model is trained, aiming to create an "ideal world dataset" [22]. *In-processing* strategies consist of changing the functioning of an existing method to make it fair. *Post-processing* mechanisms correct the decisions issued by models so that they conform to fairness criteria. Pre-processing techniques are the least computationally expensive approaches, although improving the fairness of the training set might result in a performance loss [3]. In this regard, sampling techniques such as Preferential Sampling (PS) [10] proved to be the least invasive and minimized the trade-off. Moreover, they are also flexible, tackling edge cases or tangentially-related problems. For example, [2] adapts PS to detect racial dialect bias, while Fair Oversampling (FOS) [7] also takes into account imbalanced learning. Most pre-processing algorithms only work with a single and binary sensitive attribute, although some contributions deal with more complex cases. For example, FAIR-SMOTE [5] generates instances by interpolation and can concurrently balance up to two sensitive attributes, while FAWOS [21] also works with non-binary attributes, generating data exploiting Generative Adversarial Networks (GANs).

Genetic Algorithms. Genetic algorithms (GA), often used to solve optimization problems, have already been employed in the context of fairness. The post-processing method AuFair [26] employs GA to find fair decision rules to enhance the output of a human decision-maker, replacing some decisions. The in-processing technique described in [4] focuses on a classifier that has learned to be fair through genetic programming. ExpGA [8] is an explainability algorithm that employs GAs to create synthetic counterfactual instances.

Our proposal differs from the current literature since it is able to balance datasets w.r.t. multiple sensitive attributes, also supporting non-binary ones. To create syntethic data, GENFAIR leverages a genetic algorithm, which avoids duplicating existing instances as in existing sampling techniques, creating new ones simply by interpolation as FAIR-SMOTE, or employing costly GANs as FAWOS.

3   Background and Problem Statement

In the paper, we make use of the following notation. Let us consider a dataset $D = \langle X, Y \rangle$, where X consists of instances $\{x_1, \ldots, x_l\}$ and Y consists of class labels $\{y_1, \ldots y_l\}$. An instance x_i is a vector of values $\langle x_{i,1}, \ldots, x_{i,m} \rangle$, one value for each of m attributes in $\mathbb{A} = \{A_1, \ldots, A_m\}$. A class label y_i is an element of the binary set $C = \{+, -\}$. We consider the positive class $+$ as the favorable decision, e.g., granting credit, admitting to university, etc. Moreover, we denote by SA a sensitive attribute from $\mathbb{A}$, e.g., the gender or race of an individual represented by an instance. For an instance x, we write $x[A_i]$ to refer to the value x_i of attribute A_i.

3.1   Preferential Sampling

Preferential Sampling (PS) [10] is a pre-processing mitigation technique assuming that the sensitive attribute SA is binary, taking values s and $\bar{s}$, where s is a (potentially) discriminated group and $\bar{s}$ is a (potentially) privileged group (e.g., female and male, respectively). Note that the algorithm assumes the user has an *a priori* knowledge of which group is potentially discriminated. The degree of discrimination in the dataset is measured by the distance from the statistical parity condition $P(Y = +|SA = \bar{s}) = P(Y = +|SA = s)$, estimated by a *dataset discrimination score* given by the difference:

$$disc(D, SA) = |PP|/|PP \cup PN| - |DP|/|DP \cup DN| \tag{1}$$

where $DP = \{x \in D : x[SA] = s \wedge x[C] = +\}$, $DN = \{x \in D : x[SA] = s \wedge x[C] = -\}$, $PP = \{x \in D : x[SA] = \bar{s} \wedge x[C] = +\}$ and $PN = \{x \in D : x[SA] = \bar{s} \wedge x[C] = -\}$, and $|PP|$ (respectively, $|PN|$) counts the instances from the **P**rivileged groups with a **P**ositive (respectively, **N**egative) label, and $|DP|$ (respectively, $|DN|$) counts the instances from the **D**iscriminated group with a **P**ositive (respectively, **N**egative) label.

The objective of PS is to pre-process a dataset for which $disc > 0$ to achieve $disc \approx 0$ by removing instances contributing to PP and DN, and by duplicating instances contributing to DP and PN. PS selects the instances using the *confidence* of a classifier's prediction as a rank, i.e., the predicted probability of

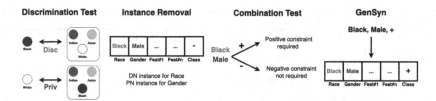

Fig. 1. Workflow of GENFAIR's steps.

the class. Hence, the closer the confidence to 0.5, the closer the instance to the decision border. The classifier used as a ranker is trained on the dataset D.

Although offering overall good results compared to other pre-processing approaches [10], PS has two significant drawbacks. First, PS cannot deal with non-binary sensitive attributes. In [10], authors explicitly suggest binarization as a possible solution. However, we reckon this solution leads to a loss of information. Second, PS cannot deal with multiple sensitive attributes simultaneously. Indeed, while balancing *disc* w.r.t. a sensitive attribute SA_1, PS could worsen it w.r.t. another sensitive attribute SA_2. For example, a *white woman positive* instance is counted in DP for the *Gender* attribute and in PP for the *Race* attribute. While PS only duplicates DP or PN instances, a multi-attribute extension should be able to add instances belonging to a DP or PN group for a SA_1 and to a DN or PP group for another SA_2. We aim at designing a pre-processing method to address these limitations by *balancing the discrimination scores of multiple multi-valued sensitive attributes*.

3.2 Genetic Algorithms

Genetic Algorithms (GA) are metaheuristic algorithms inspired by Darwin's theory of "survival of the fittest", often used to solve an optimization problem. Each possible solution is known as a *chromosome* (or as a *individual*), which includes different "genes" [11]. Initially, a GA generates a random population of chromosomes, which are subsequently evaluated using a *fitness function* that is closely tied to the problem being addressed.

Then, GA's *selection operator* selects the best individual in the population, with the objective of optimizing a fitness value, i.e., only the "fittest" individuals survive. A popular selection operator is *Tournament selection*, which chooses a random number of chromosomes as tournament participants. The chromosomes with the highest fitness are declared the winners. *NSGA2* provides a *multi-objective* selection operator, supporting multiple fitness criteria through a Pareto front. It also considers the *diversity* of chromosomes, as each of them has a "crowding distance value" estimating how dense the area around them is [20,24].

The selected chromosomes have a probability of being mixed (typically in pairs of two) by the *crossover operator*. Selected genes are "shuffled" between the chromosomes, resulting in new "children" individuals inheriting their genes from their "parents'. Chromosomes from this new generation might be selected by a *mutation operator*, mutating the value of at least one of their genes [12].

Algorithm 1: GENFAIR(D, $\mathbb{SA}$, C)

Input : D - labeled dataset, $\mathbb{SA}$ - set of sensitive attributes
Output: D' - balanced dataset
1   $\Delta \leftarrow discrimination\_test(D, \mathbb{SA})$;
2   $\Delta', D', \Pi \leftarrow instance\_removal(\Delta, \mathbb{SA}, D)$;
3   $K, \mathcal{N} \leftarrow combination\_test(\Delta', \Pi)$;
4   **for** $\kappa, n \in K, \mathcal{N}$ **do**
5     $S \leftarrow GenSyn(\kappa, n, D, DS)$; $\left.\begin{array}{c} \\ \\ \end{array}\right\}$ *synthetic data generation*
6     $D' \leftarrow D' \cup S$;
7   **return** D'

The process is repeated for a set number of generations. As the algorithm keeps only the fittest individuals, the overall fitness increases generation after generation. Finally, the algorithm chooses the individual(s) with the best fitness value as the best solution(s) to the given problem.

4   GenFair

In the following, we introduce the GENFAIR method to address the limitations of PS (see Sect. 3). To this end, we extend the notation introduced so far by considering a set $\mathbb{SA} = \{SA_1, SA_2, \ldots, SA_k\}$ of sensitive attributes, each of them possibly being multivalued, i.e., $SA_i = \{s_{i,1}, \ldots, s_{i,n_i}\}$, with $n_i \geq 2$. GENFAIR returns a dataset D' balanced w.r.t. a conservative extension of the discrimination score to multiple multivalued sensitive attributes.

In line with PS, GENFAIR first assesses the discrimination w.r.t. each $SA \in \mathbb{SA}$ in the *Discrimination Test*. Then, instances close to the decision boundary are removed in the *Instance Removal* phase. In order to create a "balanced" dataset D' with the same amount of instances as the original dataset D, the *Combination Test* computes the number of instances with a given combination of sensitive attribute values and class values. Such instances are generated by a genetic algorithm named *GenSyn*, short for *Genetic Synthesizer*, and added to D'. The four steps are depicted in Fig. 1, while Algorithm 1 reports the pseudo-code of GENFAIR which is described in detail in the rest of this section.

Step 1. Discrimination Test. The function $discrimination\_test$ takes as input D and $\mathbb{SA}$. We conservatively extend the discrimination score to a multivalued sensitive attribute SA and a value $s \in SA$ as follows:

$$disc(D, SA, s) = |P_s|/|P_s \cup N_s| - |P_{\neg s}|/|P_{\neg s} \cup N_{\neg s}| \tag{2}$$

where $P_s = \{x \in D : x[SA] = s \wedge x[C] = +\}$, $N_s = \{x \in D : x[SA] = s \wedge x[C] = -\}$, $P_{\neg s} = \{x \in D : x[SA] \neq s \wedge x[C] = +\}$ and $N_{\neg s} = \{x \in D : x[SA] \neq s \wedge x[C] = -\}$. When SA is binary, and s is the (potentially) discriminated group, this definition boils down to the discrimination score (Eq. 1): intuitively,

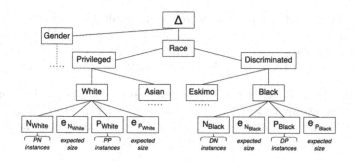

Fig. 2. Visual representation of Δ.

$x[SA] \neq s$ is equivalent to $x[SA] = \bar{s}$. When SA has more than two values, Eq. (2) compares s against all the remaining groups. Notice that the discrimination scores quantify the distance from the statistical parity condition $\forall s \in SA.P(Y = +|SA = s) = P(Y = +|SA \neq s)$, which is equivalent to independence of Y and SA (https://github.com/FedericoMz/GenFair). If $disc(D, SA, s) > 0$, the group s is considered to be discriminated. Thus, we label P_s and N_s respectively as **D**iscriminated instances with a **P**ositive class (DP) and **D**iscriminated instances with a **N**egative class (DN). Conversely, if $disc(D, SA, s) < 0$, s is considered to be privileged, and we label P_s and N_s as **P**rivileged instances with a **P**ositive class (PP) and **P**rivileged instances with a **N**egative class (PN).

GENFAIR then computes e_{P_s} and e_{N_s}, the expected size of P_s and N_s under a statistical parity condition, i.e., the expected number of instances in the group:

$$e_{P_s} = |\{x \in D : x[SA] = s\}| \cdot |\{x \in D : x[C] = +\}| / |D|$$
$$e_{N_s} = |\{x \in D : x[SA] = s\}| \cdot |\{x \in D : x[C] = -\}| / |D| \tag{3}$$

If s is discriminated, $|P_s| < e_{P_s}$ and $|N_s| > e_{N_s}$, then DP instances should be added and DN instances removed. If s is privileged, $|P_s| > e_{P_s}$ and $|N_s| < e_{N_s}$, then PP instances should be removed, and PN instances added. We notice that the absolute value of $|P_s| - e_{P_s}$ is always equal to the absolute value of $|N_s| - e_{N_s}$. Thus, for each $s \in SA$, the number of removed instances is equal to the number of added instances, and the marginal distribution of SA does not change (nor does D's size). The steps above are repeated for each $SA \in \mathbb{SA}$. Note that a discriminated instance for SA_i might be privileged for SA_j with $i \neq j$, e.g., an instance x such that $x[Race] = Black$, $x[Gender] = Male$, and $x[C] = +$ might be a DP for $Race$, but a PP for $Gender$.

After that, GENFAIR ranks every instance w.r.t. the *confidence* of a classifier trained on D. The function then returns a hierarchical dictionary Δ that maps each $SA \in \mathbb{SA}$ and each value $s \in SA$ to the instances in D, grouped as DN, DP, PP or PN, and ordered w.r.t. the classifier's confidence score either in ascending (DP, PP) or descending order (DN, PN). Δ also includes the expected size of each group. Figure 2 visualizes an example of Δ structure w.r.t. the *Race* attribute.

The Discrimination Test step addresses the first issue mentioned in Sect. 3, i.e., dealing with non-binary sensitive attributes. The vector of $disc(D, SA, s)$

values, for all $s \in SA$, defines the degree of discrimination (positive values) or privilege (negative values) for the groups in SA. For a binary SA, the two values are the opposite of each other (e.g., .20 and $-.20$). As a benefit of this approach, the user is no longer required to define *a priori* the discriminated group.

Step 2. Instance Removal. The function *instance_removal* (line 2 in Algorithm 1) takes as input Δ and D. For each DN or PP group G in Δ (such as $\{x \in D : x[Gender] = Male \wedge x[C] = +\}$), the function computes the difference between the cardinality of G and the expected size:

$$d = |G| - e_G \tag{4}$$

where e_G is the expected number of instances in G (i.e., either e_{P_s} from Eq. 3 if G is a PP group, or e_{N_s} if G is a DN group).

If $d > 0$, the function removes from G the top d instances from D and the various lists in Δ, balancing PP and DN groups. Instances are removed based on the ranker's order; in other words, GENFAIR prioritizes instances closer to the decision border – which, as in PS, are deemed the most discriminating (Sect. 3.1). If SA includes more than one sensitive attribute, removing instances from a PP or DN group of a given $SA_i \in SA$ also affects the groups (not necessarily PP or DN) of another SA_j (with $i \neq j$). For example, an instance removed from a PP group of SA_i might also belong to a DP group of SA_j; this might be the case for an instance x such that $x[Race] = Black$ (DP group), $x[Gender] = Male$ (PP group), and $x[C] = +$. In the next step, GENFAIR will re-create the instance for the DP group, but following another combination of sensitive attribute values (e.g., an instance x such that $x[Race] = Black$ (DP group), $x[Gender] = Female$ (DP group), and $x[C] = +$). We underline that removing instances to balance a sensitive attribute might also *positively* affects other sensitive attributes, if the instance belongs to a DN or a PP group for multiple sensitive attributes (e.g., an instance x such as $x[Race] = White$, $x[Gender] = Male$ and $x[C] = +$)[2].

The *instance_removal* function outputs Δ' and D', respectively obtained from Δ and D after removing instances as described above, and a list Π, which includes the unique *combinations* of removed sensitive attribute values, ordered by decreasing frequency, including all other combinations of sensitive attribute values ordered by decreasing frequency (possibly zero frequency if not in D).

Step 3. Combination Test. The function *combination_test* (Algorithm 1, line 3) takes as input Π and Δ', and computes the *constraints* that must be met while generating instances in the next step. We define a constraint as *the set of sensitive attribute values and the class value representative of an instance needed to balance the dataset*, such as *Black Female, Positive*.

---

[2] The order of sensitive attributes considered may affect the set of instances removed. GENFAIR guarantees to remove instances close to the decision boundary for the first sensitive attribute given as input. For the following ones, instances already removed might not be the closest to the decision boundary. However, the user can specify the order of sensitive attributes to be considered.

The *combination_test* checks, for each combination in Π of sensitive attribute values $s_{\pi_1,1}, s_{\pi_2,2}, \ldots, s_{\pi_q,q}$, whether one or more instances with $SA_{\pi_i} = s_{\pi_i,i}$ for $i = 1, \ldots, q$ must be generated, belonging either to the positive or negative class. For each sensitive attribute value in the combination, it extracts from Δ' the corresponding group $P_i = \{x \in D : x[SA_{\pi_i}] = s_{\pi_i,j} \wedge x[C] = +\}$ and the expected number of instances e_{P_i}, e.g., P_{White} and $e_{P_{White}}$ from Fig. 2, computed with Eq. (3). The function then computes c, the number of instances to be generated which follows the sensitive attribute values in the combination in Π, without leading any group P_i to exceed its expected size e_{P_i}.

$$c = \min(e_{P_1} - |P_1|, e_{P_2} - |P_2|, \ldots, e_{P_k} - |P_k|) \tag{5}$$

If $c > 0$, the *positive combination test* is passed, i.e., instances with a positive combination of values are needed to balance the dataset. A constraint with that combination of values and a positive class is created with cardinality c, and c placeholder instances are added to each group. The procedure is repeated for the negative class (*negative combination test*), extracting from Δ' both $N_i = \{x \in D : x[SA_{\pi_i}] = s_{\pi_i,j} \wedge x[C] = -\}$ and the expected size e_{N_i}, for each value in the combination. If both tests fail, i.e., no instances with the given combination of values are required, GENFAIR moves to the next combination in Π.

For example, suppose a combination in Π is *Race: White, Gender: Female*. If $P_{White} = \{x \in D : x[Race] = White \wedge x[C] = +\}$ and $P_{Female} = \{x \in D : x[Gender] = Female \wedge x[C] = +\}$, GENFAIR computes $e_{P_{White}} - |P_{White}|$ and $e_{P_{Female}} - |P_{Female}|$. Suppose $e_{P_{Female}} - |P_{Female}| = 3$ and $e_{P_{White}} - |P_{White}| = 5$, the constraint *Race: White, Gender: Female, C: +* is created with cardinality 3, and 3 placeholder instances are added both to P_{Female} and P_{White} in Δ. After that, the same combination is then tested with the negative class, i.e., with the groups $N_{White} = \{x \in D : x[Race] = White \wedge x[C] = -\}$ and $N_{Female} = \{x \in D : x[Gender] = Female \wedge x[C] = -\}$.

The function *combination_test* returns a list K of unique constraints, with their cardinality in the list $\mathcal{N}$. With this step, we solve the second issue mentioned in Sect. 3. Indeed, in the next step, instances are generated following the constraints, concurrently balancing multiple sensitive attributes. We also aim to maintain a plausible distribution of the combination of sensitive attributes' values. Hence, constraints are created following the order of combinations in Π and sorted by their frequency among the removed instances. GENFAIR prioritizes creating constraints and thus synthetic instances with more common combinations of sensitive attribute values, mitigating a possible distribution disruption.

Step 4. GenSyn. In the previous step, GENFAIR computed the characteristics that synthetic instances must have for a fair dataset, i.e., the constraints, and their quantity. In this last step, GENFAIR generates these synthetic instances and ensures their plausibility with the genetic algorithm *GenSyn* (Algorithm 1, lines 4–5). *GenSyn* takes as input D', the original dataset D and each constraint $\kappa \in K$ with its cardinality $n \in \mathcal{N}$, and returns a new dataset S of n instances following the values of sensitive attributes and class in κ. S that is concatenated to D' (Algorithm 1, line 6). This is iterated over all $\kappa \in K$ and $n \in \mathcal{N}$.

As said in Sect. 3, a GA includes the generation of the initial population of chromosomes, the evaluation and selection of the best ones, their crossover, and the possible mutation of their "children". With *GenSyn*, each chromosome is an instance following a given constraint, which is evaluated w.r.t. its plausibility. *combination_test* analyzes the dataset D to create the initial population of *pop* individuals (by default, 150). If instances in D have m attributes, each chromosome has m genes. The values of the genes representing the sensitive attributes and the class are given by κ; this ensures that each synthetic instance follows the values in the constraint κ. For the other attributes, the user can specify how values are generated and, later in the mutation phase, mutated (different strategies can be chosen for different attributes). With *Float Strategy*, to be used for continuous float values, a random float is selected between the minimum and maximum the attribute assume in the dataset. *Integer Strategy* is similar, but an integer is selected instead and it is meant for continuous integer values. With *Equal Probability Strategy*, a value is randomly selected among the values the attribute assumes in the dataset, and can be used for all kinds of values, including categorical and boolean. With *Weighted Probability Strategy*, each value has a probability of being selected equal to its frequency in the dataset.

Each generation strategy ensures that, for each attribute, its value in the synthetic instances belongs to the observable domain in the original dataset, since no value larger (or smaller) than the largest (or smallest) observable value can be generated. Certain use cases and the semantics of each attribute might lead to choose one strategy instead of another. By its nature, the latest strategy creates the most statistically-plausible results, but the others might increase the diversity of the data. As an example, assume a dataset D with 6 instances. *Age* has the following values: 18, 20, 23, 23, 23, 30. With the first strategy outlined, GENFAIR picks as *Age* value a random float between 18.0 and 30.0, which is wrong on a semantic level and strongly impacts the plausibility of the resulting instance. The second strategy picks a random integer in the $[18, 30]$ interval, including values not featured in D. With the third strategy, GenSyn only selects values among those in the original dataset, each having the same probability. Finally, with the fourth strategy, more common values have a higher probability (in the example, 23 has a 50% probability of being picked).

As a fitness function, inspired by [22], we consider the distance between the synthetic instance and a representative instance, i.e., the *medoid* of instances in D sharing the combination κ of sensitive attributes and class under consideration[3]. The medoid is computed on a normalized D. While compared to the medoid, the synthetic instance is also normalized. As a distance function, we adopt the cosine distance [23]. The selection operator employed by GenSyn is NSGA2, as it guarantees good diversity among the set of selected instances.

As a crossover operator, GenSyn uses a modified uniform crossover: two of the best instances of each generation are randomly selected with probability p_s (default: 0.50) to generate two "children" instances by shuffling the values of

---

[3] In the extreme case where no instances with the combination κ are featured in D, such a medoid does not exist; the algorithm fallbacks to the medoid representative of the entire dataset D.

Table 1. Number of instances and attributes, Class, Sensitive Attributes (SA), SA Values (discrimination score within parenthesis; most privileged **bolded**), Systemic Bias (SB). Values refer to the training set.

Dataset	N. Inst	N. Attrs	Class	SA	SA Values	SB
adult	36177	13	Income	Sex	**Male** (-0.197) Female (0.197)	3836
				Race	**White** (-0.103) Black (0.131) Other (0.133)	
					Asian (-0.039) Eskimo (0.143)	
german	800	10	Risk	Sex	**Male** (-0.054) Female (0.054)	0
				Age	**> 25** (-0.141) $<= 25$ (0.141)	
compas	4937	12	Recid	Sex	Male (0.137) **Female** (-0.137)	2180
				Race	**Caucasian** (-0.105) Hispanic (-0.082)	
					Native Amer. (0.042) Asian (-0.172)	
					African Amer. (0.146) Other (-0.108)	

some random chromosomes with probability p_c (default: 0.34). Each child has a probability p_m of being selected by the mutation operator (default: 0.15), which changes the value of one of its genes (following the strategy chosen for that gene), selected randomly. However, the crossover or mutation operator can not select genes representing the class or sensitive attributes. These custom operators preserve what we outlined for the initial population generation – the "children" synthetic instance follows the values in the constraint κ, and the values of other attributes belong to their respective domains in D. The evaluation, selection, crossover, and mutation steps are repeated for *gen* generations (default: 50), after which, among all the instances generated through every generation, GenSyn returns the best n instances, which GENFAIR adds to D'.

5   Experiments

In this section, we report the experiments showing that GENFAIR[4] outperforms state-of-the-art competitors regarding models fairness and the plausibility of synthetic data generation. We experimented on three datasets[5] described in Table 1. **adult** and **compas** share the same sensitive attributes, *Sex* and *Race*. *Race* is non-binary, and it is often binarized in the literature as *White* and *Non-White* [5,19]. However, certain Non-White values are privileged in both datasets (e.g., *Asian*). **german** has a high discrimination for *Age*, whereas *Sex* is relatively balanced. *Male* is always *Sex*'s privileged values, except for **compas**. For every dataset, we performed an 80–20 split in train and test respectively, and we experimented with a Random Forest (RF) with default parameters (as implemented by *scikit-learn*) as classification model.

As fairness-enhancing competitors, we adopt Preferential Sampling for balancing a single binary sensitive attribute and FAIR-SMOTE for single and multiple

---

[4] GitHub repository: https://github.com/FedericoMz/GenFair.

[5] Datasets from Kaggle. **adult** and **compas** pre-processed as in [19]. For **german**, categorical attributes are label-encoded while *Age* is binarized.

binary sensitive attributes. We also conducted some initial testing with FAWOS, however, the lack of an established heuristic for determining the optimal values of its hyperparameters (as noted by the authors themselves [21]) prompted us to cease further use of this approach. In our tests, we also included a random method, replacing the same number of instances as GENFAIR balancing multiple sensitive attributes in the original training set (refer to the beginning of Sect. 5.2). The random method generates synthetic instances by simply picking a random value for each attribute, without following any fairness constraint. Generally, it is noteworthy that while Preferential Sampling (PS), GENFAIR, and the random method output a final dataset with the same size as the original one, FAIR-SMOTE increases it. Lastly, while experimenting with FAIR-SMOTE, some features were removed from the training and test datasets as suggested in [5]. For GENFAIR and the random method, we employed the weighted probability generation strategy for all the attributes, except *Age* in `compas` and `adult` for which we used the integer generation strategy instead.

5.1   Evaluation Metrics

We evaluate the different approaches in terms of *effectiveness in terms of debiasing*, *plausibility* of the synthetic instances, and classification *performance*[6].

For the evaluation of debiasing, we computed the discrimination scores with GENFAIR's Discrimination Test, and FAT Forensics' *Systemic Bias* (SB)[7] of the training set before and after applying debiasing methods. For evaluating the fairness of the trained models, we adopted *SPD*, *DI*, *EOD*, and *AOD*, as described in Sect. 2. With these metrics, for each sensitive attribute, we compared each value to the most privileged one as identified in the training set. Negative results imply that the assumed discriminated attribute value is privileged and vice-versa.

For plausibility, we adopted a set of metrics proposed in [16], which presents the Synthetic Data Vault framework (SDV)[8], including functionalities for evaluating the quality of synthetic datasets. Among those available, we selected two statistical approaches, Kolmogorov-Smirnov Test (*KST*) and ContinuousKL-Divergence (*CKLD*), and two methods which train a model to detect synthetic data, LogisticDetection (*LD*) and SVCDetection (*SVCD*). We also computed the Inlier Score Difference (*ISD*) comparing the difference between the average inlier score of the original and balanced datasets w.r.t. the scores returned by an Isolation Forest [13], and the Average Minimum Distance (*AMD*) [6], computing the distance between each synthetic and original instance and returning the average of the minimum distances of each synthetic instance. Finally, we evaluated the performance of the classifiers w.r.t. Accuracy (Acc), Precision (Pre), Recall (Rec), and False Alarm (FA) [23].

---

[6] In tables, the best results are in **bold**, second-best in *italics*. ↑ and ↓ indicate if the measure should be maximized or minimized, while → 0 and → 1 if the ideal value is close to 0 or 1.

[7] https://fat-forensics.org/generated/fatf.fairness.data.measures.systemic_bias.html.

[8] https://sdv.dev/.

5.2 Fairness and Performance Evaluation

In this section, we report the results obtained regarding the debiasing effect of the compared methods. At first, we checked if GENFAIR can reduce the training dataset's discrimination in terms of *disc* and of SB. The former is always brought to 0, while SB of `adult` and `compas` is mitigated, respectively, to 1502 and 1975, w.r.t. the initial values reported in Table 1. `german` SB remained 0. GENFAIR achieved these results by replacing only a small number of instances: 3945 for `adult` (10.60%), 38 for `german` (4.75%), and 448 (9.07%) for `compas`.

We then checked the discrimination and the performance of a classifier trained on the dataset balanced by GENFAIR. We envisioned three scenarios: GENFAIR balancing only a single binary sensitive attribute, multiple binary sensitive attributes, and a binary and non-binary sensitive attribute. Table 2 encompasses the first two cases. For `adult` and `compas`, we binarized *Race* respectively as *White - Non-White* and *Caucasian - Non-Caucasian*. Due to the random nature of the tested methods, we ran each pre-processing algorithm on the training set 5 times, resulting in 5 balanced training sets for each dataset. Each of these training sets was used with a Random Forest. A classifier with the original train set was also trained 5 times. The algorithms balanced either one of the sensitive attributes or both of them. While balancing both, we tested the two possible orders for GENFAIR (i.e., prioritizing *Race* or, for `german` *Age*, over *Sex*, and vice-versa). Table 2 reports the average of the 5 results (the various methods offered consistent results with a low standard deviation, which is therefore

Table 2. Performance and fairness evaluation with binary sensitive attributes.

Data	SA	Algo	Acc ↑	Prec ↑	Rec ↑	FA ↓	AOD → 0		EOD → 0		SPD → 0		DI → 1	
adult		*Original Train*	.852	.738	.648	.078	.060	.033	.044	.028	.184	.097	.35	.592
		Random Baseline	.852	.742	.64	.076	.057	.036	.039	.032	.182	.098	.348	.581
	Balancing Sex	PS	**.842**	**.720**	.619	.082	-.108	.092	-.194	*.130*	*.064*	**.123**	*.733*	*.479*
		FAIR-SMOTE	.786	.568	**.663**	.172	.141	.110	**.182**	.112	.208	.159	.431	**.502**
		GENFAIR	*.837*	.702	*.624*	.090	-.108	*.103*	-.*183*	.138	**.056**	*.136*	**.772**	.445
	Balancing Race	PS	*.849*	**.733**	.642	.080	.117	-.*067*	.137	-.*119*	.214	**.027**	**.272**	**.879**
		FAIR-SMOTE	.789	.571	**.683**	.175	.232	.129	.271	.160	.299	.162	.258	.505
		GENFAIR	**.850**	**.733**	*.646*	.080	*.120*	-.065	*.140*	-.118	.217	*.030*	.265	*.869*
	Balancing Both	FAIR-SMOTE	.789	.574	**.653**	.165	.129	.111	*.162*	.133	.200	.147	.436	.526
		GENFAIR(S, R)	*.840*	**.706**	*.636*	.090	-.*093*	*.033*	-.*162*	*.037*	**.070**	*.087*	*.724*	*.638*
		GENFAIR(R, S)	*.842*	**.716**	.627	.085	-.*043*	.008	-.*079*	-.002	*.094*	*.072*	*.629*	**.690**
german		*Original Train*	.684	.760	.817	.641	-.020	.197	-.009	.068	.011	.200	.986	.750
		Random Baseline	.672	.751	.804	.652	-.037	.191	-.002	.078	-.002	.193	1.00	.757
	Balancing Sex	PS	*.670*	**.749**	**.806**	.662	.054	.141	-.*023*	**.019**	-.*021*	.141	1.02	.821
		FAIR-SMOTE	.424	*.681*	.355	**.407**	.310	**.009**	.320	-.*028*	.297	**-.008**	.456	**.978**
		GENFAIR	*.671*	.576	*.792*	*.628*	-.*064*	*.038*	-.001	-.032	-.018	.054	1.02	*.929*
	Balancing Age	PS	.660	.745	.791	.662	-.065	**.009**	-.*003*	-.119	-.025	**.008**	1.03	**.989**
		FAIR-SMOTE	**.684**	**.753**	**.823**	*.658*	-.056	.167	-.038	*.040*	-.023	.168	*1.03*	.790
		GENFAIR	*.671*	**.756**	*.792*	**.627**	-.*064*	*.038*	-.001	-.032	-.018	.054	1.02	*.928*
	Balancing Both	FAIR-SMOTE	.660	**.757**	.768	**.603**	-.100	.170	-.104	**.011**	-.073	.166	1.10	.779
		GENFAIR(S, A)	*.670*	*.754*	**.794**	*.634*	-.*084*	.068	-.038	.018	-.*043*	.086	1.05	.887
		GENFAIR(A, S)	*.664*	.749	*.792*	*.648*	-.060	**.039**	-.027	-.058	-.026	**.046**	1.03	**.939**
compas		*Original Train*	.681	.703	.733	.383	.193	.126	.115	.082	.207	.143	.722	.786
		Random Baseline	.678	.703	.721	.376	.201	.137	.127	.096	.215	.153	.709	.77
	Balancing Sex	PS	*.684*	*.710*	.725	*.367*	**.010**	.167	-.*054*	*.138*	**.029**	*.185*	*.952*	*.728*
		FAIR-SMOTE	.627	.652	.698	.461	.298	**.130**	.203	**.060**	.298	**.135**	.642	*.801*
		GENFAIR	**.691**	**.716**	*.730*	*.358*	*.021*	.182	-.028	.148	*.044*	.201	*.928*	.709
	Balancing Race	PS	*.682*	*.706*	**.728**	*.374*	.274	**.015**	.186	-.*003*	.285	**.031**	*.643*	*.948*
		FAIR-SMOTE	.628	.651	.705	.468	.225	.198	.161	.105	.231	.199	**.706**	.725
		GENFAIR	*.682*	.709	.722	*.366*	.308	**.015**	.223	-.018	.320	*.033*	.609	*.943*
	Balancing Both	FAIR-SMOTE	.621	.656	.662	.430	.216	.069	.152	**.001**	.222	**.075**	.699	**.877**
		GENFAIR(S, R)	*.694*	*.715*	*.743*	*.367*	*.173*	.094	*.104*	*.064*	*.191*	.114	*.737*	.823
		GENFAIR(R, S)	**.696**	**.716**	**.744**	.365	.160	.076	**.088**	*.043*	.178	*.096*	*.752*	*.849*

Table 3. Fairness evaluation while balancing both *Race* (non-binary) and *Sex*.

Data	Value	AOD → 0		EOD → 0		SPD → 0		DI → 1	
		Original	GENFAIR	Original	GENFAIR	Original	GENFAIR	Original	GENFAIR
adult	Black	0.048	**0.020**	0.046	**0.008**	0.131	**0.106**	0.445	**0.542**
	Asian	**-0.04**	0.171	**-0.056**	0.315	**-0.038**	0.103	**1.159**	0.557
	Eskimo	0.820	**0.021**	0.094	**-0.007**	0.122	**0.085**	0.483	**0.634**
	Other	0.139	**0.075**	0.233	**0.126**	0.127	**0.088**	0.463	**0.620**
	Female	0.059	**-0.052**	**0.042**	-0.101	0.184	**0.092**	0.352	**0.731**
compas	African Amer.	0.157	**0.034**	0.123	**-0.006**	0.185	**0.065**	0.725	**0.893**
	Hispanic	**0.057**	0.077	**-0.004**	0.041	**0.015**	0.042	**0.978**	0.931
	Other	-0.021	**0.016**	-0.080	**0.008**	-0.037	**0.011**	1.054	**0.983**
	Asian	0.415	**0.375**	0.790	**0.727**	0.675	**0.604**	0.000	0.000
	Native Amer.	0.352	**0.075**	0.210	**0.273**	0.325	**0.063**	1.482	**1.104**
	Male	0.186	**0.153**	0.115	**0.100**	0.201	**0.173**	0.731	**0.756**

not reported). For the fairness metrics, the first value refers to *Sex*, the second either to *Race* (`compas`, `adult`) or *Age* (`german`), to check whether balancing one sensitive attribute affects the other.

Generally, removing and creating instances randomly hardly impacts the model. GENFAIR maintains a good level of performance while at the same time reducing discrimination, often offering the best or the second-best results. With a single sensitive attribute, in terms of performance, GENFAIR is only sometimes behind PS, while consistently getting better results than FAIR-SMOTE. As for fairness, targeting *Sex* GENFAIR always achieves the best results with SPD and DI (with the only exception of `compas` DI). With *Race* or *Age*, the results are very slightly below those of PS (for example, the SPD of `adult` is lowered to 0.217 by PS and to 0.124 by GENFAIR; for `compas`, it is lowered respectively to 0.033 and 0.031). While balancing multiple attributes, GENFAIR provides better results than FAIR-SMOTE both in terms of performance and fairness. It can be seen that the order has a significant impact (as discussed in Sect. 4), although in all the datasets prioritizing *Race* (or *Age*, for `german`) over *Sex* gives better results even for *Sex*. It can also be seen that balancing both attributes independently achieves for them better fairness results (e.g., `adult` *Sex* SPD, alone: 0.56; with *Race*: 0.94). However, targeting only one attribute worsens the fairness of the other w.r.t. the original train. Balancing both attributes is a good trade-off, as it still provides an improvement for both.

Further tests were carried out balancing both *Race* (not-binarized) and *Sex* in the `adult` and `compas` datasets. For these tests, we employed GENFAIR with the most effective order found in the binary scenario, i.e., prioritizing *Race* over *Sex*. Neither Preferential Sampling nor FAIR-SMOTE are compatible with this test case, and the random method did not offer interesting results. Table 3 focuses on the results in terms of fairness (the impact on the performance was minimal, and therefore not reported). As it can be seen, GENFAIR successfully improves the fairness for all values, with the only exceptions of *Asian* (`adult`) and *Hispanic* (`compas`). These values were already almost balanced, but they are considered discriminated with the models trained on the new training sets.

5.3   Plausibility Evaluation

The plausibility of the synthetic data was assessed by benchmarking GENFAIR against the data generator of the SDV library. We created SDV-FAIR, a version of GENFAIR, replacing GenSyn with SDV and its *Fast_ML* method[9], which offers a good compromise between data quality and computation time. We also tested the random method outlined above. For `adult` and `compas`, we used the non-binary training set. In Table 4, we report the results w.r.t. the various metrics. GENFAIR and SDV-FAIR are roughly on par in terms of KST, CKLD, and AMD, with the exception of the `german` dataset. GENFAIR poor performance might be due to the low number of instances generated (38), which might have converged too close to the respective medoid. GENFAIR is often the best method with ISD, reducing the average outlier score of the dataset. However, SDV-FAIR is better w.r.t. SVCD and LD, implying that the instances have some traits revealing their synthetic nature, e.g., they might feature uncommon combination of values, despite being close to the medoid. To summarize, GENFAIR is better than the random method and close to the state of the art regarding data quality.

Table 4. Comparison of the random method, GENFAIR and SDV w.r.t. data quality.

Data	Method	KST ↑	CKLD ↑	LD ↑	SVCD ↑	ISD ↓	AMD ↓
`adult`	Random	0.543	0.453	0.010	0.003	0.025	0.732
	GENFAIR	**0.889**	**0.716**	0.498	0.080	**−0.005**	**0.219**
	SDV-FAIR	0.831	0.706	**0.714**	**0.175**	0.019	0.240
`german`	Random	0.726	**0.474**	0.316	0.208	−0.02	0.659
	GENFAIR	0.750	0.250	0.274	0.034	**−0.021**	0.037
	SDV-FAIR	**0.838**	0.427	**0.518**	**0.396**	−0.009	**0.019**
`compas`	Random	0.496	0.438	0.003	0.007	−0.011	0.843
	GENFAIR	**0.876**	0.728	0.321	0.111	**0.008**	**0.017**
	SDV-FAIR	0.852	**0.754**	**0.513**	**0.220**	0.019	0.082

6   Conclusions

We have presented GENFAIR, a fairness-enhancing pre-processing method for tabular data. Our experiments have shown that GENFAIR creates plausible data and is on par with the state of the art while balancing dataset with one sensitive attribute. Models trained on processed datasets showed improved fairness and maintained good levels of performance. The same holds for datasets with multiple sensitive attributes, supported only by a limited number of competitors. However, the issue of intersectional fairness, i.e., discrimination on groups characterized by combinations of sensitive attribute values, is not addressed by

---

[9] See: https://sdv.dev/SDV/user_guides/single_table/tabular_preset.html.

GenFair. Future works will study thi aspect as well as how GenFair can deal with three or more sensitive attributes, and extend it to privacy-masking, leveraging the flexibility of the genetic algorithm. In addition, the data generation and mutation could be further enhanced by inferring and exploiting causal relationships among features while creating synthetic instances.

Acknowledgment. This work is partially supported by the EU NextGenerationEU programme under the funding schemes PNRR-PE-AI FAIR (Future Artificial Intelligence Research), PNRR-SoBigData.it - Prot. IR0000013, H2020-INFRAIA-2019-1: and Res. Infr. G.A. 871042 *SoBigData++*.

References

1. Agarwal, A., et al.: A reductions approach to fair classification. In: ICML. Proceedings of Machine Learning Research, vol. 80, pp. 60–69. PMLR (2018)
2. Ball-Burack, A., et al.: Differential tweetment: mitigating racial dialect bias in harmful tweet detection. In: FAccT, pp. 116–128. ACM (2021)
3. Berk, R., et al.: Fairness in criminal justice risk assessments: the state of the art. Sociol. Methods Res. **50**(1), 3–44 (2021)
4. Cava, L., et al.: Genetic programming approaches to learning fair classifiers. In: GECCO, pp. 967–975 (2020)
5. Chakraborty, J., et al.: Bias in machine learning software: why? How? What to do? In: ESEC/SIGSOFT FSE, pp. 429–440. ACM (2021)
6. Cinquini, M., Guidotti, R.: CALIME: causality-aware local interpretable model-agnostic explanations. CoRR abs/2212.05256 (2022)
7. Dablain, D., et al.: Towards a holistic view of bias in machine learning: bridging algorithmic fairness and imbalanced learning. CoRR abs/2207.06084 (2022)
8. Fan, et al.: Explanation-guided fairness testing through genetic algorithm. In: ICSE, pp. 871–882 (2022)
9. Hardt, M., et al.: Equality of opportunity in supervised learning. In: NIPS, pp. 3315–3323 (2016)
10. Kamiran, F., et al.: Classification with no discrimination by preferential sampling. In: Proceedings of the 19th ML Conference Belgium and The Netherlands, vol. 1. Citeseer (2010)
11. Katoch, et al.: A review on genetic algorithm: past, present, and future. Multimed. Tools Appl. **80**, 8091–8126 (2021)
12. Lim, S.M., et al.: Crossover and mutation operators of genetic algorithms. Int. J. Mach. Learn. Comput. **7**(1), 9–12 (2017)
13. Liu, F.T., et al.: Isolation forest. In: ICDM, pp. 413–422. IEEE CS (2008)
14. Mehrabi, N., Fothers: a survey on bias and fairness in machine learning. ACM Comput. Surv. **54**(6), 115:1–115:35 (2021)
15. Ntoutsi, E.: Bias in AI-systems: a multi-step approach. In: NL4XAI. ACL (2020)
16. Patki, N., et al.: The synthetic data vault. In: DSAA, pp. 399–410. IEEE (2016)
17. Pedreschi, D., Ruggieri, S., Turini, F.: Discrimination-aware data mining. In: KDD, pp. 560–568. ACM (2008)
18. Pessach, D., et al.: A review on fairness in machine learning. ACM Comput. Surv. (CSUR) **55**(3), 1–44 (2022)
19. Quy, T.L., Roy, A., Iosifidis, V., Zhang, W., Ntoutsi, E.: A survey on datasets for fairness-aware machine learning. WIREs Data Min. Knowl. Discov. **12**(3) (2022)

20. Raquel, C.R., et al.: An effective use of crowding distance in multiobjective particle swarm optimization. In: GECCO, pp. 257–264. ACM (2005)
21. Salazar, et al.: Fawos: fairness-aware oversampling algorithm based on distributions of sensitive attributes. IEEE Access **9**, 81370–81379 (2021)
22. Sharma, S., et al.: Data augmentation for discrimination prevention and bias disambiguation. In: AIES, pp. 358–364. ACM (2020)
23. Tan, P.N., et al.: Introduction to data mining. Pearson Education India (2016)
24. Verma, et al.: A comprehensive review on NSGA-II for multi-objective combinatorial optimization problems. IEEE Access **9**, 57757–57791 (2021)
25. Verma, S., et al.: Fairness definitions explained. In: FairWare, pp. 1–7. ACM (2018)
26. Wang, et al.: Augmented fairness: an interpretable model augmenting decision-makers' fairness. arXiv preprint arXiv:2011.08398 (2020)

Privacy-Preserving Learning of Random Forests Without Revealing the Trees

Lukas-Malte Bammert, Stefan Kramer[⊠], Mattia Cerrato, and Ernst Althaus

Institut für Informatik, Johannes Gutenberg Universität Mainz,
Saarstraße 21, 55112 Mainz, Germany
kramerst@uni-mainz.de

Abstract. The paper presents a method for the privacy-preserving learning of random forests from private data of three parties, where not even the decision trees, i.e., neither the tree structures nor their parameters (the annotations of attributes and attribute values), are disclosed to any of the parties. To make this practical for realistically size data, a custom protocol is needed for the private comparison of two numbers, such that the numbers themselves are only available in shares and are not known to either party. Experiments with five datasets indicate that the overall protocol matches classical random forests in accuracy and can handle datasets of realistic size.

Keywords: Machine learning · Privacy · Random forest

1 Introduction

The current age of machine learning, enabled by very large data collections, high computational resources, highly effective and efficient algorithms and industrial-strength implementations, has increased the awareness both in the public discourse and on the side of policymakers. This has resulted in the enactment of robust data security regulations such as the EU General Data Protection Regulation (EU GDPR). These regulations significantly impact data processing, storage, and analysis, as data must not be transferred to third parties, even for analytical purposes. One common use case that is seriously aggravated is the collaborative analysis of data held by multiple parties. Thus, privacy-preserving solutions for learning from distributed data, without disclosing any private information, are called for.

This paper presents a protocol for the privacy-preserving training and testing of a random forest [4] with three parties, each holding a unique secret dataset. The major innovation is that we propose a custom protocol that also keeps the tree (i.e., the tree topology as well as the annotation by the attributes) secret. Existing approaches either propose custom protocols for private decision tree learning such that the tree is disclosed or employ general-purpose protocols, e.g., based on garbled circuits or homomorphic encryption, such that the tree

A. Bifet et al. (Eds.): DS 2023, LNAI 14276, pp. 372–386, 2023.
https://doi.org/10.1007/978-3-031-45275-8_25

remains undisclosed. The most important component is the private comparison of two numbers, whereby the numbers themselves are only available in shares and are not known to either party. The previous approach for a comparison of two numbers using private intersection and the evaluation only via the cardinality was for the situation in which two parties each know a number and want to compare them privately. The innovation here is the adaptation of this approach that the two numbers themselves are not known to anyone, but are only available as secret shares. The remaining steps of the approach are essentially the reduction of all steps of an ordinary decision tree learning algorithm to this comparison. Experimental results on several standard datasets demonstrate that this protocol maintains the performance of a traditional random forest while meeting privacy-preserving requirements. Importantly, the custom protocols enable running times that scale favorably in various dimensions.

This paper is structured as follows. After establishing the context in Sect. 2 with a review of existing work on privacy-preserving machine learning, Sect. 3 discusses the fundamental concepts of secure computations between multiple parties. Section 4 introduces our protocol for privacy-preserving training and testing of a random forest, where we simplify the essential steps of the algorithm to operations compatible with privacy-preserving solutions. Section 5 shares the results from test runs of a Python implementation of the protocol for selected datasets, before we conclude in Sect. 6.

2   Related Work

The concept of secure computations addresses the scenario where multiple parties need to evaluate a known function using their private inputs, all while maintaining data confidentiality. This field has been notably advanced through the foundational work of Yao and Goldreich *et al.*, both of which have contributed significantly to its progression and understanding [9,21].

These protocols guarantee two fundamental aspects: security, meaning they do not divulge any details about the private data, and correctness, which ensures accurate computation of the function's output. The security of these protocols is safeguarded against different types of adversarial behaviors, notably the 'semi-honest' or 'honest but curious' model and the 'malicious' model [3].

The semi-honest model deals with passive attackers, wherein participants fundamentally adhere to the protocol's instructions, but they might attempt to glean more information than permitted from received messages during the protocol. In contrast, the malicious model represents an active attacker, allowing for any polynomial-time attack strategy. In addition to these models, security can be divided into two categories based on their capabilities – the 'information-theoretic' model and the computational model. In the information-theoretic model, security is unconditionally guaranteed even against attackers not restricted by computability. In contrast, the computational model only promises security against polynomial-time attackers [3].

These theoretical models of security have practical applications, especially in machine learning. However, the general ability to compute an arbitrary function secretly was often impractical for many machine learning algorithms when dealing with realistic datasets. This led to the development of more specialized techniques for different algorithms.

Privacy Preserving Linear Regression for two parties has been initially discussed for horizontally distributed data [15] and later for vertically distributed data [8]. However, these models' reliance on a combination of homomorphic encryption and garbled circuits made them impractical for large datasets. In response, a model was presented based on a combination of arithmetic, binary, and Yao secret sharing that significantly improved runtime efficiency [14]. Further, an extension of this model was proposed for three parties [13].

As for privacy-preserving training of artificial neural networks (ANNs), the body of work remains limited. Models based on arithmetic, binary, and Yao secret sharing have been used for this purpose for two and three parties, respectively [13,14]. On the other hand, there are more developed approaches for privacy preserving prediction using ANNs. Notably, an ABY approach was employed for privacy-preserving prediction of an ANN [16]. Furthermore, an innovative approach for the privacy preserving learning of the weights of a sum-product network for n parties was presented recently [2].

There is also a significant amount of work on privacy-preserving training of decision trees. Most of this work [1,6,7,11,12,17,19,20] involves multi-party computational approaches where multiple parties jointly train a decision tree based on the union of their individual datasets. However, these approaches often communicate the decision tree's structure to all parties, potentially exposing information about each other's data indirectly. In contrast, a more recent study [1] presents a client-server solution for privacy-preserving training of a decision tree using homomorphic encryption in a turn-based protocol. Here, the server, unaware of the client's data, trains the encrypted decision tree on encrypted data received from the client, ensuring data privacy and protecting the decision tree's structure. Our contribution advances the state of the art by (i) proposing a secure multi-party protocol for random forest learning, which is (ii) based on secret sharing rather than homomorphic encryption.

3   Preliminaries

3.1   Secure Multi-party Computation

In this section, we aim to establish a fundamental understanding of security as applied to this research, drawing from Araki *et al.*'s book [3]. This work primarily leverages the semi-honest model. Essentially, this involves three parties who jointly execute a known function using their confidential input data. These parties are expected to adhere to the protocol guidelines, however, in doing so, they attempt to glean additional insights about each other's data from the messages exchanged in the protocol. The semi-honest model prioritizes maintaining the confidentiality of each party's input data, hence it is often termed as privacy-preserving rather than broadly secure computation. This model is suitable when the parties have a certain level of mutual trust, but have concerns about inadvertent data leaks or are bound by privacy regulations that prohibit data sharing.

First, let us briefly clarify some notation for the remainder of the section. In our scenarios, we are dealing with 3 parties. We will use subscript i to represent party $i \in 1, 2, 3$ and $i - 1$, $i + 1$ to refer to the 'preceding' and 'following' parties, respectively. Here, $i - 1 = 3$ when $i = 1$ and $i + 1 = 1$ when $i = 3$.

Consider a function $\mu : \mathbb{N} \to \mathbb{R} > 0$ to be negligible if for every polynomial $p :$ $\mathbb{N} \to \mathbb{R} > 0$ and sufficiently large n, we have $\mu(n) < \frac{1}{p(n)}$. An indexed probability sequence $X == \left\{ X(a, n) \right\}_{a \in \{0,1\}^*, n \in \mathbb{N}}$ is a series of random variables indexed by $a \in 0, 1^*$ and $n \in \mathbb{N}$. Two such sequences X, Y are deemed computationally indistinguishable ($X \overset{c}{\equiv} Y$) if for every polynomial time algorithm $\mathcal{D}$, there exists a negligible function μ such that for all $a \in 0, 1^*$ and $n \in \mathbb{N}$ we have

$$\left| Pr\left[\mathcal{D}(X(a, n)) = 1 \right] - Pr\left[\mathcal{D}(Y(a, n)) = 1 \right] \right| \leq \mu(n).$$

A 3-party protocol, denoted as π, represents three interacting algorithms. Each algorithm i has a private input $x_i \in \{0, 1\}$ (see e.g. [5] p. 13). In the context of the protocol, parties are linked to the algorithms they are executing. Party i's perspective, or 'view', during a π execution, denoted $view_i^\pi(x)$, is defined as $(x_i, r, m_1, ..., m_k)$, where r signifies the results of party i's internal random experiments, and m_j represents the j-th message received during protocol execution. Each of the three algorithms produces an output $out_i \in \{0, 1\}$ (potentially empty). The protocol's output is denoted as $output^\pi(x) = \left(out_1(x), out_2(x), out_3(x) \right)$.

Consider the following definition (from [10] p. 696): Let $f : \left(0, 1 \right)^3 \to \left(0, 1 \right)^3$ be a deterministic function with three inputs and $f_i(x_1, x_2, x_3)$ represent the i-th element of $f(x_1, x_2, x_3)$. Given π as a 3-party protocol for computing f, we assert that π computes the function f with computational assurance under the semi-honest model $i \in 1, 2, 3$, if (i) for each $x \in \left(0, 1 \right)^3$, the protocol's output $output^\pi(x)$ is such that $output^\pi(x) = f(x)$; and (ii) there exists a probabilistic polynomial time algorithm $\mathcal{S}$ such that for each i we have

$$\left\{ \mathcal{S}\left(x_i, f_i(x) \right) \right\}_{x \in \left(\{0,1\}^* \right)^3} \overset{c}{\equiv} \left\{ view_i^\pi(x) \right\}_{x \in \left(\{0,1\}^* \right)^3}.$$

Informally, in the presence of semi-honest parties, security implies that a semi-honest party's perspective/view of an execution of the protocol can be reconstructed by a simulator given only the party's input and output.

Lastly, it is important to note that while the above definition underlines the intuitive understanding of security with semi-honest parties, utilizing this definition alone to prove the security of more complex protocols can be impractical. In more complex scenarios, sub-protocols may be used to delegate individual tasks. To demonstrate the parent protocol's security would necessitate a composition theorem for protocols, as given by Canetti [5]. However, as this framework of definitions is fairly extensive, it is not adopted in this research. Therefore, it must be understood that the security of the protocols introduced in this paper is not formally proven, but rather plausibly argued. A formal proof remains the subject of future work.

3.2   Arithmetic Secret Shares

Arithmetic Secret Shares, as defined in the works of Mohassel *et al.* [13], serve as the primary encryption method in our protocol for private learning of random forests. They represent the secrets of the participating parties as integers within a prime number field, denoted as $\mathbb{Z}_p$. This choice of representation enables addition and multiplication operations to be carried out with minimal communication overhead.

This method begins with the generation of a 'distributed zero' amongst the parties. This distributed zero is essentially a set of numbers that, when added together, equal zero. Although the process of creating this set of numbers may seem complex, its validity can be straightforwardly confirmed because their sum is, indeed, equal to zero.

Once this set of numbers has been created, we can apply it to encrypt a secret. Consider a scenario where three parties want to encrypt a secret x within the field $\mathbb{Z}$. The secret x falls within the range of $[-M, M]$, and each party owns a unique part of it such that the sum of their parts equals x. Notably, it is permissible for one or two parties to possess no knowledge of x, meaning their individual part of the secret is zero.

To encrypt the secret, the three parties use the previously generated 'distributed zero' to create a set of Secret Shares. Each party then adds its part of the secret to its corresponding element in the distributed zero. This process yields three encrypted values, and their sum is equal to x modulo p. This sharing method ensures that each party needs all three shares to reveal the secret x.

Furthermore, the multiplication operation is handled in a unique way. Each party holds not only its share but also the previous party's share. Therefore, every party possesses a pair of shares, making this sharing scheme secure under the assumption that the parties will not collude.

Using Arithmetic Secret Shares, parties can also perform addition and multiplication with a publicly known constant without any communication overhead. This flexibility is due to the operations being performed modulo p.

Protocols for executing addition and multiplication operations using Arithmetic Secret Shares are described elsewhere [13], and we have employed those protocols in this work. These protocols are secure because no communication occurs between the parties during their execution.

In the case of the multiplication protocol, we can demonstrate its validity through algebraic manipulation. After each party locally computes a specific equation, the sum of the outcomes equals $x * y$ modulo p.

Just as with the creation of the distributed zero, the security of these protocols is confirmed by simulating a party's point of view. This simulation can compute all the information a party would see during the protocol's execution, demonstrating that the secret x remains secure.

In the remainder of this work, especially in the sections on protocols for privacy-preserving computations, we will maintain the convention that any computational operations between values are performed by the parties using the Arithmetic Secret Shares method.

4   Method

In this section, we describe our contribution for privacy-preserving learning of random forest models. As established in Sect. 2, some methodologies for private learning of decision trees have already been contributed to the literature (e.g. [6,12,20]). However, we argue that while the learning protocols previously developed are in itself private, the act of releasing the tree structure may itself break privacy by allowing for the reconstruction of private training data. We motivate our argument via a simple example. Let the following datasets D_1, D_2 be owned by two separate parties:

$$D_1 = \Big\{(0,0,0,0,0),(0,1,0,0,0),(1,0,0,0,0),(1,1,0,0,0)\Big\}$$
$$D_2 = \Big\{(0,0,0,0,0),(0,0,0,1,0),(0,0,1,0,0),(0,0,1,1,1)\Big\}.$$

In a data point $d \in D$ the first 4 components $a_1, \dots, a_4$ denote the values of the 4 attributes $A_1, \dots, A_4$ and the fifth component c represents the value of the class C. Then, a decision tree which classifies the joint dataset $D = D_1 \bigcup D_2$ may be found in Fig. 1.

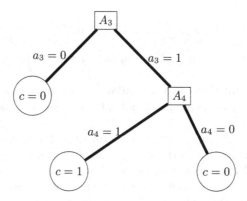

Fig. 1. Example of a decision tree that correctly classifies all data points of the joint dataset $D = D_1 \bigcup D_2$.

Any protocol for training the decision tree used to construct the decision tree would be privacy-preserving in the sense that the parties do not learn about each other's data in a direct way during this protocol. However, after the protocol is complete, an issue arises if both parties are given the final resulting decision tree. In this case, just from the structure and knowledge of its own data, any party could infer knowledge about others' data by observing the tree. In our example, party 1 could infer from the branching at the 2nd level that data points $(0,0,0,1,0)$ and $(0,0,1,1,1)$ must be contained in data set D_2. Our proposal for privacy-preserving learning of random forests, in contrast, keeps the structure of the underlying decision trees a secret from all parties. This may be achieved via arithmetic secret shares (see Sect. 3.2) and is described in the following.

4.1   Privacy-Preserving Determination of Maximum Value

This section presents a protocol for the privacy-preserving comparison of two numbers x and y. The goal is to establish if $y \leq x$. This approach will be crucial for the privacy-preserving random forest.

We base our method on previous work that reduces the private comparison problem to the Private Set Intersection Problem, which identifies if two sets A and B have a non-empty intersection $A \cap B$ in a privacy-preserving manner. Our protocol handles three semi-honest parties 1, 2, and 3 wishing to determine the maximum between two numbers x and y from $\mathbb{Z}$, given a predefined bound M. The numbers x and y exist as arithmetic secret shares with respect to a prime p, distributed among the three parties. Determining the maximum of x and y corresponds to finding the sign of $z = 2(y - x) + 1$, which is estimated through a probabilistic process with a success rate of at least $1 - 2 \cdot \frac{M}{p - 4M - 4}$.

We also define two sets, S_h^1 and S_r^0. If $s = s_n s_{n-1} \ldots s_1 \in {0,1}^n$ represents the binary representation of a positive integer, we can establish S_s^0 and S_s^1 accordingly:

$$S_s^0 := \left\{ s_n s_{n-1} \ldots s_{i+1} 1 \mid s_i = 0,\ 1 \leq i \leq n \right\},$$
$$S_s^1 := \left\{ s_n s_{n-1} \ldots s_i \mid s_i = 1,\ 1 \leq i \leq n \right\}.$$

The intersection of these two sets allows us to determine if $x > y$. If the sets share exactly one element, $x > y$ holds true; otherwise, their intersection is empty.

Due to space limitations, we cannot provide the full details and security proofs of the algorithm here. However, it involves a series of steps including random number selection, calculations based on the secret shares of x, determination of set intersections, and final calculation of the output sign of y as secret shares. We emphasize that this procedure ensures privacy as only specific parties have access to clear values of particular variables at any given time.

4.2   Privacy-Preserving Decision Tree Learning

In this section, we give an overview of our novel multi-party computation method where the decision tree's structure is concealed from all involved parties. As previously established, we envisage a scenario where three parties aim to construct a decision tree from a dataset D divided amongst them, assuming the parties are semi-honest, adhering to a protocol but trying to extract knowledge from each communication.

After preparatory steps, every data point $d \in D$ becomes a vector of $\{0,1\}^{m+1}$, implying binary-valued attributes and class. However, non-binary attributes can be converted using One-Hot-Encoding, and multiple binary classes can be learned separately. The data points are stored as additive Secret Shares modulo a large prime p, unaffected by whether the original data was horizontally or vertically distributed.

Our approach to privately training a decision tree is the same as in the core of many classical decision tree induction algorithms, where three key steps are required:

1. Establish the subset $T \subset D$ of data that is viable for the current decision tree node.
2. Determine the attribute A_j with the highest reduction of class impurity relative to T based on a random subset of attributes $A_{j_1}, \ldots, A_{j_s}$.
3. Generate the node's children as branches with respect to attribute A_j.

Our strategy to design a secure multi-party protocol which can deal with the distribution of D over multiple parties is to map these three high-level steps to arithmetical operations to be performed over secret shares of the data. We briefly elaborate on these in the following, avoiding the mathematical details and security proofs due to space limitations.

Selecting $T \subset D$. To define whether a decision tree node will branch (inner node) or select a label (leaf node), we must select the subset T from the total data D that is pertinent to this node. Keeping node composition covert from all parties, we express each node with a $2m$-length binary vector x (where m signifies attributes), distributed as secret shares among the three parties. Elements of x (i.e., x_i^0 and x_i^1) indicate permissible data point values (0 or 1) for attribute A_i in the respective node. The root node's vector permits all attribute expressions.

To identify if a data point d from D should be evaluated at node x, we calculate Z_d:

$$Z_d = 2 \sum_{i=1}^{m} z_i^{(d)}, \text{ where } z_i^{(d)} = d_i(x_i^1 - x_i^0) + x_i^0.$$

When data point d satisfies the node's attribute A_i requirement, $z_i^{(d)}$ equals 1, otherwise 0. So, d is pertinent to the node if $Z_d = 2m$. If not, Z_d becomes less than or equal to $2m - 2$, hence a comparison is necessary to determine whether the data point d is pertinent to the node or not:

$$\gamma_d = \arg\max\left\{2m - 1,\ Z_d\right\}.$$

To determine $t \subset D$, one could naively determine γ_d for each $d \in D$ using the maximum protocol. However, since the maximum protocol would have to be executed for each individual node x_j of the decision tree for all data $d \in D$ and this is significantly more communication-intensive (and therefore slower) than the arithmetic operations on the secret shares, we use the following alternative strategy. This strategy allows for all nodes at a certain level of the decision tree to determine at once which data points are residing in which node, and requires only a comparable number of communications between the parties as the maximum protocol. The idea of the method is based on exploiting the fact that for a certain data point d the distribution of the values $\sum_{i=1}^{m} z_l^{(d)}$, for a node x of the considered tree level $depth\_index$ is determined a priori. For k with $0 \le m - k \le depth$, the number of nodes on the tree level $depth\_index$ for which $\sum_{i=1}^{m} z_l^{(d)} = k$ is given by $\binom{depth\_index}{m - k}$, which can be shown by induction. In particular, this means that there is only exactly one node on each tree level in which a data point is pertinent ($k = m$). This allows the following procedure. Parties 1 and 2 permute the secret shares of the three parties of $\left(Z_d^{j1}, \ldots, Z_d^{jn}\right)$, for a datapoint $d \in D$ and all nodes $x_{j_1}, \ldots, x_{j_n}$ of the given tree level. Then party 3 can determine for all $d \in D$ the indicator vector for which permuted node j the value Z_d^j equal $2m$ is and distribute this again as secret shares between the parties. Parties 1 and 2 then repermute the shares of the indicator vector.

Perform Attribute Splits. Classical decision tree algorithms pick the next attribute for branching in a decision tree using a strategy that focuses on maximizing the reduction of class impurity, which is in our case calculated using the Gini index. To calculate this, we only consider binary attributes, simplifying the process.

We then make certain computations for each attribute that account for the size and distribution of the dataset. To simplify the process, we modify the Gini index to remove divisions and make the process more efficient, following an approach previously suggested by Hori in 2008 [17]. This approach allows us to compare the evaluation of two different attributes without requiring any division. This is crucial when we want to perform calculations privately, such as in secure multi-party computation scenarios. Private computation of division is a complex operation when performed over arithmetic secret shares (for a discussion see, e.g., a recent paper on SPN learning [2]).

To apply this process in practice, we follow a specific protocol. This protocol involves iteratively calculating and comparing the modified Gini index for each attribute. After multiple rounds of these comparisons, the protocol will indicate the attribute that should be used for branching, maximizing the reduction of class impurity based on the Gini index. The security of this protocol may be proven leveraging the security of the maximum protocol introduced in Sect. 4.1.

The correctness of this protocol – that is, whether it truly picks the attribute with maximum reduction of class impurity – can be shown through an induction argument. This argument essentially shows that after a given number of rounds, the selected attribute will indeed be the one that maximizes the evaluation function. We also make sure that the protocol does not pick an attribute that has already been branched. This is done by setting the Gini index of already-branched attributes to zero, so they will not be picked if there is at least one other attribute that is still viable. If all previous attributes are invalid, the new attribute will always be chosen.

In essence, we have transformed the original computation into a more efficient version that can be computed securely and privately, and we have shown that this approach can correctly select the attribute that maximizes reduction of class impurity in a decision tree.

Branching the Tree. Creating branches from a node in the decision tree is relatively straightforward. Since all attributes are binary, every node has two children (left and right) no matter the attribute chosen for branching. For the left (right) child, data is included only if its corresponding attribute A_j value is 0 (1).

This protocol's input includes the attribute requirements of the node and the attribute branching indicator vector. For the left child, x^0_{left} is set as x^0 and x^1_{left} as $x^1 - e$. Similarly, for the right child, x^0_{right} is set as $x^0 - e$ and x^1_{right} as x^1. The output is the two child nodes.

The correctness of the protocol stems from the definition of x, which encodes the node's attribute requirements, and e, the attribute branching indicator vector. The protocol's security relies on the security of the secret shares and the fact that no matter the attribute selected, two child nodes are always created, preventing information about the decision tree's structure from being leaked to the parties. The protocol's overhead is constant.

The previous section has provided all the essential building blocks for privacy preserving training of a decision tree. The correctness of this protocol is built upon the correctness of the previously mentioned protocols and the leaf node's label definition. The protocol's safety is ensured as the parties cannot learn anything directly or indirectly about the data. This is achieved by using secure arithmetic secret shares, and the fact that the learned decision tree's structure is always a complete binary tree of a specified depth. The total number of calls to the maximum log protocol depends on the number of interior nodes and leaf nodes, which are calculated based on the tree's depth and the data set size.

4.3 Training and Inference for Privacy-Preserving Random Forests

In the previous section, we have established a privacy-preserving protocol for training a single decision tree. Now, we extend this concept to train a random forest, which consists of multiple decision trees, where randomness is injected, in different ways, to obtain different trees from originally identical training sets.[1]

---

[1] Notice that we follow the original definition of random forests by Breiman (2001).

In our case, we obtain multiple decision trees, as in the original publication, by sampling the dataset with replacement (bootstrap sampling). However, this is the only way we inject randomness: We do not work with random subsets of features, as in the original publication and in current random forest implementations. In the following, we modify the overall process to ensure privacy.

To maintain privacy while creating these sample datasets, we use an indicator vector approach. For each draw from the original dataset, an indicator vector is generated where the corresponding index is set to 1 and all other indices are set to 0. By adding these vectors together, we obtain a weight vector that essentially counts how many times each data point is included in a sample. Once these data samples are generated independently by the respective parties, we can then proceed to train individual decision trees using the privacy-preserving protocol we previously outlined. The difference is that the computation of the protocol will be slightly adjusted to account for the weights from the weight vector. Once all trees in the forest have been generated and trained, they may be used to make predictions on new data points. This process is similar to testing a single decision tree, except that we must gather predictions from all trees in the forest and then determine the overall prediction based on a majority vote.

The communication overhead for these protocols is relatively small, primarily due to the extra communications required for the data draw process in the training of the random forest, and the accumulation of predictions in the testing process. In the next section, we will empirically test these protocols through implementation and evaluate their performance in terms of accuracy and computation time.

5   Experiments

The goal of the experiments was two-fold: First, to compare the privacy-preserving protocol to a non-privacy preserving variant in terms of prediction performance. Second, to test the running time of the privacy-preserving protocol depending on a variety of different parameters, to make sure the approach is suitable for realistically sized datasets.

All experiments were run of five datasets from the UCI repository[2]: *Breast Cancer, Balance Scale, Tic-Tac-Toe Endgame, Car Evaluation, Molecular Biology* (Splice-junction Gene Sequences). All attributes are nominal, and the task is either binary or multi-class classification. The nominal attributes were transformed to binary via one-hot encoding, employing scikit-learn. For the multiclass problems (*Balance Scale, Car Evaluation, Molecular Biology*), we learn one separate random forest for each class in turn (one-versus-all).

The Python implementation of the protocol for training and testing a random forest is based on a publicly available Python implementation for private learning of SPNs [2]. The implementation consists of a manager server that schedules individual tasks and several member servers (in our case, 3) that execute these

---

[2] http://archive.ics.uci.edu/ml.

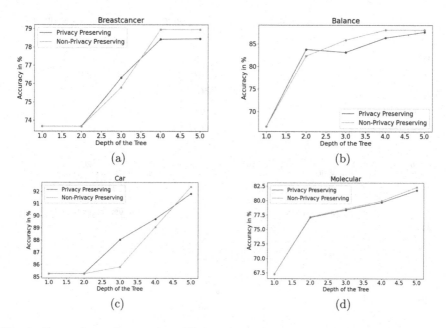

Fig. 2. Comparison of average prediction performance of privacy-preserving and non-privacy-preserving random forest for different datasets and tree depths. The curve for *Tic Tac Toe* is similar to the one of *Breast Cancer* and not shown here.

tasks. All servers communicate with each other via a WebSocket framework[3], where each member and the manager have a unique network ID.

We compared the performance of our privacy-preserving protocol with a random forest learned without any privacy-preserving restrictions. In each experiment, we used a Python package from scikit-learn [18] to learn a random forest based on Breiman's approach [4]. However, the individual decision trees were built on the basis of the Gini index.

For both the privacy-preserving and non-privacy-preserving approaches, we trained and tested a random forest consisting of 50 decision trees at tree depths $tree\_depth = 1, 2, 3, 4, 5$. We randomly divided the dataset into $\frac{2}{3}$ training data and $\frac{1}{3}$ test data for this purpose. For all learned decision trees, $\max\{tree\_depth, \log(\text{number of attributes})\}$ random attributes were considered in the attribute selection. The accuracy, i.e., the proportion of instances correctly classified by the random forest in the test data, was used as a measure to evaluate the prediction quality. The experiments were conducted on a Windows 8 machine with 16 GB of memory, AMD A10-6700 APU processor, and no internal network latency was used for inter-party communication.

Figure 2 presents the prediction performance of each random forest for different datasets and tree depths. The prediction quality of a random forest for each tree depth is averaged over the total number of random forests learned

---

[3] https://websockets.readthedocs.io/en/stable/.

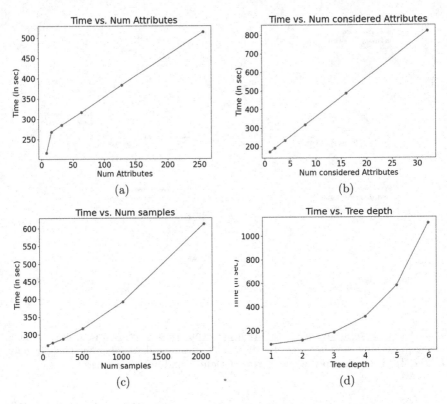

Fig. 3. Comparison of runtime of privacy-preserving random forest while varying number of attributes, tree depth and number of samples in the dataset.

for each class (in case of multi-class datasets). For all tree depths, the average prediction quality of the privacy-preserving random forest comes close to that of a non-privacy-preserving random forest, with the gap between the two being at most around 5 %. In Fig. 3, we illustrate the runtime behavior of the privacy-preserving protocol. We observe that, as the number of overall and randomly chosen attributes grows, the running time grows linearly with it. The dependency of the running time on the number of training instances also appears as linear, as one would expect. The slight bend in the curve is not easy to explain by expectations and could be due to statistical variation. The worst behavior is, as expected, in the tree depth, where the running time appears to depend a bit worse than quadratic on the depth of the tree.

To summarize, our experimental results suggest that a privacy-preserving random forest can achieve nearly similar prediction performance as a non-privacy-preserving random forest, with a manageable increase in runtime. However, there is a trade-off between privacy, performance, and runtime, which needs to be carefully considered depending on the specific requirements and constraints of a given application.

6   Conclusion and Future Work

The paper presented a privacy-preserving approach to the computation of random forests. A key innovation in our approach lies in maintaining the decision tree's structure and parameters (the annotation) as information hidden from all parties. This ensures that no party can infer any knowledge about the data of others, further strengthening the privacy protection in our methodology. Along the way, we reduce all operations to work with arithmetic secret shares and establish a protocol for securely determining the maximum of two integers. Experimental results show that prediction quality is on par with its non-privacy-preserving counterpart, and the runtime, while influenced by several variables, is manageable even for realistic datasets.

While the protocol is composed of secure primitives, a formal proof of its security would be highly desirable, potentially achieved through the construction of a framework that enables secure protocol composition. Moreover, from a practical perspective, developing an efficient protocol for finding the maximum of two numbers without any error probability would be an improvement in terms of quality. Simultaneously, reducing the protocol's runtime through parallelized operations, such as a component-wise maximum comparison of two vectors, could enhance its practical applicability.

Acknowledgements. This work was partly funded by the Carl-Zeiss-Stiftung as part of the CZS Durchbrueche project under grant number [P2021-02-014].

References

1. Akavia, A., Leibovich, M., Resheff, Y.S., Ron, R., Shahar, M., Vald, M.: Privacy-preserving decision trees training and prediction. Cryptology ePrint Archive, Paper 2021/768 (2021). https://eprint.iacr.org/2021/768
2. Althaus, E., Dousti, M.S., Kramer, S., Rassau, N.J.P.: Fast private parameter learning and evaluation for sum-product networks. CoRR abs/2104.07353 (2021). https://arxiv.org/abs/2104.07353
3. Araki, T., Furukawa, J., Lindell, Y., Nof, A., Ohara, K.: High-throughput semi-honest secure three-party computation with an honest majority. In: Proceedings of the 2016 ACM SIGSAC Conference on Computer and Communications Security, pp. 805–817. Association for Computing Machinery, New York, NY, USA (2016). https://doi.org/10.1145/2976749.2978331
4. Breiman, L.: Random forests. Mach. Learn. **45**, 5–32 (2001)
5. Canetti, R.: Security and composition of multiparty cryptographic protocols. J. Cryptol. **13**(1), 143–202 (2000). https://doi.org/10.1007/s001459910006
6. Du, W., Zhan, Z.: Building decision tree classifier on private data. In: Proceedings of the IEEE International Conference on Privacy, Security and Data Mining, vol. 14, pp. 1–8. Australian Computer Society Inc. (2002)
7. Emekci, F., Sahin, O., Agrawal, D., El Abbadi, A.: Privacy preserving decision tree learning over multiple parties. Data Knowl. Eng. **63**(2), 348–361 (2007). https://www.sciencedirect.com/science/article/pii/S0169023X07000365

8. Giacomelli, I., Jha, S., Joye, M., Page, C.D., Yoon, K.: Privacy-preserving ridge regression over distributed data from lhe. Cryptology ePrint Archive, Report 2017/979 (2017). https://eprint.iacr.org/2017/979

9. Goldreich, O., Micali, S., Wigderson, A.: How to play any mental game. In: Proceedings of the Nineteenth Annual ACM Symposium on Theory of Computing, pp. 218–229. Association for Computing Machinery, New York, NY, USA (1987). https://doi.org/10.1145/28395.28420

10. Goldreich, O.: Foundations of Cryptography - Basic Applications, vol. 2. Cambridge University Press, Cambridge (2004)

11. de Hoogh, S., Schoenmakers, B., Chen, P., op den Akker, H.: Practical secure decision tree learning in a teletreatment application. In: Christin, N., Safavi-Naini, R. (eds.) FC 2014. LNCS, vol. 8437, pp. 179–194. Springer, Heidelberg (2014). https://doi.org/10.1007/978-3-662-45472-5_12

12. Lindell, Y., Pinkas, B.: Privacy preserving data mining. In: Bellare, M. (ed.) CRYPTO 2000. LNCS, vol. 1880, pp. 36–54. Springer, Heidelberg (2000). https://doi.org/10.1007/3-540-44598-6_3

13. Mohassel, P., Rindal, P.: Aby 3: a mixed protocol framework for machine learning. In: Proceedings of the 2018 ACM SIGSAC Conference on Computer and Communications Security, pp. 35–52, October 2018

14. Mohassel, P., Zhang, Y.: SecureML: a system for scalable privacy-preserving machine learning. In: 2017 IEEE Symposium on Security and Privacy (SP), pp. 19–38 (2017). https://doi.org/10.1109/SP.2017.12

15. Nikolaenko, V., Weinsberg, U., Ioannidis, S., Joye, M., Boneh, D., Taft, N.: Privacy-preserving ridge regression on hundreds of millions of records. In: 2013 IEEE Symposium on Security and Privacy, pp. 334–348 (2013). https://doi.org/10.1109/SP.2013.30

16. Riazi, M.S., Weinert, C., Tkachenko, O., Songhori, E.M., Schneider, T., Koushanfar, F.: Chameleon: a hybrid secure computation framework for machine learning applications. CoRR abs/1801.03239 (2018). http://arxiv.org/abs/1801.03239

17. Samet, S., Miri, A.: Privacy preserving ID3 using Gini index over horizontally partitioned data. In: 2008 IEEE/ACS International Conference on Computer Systems and Applications, pp. 645–651 (2008). https://doi.org/10.1109/AICCSA.2008.4493598

18. Scikit-learn: random forest classifier. https://scikit-learn.org/stable/modules/generated/sklearn.ensemble.RandomForestClassifier.html

19. Vaidya, J., Clifton, C., Kantarcioglu, M., Patterson, A.S.: Privacy-preserving decision trees over vertically partitioned data **2**(3) (2008). https://doi.org/10.1145/1409620.1409624

20. Wang, K., Xu, Y., She, R., Yu, P.S.: Classification spanning private databases. In: Proceedings of the 21st National Conference on Artificial Intelligence, vol. 1, p. 293–298. AAAI'06, AAAI Press (2006)

21. Yao, A.C.: Protocols for secure computations. In: Proceedings of the 23rd Annual Symposium on Foundations of Computer Science, pp. 160–164. IEEE Computer Society, USA (1982)

Unlearning Spurious Correlations in Chest X-Ray Classification

Misgina Tsighe Hagos[1,2](✉)(iD), Kathleen M. Curran[1,3](iD),
and Brian Mac Namee[1,2](iD)

[1] Science Foundation Ireland Centre for Research Training in Machine Learning,
Dublin, Ireland
{kathleen.curran,brian.macnamee}@ucd.ie
[2] School of Computer Science, University College Dublin, Dublin, Ireland
misgina.hagos@ucdconnect.ie
[3] School of Medicine, University College Dublin, Dublin, Ireland

Abstract. Medical image classification models are frequently trained using training datasets derived from multiple data sources. While leveraging multiple data sources is crucial for achieving model generalization, it is important to acknowledge that the diverse nature of these sources inherently introduces unintended confounders and other challenges that can impact both model accuracy and transparency. A notable confounding factor in medical image classification, particularly in musculoskeletal image classification, is skeletal maturation-induced bone growth observed during adolescence. We train a deep learning model using a Covid-19 chest X-ray dataset and we showcase how this dataset can lead to spurious correlations due to unintended confounding regions. eXplanation Based Learning (XBL) is a deep learning approach that goes beyond interpretability by utilizing model explanations to interactively unlearn spurious correlations. This is achieved by integrating interactive user feedback, specifically feature annotations. In our study, we employed two non-demanding manual feedback mechanisms to implement an XBL-based approach for effectively eliminating these spurious correlations. Our results underscore the promising potential of XBL in constructing robust models even in the presence of confounding factors.

Keywords: Interactive Machine Learning · eXplanation Based Learning · Medical Image Classification · Chest X-ray

1 Introduction

While Computer-Assisted Diagnosis (CAD) holds promise in terms of cost and time savings, the performance of models trained on datasets with undetected biases is compromised when applied to new and external datasets. This limitation hinders the widespread adoption of CAD in clinical practice [16,21]. Therefore, it is crucial to identify biases within training datasets and mitigate their impact on trained models to ensure model effectiveness.

A. Bifet et al. (Eds.): DS 2023, LNAI 14276, pp. 387–397, 2023.
https://doi.org/10.1007/978-3-031-45275-8_26

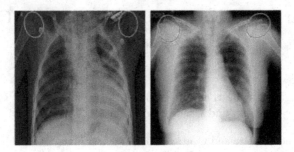

Fig. 1. In the left image, representing a child diagnosed with Viral pneumonia, the presence of Epiphyses on the humerus heads is evident, highlighted with red ellipses. Conversely, the right image portrays an adult patient with Covid-19, where the Epiphyses are replaced by Metaphyses, also highlighted with red ellipses. (Color figure online)

For example, when building models for the differential diagnosis of pathology on chest X-rays (CXR) it is important to consider skeletal growth or ageing as a confounding factor. This factor can introduce bias into the dataset and potentially mislead trained models to prioritize age classification instead of accurately distinguishing between specific pathologies. The effect of skeletal growth on the appearance of bones necessitates careful consideration to ensure that a model focuses on the intended classification task rather than being influenced by age-related features.

An illustrative example of this scenario can be found in a recent study by Pfeuffer et al. [12]. In their research, they utilized the Covid-19 CXR dataset [4], which includes a category comprising CXR images of children. This dataset serves as a pertinent example to demonstrate the potential influence of age-related confounders, given the presence of images from pediatric patients. It comprises CXR images categorized into four groups: Normal, Covid, Lung opacity, and Viral pneumonia. However, a notable bias is introduced into the dataset due to the specific inclusion of the Viral pneumonia cases collected exclusively from children aged one to five years old [9]. This is illustrated in Fig. 1 where confounding regions introduced due to anatomical differences between a child and an adult in CXR images are highlighted. Notably, the presence of Epiphyses in images from the Viral pneumonia category (which are all from children) is a confounding factor, as it is not inherently associated with the disease but can potentially mislead a model into erroneously associating it with the category. Addressing these anatomical differences is crucial to mitigate potential bias and ensure accurate analysis and classification in pediatric and adult populations.

Biases like this one pose a challenge to constructing transparent and robust models capable of avoiding spurious correlations. Spurious correlations refer to image regions that are mistakenly believed by the model to be associated with a specific category, despite lacking a genuine association.

While the exact extent of affected images remains unknown, it is important to note that the dataset also encompasses other confounding regions, such as texts and timestamps. However, it is worth mentioning that these confounding regions are uniformly present across all categories, indicating that their impact is consistent throughout. For the purpose of this study, we specifically concentrate on understanding and mitigating the influence of musculoskeletal age in the dataset.

eXplanation Based Learning (XBL) represents a branch of Interactive Machine Learning (IML) that incorporates user feedback in the form of feature annotation during the training process to mitigate the influence of confounding regions [17]. By integrating user feedback into the training loop, XBL enables the model to progressively improve its performance and enhance its ability to differentiate between relevant and confounding features [6]. In addition to unlearning spurious correlations, XBL has the potential to enhance users' trust in a model [5]. By actively engaging users and incorporating their expertise, XBL promotes a collaborative learning environment, leading to increased trust in the model's outputs. This enhanced trust is crucial for the adoption and acceptance of models in real-world applications, particularly in domains where decisions have significant consequences, such as medical diagnosis.

XBL approaches typically add regularization to the loss function used when training a model, enabling it to disregard the impact of confounding regions. A typical XBL loss can be expressed as:

$$L = L_{CE} + L_{expl} + \lambda \sum_{i=0} \theta_i^2 \,, \tag{1}$$

where L_{CE} is categorical cross entropy loss that measures the discrepancy between the model's predictions and ground-truth labels; λ is a regularization term; θ refers to network parameters; and L_{expl} is an explanation loss. Explanation loss can be formulated as:

$$L_{expl} = \sum_{i=0}^{N} M_i \odot Exp(x_i) \,, \tag{2}$$

where N is the number of training instances, $x \in X$; M_i is a manual annotation of confounding regions in the input instance x_i; and $Exp(x_i)$ is a saliency-based model explanation for instance x_i, for example generated using Gradient weighted Class Activation Mapping (GradCAM) [17]. GradCAM is a feature attribution based model explanation that computes the attention of the learner model on different regions of an input image, indicating the regions that significantly contribute to the model's predictions [18]. This attention serves as a measure of the model's reliance on these regions when making predictions. The loss function, L_{expl}, is designed to increase as the learner's attention to the confounding regions increases. Overall, by leveraging GradCAM-based attention and the associated L_{expl} loss, XBL provides a mechanism for reducing a model's attention to confounding regions, enhancing the interpretability and transparency of a model's predictions.

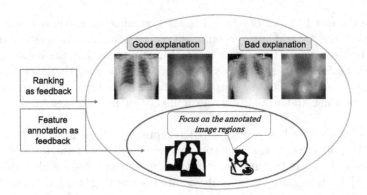

Fig. 2. The inner ellipse shows the typical mode of feedback collection where users annotate image features. The outer ellipse shows how our proposed approach requires only identification of one good and one bad explanation.

As is seen in the inner ellipse of Fig. 2, in XBL, the most common mode of user interaction is image feature annotation. This requires user engagement that is considerably more demanding than the simple instance labeling that most IML techniques require [22] and increases the time and cost of feedback collection. As can be seen in the outer ellipse of Fig. 2, we are interested in lifting this pressure from users (feedback providers) and simplifying the interaction to ask for identification of two explanations as exemplary explanations and ranking them as good and bad explanations. This makes collecting feedback cheaper and faster. This kind of user interaction where users are asked for a ranking instead of category labels has also been found to increase inter-rater reliability and data collection efficiency [11]. We incorporate this feedback into model training through a contrastive triplet loss [3].

The main contributions of this paper are:

1. We propose the first type of eXplanation Based Learning (XBL) that can learn from only two exemplary explanations of two training images;
2. We present an approach to adopt triplet loss for XBL to incorporate the two exemplary explanations into an explanation loss;
3. Our experiments demonstrate that the proposed method achieves improved explanations and comparable classification performance when compared against a baseline model.

2　Related Work

2.1　Chest X-Ray Classification

A number of Covid-19 related datasets have been collated and deep learning based diagnosis solutions have been proposed due to the health emergency caused by Covid-19 and due to an urgent need for computer-aided diagnosis (CAD) of the disease [8]. In addition to training deep learning models from scratch, transfer learning, where parameters of a pre-trained model are further trained to identify Covid-19, have been utilized [20]. Even though the array of datasets and deep learning models show promise in implementing CAD, care needs to be taken when the datasets are sourced from multiple imaging centers and/or the models are only validated on internal datasets. The Covid-19 CXR dataset, for example, has six sources at the time of writing this paper. This can result in unintended confounding regions in images in the dataset and subsequently spurious correlations in trained models [16].

2.2   eXplanation Based Learning

XBL can generally be categorized based on how feedback is used: (1) augmenting loss functions; and (2) augmenting training datasets.

Augmenting Loss Functions. As shown in Eq. 1, approaches in this category add an explanation loss, L_{expl}, during model training to encourage focus on image regions that are considered relevant by user(s), or to ignore confounding regions [7]. Ross et al. [14] use an L_{expl} that penalizes a model with high input gradient model explanations on the wrong image regions based on user annotation,

$$L_{expl} = \sum_{n}^{N} \left[M_n \odot \frac{\partial}{\partial x_n} \sum_{k=1}^{K} \log \hat{y}_{nk} \right]^2 , \qquad (3)$$

for a function $f(X|\theta) = \hat{y} \in R^{N \times K}$ trained on N images, x_n, with K categories, where $M_n \in \{0, 1\}$ is user annotation of confounding image regions. Similarly, Shao et al. [19] use influence functions in place of input gradients to correct a model's behavior

Augmenting Training Dataset. In this category, a confounder-free dataset is added to an existing confounded training dataset to train models to avoid learning spurious correlations. In order to unlearn spurious correlations from a classifier that was trained on the Covid-19 dataset, Pfeuffer et al. [12] collected feature annotation on 3,000 chest x-ray images and augmented their training dataset. This approach, however, doesn't target unlearning or removing spurious correlations, but rather adds a new variety of data. This means models are being trained on a combination of the existing confounded training dataset and the their new dataset.

One thing all approaches to XBL described above have in common is the assumption that users will provide feature annotation for all training instances to refine or train a model. We believe that this level of user engagement hinders practical deployment of XBL because of the demanding nature and expense of feature annotation that is required [22]. It is, therefore, important to build an XBL method that can refine a trained model using a limited amount of user interaction and we propose eXemplary eXplanation Based Learning to achieve this.

3   eXemplary eXplanation Based Learning

User annotation of image features, or M, is an important prerequisite for typical XBL approaches (illustrated in Eq. 1). We use eXemplary eXplanation Based Learning (eXBL) to reduce the time and resource complexity caused by the need for M. eXBL simplifies the expensive feature annotation requirement by replacing it with identification of just two exemplary explanations: a *Good explanation* (C_{good_i}) and a *Bad explanation* (C_{bad_j}) of two different instances, x_i and x_j. We pick the two exemplary explanations manually based on how much attention a model's explanation output gives to relevant image regions. A good explanation would be one that gives more focus to the lung and chest area rather than the irrelevant regions such as the Epiphyses, humerus head, and image backgrounds, while a bad explanation does the opposite.

We choose to use GradCAM model explanations because they have been found to be more sensitive to training label reshuffling and model parameter randomization than other saliency based explanations [1]; and they provide accurate explanations in medical image classifications [10].

We then compute product of the input instances and the Grad-CAM explanation in order to propagate input image information towards computing the loss and to avoid a bias that may be caused by only using a model's GradCAM explanation,

$$C_{good} := x_i \odot C_{good_i} \tag{4}$$

$$C_{bad} := x_j \odot C_{bad_j} \tag{5}$$

We then take inspiration from triplet loss [3] to incorporate C_{good} and C_{bad} into our explanation loss, L_{expl}. The main purpose of L_{expl} is to penalize a trainer according to similarity of model explanations of instance x to C_{good} and its difference from C_{bad}. We use Euclidean distance as a loss to compute the measure of dissimilarity, d (loss decreases as similarity to C_{good} is high and to C_{bad} is low).

$$d_{xg} := d(x \odot GradCAM(x), C_{good}) \tag{6}$$

$$d_{xb} := d(x \odot GradCAM(x), C_{bad}) \tag{7}$$

We train the model f to achieve $d_{xg} \ll d_{xb}$ for all x. We do this by adding a $margin = 1.0$ and translating it to: $d_{xg} < d_{xb} + margin$. We then compute the explanation loss as:

$$L_{expl} = \sum_i^N \max(d_{x_ig} - d_{x_ib} + margin, 0) \tag{8}$$

In addition to correctly classifying X, which is achieved through L_{CE}, this L_{expl} (Eq. 8) trains f to output GradCAM values that resemble the good explanations and that differ from the bad explanations, thereby refining the model to focus on the relevant regions and to ignore confounding regions. L_{expl} is zero, for a given sample x, unless $x \odot GradCAM(x)$ is much more similar to C_{bad} than it is to C_{good}—meaning $d_{xg} > d_{xb} + margin$.

4 Experiments

4.1 Data Collection and Preparation

To demonstrate eXBL we use the Covid-19 CXR dataset [4,13] described in Sect. 1. For model training we subsample 800 x-ray images per category to mitigate class imbalance, totaling 3,200 images. For validation and testing, we use 1,200 and 800 images respectively. We resize all images to 224×224 pixels. The dataset is also accompanied with feature annotation masks that show the lungs in each of the x-ray images collected from radiologists [13].

4.2 Model Training

We followed a transfer learning approach using a pre-trained MobileNetV2 model [15]. We chose to use MobileNetV2 because it achieved better performance at the CXR images classification task at a reduced computational cost after comparison among pre-trained models. In order for the training process to affect the GradCAM explanation outputs, we only freeze and reuse the first 50 layers of MobileNetV2 and retrain the rest of the convolutional layers with a custom classifier layer that we added (256 nodes with a ReLu activation with a 50% dropout followed by a Softmax layer with 4 nodes).

We first trained the MobileNetV2 to categorize the training set into the four classes using categorical cross entropy loss. It was trained for 60 epochs[1]. We refer to this model as the Unrefined model. We then use the Unrefined model to select the good and bad explanations displayed in Fig. 2. Next, we employ our eXBL algorithm using the good and bad explanations to teach the Unrefined model to focus on relevant image regions by tuning its explanations to look like the good explanations and to differ from the bad explanations as much as possible. We use Euclidean distance to compute dissimilarity in adopting a version of the triplet loss for XBL. We refer to this model as the eXBL$_{EUC}$ model and it was trained for 100 epochs using the same early stopping, learning rate, and optimizer as the Unrefined model.

For model evaluation, in addition to classification performance, we compute an objective explanation evaluation using Activation Precision [2] that measures how many of the pixels predicted as relevant by a model are actually relevant using existing feature annotation of the lungs in the employed dataset,

$$AP = \frac{1}{N} \sum_{n}^{N} \frac{\sum (T_\tau(GradCAM_\theta(x_n)) \odot A_{x_n})}{\sum (T_\tau(GradCAM_\theta(x_n)))} , \qquad (9)$$

---

[1] The model was trained with an early stop monitoring the validation loss at a patience of five epochs and a decaying learning rate $= 1e-04$ using an Adam optimizer.

where x_n is a test instance, A_{x_n} is feature annotation of lungs in the dataset, $GradCAM_\theta(x_n)$ holds the GradCAM explanation of x_n generated from a trained model, and T_τ is a threshold function that finds the $(100\text{-}\tau)$ percentile value and sets elements of the explanation, $GradCAM_\theta(x_n)$, below this value to zero and the remaining elements to one. In our experiments, we use $\tau = 5\%$.

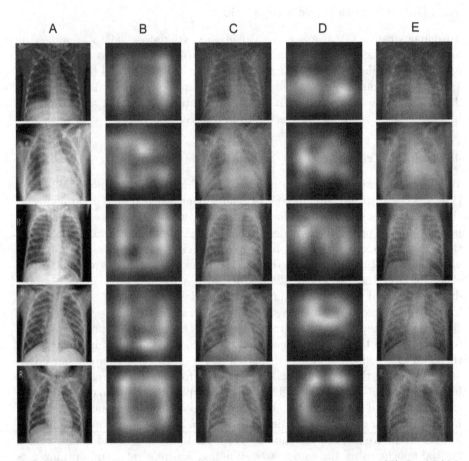

Fig. 3. Sample outputs of Viral Pneumonia category. (A) Input images; (B) GradCAM outputs for Unrefined model and (C) their overlay over input images; (D) GradCAM outputs for eXBL$_{EUC}$ and (E) their overlay over input images.

Table 1. Classification and explanation performance.

Models	Accuracy		Activation Precision	
	Validation	Test	Validation	Test
Unrefined	0.94	0.95	0.32	0.32
eXBL$_{EUC}$	0.89	0.90	0.34	0.35

5 Results

Table 1 shows classification and explanation performance of the Unrefined and eXBL$_{EUC}$ models. Sample test images, GradCAM outputs, and overlaid Grad-CAM visualizations of x-ray images with Viral pneumonia category are displayed in Fig. 3. From the sample GradCAM outputs and Table 1, we observe that the eXBL$_{EUC}$ model was able to produce more accurate explanations that avoid focusing on irrelevant image regions such as the Epiphyses and background regions. This is demonstrated by how GradCAM explanations of the eXBL$_{EUC}$ model tend to focus on the central image regions of the input images focusing around the chest that is relevant for the classification task, while the GradCAM explanations generated using the Unrefined model give too much attention to areas around the shoulder joint (humerus head) and appear angular shaped giving attention to areas that are not related with the disease categories.

6 Conclusion

In this work, we have presented an approach to debug a spurious correlation learned by a model and to remove it with just two exemplary explanations in eXBL$_{EUC}$. We present a way to adopt the triplet loss for unlearning spurious correlations. Our approach can tune a model's attention to focus on relevant image regions, thereby improving the saliency-based model explanations. We believe it could be easily adopted to other medical or non-medical datasets because it only needs two non-demanding exemplary explanations as user feedback.

Even though the eXBL$_{EUC}$ model achieved improved explanation performances when compared to the Unrefined model, we observed that there is a classification performance loss when retraining the Unrefined model with eXBL to produce good explanations. This could mean that the initial model was exploiting the confounding regions for better classification performance. It could also mean that our selection of good and bad explanations may not have been optimal and that the two exemplary explanations may be degrading model performance.

Since our main aim in this study was to demonstrate effectiveness of eXBL$_{EUC}$ based on just two ranked feedback, the generated explanations were evaluated using masks of lung because it is the only body part with pixel-level annotation in the employed dataset. However, in addition to the lung, the disease categories might be associated with other areas of the body such as the throat and torso. For this reason, and to ensure transparency in practical deployment of such systems in clinical practice, future work should involve expert end users for evaluation of the classification and model explanations.

Acknowledgements. This publication has emanated from research conducted with the financial support of Science Foundation Ireland under Grant number 18/CRT/6183. For the purpose of Open Access, the author has applied a CC BY public copyright licence to any Author Accepted Manuscript version arising from this submission.

References

1. Adebayo, J., Gilmer, J., Muelly, M., Goodfellow, I., Hardt, M., Kim, B.: Sanity checks for saliency maps. arXiv preprint arXiv:1810.03292 (2018)
2. Barnett, A.J., et al.: A case-based interpretable deep learning model for classification of mass lesions in digital mammography. Nat. Mach. Intell. **3**(12), 1061–1070 (2021)
3. Chechik, G., Sharma, V., Shalit, U., Bengio, S.: Large scale online learning of image similarity through ranking. J. Mach. Learn. Res. **11**(3) (2010)
4. Chowdhury, M.E., et al.: Can AI help in screening viral and COVID-19 pneumonia? IEEE Access **8**, 132665–132676 (2020)
5. Dietvorst, B.J., Simmons, J.P., Massey, C.: Overcoming algorithm aversion: people will use imperfect algorithms if they can (even slightly) modify them. Manage. Sci. **64**(3), 1155–1170 (2018)
6. Hagos, M.T., Curran, K.M., Mac Namee, B.: Identifying spurious correlations and correcting them with an explanation-based learning. arXiv preprint arXiv:2211.08285 (2022)
7. Hagos, M.T., Curran, K.M., Mac Namee, B.: Impact of feedback type on explanatory interactive learning. In: Ceci, M., Flesca, S., Masciari, E., Manco, G., Ras, Z.W. (eds.) ISMIS 2022. LNCS, vol. 13515, pp. 127–137. Springer, Cham (2022). https://doi.org/10.1007/978-3-031-16564-1_13
8. Islam, M.M., Karray, F., Alhajj, R., Zeng, J.: A review on deep learning techniques for the diagnosis of novel coronavirus (COVID-19). IEEE Access **9**, 30551–30572 (2021)
9. Kermany, D.S., et al.: Identifying medical diagnoses and treatable diseases by image-based deep learning. Cell **172**(5), 1122–1131 (2018)
10. Marmolejo-Saucedo, J.A., Kose, U.: Numerical grad-cam based explainable convolutional neural network for brain tumor diagnosis. Mob. Netw. Appl., 1–10 (2022)
11. O'Neill, J., Delany, S.J., Mac Namee, B.: Rating by ranking: an improved scale for judgement-based labels. In: IntRS@ RecSys, pp. 24–29 (2017)
12. Pfeuffer, N., et al.: Explanatory interactive machine learning. Bus. Inf. Syst. Eng., 1–25 (2023)
13. Rahman, T., et al.: Exploring the effect of image enhancement techniques on COVID-19 detection using chest X-ray images. Comput. Biol. Med. **132**, 104319 (2021)
14. Ross, A.S., Hughes, M.C., Doshi-Velez, F.: Right for the right reasons: Training differentiable models by constraining their explanations. arXiv preprint arXiv:1703.03717 (2017)
15. Sandler, M., Howard, A., Zhu, M., Zhmoginov, A., Chen, L.C.: Mobilenetv 2: Inverted residuals and linear bottlenecks. In: Proceedings of the IEEE Conference on Computer Vision and Pattern Recognition, pp. 4510–4520 (2018)

16. Santa Cruz, B.G., Bossa, M.N., Sölter, J., Husch, A.D.: Public COVID-19 X-ray datasets and their impact on model bias-a systematic review of a significant problem. Med. Image Anal. **74**, 102225 (2021)
17. Schramowski, P., et al.: Making deep neural networks right for the right scientific reasons by interacting with their explanations. Nat. Mach. Intell. **2**(8), 476–486 (2020)
18. Selvaraju, R.R., Cogswell, M., Das, A., Vedantam, R., Parikh, D., Batra, D.: Grad-cam: visual explanations from deep networks via gradient-based localization. In: Proceedings of the IEEE International Conference on Computer Vision, pp. 618–626 (2017)
19. Shao, X., Skryagin, A., Stammer, W., Schramowski, P., Kersting, K.: Right for better reasons: training differentiable models by constraining their influence functions. In: Proceedings of the AAAI Conference on Artificial Intelligence, vol. 35, pp. 9533–9540 (2021)
20. Yousefzadeh, M., et al.: AI-corona: radiologist-assistant deep learning framework for COVID-19 diagnosis in chest CT scans. PLoS ONE **16**(5), e0250952 (2021)
21. Zech, J.R., Badgeley, M.A., Liu, M., Costa, A.B., Titano, J.J., Oermann, E.K.: Variable generalization performance of a deep learning model to detect pneumonia in chest radiographs: a cross-sectional study. PLoS Med. **15**(11), e1002683 (2018)
22. Zlateski, A., Jaroensri, R., Sharma, P., Durand, F.: On the importance of label quality for semantic segmentation. In: Proceedings of the IEEE Conference on Computer Vision and Pattern Recognition, pp. 1479–1487 (2018)

Control and Spatio-Temporal Modeling

Explaining the Chronological Attribution of Greek Papyri Images

John Pavlopoulos[1,2]([⊠]) [iD], Maria Konstantinidou[3], Georgios Vardakas[4][iD],
Isabelle Marthot-Santaniello[5], Elpida Perdiki[3], Dimitris Koutsianos[1],
Aristidis Likas[4][iD], and Holger Essler[6]

[1] Athens University of Economics and Business, Athens, Greece
[2] Stockholm University, Stockholm, Sweden
ioannis@dsv.su.se
[3] Democritus University of Thrace, Komotini, Greece
{mkonst,eperdiki}@helit.duth.gr
[4] University of Ioannina, Ioannina, Greece
g.vardakas@uoi.gr, arly@cs.uoi.gr
[5] University of Basel, Basel, Switzerland
i.marthot-santaniello@unibas.ch
[6] Ca'Foscari University of Venice, Venice, Italy
holger.essler@unive.it

Abstract. Greek literary papyri, which are unique witnesses of antique literature, do not usually bear a date. They are thus currently dated based on palaeographical methods, with broad approximations which often span more than a century. We created a dataset of 242 images of papyri written in "bookhand" scripts whose date can be securely assigned, and we used it to train machine and deep learning algorithms for the task of dating, showing its challenging nature. To address the data scarcity problem, we extended our dataset by segmenting each image to the respective text lines. By using the line-based version of our dataset, we trained a Convolutional Neural Network, equipped with a fragmentation-based augmentation strategy, and we achieved a mean absolute error of 54 years. The results improve further when the task is cast as a multiclass classification problem, predicting the century. Using our network, we computed and provided precise date estimations for papyri whose date is disputed or vaguely defined and we undertake an explainability-based analysis to facilitate future attribution.

Keywords: Chronology Attribution · Computer Vision · Greek Papyri

1 Introduction

No autographs of classical Greek authors survive today. Our knowledge of such works (along with post-classical literature and the first Christian works including the New Testament) relies on manuscripts postdating the original compositions. Of these, the most chronologically proximal are a few thousand papyri excavated

© The Author(s), under exclusive license to Springer Nature Switzerland AG 2023
A. Bifet et al. (Eds.): DS 2023, LNAI 14276, pp. 401–415, 2023.
https://doi.org/10.1007/978-3-031-45275-8_27

mainly in Egypt in the last two centuries. Due to physical damage, these papyri usually preserve only small portions of the texts in question unlike medieval manuscripts which tend to transmit them in full-length, but both papyri and manuscripts represent copies of copies of the original works.

1.1 Background

Despite their fragmentary nature, papyri are crucial witnesses for innumerable texts, not to mention that they occasionally preserve literary works that would be otherwise lost. They are also invaluable evidence for our understanding of book culture in Antiquity, as well as for philology, the evolution of writing scripts and book production. One of the most important aspects of such research is to determine the date of the papyri involved.

Unlike their documentary counterparts (i.e. papyri preserving official and everyday documents), literary papyri bear no date before the introduction of colophons in the Middle Ages (9th century CE). We customarily employ palaeographical methods to assign an approximate and broad (often spanning more than a century) date for their production. Apart from their content, the two categories, documentary and literary papyri, are also usually written in distinctly different scripts: unformal cursive writings for the former opposed to elegant bookhand for the latter. There are some exceptions on both sides, i.e. literary texts written in cursive and documentary texts with surprisingly elegant scripts. To this day, we lack an exhaustive list of the first category (literary texts in cursive script) which does not allow us to use the numerous dated documents to date these literary papyri by script comparison. However, a few specimen of the second group (documentary texts written in bookhand) have been collected in the CDDGB (see below). Palaeographers rely on the evidence-backed assumption that handwriting styles are typical of certain periods and change over time, much like fashions and trends in anything else. The subjectivity and authoritativeness of these methods are increasingly acknowledged among scholars [3,12,15,16] and further assistance for more reliable and/or accurate ones is highly desirable.

In traditional dating, papyrologists employ comparative dating. They use the—admittedly very few—objectively dateable papyri specimens to draw comparisons with non-dated ones and estimate the latter's place on a notional timeline. The comparison is performed on the basis of the form and features of single letters, or the script overall, also used for other palaeographical tasks such as identifying scribes, or classifying styles and types of scripts. The characteristics used for such studies may focus on size (small/large, short/long), shape (round/angular), specific parts of letters (arches/loops/serifs/decorations), speed of writing, *ductus* (the number, directions and sequence of strokes required to draw a letter), formality etc. Although the same features are regularly invoked by many palaeographers, each researcher is free to focus (and they often do so) on every conceivable aspect of the writing. Hence, there is no formally established methodology, set of features to be taken into account, or even terminology that managed to reach consensus. [22,23] Even for the commonly used and agreed upon features, it is rarely possible for scholars to measure them or objectively calculate their significance towards a conclusion. Research in digital palaeography quantifying script features such as angle and direction of writing (for instance [1,2]) usually provides one such feature as the base for performing computationally palaeographical tasks. In our study, we aim at performing such a task (in this case dating) without any input in the form of human-perceived features. Instead, we attempt to identify any clues or features that lead our models to a specific date for a papyrus image.

The computer can pinpoint areas of the images which push predictions towards either extreme and/or alter these images (and predict the corresponding date) in a controlled manner. Nevertheless, it cannot provide explanations in real-life terms, nor identify features perceivable by humans. At the same time, human experts instinctively date scripts in terms of certain characteristics, however subjective, but are unable to measure each such feature's significance towards assigning a date. In this preliminary examination, our aim is to detect patterns (not necessarily semantically clear at this stage) in the application of saliency maps.

1.2 The Contributions of This Work

C1. We developed two datasets of images of Greek papyri from Egypt, along with the dates assigned to them by experts: one with whole papyri fragments; the other with lines of writing extracted from the full-size images.

C2. We proposed a Convolutional Neural Network (CNN), which we call fCNN, that is based on a fragmentation-based augmentation strategy and which predicts the date of text-line images with a mean absolute error of 54 years, using a regression head, and a macro-average F1 of 61.5%, using a classification head, setting the state of the art for Greek papyri image dating.

C3. We used fCNN to precise the dating of the lines of eleven papyri, whose previous dates based on objective criteria are ranging across two centuries, and we share our predictions: https://github.com/ipavlopoulos/palit

2 Related Work

Although researchers have suggested algorithms for the automated segmentation of papyri images to text-lines [19], and although the benefits of text-line segmentation are already known in the field of writer identification [4], no published work to date has investigated dating computationally Greek literary papyri by focusing on text line images. The baseline is set by a CNN that is fed with whole Greek literary papyri images, which achieved a mean absolute error of more than a century [18]. Our study shows that data segmentation to text lines leads to a much smaller error, with augmentation-enhanced CNNs providing the best-performing solution. In the absence of other related work for Greek papyri image dating, we summarise, next, the published work regarding dating in general [17].

Image-Based Regression

CNNs outperform approaches based on feature engineering on writer identification [6,14] and similar findings are reported in dating. In [7], the authors used pre-trained CNNs to date images of medieval Dutch charters from 14CE to 16CE, by focusing on image crops. The authors reported a mean absolute error of 10 years, a number beyond our reach with papyrus data where an approximation of 50 years is accepted. Regression using pre-trained CNNs on random crops was also suggested in [25], for the dating of medieval Swedish charters. Besides feature extraction with deep learning, earlier work approached the task with regression on top of extracted features, such as scale-invariant [8] or hinge and fraglets [9].

Dating From Other Modalities

Besides images, other modalities have also been used as input. In [11], for example, textual features were used to infer the date. Although reasonable in general, this is not a feasible approach for Greek literary papyri and manuscripts, the text of which may be of much older authors, such as Homer. A different approach was suggested in [20], where ordinal classification was combined with multispectral imaging, tracking spectral responses of iron-gall ink (of historical letters, 17-20CE) at different wavelengths. Although rich, this data representation is very expensive in time and resources to establish, which also explains why datasets in this form are very rare. Besides, papyri are mostly written with carbon-based and not iron-gall ink, which is to the present more difficult to date.

3   Data

3.1   The Nature of the Papyri

As already mentioned, papyri bearing literary texts do not carry a date and for the vast majority of them papyrologists assign a date based on the affinity of their script with objectively (not palaeographically) dated specimens. These specimens, referred to as 'objectively dated' ones, are dated using external indications (not contained in the literary text on the papyrus) [24]. Occasionally, it is archaeological evidence or even radiocarbon dating suggesting a more secure date, but most importantly, papyri were often re-used after they exceeded their lifespan and literary texts are often found on papyri that have dated documents on the opposite side.

3.2   Digitised Papyri

The images included in our dataset come from a number of collections and online resources, whereas five or six of them were scanned from images in printed volumes. Their digitisation took place during a period of more than two decades, under substantially different imaging protocols. As a result, they vary greatly in their properties, most importantly in scaling to actual size, colour capturing, resolution and bit depth. For a few of them it was not possible to extract text lines, due to very low resolution, and they returned empty files during the segmentation stage.

3.3 Our New Dataset

Our dataset comprises images of Greek papyri from Egypt and their respective dates, from the 1st to the 4th CE.[1] Images of papyri from other centuries were few, hence we did not consider them in this study. The papyri included were selected from CDDGB, the only available collection of (somewhat) securely dated literary papyri available, which includes also a few documentary texts in bookhand. The data it contains can be dated based on various objective dating criteria, such as the presence of a document that contains a date on the reverse side, internal evidence in the text (mostly for the few documentary ones and the 9th c. manuscripts having colophons), radiocarbon dating, or a dateable archaeological context associated with the manuscripts. In the CDDGB database, most records contain sampled images and we had to manually trace full-sized ones from the respective collections. We release our dataset in two forms, one where images contain whole fragments and one where they contain text lines.

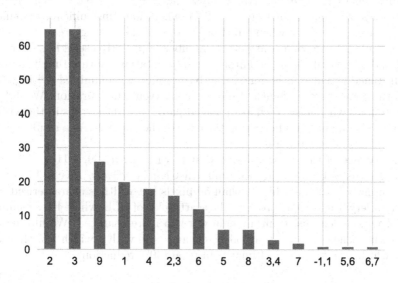

Fig. 1. The number of PLF images (vertically) per century (horizontally) or century range (when the date ranges between centuries), sorted by frequency.

---

[1] There is a very small number of exceptions which reflect the complexity of our documentation: one text is in Coptic, a few don't come from Egypt but the Near-East and another few are written on parchment, not papyrus. In this study, we collectively call them 'papyri'.

The Papyri Literary Fragments (PLF) dataset consists of 242 images of publicly available papyri fragments, from 1BCE to 9CE. As shown in Fig. 1, most fragments come from the 2nd or the 3rd CE, followed by the 9th and the 1st CE. When multiple fragments of the same manuscript were available, we included all of them. The date provided for most fragments is not specific. Typically, the minimum date range assigned to a literary papyrus spans 50 years, but it may reach up to two centuries. Most often, the latter cases concern a date between the two most frequent centuries (noted in Fig. 1 as '2,3' CE). Our study focused on the four first centuries, from 1 to 4 CE, comprising 168 images of literary papyri. Nine images were empty, which led us to 159 images in total. The final distribution across the four centuries (1-4CE) was 20, 61, 60, and 18 respectively. We converted our images to grayscale to reduce the dimensionality and to facilitate machine learning experiments.

The Papyri Literary Lines (PLL) dataset extends PLF so that images of the text lines of the fragments are provided instead of images of the whole fragments. The 159 images were segmented automatically using the Transkribus HTR platform,[2] yielding 4,655 line images. For this segmentation step, we used the default settings in Transkribus and did not train a specific baseline model, due to the multiformity of our material. We interfered minimally, by manually correcting text regions where none or very few lines were captured in the automatically generated segmentation. We also manually corrected a small number of base lines and line regions (appr. 1–2%), when no or insignificant amount of writing was captured, or when substantial and useful writing areas were obviously excluded. Even so, a considerable number of possibly useful lines were not added and in several cases the automatic segmentation captured multiple lines in an instance, or substantial amount of background with minimal writing. As a result, the dataset would benefit from more interventional curation. We did not eliminate lines with noise, such as damaged papyrus surface, gaps in the writing material (holes), and lines bordering the edge of the papyrus. As a result, several line images still contain noise.

The balance of the dates followed that of PLF, with 439, 2,116, 1,797, and 303 images from the 1st, 2nd, 3rd, and 4th CE respectively. As can be seen in Fig. 2(b), most images are higher than 50 pixels but width is characterised by a greater variety. Figure 2(a) presents the scatterplot of PLL, where lines comprise texts of various lengths, from a single word to more than ten. We filtered out images with a height lower than 50 pixels and ones with a width less than 300 pixels, which resulted in 2,774 images in total (40% reduction).

4   Method

Our method, called fCNN, is a 43m-parameter CNN that exploits augmentation so that it is robust to fragmented input, often met in papyri.

---

[2] https://readcoop.eu/transkribus/.

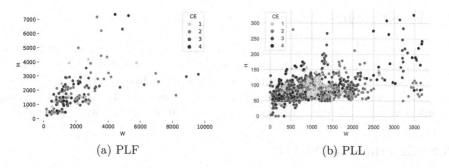

(a) PLF   (b) PLL

Fig. 2. Scatter plot of the width (shown horizontally) and height of the images.

4.1 fCNN

The network consists of two Conv2D layers to represent the image of each text line, of 32 and 64 channels respectively, followed by a 3-layer feed-forward neural network (FFNN) with a single output neuron to yield the date. We used a convolutional kernel of size 5, single stride, zero padding, and max-pooling (2×2). The FFNN receives a flat representation from the Conv2D which is reduced to 1024 and then to 512 neurons before the date is estimated. A ReLU activation function is used per layer.

Synthetic Fragmentation is a possible augmentation channel during training. Papyri are very often fragmented, leading to partial information in the image to be dated. We exploited this pattern as part of our augmentation strategy, by erasing randomly (0.5 probability) image fragments, setting their pixel values to 0.5. Images were transformed with Gaussian blur (kernel size of 3) and random affine (up to $3°$). The actual letter size as well as the image ratio to actual size in our dataset greatly varies, hence, to assist the network's robustness, we also randomly cropped and resized each image by keeping the 1:6 aspect ratio.

4.2 The Baseline

We used the state-of-the-art in regression, which is achieved by ensembles [5], including Extremely Randomized Trees (XTR) and XGBoost (XGB). We experimented with both these regressors, using patches of 50*300 windows cropped from the center of each image, which was also represented with PCA-extracted 500-dimensional features. In our preliminary experiments, PCA led to better results compared to image binarisation using Canny edge detection and Otsu, which have been reported beneficial in writer identification [13]. We used the implementation provided by SKLEARN setting all hyper-parameters to default values, besides the objective of XGB, which was set to the squared error.

5 Experiments

We approached dating, which is a regression task, with fCNN, a CNN that uses a fragmentation-based augmentation strategy. We experimented with algorithms on the PLL dataset, using as input the images of the lines of the papyri and as output the date of the respective papyri. We also show results when we cast the problem as a classification task, predicting the century as one out of four labels.

5.1 Experimental Details

We used Adam optimisation [10] with a learning rate of $1e-3$, batch size of 16, 200 epochs, early stopping with patience of 20 epochs. The regression variant was trained with a mean squared error loss and the classification variant with a cross entropy loss. We used PyTorch and we release our code in our repository.[3]

The benchmark

A majority baseline (BLM), which always predicts the 2nd CE, achieved an MAE of approx. 0.632 and an MSE of 0.772. XTR and XGB perform better than this weak baseline, with a considerable difference when looking at MSE. The latter penalizes greater distances more, which means that papyri of the 1st and 4th CE were better handled by XGB and XTR. Our fCNNr performs considerably better than all the baselines, achieving an average absolute error of 54 years.

Table 1. Mean absolute and squared error of dating along with their standard error of the mean in parenthesis.

	MAE ↓	MSE ↓
BLM	0.632 (0.032)	0.772 (0.050)
XGB	0.612 (0.005)	0.558 (0.012)
XTR	0.610 (0.006)	0.544 (0.012)
fCNNr	**0.540 (0.001)**	**0.511 (0.009)**

From Regression to Classification

By rounding the predictions of our fCNNr, we created a confusion matrix, which is shown in Fig. 3(a). Confusion regards mainly neighboring centuries. The model correctly detects images from the 2nd and 3rd while images from the 3rd may be predicted close to the 2nd, and vise versa. Difficulties in dating regard the two edges, because the 1st and 4th CE are more often predicted as of the 2nd and 3rd CE respectively.

---

[3] https://github.com/ipavlopoulos/palit.

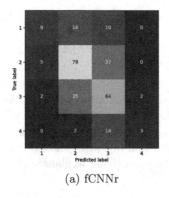
(a) fCNNr

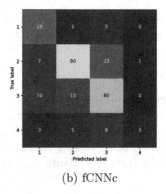
(b) fCNNc

Fig. 3. Confusion matrices of fCNNr (rounded predictions) and fCNNc.

Although our task in hand is a regression one in nature, we also trained and assessed a classification variant (fCNNc), which learns to disregard the order of centuries and simply treat them as labels. In Fig. 3(b), we observe that results improve across all centuries except from the 4th CE, where the difficulty remains approximately the same. Table 2) shows the F1 per century per fCNN variant, along with the benefit in absolute number when using the classification head instead of the regression one. We also trained an XGB and an XTR classifier, with the former performing better yet much worse than fCNNc.

Table 2. F1 per century of fCNNr (predictions are rounded) and fCNNc, the absolute difference between the two is shown in parenthesis.

	1CE	2CE	3CE	4CE
XGB	0.38	0.69	0.58	0.09
XTR	0.00	0.56	0.44	0.00
fCNNr	0.35	0.62	0.56	0.25
fCNNc	**0.69** (+0.34)	**0.78** (+0.16)	**0.73** (+0.17)	**0.26** (+0.01)

Despite the fact that both fCNN variants are trained on the same data, we note that we do not consider them as competitors. The regression-based fCNNr suggests a date, which can provide a very rough estimation of when the papyrus was written. If the predicted date was 280CE, then this is an indication that the papyrus is dated between the 3rd and the 4th CE, and that a year close to the latter is likelier. On the other hand, the classification-based fCNNc suggests a century and yields a score to indicate its confidence. If the predicted century was the 4th CE and the confidence was 80%, then this means that the network is confident that the date is 4th and no other. Although our task in hand is one of regression, both can generate useful explanations. Therefore, since our end goal is to assist and not supplement the expert, we used them both in our explainability study, discussed next.

Explainability

Saliency maps [21] reveal the parts of the image which are responsible for the network's prediction. We experimented with both variants, fCNNr and fCNNc, and we used both, gradient- and perturbation-based attribution. In this study, we opted for fCNNc using gradient-based attribution, but we observe that explanations by the two variants can be combined to yield richer explanations.

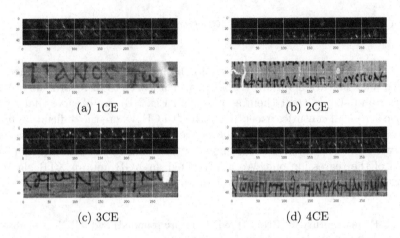

(a) 1CE   (b) 2CE

(c) 3CE   (d) 4CE

Fig. 4. Saliency maps for lines of papyri per century

We computed one heatmap per predicted line and we present a random sample of lines in Fig. 4. The heated colours show that the network consistently focuses on the letters in order to yield its predictions for the date. This means that the model is basing its prediction on the shape of specific letters, the distance between them, the size, or the intensity of the ink. By contrast, it seems invariant from background noise and other attributes which may be often present in Greek literary papyri. For example, gaps (holes in the papyrus) such as those in Figs. 4(a) and 4(c), do not get any attention by the model.

6   Assessing Data Sources Limitations

CDDGB is not a product of targeted research on securely dated papyri, but rather a compilation of such examples mentioned in other papyrological works.[4]

---

[4] More reliable compilations are promised by current projects, but are still work-in-progress for the time being.

Hence, the collection is not comprehensive and the data included is not meticulously assessed by the compilers. Shortcomings concern the accuracy of some dates. Still and all, it is the same data of objectively dated papyri that papyrologists use as reference for palaeographical dating. In this study, we introduce the computational factor in assessing scripts in connection with their assigned dates. Also, by focusing on the explainability of dating images of handwritten text, we do not consider these shortcomings detrimental. The possible inaccuracies in dating and the wide-range of the assigned dates does not affect the explanations, which aim to provide pointers on features of the script.

The imbalance in the size of the fragments and quantity of lines is an inherent issue owing to the nature of the available material. A papyrus may contain three or four usable lines, whereas others may have more than fifty. This does not affect dating significantly because, although test lines may come from a manuscript not hidden during training, each line constitutes a completely different image pattern. The same issue could be an advantage regarding explainability, because possible features are brought out in a more controlled manner when multiple lines of the same manuscripts are involved. While some features, especially palaeographically insignificant characteristics, remain consistent (such as colour/intensity of the ink, texture and colour of the background, general size of script, scale, etc.), explanations can focus on pivotal ones.

Our train and validation subsets are mutually exclusive at the line level but not necessarily at the fragment level. Although the former is straight forward, the latter is not due to the diversity of lines in the fragments. To experiment with the latter, we kept lines from papyri whose index modulo a value (13) was zero for validation and testing (in half), keeping the rest for training. Although introducing a distribution drift, presenting relatively fewer lines from 1CE during testing, this split met our restrictions. The error of fCNNr is slightly higher (0.612 ± 0.002), but remains the best. The F1 score remains approx. the same in classification, except from the 1st CE that drops to 0.4 but whose support in the test set is only 3 (out of 80) images. Future work will carefully compile more train, development, and test subsets, to investigate this issue further.

7   Error Analysis

To go further in our understanding of the relevance of our experiment, we provide in this section an error analysis, followed by an experiment on the way the model handles the damages on the papyri by ablating input images before dating.

Analysis. By studying fCNN's deviations from the ground truth, we observe that these concerned predictions toward the neighbouring century. Images from the two edge centuries, 1st and 4th CE, are scored up to the 2nd and 3rd CE respectively, the two most frequent centuries (Fig. 1). Images from the 2nd and 3rd CE, on the other hand, were scored not far from each other, most often to the 3rd and the 2nd respectively. By looking at the saliency maps of the misclassifications, we observed that letter-shaped noise, present in the source images, received the model's focus.

Ablation. Our error analysis revealed that fragments may deceive our model. In order to investigate the model's sensitivity, we fed fCNNc with test images, augmented with randomly-shaped black and white patches. We observe that the model's focus changes according to the colour of the patch. White boxes appear to be disregarded by our model, by contrast to black boxes, which are receiving attention. An example is shown in Fig. 5, where the same line from a papyrus of the 3rdCE is altered in two ways. In Fig. 5(b), the focus is everywhere except from the white patch. This is in line with our findings about the breaks, which are also depicted in white in the images (Fig. 4). By contrast, the black patch of Fig. 5(a) affects the prediction as if the model is guessing what character was missing and as if the black colour of the patch was ink.

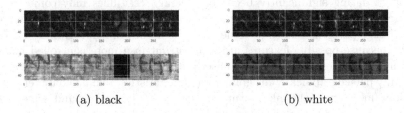

(a) black   (b) white

Fig. 5. Saliency maps of the same test line, from a papyrus of the 3rd CE, whose source image was transformed either with a black or a white patch before dating.

8   Dates in Doubt: A Computational Estimate

fCNN can accurately predict the date of a text line image (Table 1) and, when the task is simply to predict the century and not an exact date, a classification variant that ignores the temporal relation of the labels yields even better results (Table 2). As was shown from our study of saliency-based explanations, fCNN focuses on the letters, that is the foreground and not the background (e.g., the blank parts of the papyrus sheet, the fibres, the holes and damages). In order to provide the experts with suggestions that could possibly improve the current dating,[5] we apply this network to loosely dated texts (across two centuries).

In our primary source, 11 papyri are dated either to the 2nd or the 3rd CE. Using fCNNc, we found that 87% of the lines are classified to the 2nd or 3rd CE. Exceptions were from 16 which were classified to the 1st and 1 which was classified to the 4th. Figure 6 presents the analytical results. Using fCNNr, we attempted then to estimate a more precise chronology for the lines in these papyri. Despite the fact that our regressor was trained on ground truth at the century level, our expectation is that it will have learned to yield a chronology

---

[5] Datings usually come from one expert, the editor of the text. Sometimes another expert makes a case that the dating should be modified and the correction may be accepted or provided as alternative dating.

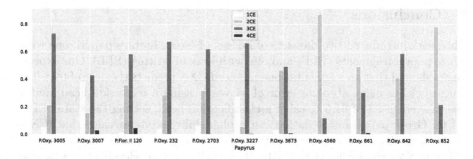

Fig. 6. Chronological attribution of fCNNc of lines in fragments dated between 2-3CE

that is closer to the objective date. Figure 7 presents our predictions, organised per papyrus. The predicted dates for the lines of P.Oxy 3005, which was classified by fCNNc on the 3rd, are diverse, with the majority falling on the late 2nd and early 3rd. Overall, our network's estimations agree with the range provided by the experts. The earliest prediction was 98CE, for a line in P.Oxy. 661. This papyrus comprises parts of a poem by Callimachus and is dated from 150 to 250 CE,[6] with the first editor arguing that it is the late 2nd CE.[7] On average, our predictions suggest 200CE, but some lines are predicted as early as 100CE while others as 250CE. The latest prediction is 270CE for a line in P.Flor. II 120,[8] dated from 250 to 261CE. In this papyrus, in very few lines our predictions agree with the experts, because on average our network dates it before the 200CE. In P. Oxy. 4560, only one line is used, and date is 100CE. In P. Oxy. 232, although lines are few, all our predictions date the papyrus between 100 and 150CE.

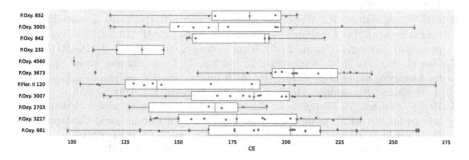

Fig. 7. Chronological attribution of fCNNr of lines in fragments dated between 2-3CE

---

[6] https://www.trismegistos.org/text/59375 (accessed: May 25, 2023).

[7] The Photographic Archive of Papyri in the Cairo Museum (accessed: May 25, 2023).

[8] https://papyri.info/ddbdp/p.flor;2;120 (accessed: May 25, 2023).

9   Conclusions

This work introduced two datasets of images of Greek literary papyri, one with whole papyri fragments (PLF) and one with lines of writing (PLL). Our experiments showed that an augmentation-enhanced CNN predicts the date of text-line images with a mean absolute error of 54 years, using a regression head, and a macro-average F1 of 61.5%, using a classification head, setting the state of the art for Greek papyri image dating. An explainability study revealed that fCNN clearly focuses on letters to predict the date, following the palaeographer's path. Using fCNN, we predicted the date of the text lines in eleven papyri, whose objective date is ranging across two centuries, and we discussed our findings.

References

1. Brink, A.A., Smit, J., Bulacu, M.L., Schomaker, L.R.B.: Writer identification using directional ink-trace width measurements. Pattern Recogn. **45**(1), 162–171 (2012). https://doi.org/10.1016/j.patcog.2011.07.005
2. Bulacu, M., Schomaker, L., Vuurpijl, L.: Writer identification using edge-based directional features. In: Proceedings of the Seventh International Conference on Document Analysis and Recognition 2003, pp. 937–941 (2003). https://doi.org/10.1109/ICDAR.2003.1227797
3. Choat, M.: Dating papyri: familiarity, instinct and guesswork. J. Study New Testament **42**(1), 58–83 (2019). https://doi.org/10.1177/0142064X19855580
4. Cilia, N.D., De Stefano, C., Fontanella, F., Marthot-Santaniello, I., Scotto di Freca, A.: PapyRow: a dataset of row images from ancient Greek papyri for writers identification. In: Del Bimbo, A., et al. (eds.) ICPR 2021. LNCS, vol. 12667, pp. 223–234. Springer, Cham (2021). https://doi.org/10.1007/978-3-030-68787-8_16
5. Fernández-Delgado, M., Sirsat, M.S., Cernadas, E., Alawadi, S., Barro, S., Febrero-Bande, M.: An extensive experimental survey of regression methods. Neural Netw. **111**, 11–34 (2019)
6. Fiel, S., Sablatnig, R.: Writer identification and retrieval using a convolutional neural network. In: Azzopardi, G., Petkov, N. (eds.) CAIP 2015. LNCS, vol. 9257, pp. 26–37. Springer, Cham (2015). https://doi.org/10.1007/978-3-319-23117-4_3
7. Hamid, A., Bibi, M., Moetesum, M., Siddiqi, I.: Deep learning based approach for historical manuscript dating. In: 2019 International Conference on Document Analysis and Recognition (ICDAR), pp. 967–972. IEEE (2019)
8. He, S., Samara, P., Burgers, J., Schomaker, L.: Discovering visual element evolutions for historical document dating. In: 2016 15th International Conference on Frontiers in Handwriting Recognition (ICFHR), pp. 7–12. IEEE (2016)
9. He, S., Sammara, P., Burgers, J., Schomaker, L.: Towards style-based dating of historical documents. In: 2014 14th International Conference on Frontiers in Handwriting Recognition, pp. 265–270. IEEE (2014)
10. Kingma, D.P., Ba, J.: Adam: A method for stochastic optimization. arXiv preprint arXiv:1412.6980 (2014)
11. Kumar, A., Baldridge, J., Lease, M., Ghosh, J.: Dating texts without explicit temporal cues. arXiv preprint arXiv:1211.2290 (2012)
12. Mazza, R.: Dating early Christian papyri: old and new methods -introduction. J. Study New Testament **42**(1), 46–57 (2019). https://doi.org/10.1177/0142064X19855579

13. Nasir, S., Siddiqi, I.: Learning features for writer identification from handwriting on papyri. In: Djeddi, C., Kessentini, Y., Siddiqi, I., Jmaiel, M. (eds.) MedPRAI 2020. CCIS, vol. 1322, pp. 229–241. Springer, Cham (2021). https://doi.org/10.1007/978-3-030-71804-6_17

14. Nguyen, H.T., Nguyen, C.T., Ino, T., Indurkhya, B., Nakagawa, M.: Text-independent writer identification using convolutional neural network. Pattern Recogn. Lett. **121**, 104–112 (2019)

15. Nongbri, B.: The limits of palaeographic dating of literary papyri: Some observations on the date and provenance of P.Bodmer II (P66). Mus. Helveticum **71**(1), 1–35 (2014)

16. Nongbri, B.: Palaeographic analysis of codices from the early Christian period: a point of method. J. Study New Testament **42**(1), 84–97 (2019). https://doi.org/10.1177/0142064X19855582

17. Omayio, E.O., Indu, S., Panda, J.: Historical manuscript dating: traditional and current trends. Multimed. Tools Appl. **81**(22), 31573–31602 (2022)

18. Paparrigopoulou, A., Kougia, V., Konstantinidou, M., Pavlopoulos, J.: Greek literary papyri dating benchmark (2023). https://doi.org/10.21203/rs.3.rs-2272076/v2

19. Papavassiliou, V., Stafylakis, T., Katsouros, V., Carayannis, G.: Handwritten document image segmentation into text lines and words. Pattern Recogn. **43**(1), 369–377 (2010)

20. Rahiche, A., Hedjam, R., Al-Maadeed, S., Cheriet, M.: Historical documents dating using multispectral imaging and ordinal classification. J. Cult. Herit. **45**, 71–80 (2020)

21. Simonyan, K., Vedaldi, A., Zisserman, A.: Deep inside convolutional networks: Visualising image classification models and saliency maps. arXiv preprint arXiv:1312.6034 (2013)

22. Stokes, P.A.: Computer-aided palaeography, present and future. In: Rehbein, M., Schaßan, T., Sahle, P. (eds.) Kodikologie und Paläographie im digitalen Zeitalter - Codicology and Palaeography in the Digital Age, vol. 2, pp. 309–338. BoD, Norderstedt (2009). https://kups.ub.uni-koeln.de/2978/

23. Stokes, P.A.: Scribal attribution across multiple scripts: a digitally aided approach. Speculum **92**(S1), S65–S85 (2017). https://doi.org/10.1086/693968

24. Turner, E.G.: Greek Manuscripts of the Ancient World. (P. J. Parsons, Ed.; Revised and Enlarged ed.). Institute of Classical Studies, London (1987)

25. Wahlberg, F., Wilkinson, T., Brun, A.: Historical manuscript production date estimation using deep convolutional neural networks. In: 2016 15th International Conference on Frontiers in Handwriting Recognition (ICFHR), pp. 205–210. IEEE (2016)

Leveraging the Spatiotemporal Analysis of *Meisho-e* Landscapes

Konstantina Liagkou[1], John Pavlopoulos[2,3]([✉]) [iD], and Ewa Machotka[3]

[1] Uni Systems, Kallithea, Greece
[2] Athens University of Economics and Business, Athens, Greece
[3] Stockholm University, Stockholm, Sweden
`ioannis@dsv.su.se, ewa.machotka@su.se`

Abstract. Japanese early-modern woodblock prints depicting pastoral views of the countryside, so-called *meisho-e* (images of famous places), are often defined today as landscapes (*fūkei*). However, the notion of *fūkei* is a modern cultural translation, which obscures specificities of Japanese visual culture, and intricacies of early modern spatiality or a socially produced space. To uncover these characteristics and provide a more nuanced understanding of *meisho-e* prints, we have engaged in a macroanalytical study of relationships between places depicted in prints and actual topography, aided by computational technologies rooted in Natural Language Processing (NLP). In our prior work, we experimented with automated harvesting of geospatial data from image-content-related inscriptions on two hundred prints. In this follow-up work, we undertake a large-scale automated mapping of *meisho* and we study the geographical distribution of sites featured in these prints. We explore two different computational paths, one using deep learning and one based on digital gazetteers, and reflect on the challenges and benefits of the applied computational approaches. We improve the former, which was the state-of-the-art, using pre-training, and we show that the latter is beneficial in terms of mapping. Finally, by using automatically extracted place-name entities, we undertake an analysis of prints over space and time. We release our code and the dataset for public use: https://github.com/Connalia/ai-jan-art.

Keywords: NLP · Spatiotemporal Analysis · Art History

1 Introduction

Among the most globally recognised artifacts kept and displayed in museums worldwide are Japanese landscape prints designed by some iconic artists such as Katsushika Hokusai (1760–1849) or Utagawa Hiroshige (1897–1858). These images depicting peaceful views of the countryside featuring mountains, rivers,

---

The first author wrote the initial draft and conducted the experiments, while all other authors contributed equally through supervision and co-authorship.

© The Author(s), under exclusive license to Springer Nature Switzerland AG 2023
A. Bifet et al. (Eds.): DS 2023, LNAI 14276, pp. 416–430, 2023.
https://doi.org/10.1007/978-3-031-45275-8_28

and rural dwellings, are commonly assumed to depict realities and topographies of pre-modern Japan, and as such defined *fūkei* or 'landscapes'. However, *fūkei* is a modern term developed in the process of cultural translation shaped by Western modern art epistemologies. In fact, these prints are more appropriately described as *meisho-e* or 'images of famous places'. Initially, these 'famous places' (*meisho*) were not depicting actual sites, which could be geolocated on the map of Japan, but poetic rhetorical figures rooted in classical poetry. These so-called *utamakura* (lit. poem pillows) tied seasonal images and symbolic motives, with particular places [9].

In the context described above, the actual topography of *meisho* places was not the guiding principle for their visual depictions [4]. In fact, *meisho-e* prints such as those designed by Hokusai or Hiroshige curate geographical reality in multiple ways. They maintain links to this reality mainly through printed inscriptions that feature a wide variety of place-names. These characteristics changed in time, especially in the first half of the ninetieth century, when many new toponyms entered the world of printed culture, and *meisho* strengthened their relationships with physical reality. Nonetheless, identification of the depicted places and their geolocation is far from being straightforward, which challenges interpretation of *meisho-e* prints as landscapes, and hinders understanding of their social function in general.

The issues, which were only briefly presented above, have not been comprehensively addressed to date. The research on *meisho-e* prints remains fragmented and often focuses on specific print series or individual designers rather than attempting to look at the genre and its epistemology at large. This is mainly due to the richness and diversity of the visual material that escapes traditional analytical methods based on close reading or interpretation of selected individual images. In this context, Natural Language Processing (NLP) can facilitate the discovery of new knowledge, through the analysis of large cultural datasets of digitised objects, and offer a possibility to rectify this situation.

1.1 Research Aims

This study provides a macro-analytical exploration of *meisho-e* prints through Named Entity Recognition (NER). The contributions of this work are as follows:

- We benchmark NER on inscriptions of *meisho-e* prints, comparing the state-of-the art [12], a fine-tuned Japanese BERT [6], with simpler gazetteer-based approaches. We show (i) that the former is better, and (ii) that further pre-training of BERT leads to better results, setting the **new state-of-the-art**.
- Despite its superior performance overall, we show that BERT-based approaches still fall short compared to the much simpler gazetteers, when geolocation of places is the objective.
- We extract place-name entities for approx. 20k *meisho-e* prints, which we use to perform **a large-scale spatiotemporal analysis** of place-names distribution across time, aiming to discover which *meisho* are the most popular per time period, and how these preferences were distributed in space; i.e., which areas were considered culturally significant and at which times.

As spatiotemporal analyses are not limited to the mapping of depicted places, we also conduct the analysis of formal aspects of prints aiming at developing an understanding of how space is represented in prints. More specifically, we take a pivotal step to study how the colour schemes in prints changed in time and in relation to the depiction of different types of places. This new experiment opens up new analytical venues that we plan to explore in the future.

2   Related Work

Recent years saw an advancement in the field of Spatial Humanities and Spatial Art History building on Geographical Information Systems (GIS), Natural Language Processing (NLP) and Corpus Linguistics [17]. However, although these analytical paths bring good results in the study of contemporary datasets it is not always the same for historical materials. Also, spatial analysis of Japanese pre-modern materials such as *meisho-e* prints, remains especially challenging, among others due to their formal characteristics (e.g., different perspective principles), difficulties with place identification (e.g, ambiguity of the depicted visual motives, problems with accurate transcription and lack of textual metadata) etc. Important contributions have been made recently e.g. the digitization of Japanese prints collections and the development of print databases worldwide [1], and among others, the initiation of computational analysis of Japanese prints targeting the questions of style [16], attribution [8] and content e.g. images featuring figures of kabuki actors [19] etc. or even geolocating of the selected print series depicting Edo city [15]. However, these important efforts study relatively small datasets, focus on testing technical solutions rather than large theoretical questions and do not target 'landscape' images at a scale and depth that would enable the development of an entirely new epistemology of this art genre.

Large-scale spatial analysis and the mapping of *meisho-e* prints with geographical information systems remains challenging as place identification is facilitated by the image-content related inscriptions printed in the images that often feature place names. But due to the high complexity of the task, reading early-modern inscriptions has so far been conducted by experts in Japanese pre-modern art history and literature. The task has been challenging for both humans and machines and it is estimated that only 1% of pre-modern textual sources have been transcribed [13].

Computational tools are expected to improve this situation. Automated text recognition technologies, e.g., Optical Character Recognition (OCR) and Handwritten Text Recognition (HTR), can yield the text in an image in a machine readable form. However, conducting OCR and HTR analysis on pre-modern Japanese texts is a challenging task due to the intricacies of Japanese writing systems. There are several reasons for this. First, this is due to a large number of characters used in pre-modern texts (ca. 4,500 characters are used [11]), of which many appear only once or twice in a given dataset. Second, the texts use so-called *hentaigana* or 'kanji variations' in which kanji characters could be used alternately depending on their phonetic value, and in multiple ways or forms. As a result, the same word could be written in different characters [18] and in different forms, or they could be written in a phonetic alphabet. Third, methodological challenges also include a lack of training data and applicability of software developed for the analysis of Western materials and texts in the study of non-Western materials. Fourth, the layout of texts is not always sequential and is characterised by a great variety of spatial distribution across the page and integration with the illustrations (both in single prints and printed books).

Therefore, the application of OCR software for automated transcription of text has been developing slowly in the case of East Asian languages using logo-graphic writing systems [3]. Only recently, the field noted considerable progress in the development of OCR instruments. The new tools for automated reading of early-modern texts such as KuLA (Kuzushiji Learning Application) [7], KuroNet and Miwo [5,11] and newly established databases, among others, the National Institute for Japanese Literature, the National Institute for Japanese Language and Linguistics, the National Diet Library Digital Collections, the Waseda University's database, the Ritsumeikan University's ARC Portal Databases as well as datasets created, among others, by the National Institute of Informatics and the Center for Open Data in the Humanities (CODH) provided an excellent incentive and facilitated progress in the field of automated data harvesting from Japanese pre-modern textual sources. However, it is not feasible to develop one computational transcription tool for all pre-modern manuscripts, printed texts in books and visual images.

These and other challenges are also relevant to the automated reading of place-names inscriptions on visual images, even the prints produced at the same period. Therefore, transcription of inscriptions is one of the main obstacles not only for historians and literary scholars but also for art historians interested in a large-scale spatiotemporal analysis of *meisho-e* prints. Therefore, to expand the existing analysis of prints, and achieve our research goals we engage the so-called 'distant viewing' approach [2] and explore technological solutions facilitating a large-scale automated mapping of *meisho* or famous places depicted in prints produced between ca. 1750 and 1850, and we study the geographical distribution of sites featured in these prints. We explore two different computational paths, one using deep learning and one based on digital gazetteers, and reflect on the challenges and benefits of the applied computational approaches.

3 The Corpus of Digitised *Meisho-e* Prints

The access to the data for this work was facilitated by the database hosted at the Art Research Centre at Ritsumeikan University, Kyoto. The Centre's digital databases of Japanese printed culture host approx. 700,000 (678,429) digitised objects kept at 28 institutions in Japan and abroad. We use the 200 train- and test-annotated data with their place-names from [12]. Our study used 22,959 inscriptions that related to keyword *meisho* (famous place). These digitised prints depict mainly natural environments. Only 10,421 of them have meta-data specifying production dates important for our study. Statistical overview of textual information in the 200 inscriptions studied previously [12] as compared to our new set of 22,959 is presented in Table 1.

Table 1. Statistics of the inscription texts in our study and in [12], including the number of texts, the average, the min., the 1st and 3rd quartile, the max. length in characters.

	Num.	Avg	St.dev.	Min	25%	75%	Max
Ours	22,959	11	7.9	2	9	13	444
[12]	200	20	8.5	5	14	23	59

4 Methods

In this study, we used NER to conduct a large-scale automated mapping of *meisho* or famous places depicted in prints produced between ca. 1697 and 1978. BERT-based NER has been found to be accurate for the task in hand [12]. On the other hand, gazetteer-based approaches hold valid latitude and longitude coordinates while not requiring any labelled data, as is the case of deep learning.

BERT-FT. In [12], the authors used a Japanese pretrained BERT.[1] The model was pre-trained on Japanese Wikipedia articles and was fine-tuned for place-name entity recognition. The authors showed that merging location and geopolitical entities into a single place-name one leads to improved inter-annotator agreement and results. We used this model as a strong baseline, geolocating the recognised entities on a map, in order to elaborate on the limitations of BERT-based models. We call this fine-tuned BERT as BERT-FT.

GOJ. The Gazetteer of Japan (GOJ) issued by the Government of Japan,[2] includes more than 4,000 modern place-names. It does not include any historical place names. This is an important limitation, because the inscriptions of this study refer to such historical places, some of which may have been altered over time. Edo, for example, is now called Tokyo, and as such is found in GOJ.

GeoLOD. In [10], the authors introduced a gazetteer of Japanese toponyms that comes bundled with a geotagging algorithm. The tool is based on data from three databases: "Prefectures of Japan", "Historical Administrative Area Data Set Beta Dictionary of Place Names", "Railroad Stations in Japan (2019)".

---

[1] https://huggingface.co/cl-tohoku/bert-base-japanese.
[2] https://www.gsi.go.jp/ENGLISH/pape_e300284.html.

BERT-FP-FT. The Wikipedia articles which were used to pre-train BERT-FT do not cover the language used in our inscriptions, which were produced during the Edo period (1600–1868). To address this issue, in this study, we used 20,346 inscriptions to further pre-train Japanese BERT, before fine-tuning for NER. We used a masked language modelling objective, which is a language modelling task where the model is trained to predict the missing token(s) in a text. Our hypothesis is that this objective will allow the model to learn the language and the context of inscriptions from the Edo period, used in our dataset.

5   Experiments

Masked language modelling, or MLM in short, was used to further pre-train the Japanese BERT. We used a batch size of 64, a max length of 128 tokens, a learning rate of $2e-5$, 20 epochs, 500 warm-up steps, and a weight decay of 0.01. To measure the model's ability, we measured the accuracy of the predictions for masked tokens. Further pre-training improves the average negative log likelihood of masked tokens (the same for the two models), from 2.22 to 1.27 (-43%).

We used MLM to yield our BERT-FP-FT model, which we compared with GOJ, GeoLOD, and BERT-FT [12]. As shown in Table 2, GeoLOD was better than GOJ, achieving 39% in F1. Both gazetteer-based approaches achieved high precision but low recall. A preliminary error analysis revealed that they could not detect all place names from the Edo period, but more experiments are needed to verify this. BERT-FT achieved 77% in F1. BERT-FP-FT outperformed its competitors in all the evaluation metrics and its difference in F1 (+4) from BERT-FT, which we consider as the previous state of the art, is also robust, as we can see in Table 3, where we repeated the experiment three times.

Table 2. Evaluation of GOJ, GeoLod, BERT-FT [12] and BERT-FP-FT.

	Precision	Recall	F1
GOJ	0.92	0.07	0.13
Geolod	0.97	0.25	0.39
BERT-FT	0.76	0.78	0.77
BERT-FP-FT	**0.79**	**0.82**	**0.81**

Table 3. F1 across the 3 folds used for Monte Carlo Cross Validation

	#1	#2	#3	AVG
BERT-FT	0.80	0.82	0.85	0.82
BERT-FP-FT	**0.82**	**0.85**	**0.90**	**0.86**

6 Empirical Analysis

6.1 Inscription Text Restoration

One of the major problems in the transcription of inscriptions on prints, which facilitates their spatiotemporal mapping, is the quality and readability of historical material. In time, the material quality of prints deteriorates e.g., as the result of light exposure the paper and pigments undergo discolouring, which hinders the readability of the texts. Prints also may be damaged in other ways (via tearing, insect activity), which diminishes their readability by the public (both experts and the wider public). In this context, MLM can help restore fragmented inscriptions, by replacing with [MASK] the token to be restored.

We present a use-case of this method by testing its applicability in the restoration of an inscription on a selected print by Utagawa Hiroshige Fig. 1. The print presents a view of the famous Seta Bridge in the southeast part of Lake Biwa with Mt. Mikami in the background. The upper left cartouche comprises the inscription (also shown in Fig. 1). We masked the second kanji character in this inscription assuming that it was not readable (e.g., destroyed) Fig. 1. We fed our MLM with "瀬[MASK]夕照" and the model correctly predicted that the missing character is 田 (Table 4, first two columns of the first row). By masking and restoring the first kanji character, however, the model had trouble identifying the correct character (Table 4, last two columns). This is probably due to the fact that the first character refers to the name of a specific place, while the second is a generic term 田 or 'rice field' often used in different place names in Japan.

Table 4. Two masked tokens predicted by BERT w/o and w/further pre-training.

瀬[mask]夕照		[mask]田夕照	
w/o	w/	w/o	w/
田 0.15	田 0.79	UNK 0.13	嶋 0.04
野 0.06	川 0.03	都 0.02	隠 0.03
下 0.05	崎 0.02	狩 0.01	茨 0.02

Similarly, in other prints featuring the inscription with the place-name 浅間山 or 'Mt. Asama' or 'Asama Mountai' (which is, in fact, a volcano), we observe that the word 山 or 'mountain' is correctly restored, but this is not the case when we mask the first word 浅間 or 'Asama' instead. This is because the word is a generic term describing a topographical formation (ie., mountain) used in many place-names (e.g. mountains) in Japan. Nonetheless, this experiment has shown both opportunities and challenges related to the automated transcription of inscriptions in *meisho-e* prints.

Fig. 1. The print by Utagawa Hiroshige (1797–1858) entitled "Sunset Glow at Seta" (瀬田夕照), from the series "The Eight Views of Ōmi" (近江八景), 1834–35, woodblock print, MET (OA). On the left we see two cartouches with inscriptions, the first one with the second kanji character masked, and the second one, the original version with no character masked.

6.2 Data Augmentation with OCR

The next analytical step towards large-scale mapping of places depicted in print is experimenting with computational tools with automated transcription of inscriptions on prints, which feature place-names. The corpus of Japanese pre-modern printed books and single prints is extremely rich. For example, Kokusho Sōmokuroku (General Catalog of National Books) alone includes more than 450,000 pre-modern books [14], 90% dated to the Edo period (1600–1868). However, only less than 1% of these books have been transcribed to date. OCR could provide a solution to this problem, presuming that historical books, prints, and documents are available in digital format. In this work, we hypothesise that NER can be performed on the OCR-recognised text, and information about the place-names can be extracted despite the noise generated during the automated recognition phase.

We used a pretrained Japanese OCR model,[3] in order to extract the text inscribed on *meisho-e* print images. Then, we applied NER on the OCRed output, investigating the possibility of enriching our primary source of data before moving on to exploration at a larger scale. For example, as is shown in Fig. 2, we analysed Hokusai's print entitled "Inume Pass in Kai Province" (甲州犬目峠) from the series "Thirty-six Views of Mount Fuji" (富嶽三十六景, ca. 1830–31). We extracted the white rectangular cartouche with the inscription, located in the upper left corner of the image, and we experimented with different OCR models to transcribe the text. Then, we applied NER on the OCRed text. The recognition tool performed relatively well, with the model recognising 富嶽 or

---

[3] https://huggingface.co/kha-white/manga-ocr-base.

"Fugaku" (marked in red colour) which is an alternative name for Mount Fuji. This is a promising result given that our model was not trained on OCRed input. If successful, this application will not only allow large-scale exploration, but can also unlock related applications for the study of early-modern printed books.

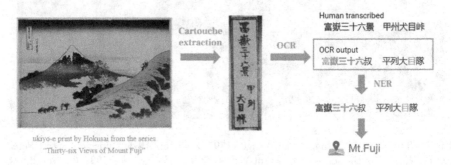

Fig. 2. Brown colour indicates the correct Japanese characters while red colour indicates the place-name extracted by the Bert NER model (Color figure online)

NER on OCRed transcriptions is indeed promising, but challenges exist. When it comes to longer texts such as poems, due to the characteristics of the Japanese pre-modern writing system and the formal specificities of the design (e.g., multicolour cartouches), the recognition is largely distorted, for example, when the model delivered only four correct characters scattered across the inscription. The following question arises, then: is this error affecting the performance of NER models, and if so, can we learn to bypass it? To address this question, we plan to assess NER on OCR output, in order to quantify the error that is propagated from the recognition of written text to the recognition of named entities. Furthermore, we also plan to improve the written text recognition outcome. For example, one thing we are considering is joining forces with OCR error correction challenges.

6.3   Geolocating Recognised Place-Name Entities

Finally, we experimented with geolocating the recognised places, by applying several methods (as described earlier in Sect. 4 (Analytical Tools). The scope of this experiment was to assess if the reported evaluation results are reflected in a use case. In principle, the map featuring places depicted in prints, which would result from our spatiotemporal mapping experiments, could be envisioned as a tool for the study of spatial relationships between geographical places and their representation in prints and their changes across time. Figure 3 provides such an example, by visualising the frequency of appearance of recognised place-names across our dataset.

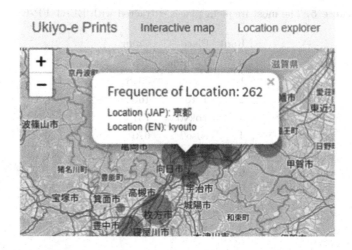

Fig. 3. Visualisation of a publicly available interactive map we created, serving exploratory purposes of the recognised (using GeoLOD) locations which *meisho-e* prints depict (the disc radius reflects the frequency)

BERT-Based Methods. By using BERT-FP-FT, we observe that the model included different kinds of entities and not only place-names, yielding a faulty geolocation. In Table 5, which presents the ten most frequently recognised place-names across our data, we observe the presence of generic categories of landforms or human-made objects, e.g., 'mountain' or 'bridge', as well as grammatical forms such as the preposition 'of'. We also observe two more error sources. First, a few recognised places were geolocated outside of Japan, like Jiang, and Sichuan. Second, historical place-names extracted with NER from *meisho-e* inscriptions are not easily geolocated on contemporary maps of Japan. This is due to the historical transformation of Japanese writing systems, and historical changes in administrative geography in Japan as the retrieval algorithms are trained on contemporary datasets and gazetteers of toponyms (place names). For example, the tool could not geolocate the names of roads such as 'Tōkaidō' that are not pinnable on a map by a single pin. 'Tōkaidō' was represented as a larger grouping of pins and was located in Aichi prefecture. Also, Tokyo, Kawasaki, and Fujisawa were located on the Izu peninsula, which is not geographically correct.[4]

Gazetteers. When tested, gazetteer-based approaches, which already hold valid latitude and longitude coordinates, were overall more precise compared to BERT-based ones. The Gazetteer of Japan (GOJ) did not detect many place-names, mainly because it does not cover historical toponyms from the pre-modern period. GeoNLP was more precise but it could also not detect all place-names from the period. Therefore, despite the better performance of BERT-based mod-

---

[4] We observed similar findings for BERT-FT.

Table 5. The most frequent places extracted with BERT-FP-FT

Rank	Place	Count
0	Tōkaidō	3657
1	Edo	3328
2	Toto	1602
3	Tokyo	751
4	Kiso	670
5	of	545
6	bridge	478
7	Jiang	476
8	Mountain	462
9	Sichuan	452

els, we find that gazetter-based approaches still hold the advantage when the goal is geolocation and mapping of *meisho-e* prints.

6.4 Spatiotemporal Analysis

Following our findings, we conducted spatiotemporal mapping of *meisho* depicted in prints aimed at the identification of the most popular places at different historical periods. In Fig. 4, we present the frequency of recognised place-name entities over time. We used the GeoLOD for our purposes, to opt for high precision (Table 2). We have found out that the depiction of places (including both natural and man-made formations) flourished especially in the 19th century, especially between the 1830 and 1860s, and that different places were popular in different historical eras.

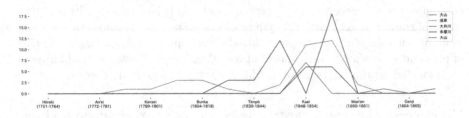

Fig. 4. Frequency of place name entities, recognised with GeoLOD, over time

We also experimented with analysis of formal aspects of the prints and how certain places were depicted, namely what colour schemes were used in their depiction across time. We analysed the colour of 3,505 *meisho-e* images printed during 16 historical eras, from 1751 to 1868. We transformed images from RGB (i.e., red, green, blue) to HSV (i.e., hue, saturation, lightness) and each pixel was classified to the colour it depicts, by using the classification presented in Table 6. This classification allows us to compute the percentage of each colour for an image (i.e., how many pixels are classified to the respective colour out of all in the image). Figure 6 presents the colour percentages averaged across the images of each time period. We observe that yellow is the most frequently-used colour over time. Red and green follow, with the former being most dominating during early time periods (1751–1772) while the latter dominated after the Kaei era (after 1854). Blue and cyan followed, with low percentages over time. These analytical results may indicate historical changes in colour preferences among the print producers (designers, artists) and consumers as well as technological developments e.g. availability of certain pigments etc. They also may be correlated with certain types of motives and topographical formations (e.g., mountain, sea) and to a lesser degree even specific places (e.g., colour schemes linked to the depiction of the Edo city).

Table 6. HSV ([hue, saturation, value]) range (from, to) per colour

Colour	HSV: from	HSV: to
BLUE	[110, 50, 0]	[130, 255, 255]
CYAN	[80, 100, 100]	[100, 255, 255]
GREEN	[36, 0, 0]	[70, 255, 255]
RED	[0, 25, 0]	[15, 255, 255]
YELLOW	[16, 25, 0]	[35, 255, 255]

Figure 5 for example, we observe that brown and blue dominate, which reflects the type of depicted motive, i.e. the bridge, representing both water and land. Beige colouring also results in discolouration of the paper which prints undergo in time under the influence of the exposure to light. The dominance of brown and blue can also be seen in the 3D representation of the colours of that print (Fig. 5).

We investigated the correlation of specific colour schemes and historical periods and specific places. We focused on the two most frequent depicted places 東海道 or 'Tōkaidō' and 江戸 or 'Edo', which generated 468 and 575 hits respectively. We calculated the amount of blue and cyan per image and historicised the results Fig. 7. We observed that prints depicting stations along the 'Tōkaidō' road have more cyan colour than images of Edo city. This is not surprising considering that the 'Tōkaidō' road linking Kyoto and Edo city crosses large areas with rich waterscapes. Moreover, we observe that the colour is not directly linked to the place that is depicted, but rather follows the colour trends specific for a given era characteristic for the majority of prints produced at a given period. This finding underlines a general understanding of the prints and their material aspects in Japanese art history and indicates that to contribute to the state-of-

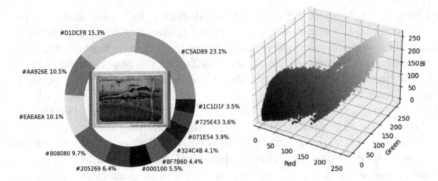

Fig. 5. Visualisation of the colour scheme in a selected print (title: 「東海道五十三次之內 京師 三条大橋」) focusing on the twelve most frequent colours (left) and a 3d RBG representation of all the colours in the image (right) (Color figure online)

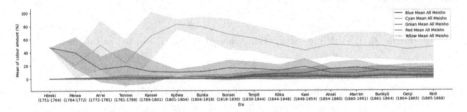

Fig. 6. Visualisation of the temporal distribution of the frequency of basic colours used in prints in different historical epochs between ca. 1750s and 1850s with the standard error of the mean shadowing each respective time series (Color figure online)

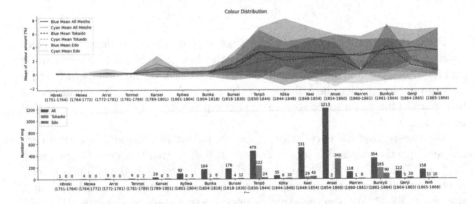

Fig. 7. On top is the average (and the st. error of the mean) percentage of blue and cyan in prints of specific time periods is shown, in the two most frequent places versus all images. The barchart shows the support (# prints) per place per era (i.e., aligned). (Color figure online)

the-art in the field in a meaningful more fine-tuned analysis needs to follow (e.g., on the level of a print designer, publisher, print format, topic).

Analytical Limitations in the Study of Colour Schemes. The computational study of colour schemes in *meisho-e* prints has several analytical limitations. Most importantly, the colour schemes depend on the material characteristics and quality of the photographs of prints (subsequently digitised). Also, the results indicate a strong presence of silver and grey colour, which most probably is not included in the compositions but in the composition frames, which are also part of the image and its photo. These limitations need to be accounted for in the future interpretation.

7 Conclusion

In this study, we benchmarked NER on inscriptions of *meisho-e* prints, comparing deep learning with simpler gazetteer-based approaches. We improved the former with further pre-training and we showed that BERT still falls short compared to the latter, when geolocation is the goal. By applying NER on approx. 20k images, we undertook a large-scale spatiotemporal analysis of recognised place-names, discovering popular *meisho* per historical era. By focusing on the two most frequently recognised *meisho*, we showed that colour reveals characteristics of the landscape, in the case of the waterscapes, and that it is not directly linked to the place depicted but may follow other trends. Future work will explore more *meisho* and we will interpret our analytical results, attempting to contextualise them in relation to sociopolitical changes and technological advancements (e.g. introduction of new printing materials), as well as correlate them with other factors such as print designer, publisher, and print format that may have played the role in their production and popularity at a given time.

References

1. Akama, R.: 赤間 亮, 立命館大学アート・リサーチセンターの古典籍デジタル化：Arc国際モデルについて(¡特集¿古典籍資料の最前線), 情報の科学と技術 J. Inf. Sci. Technol. Assoc. (2015)
2. Arnold, T., Tilton, L.: Distant viewing: analyzing large visual corpora. Digit. Scholarsh. Humanit. **34**(1), i3–i16 (2019)
3. Catalinac, A.: Quantitative text analysis with Asian languages: some problems and solutions. Polimetrics I (1) (2014)
4. Chino, K.: The emergence and development of famous place painting as a genre. Rev. Jpn. Cult. Soc. **15**, 39–61 (2003)
5. Clanuwat, T., Bober-Irizar, M., Kitamoto, A., Lamb, A., Yamamoto, K., Ha, D.: Deep learning for classical Japanese literature. In: Neural Information Processing Systems (NeurIPS) Creativity Workshop (2018)
6. Devlin, J., Chang, M.W., Lee, K., Toutanova, K.: BERT: pre-training of deep bidirectional transformers for language understanding. arXiv preprint arXiv:1810.04805 (2018)

7. Hashimoto, Y., Iikura, Y., Hisada, Y., Kang, S., Arisawa, T., Kobayashi-Better, D.: The Kuzushiji project: developing a mobile learning application for reading early modern Japanese texts. Digit. Humanit. Q. **11**, 1–13 (2016)

8. Hirose, S., Yoshimura, M., Hachimura, K., Akama, R.: Authorship identification of ukiyoe by using rakkan image. In: The Eighth IAPR International Workshop on Document Analysis System (2008)

9. Kamens, E.: Utamakura, allusion, and intertextuality in traditional Japanese poetry (1997)

10. Kitamoto, A.: Toponym information platform geolod and its cooperation with rekiske and rekiroku (2022)

11. Lamb, A., Clanuwat, T., Kitamoto, A.: KuroNet: regularized residual U-Nets for end-to-end Kuzushiji character recognition. SN Comput. Sci. **1**, 177 (2020)

12. Liagkou, K., Pavlopoulos, J., Machotka, E.: A study of distant viewing of ukiyo-e prints. In: Proceedings of the Thirteenth Language Resources and Evaluation Conference, pp. 5879–5888 (2022)

13. Mitsutoshi, N., Susume, W.: 和本のすすめ：江戸を読み解くために (2011). https://cir.nii.ac.jp/crid/1130000795601158272

14. Somokuroku, K.: Iwanami shoten, vol. 8 (1963)

15. Suzuki, C., Kitamoto, A.: Pre-modern Japanese books as data of humanities: finding image of edo famous place from meisho-ki 名所記 and 名所 meisho-zue using iiif curation platform. Keynote Session (2019)

16. Tian, Y., Clanuwat, T., Suzuki, C., Kitamoto, A.: P2 - ukiyo-e analysis and creativity with attribute and geometry annotation. In: International Conference on Computational Creativity, Underline Science Inc (2021)

17. Won, M., Murrieta-Flores, P., Martins, B.: Ensemble named entity recognition (NER): evaluating NER tools in the identification of place names in historical corpora (2018)

18. Yada, T.: 矢田勉. 国語文字●表記史の研究 (a historical study of Japanese writing system). kyūko shoin (2012)

19. Yin, X., Xu, W., Akama, R., Tanaka, H.: A synthesis of 3-D Kabuki face from ancient 2-D images using multilevel radial basis function. J. Soc. Art Sci. **7**(1), 14–21 (2008)

Predictive Inference Model of the Physical Environment that Emulates Predictive Coding

Eri Kuroda[✉][iD] and Ichiro Kobayashi[iD]

Ochanomizu University, Tokyo, Japan
{kuroda.eri,koba}@is.ocha.ac.jp

Abstract. In recent years, the significance of artificial intelligence in comprehending the real-world has increased, by leveraging the inherent ability of humans to process intuitive physics on a computer. Prior investigations on real-world understanding have mainly relied on image inference to recognize the physical environment. In contrast, we propose an inference model that can predict the observed environment using both visual and physical features, emulating the predictive coding hypothesized to occur in the human brain, and detects change points in response to predictive events. Additionally, the model verifies the correctness of the timing of important physical events of objects, such as object collisions and disappearances. Furthermore, the results of the physical information prediction are also described as natural language sentences to confirm whether the model accurately recognizes the real-world and predicts the next behavior based on the physical information.

Keywords: physical characteristics · latent hierarchical structure of physical relationships · prediction

1 Introduction

When faced with a specific circumstance, humans possess the innate capacity to swiftly comprehend environmental cues, predominantly through visual perception. This capability is believed to rely on the mental construction and simulation of the environment within the brain, contingent upon perceived stimuli [9]. Concurrently, humans are able to apprehend and anticipate the actions of objects in the environment, founded on the environmental framework constructed within their brain. At this point, humans generate predictions concerning both the physical and visual aspects of the perceived objects. It is believed that physical prediction pertains to significant events in the object, rather than forecasting all possible states of the environment. Considerable research has been conducted to achieve the human capacity to identify and forecast environmental information on a computer [1,6,8,9,13,18,22,26,30]. Nonetheless, the majority of real-world prediction studies have produced results based on either visual predictions via pixel alterations, or physical predictions via numerical variations in simulators,

A. Bifet et al. (Eds.): DS 2023, LNAI 14276, pp. 431–445, 2023.
https://doi.org/10.1007/978-3-031-45275-8_29

and no prediction model that can simultaneously generate both visual and physical predictions has been put forward, as humans are capable of doing. In this investigation, we present a novel model capable of producing both visual and physical predictions regarding objects in the environment, whilst simultaneously extracting the timing of important events amongst the predicted events. The model is constructed through a combination of PredNet [20], a prediction model that replicates the top-down and bottom-up hierarchical information processing in the human brain, and the Variational Temporal Abstraction (VTA) [14] mechanism, which retrieves change points within the observed environment based on the visual information's image characteristics. The proposed model is rigorously evaluated to confirm its efficacy, wherein the timing of predicted object collisions within the event is ascertained using CLEVRER [33], representing physical phenomena such as object collisions, and the model's accuracy is verified by computing the correct timing. Furthermore, the physical prediction results are generated as sentences to facilitate interpretation and validate whether the model accurately forecasts the next action based on physical information.

2   Related Work

Real-World Cognition. Real-world cognition refers to the study of machine learning and artificial intelligence for recognizing and interpreting the real-world. Ha et al [9] proposed the concept of world models as a mechanism by which humans perceive and understand the environment. When humans visually observe the environment, they can quickly recognize the objects and their behavior in the environment. This is made possible by modeling and simulating the environment in the brain based on the sensory input. LeCun [16] identified one of the three challenges that AI research must address in the future: "How can machines learn to represent, predict, and act on the world from observation?" Humans and animals can gain insight into how the world works and acquires background knowledge through limited interaction and observation. This is considered the basis of common sense, which not only predicts future outcomes but also fills in information gaps in time and space. Common sense consists of models of the world that inform us about what is probable and what is improbable. This allows humans and animals to predict, reason, plan, explore sequences of actions, and imagine novel solutions to problems. The study of real-world cognition is therefore crucial.

Prediction. Research on real-world cognition often focuses on visual prediction and commonly employs Recurrent Neural Networks (RNN) and Long Short Term Memory (LSTM) methods [11]. However, while Chang et al. [2] proposed a new model, STIP, to address the problem of generating high-resolution predictions due to a loss function based on information loss and mean squared error, the emphasis of many studies is solely on capturing temporal dependencies between frames, with little discussion of the spatial features within frames. To rectify this, Wang et al. introduced a spatio-temporal LSTM (ST-LSTM) structure to predict high-quality videos and proposed novel prediction models such as PredRNN++ [28] and PredRNN [29]. Additionally, to enable long-term prediction, Lin et al. [19] integrated a self-attention mechanism into ST-LSTM to store long-range spatial features, while Lee et al. [17] introduced memory alignment learning to store long-term temporal dependencies. Other proposed models include Iso-Dream [23], an improved version of Dreamer [10], which separately learns controllable and uncontrollable state transitions and combines them with prediction, and Gao et al.'s SimVP [7], a prediction model that merges image recognition with Transformer technology and uses Vision Transformer [5]. These studies aim to produce highly accurate prediction results and expand research on models that can make long-term predictions, such as for humans.

Physical Reasoning, Intuitive Physics. The field of common sense or intuitive physics, which involves computational understanding of the physical world, has been studied extensively in recent years. Representing intuitive physics is crucial for modeling object interactions and predicting their dynamics, and has received considerable attention [3,4,26]. Tang et al. [26] proposed PHYCINE, a hierarchical prediction model that focuses not only on first-order features such as object position and shape, but also on hidden behaviors of objects such as mass and charge, by discovering physical concepts of objects from low-level (color, shape) to high-level abstract (mass, charge) from video images. Ye et al. [31] and Piloto et al. [25] focus on learning intuitive physical properties that can be interpreted. In addition, many studies have attempted to learn intuitive physical properties from a few frames of a video image. Yi et al. [32] have focused on the complex temporal and causal structures underlying object interactions, using the image reasoning dataset CLEVR (Compositional Language and Elementary Visual Reasoning diagnostics dataset) [12] with CLEVRER (CoLlision Events for Video REpresentation and Reasoning) [33]. They also extended CLEVRER and proposed CLEVRER-Humans as a video inference dataset for human-labeled causality inference [22].

3   PredNet

PredNet [21] is a deep prediction neural network construct to mimic the concept of predictive coding. An overview of the model is shown in Fig. 1. Each module has four internal components: an input convolutional layer (x_{t_k}), a recurrent convolutional representation layer (R), a convolutional prediction layer (A), and an error representation (E). The representation layer in each module captures the state for prediction, while the input layer processes the input information. The prediction layer generates the internal prediction state, and the error layer outputs the error representation by taking the difference between the prediction state and the input. PredNet utilizes a bidirectional process to generate predictions, where predictions made in the upper layers of the network are conveyed to the lower layers via the representation module, and errors detected in the lower layers are transmitted to the upper layers. This mechanism mimics the operation of a generalized state equation, which enables accurate predictions to be made.

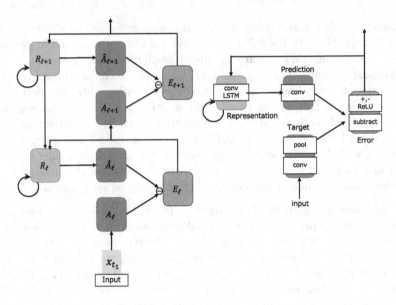

Fig. 1. Schematic diagram of PredNet.

4   Variational Temporal Abstraction (VTA)

In Variational Temporal Abstraction (VTA) [14], a state-space model is proposed to extract hierarchical abstractions from series data and detect change points. Figure 2 (right) is the graphical model of the hierarchical state-space model obtained by VTA. In this figure, X is the input, S is the observation abstraction, and Z is the temporal abstraction. X is the lowest layer closest to the input, and S and Z are the upper layers in that order. By processing the input series information and obtaining the hierarchical structure, VTA enables the acquisition of the upper Z representation, which indicates the transition of the environment. However, as for the state space models that handle sequential series information, in general, it is difficult to determine when to transition to the upper layer Z, taking into account temporal transitions, as depicted in Fig. 2 (left) to (right). To address this issue, VTA introduces a binary latent variable m that determines the timing, as shown in Fig. 2(left). The boundary indicator $M = m_{1:T}$ takes the value 0 or 1. When the change in the observed or temporal abstraction is significant, m becomes 1, and the upper layer Z transitions accordingly.

5   Proposed Model

5.1   Mechanism of the Change Point Prediction Model

PredNet [20] and VTA mechanisms are integrated to construct an change point prediction model. This model mimics predictive coding which is a hypothesized

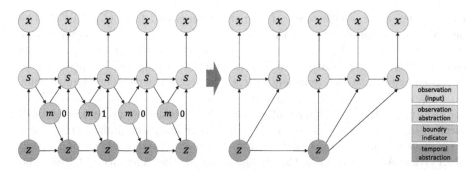

Fig. 2. Schematic diagram of VTA. (left) Model with boundary index $M = \{0, 1, 0, 0\}$. (right) Model with time structure obtained from the boundary index M.

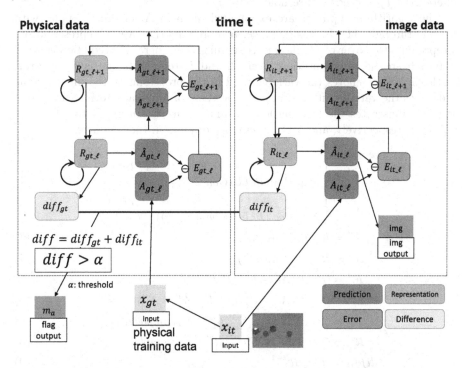

Fig. 3. Schematic diagram of a change point prediction model.

function in the human brain. The model architecture is presented in Fig. 3. The proposed model is a parallel hierarchical structure of two PredNet models, one of which predicts physical phenomena of the environment by representing them as graphs, and the other of which predicts them by visual information of the environment. The proposed model also incorporates the change point discrimination flag m, which is a mechanism of VTA. The input information consists of two datasets: the CLEVRER dataset with image information x_{it} and the physical training dataset x_{gt} with physical properties generated from CLEVRER. The output information consists of two pieces of information: the predicted image ("img output" in Fig. 3), which was sequentially predicted for the image, and the change point m_a ("flag output" in Fig. 3), which was computed by the inference of the embedding vector representing the physical property. The change point m_a serves as an indicator flag, signifying when the cumulative value of physical and image data has significantly changed and takes the value 0 or 1. Both mechanisms learn by error propagation to higher levels, minimizing the differences between prediction $\hat{A}$ derived from the representation tier R and actual observation A. To determine the change point m, the difference $diff$ between the representation layer R at time $t-1$ and time t is calculated for the physical and image data, respectively, such that the change point m_a becomes 1 if the difference $diff$ exceeds a threshold value α.

The algorithmic updates are expounded upon in Algorithm 1, along with Eqs. (1) through (11). In this instance, R represents the layer of representation, A represents the layer of prediction, $\hat{A}$ signifies the generated prediction content derived from the representation layer R, and E represents the layer of error. Furthermore, it denotes the variable used in the processing of the image, while gt denotes the variable utilized in the processing of physical information. Equation (12) illustrates the training loss. λ_t and λ_ℓ are weighting factors for time and layer, respectively, and n is the number of units in the ℓ-th layer.

$$A_l^{it} = \begin{cases} x_{it} & \text{if } l = 0 \\ MaxPool(ReLU(Conv(E_{l-1}^{it}))) & l > 0 \end{cases} \tag{1}$$

$$A_l^{gt} = \begin{cases} x_{gt} & \text{if } l = 0 \\ MaxPool(ReLU(Conv(E_{l-1}^{gt}))) & l > 0 \end{cases} \tag{2}$$

$$\hat{A}_l^{it} = ReLU(Conv(R_l^{it})) \tag{3}$$

$$\hat{A}_l^{gt} = ReLU(Conv(R_l^{gt})) \tag{4}$$

$$E_l^{it} = [ReLU(A_l^{it} - \hat{A}_l^{it}); ReLU(\hat{A}_l^{it} - A_l^{it})] \tag{5}$$

$$E_l^{gt} = [ReLU(A_l^{gt} - \hat{A}_l^{gt}); ReLU(\hat{A}_l^{gt} - A_l^{gt})] \tag{6}$$

$$R_l^{it} = ConvLSTM(E_l^{it-1}, R_l^{it-1}, Upsample(R_{l+1}^{it})) \tag{7}$$

$$R_l^{gt} = ConvLSTM(E_l^{gt-1}, R_l^{gt-1}, Upsample(R_{l+1}^{gt})) \tag{8}$$

$$diff_{it} = R_l^{it} - R_l^{it-1} \tag{9}$$

$$diff_{gt} = R_l^{gt} - R_l^{gt-1} \tag{10}$$

$$diff = diff_{it} + diff_{gt} \tag{11}$$

$$L_{train} = \sum_t \lambda_t \sum_l \frac{\lambda_l}{n_l} \sum_{n_l} E_l^t \tag{12}$$

Algorithm 1. Calculation of change point prediction model

Require: x_{it}, x_{gt}
$\quad A_0^{it} \leftarrow x_{it}, A_0^{gt} \leftarrow x_{gt}$
$\quad E_l^0, R_l^0 \leftarrow 0$
$\quad$ **for** $t = 1$ **to** T **do**
$\quad\quad$ **for** $l = L$ **to** 0 **do**
$\quad\quad\quad$ **if** $l = L$ **then**
$\quad\quad\quad\quad R_L^{it} = \text{ConvLSTM}(E_L^{it-1}, R_L^{it-1})$
$\quad\quad\quad\quad R_L^{gt} = \text{ConvLSTM}(E_L^{gt-1}, R_L^{gt-1})$
$\quad\quad\quad$ **else**
$\quad\quad\quad\quad R_l^{it} = \text{ConvLSTM}(E_l^{it-1}, R_l^{it-1}, \text{Upsample}(R_{l+1}^{it}))$
$\quad\quad\quad\quad R_l^{gt} = \text{ConvLSTM}(E_l^{gt-1}, R_l^{gt-1}, \text{Upsample}(R_{l+1}^{gt}))$
$\quad\quad\quad$ **end if**
$\quad\quad$ **end for**
$\quad\quad$ **for** $l = 0$ **to** L **do**
$\quad\quad\quad$ **if** $l = 0$ **then**
$\quad\quad\quad\quad \hat{A}_0^{it} = \text{SatLU}(\text{ReLU}(\text{Conv}R_0^{it}))$
$\quad\quad\quad\quad \hat{A}_0^{gt} = \text{SatLU}(\text{ReLU}(\text{Conv}R_0^{gt}))$
$\quad\quad\quad$ **else**
$\quad\quad\quad\quad \hat{A}_l^{it} = \text{ReLU}(\text{Conv}R_l^{it})$
$\quad\quad\quad\quad \hat{A}_l^{gt} = \text{ReLU}(\text{Conv}R_l^{gt})$
$\quad\quad\quad$ **end if**
$\quad\quad\quad E_l^{it} = [\text{ReLU}(A_l^{it} - \hat{A}_l^{it}); \text{ReLU}(\hat{A}_l^{it} - Ait_l)]$
$\quad\quad\quad E_l^{gt} = [\text{ReLU}(A_l^{gt} - \hat{A}_l^{gt}); \text{ReLU}(\hat{A}_l^{gt} - A_l^{gt})]$
$\quad\quad\quad$ **if** $l < L$ **then**
$\quad\quad\quad\quad A_{l+1}^{it} = \text{MaxPool}(\text{Conv}(E_{it}^l))$
$\quad\quad\quad\quad A_{l+1}^{gt} = \text{MaxPool}(\text{Conv}(E_{gt}^l))$
$\quad\quad\quad$ **end if**
$\quad\quad\quad diff_{it} = R_l^{it} - R_l^{it-1}$
$\quad\quad\quad diff_{gt} = R_l^{gt} - R_l^{gt-1}$
$\quad\quad\quad diff = diff_{it} + diff_{gt}$
$\quad\quad\quad$ **if** $diff > \alpha$ **then**
$\quad\quad\quad\quad m_a = 1$
$\quad\quad\quad$ **else**
$\quad\quad\quad\quad m_a = 0$
$\quad\quad\quad$ **end if**
$\quad\quad$ **end for**
$\quad$ **end for**

---

6   Experiment

6.1   Change Point Extraction in Predictive Inference

To verify the effectiveness of our proposed model, an experiment was conducted to see if the model can correctly extract the change point of the next step state. The dataset we used was the CLEVRER dataset and physical training data generated from CLEVRER.

Physical Training Dataset. The two datasets we used were CLEVRER dataset [33] and a dataset representing physical properties of real-world objects – the procedure for creating the physical properties dataset is shown in Fig. 4.

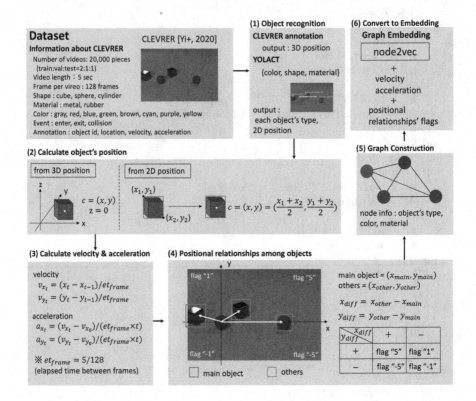

Fig. 4. Steps to create physical training dataset.

Table 1 shows the experimental settings of the model, which is the same as used in the previous study [20].

Results and Discussion. The results of the change point prediction accuracy in the proposed model are shown in Table 2. The physical data in Table 2 shows the results obtained from the data set created in Fig. 4, and the annotation data in the table shows the results obtained from the CLEVRER annotation dataset.

The results show that the accuracy of the physical data is equivalent to that of the annotation data, which is the supervised data, in predicting the change points. As a result example, the predicted images and flags for region i are shown in Fig. 5. As the predicted image is also accurately generated, it can be said that this model is proficient in generating both the predicted image and flag of the next time's change point.

Table 1. Experimental Settings.

Number of training data	600,000
Number of validation data	60,000
Number of times studied	500,000
#Layers	4
Size of convolutional filter	3 × 3 (for all conventions)
#Channels	From lower module, 3, 48, 96, 192
Optimization	Adam [15]
Learning rate decay	0.0001
α	5

Table 2. Accuracy of the proposed prediction model.

Validation range	i	ii	iii	iv	v	vi
physical data	33.3	**50**	50	33.3	**66.7**	**50**
annotation data	**66.7**	**50**	**66.7**	40	50	**50**

6.2 Text Generation of Prediction Results

The proposed model made two predictions, one for physical data and one for visual data. Humans apprehend and acquire knowledge of the real-world by perceiving it and engaging in predictions and inferences. Furthermore, linking language to the physical world enables us to gain a more profound comprehension of reality and our prior experiences. Put differently, human intelligence can be conveyed through symbol manipulation using language that pertains to the real-world. Therefore, research on comprehending the physical world through machine learning technology should express reasoning as a language, with the aim of linking the recognition of real-world objects, understanding of physical properties, and prediction using language. This study generated embedded vectors, extracted as change points in physical data, as a form of language information. Additionally, only collisions were used as change points for the generation.

Dataset. To generate language from the embedded vectors predicted by the change-point prediction model, it is necessary to learn new linguistic information. For this purpose, we developed a language dataset consisting of a pair of data: an embedded vector of graph representations representing physical properties and a sentence describing the state of the graph. Although the experiment was conducted in Japanese, this paper covers both English and Japanese. The graph's embedding vector representation was created from the CLEVRER annotation data using the procedure illustrated in Fig. 4. The paired sentences were devised to fit into nine templates of three (before collision, collision, and after collision) × three (type of sentences). The correct answer for each image was three sentences.

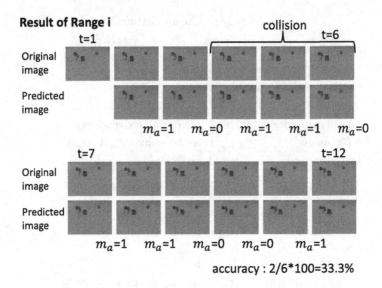

Fig. 5. Predicted change point extraction results in range i.

The details of the templates are as follows: Two objects A and B collide with each other, and A and B are "{gray, red, blue, green, brown, water, purple, yellow} {sphere, cylinder, cube}." For example, "Red sphere" and "Blue cylinder" are now included. In addition to the collision data, we also created a dataset for when the objects were approaching before the collision and when they were leaving after the collision. The approaching time was five frames before the collision, and the leaving time was five frames after the collision. An example of the generated pair dataset is shown in Fig. 6.

Text Generation Model. The text generation model utilized only the decoder component of the Transformer [27]. The decoder architecture is depicted in Fig. 7. Although conventional transformers are based on an encoder-decoder model, this study adopts the embedding vector prediction result of the graph in the change-point prediction model of the proposed model as the encoder output. This prediction result is employed as input from the encoder to the decoder. The paired data generated in Fig. 4 was utilized to train the decoder, with the number of paired sentence data set at 219,303 (nine sentences × 24,367 collisions) and the predictive graph embeddings for test data set at 10,965. The training settings are detailed in Table 3.

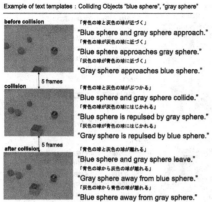

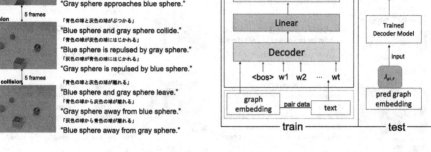

Fig. 6. Example of text templates.

Fig. 7. Schematic diagram of the text generation model.

Table 3. Experiment Setting.

batch size	8
Embedding	128
hidden layer	512
Optimization	Adam [15]

Results and Discussion. We confirmed that the embedded representations of the predicted graphs made correct predictions about the real-world by generating a language sentence describing the observed real-world situation. The four ranges that were examined for description were those shown in Fig. 2, i, ii, iv, and vi, which indicate the time of the collision.

Range of i. In range i, a green sphere collides with a red cylinder and the assumed correct statement is shown in Fig. 8. The generated sentence was "A green cylinder is repelled by a red cylinder." The sentence was correct about the color of the object, but incorrect about its shape.

Range of ii. In range ii, a green cylinder collides with a brown cube and the assumed correct statement is shown in Fig. 8. The sentence generated was "A green cylinder collides with a brown cube." The sentence was correct for both color and shape of the objects.

Range of iv. In range iv, a gray sphere collides with a blue cube and the assumed correct statement is shown in Fig. 8. The generated sentence was "A grey sphere is repelled by blue cube." The sentence was correct for the color of the object, but incorrect for the shape.

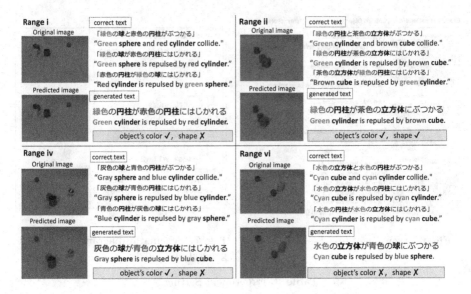

Fig. 8. Example of text generation for prediction results.

Range of vi. In range vi, a cyan cube collides with a blue sphere and the assumed correct statement is shown in Fig. 8. The generated sentence was "A cyan cube collides with a blue sphere," which was incorrect for the object's color and shape. Unlike the other results, range vi produced incorrect judgments for both color and shape of the object. Figure 9 depicts the objects' transition up to the collision in range vi, which includes the "cyan cube" and "cyan cylinder" colliding objects. It is noticeable that the "cyan cube" passed through the "blue sphere" without collision. The infinitesimally small distance between the cyan cube and the blue sphere led to the incorrect prediction of their collision. It is likely that considering the cyan cylinder hidden behind the cyan cube, and both objects being of the same color, contributed to the failure to generate a description accurately. To improve the text generation accuracy, it is necessary to improve the points where objects of the same color are regarded as the same object and where incorrect collision predictions are made.

Accuracy Verification with BLEU

The accuracy of the generated text is evaluated by BLEU [24]. BLEU@n is a measure of how well each correct and generated sentence matches in the n-gram. The evaluation results of the generated sentences using the BLEU evaluation metric are presented in Table 4. Since there were three correct answers for each generated sentence, the average of each score was used as the BLEU score for the generated sentence. The BLEU scores were computed for Japanese sentences, and the generated sentences achieved scores of 80 for the 2-g, 75 for the 3-g, and 69 for the 4-g, indicating that they were able to generate informative and accurate sentences about the observed environment to a certain extent.

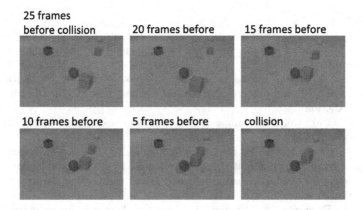

Fig. 9. Object transition status of range vi.

Table 4. BLEU evaluation.

	BLEU@2	BLEU@3	BLEU@4
score	79.7	74.5	68.8

7   Conclusions

In this study, we constructed a model that emulates the structures in the human brain, which can predict the observed environment visually and physically. The predictive model was able to appropriately retrieve change points occurring in the next step, such as object collisions in the environment. Moreover, we generated descriptions from the predicted physical attributes of the environment and calculated the BLEU score, resulting in a language generation capability with a certain degree of accuracy. Based on this outcome, we assert that this model is capable of not only visual prediction but also physical prediction. The outputs of this model and language generation have allowed us to establish a link between the recognition of real-world objects and the understanding and prediction of their physical properties, mediated through the use of language. On the other hand, we believe that there is still room for improvement in both prediction model and language generator since the target dataset is less complex than the actual environment perceived by humans. As future work, we aim to enhance the model and expand the number of language datasets to allow language generation for various physical properties other than collisions.

Acknowledgement. This work was supported by the Japan Society for the Promotion of Science KAKENHI Grant Numbers JP22J21786, JP22KJ1355, 23H03453 and JSPS Bilateral Program Number JPJSBP120213504.

References

1. Bear, D.M., et al.: Physion: evaluating physical prediction from vision in humans and machines (2021)
2. Chang, Z., Zhang, X., Wang, S., Ma, S., Gao, W.: STIP: A SpatioTemporal Information-Preserving and Perception-Augmented model for High-Resolution video prediction (2022)
3. Chen, Z., et al.: ComPhy: compositional physical reasoning of objects and events from videos (2022)
4. Ding, M., Chen, Z., Du, T., Luo, P., Tenenbaum, J.B., Gan, C.: Dynamic visual reasoning by learning differentiable physics models from video and language (2021)
5. Dosovitskiy, A., et al.: An image is worth 16x16 words: Transformers for image recognition at scale (2020)
6. Duan, J., Dasgupta, A., Fischer, J., Tan, C.: A survey on machine learning approaches for modelling intuitive physics (2022)
7. Gao, Z., Tan, C., Wu, L., Li, S.Z.: SimVP: Simpler yet better video prediction (2022)
8. Ge, J., et al.: Learning the relation between similarity loss and clustering loss in Self-Supervised learning (2023)
9. Ha, D., Schmidhuber, J.: World models (2018)
10. Hafner, D., Lillicrap, T., Ba, J., Norouzi, M.: Dream to control: Learning behaviors by latent imagination (2019)
11. Hochreiter, S., Schmidhuber, J.: Long short-term memory. Neural Comput. $9(8)$, 1735–1780 (1997)
12. Johnson, J., Hariharan, B., van der Maaten, L., Fei-Fei, L., Zitnick, C.L., Girshick, R.B.: CLEVR: A diagnostic dataset for compositional language and elementary visual reasoning. CoRR abs/1612.06890 (2016), http://arxiv.org/abs/1612.06890
13. Kandukuri, R.K., Achterhold, J., Moeller, M., Stueckler, J.: Physical representation learning and parameter identification from video using differentiable physics. Int. J. Comput. Vis. $130(1)$, 3–16 (2022)
14. Kim, T., Ahn, S., Bengio, Y.: Variational temporal abstraction. CoRR abs/1910.00775 (2019), http://arxiv.org/abs/1910.00775
15. Kingma, Ba: Adam: A method for stochastic optimization. arXiv:1412.6980 (2017)
16. LeCun, Y.: A path towards autonomous machine intelligence
17. Lee, S., Kim, H.G., Choi, D.H., Kim, H.I., Ro, Y.M.: Video prediction recalling long-term motion context via memory alignment learning (2021)
18. Li, Z., Zhu, X., Lei, Z., Zhang, Z.: Deconfounding physical dynamics with global causal relation and confounder transmission for counterfactual prediction. AAAI $36(2)$, 1536–1545 (2022)
19. Lin, Z., Li, M., Zheng, Z., Cheng, Y., Yuan, C.: Self-Attention ConvLSTM for spatiotemporal prediction. AAAI $34(07)$, 11531–11538 (2020)
20. Lotter, Kreiman, Cox: Deep predictive coding networks for video prediction and unsupervised learning. arXiv:1605.08104 (2017)
21. Lotter, W., Kreiman, G., Cox, D.: A neural network trained to predict future video frames mimics critical properties of biological neuronal responses and perception (2018)
22. Mao, J., Yang, X., Zhang, X., Goodman, N., Wu, J.: CLEVRER-Humans: Describing physical and causal events the human way (2022)
23. Pan, M., Zhu, X., Wang, Y., Yang, X.: Iso-Dream: Isolating and leveraging non-controllable visual dynamics in world models (2022)

24. Papineni, K., Roukos, S., Ward, T., Zhu, W.J.: BLEU: a method for automatic evaluation of machine translation. In: Proceedings of the 40th Annual Meeting on Association for Computational Linguistics, ACL 2002, pp. 311–318. Association for Computational Linguistics, USA (2002)

25. Piloto, L.S., Weinstein, A., Battaglia, P., Botvinick, M.: Intuitive physics learning in a deep-learning model inspired by developmental psychology. Nat. Hum. Behav. **6**(9), 1257–1267 (2022). https://doi.org/10.1038/s41562-022-01394-8

26. Tang, Q., Zhu, X., Lei, Z., Zhang, Z.: Intrinsic physical concepts discovery with Object-Centric predictive models (2023)

27. Vaswani, A., et al.: Attention is all you need. CoRR abs/1706.03762 (2017). http://arxiv.org/abs/1706.03762

28. Wang, Y., Gao, Z., Long, M., Wang, J., Yu, P.S.: PredRNN++: Towards a resolution of the Deep-in-Time dilemma in spatiotemporal predictive learning (2018)

29. Wang, Y., et al.: PredRNN: a recurrent neural network for spatiotemporal predictive learning (2021)

30. Wu, B., Yu, S., Chen, Z., Tenenbaum, J.B., Gan, C.: STAR: a benchmark for situated reasoning in Real-World videos (2022)

31. Ye, T., Wang, X., Davidson, J., Gupta, A.: Interpretable intuitive physics model. In: Proceedings of (ECCV) European Conference on Computer Vision, pp. 89–105 (2018)

32. Yi, K., et al.: CLEVRER: CoLlision events for video REpresentation and reasoning. arXiv:1910.01442 (2020)

33. Yi, K., et al.: Clevrer: collision events for video representation and reasoning. In: ICLR (2020)

Transferring a Learned Qualitative Cart-Pole Control Model to Uneven Terrains

Domen Šoberl[1]([⊠]) and Ivan Bratko[2]

[1] Faculty of Mathematics, Natural Sciences and Information Technologies, University of Primorska, Koper, Slovenia
domen.soberl@famnit.upr.si
[2] Faculty of Computer and Information Science, University of Ljubljana, Ljubljana, Slovenia
ivan.bratko@fri.uni-lj.si

Abstract. Qualitative modeling can be applied to the control of dynamic systems by following these steps: (1) learning a qualitative model of the controlled dynamic system from the system's behaviors in time, (2) using the learned model to derive a qualitative plan for the control task, and (3) executing the qualitative plan on an actual dynamic system. This approach has been demonstrated in the usual cart-pole control domain as significantly more sample efficient than the usual variants of reinforcement learning, by at least two orders of magnitude. The qualitative approach also enables better explanation of the learned control strategy through symbolic planning. In this paper, we generalize the cart-pole problem to uneven terrains, such as driving over a crater or a hill. We study whether the learned flat-surface qualitative controller can be successfully transferred to the tasks of negotiating uneven terrain. Experiments show that the flat-surface qualitative controller is remarkably robust on new, more difficult tasks.

Keywords: Qualitative Modeling · Qualitative Reasoning · Qualitative Control · Transfer Learning

1 Introduction

The problem of controlling the cart and pole system also called the inverted pendulum, is a popular benchmark problem for control learning methods. Michie and Chambers [9] were among the first to study adaptive control on this dynamic system. They implemented a reinforcement learning algorithm called BOXES, which discretized the continuous domain into 'boxes' and kept a record on how actions are performed within each 'box'. Later experiments involved various types of neural networks [1,2,5], policy gradient learning [12], and Q-learning [5,8,10]. Ramamoorthy and Kuipers [11] used qualitative modeling to design a controller for the pole-and-cart system. Their control policy, which was derived manually, was robust enough to accommodate a large amount of abuse from the user.

In [15], experiments with cart-pole control are presented using an approach in which a qualitative model is learned through experimentation. Then a control policy is derived automatically by planning with the learned model. The result of planning with a qualitative model is called a qualitative plan. Such a qualitative plan cannot be directly executed on the actual dynamic system or its simulator because quantitative actions are needed for that. Therefore, qualitative actions in a qualitative plan have to be first reified into numerical values, which is done by a 'reactive executor' [13]. The reactive executor is also capable of adaptation to concrete numerical system models that share the same qualitative model. This adaptation ability was experimentally shown to work rather well over several examples of dynamic systems, including quadcopter flying [16], moving complicated objects by pushing [17], robotic walk [13], and standard cart-pole control [15]. It should be noted that qualitative plans enable interpretable strategies for the control of dynamic systems, which was discussed in particular in [15].

In this paper, we generalize the standard cart-pole problem to driving over uneven terrain. Our approach is based on learning a *qualitative model* and using this model to derive a controller through planning and learning. The approach has the advantage in comparison with the usual reinforcement learning in that (1) learning a qualitative model is easier than a quantitative model, and (2) a qualitative model enables symbolic planning of the control task. For these reasons, learning with this approach can be significantly faster (in terms of required sample size) than standard reinforcement learning [4,12].

We investigate the question, how robust is the qualitative controller trained on the usual, flat surface, when transferred to the more difficult problem of driving over uneven terrain. We present the results of experiments carried out in a simulated environment.

In Sect. 2 we introduce the cart-pole problem on uneven terrain. In Sect. 3 we give details of qualitative learning and planning for the flat-surface version of the problem. In Sect. 4 we explain how qualitative plans are executed. In Sect. 5 we present experiments in the transfer of the flat-surface qualitative controller to the tasks on uneven terrain.

2   The Cart-Pole System on Uneven Terrain

In the usual version of the cart-pole system, a pole is freely hinged on the top of a wheeled cart that moves along a one-dimensional track on a flat surface. It is assumed that there is no friction between the cart and the track or the pole and the cart. The controller can apply force F of a fixed magnitude at discrete time intervals, pushing the cart in either direction, left or right. This is also known as bang-bang control. In this paper, we consider the control tasks when the goal of control is for the cart to reach a given goal position from a given starting position while preventing the pole from falling. It should be noted that this version of the cart-pole task is considerably more difficult than just balancing the pole, which is the task usually tackled in experiments with reinforcement learning.

An example of extending the standard control task to uneven terrain is shown in Fig. 1. The terrain has the shape of a crater, the cart's starting position is on

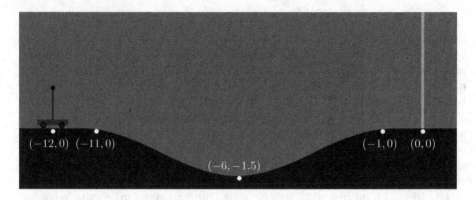

Fig. 1. The task is to drive the cart from the start position at $(-12\,\text{m},\,0\,\text{m})$ through the crater to reach the goal position at $(0\,\text{m},\,0\,\text{m})$ indicated by the yellow pole. Three white dots at $(-11\,\text{m},\,0\,\text{m})$, $(-6\,\text{m},\,-1.5\,\text{m})$ and $(-1\,\text{m},\,0\,\text{m})$ indicate the left edge, the bottom and the right edge of the crater, respectively. The shape of the crater is modeled by a Bezier curve of 2nd degree interpolated between these three dots.

the left of the crater, and the goal is to reach a given position to the right of the crater.

To simulate the cart-pole on uneven terrain, we have to generalize the usual mathematical model of cart-pole, which assumes a flat surface. Figure 2 shows the variables in the generalized model. We derived the following differential equations for pole-and-cart on a slope:

$$\ddot{x} = \frac{F + ml(\dot{\theta}^2 \sin\theta - \ddot{\theta}\cos\theta)}{M + m} - g\sin\varphi \tag{1}$$

$$\ddot{\theta} = \frac{(M+m)g\sin(\theta - \varphi)}{(M+m)l - ml\cos^2\theta} -$$
$$- \frac{\cos\theta\left(F + ml\dot{\theta}^2\sin\theta - (M+m)g\sin\varphi\right)}{(M+m)l - ml\cos^2\theta} \tag{2}$$

The parameters in our experiments were:

Cart mass: $M = 1\,\text{kg}$
Pole mass: $m = 0.1\,\text{kg}$
Pole length: $l = 1\,\text{m}$
Gravity: $g = 9.81\,\text{m/s}^2$

The amount of force F varied between experiments so that $|F|$ was either 5 N or 10 N. These amounts correspond to a "weak" or a "strong" motor, which produced different control behaviors, with the stronger motor being more effective. Action frequency was 50 Hz, which means that the controller determined the next force direction every 0.02 s.

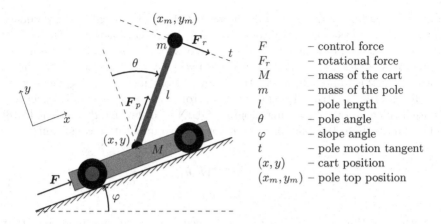

Fig. 2. Cart and pole on a slope, the slope angle is denoted by φ.

3 Learning a Qualitative Controller for Cart-Pole on Flat Surface

The approach of learning a qualitative controller from observations we used in this paper is as described in detail in [13,16]. The approach consists of the following:

1) Collect learning data by observing the target dynamic system when sequences of random actions are executed.
2) Use the collected observations as examples for learning a qualitative model of QSIM-type of the dynamic system.
3) Use QSIM-like qualitative simulation [6,7] to generate a search space for qualitative plans that possibly solve the control task. Note that successful execution of these plans on the actual (quantitative) system is not guaranteed, due to the ambiguous nature of qualitative simulation, and success depends on the numerical parameters of our concrete dynamic system. So these plans can be viewed as a source of plausible ideas for solving the control task.
4) Try to execute a qualitative plan by the 'reactive executor' on the dynamic system. The reactive executor observes the actual (quantitative) state of the controlled system and finds appropriate numerical actions according to the qualitative plan.

3.1 Learning a Qualitative Model

We here summarize the results of applying the above procedure to learning to control the cart-pole system on a flat surface as conducted in [15]. To collect numerical data needed to learn a qualitative model of cart-pole, we acquired samples of the form $(F, x, \dot{x}, \theta, \dot{\theta})$ at the rate of 50 Hz of the running experimentation with the system controlled by random bang-bang actions. After 3 s of experimentation, two behaviors in time were observed, both ending with a pole crash. 150 samples were collected in these two traces. Note that in learning a controller in this approach, unlike the usual reinforcement learning, there is no need to observe any trials in which the control task was successfully executed. We used the qualitative learning program QUIN (QUalitative INduction) [3, 18] to learn from these 150 examples the following two qualitative constraints:

$$\ddot{\theta} = M^{-,+}(F, \theta)$$
$$\ddot{x} = M^{+}(F)$$

$$(3)$$

Here, the notation $y = M^{+}(x)$ means, as usual in qualitative modeling, that y is a monotonically increasing function of x. A multivariate constraint like $z = M^{-,+}(x, y)$ means that z monotonically decreases with x and monotonically increases with y. More formally: $\partial z/\partial x < 0$, and $\partial z/\partial y > 0$. Note that these learned constraints are an approximation to the qualitative abstraction of the differential equation model of cart-pole for the flat surface ($\varphi = 0$). A completely correct qualitative model would be considerably more complex.

3.2   Finding a Qualitative Plan

Given the above qualitative model (as two qualitative constraints), we can use a QSIM-like algorithm [6, 7] to generate the search space of possible qualitative behaviors for all possible action sequences (force F in the case of the cart-pole system). The planning consists of searching for such action sequences that, starting in the given start state of the system, possibly result in a given goal state. For example, starting at the start state with $x = x_0$ and the pole upright, with all the variables being steady, we want to reach a goal state with $x = x_1$ ($x_1 > x_0$) and the pole is upright again, and all the variables steady. This means moving the cart to the right to some position x_1.

The search space consists of all possible qualitative states of the system, connected by state transitions, allowed under the assumption of smoothness, that comply with the given qualitative model. Qualitative states are obtained by discretizing the domain at certain landmarks. Typically, the initial and goal values are taken as landmarks, while the learned qualitative model may additionally introduce it own landmarks, e.g. QUIN may learn that a certain qualitative constraint holds for all $x \leq l$ and another for all $x > l$, hence l in considered a landmark within the real-valued domain of x. In our cart-pole domain, the landmarks for x are the initial and the goal value x_0 and x_1, and for θ the goal value 0 and the interval end-points -180 and 180. Time derivatives $\dot{x}$ and $\dot{\theta}$ contain only the landmark 0 and are thus qualitatively abstracted as *inc*, *dec* and *std*, respectively for positive values, negative values and 0. Suppose $x_0 = 0$ and $x_1 = 1$. The numerical state ($x = 0.5, \dot{x} = 0, \theta = 45, \dot{\theta} = -0.1$) is therefore qualitatively abstracted as ($x = x_0..x_1/std, \theta = 0..180/dec$).

The *assumption of smoothness* means that the transition between two qualitative states is possible only if all variables and their time derivatives transition in a numerically continuous manner. Considering our variable θ, transition $(\theta = -180..0/inc) \rightarrow (\theta = 0/inc)$ is possible, while $(\theta = -180..0/inc) \rightarrow (\theta = 0..180/inc)$ violates the continuity of θ. Transition $(\theta = 0..180/inc) \rightarrow (\theta = 0..180/dec)$ is not possible, because it violates the continuity of $\dot{\theta}$, while the sequence of transitions $(\theta = 0..180/inc) \rightarrow (\theta = 0..180/std) \rightarrow (\theta = 0..180/dec)$ is possible. See [6,7,13] for more details.

Table 1. An executable plan to move the cart-pole from x_0 to x_1 found by the qualitative planner. The symbols min and max stand for different minimal and maximal values for different variables F, x and θ.

Step	F	$x/\dot{x}$	$\theta/\dot{\theta}$
0	0	x_0/std	0/std
1	min..0	min..x_0/dec	0..max/inc
2	0	min..x_0/dec	0..max/inc
3	0..max	min..x_0/dec	0..max/inc
4	0..max	min..x_0/std	0..max/std
5	0..max	min..x_0/inc	0..max/dec
6	0..max	x_0/inc	0..max/dec
7	0..max	$x_0..x_1$/inc	0..max/dec
8	0	$x_0..x_1$/inc	0..max/dec
9	min..0	$x_0..x_1$/inc	0..max/dec
10	min..0	x_1/std	0/std

The planner may find many different plans of various sizes. Some may qualitatively be correct, but may not be executable under some specific numerical properties of the system. This can only be verified by actually executing a plan and discarding it if the execution is unsuccessful. To find plans that have the best practical chances to succeed when executed, we use the following heuristics:

1) Favor short solutions. Short plans offer simpler explanations than long ones and hopefully take less time to find. They are therefore first to be tested. It should be noted that there is no guarantee that the shortest qualitative plan will be in fact the quickest to achieve the goal on the physical system.
2) Favor solutions with effective actions. An action A in some qualitative state S is *effective* if it causes a deterministic transition to the next state. Plans with effective actions are more likely to succeed.

Using these heuristics, the shortest executable plan found by the planner to move the cart-pole from position x_0 to position x_1 is shown in Table 1 and visualized in Fig. 3.

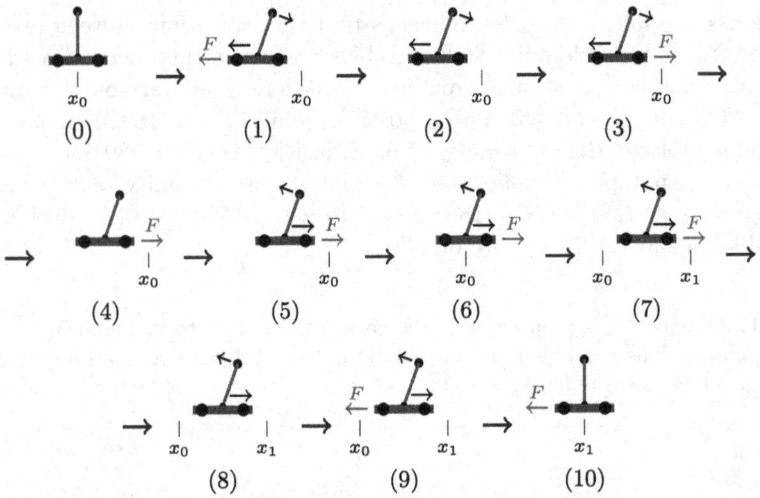

Fig. 3. A visualization of the qualitative plan from Table 1. Position x is depicted relative to landmarks x_0 and x_1. Blue arrows show transitions between system's states. Black arrows depict the directions of motion of the cart and the pole. The directions of actions F are respectively shown by red arrows.

A possible explanation of the plan is the following: In the initial state (state 0) apply a negative force F, which will cause the cart to move backward and the pole to lean forward. The cart will move left from its initial position x_0 (state 1). Stop applying the negative force. The cart and the pole will keep their momenta (state 2). Apply a positive force F, the cart and the pole are expected to keep moving in the same direction for a while (state 3). Eventually, they would both come to a stop, while the positive force is still being applied (state 4), and change their direction of motion (state 5). The cart will assume its initial position x_0, while the cart and the pole will maintain their previous qualitative directions (state 6). The cart will then move past x_0, between the points x_0 and x_1 (state 7). Stop applying the positive force. The cart and the pole will keep their momenta (state 8). Apply a negative force F. The cart and the pole will for a while keep moving in the same direction (state 9), but with the right amount of force at the right timing, the cart and the pole would stop precisely at their goal positions (state 10).

Such an ideal execution is obviously not feasible in practice and the cart is likely to overshoot the goal position x_1 by a certain amount. Finding itself in some qualitative state with $x = x_1..max$, the system will devise a similar but mirrored plan to move *left* towards x_1. Finally, the execution would end up continuously balancing the pole near the goal position, as seen in Fig. 6.

4   Reactive Execution

The role of our reactive executor is to implement in real-time a numerical transition $S_i \rightarrow S_{i+1}$ between two consecutive qualitative states in a given qualitative plan $S_0 \rightarrow \cdots \rightarrow S_n$. The main challenge of such an execution are unknown numerical properties of the system, which makes it difficult to predict what numerical state the system will transition to after some continuous action has been executed for a certain amount of time and with a certain magnitude. We approach this problem reactively, which means that we consider only one action ahead after observing the current numerical state. In our case, a decision about action is made 50 times per second. Every time, the executor performs the following steps:

1) Observe the current numerical state, the value, speed and acceleration of each domain variable.
2) Estimate how far each variable is from its next goal value defined by the next qualitative state S_{i+1}. This 'distance' to the next goal value is estimated as the minimal time needed to reach the goal value, according to the previously observed speeds and accelerations. Classical kinematic equations are used independently on each variable. We denote by e_i the time estimate for variable x_i.
3) Construct an n-dimensional hypersurface H, where n is the number of observed variables, and embed the surface into an $(n+1)$-dimensional hyperspace, so that the next immediate goal state S_{i+1} is at the global minimum, and the current state S at a certain distance from S_{i+1}, as depicted in Fig. 5. The hypersurface is constructed in such a way that the steepness of the slope increases with the goal distance.
4) Use the (learned) qualitative model to resolve the qualitative effect of each available action. A qualitative effect of an action merely determines in which direction a variable would change, but not its rate of change.
5) Choose and execute the action whose effect follows the steepest descent along the surface H (defined by formula 4).

The plan that the executor attempts to execute is a *reduced* version of the planner's output. Consider again the plan from Fig. 3. This detailed plan depicts not only the positions of individual variables, but also the directions of their change (qualitative velocities). These, however, can be inferred by the executor from consecutive positions. E.g., if the pole transitions from $\theta = 0$ to $\theta = 0..90$, the qualitative direction of θ must obviously be *increasing*. The executor, therefore, follows a succession of qualitative states in the plan that differ only in the qualitative values of the state variables, but not in actions. The plan from Fig. 3 is thus reduced by the executor as shown in Fig. 4. The crucial information passed from the planner to the executor is here to tilt the pole in the direction of the goal before moving forward towards x_1.

Besides the plan, the executor can also be given a set of numerical constraints that determine fail states. In our case, we allow the pole to move within a

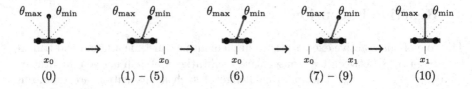

Fig. 4. A reduced sequence of qualitative states executed by the reactive executor when following the qualitative plan from Fig. 3. A numerical constraint $\theta \in [\theta_{\min}, \theta_{\max}]$ is given additionally with the qualitative model.

predetermined interval $\theta \in (\theta_{\min}, \theta_{\max})$, otherwise the execution fails. While the executor will aim to bring the variables to their goal values, it will simultaneously try to keep numerically constrained variables within their constraints, which is done by targeting the midpoint of the constraining interval. The idea behind this type of numerical constraints is to specify a fail-safe ϵ-neighborhood around a 'safe point'. However, other types of numerical constraints may also be considered in the future.

For a numerically constrained variable x_i, the time to reach its interval's midpoint from the current state is estimated (denoted as e_i), as well as the time to reach the midpoint from the farthest point within that interval (denoted as $e_i^{\max}$). The conversion of the current numerical state to such time estimates allows a comparison between variables of different measures and kinematic properties, either pursuing a goal value or being numerically constrained to an interval.

Prioritization of actions is done by determining the steepness of the descent along the hypersurface H, which is constructed in the following way. Let x_i for all $i \leq k$ have a goal value, while x_i for all $i > k$ are numerically constrained. Define the function:

$$f(e_1, \ldots, e_n) = \sum_{i=1}^{k} \frac{e_i^2}{2} + \sum_{i=k+1}^{n} \left(\left(1 + \frac{e_i}{e_i^{\max}}\right)^{-1} \cdot \left(1 - \frac{e_i}{e_i^{\max}}\right)^{-1} - 1 \right), \quad (4)$$

where $(e_1, \ldots, e_n) \in H$ is a point on the surface H and F the embedding of H into a higher dimensional space. An example of such an embedding for variables x and θ is shown in Fig. 5. Note that the goal is reached when $e_i = 0$ for all $i \leq k$, while the executor also targets $e_i = 0$ for all $i > k$. The gradient

$$\nabla f = \left(\frac{\partial f}{\partial e_1}, \ldots, \frac{\partial f}{\partial e_n} \right) \quad (5)$$

at the current time estimates $(e_1, \ldots, e_n)$ represents the direction of the steepest ascent, while the executor follows the direction of the steepest descent $-\nabla f$ towards the goal.

After the gradient is computed, the executor must resolve, which qualitative action works in the direction of the steepest descent, or at least closest to it. Qualitative action $A = [\mathrm{dir}_{c_0}, \ldots, \mathrm{dir}_{c_m}]$ is defined as a vector of qualitative directions $\mathrm{dir}_i \in \{\mathrm{inc}, \mathrm{std}, \mathrm{dec}\}$ of control variables $c_0, \ldots, c_m$. The effect of action

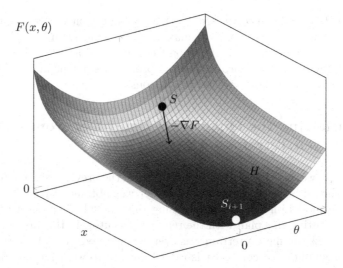

Fig. 5. Executor's evaluation of the current system's state S relative to the next immediate goal state S_{i+1} in the form of a hypersurface H. The shape of H is determined by the executor's observations of speeds and accelerations for each variable x_i. Action with the steepest descending effect is chosen for execution.

A on other variables is deduced from the given qualitative model. The qualitative algebra for such deduction is theoretically analyzed in [13,14]. It is, however, not difficult to understand intuitively. Consider again the learned qualitative model (3) and the learned qualitative constraint $\ddot{\theta} = M^{-,+}(F, \theta)$. Note that in our domain, F is the sole control variable. Qualitative action $\text{dir}_F = dec$ implies $\text{dir}_{\ddot{\theta}} = inc$, if $\text{dir}_\theta \in \{std, inc\}$. However, if $\text{dir}_\theta = dec$, the effect of F on $\ddot{\theta}$ is qualitatively non-deterministic. The qualitative effects of time derivatives are further propagated to their time integrals, e.g., from $\ddot{\theta}$ to $\dot{\theta}$ and to θ. A qualitative vector $E = [\text{dir}_{x_0}, \ldots, \text{dir}_{x_n}]$ of qualitative effects of action A on every variable x_i is finally obtained. Deterministic qualitative effects *inc* and *dec* are interpreted as real values 1 and -1 respectively, while non-deterministic effects and *std* are interpreted as value 0. Action A is then quantified as

$$Q(A) = -\nabla f \cdot E, \tag{6}$$

and a qualitative action A_i with the maximum $Q(A_i)$ is chosen for execution. Interpreting non-deterministic effects as 0 results in prioritizing actions with a higher level of determinism. Therefore, if two actions could achieve the same qualitative effect, the one with a higher certainty is chosen over the other. Suppose the qualitative effect of action A_1 on variable x is *inc* and the qualitative effect of action A_2 is non-deterministic. None of the two actions exclude the possibility of increasing x, but action A_1 is certain to do it.

The chosen qualitative action A determines for each output variable (signal) whether to increase or decrease its current value. In our domain, this is either

F = inc or F = dec. Since we are using bang-bang control, this immediately translates to numerical actions F = max and F = min. Even though this type of control may seem crude, a high action frequency still allows the controller to successfully regulate any critical numerical property, e.g., the angle of the pole, by issuing patterns of short *inc* and *dec* actions.

5   Transferring the Qualitative Controller from Flat to Uneven Terrains

In this section, we test the qualitative controller of the previous section on uneven terrain. It should be noted that this controller was obtained by merely learning a qualitative model on a flat surface, using that model to derive a qualitative plan, and executing this plan with the reactive executor. In this process, no case of a control task on uneven terrain was ever encountered. It is of interest to see, whether the so-obtained controller is robust enough to work on modified, more difficult tasks such as driving over a crater (Fig. 1).

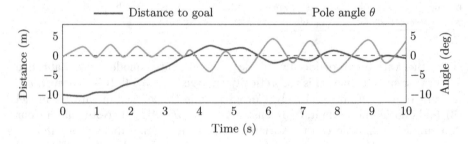

Fig. 6. Reactive execution of the qualitative plan for cart-pole on a flat surface.

The qualitative model (3), which was learned and previously successfully used on flat terrain, was transferred to the uneven terrain environment unchanged. However, as seen in Fig. 2, the new environment introduces the variable φ, which indicates the steepness of the slope. This changes the notion of the pole's 'upwards' position from $\theta = 0$ to $(\theta - \varphi) = 0$. Hence the goal conditions and constraints that involve pole angle were altered accordingly.

We experimented with three types of tasks:

1) Driving over a crater with start and goal as in Fig. 1: start left of the crater and drive to the goal on the right of the crater;
2) Driving out of the crater: start at the bottom and drive to the goal on the right of the crater;
3) Driving over a hill with the shape complementary to the crater of Fig. 1; start on the left of the hill and drive to the goal to the right of the hill.

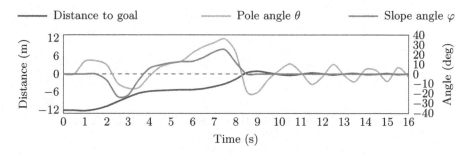

Fig. 7. Trace in time of driving over the crater with bang-bang force 10 N. The angles are in degrees, and the distance to the goal (blue curve) is in meters.

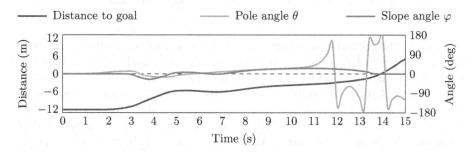

Fig. 8. Unsuccessful control in driving over the crater with bang-bang force 5 N. The pole collapses soon after 11 s (red curve) at the steep section of the crater when the motor is too weak to control the pole. After that, the pole rotates in full circles a few times.

In the experiments, we varied the amount of bang-bang control force. We started with the usual $|F| = 10$ N. Figure 7 shows the control trace for successfully driving over the crater (task 1). Similarly, the flat surface controller completed the remaining two "non-flat" tasks above without problems. Then we reduced the motor strength $|F|$ to make the control task harder. A weaker motor with $|F| = 5$ N then indeed failed on all three tasks. Figure 8 shows the control trace with 5 N on task 1 (driving over crater). The failure occurred at about the 11th second of the execution when the pole crashed while the cart was driving out of the crater over the steep section close to the goal. We then analyzed this control behavior to find the reason for failure. The conclusion of this analysis was that the task cannot be completed, even theoretically, with the amount of force limited to 5 N. This can be explained with the following reasoning. The balancing of the pole is done by the pole's angle oscillating about the vertical. For example, if the pole is leaning to the right, the cart has to be accelerated to the right to cause negative angular acceleration. Accelerating the cart to the right is normally done by a positive force F, pushing to the right. When the cart is climbing out of the crater, the control force has to overcome the opposing gravitational force. If the motor force is too weak then the pole cannot be

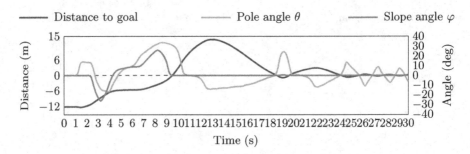

Fig. 9. Control trace of driving over the crater with bang-bang force 6.9 N, which is just enough to catch the pole, but at the price of largely overshooting the goal $x = 0$ and only then returning to the goal (blue curve). (Color figure online)

controlled. The critical situation arises at the steepest section of the slope where the slope angle is largest, about 0.45 rad and the pole angle is about 0.61 rad. For the pole angular acceleration to be negative, the required force $F > 6.83$ N, which shows that control cannot succeed with $|F| = 5$ N. Interestingly, setting $|F| = 6.9$ N (just above the required threshold) is enough for the qualitative controller to complete the task (Fig. 9). The trace shows that this is achieved at a narrow margin, when the cart overshoots the goal $x = 0$ by first moving far to the right, before eventually stabilizing close to the goal.

6   Conclusions

The approach to the learning to control dynamic systems experimented with in this paper consists of: learning a QDE qualitative model (Qualitative Differential Equations) from examples of random behaviors of the controlled system, finding a qualitative plan for the control task by searching the qualitative state space generated by the learned model, and reactively executing a qualitative plan on the actual dynamic system. This approach was applied to a standard version of the cart-pole control task in [15], in which the cart moves on a flat surface. In this paper, we generalized the application problem to cart-pole on uneven surfaces, such as the task of driving over a crater or a hill. We investigated whether the flat-surface controller can be usefully transferred to the generalized control problem. We showed in the experiments that the flat-surface controller can handle the new control tasks (never encountered by the controller before) remarkably robustly. The key seems to be the generality of qualitative plans learned for the simpler control problem, and the robustness of reactive execution of qualitative plans, even if this may be suboptimal. These features make the sample efficiency of learning qualitative controllers much better than that usually reported in reinforcement learning approaches, e.g. [4,12]. As experiments indicate, typical reinforcement learning methods require at least two orders of magnitude larger samples than our approach based on qualitative learning and planning.

References

1. Anderson, C.W.: Learning to control an inverted pendulum using neural networks. IEEE Control Syst. Mag. **9**(3), 31–37 (1989)
2. Barto, A.G., Sutton, R.S., Anderson, C.W.: Neuronlike adaptive elements that can solve difficult learning control problems. IEEE Trans. Syst. Man Cybern. SMC **13**(5), 834–846 (1983)
3. Bratko, I., Šuc, D.: Learning qualitative models. AI Mag. **24**(4), 107–119 (2003)
4. Hein, D., Limmer, S., Runkler, T.A.: Interpretable control by reinforcement learning. IFAC-PapersOnLine **53**(2), 8082–8089 (2020). https://doi.org/10.1016/j.ifacol.2020.12.2277
5. Hosokawa, S., Kato, J., Nakano, K.: A reward allocation method for reinforcement learning in stabilizing control tasks. Artif. Life Rob. **19**(2), 109–114 (2014). https://doi.org/10.1007/s10015-014-0146-0
6. Kuipers, B.: Qualitative simulation. Artif. Intell. **29**(3), 289–338 (1986)
7. Kuipers, B.: Qualitative Reasoning: Modeling and Simulation with Incomplete Knowledge. MIT Press, Cambridge, MA, USA (1994)
8. Linglin, W., Yongxin, L., Xiaoke, Z.: Design of reinforce learning control algorithm and verified in inverted pendulum. In: 2015 34th Chinese Control Conference (CCC), pp. 3164–3168 (2015)
9. Michie, D., Chambers, R.A.: BOXES: an experiment in adaptive control. Mach. Intell. **2**, 125–133 (1968)
10. Puriel-Gil, G., Yu, W., Sossa, H.: Reinforcement learning compensation based PD control for inverted pendulum. In: 15th International Conference on Electrical Engineering, Computing Science and Automatic Control (CCE), pp. 1–6 (2018)
11. Ramamoorthy, S., Kuipers, B.: Qualitative heterogeneous control of higher order systems. In: Maler, O., Pnueli, A. (eds.) HSCC 2003. LNCS, vol. 2623, pp. 417–434. Springer, Heidelberg (2003). https://doi.org/10.1007/3-540-36580-X_31
12. Riedmiller, M., Peters, J., Schaal, S.: Evaluation of policy gradient methods and variants on the cart-pole benchmark. In: 2007 IEEE International Symposium on Approximate Dynamic Programming and Reinforcement Learning, pp. 254–261 (2007)
13. Šoberl, D.: Automated planning with induced qualitative models in dynamic robotic domains. Ph.D. thesis, University of Ljubljana (2021). https://repozitorij.uni-lj.si/IzpisGradiva.php?id=126285
14. Šoberl, D., Bratko, I.: Reactive motion planning with qualitative constraints. In: Benferhat, S., Tabia, K., Ali, M. (eds.) IEA/AIE 2017. LNCS (LNAI), vol. 10350, pp. 41–50. Springer, Cham (2017). https://doi.org/10.1007/978-3-319-60042-0_5
15. Šoberl, D., Bratko, I.: Learning explainable control strategies demonstrated on the pole-and-cart system. In: Wotawa, F., Friedrich, G., Pill, I., Koitz-Hristov, R., Ali, M. (eds.) IEA/AIE 2019. LNCS (LNAI), vol. 11606, pp. 483–494. Springer, Cham (2019). https://doi.org/10.1007/978-3-030-22999-3_42
16. Šoberl, D., Bratko, I.: Learning to control a quadcopter qualitatively. J. Intell. Rob. Syst. (2020). https://doi.org/10.1007/s10846-020-01228-7
17. Šoberl, D., Žabkar, J., Bratko, I.: Qualitative planning of object pushing by a robot. In: Esposito, F., Pivert, O., Hacid, M.-S., Raś, Z.W., Ferilli, S. (eds.) ISMIS 2015. LNCS (LNAI), vol. 9384, pp. 410–419. Springer, Cham (2015). https://doi.org/10.1007/978-3-319-25252-0_44
18. Šuc, D.: Machine Reconstruction of Human Control Strategies. Frontiers in Artificial Intelligence and Applications, IOS Press, Inc. (2003)

Which Way to Go - Finding Frequent Trajectories Through Clustering

Thiago Andrade[1,2]([✉]) [iD] and João Gama[1,3] [iD]

[1] INESC TEC, Faculdade de Engenharia da Universidade do Porto, Porto, Portugal
thiago.a.silva@inesctec.pt
[2] Faculdade de Ciências, Universidade do Porto, Porto, Portugal
[3] Faculty of Economics, University of Porto, 4200-464 Porto, Portugal

Abstract. Trajectory clustering is one of the most important issues in mobility patterns data mining. It is applied in several cases such as hot-spots detection, urban transportation control, animal migration movements, and tourist visiting routes among others. In this paper, we describe how to identify the most frequent trajectories from raw GPS data. By making use of the Ramer-Douglas-Peucker (RDP) mechanism we simplify the trajectories in order to obtain fewer points to check without losing information. We construct a similarity matrix by using the Fréchet distance metric and then employ density-based clustering to find the most similar trajectories. We perform experiments over three real-world datasets collected in the city of Porto, Portugal, and in Beijing China, and check the results of the most frequent trajectories for the top-k origins x destinations for the moves.

Keywords: Clustering · Fréchet Distance · Transportation · Frequent Trajectories · Mobility Patterns

1 Introduction

With the development of location-based positioning devices and the advent of the Internet-of-Things (IoT), more and more moving objects are traced and their trajectories are recorded joining diversified information about their carriers and equipment. These data comprise a rich source of spatial and temporal semantic information. Therefore, moving object trajectory clustering undoubtedly becomes the focus of the study in moving object data mining [1].

Many areas can leverage the similarity of trajectory analysis such as policy-makers/government, transportation companies/authorities, last-mile parcel carriers, biologists, and retail and marketing companies, among others. In the public sector, the managers can analyze the moves of people at different hours of the day and week to promote changes in the infrastructure of a region, change the bus routes, increase the number of trains or metro cars, and take measures to diminish the bottlenecks of traffic hot-spots and try to diminish the vehicle emissions. The private sector can also make use of these studies to target advertisements to specific groups of users that travel along some routes or visit some

A. Bifet et al. (Eds.): DS 2023, LNAI 14276, pp. 460–473, 2023.
https://doi.org/10.1007/978-3-031-45275-8_31

points of interest. Biologists can use these techniques to help to understand the whereabouts of animals such as birds and fish, where they go, and which are the recurrent routes taken. Tourism in both public and private sectors can also make good use of trajectory analysis by recommending tourist routes or using these routes to improve or even deploy services along the path.

A common approach to performing trajectory analysis is by making use of clustering techniques where the process assigns a set of similar trajectories into groups (the clusters) having highly similar trajectories within each cluster and low similar trajectories among the different cluster sets [1,2]. Among the clustering approaches, one, in particular, has shown to be more suitable for trajectory analysis due to its possibility of forming clusters of arbitrary shapes in Euclidean space: the density-based approach. One of the most popular algorithms in this group is DBSCAN [3]. Still, one key component of good-quality trajectory analysis is how to calculate the similarity between trajectories in a group. Different similarity measures can be used but not all of them take into consideration the order of the data points in the trajectory set and this is paramount for a good-quality cluster of similar trajectories. Fréchet distance is one of the metrics that can be used to solve the problem.

In this paper, we show how to identify frequent spatiotemporal trajectories using density-based clustering techniques with Fréchet distance and experiment with the proposal in three real-world trajectory datasets.

The paper is organized as follows. Section 2 briefly reviews the related works in this area and the common trajectory clustering methods. Section 3 explains the data set and the methods used in detail. Section 4, describes the experimental setup and discusses the results for the study. Conclusions and suggestions for further research are presented in Sect. 5.

2 Related Work

Nowadays, due to the massive volume of data that is being generated the need for methods to analyze mobility data has also grown. Many researchers have been applying data-mining techniques over many scenarios related to the identification of meaningful locations, habits, and common paths users and objects take for diverse goals and the use of clustering techniques is very common when dealing with these problems. We can view the clustering methods developed for handling data into five categories: partition-based methods, hierarchy-based methods, density-based methods, grid-based methods, and model-based methods [2]. In this section, we review some relevant works which leverage the information contained in these data for a multitude of different applications and show how it can be effectively used for analyzing the moves of objects.

In [4,5] Andrade et al. propose a method for discovering common pathways across users' habits without any a priori or external knowledge. First, they perform density-based clustering for spatiotemporal data to obtain the places the user visits the most. Secondly, a Gaussian Mixture Model (GMM) is applied over the clustered origin x destination (OD) places to automatically separate the trajectories into habits. Finally, they apply an extension of the Longest Common

Sub-sequence (LCSS) algorithm over the habits to find the trips that are more similar. Gudmundsson et al. [6] proposed a sub-trajectory clustering approach based on Fréchet distance, considering a trajectory as a directed curve in 2D. Dynamic Time Warping (DTW) was used by Sanchez et al. [7] for fast trajectory similarity. They proposed hashing techniques named Distance-Based Hashing (DBH) and Locality Sensitive Hashing (LSH) by clustering the trajectories using the k-means algorithm. Hung et al. [8] proposed a "clue-aware" trajectory clustering algorithm to cluster similar trajectories into groups by similarity, and then aggregate trajectories in each group to derive the corresponding trajectory centroids. Brankovic et al. [9] consider a center-based trajectory clustering algorithm to compute the ℓ-simplifications of the center trajectories, where a centroid trajectory has only ℓ points.

3   Methodology

3.1   Definitions

Here we introduce the definition of a point, trajectory, and frequent trajectory.

Point: A point is a triple of the form p = *(latitude, longitude, time)* that represents a latitude-longitude location and a time-stamp.

Trajectory: A trajectory Tr is a sequence of ordered point triples $Tr = (p_1, p_2, \ldots, p_n)$ where p_i is a point and $p_1.time < p_2.time < \ldots < p_n.time$. From this sequence of points, we can derive the trajectory length. This property is the sum of the distances between each of the sequence points often represented in meters or kilometers. Direction is also another property of the trajectory. In our study, it is denoted by the direction of the start x endpoint which means from where the individual is departing to the destination or final observation. For this study, we also observe the cardinality of the trajectory. This property is denoted by the number of points in the sequence.

Frequent Trajectory: A frequent trajectory is described as a regular route an individual tends to follow when traveling/moving between two locations (origin and destination) [10]. In a real-life example, this can be a street or highway one can take to drive from home to work, a metro line used to commute, a sidewalk that is taken to walk to the user's preferred restaurant, to go to a shopping mall, etc. In this study, we have focused on the discovery of the most frequent places that individuals visit and the common routes that are related to these displacements.

Other characteristics are also important to mention:

- Trajectories may have different lengths as individuals tend to move accordingly to their needs and singularities (e.g., N and M can be different for $Tr_i = (p_1, p_2, \ldots, p_N)$ and $Tr_j = (p_1, p_2, \ldots, p_M)$).
- Trajectories may have different directions. In the context of individuals' movement, the direction of each trajectory is an essential condition for the similarity of trajectories. As we propose the discovery of frequent routes, two

trajectories moving in opposite directions should be considered as different moves despite their close proximity to each other as they may represent different habits (e.g., going to work from home and going back home from work).

3.2 Trajectories Simplification with Ramer-Douglas-Peucker (RDP)

In some cases, GPS raw data can be very densely represented. The three datasets used in this work have different granularities and for the frequent trajectories discovery/clustering step, we end up not needing so much detail. Many of these points can be removed as they are somehow redundant whereas other key positions need to be kept. A good way to avoid unnecessary processing is by using compression techniques. For simplifying the trajectories we use the Ramer-Douglas-Peucker algorithm [11,12]. The aim of the algorithm is to produce a simplified poly-line that has fewer points than the original but still keeps the original's characteristics/shape. The method takes one threshold parameter ϵ and starts by connecting the first and the last point of the original line with a reference OD pair. It then finds the point that is furthest away from that baseline reference and checks if it's greater than ϵ. If true, it keeps the point and the function continues to recursively split the line into two segments creating new reference points and repeating the procedure. If the point is nearer to the baseline reference than ϵ it discards all the points between these reference points simplifying the trajectory.

Figure 1 (a) shows an example of a trajectory being split by the RDP algorithm.

3.3 Clustering Algorithm and Similarity Measures

Clustering is an efficient way to group data into different classes on the basis of the internal and previously unknown schemes inherent in the data and trajectory clustering is the most popular topic in current trajectory data mining. The aim is to discover the similarity (distance) in moving object databases, grouping similar trajectories into the same cluster, and finding the most common patterns [2].

Density-based clustering techniques are very popular methods for location detection, as they have the ability to detect clusters of arbitrary shapes without specifying the number of clusters in the data apriori. Furthermore, they are tolerant of outliers (noise). Some recent studies have addressed the location detection techniques in order to improve the quality of the discoveries [4,5,10, 13,14].

In this study, we apply the clustering method proposed by [10] which is a variation of DBSCAN [3] to form the clusters of trajectories between the start (origin) and end (destination) points of all the trajectories.

One of the most important parts of a clustering algorithm is the similarity measure of two items. This is the step where the distance of two points is calculated before the algorithm decides to group these items or not. Different comparison strategies must be taken accordingly to the purpose of the clustering task. Some of the most common distances are Euclidean, Hausdorff [1], Longest

Common Sub-Sequence [15], Dynamic Time Warping [1], and Fréchet distance [16].

For the Euclidean distance (ED), the similarity between two trajectories is simple and intuitive, because it is parameter-free. In addition, its time complexity is linear which means that it can handle a large dataset. However, noise existing in trajectory data will have a great influence on the result. Another main disadvantage of using the ED for measuring the similarity between trajectories is that the sampling points must be in corresponding positions (at the same time) and the trajectories must have the same length. In real-world scenarios, this is not true even though the origin and destination are the same.

Hausdorff distance (HD) between trajectory segments A and B selects the maximum unidirectional HD from A to B and from B to A. It measures the maximum mismatching degree between two trajectory segments. HD tolerates the influence caused by the disturbance of points but is sensitive to noise data. This last point is also an issue in real-world scenarios when dealing with GPS data due to the signal interference caused by objects. For this reason, we avoid using this distance function.

For the Longest Common Sub-Sequence (LCSS), as the name suggests, the idea is to get the longest list of common items in sequence between two trajectories. It uses a distance function (that can be ED or any other) to compare if the combination of pair of points is less than a threshold ϵ. Having the distance value less than the expected threshold, the value of LCSS is increased by 1. One advantage of LCSS is that it allows certain deviations existing in the sampling data (which is common in the real world). The advantages are the distance measure choice and parameter specification as well as the time complexity.

In order to find an optimal alignment between two given (time-dependent) sequences under certain restrictions the Dynamic Time Warping (DTW) algorithm was proposed. This method can match trajectories even if their lengths are different. The goal is to minimize the warping cost of finding similar paths between two trajectories. It is also sensitive to noise. The disadvantages are that when two trajectories are completely dissimilar in a small range, the DTW distance cannot be found and time cost and complexity are higher than the previous techniques.

Finally, the discrete Fréchet Distance (DFD) considers both the sequential relationship as well as the location of the points in the trajectories while measuring their similarity. It also relies on ED to calculate the distance in a point-wise fashion as shown in Eq. 1.

$$DFD(x, y) = \max \left(\|p_{i(t)} - q_{i(t)}\|, \min \left(DFD(x - 1, y), DFD(x, y - 1) \right) \right) \quad (1)$$

where given two sequences of points $p = (p_1, p_2, p_3, \ldots, p_n)$ and $q = (q_1, q_2, q_3, \ldots, q_m)$, Fréchet distance represented by $DFD(x, y)$ is the maximum of the minimum distances between points p_i and q_i. Figure 1 (b) shows an example of the Fréchet distance between two trajectories.

One of the key issues previous studies have shown is that the Fréchet distance contains the temporal relationship between the points. More particularly, the

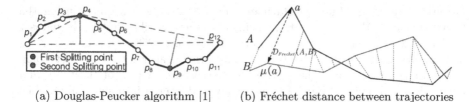

(a) Douglas-Peucker algorithm [1]   (b) Fréchet distance between trajectories

Fig. 1. RDP and Fréchet distance examples

structure of the nodes inside the trajectory is taken into consideration in the computation process, which can more accurately describe the similarity between the trajectories yelling better results [17]. In some scenarios where there exists a backward direction, ring, or crisscross in a trajectory, the Fréchet distance value doesn't show more distortion than other distance measurements. Due to these characteristics, this metric is more descriptive and more suitable as a measure of the similarity between trajectories. The time complexity is also similar to the other mentioned metrics.

For this work, we decided to use the DFD to cluster the trajectories. To obtain the clusters of frequent trajectories between the origin and destination pairs we first construct a symmetric distance matrix of each of the pairs in the trajectories connecting the given OxD using the DFD. We then fit the symmetric distance matrix to the DBMeans [10] method to obtain the different groups of trajectories.

4 Experiments and Results

4.1 Datasets

Here we describe the data we use in the paper as well as the preprocessing steps taken to handle the raw data and transform it to perform the clustering.

Porto Taxi Dataset. The dataset comprises 442 taxis running in the city of Porto, in Portugal, for an entire year (from 2013–07–01 to 2014–06–30) [18]. These taxis operate through a taxi dispatch central, using mobile data terminals installed in the vehicles. Each data sample corresponds to one completed trip. It contains a total of 9 (nine) features and for this study following were used: TRIP ID: a unique identifier for each trip; TAXI ID: a unique identifier for the taxi that performed each trip; TIMESTAMP: Unix Timestamp (in seconds) that identifies the trip's start; POLYLINE: It contains a list of GPS coordinates between brackets organized in pairs as [LONGITUDE, LATITUDE]. This list contains one pair of values for every 15 s. The last list item corresponds to the trip's destination while the first one represents its start. The dataset has 1,710,670 instances in total.

Google Location History Dataset. This dataset was acquired using the Google Location History Data from a single user as described in [4]. The dataset contains 120.847 instances from a period of 9 months or 253 unique days starting in February 2019 to October 2019. Among other features that are not going to be used in this study, the dataset is composed of a pair of (latitude, and longitude), and a timestamp. All the data was delivered in a single JSON file. As the locations of this dataset are well known by the researchers that published the files in [4], this dataset is going to be used as ground truth.

T-Drive Dataset. This dataset contains the GPS trajectories of 10,357 taxis during the period of one week from 02 to 08, February 2008, within Beijing, China [19]. The total number of points in this dataset is about 16,3 million and the total distance of the trajectories reaches 9 million kilometers. The average sampling interval is about 177 s with a distance of about 623 m. Each file of this dataset, which is named by the taxi ID, contains the trajectories of one taxi.

4.2 Data Filtering and Preprocessing

The first step is the preprocessing task that is including among other activities, the data cleaning process where we perform outliers and noise removal [20,21]. First of all, we need to look for duplicate data in the dataset and remove it. We also look for null data in the points where we cannot use the latitude or longitude to create new features in the next step.

Due to the influence of GPS signal loss and data drift, there are a number of unwanted points in the trajectories set during the data acquisition. Hence, cleaning tasks need to be performed in order to have more trustworthy data. This inconsistent data must be deleted. We apply a smoothing median filter to each set of 5 of GPS points to remove the noise as it is more robust to outliers [4,10]. We also perform filtering accordingly to the speed and acceleration: points with speed greater than 150 km/h and acceleration greater than 10 m/s^2 are removed [22].

For the Porto Taxi dataset, $1,290,226$ rows and $28,599$ trajectories without missing values were kept. The Google Location History (GLH) dataset was kept with $209,038$ rows for 333 distinct trajectories while for T-Drive dataset was kept with $16,325,487$ rows and $146,749$ unique trajectories.

4.3 Experimental Setup

For the experiments, we have extracted a subset of the Porto Taxi dataset. We used a sub-sample of one week of data between 2013–08–01 and 2013–08–07. For the GLH we used the dataset as a whole due to the data being collected from a single user while for the T-Drive Taxi dataset, we used a single day of data (2008–02–03) chosen by the mean value of the 7 days of data.

For the clustering step in the GLH dataset, we set the ϵ to 100 meters and the *MinPts* to 5 as it comprises data from single user mobility while for the Porto

Taxi dataset, we used a more strict value for ϵ: 50 meters and $MinPts$ was set to 10 due to the nature of the taxi business, the same values were applied to the T-Drive taxi dataset. For the simplification of the trajectories using the RDP algorithm, we set the ϵ to 50 for the GLH dataset due to the more flexible moves the individual may perform freely walking, running, cycling, or other activities related to vehicles. Regarding the Porto Taxi and the T-Drive taxi datasets, we set the ϵ to 25 due to the constraints for the vehicles in the road network. Table 1 shows the results for the top 10 trajectories with more points. Likewise, Table 2 shows the results for the top 5 trajectories with more points for the GLH dataset while Table 3 shows the values for the T-Drive dataset. In many cases (more evident in Tables 1 and 2) the reduction is greater than 50% and in some situations, the values reach up to 80%.

All the experiments were performed on a 64-bit Intel(R) Core(TM) i7-6700K CPU 4.00 GHz and 32 GB memory RAM machine over Python 3.6.7 environment.

Table 1. Porto Taxi RDP

Traj. Id	Raw	RDP
6	111	34
35	102	30
4	101	46
5	101	31
27	98	35
42	95	27
41	92	30
38	92	29
8	90	32
24	88	42

Table 2. GLH RDP

Traj. Id	Raw	RDP
0	38	10
1	22	4
2	24	5
3	30	11
4	53	13

Table 3. T-Drive RDP

Traj. Id	Raw	RDP
1	39	30
2	12	7
8	8	6
12	8	5
11	7	5
14	7	4
7	6	4
4	5	4
6	4	3
13	4	3

4.4   Evaluation

In this subsection, we compare our proposal with the one made by [4] where the authors use the LCSS to identify common pathways (frequent trajectories in our case).

Porto Taxi Dataset Results. In Fig. 2 one can see the most frequent trajectories between the top origin x destination pair in the great Porto area. As expected, this is a set of options of routes from the city center (start green marker) at the bottom to the airport area (end red marker) at the top of the figure. The strong blue cluster on the right side seems to be the longest path while all the other colors merge with the strong green cluster before reaching the end of the trajectory.

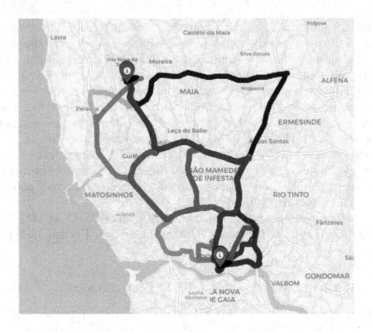

Fig. 2. Porto Taxi frequent trajectories. (Color figure online)

The most frequent trajectories that each model yields are different. One can notice that in our case the frequent trajectories group contains 22 trajectories while in the work LCSS method, it returns only 2. It is also possible to identify that they are different routes. For our approach (in purple on the left side) the most frequent trajectories are departing from the green marker and then going right before going up and then to the left. For the LCSS (in blue on the right side) it goes straight up and then to the left and then up again. Figure 3 shows the comparison between the two methods.

Google Location History Dataset Results. For the Google Location History dataset, by using the ground truth information given in [4], we can identify the clusters of trips starting at the individual's workplace and ending at the individual's home place depicted in Fig. 4. One interesting pattern to observe is the cluster in the lime color that is very different from the other pathways. The user takes a very distinct route from work to home. Further investigation on the location on the top of the image can help us to understand why.

The most frequent trajectories in this dataset are the same where both groups contain 2 trajectories. Figure 5 shows the comparison between the two methods.

(a) Fréchet distance trajectories (purple)   (b) LCSS trajectories (blue)

Fig. 3. Porto taxi frequent trajectories with Fréchet and LCSS results. (Color figure online)

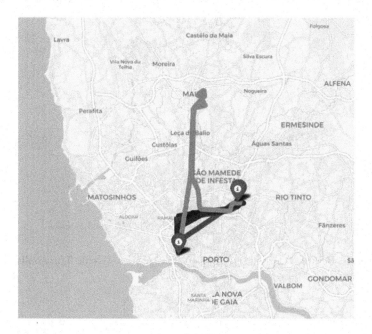

Fig. 4. Google Location History frequent trajectories. (Color figure online)

T-Drive Taxi Dataset Results. In Fig. 6 one can see the most frequent trajectories between the top origin x destination pair in the great Beijing area. The group has 15 trajectories in total. Unlike the results from the Porto Taxi dataset, the most frequent trajectories, in this case, are very short ones connecting the Capital Airport and a place nearby that we suggest being a hotel. The only exception to this rule is the long trajectory (in red) that is spread around the northeastern part of the city (obviously an outlier).

The most frequent trajectories in this dataset are the same where both groups contain 11 trajectories. Figure 7 shows the comparison between the two methods.

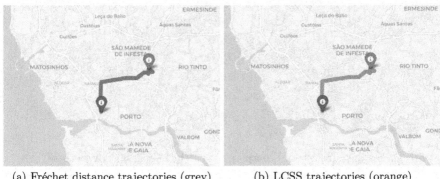

(a) Fréchet distance trajectories (grey)   (b) LCSS trajectories (orange)

Fig. 5. GLH frequent trajectories with Fréchet and LCSS results. (Color figure online)

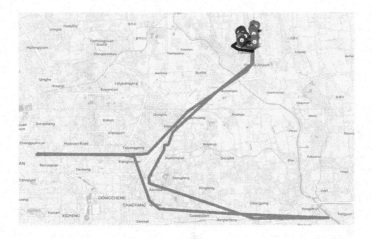

Fig. 6. Beijing T-Drive Taxi frequent trajectories. (Color figure online)

(a) Fréchet distance trajectories (blue)   (b) LCSS trajectories (brown)

Fig. 7. T-Drive taxi frequent trajectories with Fréchet and LCSS results. (Color figure online)

4.5   Discussion

Regarding the GLH dataset, as it has mixed data logs, sometimes the trajectories are not that rich and the precision of the points is low. Still, we can obtain clear trajectory groups from our method. It is important to mention that in [10] the authors already classified the locations turning this dataset into a labeled version of the raw data. In this case, is possible to use it as a ground truth for identifying the frequent trajectories between origins and destinations. By performing a visual inspection of the outputs for the two datasets one can clearly see the distinct groups between the same origin and destination. As this dataset is from individual user mobility is expected that we find two main clusters that connect the user's home and work and the opposite trips going back from work to home.

For the Porto Taxi dataset, the results are more clear as the GPS logs are from vehicles that are constrained into the road network, and the trips are started and ended in clear places usually outdoors which increases the quality of data. Also, it is important to remark that the logs are constant in intervals of 15 s apart.

The results for the T-Drive Taxi dataset are also interesting as we found out that the most frequent trajectories are short and close to the airport. One can reason for not finding many trajectories connecting regions closer to the city center due to Beijing being a large city (much larger than Porto) and the users could be using the metro, bus, or train to perform these moves.

5   Conclusions and Future Work

Frequent trajectory discovery is an important research topic in mobility patterns and clustering analysis has the tools to help us to understand the movements of individuals and objects with several application scenarios. Still, the problem is challenging due to the ubiquitous shifting in trajectory data and variable logging rates of different objects. In this paper, we showed how to handle raw GPS data and identify the most frequent trajectories between origin and destination pairs by using density-based clustering techniques, polyline simplification, and Fréchet distance. The experiments over three real-world datasets show the usefulness of our proposal in different scenarios.

For future work, we intend to perform more extensive experiments on different real-world datasets and scenarios. We also plan to improve the clustering algorithm for streaming processing.

Acknowledgement. This work was developed under the project "City Analyser" (POCI-01-0247-FEDER-039924), financed by European Regional Development Fund (ERDF), through the Research and Technological Development Incentive System, within the Portugal2020 Competitiveness and Internationalization Operational Program.

This work is also financed by National Funds through the Portuguese funding agency, FCT - Fundação para a Ciência e a Tecnologia within the grant: UI/BD/152697/2022.

References

1. Zheng, Y.: Trajectory data mining: an overview. ACM Trans. Intell. Syst. Technol. (TIST) **6**(3), 1–41 (2015)
2. Yuan, G., et al.: A review of moving object trajectory clustering algorithms. Artif. Intell. Rev. **47**, 123–144 (2017)
3. Ester, M., et al.: A density-based algorithm for discovering clusters in large spatial databases with noise. In: Kdd, vol. 96, no. 34, pp. 226–231 (1996)
4. Andrade, T., Cancela, B., Gama, J.: From mobility data to habits and common pathways. Expert. Syst. **37**(6), e12627 (2020)
5. Andrade, T., Cancela, B., Gama, J.: Discovering common pathways across users' habits in mobility data. In: Moura Oliveira, P., Novais, P., Reis, L.P. (eds.) EPIA 2019. LNCS (LNAI), vol. 11805, pp. 410–421. Springer, Cham (2019). https://doi.org/10.1007/978-3-030-30244-3_34
6. Gudmundsson, J., Valladares, N.: A GPU approach to subtrajectory clustering using the Fréchet distance. In: Proceedings of the 20th International Conference on Advances in Geographic Information Systems, pp. 259–268 (2012)
7. Sanchez, I., et al.: Fast trajectory clustering using hashing methods. In: 2016 International Joint Conference on Neural Networks (IJCNN), pp. 3689–3696. IEEE (2016)
8. Hung, C.-C., Peng, W.-C., Lee, W.-C.: Clustering and aggregating clues of trajectories for mining trajectory patterns and routes. VLDB J. **24**, 169–192 (2015)
9. Brankovic, M., et al.: (k, l)-medians clustering of trajectories using continuous dynamic time warping. In: Proceedings of the 28th International Conference on Advances in Geographic Information Systems, pp. 99–110 (2020)
10. Andrade, T., Cancela, B., Gama, J.: Discovering locations and habits from human mobility data. Ann. Telecommun. **75**, 505–521 (2020)
11. Ramer, U.: An iterative procedure for the polygonal approximation of plane curves. Comput. Graph. Image Process. **1**(3), 244–256 (1972)
12. Douglas, D.H., Peucker, T.K.: Algorithms for the reduction of the number of points required to represent a digitized line or its caricature. Cartographica: Int. J. Geogr. Inf. Geovisualization **10**(2) 112–122 (1973)
13. Andrade, T., Gama, J.: Identifying points of interest and similar individuals from raw GPS data. In: Cagáñová, D., Horňáková, N. (eds.) Mobility IoT 2018. EICC, pp. 293–305. Springer, Cham (2020). https://doi.org/10.1007/978-3-030-30911-4_21
14. Andrade, T., Cancela, B., Gama, J.: Mining human mobility data to discover locations and habits. In: Cellier, P., Driessens, K. (eds.) ECML PKDD 2019. CCIS, vol. 1168, pp. 390–401. Springer, Cham (2020). https://doi.org/10.1007/978-3-030-43887-6_32
15. Vlachos, M., Kollios, G., Gunopulos, D.: Discovering similar multidimensional trajectories. In: Proceedings of the 18th International Conference on Data Engineering, ICDE 2002, p. 673. IEEE Computer Society, Washington, DC, USA (2002). http://dl.acm.org/citation.cfm?id=876875.878994. Accessed 16 June 2019
16. Eiter, T., Mannila, H.: Computing discrete Fréchet distance (1994)
17. Magdy, N., et al.: Review on trajectory similarity measures. In: 2015 IEEE Seventh International Conference on Intelligent Computing and Information Systems (ICICIS), pp. 613-619. IEEE (2015)
18. Moreira-Matias, L., et al.: Predicting taxi-passenger demand using streaming data. IEEE Trans. Intell. Transp. Syst. **14**(3), 1393–1402 (2013)

19. Yuan, J., et al.: Driving with knowledge from the physical world. In: Proceedings of the 17th ACM SIGKDD International Conference on Knowledge Discovery and Data Mining, pp. 316–324 (2011)
20. Andrade, T., et al.: Anomaly detection in sequential data: principles and case studies. Wiley Encycl. Electr. Electron. Eng., 1–14 (2019)
21. Gama, J., et al.: Extração de conhecimento de dados: data mining, 3rd edn., p. 428. Edições Sílabo, Lisboa (2017)
22. Andrade, T., Gama, J.: Estimating instantaneous vehicle emissions. In: Proceedings of the 38th ACM/SIGAPP Symposium on Applied Computing, SAC 2023, Tallinn, Estonia, pp. 422–424 (2023)

Graph Theory and Network Analysis

Boosting-Based Construction of BDDs for Linear Threshold Functions and Its Application to Verification of Neural Networks

Yiping Tang[1,2] , Kohei Hatano[1,2(✉)] , and Eiji Takimoto[1]

[1] Kyushu University, Fukuoka, Japan
tang.yiping.641@s.kyushu-u.ac.jp,
{hatano,eiji}@inf.kyushu-u.ac.jp
[2] RIKEN AIP, Tokyo, Japan

Abstract. Understanding the characteristics of neural networks is important but difficult due to their complex structures and behaviors. Some previous work proposes to transform neural networks into equivalent Boolean expressions and apply verification techniques for characteristics of interest. This approach is promising since rich results of verification techniques for circuits and other Boolean expressions can be readily applied. The bottleneck is the time complexity of the transformation. More precisely, (i) each neuron of the network, i.e., a linear threshold function, is converted to a Binary Decision Diagram (BDD), and (ii) they are further combined into some final form, such as Boolean circuits. For a linear threshold function with n variables, an existing method takes $O(n2^{\frac{n}{2}})$ time to construct an ordered BDD of size $O(2^{\frac{n}{2}})$ consistent with some variable ordering. However, it is non-trivial to choose a variable ordering producing a small BDD among $n!$ candidates.

We propose a method to convert a linear threshold function to a specific form of a BDD based on the boosting approach in the machine learning literature. Our method takes $O(2^n \text{poly}(1/\rho))$ time and outputs BDD of size $O(\frac{n^2}{\rho^4} \ln \frac{1}{\rho})$, where ρ is the margin of some consistent linear threshold function. Our method does not need to search for good variable orderings and produces a smaller expression when the margin of the linear threshold function is large. More precisely, our method is based on our new boosting algorithm, which is of independent interest. We also propose a method to combine them into the final Boolean expression representing the neural network. In our experiments on verification tasks of neural networks, our methods produce smaller final Boolean expressions, on which the verification tasks are done more efficiently.

Keywords: Convolutional Neural Network · Binary decision diagram · Boosting · Verification

1 Introduction

Interpretability of Neural Networks (NNs) has been relevant since their behaviors are complex to understand. Among many approaches to improve interpretability, some results apply verification techniques of Boolean functions to understand NNs, where

A. Bifet et al. (Eds.): DS 2023, LNAI 14276, pp. 477–491, 2023.
https://doi.org/10.1007/978-3-031-45275-8_32

NNs are represented as an equivalent Boolean function and then various verification methods are used to check criteria such as robustness [6,7,12–14]. This approach is promising in that rich results of Boolean function verification can be readily applied. The bottleneck, however, is to transform a NN into some representation of the equivalent Boolean function.

A structured way of transforming NNs to Boolean function representations is proposed by [10]. They proposed (i) to transform each neuron, i.e., a linear threshold function, into a Binary Decision Diagram (BDD) and then (ii) to combine BDDs into a final Boolean function representations such as Boolean circuits. In particular, the bottleneck is the transformation of a linear threshold function to a BDD. To do this, they use the transformation method of [3]. The method is based on dynamic programming, and its time complexity is $O(n2^{\frac{n}{2}})$ and the size of resulting BDD is $O(2^{\frac{n}{2}})$, where n is the number of the variables. In addition, the method requires a fixed order of n variables as an input and outputs the minimum BDD consistent with the order. Thus, to obtain the minimum BDD, it takes $O(n!n2^{\frac{n}{2}})$ time by examining $n!$ possible orderings. Even if we avoid the exhaustive search of orderings, it is non-trivial to choose a good ordering.

In this paper, we propose an alternative method to obtain a specific form of BDD representation (named Aligned Binary Decision Diagram, ABDD) of a linear threshold function. Our approach is based on *Boosting*, a framework of machine learning which combines base classifiers into a better one. More precisely, our method is a modification of the boosting algorithm of Mansour and McAllester [8]. Given a set of labeled instances of a linear threshold function, their algorithm constructs a BDD that is consistent with the instances in a top-down greedy way. The algorithm can be viewed as a combination of greedy decision tree learning and a process of merging nodes. Given a linear threshold function $f(x) = \sigma(w \cdot x + b)$ where σ is the step function, we can apply the algorithm of Mansour and McAllester by feeding all 2^n possible labeled instances of f and obtain a BDD representation of f of the size $O(\frac{n^2}{\rho^4} \ln \frac{1}{\rho})$ in time $O(2^n \text{poly}(1/\rho))$, where ρ is the margin of f, defined as $\rho = \min_{x \in \{-1,1\}^n} |w \cdot x + b|/\|w\|_1$. An advantage of the method is that the resulting BDD is small if the linear threshold function has a large margin. Another merit is that the method does not require a variable ordering as an input. However, in our initial investigation, we observe that the algorithm is not efficient enough in practice. Our algorithm, in fact, a boosting algorithm, is obtained by modifying their algorithm so that we only use one variable (base classifier) in each layer. We show that our modification still inherits the same theoretical guarantees as Mansour and McAllester's. Furthermore, surprisingly, the small change makes the merging process more effective and produces much smaller BDDs in practice. Our modification might look easy but is non-trivial in a theoretical sense. To achieve the same theoretical guarantee, we introduce a new information-theoretic criterion to choose variables that is different from the previous work. That is one of our technical contributions[1].

In our experiments on verification tasks of Convolutional Neural Networks (CNNs), by following the same procedures as [10], we construct smaller BDDs and resulting Boolean representations of CNNs faster than in previous work, thus contributing to more efficient verification.

---

[1] All proofs will be found in the arXiv version http://arxiv.org/abs/2306.05211.

Table 1. Time and size for several methods to convert to DDs from a given linear threshold function (LTF, for short) of margin ρ. The fourth result is only for LTFs with integer weights whose $L2$-norm is W.

Method	DD type	Size	Time
[3]	OBDD	$O(2^{\frac{n}{2}})$	$O(n!n2^{\frac{n}{2}})$
[8]	BDD	$O(\frac{n^2}{\rho^4}\ln\frac{1}{\rho})$	$O(2^n\text{poly}(1/\rho))$
Ours	ABDD	$O(\frac{n^2}{\rho^4}\ln\frac{1}{\rho})$	$O(2^n\text{poly}(1/\rho))$
(cf. [10])	OBDD	$O(nW)$	$O(nW)$

This paper is organized as follows: Sect. 2 overviews the preliminaries of binary NN (BNN), BDD, Ordered BDD (OBDD), and Aligned BDD (ABDD). Sections 3 and 4 detail our proposed method to construct ABDD. Section 5 details the construction of the Boolean circuit and SDD. Section 6 handles the experimental results with analysis, followed by the conclusion in Sect. 7.

Related Work. The work in [9] proposed a precise Boolean encoding of BNNs that allows easy network verification. However, it only works with small-sized networks. [11] leveraged the Angluin-style learning algorithm to convert the BNN (the weights and input are binarized as $\{-1,1\}$) and OBDD into Conjunctive Normal Form (CNF) and then used the Boolean Satisfiability (SAT) solver to verify the equivalence of its produced CNF. However, they modified the OBDD several times and utilized limited binary network weights. [3] suggested a method to convert a linear threshold classifier with real-valued weight into OBDD. However, their approach owns time complexity of $O(n!n2^{\frac{n}{2}})$ and OBDD size complexity of $O(2^{\frac{n}{2}})$ via searching the full ordering, which increases exponentially when n becomes larger. Still, [3,9,11] can only handle small dimension NN weight, and the large Boolean expression was represented as Sentential Decision Diagram (SDD), which owns enormous time complexity. Moreover, [4] proposed a rule extraction method inspired by the 'rule extraction as learning' approach to express NN into a Reduced Ordered DD (RODD), which has the time complexity of $O(n2^{2n})$.

2   Preliminaries

2.1   Binary Neural Network

A binary neural network (BNN) is a variant of the standard NN with binary inputs and outputs [2]. In this paper, each neural unit, with a step activation function σ, is formulated as follows:

$$\sigma\left(\sum_i x_i w_i + b\right) = \begin{cases} 1, & \sum_i x_i w_i + b \geq 0 \\ -1, & \text{otherwise} \end{cases} \tag{1}$$

where $x \in \{-1,1\}^n$, $w \in \mathbb{R}^n$ and $b \in \mathbb{R}$ are the input, the weight vector and the bias of this neural unit, respectively.

2.2   Definition of BDD, OBDD and ABDD

A binary decision diagram (BDD) T is defined as a tuple $T = (V, E, l)$ with the following properties: (1) (V, E) is a directed acyclic graph with a root and two leaves, where V is the set of nodes, E is the set of edges such that $E = E_- \cup E_+$, $E_- \cap E_+ = \varnothing$. Elements of E_+ and E_-'s are called $+$-edges, and $-$-edges, respectively. Let $L = \{0\text{-leaf}, 1\text{-leaf}\} \subset V$ be the set of leaves. For each $v \in V$, there are two child nodes $v^-, v^+ \in V$ such that $(v, v^-) \in E_-$ and $(v, v^+) \in E_+$. (2) l is a function from $V \setminus L$ to $[n]$.

Given an instance $x \in \{-1, 1\}^n$ and a BDD T, we define the corresponding path $P(x) = (v_0, v_1, \ldots, v_{k-1}, v_k) \in V^*$ over T from the root to a leaf as follows: (1) v_0 is the root. (2) for any $j = 0, \ldots k - 1$, we have $(v_j, v_{j+1}) \in E_+ \Leftrightarrow x_{l(v_j)} = 1$ and $(v_j, v_{j+1}) \in E_- \Leftrightarrow x_{l(v_j)} = -1$. (3) v_k is a leaf node. We say that an instance $x \in \{-1, 1\}^n$ reaches node u in T, if $P(x)$ contains u. Then, a BDD T naturally defines the following function $h_T : \{-1, 1\}^n \to \{-1, 1\}$ such that

$$h_T(x) \triangleq \begin{cases} -1, & x \text{ reaches 0-leaf} \\ 1, & x \text{ reaches 1-leaf.} \end{cases} \tag{2}$$

Given a BDD $T = (V, E, l)$, we define the depth of a node $u \in V$ as the length of the longest path from the root to u. An ordered BDD (OBDD) $T = (V, E, l)$ is a BDD satisfying an additional property: There is a strict total order $<_{[n]}$ on $[n]$ such that for any path $P = (v_0, \ldots, v_k)$ on from the root to a leaf, and any nodes v_i and v_j $(i < j < k)$, $l(v_i) <_{[n]} l(v_j)$. An Aligned BDD (ABDD) $T = (V, E, l)$ is defined as a BDD satisfying that for any nodes $u, v \in V \setminus L$ with the same depth, $l(u) = l(v)$. We employ $v_{i,j}$ to appear the positional information of a node in the BDD, where j represents the depth of the node, and i represents the position of the node at depth j.

BDD, OBDD and ABDD are illustrated in Fig. 1.

2.3   Instance-Based Robustness (IR), Model-Based Robustness (MR) and Sample-Based Robustness (SR)

Robustness is a fundamental property of the neural network, which represents the tolerance of the network to noise or white attacks. For binary input images, the robustness k represents that as long as at least k pixels are flipped from 0 to 1 or 1 to 0, and the neural network's output will be changed.

We define the IR and MR of a network as follows.

Definition 1. *(Instance-based Robustness) [10]*
Consider a classification function $f : \{-1, 1\}^n \to \{-1, 1\}$ and a given instance x. The robustness of the classification of x by f, denoted by $r_f(x)$. If f is not a trivial function (always $True$ or $False$),

$$r_f(x) = \min_{x': f(x) \neq f(x')} dis(x, x') \tag{3}$$

where $dis(x, x')$ denotes the Hamming distance between x and x'.

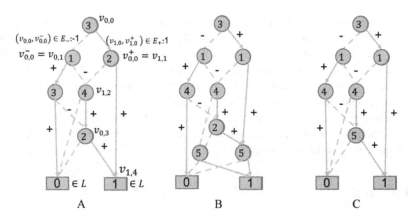

Fig. 1. Examples of BDD (A), OBDD (B) and ABDD (C). To express the same linear threshold function in BDD form: nodes at the same depth can be labeled by different variables, which means that the number of variables does not limit the depth of BDD; in OBDD form: the nodes at the same depth are all labeled by the same variables, which results in the depth of OBDD is smaller than the number of variables; in ABDD form: nodes at the same depth are labeled by the same variables, and the depth of ABDD only depends on the reduction of entropy to 0 in our algorithm.

Definition 2. *(Model-based Robustness) [10]*
Consider a classification function $f : \{-1, 1\}^n \rightarrow \{-1, 1\}$. The Model-based Robustness of f is defined as:

$$MR(f) = \frac{1}{2^n} \sum_x r_f(x). \tag{4}$$

However, we consider that computing MR on full-size data is not practically meaningful. In practical applications, robustness validation based on sample data is common. Here, we regard the samples in the dataset as instances randomly extracted from the full-size data under the uniform distribution. Then, we have the following definition.

Definition 3. *(Sample-based Robustness, SR)*
Consider a classification function $f : \{-1, 1\}^n \rightarrow \{-1, 1\}$. Given a sample S under uniform distribution from $\{-1, 1\}^n$. The Sample-based Robustness of f is defined as:

$$SR(f) = \frac{1}{|S|} \sum_{x \in S} r_f(x). \tag{5}$$

2.4   Overview of Our Method

In Sect. 3, we propose an algorithm that constructs an ABDD whose training error is small with respect to a given sample of some target Boolean function. Our algorithm is based on boosting, which is an effective approach in machine learning that constructs a more accurate classifier by combining "slightly accurate" classifiers. In Sect. 4, we apply our boosting algorithm for finding an equivalent ABDD with a given linear thresh-old function. Under a natural assumption that the linear threshold function has a large "margin", we show the size of the resulting ABDD is small. In Sect. 5, we show how to convert a given BNN to an equivalent Boolean expression suitable for verification tasks. More precisely, (i) Each neural unit is converted to an equivalent ABDD by applying our boosting algorithm, (ii) each ABDD is further converted to a Boolean circuit, and (iii) all circuits are combined into the final circuit, which is equivalent to the given BNN. Furthermore, for a particular verification task, we convert the final circuit to an equivalent sentential decision diagram (SDD).

3 Boosting

3.1 Problem Setting

Boosting is an approach to constructing a strongly accurate classifier by combining weakly accurate classifiers. We assume some unknown target function $f : \{-1, 1\}^n \to \{-1, 1\}$. Given a sample $S = ((x_1, f(x_1)), \ldots, (x_m, f(x_m))) \in (\{-1, 1\}^n \times \{-1, 1\})^m$ of m instances labeled by f and a precision parameter ε, we want to find a classifier $g : \{-1, 1\}^n \to \{-1, 1\}$ such that its training error $\Pr_U\{g(x) \neq f(x)\} \leq \varepsilon$, where U is the uniform distribution over S. We are also given a set $\mathcal{H}$ of base classifiers from $\{-1, 1\}^n$ to $\{-1, 1\}$. We assume the following assumption which is standard in the boosting literature [8].

Definition 4. *(Weak Hypotheses Assumption (WHA))*
A hypothesis set $\mathcal{H}$ satisfies γ-Weak Hypothesis Assumption (WHA) for the target function $f : \{-1, 1\}^n \to \{-1, 1\}$ if for any distribution d over $\{-1, 1\}^n$, there exists $h \in \mathcal{H}$ such that $edge_{d,f}(h) \triangleq \sum_{x \in \{-1,1\}^n} d_x f(x) h(x) \geq \gamma$.

Intuitively, WHA ensures the set of $\mathcal{H}$ of hypotheses and f are "weakly" related to each other. The edge function $edge_{d,f}(h)$ takes values in $[-1, 1]$ and equals to 1 if $f = h$. Under WHA, we combine hypotheses of $\mathcal{H}$ into a final hypothesis h_T represented by an ABDD T.

Our analysis is based on the conditional entropy of f given an ABDD without leaves, where the entropy is measured by a variant of entropy function $G : [0, 1] \to [0, 1]$, defined as $G(q) \triangleq 2\sqrt{q(1-q)}$. Like the Shannon entropy, G is concave and $G(1/2) = 1$, and $G(0) = G(1) = 0$. In particular, $\min(q, 1-q) \leq G(q)$. An ABDD T without leaves is an ABDD that has no 0-leaf and 1-leaf. For each node u in T, let $p_u \triangleq \Pr_U\{x \text{ reaches } u\}$ and $q_u \triangleq \Pr_U\{f(x) = 1 \mid x \text{ reaches } u\}$, respectively. Let $N(T)$ be the set of nodes in T with no outgoing edges. Now we define the conditional entropy of f given T as

$$H_U(f|T) = \sum_{u \in N(T)} p_u G(q_u). \tag{6}$$

We can construct a standard ABDD $\hat{T}$ (i.e. with 0-leaf and 1-leaf) by merging those nodes u in $N(T)$ with $q_u < 1/2$ into 0-leaf and those nodes u in $N(T)$

with $q_u \geq 1/2$ into 1-leaf. Then, it is easy to see that $\mathrm{Pr}_U\{f(x) \neq h_{\hat{T}}(x)\} = \sum_{u \in N(T)} p_u \min(q_u, 1 - q_u) \leq \sum_{u \in N(T)} p_u G(q_u)$. Thus, minimizing $H_U(f|T)$ implies minimizing $\mathrm{Pr}_U\{h_{\hat{T}}(x) \neq f(x)\}$, and hence in what follows we consider ABDDs without leaves and simply call them ABDDs.

Proposition 1. $\mathrm{Pr}_U\{h_{\hat{T}}(x) \neq f(x)\} \leq H_U(f|T)$.

Therefore, it suffices to find an ABDD T whose conditional entropy $H_U(f|T)$ is less than ε.

We will further use the following notations and definitions. Given an ABDD T, S and $u \in N(T)$, let $S_u = \{(x, y) \in S \mid x \text{ reaches } u\}$. The entropy $H_d(f)$ of $f : \{-1, 1\}^n \rightarrow \{-1, 1\}$ with respect to a distribution d over $\{-1, 1\}^n$ is defined as $H_d(f) \triangleq G(q)$, where $q = \mathrm{Pr}_d\{f(x) = 1\}$. The conditional entropy $H_d(f|h)$ of f given $h : \{-1, 1\}^n \rightarrow \{-1, 1\}$ with respect to d is defined as $H_d(f|h) = \mathrm{Pr}_d\{h(x) = 1\}G(q^+) + \mathrm{Pr}_d\{h(x) = -1\}G(q^-)$, where $q^{\pm} = \mathrm{Pr}_d\{f(x) = 1 \mid h(x) = \pm 1\}$, repetively.

3.2 Our Boosting Algorithm

Our algorithm is a modification of the boosting algorithm proposed by Mansour and McAllester [8]. Both algorithms learn Boolean functions in the form of BDDs in a top-down manner. The difference between our algorithm and Mansour and McAllester's algorithm lies in the construction of the final Boolean function, where ours utilizes ABDDs, while Mansour and McAllester's algorithm does not. Although this change may appear subtle, it necessitates a new criterion for selecting hypotheses in $\mathcal{H}$ and demonstrates improved results in our experiments.

Our boosting algorithm iteratively grows an ABDD by adding a new layer at the bottom. More precisely, at each iteration k, given the current ABDD T_k, the algorithm performs the following two consecutive processes (as illustrated in Fig. 2).

Split: It chooses a hypothesis $h_k \in \mathcal{H}$ using some criterion and adds two child nodes for each node in $N(T_k)$ in the next layer, where each child corresponds to ± 1 values of h_k. Let T'_k be the resulting DD.

Merge: It merges nodes in $N(T'_k)$ according to some rule and let T_{k+1} be the ABDD after the merge process.

The full description of the algorithm is given in Algorithm 1 and 2, respectively. For the split process, it chooses the hypothesis h_k maximizing the edge $edge_{\hat{d}, f}(h)$ with respect to the distribution $\hat{d}$ specified in (8). For the merge process, we use the same way in the algorithm of Mansour and McAllester [8].

Definition 5. [8] For δ and λ $(0 < \delta, \lambda < 1)$, a (δ, λ)-net $\mathcal{I}$ is defined as a set of intervals $[v_0, v_1], [v_1, v_2], \ldots, [v_{w-1}, v_w]$ such that (i) $v_0 = 0$, $v_w = 1$, (ii) for any $I_k = [v_{k-1}, v_k]$ and $q \in I_k$, $\max_{q' \in I_k} G(q') \leq \max\{\delta, (1 + \lambda)G(q)\}$.

Mansour and McAllester showed a simple construction of (δ, λ)-net with length $w = O((1/\lambda) \ln(1/\delta))$ [8] and we omit the details. Our algorithm uses particular (δ, λ)-nets for merging nodes.

3.3 Analyses

The conditional entropy of a function $f : \{-1,1\}^n \to \{-1,1\}$ given an ABDD T and a hypothesis $h : \{-1,1\}^n \to \{-1,1\}$ with respect to distribution d over $\{-1,1\}^n$ is defined as follows:

$$H_d(f|T,h) = \sum_{u \in N(T)} (p_{u+}G(q_{u+}) + p_{u-}G(q_{u-})) \tag{7}$$

where $p_{u+} = Pr_d\{x$ reaches u and $h(x) = 1\}$ and $q_{u+} = Pr_d\{f(x) = 1 | x$ reaches $u, h(x) = 1\}$. p_{u-} and q_{u-} are defined similarly. The conditional entropy in (7) is about the nodes in $N(T)$ after split by hypothesis h.

Then, we establish the connection between γ and the entropy function and have a $\gamma \in (0,1)$ at each depth to reflect the entropy change under our algorithm as shown in Lemma 1 and 2.

Lemma 1. *Let $\hat{d}$ be the distribution over S specified in (8) when T_k, $\mathcal{H}$ and S is given by Algorithm 2 and let h_k be the output. If $\mathcal{H}$ satisfies γ-WHA, then the conditional entropy of f with respect to the distribution U over S given T_k and h_k is bounded as $H_U(f|T_k, h_k) \leqslant (1 - \gamma^2/2)H_U(f|T_k)$.*

Lemma 2. *([8]) Assume that, before the merge process in Algorithm 1, $H_U(f|T_k, h_k) \leq (1 - \lambda)H_U(f|T_k)$ for some λ ($0 < \lambda < 1$). Then, by merging based on the (δ, η)-net with $\delta = (\lambda/6)H_U(f|T_k)$ and $\eta = \lambda/3$, the conditional entropy of f with respect to the distribution U over S given T_{k+1} is bounded as $H_U(f|T_{k+1}) \leqslant (1-\lambda/2)H_U(f|T_k)$, where the width of T_{k+1} is $O((1/\lambda)(\ln(1/\lambda) + \ln(1/\varepsilon)))$, provided that $H_U(f|T_k) > \varepsilon$.*

Now we are ready to show our main theorem.

Theorem 1. *Given a sample S of m instances labeled by f, and a set $\mathcal{H}$ of hypotheses satisfying γ-WHA, Algorithm 1 outputs an ABDD T such that $\mathrm{Pr}_U\{h_T(x) \neq f(x)\} \leq \varepsilon$. The size of T is $O((\ln(1/\varepsilon)/\gamma^4)(\ln(1/\varepsilon) + \ln(1/\gamma)))$ and the running time of the algorithm is $poly(1/\gamma, n)m$.*

4 ABDD Construction

We now apply the ABDD Boosting algorithm developed in the previous section to a given linear threshold function f to obtain an ABDD representation T for f. In particular, we show that the size of T is small when f has a large margin.

To be more specific, assume that we are given a linear threshold function $f : \{-1,1\}^n \to \{-1,1\}$ of the form

$$f(x) = \sigma(w \cdot x + b)$$

for some weight vector $w \in \mathbb{R}^n$ and bias $b \in \mathbb{R}$, where σ is the step function, i.e., $\sigma(z)$ is 1 if $z \geq 0$ and -1 otherwise. Note that there are infinitely many (w, b) inducing the same function f. We define the margin ρ of f as the maximum margin $f(x)(w \cdot x +$

Algorithm 1. ABDD Boosting

Input: a sample $S \in (\{-1,1\}^n \times (-1,1))^m$ of m instances by f, and a set $\mathcal{H}$ of hypotheses, and precision parameter ε ($0 < \varepsilon < 1$);

Output: ABDD T;

1: initialization: T_1 is the ABDD with a root and $0, 1$-leaves, $k = 1$.
2: **repeat**
3:     (**Split**) Let $h_k = Split(T, \mathcal{H}, S)$ and add child nodes with each node in $N(T_k)$. Let T'_k be the resulting ABDD.
4:     **for** $u \in N(T'_k)$ **do**
5:         merge u to the 0-leaf (the 1 leaf) if $q_u = 0$ ($q_u = 1$, resp.).
6:     **end for**
7:     (**Merge**) Construct a $(\hat{\delta}, \hat{\lambda}/3)$-net $\mathcal{I}_k$ with
8:     $\hat{\lambda} = 1 - \frac{H_U(f|T_k, h_k)}{H_U(f|T_k)}$, and $\hat{\delta} = \frac{\hat{\lambda} H_U(f|T_k)}{6}$.
9:     **for** $I \in \mathcal{I}_k$ **do**
10:        merge all nodes $u \in N(T'_k)$ such that $q_u \in I$.
11:     **end for**
12:     Let T_{k+1} be the resulting ABDD and update $k \leftarrow k + 1$.
13: **until** $H_U(f|T_k) < \varepsilon$
14: Output $T = T_k$.

$b)/\|w\|_1$ over all (w, b). We let our hypothesis set $\mathcal{H}$ consist of projection functions, namely, $\mathcal{H} = \{h_1, h_2, \ldots, h_n, h_{n+1}, h_{n+2}, \ldots, h_{2n}\}$, where $h_i : x \mapsto x_i$ if $i \leq n$ and $h_i : x \mapsto -x_i$ otherwise, so that we can represent f as $f(x) = \sigma(\sum_{i=1}^{2n} w_i h_i(x) + b)$ for some non-negative $2n$-dimensional weight vector $w \geq 0$ and bias b. Then, we can represent the margin ρ of f as the solution of the optimal solution for the following LP problem:

$$\max_{w,b,\rho} \rho \tag{9}$$

$$\text{s.t.} \quad f(x)(\sum_{i=1}^{2n} w_i h_i(x) + b) \geq \rho \text{ for any } x \in \{-1, 1\}^n,$$

$$w \geq 0,$$

$$\sum_i w_i = 1.$$

Now we show that $\mathcal{H}$ actually satisfies ρ-WHA for f.

Lemma 3. *Let f be a linear threshold function with margin ρ. Then $\mathcal{H}$ satisfies ρ-WHA.*

By the lemma and Theorem 1, we immediately have the following corollary.

Corollary 1. *Let f be a linear threshold function with margin ρ. Applying the ABDD Boosting algorithm with the sample $S = \{(x, f(x)) \mid x \in \{-1, 1\}^n\}$ of all (2^n) instances, our hypothesis set $\mathcal{H}$, and the precision parameter $\varepsilon = 1/2^n$, we obtain an ABDD T, which is equivalent to f of size $O\left(\frac{n}{\rho^4}(n + \ln(1/\rho))\right)$.*

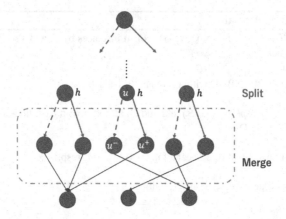

Fig. 2. Illustration of our boosting algorithm. The blue dotted line part represents the process of merging the split temporary nodes into new nodes by searching the equivalence space that does not appear in the ABDD. Following algorithm 1, we find some hypothesis h to construct child nodes in the new layer and then merge nodes afterward. (Color figure online)

5   Circuit and SDD Construction

Since we have a method to convert linear threshold functions into a DD representation, the next step is to connect them according to the structure of the NN to form an equivalent $(\lor, \land, \neg)$-circuit, which is used to verify SR. Subsequently, we can convert the circuit into an SDD, which is used to verify robustness.

The conversion process from a DD to a circuit is performed in a top-down manner. As shown in Fig. 3, the region enclosed by the red dashed box represents a conversion unit. This unit converts a node x_1 and its two edges in the DD into four gates and a variable in the circuit. As the next node x_2 is connected to a different edge of node x_1, we generate a new circuit beginning with an'or'-gate and connect it to the'and'-gate obtained from the conversion of an edge associated with x_1. This process is repeated until all edges of the nodes reach either the 0-leaf or the 1-leaf. The resulting circuit is equivalent to the DD and its corresponding neural unit. To construct the NN's equivalent circuit, we utilize Algorithm 1 to generate DDs for each neuron in the NN. These DDs are then converted into circuit form. We combine these equivalent circuits based on the structure of the NN, specifically establishing a one-to-one correspondence between the inputs and outputs of each neuron in the NN and the inputs and outputs of the circuit. This completes the construction of the NN's equivalent circuit. Such a method is also described in [10].

Once we have the equivalent circuit of a neural network (NN), the subsequent step is to convert it into an SDD. SDD is a subclass of deterministic Decomposable Negation Normal Form (d-DNNF) circuits that assert a stronger decomposability and a more robust form of determinism [5]. The class of SDDs generalizes that of OBDDs in that, every OBDD can be turned into an SDD in linear time. In contrast, some Boolean functions have polynomial-size SDD representations but only exponential-size OBDD

Algorithm 2. *Split*

Input: ABDD T, a set $\mathcal{H}$ of hypotheses, a set S of m instances labeled by f;
Output: $h \in \mathcal{H}$;

1: **for** $u \in N(T)$ **do**
2:     Let $\bar{d}^u$ be the distribution over S_u defined as follows:

$$\bar{d}^u_{(x,y)} = \begin{cases} \frac{1}{2 \times |\{(x',y') \in S_u | y=1\}|}, & \text{if } y = 1 \ \& \ (x,y) \in S_u \\ \frac{1}{2 \times |\{(x',y') \in S_u | y=-1\}|}, & \text{if } y = -1 \ \& \ (x,y) \in S_u, \text{ for } (x,y) \in S. \\ 0, & \text{if } (x,y) \notin S_u \end{cases}$$

3:     Let $p'_u = \frac{p_u G(q_u)}{\sum_{u \in N(T)} p_u G(q_u)}$,
4: **end for**
5: Let $\hat{d}$ be the distribution over S given follows:

$$\hat{d}_{(x,y)} = \sum_{u \in N(T)} p'_u \bar{d}^u_{(x,y)}, \tag{8}$$

    for $(x,y) \in S$.
6: Output $h = \arg\max_{h' \in \mathcal{H}} edge_{\hat{d}, f}(h')$.

---

representations [1]. In SDD, Boolean functions are represented through the introduction of"decision ($\vee$) nodes" and "conjunction ($\wedge$) nodes."

Indeed, the size complexity of each *apply* operation between two SDD nodes is proportional to the product of their internal nodes. Consequently, as the complexity of the circuit increases, the construction of the corresponding SDD requires more space. Considering the conversion process according to ABDD, if the resulting circuit is smaller, it follows that the corresponding SDD constructed from it will also be smaller in size. The size reduction in the circuit conversion directly influences the size of the resulting SDD. Therefore, by optimizing the BDD/circuit representation, we can achieve a more compact SDD.

6   Experiments

6.1   Experimental Setup

We use the USPS digits dataset of hand-written digits, consisting of 16×16 binary pixel images. We use the data with labels 0 and 1 for verification of SR. The NN design we used is similar to [10], which has two convolution layers and a full-connected layer. In training, we use two real-valued-weight convolutional layers (kernel size 3, stride 2; kernel size 2, stride 2) with a sigmoid function and a real-valued-weight fully-connected layer in the NN. In testing, the sigmoid function is replaced with the step activation function mentioned before. The experiments are conducted using a CPU of Intel(R) Xeon(R) Gold 2.60GHz. The batch size equals 32 and the utilized learning rate is 0.01 decaying to 10% at half and three-quarters of all learning epochs. Here, the Stochastic Gradient Descent (SGD) optimizer with a momentum of 0.9 and a weight decay of 0.0001 is used.

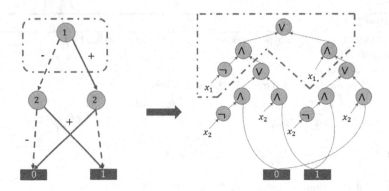

Fig. 3. Example of converting BDD/OBDD/ABDD (Left) to the circuit (Right).

6.2 Sample-Based Robustness (SR) Validation

The SR of a standard NN whose output is real-valued is achieved simply by querying the pixels that affect the recognition the most and flipping them until the recognition result changes. Nevertheless, on the standard methods, for a BNN that has input size $h \times w$, the time cost to compute robustness k is $O((hw)^k)$, which is difficult to use in BNNs.

Since BDD, OBDD, and ABDD are easily represented as a circuit, as shown in Fig. 3. The circuits of several neural units are linked into a circuit f that represents the entire NN according to its network structure. Note that we separately verify the SR of the OBDDs generated by the methods based on Theorem 1 and Theorem 2 in [10] using a circuit representation. The integration of weights and bias in Theorem 2 is performed as follows.

For each w in weight W and bias b, We set $\alpha = \max\{|w_1|, \dots, |w_n|, |b|\}$, then we turn them to integer weight $\hat{w}_i = \lfloor \frac{10^p}{\alpha} w_i \rfloor$ and bias $\hat{b} = \lfloor \frac{10^p}{\alpha} b \rfloor$ where p is the number of digits of precision.

As for $dis()$ in Definition 1, it's easy to express the $k \leq dis(x, x')$ between x and x' in a circuit form and likewise convert it to circuit $g_{k,x}$ denoted as follows:

$$g_{k,x}(x') = \begin{cases} 1, & if \mid x \oplus x' \mid \leq k \\ -1, & otherwise \end{cases} \tag{10}$$

We calculate the SR of f on (positive and negative) instance x, where have $f(x) = 1$ and $f(x) = -1$, by running Algorithm 3 of BDD, OBDD and ABDD on 10 CNNs, the results as shown in Table 2. Note that the SR of negative instances can be computed by invoking Algorithm 3 on function $\neg f$.

In table 2, (Shi.1) circuits are generated using the Theorem 1 proposed in [10], while (MM.) circuits are generated using the method described in [8], and (Shi.2 p) circuits are generated using the Theorem 2 proposed in [10] under the number of digits of precision $p = 2, 3, 4$. As observed, the number of gates of (Shi.2 2, 3, 4) are significantly larger than the others, it also consumes more time during SR validation (over 10 h).

Algorithm 3. SR

Input: circuit f, (positive) instance $x \in \{-1, 1\}^n$;
Output: $r_f(x)$;
1: initial: $r_f = 0$;
2: **for** $k = 1$ to n **do**
3:     **if** $g_{k,x} \wedge \overline{f}$ is satisfiable **then**
4:         break
5:     **end if**
6: **end for**
7: **return** k

Table 2. Summary of experimental results for SR calculations.

ID	SR	Time for SR (s)			Num of gates					
		Ours	(Shi.1)	(MM.)	Ours	(Shi.1)	(MM.)	(Shi.2 2)	(Shi.2 3)	(Shi.2 4)
1	1.83	3161	3994	4709	3737	5536	4433	48103	57079	57511
2	7.87	7142	9718	10626	1595	4039	2300	46972	57511	57505
3	3.79	3788	5050	5425	2384	4129	3146	51952	56611	54883
4	2.9	3970	4781	7655	3464	5128	5300	37474	57487	57493
5	4.94	5035	5902	7353	3235	3994	4249	43501	53971	57511
6	3.53	5229	7402	9731	3479	6049	4709	47083	55021	57511
7	2.84	4714	5839	7685	4963	5824	6613	47227	57511	57511
8	6.08	6559	8740	9334	3227	5233	3902	38983	57487	57511
9	4.01	5774	7323	8633	3659	5500	4577	47767	57487	57511
10	2.82	3711	5373	5393	2520	5497	3600	42439	57511	57511

6.3 Analysis

We train a standard CNN and show 99.73% accuracy on the test set. Then we replace the sigmoid function with the step activation function, and the accuracy of the CNN dropped to 99.22%. To represent BCNN, compare to BDD and OBDD, our algorithm can generate smaller ABDD while keeping the same recognition accuracy. It provides certain advantages in later work. Note that since [10]'s strategy is to randomly select multiple orders to generate many OBDDs and select one with the smallest size. Although we set the number of random orders to 100, the size of OBDD is still greater than ABDD.

Comparing these methods in Table 1, we observe that the BDD algorithm proposed by Mansour and McAllester [8] aims to identify the hypothesis h that yields the most significant reduction in entropy at each node. This approach initially appears to generate a smaller DD due to the intuitive advantage of entropy reduction. However, our observations reveal that this advantage diminishes when the number of samples within a node changes during the merging process. Specifically, if the number of samples remains unchanged, i.e., the entropy of the child node does not decrease to zero, the entropy of both BDD and our algorithm (ABDD) at the same depth remains constant. However,

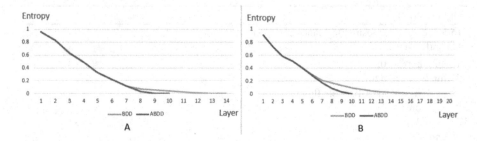

Fig. 4. Illustration of the entropy change of ABDD and BDD at each depth. We show the change in the total entropy in each depth of the DD generated by the linear threshold functions of the 3 × 3 convolution layer (A) and the fully connected layer (B) when using the ABDD and BDD algorithms, respectively.

once this phenomenon occurs, the entropy of ABDD decreases at a faster rate compared to that of BDD, as depicted in Fig. 4. We attribute this phenomenon to the fact that our approach ensures that nodes at the same depth share the same variables, thereby granting us an advantage when merging based on the energy of each node. When comparing ABDD with the OBDD [10], we contend that ABDD offers a precise algorithm for finding the DD with the smallest size, instead of relying on multiple randomizations. In contrast, when [10] applies OBDD based on neural unit weights, the time complexity increases exponentially with the size of the weights. In contrast, our algorithm exhibits minimal time consumption, unaffected by changes in weight size. Undoubtedly, this represents a notable advantage of our approach.

7   Conclusion

This paper introduced a Boosting-aided method that generates DD with smaller size and time complexities than conventional works. Our proposed method's resulting diagram is named ABDD, a variant of standard DD, in which the same variable labels the nodes at the same depth, and the depth is not limited to the number of dimensions of the variable. Experimental results show that ABDD can be connected in various forms to express NN and can be used to implement various verification tasks efficiently. Since our method can be used to generate a smaller equivalent circuit of the NN, it can be applied in tasks such as hardware-based transformations of NNs. In the future, we aim to extend our method to more complex NNs.

Acknowledgement. We thank Sherief Hashima of RIKEN AIP and the reviewers for their help-ful comments. This work was supported by JSPS KAKENHI Grant Numbers JP23H03348, JP20H05967, JP19H04174 and JP22H03649, respectively.

References

1. Bova, S.: SDDs are exponentially more succinct than OBDDs. In: Proceedings of the Thirtieth AAAI Conference on Artificial Intelligence, 12–17 February 2016, Phoenix, Arizona, USA, pp. 929–935. AAAI Press (2016)
2. Bshouty, N.H., Tamon, C., Wilson, D.K.: On learning width two branching programs. Inf. Process. Lett. **65**(4), 217–222 (1998)
3. Chan, H., Darwiche, A.: Reasoning about Bayesian network classifiers. In: UAI 2003, Proceedings of the 19th Conference in Uncertainty in Artificial Intelligence, Acapulco, Mexico, 7–10 August 2003, pp. 107–115. Morgan Kaufmann (2003)
4. Chorowski, J., Zurada, J.M.: Top-down induction of reduced ordered decision diagrams from neural networks. In: Honkela, T., Duch, W., Girolami, M., Kaski, S. (eds.) ICANN 2011. LNCS, vol. 6792, pp. 309–316. Springer, Heidelberg (2011). https://doi.org/10.1007/978-3-642-21738-8_40
5. Darwiche, A.: SDD: A new canonical representation of propositional knowledge bases. In: Walsh, T. (ed.) IJCAI 2011, Proceedings of the 22nd International Joint Conference on Artificial Intelligence, Barcelona, Catalonia, Spain, 16–22 July 2011, pp. 819–826. IJCAI/AAAI (2011)
6. Liu, B., Malon, C., Xue, L., Kruus, E.: Improving neural network robustness through neighborhood preserving layers. In: Pattern Recognition. ICPR International Workshops and Challenges, pp. 179–195 (2020)
7. Mangal, R., Nori, A.V., Orso, A.: Robustness of neural networks: a probabilistic and practical approach. In: Proceedings of the 41st International Conference on Software Engineering: New Ideas and Emerging Results, ICSE (NIER), pp. 93–96. IEEE/ACM (2019)
8. Mansour, Y., McAllester, D.: Boosting using branching programs. J. Comput. Syst. Sci. **64**(1), 103–112 (2002)
9. Narodytska, N., Kasiviswanathan, S.P., Ryzhyk, L., Sagiv, M., Walsh, T.: Verifying properties of binarized deep neural networks. In: Proceedings of the AAAI Conference on Artificial Intelligence, vol. 32(AAAI-18) (2018)
10. Shi, W., Shih, A., Darwiche, A., Choi, A.: On tractable representations of binary neural networks. In: Proceedings of the 17th International Conference on Principles of Knowledge Representation and Reasoning, KR 2020, Rhodes, Greece, 12–18 September 2020, pp. 882–892 (2020)
11. Shih, A., Darwiche, A., Choi, A.: Verifying binarized neural networks by angluin-style learning. In: Janota, M., Lynce, I. (eds.) SAT 2019. LNCS, vol. 11628, pp. 354–370. Springer, Cham (2019). https://doi.org/10.1007/978-3-030-24258-9_25
12. Weng, T., et al.: Evaluating the robustness of neural networks: an extreme value theory approach. In: 6th International Conference on Learning Representations, ICLR 2018 (2018)
13. Yu, F., Qin, Z., Liu, C., Zhao, L., Wang, Y., Chen, X.: Interpreting and evaluating neural network robustness. In: Proceedings of the Twenty-Eighth International Joint Conference on Artificial Intelligence, IJCAI 2019, pp. 4199–4205 (2019)
14. Zheng, S., Song, Y., Leung, T., Goodfellow, I.J.: Improving the robustness of deep neural networks via stability training. In: 2016 IEEE Conference on Computer Vision and Pattern Recognition, CVPR 2016, pp. 4480–4488 (2016)

Interpretable Data Partitioning Through Tree-Based Clustering Methods

Riccardo Guidotti[1,2]([ID]), Cristiano Landi[1]([ID]), Andrea Beretta[2]([ID]),
Daniele Fadda[2]([ID]), and Mirco Nanni[2]([ID])

[1] University of Pisa, Pisa, Italy
{riccardo.guidotti,cristiano.landi}@unipi.it
[2] ISTI-CNR Pisa, Pisa, Italy
{riccardo.guidotti,andrea.beretta,daniele.fadda,
mirco.nanni}@isti.cnr.it

Abstract. The growing interpretable machine learning research field is mainly focusing on the explanation of supervised approaches. However, also unsupervised approaches might benefit from considering interpretability aspects. While existing clustering methods only provide the assignment of records to clusters without justifying the partitioning, we propose tree-based clustering methods that offer interpretable data partitioning through a shallow decision tree. These decision trees enable easy-to-understand explanations of cluster assignments through short and understandable split conditions. The proposed methods are evaluated through experiments on synthetic and real datasets and proved to be more effective than traditional clustering approaches and interpretable ones in terms of standard evaluation measures and runtime. Finally, a case study involving human participation demonstrates the effectiveness of the interpretable clustering trees returned by the proposed method.

Keywords: Interpretable Clustering · Tree-based Clustering ·
Interpretable Data Partitioning · Explainable Unsupervised Learning

1 Introduction

The growing interest in eXplainable Artificial Intelligence (XAI) led to the design of a huge amount of methods to explain supervised learning approaches [22]. Our objective is to enhance the interpretability of unsupervised learning algorithms. In particular, we focus on the interpretability of clustering algorithms considering two aspects: first, understanding the insights gained from the record-to-cluster assignment, and second, understanding the logic used by the algorithm to partition the data. The results returned by traditional clustering algorithms such as k-Means [39] and the hierarchical Complete-Linkage algorithm [39] do not provide results that are easy to understand. Indeed, the centroids used by k-Means to characterize the different groups may not be suitable when there are many features or similar feature-values in the centroids. Similarly, the dendrogram returned by Complete-Linkage only provides an idea of at which distance

A. Bifet et al. (Eds.): DS 2023, LNAI 14276, pp. 492–507, 2023.
https://doi.org/10.1007/978-3-031-45275-8_33

the clusters were joined, but not why. Moreover, it is difficult for a human user to understand which branch of the dendrogram should be followed to assign a record to a cluster. These considerations are in line with recent studies on interpretable clustering [1,2,7,15,29], which argue that clustering should be derived from unsupervised binary trees. In a binary tree, each node is associated with a feature-threshold pair that recursively splits the dataset, and leaves have labels corresponding to clusters. If shallow trees are considered, any cluster assignment can be explained with a small number of thresholds [27]. Indeed, in supervised problems requiring interpretability, decision trees are typically adopted [4,22,37].

We advance the state-of-the-art by designing a tree-based clustering method that returns an unsupervised binary tree to interpret the data partitioning. The method simultaneously creates the tree and the data partitioning by searching for the best splits along subsequent iterations. This is different from state-of-the-art approaches, which first apply a traditional clustering algorithm and then try to infer a tree that approximates the clustering [2,7,15,29]. We name our proposal Partitioning Tree (PARTREE), and we show its effectiveness by implementing three versions that differ w.r.t. the criterion adopted to perform the data split. PARTREE can be applied to numerical, categorical, and mixed datasets. The experiments conducted on synthetic and real datasets show that PARTREE is as effective as traditional clustering methods and outperforms tree-based state-of-the-art approaches. Additionally, we conducted a case study where participants were asked to assign a record to a cluster on the basis of the clustering explanations returned by different algorithms. The results of the survey show that the usage of tree-based clustering methods improves user performance.

2   Related Works

We provide here a self-contained literature review of tree-based unsupervised approaches and of interpretable clustering methods adopting tree-based models.

We consider in the first group trees that do not account for any loss function such as k-d-trees [39], Random Projection tree [10], PCA trees [16,42], Approximate Principal Direction trees [31]. Their aim is is to create balanced trees and reduce the distance between records in the same partition.

The second group of approaches determine the partitioning features by optimizing different metrics, either locally or globally. TIC [3] uses maximum distance between prototypes as a splitting criterion and the F-test as a stopping condition. CLTree [29] is based on a supervised decision tree construction that uses uniformly at random synthetic points to capture the natural distribution of the data. In [41] is described a simplified solution of CLTree: k-Means is executed, and the clusters assignment labels are used as classes for a decision tree classifier. In [1] is proposed a method for hierarchical clustering based on trees where the selection of the splitting attribute is established w.r.t. four measures based on heterogeneity. In [6] additional split criteria and agglomeration measures are developed. TASC [8] partitions data space by constructing a decision tree using one attribute set and measures the degree of similarity through another one. INCONCO [36] finds clusters that minimize minimum description length

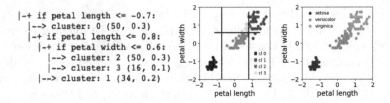

Fig. 1. PARTREE application on the `iris` dataset. Partitioning logic as rules (1st plot, numbers within parenthesis are the absolute and relative number of records in a cluster); partitioning logic as scatter plots (2nd), `iris` classes (3th plot).

and retrieves rules by assuming a multivariate normal distribution. CUBT [15] retrieves interpretable Clustering using Unsupervised Binary Trees. A problem with CUBT is that it joins similar clusters even if they do not share the same parent in the tree. Since this last step decreases the interpretability of the clustering destroying the tree, in our proposal we consider only "valid" partitioning trees. In [19] is proposed an extension of CUBT to nominal data through heterogeneity criteria and dissimilarity measures. PCN [23] extracts a subset of patterns to cluster the data and uses multiple unsupervised trees similar to TIC to derive these patterns. UD3.5 [30] extends PCN by introducing two quality measures to control the split and does not require any empirical parameter to control the depth of the trees. Recent approaches returning clustering trees focus on interpretability to address the need for XAI [22]. DReaM [7] uses a probabilistic discriminative model to learn a rectangular decision rule for each cluster. Iterative Mistake Minimization (IMM) [33] recursively builds a binary tree to minimize mistakes in k-Means. ExKMC [11,17] extends IMM to allow for more accurate but less interpretable clustering trees. Ex-Greedy [26] additionally boosts ExKMC, while ExShallow [27] favors the construction of shallow trees by accounting for the depth through a penalty term. K-Means Tree (KMT) [40] extends k-Means optimizing tree and centroids jointly for fast clustering. ICOT [2] uses Mixed Integer Optimization (MIO) to generate an optimal tree-based clustering model. Also in [28] is used MIO to jointly find clusters and define polytopes explaining the clusters. In [18] is created a tree using oblique trees on top of any clustering method defined by optimizing a cost function.

We place our proposal between these two groups. Indeed, as for the second group, our objective is to *(i)* cluster a dataset effectively and efficiently and *(ii)* to obtain an interpretable tree explaining the logic to partition the data. However, similarly to the first group, we aim at *(iii)* inducing the tree directly on the dataset without relying on other clustering approaches but *(iv)* by following local heuristic splitting criteria. Furthermore, our proposal aims at handling continuous, categorical, and mixed data.

3   Partitioning Tree Methods

Given a set of n records $X = \{x_1, \ldots, x_n\}$, where each observation is a d-dimensional vector, the clustering problem consists in partitioning X into $k < n$

Algorithm 1: PARTREE(X, $max\_clusters$, max_depth, min_sample, ε)

Input : X - dataset, $max\_clusters$ - max number of clusters, max_depth - max tree depth,
min_sample - min cluster size, ε - percentage of BIC parent discount
Param : qs - queue score function
Output: C - clustering, R - clustering tree

1 $C \leftarrow \emptyset$;   // init. clustering result
2 $\mathscr{R} \leftarrow \text{make\_node}(X)$;   // init. tree root
3 $\mathcal{Q} \leftarrow push(\mathcal{Q}, \text{qs}(X), \langle X, \mathscr{R} \rangle)$;   // init. priority queue
4 **while** $|\mathcal{Q}| > 0 \wedge |C| + |\mathcal{Q}| < max\_clusters$ **do**
5 $\quad$ $\langle C, \mathscr{N} \rangle \leftarrow pop(\mathcal{Q})$;   // extract tree node from queue
6 $\quad$ **if** $|C| < min\_sample \vee depth(\mathscr{N}) > max\_depth$ **then**
7 $\quad\quad$ $C \leftarrow C \cup \{C\}$; $\mathscr{L} \leftarrow \text{make\_leaf}(\mathscr{N})$;   // add cluster and make leaf
8 $\quad\quad$ **continue**;   // go next iteration
9 $\quad$ $f, C_1, C_2 \leftarrow \text{make\_split}(C)$ // make split
10 $\quad$ **if** $bic(C) < bic([C_1, C_2]) - \varepsilon|bic(C)|$ **then**
11 $\quad\quad$ $C \leftarrow C \cup \{C\}$; $\mathscr{L} \leftarrow \text{make\_leaf}(\mathscr{N})$;   // add cluster and make leaf
12 $\quad\quad$ **continue**;   // go next iteration
13 $\quad$ $\mathscr{N}_l \leftarrow \text{make\_node}(C_1)$; $\mathscr{N}_r \leftarrow \text{make\_node}(C_2)$;   // make left and right node
14 $\quad$ $\mathscr{N} \leftarrow \text{update\_node}(f, \mathscr{N}_l, \mathscr{N}_r)$ // update tree
15 $\quad$ $\mathcal{Q} \leftarrow push(\mathcal{Q}, \text{qs}(C_1), \langle C_1, \mathscr{N}_l \rangle)$; $\mathcal{Q} \leftarrow push(\mathcal{Q}, \text{qs}(C_2), \langle C_2, \mathscr{N}_r \rangle)$;   // update queue
16 **while** $|\mathcal{Q}| > 0$ **do**
17 $\quad$ $\langle C, \mathscr{N} \rangle \leftarrow pop(\mathcal{Q})$;   // extract tree node from queue
18 $\quad$ $C \leftarrow C \cup \{C\}$; $\mathscr{L} \leftarrow \text{make\_leaf}(\mathscr{N})$;   // add cluster and make leaf
19 **return** $C, \mathscr{R}$;

---

disjoint sets (or clusters) $C = \{C_1, \ldots, C_k\}$, such that C is optimal in terms of homogeneity and simplicity, i.e., similar records belong to the same clusters and the number of clusters is small. Our objective is to define an algorithm that, given the dataset X, not only returns the clustering C, but also a way to humanly understand the logic used by the algorithm to obtain the clustering.

To this aim, we define Partitioning Tree (PARTREE), an interpretable tree-based clustering method that, besides the record to cluster assignments, returns an unsupervised binary tree describing the partitioning logic adopted. PARTREE is based on hierarchical top-down iterative bisections to find the best feature to partition the data. We highlight that, differently from bisecting k-Means [39], the crucial point of PARTREE is that the data partitioning at every tree level is done by selecting a unique feature-value to separate the data to guarantee cohesion among the records in the same partition. Indeed, in short, PARTREE aims at guaranteeing interpretability without sacrificing the clustering quality.

Figure 1 shows the clustering of PARTREE on the iris dataset. On the left is reported the partitioning logic learned by PARTREE. The central plot shows the effect of the axis-parallel partitioning w.r.t. the features in the rules. By comparing this plot with the right one reporting the iris classes we can appreciate how PARTREE, without any knowledge of the class, retrieved that setosa flowers can be recognized by the smallest *petal lengths* (cl. 0), while the remaining flowers are partitioned into versicolor (cl. 2) and "normal"-virginica (cl. 1), still w.r.t *petal length*. The "anomalous"-virginica, in cluster 3, are separated w.r.t. *petal width* from the rest. In the following, we describe the overall algorithm.

ParTree Algorithm. In line with [21,35], PARTREE adopts a top-down, divide-and-conquer strategy. It starts from a set containing a single cluster, then, iteratively tries to partition a cluster in two sub-clusters. The general schema of PARTREE is illustrated in Algorithm 1. It starts by initializing an empty set of sets $\mathcal{C}$ containing the clustering result (line 1) and by building the root of the tree $\mathcal{R}$ containing the whole dataset X (line 2). Then, it pushes both $\mathcal{C}$ and $\mathcal{R}$ into a priority queue $\mathcal{Q}$ that keeps track of the set of candidate clusters to be considered for partitioning (line 3). The priority of the objects in the queue is given by a qs$(\cdot)$ function that can be implemented in different ways. In our experiments, we considered the size of the cluster.

At each iteration, a candidate cluster C and the tree node $\mathcal{N}$ modeling C are extracted from $\mathcal{Q}$ (line 5). If C is too small (min_sample) to be partitioned or the tree node $\mathcal{N}$ is too deep (max_depth), then C is added to the clustering result $\mathcal{C}$, $\mathcal{N}$ is turned into a leaf $\mathcal{L}$, and the control passes to the next iteration (lines 6–8). On the other hand, the partitioning of C is performed by make_split$(\cdot)$ (line 9) that returns a binary partitioning function f, as well as the partitioning obtained by f on C, i.e., C_1, C_2, such that $C = C_1 \cup C_2$. Alternative ways to implement make_split$(\cdot)$ are illustrated in the remainder of this section. After that, PARTREE calculates the Bayesian Information Criterion (BIC) as in [35] on C and on the two sub-clusters C_1, C_2 (lines 10–11). If the *data partitioning* of C into C_1 and C_2 *is not advantageous enough* than keeping C united w.r.t. the BIC score, then C is added to the clustering result $\mathcal{C}$, $\mathcal{N}$ is turned into a leaf $\mathcal{L}$, and the control passes to the next iteration (lines 10–12). Otherwise, if the *data partitioning is advantageous*, the tree nodes $\mathcal{N}_l$ and $\mathcal{N}_r$ are created for sub-clusters C_1 and C_2 (line 13), and linked to the parent node together with the partitioning function f (line 14). After that, the novel candidate clusters C_1, C_2 with $\mathcal{N}_l, \mathcal{N}_r$ are pushed into (line 15). The $\varepsilon \in [0, 1]$ parameter controls to which extent the BIC of the children must be lower than the BIC of the parent w.r.t. the absolute BIC of the parent. If $\varepsilon > 0$ the condition to stop the partitioning is relaxed and allows to obtain more clusters.

PARTREE termination is controlled by three conditions: *(i)* the maximum number of admissible clusters ($max\_clusters$), *(ii)* the maximum admissible tree depth (max_depth), *(iii)* if it is worth to split the clusters obtained so far w.r.t. the BIC criterion. The while-loop (lines 4–15) ends either when $\mathcal{Q}$ is empty or when the number of candidate clusters and the clusters contained in the clustering result exceeds $max\_clusters$. If the queue is not empty after the first while-loop, a further loop completes the clustering and the tree construction (lines 16–18). Finally, it returns the clustering and the unsupervised binary partitioning tree. We highlight that the selection of the best splitting attribute (make_split$(\cdot)$) can be done in parallel due to the fact that the search is made independently among the various attributes. The computational complexity of PARTREE is $O(It \cdot n \cdot d \cdot s)$, where It is the number of iterations required to stop the algorithm, and s is the cost of the splitting strategy adopted. This leads to a theoretical worst case of $O(d^3 n^3 \log n)$, which yet in normal situations is expected to boil down to $O(d^2 n \log^2 n)$, thus almost linear in the number of records and quadratic in the number of initial features. In the following, we define three different implementations of PARTREE make_split fulfilling different partitioning principles.

Algorithm 2: CPT_split(X)

Input : X - dataset
Param : $dist$ - distance function, $get\_centroids$ - calculate centroids,
Output: f - binary partitioning function, C_1, C_2 - cluster partitions

1 $j^* \leftarrow \infty$; $t^* \leftarrow \infty$; $MSE^* \leftarrow \infty$;   // init. split feature, threshold, best MSE
2 **for** $j \in [1, d]$ **do**   // for every feature
3   **for** $t \in values(X^{(j)})$ **do**   // for every value
4          $f \leftarrow$ make_partitioner($cond(X^{(j)}, t)$);   // build data partitioner
5          $X_a, X_b \leftarrow$ split_data(f, X);   // make binary partitions
6   **if** $|X_a| = 0 \vee |X_b| = 0$ **then**
7 | continue;   // go next iteration
8          $\mu_a, \mu_b \leftarrow$ get_centroids(X_a, X_b);   // make centroid
9          $MSE \leftarrow \frac{|1|}{|X|} \left(\sum_{x \in X_a} \text{dist}^2(\mu_a, x) + \sum_{x \in X_b} \text{dist}^2(\mu_b, x) \right)$;   // calculate MSE
10   **if** $MSE < MSE^*$ **then**   // if better partitioning
11 | $j^* \leftarrow j$; $t^* \leftarrow t$; $MSE^* \leftarrow MSE$;   // update split feature, thr, MSE
12 $f \leftarrow$ make_partitioner($cond(X^{(j^*)}, t^*)$);   // build data partitioner
13 $C_1, C_2 \leftarrow$ split_data(f, X);   // make binary partitions
14 **return** f, C_1, C_2;

Center-Based Split. Inspired by bisecting k-Means [39] and similarly to [3,23], Center-based PARTREE (CPT) employs a center-based strategy. The pseudo-code of CPT is reported in Algorithm 2. CPT tests each feature j value t (lines 2–3) to find the best axis-parallel split (lines 4–5) to partition X based on the centers μ_a, μ_b of the two partitions (line 8). The goal is to optimize local compactness using a single axis-parallel split. The partitioning leading to the smallest weighted average Mean Squared Error (MSE) is selected as best split (lines 9–11).

The CPT split can be parameterized w.r.t. *(i)* the distance function $dist$ used to calculate the distance between the records and its center contributing for the MSE, and *(ii)* the function $get\_centroids$ used to calculate the centroids. These functions can be adapted to different types of data, by using appropriate distance measures and feature aggregators. For continuous features we use the Euclidean distance for $dist$ and of the mean values for every feature for $get\_centroids$ (like in k-Means), while for categorical features we can use Jaccard distance for $dist$ and the mode values for $get\_centroids$ (like in k-Modes). Thus, CPT can be used on any data type, also mixed, if the appropriate functions are provided. We assume without loss of generality that the values of each feature are normalized.

Impurity-Based Split. Inspired by Decision Tree models for classification and regression [39] and similarly to [1,19], Impurity-based PARTREE (IPT) adopts an impurity-based strategy. The pseudo-code of IPT is reported in Algorithm 3. The objective of IPT is to use impurity measures, such as Gini Index and Entropy for classification and Mean Absolute Percentage Error (MAPE) and R2 score for regression [39], to optimize the impurity w.r.t. the various features for an axis-parallel split. Since the problem is unsupervised and there is no target variable, the impurity measures cannot be directly adopted. Instead, given a partitioning of X into X_a and X_b w.r.t an axis-parallel split on feature j value t, the impurity is estimated by first computing the impurity values taking each remaining feature l $(l \neq j)$ as target, and then aggregating the results (lines 8–13). IPT solves the problem of requiring a notion of *(i)* distance, *(ii)* centroids.

Algorithm 3: IPT_split(X)

Input : X - dataset
Param : impurity - impurity function, agg_fun - impurity aggregation function
Output: f - binary partitioning function, C_1, C_2 - cluster partitions

1	$j^* \leftarrow \infty;\ t^* \leftarrow \infty;\ h^* \leftarrow \infty;$	// init. split feature, threshold, best impurity
2	**for** $j \in [1, d]$ **do**	// for every feature
3	**for** $t \in values(X^{(j)})$ **do**	// for every value
4	$f \leftarrow$ make_partitioner($cond(X^{(j)}, t)$);	// build data partitioner
5	$X_a, X_b \leftarrow$ split_data(f, X);	// make binary partitions
6	**if** $|X_a| = 0 \vee |X_b| = 0$ **then**	
7	**continue**;	// go next iteration
8	$H \leftarrow \emptyset;$	// init. impurity list
9	**for** $l \in [1, d] \wedge l \neq j$ **do**	// for every target feature
10	$h_a \leftarrow$ impurity($X_a^{(l)}$); $h_b \leftarrow$ impurity($X_b^{(l)}$);	// calculate impurities
11	$h \leftarrow \frac{|X_a|}{|X|} h_a + \frac{|X_b|}{|X|} h_b;$	// calculate total impurity
12	$H \leftarrow H \cup \{h\};$	// update impurity list
13	$h' \leftarrow$ agg_fun(H);	// aggregate impurities
14	**if** $h' < h^*$ **then**	// if better partitioning
15	$j^* \leftarrow j;\ t^* \leftarrow t;\ h^* \leftarrow h';$	// update split feature, thr, MSE
16	$f \leftarrow$ make_partitioner($cond(X^{(j^*)}, t^*)$);	// build data partitioner
17	$C_1, C_2 \leftarrow$ split_data(f, X);	// make binary partitions
18	**return** $f, C_1, C_2;$	

IPT split is parameterized w.r.t. *(i)* the *impurity* function(s), and *(ii)* *agg_fun* used to aggregate the list of impurities H (line 13). If l is a continuous feature we used MAPE or non-negative R2, i.e., R2 where values smaller than zero are considered zero, while if l is a categorical feature we used the Gini Index or the normalized Entropy, as they all take values between 0 and 1. As aggregation function *agg_fun*, similarly to Complete-Linkage [39], we experimented by using the minimum, maximum and average values. IPT does not require to normalize the data as the features are all analyzed independently from each other.

Principal Component-Based Split. Principal-based PARTREE (PPT), with pseudo-code in Algorithm 4, is inspired by [16,31,40,42], by Oblique Decision Trees [43] and Householder reflection [25]. The intuition is that principal components are capturing as much variance as possible and they are orthogonal each other. PPT first runs a dimensionality reduction algorithm on X to obtain a reduced version A. For a continuous dataset, Principal Component Analysis (PCA) [39] can be used, for a categorical dataset Multiple Correspondence Analysis (MCA) [20], and for a mixed dataset Factor Analysis of Mixed Data (FAMD) [13]. After that, the principal components $A^{(j)}$ are used as target variables of a decision tree regressor f with a single split (line 4) that is used to identify the axis-parallel split that better predicts the principal component values of the data. Depending on the data partition X, one of the first *nbr_components* principal components is used to separate X w.r.t. one of the features of X. Finally, the goodness of each partitioning is measured using the BIC[1] (lines

---

[1] After a preliminary experimentation we discarded evaluation measures for regressors as they do not consider the separation of the data but the performance of the regressor.

Algorithm 4: PPT_split(X)

Input : X - dataset
Param : $nbr\_components$ - number of components
Output: f - binary partitioning function, C_1, C_2 - cluster partitions

1 $j^* \leftarrow \infty$; $t^* \leftarrow \infty$; $bs^* \leftarrow \infty$;   // init. split feature, threshold, best BIC
2 $A \leftarrow$ get_PCA($X, nbr\_components$));   // calculate PCA
3 **for** $j \in [1, nbr\_components]$ **do**   // for every principal component
4 | $f \leftarrow$ train_tree_regressor($X, A^{(j)}$));   // build data partitioner
5 | $X_a, X_b \leftarrow$ split_data(f, X);   // make binary partitions
6 | **if** $|X_a| = 0 \lor |X_b| = 0$ **then**
7 | | **continue**;   // go next iteration
8 | $bs \leftarrow$ bic(X_a, X_b);   // calculate BIC partitions
9 | **if** $bs < bs^*$ **then**   // if better partitioning
10 | | $j^* \leftarrow j$; $t^* \leftarrow t$; $bs^* \leftarrow bs$;   // update split feature, thr, BIC
11 $f \leftarrow$ train_tree_regressor($X, A^{(j^*)}$));   // build data partitioner
12 $C_1, C_2 \leftarrow$ split_data(f, X);   // make binary partitions
13 **return** f, C_1, C_2;

8–10). The PPT split can be parameterized w.r.t. the number of components $nbr\_components$, and like CPT, it requires to normalize the dataset before usage.

4 Experiments

In this section we evaluate the performance of PARTREE, that we implemented in Python[2], on different datasets and against a wide array of competitors. Our objective is to demonstrate that PARTREE is more accurate or faster than state-of-the-art tree-based clustering algorithms, as well as a reasonable competitor of traditional (mostly non-interpretable) approaches.

Datasets. We experimented with 18 synthetic and 15 real datasets taken by UCI Machine Learning and Kaggle[3]. The synthetic datasets are all continuous and typically bi-dimensional, while some real datasets are composed also by categorical attributes. We consider the datasets grouped as follows: DS1 *continuous synthetic datasets*, DS2 *continuous real datasets*, i.e., real datasets without categorical attributes, DS3 *categorical real datasets*, i.e., real datasets with continuous attributes turned into categorical ones through equal-width binning with 20 bins, DS4 *mixed real datasets*, i.e., the datasets remain as they are. After that, in order to be processed by the clustering method implementations adopted, categorical attributes in DS3 and DS4 are turned into labels. All the datasets are normalized using z-score normalization [39]. We underline that PARTREE can always work with continuous attributes and, depending on the type of split, it can be applied on categorical and mixed dataset either by design with IPT and PPT, or through one-hot encoding or custom-defined distance functions with CPT. Table 1 summarizes the characteristics of the datasets types.

---

[2] https://github.com/cri98li/ParTree.
[3] https://github.com/deric/clustering-benchmark,   https://archive.ics.uci.edu/ml/datasets.php, https://www.kaggle.com/datasets.

Table 1. Dataset groups characteristics.

	nbr.	records			features			con. feat.			cat. feat.			clusters		
		min	avg	max	min	avg	max	min	avg	max	min	avg	max	min	avg	max
DS1	18	238	1860.9	5000	2	2.0	3	2	2.056	3	0	0	0	3	8.2	31
DS2	7	1000	10913.4	32561	14	24.2	44	6	16.8	32	1	7.4	13	2	2.0	2
DS3	14	150	15366.5	150000	4	18.2	44	4	15.2	34	0	3	13	2	2.2	4
DS4	14	150	15366.5	150000	4	18.2	44	4	15.2	34	0	3	13	2	2.2	4

Competitors. We compare PARTREE with traditional clustering algorithms and with tree-based competitors. In particular, we considered k-Means (KM), bisecting k-Means (BKM), BIRCH (BIR), DBSCAN (DBS), OPTICS (OPT) and the *Agglomerative Hierarchical Clustering* (AHC) [39] single/complete/ward linkage approaches as implemented by `scikit-learn`, x-Means (XM) [35] as implemented by `pyclustering`, and k-Modes (KMD)[4]. As tree-based competitors we considered CLTree (CLT) [29] that provides a usable Python implementation. Furthermore, we implemented k-Means-Tree (KMT) [41] by combining the `scikit-learn` k-Means and Decision Tree classifier. Finally, inspired by [3], we realize a variant of PARTREE using the maximum variance as indicator to identify the feature-values to use for partitioning (VPT).

Experimental Setting. We evaluated the performance of the algorithms by adopting a large array of measures. We report the results w.r.t. the *external* validation measures Adjusted Rand Index (ARI) [38] Normalized Mutual Information (NMI) [34], and Fowlkes-Mallows Score (FMS) [14], and the *internal* validation measure Silhouette (SIL) [39]. Also, we evaluated the running time in seconds. For each algorithm, we executed the clustering with different parameter combinations on each dataset[5] and we considered the evaluation measures corresponding to the best performance w.r.t SIL as being an internal validation measure can be used without the need of the clustering ground truth[6]. The values adopted to test the parameters of the various algorithms as well as other validation measures are available on the Github of the project. For the clustering methods requiring a distance functions we used the Euclidian distance for DS1, DS2, and DS4, while for DS3 the Cosine distance [39].

Results. Table 2 shows the average values of the measures obtained by the 14 algorithms and their average rank for the different datasets types. k-Means (KM) is the best performer overall, while the second best performer is the Agglom-

---

[4] https://scikit-learn.org/stable/index.html, https://github.com/annoviko/pyclustering/, https://github.com/nicodv/kmodes.

[5] Details for the parameter values tested are available on the repository. However, since the objective is towards interpretable clustering, we do not search for more than 12 clusters or trees deeper than 10. Also, we remark that it is outside the purpose of this study to design strategies to identify good values for $max\_clusters, max\_depth, min\_sample, \varepsilon$. We leave this task for a future study.

[6] Similar results to those reported are obtained with best parameters w.r.t other measures.

Table 2. Experimental results. For each algorithm and dataset we report the performance of the best parameter configuration w.r.t SIL. The overall highest value per measure is in bold while the highest among tree-based methods is in italic and blue.

	DS1										DS2									
	NMI ↑		ARI ↑		FMS ↑		SIL ↑		Time ↓		NMI ↑		ARI ↑		FMS ↑		SIL ↑		Time ↓	
	avg	rnk	avg	rnk	avg	rnk	avg	rnk	avg	rnk	avg	rnk	avg	rnk	avg	rnk	avg	rnk	avg	rnk
CPT	*.80*	5.5	*.72*	5.1	*.81*	5.4	.59	5.1	1.2	11.0	.54	1.7	.56	1.7	.83	4.0	.32	6.2	16.8	10.5
IPT	.48	10.8	.39	10.3	.65	9.3	.42	11.0	1.9	10.4	.30	5.0	.26	6.0	.71	6.5	.32	6.7	14.7	9.5
PPT	.77	6.6	.69	6.3	.78	6.6	.57	6.9	*0.1*	5.3	*.69*	2.0	*.70*	2.0	**.85**	2.0	.35	5.0	*0.2*	3.3
VPT	.66	8.3	.50	8.5	.66	8.7	.51	8.6	0.9	10.8	.01	6.0	.02	8.0	.65	7.6	.24	5.6	1.2	8.0
CLT	.10	13.2	.03	13.1	.46	12.5	.02	13.2	25.4	13.9	.01	3.5	<0	6.5	.69	3.0	<0	11.0	>100	11.0
KMT	.72	8.0	.58	7.1	.71	7.5	.54	7.6	0.1	5.6	.23	4.7	.24	3.1	.70	5.1	.37	5.0	0.4	5.0
KM	.83	4.2	.71	4.7	.79	5.0	.61	2.1	0.1	5.7	.23	2.7	.24	2.1	.71	4.7	.38	3.0	0.5	5.8
BKM	.79	6.2	.71	5.6	.79	5.9	.59	4.8	0.1	7.8	.23	2.7	.24	2.1	.71	4.7	.38	3.0	0.5	5.7
xM	**.85**	3.6	**.76**	4.3	**.83**	4.6	**.63**	1.6	**0.0**	2.4	.22	4.8	.21	4.8	.73	3.2	.51	2.2	**0.1**	1.8
KMD	.00	13.5	<0	13.3	.47	12.6	<0	13.7	0.1	7.7	.10	8.4	.11	6.7	.65	7.1	.05	10.8	5.9	8.1
BIR	.80	5.1	.68	5.8	.77	5.8	.58	5.6	0.1	4.5	.21	5.4	.21	6.7	.69	6.2	.32	5.8	1.0	5.2
AHC	**.85**	2.8	.74	3.3	.82	3.6	.60	3.4	0.1	4.2	.13	6.1	.10	6.4	.77	2.7	**.64**	1.0	1.6	4.2
DBS	.81	3.8	.75	3.6	.82	3.7	.54	7.2	**0.0**	2.2	.12	8.1	.10	7.5	.73	6.2	.33	7.7	1.2	3.4
OPT	.51	10.0	.36	10.1	.58	10.2	.38	10.9	4.7	12.8	.15	5.2	.10	6.7	.57	5.5	.26	7.5	90.7	10.8

	DS3										DS4									
	NMI ↑		ARI ↑		FMS ↑		SIL ↑		Time ↓		NMI ↑		ARI ↑		FMS ↑		SIL ↑		Time ↓	
	avg	rnk	avg	rnk	avg	rnk	avg	rnk	avg	rnk	avg	rnk	avg	rnk	avg	rnk	avg	rnk	avg	rnk
CPT	.13	5.4	.10	6.8	.59	7.3	.12	11.0	22.7	10.7	.13	5.3	.10	6.5	.59	7.0	.12	10.8	21.9	9.7
IPT	.11	6.0	.10	6.5	*.66*	4.5	.17	9.7	1.1	7.5	.11	6.2	.10	6.7	*.66*	4.7	.17	10.5	1.1	7.0
PPT	*.25*	3.5	*.26*	3.1	*.66*	6.3	.28	7.8	0.9	6.2	*.25*	3.7	*.26*	3.2	*.66*	6.6	.28	8.2	0.9	5.8
VPT	.10	6.4	.09	7.7	.64	5.8	.34	8.2	1.4	6.6	.13	5.5	.14	5.6	.64	6.5	.44	6.5	3.9	10.0
CLT	.11	5.1	.08	8.1	.59	7.1	<0	13.5	>100	14.0	.11	5.2	.08	8.2	.59	7.8	<0	13.7	>100	14.0
KMT	.13	6.6	.13	5.2	.63	5.5	*.48*	4.2	*0.6*	6.3	.14	6.5	.14	5.4	.63	6.1	*.48*	4.5	*0.6*	6.1
KM	.17	5.4	.17	3.0	.62	5.6	.46	2.6	0.6	8.3	.18	4.9	.17	3.1	.62	6.1	.46	3.2	0.5	6.9
BKM	.18	5.0	.17	2.9	.63	5.2	.46	2.9	1.0	8.0	.18	5.2	.17	3.0	.63	5.7	.46	3.6	0.9	7.6
xM	.18	4.9	.17	3.9	.61	6.6	.45	2.9	**0.1**	1.7	.17	6.1	.15	4.5	.64	5.2	.49	2.6	**0.1**	2.3
KMD	.08	6.7	.06	6.2	.57	8.3	.08	11.6	2.2	8.7	.09	6.5	.09	6.2	.60	6.8	.10	11.5	1.6	8.3
BIR	.16	6.3	.15	6.1	.63	6.4	.40	5.3	1.9	5.3	.16	6.3	.15	6.4	.63	6.0	.42	5.2	1.7	5.3
AHC	.07	7.6	.06	8.6	**.68**	3.3	**.60**	1.6	4.3	3.3	.07	8.0	.06	8.6	**.68**	3.2	**.61**	1.6	3.7	3.3
DBS	.09	8.4	.06	9.0	.67	5.2	.35	7.3	1.6	3.8	.09	7.7	.06	9.3	.67	5.2	.35	7.7	1.6	3.7
OPT	.15	5.1	.08	7.3	.52	7.0	.20	8.6	65.2	12.8	.16	4.8	.08	7.3	.53	7.0	.23	8.5	62.9	12.6

erative Hierarchical clustering (AHC), especially w.r.t. FMS that is meant to adequately judge the performance of hierarchical clustering algorithm. The best tree-based clustering algorithm is often a PARTREE method. Among the synthetic datasets DS1, CPT is the best algorithm for NMI, ARI, and FMS. On the other hand, among the real datasets DS2, DS3, and DS2, independently from the data type representation, PPT is the best method for NMI and ARI, and is on par with IPT for FMS. Besides the quality measures, also w.r.t. runtime, PPT is competitive with the traditional clustering algorithms and better or on par with KMT, i.e., k-Means followed by a Decision Tree. Furthermore, we highlight that PARTREE methods are considerably better than the unique tree-based method for which we found a usable implementation, i.e., CLTree (CLT).

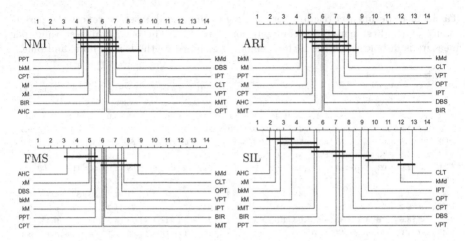

Fig. 2. Critical Difference plots with Nemenyi at 95% confidence among all datasets.

The comparison of the ranks of all methods against each other considering all datasets and types is visually represented in Fig. 2 with Critical Difference (CD) diagrams [12]. Two methods are tied if the null hypothesis that their performance is the same cannot be rejected using the Nemenyi test at $\alpha = 0.05$. We immediately notice that for NMI and ARI, the methods are compactly tied to each other but having PPT and CPT in the top-5 together with the centroids-based methods KM, BKM and XM. For FMS, we need to consider the top-7 to find PPT and CPT, which are overtaken by AHC and DBS besides KM, BKM and XM. Finally, for SIL, the performance of PARTREE methods are slightly worse. However, PPT performance is not statistically worse than KMT, the best tree-based clustering method w.r.t. SIL, as they are tied in the CD plot. Furthermore, we recall that KMT [41] first applies k-Means and then a Decision Tree that uses as target variable the cluster labels. Similarly to CUBT [15], this causes an inconsistency in the tree-based logic of the clustering structure, as two different records satisfying different conditions in the tree can be assigned to the same cluster. On the other hand, for PARTREE methods, this cannot happen by construction, making PARTREE the best interpretable-by-design tree-based clustering method. Moreover, the SIL measure, and other internal evaluation measures, are influenced by the distance function used to estimate it. In contrast, the external validation measures NMI, ARI, and FMS are more objective as they report the agreement between the returned and expected clustering assignments. Therefore, the good performance of PARTREE methods is even more remarkable, considering that the best parameter configuration has been selected w.r.t. SIL. Thus, SIL can be used for an unsupervised parameter tuning of PARTREE.

5   User Study

We describe here the user study carried out to evaluate the interpretability of different clustering algorithms. First, we aim to study if a user assigning a

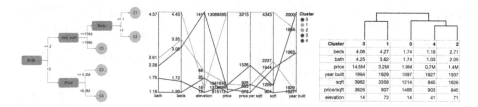

Fig. 3. Clustering visualizations for HOME. From left to right: PARTREE, KM, AHC.

record to a cluster, i.e., trying to repeat the same logic followed by a clustering algorithm, makes fewer mistakes by using a tree rather than other clustering visualizations like centroids or dendrograms. Second, we want to understand if the tree resulting from PARTREE has a lower cognitive effort and higher usability than other clustering summarizations. As competitors, we selected *(i)* the centroids returned by k-Means (KM) represented with a parallel plot, and *(ii)* the dendrogram returned by Agglomerative Hierarchical Clustering (AHC). We adopted KM and AHC because they are the most effective approaches in the benchmarking in the previous section and also because they are among the most widely adopted methods. We do not compare PARTREE against other tree-based clustering algorithms as we have assessed in the previous section that PARTREE is the best tree-based clustering algorithm, and since all methods in this family return a tree, there is small interest in comparing different trees, while we are interested in checking if a tree is more interpretable than the visualizations available from traditional non-tree-based clustering algorithms.

Our experiment consists in providing a user with a record x and a clustering visualization in terms of tree, centroids, or dendrogram and ask to the user to insert x into the right cluster. For all the visualizations, we do not provide distances between x and the clusters to avoid biasing the user toward a number that might be unable to understand and to entirely judge how much a user can really exploit the different clustering representations provided. Thus, our *hypotheses* are the following. By comparing the correctness of the assignments, i.e., correspondence of matches between users' assignments and real cluster assignments, i.e., the Success Rate (SR), we are able to evaluate the clustering interpretability and usability. *HP1: PARTREE visualization makes users perform better in terms of SR than the competitors.* We assess the Cognitive Effort by means of the NASA Task Load Index (NASA-TLX) questionnaire [5], and the clustering visualization Usability with the System Causability Scale (SCS) [24]. Our expectation is that the tree-based visualization of PARTREE illustrates the clustering logic in a clearer way than the competitors because, at the cognitive level, following precise partitioning steps on a tree is easier than simultaneously comparing a set of features on centroids or on a dendrogram. *HP2: PARTREE visualization requires a lower cognitive effort than the competitors as NASA-TLK. HP3: PARTREE visualization provides better usability than the competitors as SCS.*

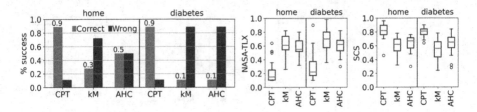

Fig. 4. User study results: (left) success rate, (center) cognitive effort, (right) usability.

Experimental Setting. We experimented with home and diabetes from which we randomly selected three records. For PARTREE, we adopted the CPT version. We set parameters of CPT, ΚM, and AHC to obtain 5 clusters for home and 10 clusters for diabetes, respectively. Figure 3 shows the visualizations for home. More details are available in the repository, including the clustering logics provided to user study participants. We ran an online experiment with the Qualtrics platform[7]. We collected data from 40 participants[8] with a background in Computer Science and Data Science with an average age of 26 ranging from 22 to 37 years old. We asked the participants to assign a record to a cluster using CPT, ΚM and AHC for home ("easy" task, with 7 features and 5 clusters) and diabetes ("difficult" task, with 8 features and 10 clusters). Then participants were asked to express their confidence in the accomplishment of the task. To prevent any *learning effect*, each participant used different yet analogous records for each algorithm. To prevent any *order effect*, participants were presented with randomized orders for the complexity and for the algorithms.

Results. Figure 4 (left) shows the *Success Rate* in the record to cluster assignment. It is clear how it is easier for a user to correctly assign a record to a cluster following the tree logic by reasoning on one feature after the other, instead of considering all the features simultaneously with parallel plots and dendrograms. A one-way ANOVA test on the ST revealed that there is a statistically significant difference in performance between at least two methods. We identify the methods through the Tukey's Honestly Significant Difference (THSD) [32] for multiple comparisons. It showed that the mean value of SR is significantly different for CPT and ΚM (p>0.01, 95% Confidence Interval (CI) = [0.47, 0.92]) and for CPT and AHC (p>0.01, 95% CI = [0.4, 0.8]). This result confirms *HP1*. Also, the THSD found that the mean value of SR is not significantly different between the easy and complex conditions (p = 0.213, 95% CI = [−14.6, 3.3]). This lack of statistical significance holds also for the Cognitive Effort and Usability. Figure 4 (center) shows the *Cognitive Effort* as NASA-TLX box-plots (the lower, the better): CPT has the lowest and more compact box-plots. The THSD found that the mean value of NASA-TLX is significantly different between CPT and ΚM (p>0.001, 95% CI = [−48.6, −30.6]), and between CPT and AHC (p>0.001,

---

[7] https://www.qualtrics.com/.

[8] All participants provided written informed consent and received no monetary rewards.

95% CI = $[-41.9, -24.0]$), while this does not happen between AHC and KM. Similarly, Fig. 4 (right) shows the *Usability* as SCS box-plots (the higher, the better): CPT has the highest and more compact box-plots. Also in this case the THSD found that the mean value of SCS is significantly different between CPT and KM (p>0.001, 95% CI = $[0.18, 0.11]$), and between CPT and AHC (p>0.001, 95% CI = $[-0.04, 0.09]$), while again this does not happen between AHC and KM. These results confirm *HP2* and *HP3*.

6   Conclusion

We have presented PARTREE, an interpretable tree-based clustering method that, besides the record to cluster assignments, returns an unsupervised binary tree describing the partitioning logic adopted. Our experiments show that PARTREE is on par with, or in some cases even better than, traditional non-tree-based clustering methods while it overcomes state-of-the-art tree-based approaches. Also, a survey involving real users has shown that the clustering tree provided by PARTREE is the most effective way for a user to replicate the algorithm's behavior in assigning a record to a cluster. We plan to repeat the user study involving more users and providing better visualizations. Furthermore, we would like to study the fair-clustering [9] problem through PARTREE and change its partitioning procedure to guarantee fair partitions, i.e., groups of records not separated w.r.t. a sensitive attribute such as sex or race.

Acknowledgment. This work is partially supported by the EU NextGenerationEU programme under the funding schemes PNRR-PE-AI FAIR (Future Artificial Intelligence Research), PNRR-SoBigData.it - Strengthening the Italian RI for Social Mining and Big Data Analytics - Prot. IR0000013, H2020-INFRAIA-2019-1: Res. Infr. G.A. 871042 *SoBigData++*, G.A. 761758 *Humane AI*, G.A. 952215 *TAILOR*, ERC-2018-ADG G.A. 834756 *XAI*, G.A. 101070416 *Green.Dat.AI* and CHIST-ERA-19-XAI-010 SAI.

References

1. Basak, J., Krishnapuram, R.: Interpretable hierarchical clustering by constructing an unsupervised decision tree. IEEE TKDE **17**(1), 121–132 (2005)
2. Bertsimas, D., Orfanoudaki, A., Wiberg, H.M.: Interpretable clustering: an optimization approach. Mach. Learn. **110**(1), 89–138 (2021)
3. Blockeel, H., Raedt, L.D., Ramon, J.: Top-down induction of clustering trees. In: ICML, pp. 55–63. Morgan Kaufmann (1998)
4. Breiman, L., Friedman, J.H., Olshen, R.A., Stone, C.J.: Classification and regression trees. Wadsworth (1984)
5. Cao, A., Chintamani, K.K., Pandya, A.K., Ellis, R.D.: NASA TLX: software for assessing subjective mental workload. Behav. Res. Meth. **41**(1), 113–117 (2009). https://doi.org/10.3758/BRM.41.1.113
6. Castin, L., Frénay, B.: Clustering with decision trees: divisive and agglomerative approach. In: ESANN, pp. 455–460 (2018)

7. Chen, J., et al.: Interpretable clustering via discriminative rectangle mixture model. In: ICDM, pp. 823–828. IEEE Computer Society (2016)
8. Chen, Y., Hsu, W., Lee, Y.: TASC: two-attribute-set clustering through decision tree construction. Eur. J. Oper. Res. **174**(2), 930–944 (2006)
9. Chierichetti, F., Kumar, R., Lattanzi, S., Vassilvitskii, S.: Fair clustering through fairlets. In: NIPS, pp. 5029–5037 (2017)
10. Dasgupta, S., Freund, Y.: Random projection trees and low dimensional manifolds. In: STOC, pp. 537–546. ACM (2008)
11. Dasgupta, S., Frost, N., Moshkovitz, M., Rashtchian, C.: Explainable k-means clustering: theory and practice. In: XXAI Workshop. ICML (2020)
12. Demsar, J.: Statistical comparisons of classifiers. JMLR **7**, 1–30 (2006)
13. Escofier, B., et al.: Analyses factorielles simples et multiples. Dunod **284** (1998)
14. Fowlkes, E.B., Mallows, C.L.: A method for comparing two hierarchical clusterings. J. Am. Stat. Assoc. **78**(383), 553–569 (1983)
15. Fraiman, R., Ghattas, B., Svarc, M.: Interpretable clustering using unsupervised binary trees. Adv. Data Anal. Classif. **7**(2), 125–145 (2013)
16. Freund, Y., et al.: Learning the structure of manifolds using random projections. In: NIPS, pp. 473–480. Curran Associates, Inc. (2007)
17. Frost, N., Moshkovitz, M., Rashtchian, C.: ExKMC: expanding explainable k-means clustering. CoRR abs/2006.02399 (2020)
18. Gabidolla, M., Carreira-Perpiñán, M.Á.: Optimal interpretable clustering using oblique decision trees. In: KDD, pp. 400–410. ACM (2022)
19. Ghattas, B., Michel, P., Boyer, L.: Clustering nominal data using unsupervised binary decision trees. Pattern Recognit. **67**, 177–185 (2017)
20. Greenacre, M., et al.: Multiple correspondence analysis. CRC (2006)
21. Guidotti, R., et al.: Clustering individual transactional data for masses of users. In: KDD, pp. 195–204. ACM (2017)
22. Guidotti, R., et al.: A survey of methods for explaining black box models. ACM CSUR **51**(5), 93:1–93:42 (2019)
23. Gutiérrez-Rodríguez, A.E., et al.: Mining patterns for clustering on numerical datasets using unsupervised decision trees. KBS **82**, 70–79 (2015)
24. Holzinger, A., et al.: Measuring the quality of explanations: the system causability scale (SCS) comparing human and machine explanations. KI **34**(2), 193–198 (2020)
25. Householder, A.S.: Unitary triangularization of a nonsymmetric matrix. J. ACM **5**(4), 339–342 (1958)
26. Laber, E.S., Murtinho, L.: On the price of explainability for some clustering problems. In: ICML, vol. 139, pp. 5915–5925. PMLR (2021)
27. Laber, E.S., Murtinho, L., Oliveira, F.: Shallow decision trees for explainable k-means clustering. Pattern Recognit. **137**, 109239 (2023)
28. Lawless, C., et al.: Interpretable clustering via multi-polytope machines. In: AAAI, pp. 7309–7316. AAAI Press (2022)
29. Liu, B., Xia, Y., Yu, P.S.: Clustering through decision tree construction. In: CIKM, pp. 20–29. ACM (2000)
30. Loyola-González, O., et al.: An explainable artificial intelligence model for clustering numerical databases. IEEE Access **8**, 52370–52384 (2020)
31. McCartin-Lim, M., McGregor, A., Wang, R.: Approximate principal direction trees. In: ICML. icml.cc/Omnipress (2012)
32. Montgomery, D.C.: Design and Analysis of Experiments. Wiley, Hoboken (2017)
33. Moshkovitz, M., Dasgupta, S., Rashtchian, C., Frost, N.: Explainable k-means and k-medians clustering. In: ICML, vol. 119, pp. 7055–7065. PMLR (2020)

34. Nguyen, X.V., et al.: Information theoretic measures for clusterings comparison: is a correction for chance necessary? In: ICML, vol. 382, pp. 73–80. ACM (2009)
35. Pelleg, D., Moore, A.W.: X-means: extending k-means with efficient estimation of the number of clusters. In: ICML, pp. 727–734. Morgan Kaufmann (2000)
36. Plant, C., Böhm, C.: INCONCO: interpretable clustering of numerical and categorical objects. In: KDD, pp. 1127–1135. ACM (2011)
37. Quinlan, J.R.: Induction of decision trees. Mach. Learn. **1**(1), 81–106 (1986)
38. Rand, W.M.: Objective criteria for the evaluation of clustering methods. J. Am. Stat. Assoc. **66**(336), 846–850 (1971)
39. Tan, P.N., et al.: Introduction to Data Mining. Pearson Education India, Noida (2016)
40. Tavallali, P., Tavallali, P., Singhal, M.: K-means tree: an optimal clustering tree for unsupervised learning. J. Supercomput. **77**(5), 5239–5266 (2021)
41. Thomassey, S., Fiordaliso, A.: A hybrid sales forecasting system based on clustering and decision trees. Decis. Support Syst. **42**(1), 408–421 (2006)
42. Verma, N., Kpotufe, S., Dasgupta, S.: Which spatial partition trees are adaptive to intrinsic dimension? In: UAI, pp. 565–574. AUAI Press (2009)
43. Wickramarachchi, D.C., Robertson, B.L., Reale, M., Price, C.J., Brown, J.: HHCART: an oblique decision tree. Comput. Stat. Data Anal. **96**, 12–23 (2016)

Jaccard-Constrained Dense Subgraph Discovery

Chamalee Wickrama Arachchi[✉] and Nikolaj Tatti

HIIT, University of Helsinki, Espoo, Finland
{chamalee.wickramaarachch,nikolaj.tatti}@helsinki.fi

Abstract. Finding dense subgraphs is a core problem in graph mining with many applications in diverse domains. At the same time many real-world networks vary over time, that is, the dataset can be represented as a sequence of graph snapshots. Hence, it is natural to consider the question of finding dense subgraphs in a temporal network that are allowed to vary over time to a certain degree. In this paper, we search for dense subgraphs that have large pairwise Jaccard similarity coefficients. More formally, given a set of graph snapshots and a weight λ, we find a collection of dense subgraphs such that the sum of densities of the induced subgraphs plus the sum of Jaccard indices, weighted by λ, is maximized. We prove that this problem is **NP**-hard. To discover dense subgraphs with good objective value, we present an iterative algorithm which runs in $\mathcal{O}\left(n^2 k^2 + m \log n + k^3 n\right)$ time per single iteration, and a greedy algorithm which runs in $\mathcal{O}\left(n^2 k^2 + m \log n + k^3 n\right)$ time, where k is the length of the graph sequence and n and m denote number of nodes and total number of edges respectively. We show experimentally that our algorithms are efficient, they can find ground truth in synthetic datasets and provide interpretable results from real-world datasets. Finally, we present a case study that shows the usefulness of our problem.

1 Introduction

Finding dense subgraphs is a core problem in graph mining with many applications in diverse domains such as social network analysis [20], temporal pattern mining in financial markets [8], and biological system analysis [10]. Often, many real-world networks vary over time, in which case a sequence of graph snapshots naturally exists. Consequently, mining dense subgraphs over time has gained an attention in data mining literature [12,15,19,20].

Our goal is to find dense subgraphs in a temporal network. In order to measure density, we will use a popular choice of the ratio between number of induced edges and nodes. This choice is popular since the densest subgraph can be found in polynomial time [13] and approximated efficiently [5].

Given a graph sequence, there are natural extremes to find the densest subgraphs: the first approach is to find a common subgraph that maximizes the sum of densities for individual snapshots, as proposed by Semertzidis et al. [20]

© The Author(s), under exclusive license to Springer Nature Switzerland AG 2023
A. Bifet et al. (Eds.): DS 2023, LNAI 14276, pp. 508–522, 2023.
https://doi.org/10.1007/978-3-031-45275-8_34

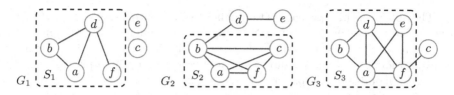

Fig. 1. Toy graphs used in Example 1

among other techniques. The other approach is to find the densest subgraphs for each snapshot individually.

In this paper, we study the problem that bridges the gap between these two extremes, namely we seek dense subgraphs in a temporal network that are allowed to vary over time to a certain degree. Our approach is to incorporate the Jaccard similarity index directly into our objective function along with the density. Here, we reward similar graphs over the snapshots. More formally, given a graph sequence $\mathcal{G}$ and parameter λ, we seek a sequence of subgraphs, such that the sum of densities plus the sum of Jaccard indices, weighted by λ, is maximized.

We demonstrate the objective in the following toy example.

Example 1. Consider a graph sequence $\{G_1, G_2, G_3\}$ shown in Fig. 1, each graph consisting of 6 vertices and varying edges. We denote the density induced by the vertex set S_i by $d(S_i)$, defined as the ratio between number of induced edges and nodes, $d(S_i) = \frac{|E(S_i)|}{|S_i|}$.

Given a weight parameter λ and a sequence of subgraphs $\{S_1, S_2, S_3\}$, we define our objective function as $\sum_{i=1}^{3} d(S_i) + \lambda \sum_{i<j} J(S_i, S_j)$. Note that Jaccard index between set S and T is defined as $J(S, T) = \frac{|S \cap T|}{|S \cup T|}$.

Assume that we set $\lambda = 0.3$ and select $S_1 = \{a, b, d, f\}$, $S_2 = \{a, b, c, f\}$, and $S_3 = \{a, b, d, e, f\}$. The sum of densities is $\frac{4}{4} + \frac{6}{4} + \frac{8}{5} = 4.1$. Here, $J(S_1, S_2) = \frac{3}{5} = 0.6$, $J(S_1, S_3) = \frac{4}{5} = 0.8$, and $J(S_2, S_3) = \frac{3}{6} = 0.5$. Therefore, our objective is equal to $4.1 + 0.3 \times (0.6 + 0.8 + 0.5) = 4.67$.

We show that our problem is **NP**-hard and consequently propose two greedy algorithms. The first approach is an iterative algorithm where we start either with the common densest subgraph or a set of the densest subgraphs for each individual snapshot and then iteratively improve each individual snapshot. The improvement step is done with a classic technique used to approximate the densest subgraph (in a single graph) [2,5]. We start with the complete snapshot, and iteratively seek out the vertex so that the remaining graph yields the largest score. We remove the vertex, and the procedure is repeated until no vertices remain; the best subgraph for that snapshot is selected. This is repeated for each snapshot until no improvement is possible. The second algorithm uses a similar approach except we consider all snapshots at the same time, and select the best subgraph once we are done.

The appeal of this approach is that, when dealing with a single graph, finding the next vertex can be done efficiently using a priority queue [2,5]. We cannot use this approach directly due to the updates in Jaccard indices. Instead we maintain a *set* of priority queues that allow us to find vertices quickly in practice.

The remainder of the paper is organized as follows. In Sect. 2, we provide preliminary notation along with the formal definitions of our optimization problem. Next, we prove **NP**-hardness of our problem in Sect. 3. All our algorithms and their running times are presented in Sect. 4. Related work is discussed in Sect. 5. Section 6 contains an experimental study both with synthetic and real-world datasets. Finally, Sect. 7 summarizes the paper and provides directions for the future work.

2   Preliminary Notation and Problem Definition

We begin by providing preliminary notation and formally defining our problem.

Our input is a sequence of graphs $\mathcal{G} = \{G_1, \ldots, G_k\}$, where each snapshot $G_i = (V, E_i)$ is defined over the same set of nodes. We often denote the number of nodes and edges by $n = |V|$ and $m_i = |E_i|$, or $m = |E|$, if i is omitted.

Given a graph $G = (V, E)$, and a set of nodes $S \subseteq V$, we define $E(S) \subseteq E$ to be the subset of edges having both endpoints in S.

As mentioned before, our goal is to find dense subgraphs in a temporal network, and for that we need to quantify the density of a subgraph. More formally, assume an unweighted graph $G = (V, E)$, and let $S \subseteq V$. We define the *density* $d(S)$ of a single node set S and extend this definition for a sequence of subgraphs $\mathcal{S} = \{S_1, \ldots, S_k\}$ by writing

$$d(S_i) = \frac{|E(S_i)|}{|S_i|} \quad \text{and} \quad d(\mathcal{S}) = \sum_{i=1}^{k} d(S_i) \quad .$$

We will use the Jaccard index in order to measure the similarity between two subgraphs. More formally, given two sets of nodes S and T, we write

$$J(S, T) = \frac{|S \cap T|}{|S \cup T|} \quad .$$

Ideally, we would like to have each subgraph to have high density, and share as many nodes as possible with each other. This leads to the following score and optimization problem. More specifically, given a weight parameter λ and a sequence of subgraphs $\mathcal{S} = \{S_1, \ldots, S_k\}$ we define a score

$$q(\mathcal{S}; \lambda) = d(\mathcal{S}) + \lambda \sum_{i=1}^{k} \sum_{j=i+1}^{k} J(S_i, S_j) \quad .$$

Problem 1. (Jaccard Weighted Densest Subgraphs (JWDS)). Given a graph sequence $\mathcal{G} = \{G_1, \ldots, G_k\}$, with $G_i = (V, E_i)$, and a real number λ, find a collection of subset of vertices $\mathcal{S} = \{S_1, \ldots, S_k\}$, such that $q(\mathcal{S})$ is maximized.

We will consider two extreme cases. The first case is when λ is very large, say $\lambda = \sum m_i$, which we refer to as *densest common subgraph* or DCS. This problem can be solved by first flattening the graph sequence into one weighted graph, where an edge weight is the number of snapshots in which an edge occurs. The problem is then a standard (weighted) densest subgraph problem that can be solved using the method given by Goldberg [13] in $\mathcal{O}(nm \log n)$ time. The other extreme case is $\lambda = 0$ which can be found by solving the densest subgraph problem for each individual snapshot.

The main difference from prior studies [15,20] is that we allow the subsets to be varied within a given margin (which is defined by Jaccard coefficient), without enforcing subsets to be fully identical.

3 Computational Complexity

The problem of finding a common subgraph which maximizes the sum of densities can be solved optimally in polynomial time. Moreover, if we set $\lambda = 0$, then we can solve the problem by finding optimal dense subgraphs for each snapshot individually. Next we show that JWDS is **NP**-hard. Note that the hardness relies on the fact that we can choose a specific λ.

Proposition 1. JWDS *is **NP**-hard.*

Proof. We will prove the hardness by reducing from a 2-bounded 3-set packing problem 3DM-2, a problem where we are given a set of items U, a family of sets $\mathcal{C}$ each of size 3 such that each item in U is included in exactly two sets, and are asked to find a maximum matching [7].

Assume that we are given an instance with $r = |U|$ items and $\ell = |\mathcal{C}|$ sets. For each set C_i we introduce two nodes v_i and v_i', and for each item u_i we introduce a node w_i. We also introduce two additional nodes z_1 and z_2. In total, we have $n = 2\ell + r + 2$ nodes.

Let u_i be an item and let C_a and C_b be two sets containing u_i. We add two snapshots: the first graph G_i contains (z_1, z_2), (w_i, z_1), (v_a, z_1) and (v_b, z_1) edges and the second graph G_i' contains (z_1, z_2), (w_i, z_1), (v_a', z_1) and (v_b', z_1) edges.

We also add $q = 2rn^4$ graphs F_j, each with one edge (z_1, z_2). We set $\lambda = 1/q$. Let $\mathcal{S}$ be the optimal solution. We claim that

$$q(\mathcal{S}) \geq T_1 + T_2 + T_3, \text{ where } T_1 = \frac{q}{2} + \lambda \binom{q}{2}, \ T_2 = 2r\frac{4}{3}, \ T_3 = \frac{\lambda}{2}\binom{2r}{2} + \lambda 2p + \lambda r,$$

(1)

if and only if there is a matching with p sets.

Assume that Eq. 1 holds. Let Q_j be the optimal subgraphs in F_j. Write S_i to be the optimal subgraph in G_i and S_i' to be the optimal subgraph in G_i'. Due to symmetry, we can safely assume that the subgraphs Q_j are all equal. Moreover, $Q_j = \{z_1, z_2\}$ since the densities and the Jaccard indices of S_i and S_i' cannot reach $q/2$. The densities and Jaccard indices of Q_j now correspond to T_1.

Next, we claim that S_i (and S_i') consists of 3 nodes and 2 edges, one of them being (z_1, z_2). Fix S_i with $y = |E(S_i)|$ edges and $x = |S_i|$ nodes. Let $z = |S_i \cap \{z_1, z_2\}|$ be the intersection with Q_j. We can show that the score is

$$q(S) = \frac{y}{x} + \frac{z}{x + 2 - z} + R + C,$$

where R contains the Jaccard terms using S_i and not Q_j, and C contains the remaining densities Jaccard terms not depending on S_i. The first two terms form a fraction with a denominator of at most n^2. Consequently, any changes to x, y, and z change the first two terms by at least n^{-4}. Note that R contains only $2r - 1$ terms, and due to λ, we have $R < n^{-4}$. In other words, S must optimize the first two terms. Since there are only 5 non-singleton nodes in G_i, $x \leq 5$. Moreover, $z \leq \min(2, x)$ and $y \leq \min(4, x - 1)$. Enumerating all the possible combinations show that $x = 3$ and $z = y = 2$ yield optimal score. This is only possible if S_i consists of 3 nodes and 2 edges, one of them being (z_1, z_2). Now, densities of S_i (and S_i') and Jaccard indices between S_i (and S_i') and Q_j correspond to T_2.

Finally, let us look now at the Jaccard terms between S_i and/or S_j'. These terms will constitute T_3. Let a be the number of nodes v_i or v_i' that are included in 3 subgraphs; any such node will yield 3 Jaccard terms of value 1. Let b be the number of nodes v_i v_i', or w_i that are included in 2 subgraphs; any such node will yield one Jaccard term of value 1. The remaining Jaccard terms between S_i and S_i' are all of value $2/4$. In summary, the terms are equal to

$$\lambda \binom{2r}{2} \frac{2}{4} + \lambda a 3 \left(1 - \frac{2}{4}\right) + \lambda b \left(1 - \frac{2}{4}\right) = \frac{\lambda}{2} \binom{2r}{2} + \lambda a + \lambda(a + b)/2 \quad .$$

Assume that $a < 2p$. Then since $a + b \leq 2r$ these terms are less than T_3, which is a contradiction. Therefore, $a \geq 2p$. Now, there are p vertices in $\{v_i\}$ or p vertices in $\{v_i'\}$ that are included in 3 subgraphs. These sets correspond to matchings, and at least one of them will have p sets.

To prove the other direction, assume there is a matching $\mathcal{M}$ with p sets. To form the subgraph sequence, we first select (z_1, z_2) in every set. For $u_i \in C_j \in \mathcal{M}$, we also select (v_j, z_1) in G_i and (v_j', z_1) in G_i'. For an item u_i not covered by $\mathcal{M}$, we select (w_i, z_1) in G_i and (w_i, z_1) in G_i'. A straightforward calculation shows that this sequence yields the score given in Eq. 1.   □

4   Algorithms

Since our problem is **NP**-hard, we need to resort to heuristics. In this section we present two algorithms that we will use to find good subgraphs.

The first algorithm is as follows. We start with an initial candidate S. This set is either the solution of the densest common subgraph or the densest subgraphs of each individual snapshots; we test both and select the best end result.

In order to improve the initial set we employ the strategy used in [2,5] when approximating the densest subgraph: Here, the algorithm starts with the whole graph and removes a node with the minimum degree, or equivalently, removes a node such that the remaining subgraph has the highest density. This is continued

---

Algorithm 1: $\text{ITR}(\mathcal{G}, \lambda, \mathcal{S})$, finds subgraphs with good $q(\cdot; \lambda)$

---

1 **while** changes **do**
2 | **foreach** $i = 1, \ldots, k$ **do**
3 | | $C \leftarrow V$;
4 | | **foreach** $j = 2, \ldots, |V|$ **do**
5 | | | $u \leftarrow \arg\max\limits_{v \in C} q(S_1, \ldots, S_{i-1}, C \setminus \{v\}, S_{i+1}, \ldots, S_k)$;
6 | | | $C \leftarrow C \setminus \{u\}$;
7 | | $S_i \leftarrow$ best tested C, if the score improves;

8 **return** $\mathcal{S}$;

Algorithm 2: $\text{GRD}(\mathcal{G}, \lambda)$, finds subgraphs with good $q(\cdot; \lambda)$

---

1 $\mathcal{S} \leftarrow S_1, \ldots, S_k$, where $S_i = V$;
2 **while** there are nodes **do**
3 | $u, j \leftarrow \arg\max\limits_{v, i | v \in S_i} q(S_1, \ldots, S_{i-1}, S_i \setminus \{v\}, S_{i+1}, \ldots, S_k)$;
4 | $S_j \leftarrow S_j \setminus \{u\}$;

5 **return** best tested $\mathcal{S}$;

---

until no nodes remained, and among the tested subgraphs the one with the highest density is selected.

We employ a similar strategy. For a snapshot G_i, we start with $S_i = V$, and then iteratively remove the vertices so that the score is maximal. After removing all vertices, we pick the subgraph for S_i which maximizes our objective $q(\mathcal{S}; \lambda)$. We iterate over all snapshots, we keep on modifying the sets until the algorithm converges. The pseudo-code for this approach is given in Algorithm 1.

Our second algorithm is similar to ITR. In Algorithm 1 we consider each snapshot separately and peel off vertices. In our second algorithm, we initialize each S_i with V. In each iteration, we find a snapshot S_i and a vertex v so that the remaining subgraph sequence is maximized. We remove the node and continue until no nodes are left. In the process, we choose the one which maximizes our objective function. The pseudo-code for this method is given in Algorithm 2.

The bottleneck in both algorithms is finding the next vertex to delete. Let us now consider how we can speed up this selection. To this end, select S_i and let $v \in S_i$. Let us write $\mathcal{S}'$ to be $\mathcal{S}$ with S_i replaced with $S_i \setminus \{v\}$. We can write the score difference between $\mathcal{S}$ and $\mathcal{S}'$ as

$$q(\mathcal{S}') - q(\mathcal{S}) = \frac{|E(S_i)| - \deg v}{|S_i| - 1} - \frac{|E(S_i)|}{|S_i|} + \sum_{j \neq i} J(S_i \setminus \{v\}, S_j) - J(S_i, S_j) \quad . \quad (2)$$

Let us first consider GRD. To find the optimal v and i, we will group the nodes in S_i such that the sum in Eq. 2 is equal for the nodes in the same group. In order to do that we group the nodes based on the following condition: if two nodes $u, v \in S_i$ belong to the exactly same S_j for each j, that is, $u \in S_j$ if and only if $v \in S_j$, then u and v belong to the same group. Let us write $\mathcal{P}$ to be the collection of all these groups (across all i).

Select $P \in \mathcal{P}$. Since the sum in Eq. 2 is constant for all nodes in P, the node in P maximizing Eq. 2 must have the smallest degree. Thus, we maintain the nodes in P in a priority queue keyed by the degree. We also maintain the difference of the Jaccard indices, the sum in Eq. 2. In order to maintain the difference, we maintain the sizes of intersection $|S_i \cap S_j|$ and the union $|S_i \cup S_j|$ for all i and j. To find the optimal v and i, we find the vertex with the smallest degree in each group, and then compare these candidates among different groups.

This data structure leads to the following running time.

Proposition 2. *Assume a graph sequence $G_1, \ldots, G_k$ with n nodes and total $m = \sum_i m_i$ edges. Let $\mathcal{P}_{ir}$ be the groups of S_i (based on the node memberships in other snapshots) when deleting rth node. Define $\Delta = \max |\mathcal{P}_{ir}|$. Then the running time of* GRD *is in*

$$\mathcal{O}\big(nk^2\Delta + m\log n + k^2 n(k + \log n)\big) \subseteq \mathcal{O}\big(n^2k^2 + m\log n + k^3 n\big) \quad .$$

Proof. Finding the best node u and the snapshot S_i requires $\mathcal{O}(\sum_i |\mathcal{P}_{ir}|) \in \mathcal{O}(k\Delta)$ time. Consider deleting u from S_i.

Deleting u from its queue requires $\mathcal{O}(\log n)$ time. Upon deletion, we update the degrees of the neighboring nodes in the corresponding queues, in *total* time of $\mathcal{O}(m\log n)$. Updating the intersection and the union sizes requires $\mathcal{O}(k)$ time.

We also need to update the gain coming from Jaccard indices for each group P. Only one term changes if P is not a subset of S_i; there are at most $k\Delta$ such groups. Otherwise, if $P \subseteq S_i$, then $k-1$ terms change; there are at most Δ such groups. In summary, we need $\mathcal{O}(k\Delta)$ time.

Node u is included in $\mathcal{O}(k)$ queues. As we remove u from S_i, these queues need to be updated by moving u to the correct queue. A single such update requires deleting u from its current queue, finding (and possibly creating) the new queue, and adding u to it. This can be done in $\mathcal{O}(k + \log n)$ time.

Combining these times proves the claim. □

We should point out that the running time depends on Δ, the number of queues in a single snapshot. This number may be as high the number of vertices, n, but ideally $\Delta \ll n$.

The same data structure can be also used ITR. The only difference is that we do not select optimal i; instead i is fixed when looking for the next vertex to delete. Trivial adjustments to the proof of Proposition 2 imply the following claim.

Proposition 3. *Assume a graph sequence $G_1, \ldots, G_k$ with n nodes and total $m = \sum_i m_i$ edges. Let $\mathcal{P}_{ir}$ be the groups of S_i (based on the node memberships in other snapshots) when deleting rth node. Define $\Delta = \max |\mathcal{P}_{ir}|$. Then the running time of a single iteration of* ITR *is in*

$$\mathcal{O}\big(nk^2\Delta + m\log n + k^2 n(k + \log n)\big) \subseteq \mathcal{O}\big(n^2k^2 + m\log n + k^3 n\big) \quad .$$

5   Related Work

In this section we discuss previous studies on discovering the densest subgraph in a single graph, the densest common subgraph over multiple graphs, overlapping densest subgraphs, and other types of density measures.

The Densest Subgraph: Given an undirected graph, finding the subgraph which maximizes density has been first studied by Goldberg [13] where an exact, polynomial time algorithm which solves a sequence of min-cut instances is presented. Asahiro et al. [2] provided a linear time, greedy algorithm proved to be an 1/2-approximation algorithm by Charikar [5]. The idea of the algorithm is that at each iteration, a vertex with minimum degree is removed, and then the densest subgraph among all the produced subgraphs is chosen.

Several variants of the densest subgraph problem constrained on the size of the subgraph $|S|$ have been studied: finding the densest k-subgraph ($|S| = k$) [2,9,16], at most k-subgraph ($|S| \leq k$) [1,17], and at least k-subgraph ($|S| \geq k$) [1,17]. Unlike the densest subgraph problem, when the size constraint is applied, the densest k-subgraph problem becomes **NP**-hard [9]. Furthermore, there is no polynomial time approximation scheme (PTAS) [16]. Approximating the problem of finding at most k-subgraph is shown as hard as the densest k-subgraph problem by Khuller and Saha [17]. To find exactly k-size densest subgraph, Bhaskara et al. [4] gave an $\mathcal{O}(n^{1/4+\epsilon})$-approximation algorithm for every $\epsilon > 0$ that runs in $n^{\mathcal{O}(1/\epsilon)}$ time. Andersen and Chellapilla [1] provided a linear time 1/3-approximation algorithm for at least k densest subgraph problem.

The Densest Common Subgraph over Multiple Graphs: Jethava and Beerenwinkel [15] extended the densest subgraph problem (DCS) for the case of multiple graph snapshots. As a measure the authors' goal was to maximize the minimum density. Moreover, Semertzidis et al. [20] introduced several variants of this problem by varying the aggregate function of the optimization problem, one variant, BFF-AA, is same as the DCS problem discussed in Sect. 2. DCS can be solved exactly through a reduction to the densest subgraph problem, and is consequently polynomial. The hardness of DCS variants has been addressed [6].

Overlapping Densest Subgraphs of a Single Graph: Finding multiple dense subgraphs in a single graph which allows graphs to be overlapped is studied by adding a hard constraint to control the overlap of subgraphs [3]. Later, Galbrun et al. [11] formulated the same problem adding a penalty in the objective function for the overlap. The difference of our problem to the works of Balalau et al. [3] and Galbrun et al. [11] is that our goal is to find a collection of dense subgraphs over multiple graph snapshots (one dense subgraph for each graph snapshot) while they discover a set of dense subgraphs within a single graph. Due to this difference, we want to reward similar subgraphs while the authors want to penalize similar subgraphs.

Other Density Measures: We use the ratio of edges over the nodes as our measures as it allows us to compute it efficiently. Alternative measures have been

Table 1. Characteristics of synthetic datasets. Here, $|V^d|$ and $|V^s|$ give initial number of dense and sparse vertices respectively, $E[|E|]$ is the expected number of edges, k is the number of snapshots, p_d, p_s, and p_c gives the dense, sparse, and cross edge probabilities, $J_{min} = \min_{i<j} J(V_i^d, V_j^d)$ is the minimum Jaccard index between ground truth dense sets of vertices, d_{true} is the ground truth density of dense components, d_{dcs} gives the density of densest common subgraph, and d_{ind} gives the sum of densities of locally densest subgraph from each graph snapshot.

Dataset	$|V^d|$	$|V^s|$	$E[|E|]$	k	p_d	p_s	p_c	d_{true}	d_{dcs}	d_{ind}	J_{min}
Syn-1	100	900	672.8	10	0.05	0.0005	0.0005	34.68	24.26	35.74	0.47
Syn-2	100	5 000	3 850.2	5	0.05	0.0001	0.0001	40.47	13.03	40.57	0.18
Syn-3	120	1 200	3 922	5	0.06	0.005	0.002	27.4	17.62	27.62	0.45
Syn-4	250	5 000	4 709	8	0.03	0.001	0.001	55.79	29.65	55.98	0.27
Syn-5	500	3 500	12 136.29	7	0.05	0.0003	0.0003	112.11	87.27	112.12	0.53
Syn-6	350	3 500	32 015	5	0.06	0.005	0.002	67.96	52.17	67.96	0.49

also considered. One option is to use the proportion of edges instead, that is, $|E|/\binom{|V|}{2}$. The issue with this measure is that a single edge yields the highest density of 1. Moreover, finding the largest graph with the edge proportion of 1 is equal to finding a clique, a classic problem that does not allow any good approximation [14]. As an alternative approach, Tsourakakis et al. [24] proposed finding subgraphs with large score $|E| - \alpha\binom{|V|}{2}$. Optimizing this measure is an **NP**-hard problem but an algorithm similar to the one given by Asahiro et al. [2] leads to an additive approximation guarantee. In similar vein, Tatti [21] considered subgraphs maximizing $|E| - \alpha|V|$ and showed that they form nested structure similar to k-core decomposition. An alternative measure called triangle-density has been proposed by Tsourakakis [23] as a ratio of triangles and the nodes, possibly producing smaller graphs. Like the density, optimizing this measure can be done in polynomial time. We leave adopting these measures as a future work.

6   Experimental Evaluation

The goal in this section is to experimentally evaluate our algorithms. We first generate several synthetic datasets and plant dense subgraph components, in each snapshot and test how well our algorithms discover the ground truth. Next we study the performance of the algorithm on real-world temporal datasets in terms of running time. We compare our results with the solutions obtained with the densest common subgraph [20] and the sum of densities of locally dense subgraphs [13]. Finally, we present results of a case study.

We implemented the algorithms in Python[1] and performed the experiments using a 2.4GHz Intel Core i5 processor and 16GB RAM.

Synthetic Datasets: Next, we explain in detail how the synthetic datasets are generated, the statistics and the related parameters are given in Table 1.

---

[1] The source code is available at https://version.helsinki.fi/dacs/. .

Table 2. Computational statistics from the experiments for synthetic datasets using ITR and GRD algorithms. Here, λ is the parameter in $q(\cdot; \lambda)$, i is the number of iterations using ITR, columns d_{dis} and q are the sum of densities and scores of the discovered sets, J_{min} provide the minimum Jaccard index between discovered sets, columns ρ give the average Jaccard index between discovered and ground truth sets, and columns $time$ give the computational time in seconds.

Data	λ	ITR						GRD				
		d_{dis}	q	J_{min}	ρ	$time$	i	d_{dis}	q	J_{min}	ρ	$time$
Syn-1	0.3	34.84	43.1	0.51	0.93	10	4	34.84	43.09	0.5	0.93	30
	0.5	33.54	48.97	0.54	0.89	11	5	33.44	48.97	0.54	0.88	33
	0.7	26.43	56.08	0.84	0.74	4	2	26.22	56.08	0.86	0.73	32
Syn-2	0.3	40.57	41.18	0.18	0.99	50	3	40.57	40.77	0.18	0.99	92
	0.4	40.57	41.39	0.18	0.99	53	3	40.57	41.39	0.18	0.99	84
	0.5	40.57	41.6	0.18	0.99	50	3	40.57	41.6	0.18	0.99	87
Syn-3	0.4	27.59	29.59	0.44	0.97	16	3	27.6	29.59	0.44	0.97	29
	0.5	27.59	30.1	0.44	0.97	18	3	27.6	30.08	0.44	0.97	26
	0.8	27.44	31.65	0.46	0.98	17	3	27.6	31.56	0.43	0.98	28
Syn-4	0.4	55.75	60.38	0.27	0.98	66	2	55.89	60.42	0.27	0.98	307
	0.5	55.82	61.56	0.27	0.98	132	4	55.83	61.55	0.27	0.98	308
	0.6	55.76	62.71	0.27	0.98	132	4	55.79	62.71	0.27	0.98	285
Syn-5	0.01	112.12	112.26	0.53	1	73	2	112.12	112.26	0.53	1	249
	0.4	112.12	117.7	0.53	1	69	2	112.12	116.3	0.53	1	251
	0.5	112.12	119.09	0.53	1	103	3	112.12	119.09	0.53	1	257
Syn-6	0.1	67.96	68.61	0.49	1	42	2	67.96	68.61	0.49	1	226
	0.8	67.96	73.18	0.49	1	67	3	67.96	73.17	0.49	1	252
	1	67.95	74.49	0.49	1	69	3	67.96	74.48	0.49	1	238

Each dataset consists of k graphs given as $\{G_1, \ldots, G_k\}$. We split the vertex set into dense and sparse components V^d and V^s. To generate the ith snapshot we create two components V_i^d and V_i^s by starting from V^d and V^s and moving nodes from V^s to V^d with a probability of η_i. The probability η_i is selected randomly for each snapshot from a uniform distribution $[0.01, 0.09]$. Once the vertices are generated, we sample the edges using a stochastic block model, with the edge probabilities being p_d, p_s, and p_c for dense component, sparse component, and cross edges, respectively. We created 6 such synthetic datasets in total to test our algorithms.

Results of Synthetic Datasets: We report our results in Table 2.

First, we observe that the discovered density values d_{dis} approximately match each other, that is, both ITR and GRD perform equally well in terms of the densities. Similar result holds also for the scores $q(\cdot; \lambda)$ and minimum Jaccard coefficients J_{min}. However, ITR runs faster than GRD. This is probably due to

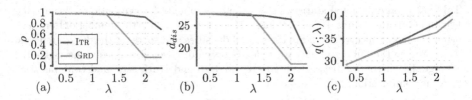

Fig. 2. Average Jaccard index to the ground truth ρ as a function of λ in Fig. 2b. Discovered density d_{dis} as a function of λ in Fig. 2b. Scores $q(\cdot; \lambda)$ as a function of λ in Fig. 2c. This experiment was performed using *Syn-3* dataset.

Table 3. Characteristics of real-world datasets. Here, $|V|$ gives the number of vertices, $|E|$ is the expected number of edges, k is the number of snapshots, d_{dcs} gives the density of densest common subgraph, and d_{ind} gives the sum of densities of locally densest subgraph from each graph snapshot.

Data	$|V|$	$|E|$	k	d_{dcs}	d_{ind}
Twitter-#	806	101.2	15	7.54	38.79
Enron	1 079	23.2	183	43.47	185.34
Facebook	4 117	83.13	104	14	88.65
Students	889	43.68	122	24.15	117.98
Twitter-user	4 605	109.19	93	23	90.63
Tumblr	1 980	65.3	89	36.67	103.98

the fact that GRD takes more time to select the next vertex to delete, which is the bottleneck in both algorithms despite of having same asymptotic time complexity per iteration in ITR and overall time complexity in GRD. Let us now compare the discovered sets to the ground truth, given in columns ρ. We can see both algorithms gives similar values which indicates equally good performance of ITR and GRD.

Our next step is to study the effect of the input parameter λ. First, we observe Fig. 2 which demonstrates ρ as a function of λ. In Fig. 2a, we see that ρ gradually decreases as we increase λ. This is due to the fact that when we increase the weight of constraint part of q, the algorithms try to find dense sets with higher Jaccard coefficients which eventually forces to deviate from its ground truth. Furthermore, if we set $\lambda = 2$, we can see a drastic change in ρ.

Let us now consider Fig. 2b which demonstrates how the discovered sum of densities change with respect to λ. We see the decreasing trend showing that second term in the objective function starts to dominate with the increase of λ.

Next, we study how the score function q behaves over λ shown in Fig. 2c. We can observe that both ITR and GRD have an increasing trend when we increase λ from 0.3 to 2.3. This is expected as larger λ should yield larger scores.

Real-World Datasets: We consider 6 publicly available, real-world datasets. The details of the datasets are shown in Table 3. *Twitter-#* [22][2] is a hash-

---

[2] https://github.com/ksemer/BestFriendsForever-BFF-.

Table 4. Computational statistics from the experiments for real-world datasets. Here, λ is the parameter in $q(\cdot; \lambda)$, i gives the number of iterations using ITR, columns d_{dis} are the discovered sum of densities, columns q are the discovered scores, and columns *time* give the computational time in seconds.

Data	λ	ITR				GRD		
		d_{dis}	q	*time*	i	d_{dis}	q	*time*
Twitter-#	0.3	37.34	40.52	2.81	5	37.65	38.13	5.79
	0.5	14.14	54.54	3.06	7	31.01	42.37	6.38
	0.7	11	74.7	2.62	5	16.8	56.18	6.23
	0.8	11	83.8	2.79	6	9.33	75.47	6.88
Enron	0.05	130.43	357.29	72.02	5	122.08	343.02	816.54
	0.1	111.88	593.87	148.9	10	62.08	589.96	826.28
	0.5	95.88	2 662.7	81.56	6	20	3 139.25	854.43
	5	88.66	25 254.81	98.87	7	20	31 212.5	774.39
Facebook	0.1	52.86	126.98	214.74	7	56.03	85.64	9 461.59
	0.5	36.4	446.09	148.94	6	25.47	344.6	10 193.96
	0.7	31.25	657.92	185.9	1 7	23.41	475.83	9 781.9
	1	27.41	922.45	136.72	6	22.48	659.32	9 034.35
Students	0	108.9	108.9	124.38	4	102.7	102.7	2 787.8
	0.2	46.99	526.19	100.17	5	33.17	567.84	2 178.28
	0.5	46.17	1 258.41	79.65	4	33.17	1 369.86	2 256.26
	0.8	44.5	2 020.81	78.61	4	33.17	2 171.88	2 125.67
Twitter-user	0.01	79.21	81.61	572.25	8	63.53	68.53	7 524.47
	0.1	12.04	269	260.31	9	13.9	212.28	6 422.46
	0.2	11.49	545.93	149.91	5	11.64	423.1	6 770.16
	0.5	11.49	1 347.59	150.08	5	10.83	1 040.05	6 847.68
Tumblr	0.1	71.33	316.07	50.23	4	66.47	302.83	1 229.98
	0.5	64.37	1 456.47	49.31	4	62.97	1 291.82	1 151.85
	0.7	59.25	2 182.29	48.96	4	62.97	1 783.35	1 217.76
	1	59.17	3 092.08	48.37	4	62.97	2 520.66	1 252.05

tag network where nodes correspond to hashtags and edges corresponds to the interactions where two hashtags appear in a tweet. This dataset contains 15 such daily graph snapshots in total. *Enron*[3] is a popular dataset which contains email communication data within senior management of Enron company. It contains 183 daily snapshots in which daily email count is at least 5. *Facebook* [25][4] is a network of Facebook users in New Orleans regional community. It contains a set of facebook wall posts among these users during 9th of June to 20th of August,

---

[3] http://www.cs.cmu.edu/~./enron/.
[4] https://networkrepository.com/fb-wosn-friends.php.

Table 5. Twitter hash tags discovered for *Twitter-8* dataset.

DCS: abudhabigp, fp1, abudhabi, guti, f1, pushpush, skyf1, hulk, allowin, bottas, kimi, fp3, fp2, unleashthehulk, density: 7.07
ITR algorithm: $\lambda = 0.8$, density: 12.23, objective: 21.76
Day 1: indiangp, skyf1, kimi, f1
Day 2: abudhabigp, skyf1, f1
Day 3: kimi, skyf1, abudhabigp, f1
Day 4: abudhabigp, skyf1, f1
Day 5: abudhabigp, english, arabic, spanish, french, danish, swedish, f1, endimpunitybh, skyf1, bahrain
Day 6: abudhabigp, fp1, abudhabi, guti, f1, skyf1, hulk, allowin, bottas, kimi, fp3, fp2
Day 7: abudhabigp, abudhabi, guti, f1, pushpush, skyf1, hulk, allowin, bottas, kimi, fp3, quali
Day 8: abudhabigp, skyf1, f1

2006. *Students*[5] is an online message network at the University of California, Irvine. It spans over 122 days. *Twitter-user* [19][6] is a network of twitter users in Helsinki 2013. It contains a set of tweets which appear each others' names. *Tumblr* [18][7] contains phrase or quote mentions appeared in blogs and news media. It contains author and meme interactions of users over 3 months from February to April in 2009.

Results of Real-World Datasets: We report the results obtained from the experiments with real-world datasets in Table 4.

First, we compare scores q obtained using ITR and GRD. As we can see, apart from few cases in *Enron* and *Students* datasets, ITR achieves greater score than GRD. Furthermore, we can observe in *time* columns, ITR runs faster than GRD. Next, let us observe column i which gives number of iterations with ITR algorithm. We can see that we have at most 10 number of iterations which is reasonable to deal with real-world datasets.

As we compare columns d_{dis} which show discovered densities, we can occasionally see approximately similar values but also deviations. Next, let us observe the effect of λ on d_{dis} and q. Based on λ, the ratio between d_{dis} and q tends to change. For example, in general, when λ is lowered, d_{dis} tends to increase while q decreases, as expected.

Case Study Using Twitter-8 Dataset: In this section, we present a case-study and analyze the result which illustrates trending twitter hash tags over a span of 8 days, under a Jaccard constrained environment.

---

[5] https://toreopsahl.com/datasets/#online_social_network.

[6] https://github.com/polinapolina/segmentation-meets-densest-subgraph/tree/master/data.

[7] http://snap.stanford.edu/data/memetracker9.html.

Twitter-8 contains a hashtag network from November, 2013. We prepared this dataset by extracting first 8 daily graph snapshots from *Twitter-#* dataset. Here, each node of the graph represents a specific hashtag. As seen in the tags from Table 5, Formula-1 racing car event which occurred on Abu Dhabi has been trending during the period. We set $\lambda = 0.8$ and find different sets of dense subgraphs for each day using ITR. On Day 1, tags *indeangp*, *skyf1*, *kimi*, and *f1* have been added to the dense hashtag collection whereas on Day 2, the tag *kimi* who is a Finnish racing legend and the tag *indeangp* have been removed from trending list. On Day 3, *kimi* has been re-entered into the trending list and the tag *indeangp* has been replaced by *abudhabigp*. On Day 6, the tags like *bottas* (another Finnish racing car driver), *fp2* (Free practice 2), *fp3* (Free practice 3), and etc. have been added which indicates that additional racing car event related tags are trending. On Day 5, we can observe more new tags like *bahrain*, *english*, *arabic*, *french*, *danish*, and *swedish* have been appeared which seems not directly related to racing car event. Moreover, new dense collection gives a higher density of 12.22 with respect to the DCS density of 7.07.

7   Concluding Remarks

We introduced a novel Jaccard weighted, dense subgraph discovery problem (JWDS) for graphs with multiple snapshots. Here, our goal was to find a dense subset of vertices from each graph snapshot such that the sum of densities and the similarity between the snapshots is maximized.

We proved that our problem is **NP**-hard, and designed an iterative algorithm which runs in $\mathcal{O}(n^2k^2 + m \log n + k^3n)$ time per iteration and a greedy algorithm which runs in $\mathcal{O}(n^2k^2 + m \log n + k^3n)$ time.

We experimentally showed that the number of iterations was low in iterative algorithm, and that the algorithms could find the ground truth using synthetic datasets and could discover dense collections in real world datasets. We also studied the effect of our user parameter λ. Finally, we performed a case study showing interpretable results.

The paper introduces several interesting directions for future work. In this paper, we enforced the pairwise Jaccard constraint between all available pairs of snapshots. However, we can relax this constraint further by letting only a portion of sets which lies within a specific window to assure the Jaccard similarity constraint which may lead to future work. Another possible direction is adopting different type of density for our problem setting.

References

1. Andersen, R., Chellapilla, K.: Finding dense subgraphs with size bounds. In: WAW, pp. 25–37 (2009)
2. Asahiro, Y., Iwama, K., Tamaki, H., Tokuyama, T.: Greedily finding a dense subgraph. J. Algorithms **34**(2), 203–221 (2000)

3. Balalau, O.D., Bonchi, F., Chan, T.H., Gullo, F., Sozio, M.: Finding subgraphs with maximum total density and limited overlap. In: WSDM, pp. 379–388 (2015)

4. Bhaskara, A., Charikar, M., Chlamtac, E., Feige, U., Vijayaraghavan, A.: Detecting high log-densities: an $O(n^{1/4})$ approximation for densest k-subgraph. In: STOC, pp. 201–210 (2010)

5. Charikar, M.: Greedy approximation algorithms for finding dense components in a graph. In: APPROX, pp. 84–95 (2000)

6. Charikar, M., Naamad, Y., Wu, J.: On finding dense common subgraphs (2018). https://arxiv.org/abs/1802.06361

7. Chlebík, M., Chlebíková, J.: Approximation hardness for small occurrence instances of np-hard problems. In: CIAC, pp. 152–164 (2003)

8. Du, X., Jin, R., Ding, L., Lee, V.E., Thornton, J.H.: Migration motif: a spatial-temporal pattern mining approach for financial markets. In: KDD, pp. 1135–1144 (2009)

9. Feige, U., Peleg, D., Kortsarz, G.: The dense k-subgraph problem. Algorithmica **29**, 410–421 (2001)

10. Fratkin, E., Naughton, B.T., Brutlag, D.L., Batzoglou, S.: Motifcut: regulatory motifs finding with maximum density subgraphs. Bioinformatics **22**(14), e150–e157 (2006)

11. Galbrun, E., Gionis, A., Tatti, N.: Top-k overlapping densest subgraphs. DMKD **30**(5), 1134–1165 (2016)

12. Galimberti, E., Bonchi, F., Gullo, F., Lanciano, T.: Core decomposition in multi-layer networks: theory, algorithms, and applications. TKDD **14**(1), 1–40 (2020)

13. Goldberg, A.V.: Finding a maximum density subgraph (1984)

14. Håstad, J.: Clique is hard to approximate within $n^{1-\epsilon}$. In: STOC, pp. 627–636 (1996)

15. Jethava, V., Beerenwinkel, N.: Finding dense subgraphs in relational graphs. In: ECMLPKDD, pp. 641–654 (2015)

16. Khot, S.: Ruling out PTAS for graph min-bisection, dense k-subgraph, and bipartite clique. SIAM J. Comput. **36**(4), 1025–1071 (2006)

17. Khuller, S., Saha, B.: On finding dense subgraphs. In: ICALP, pp. 597–608 (2009)

18. Leskovec, J., Backstrom, L., Kleinberg, J.: Meme-tracking and the dynamics of the news cycle. In: KDD, pp. 497–506 (2009)

19. Rozenshtein, P., Bonchi, F., Gionis, A., Sozio, M., Tatti, N.: Finding events in temporal networks: segmentation meets densest subgraph discovery. KAIS **62**(4), 1611–1639 (2020)

20. Semertzidis, K., Pitoura, E., Terzi, E., Tsaparas, P.: Finding lasting dense subgraphs. DMKD **33**(5), 1417–1445 (2019)

21. Tatti, N.: Density-friendly graph decomposition. TKDD **13**(5), 1–29 (2019)

22. Tsantarliotis, P., Pitoura, E.: Topic detection using a critical term graph on news-related tweets. In: EDBT/ICDT Workshops, pp. 177–182 (2015)

23. Tsourakakis, C.: The k-clique densest subgraph problem. In: WWW, pp. 1122–1132 (2015)

24. Tsourakakis, C., Bonchi, F., Gionis, A., Gullo, F., Tsiarli, M.: Denser than the densest subgraph: extracting optimal quasi-cliques with quality guarantees. In: KDD, pp. 104–112 (2013)

25. Viswanath, B., Mislove, A., Cha, M., Gummadi, K.P.: On the evolution of user interaction in facebook. In: WOSN, pp. 37–42 (2009)

RIMBO - An Ontology for Model Revision Databases

Filip Kronström[1]([✉]) [iD], Alexander H. Gower[1] [iD], Ievgeniia A. Tiukova[1,2] [iD], and Ross D. King[1,3,4] [iD]

[1] Chalmers University of Technology, Gothenburg, Sweden
filipkro@chalmers.se
[2] KTH Royal Institute of Technology, Stockholm, Sweden
[3] University of Cambridge, Cambridge, UK
[4] Alan Turing Institute, London, UK

Abstract. The use of computational models is growing throughout most scientific domains. The increased complexity of such models, as well as the increased automation of scientific research, imply that model revisions need to be systematically recorded. We present RIMBO (Revisions for Improvements of Models in Biology Ontology), which describes the changes made to computational biology models.

The ontology is intended as the foundation of a database containing and describing iterative improvements to models. By recording high level information, such as modelled phenomena, and model type, using controlled vocabularies from widely used ontologies, the same database can be used for different model types. The database aims to describe the evolution of models by recording chains of changes to them. To make this evolution transparent, emphasise has been put on recording the reasons, and descriptions, of the changes.

We demonstrate the usefulness of a database based on this ontology by modelling the update from version 8.4.1 to 8.4.2 of the genome-scale metabolic model Yeast8, a modification proposed by an abduction algorithm, as well as thousands of simulated revisions. This results in a database demonstrating that revisions can successfully be modelled in a semantically meaningful and storage efficient way. We believe such a database is necessary for performing automated model improvement at scale in systems biology, as well as being a useful tool to increase the openness and traceability for model development. With minor modifications the ontology can also be used in other scientific domains.

The ontology is made available at https://github.com/filipkro/rimbo and will be continually updated.

Keywords: Ontology · Knowledge representation · Database · Computational biology · Semantic web

1 Introduction

Computational models play a crucial part in our understanding of complex biological systems [1], and the further improvements of such models have been described as a grand challenge for the 21st century [2].

© The Author(s) 2023
A. Bifet et al. (Eds.): DS 2023, LNAI 14276, pp. 523–534, 2023.
https://doi.org/10.1007/978-3-031-45275-8_35

A promising way forward, enabled by advances in AI and automation, is to develop autonomous laboratories performing experiments and discovering knowledge. This has been demonstrated by robot scientists performing cycles of experiments to determine gene functions in yeast [3], discover drugs [4], and optimise cell culturing conditions [5]. Largely automated pipelines has been used to optimise strain engineering in both *Saccharomyces cerevisiae* and *Escherichia coli* [6], and a mobile robotic chemist has searched for photocatalysts for hydrogen production [7].

Using computational models to guide experiment design and the experimental results to improve the models in a closed loop manner has proved to be a successful and scaleable way of developing systems biology models [8].

For an AI agent to be able to autonomously reason about improvements to a model, a structured and semantically unambiguous way of storing models is required. Critically, such a store should also handle large numbers of revisions to the models. A semantically meaningful representation of these revisions will enable human researchers to gain insights from the model improvement cycles, access and use the models, as well as facilitating for computer systems to reason about previous changes to models.

Across many domains the importance and use of computational models, as well as the number of models available, is increasing [9]. No matter if the model is used in science or any other field, it needs the trust of a wider community. One way this can be achieved is by making the steps taken during development more open and transparent, regardless if it was done by humans or machines.

In this work we propose an ontology, capturing and explaining changes to different types of computational biology models, which can also be used for a model revision database. Such a database is important in an automated scientific discovery setting, where we seek improvements to computational models. We also demonstrate how this ontology can be used to model community consensus updates, as well as machine generated hypotheses about improvements for yeast metabolic models.

2   Background and Related Work

There are several repositories or databases where the computational biology community share models today. Most notably BioModels [10] with over 2 000 submitted models of different types, but also BiGG Models [11] with genome-scale metabolic models (GEMs), and the CellML Model Repository [12]. Although some repositories support version control (e.g. BioModels), they are not designed to deal with the large numbers of small revisions generated when developing and refining models.

Central to the increased sharing and reuse of computational models, and other biological information, are common and unambiguous model descriptions. Biological Pathway Exchange (BioPax) [13] is a language for exchange and integration of biological pathways. For computational models, CellML [14], and the Systems Biology Markup Language (SBML) [15] are widely used. They both enable the databases mentioned above, with BioModels and BiGG containing large amounts of SBML models and the CellML Model Repository naturally being for CellML models. Although slightly different, the three model formats are all XML-based.

These three formats share a heavy reliance on ontologies. Ontologies and controlled vocabularies provide semantic meaning to data, both to humans and machines. The Gene Ontology (GO) [16], provides structure and semantics to genes and gene products across species. The Systems Biology Ontology (SBO) [17] is closely tied to SBML, and contains vocabularies useful for computational modelling and systems biology, and the Kinetic Simulation Algorithm Ontology (KiSAO) [18] complements it with additional terms describing simulation and algorithms. Cell types and processes in cells can be found in the Cell Ontology (CL) [19] and the Ascomycete Phenotype Ontology (APO) [20] contains phenotypes for Ascomycete fungi. The EDAM ontologies [21,22] have vocabularies for data management and analysis. Provenance models, describing the provenance of both scientific experiments and general processes, has been encoded in ontologies like PROV-O [23] and REPRODUCE-ME [24].

The COMODI (COmputational MOdels DIffer) ontology [25] attempted to characterise changes to computational models in XML format. Changes to "XmlEntities" were identified along with "Reasons", "Intentions", and "Targets" for them. Such annotations provide very detailed descriptions of each change to the XML tree, which is helpful when studying single updates or for understanding the mechanics of the format the model is encoded in. However, this verbosity is not helpful when chains of revisions are studied. Instead of providing a detailed description of all changes to the encoding of the model, we argue that overarching intentions are important. There are also differences between describing a change, and providing an unambiguous and storage efficient patch that can be used to recreate the actual file, without storing a copy of it.

Apart from just offering semantic meaning, ontologies can also effectively be used as the schema for databases. By modelling data as Resource Description Framework (RDF) triples (*subject, predicate, object*) using terms from ontologies, knowledge graphs can be created. Such graphs can be queried or reasoned over, and have previously acted as the knowledge base for closed loop model improvements [8].

3 Results

We propose that model revisions are represented using the Revisions for Improvements of Models in Biology Ontology (RIMBO). It is designed to be the schema for a graph database containing iteratively improved computational biology models. In theory it can be used with any type of model, but we have focused on models that are improved by making small changes which can be described in a semantically meaningful way. Below, a non exhaustive list can be seen, illustrating the type of competency questions we want such a database to answer.

– Which model was introduced in publication p?
– Which models are derived from model M_1?
– What was the reason for the revision R?
– Which revision tried to correct the predicted essentiality of gene g in model M_2?
– Which model is a revision of model M_3, where the change affects reaction r?

The ontology is expressed in OWL2 and developed in Protégé (v. 5.5.0, https://protege.stanford.edu/). We will first, in Sect. 3.1, describe the ontology and then, in Sect. 3.2, show examples of model revisions in this format and demonstrate it can be used as a database for large numbers of revisions.

3.1   Description of RIMBO

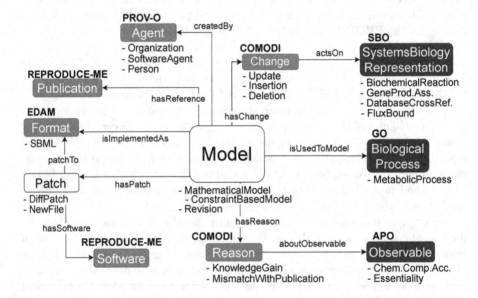

Fig. 1. Overview of RIMBO showing classes, how they are connected, and which ontologies they are from. The text under the boxes specifies subclasses used for the demonstration in Sect. 3.2. Red boxes denotes domain specific classes that would need replacing if the ontology is applied to another domain. Blue denotes classes from other foundational scientific ontologies and the white boxes are classes introduced in RIMBO. (Color figure online)

RIMBO combines classes from different ontologies and an overview illustrating how this is done can be seen in Fig. 1. To connect these classes the relations described in Table 1, along with their domains and ranges, are introduced.

The central class in RIMBO is Model, being a superclass to different modelling types, imported from ontologies such as the Mathematical Modelling

Ontology (MAMO) and EDAM. Information about this model is provided through links to other concepts. For example, `BiologicalProcess` classes from GO or CL can describe which phenomenon is being modelled, and terms from REPRODUCE-ME and PROV-O specify important metainformation, such as when and by whom it was created, as well as links to relevant publications. The model is also linked to the model file, represented by an instance of its corresponding `Format` class from EDAM. This connects either to an external reference to a filestore or an online resource, or a representation of the file in the graph. There are advantages and disadvantages to each option. External references require maintenance to ensure they point to correct locations, but are more storage-efficient. Having large files in the graph may affect query performance.

Table 1. The relations used to model revisions with RIMBO, along with domains and ranges when applicable. The namespaces specify which ontology the classes are from, when no namespace is specified the term is introduced in RIMBO. `rep-me` is short for REPRODUCE-ME.

Relation	Domain	Range	Description
aboutObservable	comodi: Reason	apo: Observable	Describes the Reason by linking it to Observable terms from APO.
actsOn	comodi: Change	sbo: Sys.Bio.Repr.	Describes what part of the model is affected by the Change.
createdBy	Model	prov: Agent	Specifies who created a Model (Organization, Person ⊑ Agent).
hasChange	Revision	comodi: Change	Connects the Revision to a Change.
hasPatch	Revision	Patch	Connects the Revision to a Patch.
hasReason	Revision	comodi: Reason	Connects the Revision to a Reason.
hasReference		rep-me: Publication	Links, e.g., a Model or Mism.W.Pub. to its Publication.
hasSoftware		rep-me: Software	Specifies software, e.g., used to find the diff-patch.
isImplementedAs	Model	edam: Format	Links the Model to Format with information about the model file.
isUsedToModel	Model		Describes which phenomenon is modelled.
ofMaterialEntity	apo: Observable	Mat.Entity	Links Observable to Mat.Entity (Gene, Chem.Entity ⊑ Mat.Entity).
patchTo	Patch	edam: Format	Links a Patch to the model file it applies to.
revisionTo	Revision	Model	Links the Revision to the Model it was revised from.

The other central class in this ontology is `Revision`, which is also a subclass of `Model`, and describes a modified version of a `Model`. An important thing to note is that the `Revision` class is not disjoint with classes describing the model type, for example `MathematicalModel` classes from MAMO. Hence, a revised model is described as the intersection of its model type and a `Revision`. Recording the reason along with descriptions of the changes made to models is important, both when improvements are generated by humans and machines. For a human generated revision, it can, for example, be used as a way of documenting the research. For a machine, it enables the system to reason about the effect of previous changes, as well as providing a way of communicating and motivating its findings with human researchers. The `Reason` class is from COMODI and has subclasses such as `MismatchWithPublication` and `KnowledgeGain`. Linking this to terms from ontologies like APO and relevant genes or chemicals gives a description of the cause of a change. As one revision might be made up of several changes, such as the addition of multiple new reactions, it is described by a `Change` collecting, possibly several, instances of `Deletions`, `Insertions`, or `Updates`, all from the COMODI ontology. The change can be described by linking these classes to subclasses of `SystemsBiologyRepresentation` from SBO and specific reactions or genes.

The actual change to the file is saved using the `Patch` class, with subclasses `DiffPatch` and `NewFile`. As iterative changes often are small, in terms of the actual changes to the files, it makes sense to just store the differences between the two files to the database. This is done with the `DiffPatch` class along with information on what software was used to find it. In some cases it might be desirable to just store a new version of the model file, for instance for binary model representations, for larger changes, or to avoid lengthy chains of patches. This is done using the `NewFile` class.

3.2   Demonstration

To demonstrate the usefulness of this ontology and a resulting database, we have generated a demonstration knowledge graph with model revisions. This example is based on revisions to the genome-scale metabolic model (GEM) Yeast8 [26] for the yeast species *Saccharomyces cerevisiae*. A GEM is a network collecting information about, for example, genes and reactions in a biological system. First, we model a part of a community update of Yeast8, from v8.4.1 to v8.4.2. Then, by expressing the model in first-order logic, an algorithm using abductive reasoning, LGEM$^+$ [27], was used to suggest modifications to the theory. Finally, we perform 31 400 random revisions.

The first update, from v8.4.1 to v8.4.2, was about improving the simulation of alcoholic fermentation conditions by adding several fatty acid ester producing reactions. The modification suggested by LGEM$^+$ was to remove the gene YJL130C as a requirement for an enzyme catalysing the reaction carbamoyl-phosphate synthase (glutamine-hydrolysing). This was suggested as a remedy to YJL130C being predicted as essential for growth, when empirical evidence showed it was not [28]. Finally, starting with this version, random revisions were generated by iteratively either removing a reaction, modifying the gene requirements for a reaction, or modifying the flux bounds for a reaction.

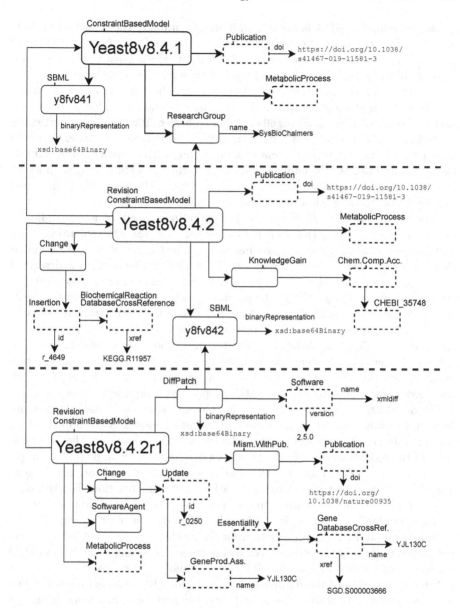

Fig. 2. The knowledge graph containing the base model, Yeast8v.8.4.1, the update to v8.4.2, and the revision changing the gene reaction rule for reaction r_0250, described in Sect. 3.2. The boxes are instances of the classes named above them, solid boxes represent named nodes and dashed correspond to blank nodes. The dashed red lines separates entries belonging to the different models/revisions. (Color figure online)

The knowledge graph with the first two revisions can be seen in Fig. 2, where the base model, Yeast8 v8.4.1, is added as an instance of a `ConstraintBasedModel` modelling a `MetabolicProcess`. It is linked to the

ResearchGroup "SysBioChalmers", who are maintaining the model on Github, as well as the corresponding Publication by Lu et al. [26]. The model file itself is represented as an instance of the SBML format which links to a compressed copy of the original model file as a literal of type xsd:byte64Binary.

Yeast8 v8.4.2 is still a ConstraintBasedModel, but also a Revision, meaning this entry is the intersection of the two classes. The reason for the this revision is modelled as a KnowledgeGain about ChemicalCompoundAccumulation of CHEBI_35748 (fatty acid ester) and it is described by Insertions of BiochemicalReactions and TransportReactions with references to the KEGG Reaction database. As the number of changes to the model file, going from v8.4.1 to v8.4.2, is rather large, we save a compressed copy of the entire file, represented as a NewFile, linked with a new instance of SBML.

The reason for the change from the abduction algorithm is contradicting results in a publication. Hence, it is modelled as a MismatchWithPublication referring to a Publication representing the work by Giaever et al. [28], as well as the predicted Essentiality of the gene, "YJL130C". The revision is described as an Update of the reaction r_0250's GeneProductAssociation associated to the aforementioned gene. Unlike the previous models, this iteration was not generated by "SysBioChalmers", instead it is linked to a SoftwareAgent referring to LGEM+. This time the model file is represented by the difference to Yeast8v8.4.2. An instance of DiffPatch links a literal of the type xsd:base64Binary, containing the patch recreating the updated model, to the revision and the previous model file. The software and version, xmldiff, v2.5.0 (https://xmldiff. readthedocs.io/), used to find the patch is specified using the Software class.

To demonstrate that a database using this ontology can handle large numbers of revisions, chains of thousands of modified models were added, along with metainformation describing the change and who made it. The modifications of the models were performed using COBRApy (v.0.26.3, https://cobrapy. readthedocs.io/). When altering the gene-reaction rule a randomly picked gene was either removed or added to the rule1of a random reaction. For the flux-bound modifications either the upper or the lower bound for some reaction was updated randomly such that it still is valid. Removing a reaction was done by chosing a random reaction to delete from the model. The different actions were not picked uniformly to better reflect real revisions, resulting in 25 793 modified gene-reaction rules, 3 857 altered flux bounds, and 1 750 removed reactions. In our implementation of the database a copy of every 100th model file was saved to reduce the sizes of the patches stored for every revision. The knowledge graph with 31 400 revisions contains 688 512 triples and the size of it, serialised as a .ttl-file, is 1.17 GB (as a reference, one uncompressed Yeast8 file in SBML format has a file size of ~10 MB).

To validate the database, iterations containing more and more data were deployed on an Apache Jena Fuseki server running on a 2021 MacBook Pro M1. The growing database was queried for the binary patch, along with the file it should be applied to, belonging to revisions updating the gene-reaction rules of specific reactions. Figure 3a shows an example of the queries executed in the experiment where the gene-reaction pairs were varied. In Fig. 3b a box-plot of the query times is shown, based on 100 queries for each database. All pairs were present in every iteration of the databases and the same series of queries were executed between the iterations. The query times increase with growing database size, but the majority of the queries show a rather small increase. The major difference between the different database iterations is the worst case queries, which is primarily explained by the number of results retrieved. The gene-reaction pairs are not necessarily unique and with a bigger database we can expect more duplicates.

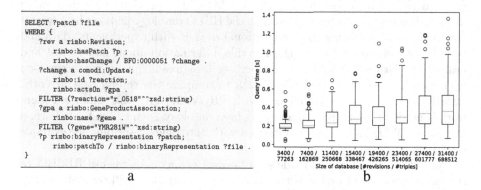

a   b

Fig. 3. (a) shows a query retrieving the patch and the file it applies to for a revision where a modification, involving the gene YPL280W, of the gene-reaction rule for reaction r_4133 is performed. This type of queries, but with varying gene-reaction pairs were used to generate (b), showing a box-plot of query times from 100 queries for databases of different size, deployed on an Apache Jena Fuseki server.

4   Discussion and Conclusion

In this work we demonstrate the usefulness of a structured and semantically sound representation for computational models, not only for sharing with the community, but also during the development process. We view RIMBO as a complement to public model repositories, such as BioModels, BiGG Models, and CellML Model Repository, providing structure and transparency to model development. One could envision revision traces, expressed in controlled vocabularies, describing the provenance published along with new models. We argue this would be useful both for automated and traditional labs, as it could greatly increase the openness and traceability of research. Along with this, RIMBO also works as a useful tool to organise the models during development in a storage efficient manner.

The ontology based graph structure allows for flexibility in the implementation of a database. For example, what level of detail to use when explaining a change might vary depending on the needs of the specific lab, and what kind of model is revised. One might be interested in recording more fine-grained descriptions of a change for a more specific model. Sometimes it could also be useful to describe the actual changes to the XML-tree using the COMODI ontology. Depending on the domain and what kind of changes are made, there might also be a need to introduce new terms to describe the revision. APO and SBO, along with some new classes describing terms connected to the SBML Level 3 Flux Balance Constraints package covers our current needs, working with yeast systems biology, but other domains most likely need other, domain specific, classes.

A planned future extension and generalisation of this work, interesting for both traditional and autonomous labs, is to also model and record hypotheses, e.g., generating the revised model. This would build on previous work attempting to formalise scientific discovery, such as the HELO ontology [29] and be a way of connecting improvements to computational models with experimental data and back to new biological knowledge. For this, information about how to test and evaluate hypotheses should be described, such as unambiguous instructions on what simulations to run and which data to compare the results to. Currently, RIMBO is, to some extent, aligned with PROV-O. With this extension more work is needed to align it with an upper level ontology, such as the Basic Formal Ontology (BFO) [30], to easier interface with ontologies describing for example experimental data.

As with computational models, ontologies change. As we use RIMBO to represent models and revisions to models in our project, it will be continuously developed and new releases will be published here: https://github.com/filipkro/rimbo.

Although this work is focused on computational biology models, the techniques and ideas presented are not domain specific. As the iterative nature of new knowledge gain is common for most fields, we think the approach of recording smaller changes to models, no matter if the improvements have been found by humans or machines, along with reasons and intentions for changes can be useful in many scientific fields.

5   Code and Availability

The code and knowledge graph for the demonstration is available here: https://github.com/filipkro/rimbo-demo. The ontology and future updates of it, is available here: https://github.com/filipkro/rimbo.

Acknowledgements. We want to thank the rest of the Ross King Group at Chalmers University for their thoughtful insights and discussions. This work was partially supported by the Wallenberg AI, Autonomous Systems and Software Program (WASP) funded by the Alice Wallenberg Foundation. Funding was also provided by the Chalmers AI Research Centre, the UK Engineering and Physical Sciences Research Council (EPSRC) grant nos: EP/R022925/2 and EP/W004801/1, as well as the Swedish Research Council Formas (2020-01690).

References

1. Noble, D.: The rise of computational biology. Nat. Rev. Mol. Cell Biol. **3**(6), 459–463 (2002)
2. Omenn, G.S.: Grand challenges and great opportunities in science, technology, and public policy. Science **314**(5806), 1696–1704 (2006)
3. King, R.D., et al.: Functional genomic hypothesis generation and experimentation by a robot scientist. Nature **427**(6971), 247–252 (2004)
4. Williams, K., et al.: Cheaper faster drug development validated by the repositioning of drugs against neglected tropical diseases. J. Roy. Soc. Interface **12**(104), 20141289 (2015)
5. Kanda, G.N., et al.: Robotic search for optimal cell culture in regenerative medicine. eLife **11**, e77007 (2022)
6. Singh, A.H., et al.: An automated scientist to design and optimize microbial strains for the industrial production of small molecules (2023)
7. Burger, B., et al.: A mobile robotic chemist. Nature **583**(7815), 237–241 (2020)
8. Coutant, A., et al.: Closed-loop cycles of experiment design, execution, and learning accelerate systems biology model development in yeast. Proc. Natl. Acad. Sci. **116**(36), 18142–18147 (2019)
9. Barton, C.M., et al.: How to make models more useful. Proc. Natl. Acad. Sci. **119**(35), e2202112119 (2022)
10. Malik-Sheriff, R.S., et al.: BioModels-15 years of sharing computational models in life science. Nucleic Acids Res. **48**, D407–D415 (2020)
11. King, Z.A., et al.: BiGG models: a platform for integrating, standardizing and sharing genome-scale models. Nucleic Acids Res. **44**, D515–D522 (2016)
12. Lloyd, C.M., et al.: The CellML model repository. Bioinformatics **24**(18), 2122–2123 (2008)
13. Demir, E., et al.: The BioPAX community standard for pathway data sharing. Nat. Biotechnol. **28**(9), 935–942 (2010)
14. Lloyd, C.M., et al.: CellML: its future, present and past. Prog. Biophys. Mol. Biol. **85**(2), 433–450 (2004)
15. Hucka, M., et al.: The systems biology markup language (SBML): a medium for representation and exchange of biochemical network models. Bioinformatics **19**(4), 524–531 (2003)
16. Ashburner, M., et al.: Gene ontology: tool for the unification of biology. Nature Genet. **25**(1), 25–29 (2000)
17. Juty, N., le Novère, N.: Systems biology ontology. In: Dubitzky, W., et al. (eds.) Encyclopedia of Systems Biology, pp. 2063–2063. Springer, New York (2013). https://doi.org/10.1007/978-1-4419-9863-7_1287
18. Zhukova, A., et al.: Kinetic simulation algorithm ontology. Nat. Proc. (2011)
19. Diehl, A.D., et al.: The cell ontology 2016: enhanced content, modularization, and ontology interoperability. J. Biomed. Semant. **7**(1), 44 (2016)
20. Costanzo, M.C., et al.: New mutant phenotype data curation system in the saccharomyces genome database. Database J. Biol. Databases Curation **2009**, bap001 (2009)
21. Black, M., et al.: EDAM: the bioscientific data analysis ontology (update 2021). F1000Research, vol. 11 (2022)
22. Kalaš, M., et al.: EDAM-bioimaging: the ontology of bioimage informatics operations, topics, data, and formats (2019 update) [version 1; not peer reviewed]. F1000Research, vol. 8(ELIXIR), p. 158 (2019)

23. Lebo, T., et al.: PROV-o: the PROV ontology. Technical report, World Wide Web Consortium (2013)
24. Samuel, S., König-Ries, B.: End-to-end provenance representation for the understandability and reproducibility of scientific experiments using a semantic approach. J. Biomed. Semant. **13**(1), 1 (2022)
25. Scharm, M., et al.: COMODI: an ontology to characterise differences in versions of computational models in biology. J. Biomed. Semant. **7**(1), 46 (2016)
26. Lu, H., et al.: A consensus s. cerevisiae metabolic model yeast8 and its ecosystem for comprehensively probing cellular metabolism. Nat. Commun. **10**(1), 3586 (2019)
27. Gower, A.H., et al.: LGEM$^+$: a first-order logic framework for automated improvement of metabolic network models through abduction. arXiv, arXiv:2306.06065 (2023)
28. Giaever, G., et al.: Functional profiling of the saccharomyces cerevisiae genome. Nature **418**(6896), 387–391 (2002)
29. Soldatova, L.N., et al.: Representation of probabilistic scientific knowledge. J. Biomed. Semant. **4**(1), S7 (2013)
30. Arp, R., et al.: Building Ontologies with Basic Formal Ontology. The MIT Press, Cambridge (2015)

Unsupervised Graph Neural Networks for Source Code Similarity Detection

Julien Cassagne[1]([⊠])(iD), Ettore Merlo[1], Paula Branco[2], Guy-Vincent Jourdan[2], and Iosif-Viorel Onut[3]

[1] Polytechnique Montreal, Montreal, Canada
{julien.cassagne,ettore.merlo}@polymtl.ca
[2] University of Ottawa, Ottawa, Canada
{pbranco,gjourdan}@uottawa.ca
[3] IBM Centre for Advanced Studies, Toronto, Canada
vioonut@ca.ibm.com

Abstract. In this paper, we propose a novel unsupervised approach for code similarity and clone detection that is based on Graph Neural Networks. We propose a hybrid approach to detect similarities within source code, using centroid distances and a Graph Auto-Encoder that uses a raw abstract syntax trees as input. When compared to $R_{TV}NN$ [33], the state-of-the-art unsupervised approach for code similarity and clone detection, our method improves significantly training and inference time efficiency, while preserving or improving precision. In our experiments, our algorithm is on average 77 times faster during training and 21 times faster during inference. This shows that using Graph Auto-Encoders in the domain of source code similarity analysis is the better option in an industrial context or in a production environment. We illustrate this by using our approach to compute source code similarity within a large dataset of phishing kits written in PHP provided by our industry partner.

Keywords: Graph neural network · Unsupervised Learning · Machine learning · Phishing kits similarity · Software similarity analysis · Static analysis

1 Introduction

The concept of a Graph Neural Network (GNN) was first introduced around 2005 [7,27] and was motivated by the non-regularity of data structures. In recent literature, GNNs have been successfully applied in a wide range of domains and have often outperformed previous approaches [39]. Recently, GNNs have been shown to be particularly effective for source code analysis [28] and have the advantage of being able work directly from graphs representing the source code. Often, programs are written using highly structured languages. In their graph representation, nodes are entities such as functions, variables, and statements, while edges represent relationships such as calls, uses, and dependencies between these entities. By learning the graph representation of source code, GNNs capture

© The Author(s), under exclusive license to Springer Nature Switzerland AG 2023
A. Bifet et al. (Eds.): DS 2023, LNAI 14276, pp. 535–549, 2023.
https://doi.org/10.1007/978-3-031-45275-8_36

the structural relationships between different parts of the code. They can use this information to perform tasks such as code classification, vulnerability or bug detection, and clone detection [8,15,18,20,21].

Recently [14], the lack of available labelled datasets for supervised Neural Clone Detection models has been flagged as a problem. In order to perform well, supervised Neural Clone Detection models require access to a large amount of datasets that have been labelled by human. However, such datasets are very hard to find and obtain. This is also true in the field of malware analysis, where the lack of labelled datasets limits the ability of models to detect and analyze malicious code.

One way around this problem is using unsupervised learning. One example is $R_{TV}NN$ [33]. This model uses AST and lexical embeddings to match clones using two different recursive neural networks and performs well on the clone detection problem. It can detect similarities in a large unlabelled data set but it is computationally intensive because of the upstream AST transformations and the presence of two different neural networks. The training time of the model is one of the problems identified by the authors of [33].

In this paper, we propose a new architecture to perform code similarity and clone detection in an unsupervised manner while improving the limitations of [33]. Our contributions are:

- We introduce a novel hybrid technique to detect similarities within source code, using centroid distances and a Graph Auto-Encoder (GAE) that represents a specialized architecture of unsupervised GNNs.
- We compare our approach to $R_{TV}NN$ [33] and show that our solution is much more efficient in training time (x77) and in inference time (x21) while preserving or even slightly improving precisions.
- We apply the proposed approach on a large dataset of 20,000+ "phishing kits" to compute similarity metrics between these kits.

The remainder of this paper is organized as follows: we first introduce the related work in Sect. 2. We then present our approach in Sect. 3. We compare our method with the state-of-the-art in unsupervised models in Sect. 4. In Sect. 5, we discuss the limitations of our experiments and future work, finally conclusions are discussed in Sects. 6.

2   Related Work

2.1   Tree-Based Code Clone Detection

Code clone detection and similarity analysis in software code have been studied extensively. Different granularity level (block, function, file) can be used for the analysis, defining the syntactic unit called code fragment. In the literature [25], clones have been divided into categories:

Type 1: Identical code fragments (except for variations in white space and comments)

Type 2: Structurally/syntactically identical code fragment except for variations in identifiers, literals, types, layout and comments.

Type 3: Copied fragments with further modifications. Statements can be changed, added or removed in addition to variations in identifiers, literals, types, layout and comments.

Type 4: Code fragments that perform the same computation but are implemented through different syntactic variants.

Each category is more difficult to detect than the previous one. The main difference between previous articles describing attempts to solve this problem comes from the representation used as input. A simple string-based approach was proposed in 2006 [3] and obtained a recall of 100% for type 1 clones on tested data sets. By adding additional transformations on identifiers and constants, they obtained a good result on type 2 clones, up to 88%.

Detection of more complex code clones requires a richer representation of code that contains semantic information like abstract syntax trees (AST). Building such trees relies on a parser that is language dependent. Then it is possible to find code clones using sub-trees comparisons [2]. The approach proposed in [32], "CDLH", uses a Word2Vec model to learn token embeddings and capture lexical information. It then trains a tree-LSTM [29] model based on an AST to combine these embeddings into a binary vector to represent a code fragment. Other AST-based deep learning approaches have been developed, such as ASTNN [38] and TBCCD [36]. However, these machine learning models require a labelled data set during the training phase, as they need to be trained in a supervised way. One way to overcome this issue is using transfer learning [35], or using unsupervised learning.

One architecture, called $R_{TV}NN$ [33], uses unsupervised learning and performed well on a clone detection problem. The model uses AST and lexical embeddings to match clones using two different recursive neural networks. This method can detect similarities in a large unlabelled data set. One drawback of this method is that it is computationally intensive because of the upstream AST transformations and the presence of two different neural networks. An architecture was proposed to overcome these limitations using weighted recursive autoencoders [37], but it still require pre-processing and linearization of the AST. In this paper, we propose to use a Variational Graph Auto-Encoder (VGAE) and directly send the AST as input to the graph neural network, to improve training and inference time.

2.2 Graph Neural Network

The point of a Graph Neural Network (GNN [7]) is to learn a state embedding for each node that contains information about itself and about its neighbourhood. Different types of layers have been proposed. The most generic and common one is the graph convolutional network (GCN) [11]. There are more specialized layers, such as gated graph neural networks (GGNN) [13] that use gated recurrent units and unroll the recurrence for a fixed number of steps. More recently, the graph matching network (GMN) was proposed [12]. It takes two graphs as input and uses an attention mechanism to compute a similarity score.

Learning a similarity metric between graphs has been a key problem that recent studies have tried to address using GNNs [16]. In [30], architecture that creates a graph embedding through different steps (sampling, encoding, embedded distribution) is presented. It obtains good results on most of the open data sets used for benchmarking and the final embedding is claimed to be close to a graph isomorphism. Aside from embedding structural information, a basic auto-encoder allows eliminating irrelevant and redundant features [4]. Different types of graph auto-encoders (GAEs) have been proposed with different objective functions [10,23].

A recent supervised architecture had been proposed to compute graph similarity directly from a neural network [31]. The authors compared the performance of a GGNN and a GMN in code clones detection based in the similarity between two graphs that aggregate AST, CFG and Dataflow information. Their results outperformed state-of-the-art approaches on the *BigCloneBench* and *Google Code Jam* data sets. However, using a deep neural network for code clone detection and fragment matching implies doing n^2 inferences for each pair, which can be an issue on large data sets. Besides, the addition of a new entry to the dataset implies running an inference between this new entry and all other elements of the data set.

In this paper, we propose a more scalable approach based on a Graph Neural Networks trained in an unsupervised manner.

3   Methodology

In this section, we first provide an overview of our approach based on an AST and an auto-encoder. We then explain how we obtain a vector representation of a set of source files and how we compute the pair-wise similarity. The following steps describe our method:

– Parse source code and build the corresponding AST
– Build a data set of fragments and train the VGAE
– Create a representation from the encoded fragments that compose the source code

Figure 1 shows an overview of our architecture. Our model creates a representation of each fragment, from which we can then compute distances. We chose a VGAE to create these representations, as it allows us to train the GNN in an unsupervised manner [34] while providing the advantage of GNNs to consider the structural information of the input graph. Details about internal VGAE architecture is provided in Fig. 2.

3.1   Parser

The extraction of program AST is done through a parser generated with JavaCC. The first pre-processing step is to identify files that contain code. The parser

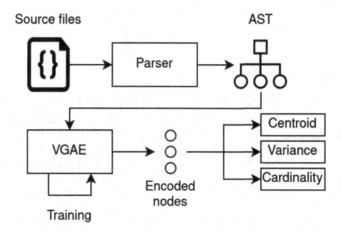

Fig. 1. Architecture overview

then extracts the AST for each file that is then decomposed into fragments at the function and method level. By representing a file as a set of AST fragments, our representation is not sensible to permutation of function.

3.2 Graph Neural Network

Each fragment $F = (V, E)$ is represented as a set of nodes V and edges E. Each node $i \in V$ is associated with a feature vector x_i of dimension H resulting from the one-hot encoding on node type. We can then represent each fragment as a node matrix X of shape $(|V|, H)$ and an adjacency matrix A of shape $(2, |E|)$, and then use these matrices as input to the first layer of our model (Fig. 3).

In our architecture, we selected the Graph Convolutional Layer (GCN) [11]. It implements the core concept of Graph Neural Network by learning the features through direct neighbors node inspection. This layer gave good result in the recent literature [26,39] and can be fit, like most GNNs, into the general framework message passing neural networks (MPNN) [6] to obtain the following node-wise formulation:

$$\mathbf{x}_i^{(k)} = \Theta \sum_{j \in \mathcal{N}(i) \cup \{i\}} \frac{e_{j,i}}{\sqrt{\hat{d}_j \hat{d}_i}} \mathbf{x}_j^{(k-1)} \tag{1}$$

with $\hat{d}_i = 1 + \sum_{j \in \mathcal{N}(i)} e_{j,i}$, where $e_{j,i}$ denotes the edge weight from source node j to target i, $\mathcal{N}(i)$ is the set of direct neighbors of i, $\mathbf{x}_i^{(k-1)} \in \mathbb{R}^F$ denotes the node features of node i in layer $(k-1)$, and Θ is a weight matrix.

In the case of an AST, all edges have the same relevance and weights. Therefore, we chose to set all weights $e_{i,j}$ to 1. It should be noted that the described layer does not consider the order of edges. In a recent article [31], the authors proposed to generate new edge types to link siblings in their order. The creation of such edge types, as well as edges bearing semantic information from CFG, is left as a future work here.

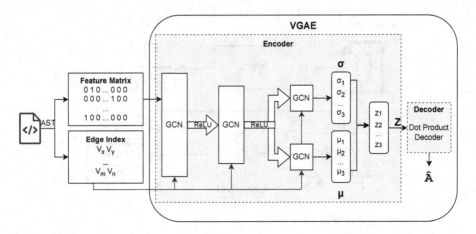

Fig. 2. VGAE overview

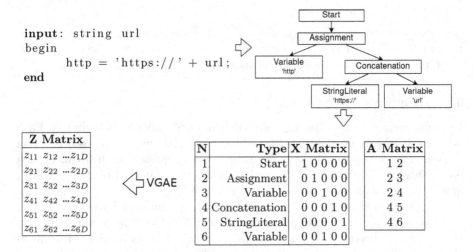

Fig. 3. Fragment Representation

We implemented our model composed of three GCN layers using the geometric deep learning extension library for PyTorch [24] named PyTorch Geometric [5]. The model produces an output matrix Z, composed of $|V|$ vectors z (one per node) of size D. The list of chosen parameters is available in the next section, Table 2.

3.3 Graph Auto-encoder

Next, the unsupervised learning begins. As presented in a recent survey [34], currently the main way to perform unsupervised learning with Graph Neural Networks is using Graph Auto-Encoder [10, 22]. They represent an unsupervised learning frameworks which encode nodes/graphs into a latent vector space and reconstruct graph data from the encoded information. GAEs learn latent node representations through reconstructing graph structural information such as the graph adjacency matrix.

In contrast, the Variational Auto-Encoder (VAE) [9] is a Bayesian model which learns the compressed representation of the Auto-Encoder, and constructs the parameters representing the probability distribution of the data. The commonly adopted evaluation method of GAE is to minimize the reconstruction errors of the input fragment or, in other words, to reconstruct the adjacency matrix A from the embedding Z. To train our model, we used the variational graph auto-encoder of [10] that proposes using the following objective function:

$$\mathcal{L} = \mathbb{E}_{q(\mathbf{Z}|\mathbf{X},\mathbf{A})}\big(\log p(\mathbf{A}|\mathbf{Z})\big) - D_{\mathrm{KL}}(q(\mathbf{Z}|\mathbf{X},\mathbf{A})\|p(\mathbf{Z})) \qquad (2)$$

where X, A and Z are the matrix defined in Sect. 3.2 and represented in Fig. 3, p and q are two discrete probability distributions, and D_{KL} is the Kullback-Leibler divergence that measures the distance between the two probability distributions.

3.4 Fragment Representation

Once the model has converged after the training phase, we can use the encoder to create our fragment representation. To this end, we work on the distribution of the encoded nodes of the AST by aggregating all nodes embedding Z returned by the VGAE and then compute the following three metrics:

– Centroid: column-wise mean of Z
– Variance: column-wise standard deviation of Z
– Cardinality: $|Z|$

These three components are our final fragment representation from which we can compute the distance between others fragments to obtain a similarity score. The cardinality is a scalar, and the centroid and variance are both vectors of dimension D Fig. 3 (dimension of the latent space). Using these three metrics, it is possible to obtain a list of similar fragments [19] by computing the distance between centroids, the variance, and the cardinality and selecting the closest ones. Note that these three components can also be computed at the file level by aggregating all encoded nodes of a file or at the overall software level by aggregating all encoded nodes from all ASTs that compose the software.

We provide in our appendix repository [1] the corresponding implementation using Pytorch [24] and Pytorch Geometric [5], which we used in our experiments.

4 Experiments

In this section, we compare our architecture to the prior art: we limited the comparison to $R_{\mathrm{TV}}NN$ [33], as it is the only approach based on unsupervised learning we found for clone detection. Comparison with supervised system would not be feasible here, as we are analysing unlabeled datasets.

We performed our experiments on the same eight datasets used in [33], and we estimated the precision of our architecture based on the test sets of code clones used in [33] as well. These datasets are based on source code of open source softwares, and were downloaded from GitHub. More information can be found in Table 1.

Table 1. Systems statistics

System	Version	Files	Lines of codes	Tokens
ANTLR	4	514	104,225	701,807
Apache Ant	1.9.6	1,218	136,352	888,424
ArgoUML	0.34	1,908	177,493	1,172,058
CAROL	2.0.5	184	12,022	80,947
DNSjava	2.0.0	196	24,660	169,219
Hibernate	2	555	51,499	365,256
JDK	1.4.2	4,129	562,120	3,512,807
JHotDraw	6	984	58,130	377,652

Table 2. Training parameters

Number of features (H)	148
Latent space size (D)	10
Layers output size	100-20-10
Optimizer	Adam
Learning rate	$1e^{-2}$
Batch size	10
Epoch	30

All the systems analyzed are written in Java. To parse the code, we used a custom made top-down parser for Java 1.7 written in JavaCC. The extracted AST is then used as input to our VGAE model. The selected hyperparameters are provided in Table 2. The first one corresponds to the number of different AST node types (parser dependent value). In this case, it is 148. The other hyperparameters were empirically chosen to obtain good results while keeping a reasonable training time.

The comparison with the results of [33] is based on three aspects: the training time, the inference time, and the precision. As mentioned earlier, we made the comparison on the same set of software, at the same version, using the same set of code clones. However, we should note that the experiments have been executed on different computers. We used a Fedora 31 server with a 4-core Intel Core i5

4670 at 3.4 GHz (released in 2013), whereas the results of [33] were obtained using a compute node serving two 8-core Intel Xeon E5-4627 v2 processors at 3.3 GHz (released in 2014).

Our experiment was limited by the speed of our HDD, leading the CPU to not be used to full capacity. We believe the our reported performance would have been even better using an SSD and a two Intel Xeon E5-4627 processors, as such CPU comes with larger caches, more cores and faster memory.

4.1   Clone Detection

To solve the clone code detection problem, we chose a file-level and method-level threshold that we used throughout the whole experiment and that we did not optimize system-wise. When the distance between two fragments is below all three thresholds, the pair of fragments is then considered a "code clone." As our representation has three components, we identified three thresholds that ensure the point representing the second fragment is inside a box centred on the first fragment: if the distance is bellow all three thresholds, the two fragments are considered clones. We also decided to tighten the threshold for method-level clone detection because methods are on average smaller than files. We provide the thresholds used for detect file-level and method-level code clones in Table 3. They were selected empirically based on the average accuracy on all datasets.

One difference between our solution and $R_{TV}NN$ is the ability to differentiate between clones of types 1 and 2. Our representation does not take into account identifiers information, meaning that parametric and identical clones get the same representation. This does not impact the following experiments, since comparisons are based on clone detection no matter what type they are. We do not believe that this limitation is significant, since clone 1 detection is an easy problem to solve and does not require an architecture as complex as the one described here.

4.2   Training and Inference

For each system, we trained our model on all the AST extracted from source code Java files. As in the original experiment, we excluded all files containing more than 4,000 lexical elements from the training set. We provide a comparison of the training time per epoch between the two modes that the authors of $R_{TV}NN$ [33] proposed (AST-based and greedy) and our approach (VGAE) in Table 4.

Table 3. Clone detection thresholds

	File-level	Method-level
Centroid distance	6e−1	4e−1
Variance distance	3e−1	3e−1
Size distance	$size \times 8e{-}1$	$size \times 5e{-}1$

Table 4. Average training/inference time per epoch number in parenthesis corresponds to the median

	Training (sec)			Inference (sec)		
CPU	Xeon E5@3.3 GHz[a]	i5@3.4 GHz[b]		Xeon E5@3.3 GHz[a]		i5@3.4 GHz[b]
System	AST-Based	Greedy	VGAE	AST-Based	Greedy	VGAE
ANTLR	443	3516	15	3.21 (1.18)	33.36 (1.96)	5.08 (0.06)
Apache Ant	813	3476	53	3.31 (1.76)	25.20 (3.10)	0.25 (0.06)
ArgoUML	1018	3868	47	2.58 (1.24)	16.35 (1.80)	0.26 (0.04)
CAROL	34	116	5	0.88 (0.48)	4.87 (0.95)	0.16 (0.07)
DNSjava	148	1169	14	3.63 (2.16)	30.67 (4.30)	0.81 (0.12)
Hibernate	277	1077	15	2.49 (1.17)	17.70 (1.70)	0.87 (0.05)
JDK	2977	14965	201	3.46 (1.19)	35.06 (1.80)	0.55 (0.05)
JHotDraw	336	792	24	1.67 (0.93)	6.40 (1.19)	0.11 (0.02)

[a]Two Intel Xeon E5-4627 v2 @ 3.3 GHz, released in 2014
[b]One Intel Core i5 4670 @ 3.4 GHz, released in 2013

On average, our approach is 16 times faster than the $R_{TV}NN$ AST-based mode and 77 times faster than the $R_{TV}NN$ greedy mode. This can be explained by two factors:

– Our model does not need to perform any pre-processing of the AST; the tree is given as is to the GCN. In contrast, $R_{TV}NN$ performs heavy pre-processing steps.
– Our architecture uses only one GNN, whereas $R_{TV}NN$ uses two different recurrent neural networks.

For the same reasons, the same effect can be observed regarding the inference time, also shown in Table 4: our approach is on average 22 times faster than $R_{TV}NN$ greedy. Note that the inference times provided for our method does includes all the steps to obtain our fragment representation, namely:

– Parsing
– Exporting the AST to a *PyTorch* object
– GNN inference
– Fragment representation from embedding

We can see that the average inference time for ANTLR is worse than $R_{TV}NN$ AST-based here, but the median time is better. This is due to a few extremely large files in ANTLR that take more time to be processed and fed to the model.

4.3 Precision

After training the model and inferring all representations, we can compare the precision. By applying the threshold defined before, we determined whether each considered pair of fragments is correctly classified as clones. The pairs of fragments from the original experiments were manually selected in each system [33]

Table 5. Precision results

System	File-level		Method-level		File-level		Method-level	
	AST-Based	VGAE	AST-Based	VGAE	Greedy	VGAE	Greedy	VGAE
ANTLR	97% (30)	97% (30)	100% (30)	100% (30)	100% (30)	100% (30)	100% (30)	100% (30)
Apache Ant	92% (24)	92% (24)	100% (30)	100% (30)	93% (30)	97% (30)	100% (30)	100% (30)
ArgoUML	90% (30)	90% (30)	100% (30)	100% (30)	100% (30)	100% (30)	100% (30)	100% (30)
CAROL	100% (1)	100% (1)	100% (30)	100% (30)	100% (10)	100% (10)	100% (30)	100% (30)
DNSjava	47% (30)	50% (30)	73% (30)	77% (30)	100% (30)	100% (30)	87% (30)	90% (30)
Hibernate	100% (13)	100% (13)	53% (30)	70% (30)	100% (20)	100% (20)	70% (30)	73% (30)
JDK	90% (30)	94% (30)	100% (30)	100% (30)	100% (30)	100% (30)	100% (30)	100% (30)
JHotDraw	100% (30)	100% (30)	100% (30)	100% (30)	100% (30)	100% (30)	100% (30)	100% (30)

and correspond to fragments that are clones (types 1 to 4) and fragments that are different. The authors of $R_{TV}NN$ showed that they obtained better precision using the "greedy" method than the "AST-based" method. We could not evaluate all methods simultaneously on the same set because, originally, AST-based and greedy were not tested on the same set of clones. We performed two evaluations, one comparing results with the AST-based method, and one comparing results with the greedy method (Table 5). In both scenarios, our architecture performed as well or better. The remaining undetected clones are mostly type 4 clones with substantially different amounts of code. However, these results do not represent a significant improvement as there was not much room for improvement.

4.4   Analysis of Results on DNSJava

To better understand the limitations of our architecture and the edge cases that were not detected, we performed deeper results analyses on DNSJava method-level clones. We started by using uniform manifold approximation and projection (UMAP) [17] to perform dimensionality reduction on the fragment representations, as illustrated on Fig. 4.

The grey lines link two code clones. The blue points correspond to fragments that are correctly classified as a clone, and the red points correspond to fragments that are not correctly classified. We showed only one of the three code clones that have not been correctly classified for easier reading.

Fig. 4. 2D representation (UMAP) of method embeddings on a subset of DNSJava code clones

The two fragments corresponding to the red points are type 4 clones, and our architecture did not consider these two pieces of code as clones mainly because of the size difference. Overall, most other undetected pairs also correspond to type 4 clones with substantially different amounts of code. At the same time, most of the type 3 clones were correctly classified. Finally, all type 1 and 2 clones were correctly classified.

We note that, by tuning the threshold for this particular system, we were able to reach 100% precision at the method and file levels.

5   Limitations and Future Research

Hyperparameters. One limitation of our approach is the choice of hyper-parameters of our neural network. We did not do a thorough fine-tuning of our model using a validation set: there is room for performance improvement that could lead to better precision or even faster training and inference time.

Large Trees. As mentioned in our first experiment, we excluded from the training set files with more than 4,000 lexical elements (to be comparable with the results obtained in $R_{TV}NN$ [33]). Including them highlighted a limitation of our architecture: large graphs significantly increase the training and inference time.

Future Research. Our first experiment was performed on the dataset of clones supplied with $R_{TV}NN$. This dataset had been manually labeled by the authors [33] and is relatively small. In addition to the second experiment on physhing kits and to directly compare our architecture to other state-of-the-art models for clone detection in conventional software, we would like to run new experiments on other public benchmark datasets and compare the performances on a larger scale. As an additional datasets for future experiments, we can mention BigCloneBench that is a large, diverse, and publicly available dataset of code clones. BigCloneBench is widely used in the research community for evaluating and benchmarking code clone detection systems.

6   Conclusions

In this paper, we present a novel approach for code similarity analysis that combines centroid distances and unsupervised training of GNN. Unlike traditional metric-based methods, our architecture is designed to automatically learn discriminating features in source code using a GAE.

In the first experiment, our model significantly improves training and inference time compared to the prior-art model, mainly because of: the advantages brought by GNNs, the simplicity of the architecture, and the absence of AST pre-processing steps. This is the first architecture to apply a GAE approach to clone detection.

In the second experiment, we have presented results about how we used our architecture on a large data set of more than 20,000 phishing kits corresponding to a total size of about 84 MLOCs. The goal was to efficiently detect kits that share high source code similarity. We believe that after training, our model may help to quickly identify or classify new incoming phishing kits by similarity to previous ones. It may also help to detect kits that may be derived from one another.

References

1. Repository. https://gitlab.com/polymtl-static-analysis/vgae-code-analysis
2. Baxter, I., Yahin, A., Moura, L., Sant'Anna, M., Bier, L.: Clone detection using abstract syntax trees. In: Proceedings of the International Conference on Software Maintenance (Cat. No. 98CB36272), pp. 368–377 (1998)
3. Ducasse, S., Nierstrasz, O., Rieger, M.: On the effectiveness of clone detection by string matching: research articles. J. Softw. Maint. Evol. **18**(1) (2006)
4. Feng, S., Duarte, M.F.: Graph autoencoder-based unsupervised feature selection with broad and local data structure preservation. Neurocomputing (2018)
5. Fey, M., Lenssen, J.E.: Fast Graph Representation Learning with PyTorch Geometric (2019)
6. Gilmer, J., Schoenholz, S.S., Riley, P.F., Vinyals, O., Dahl, G.E.: Neural message passing for quantum chemistry. CoRR abs/1704.01212 (2017)
7. Gori, M., Monfardini, G., Scarselli, F.: A new model for learning in graph domains. In: Proceedings of the 2005 IEEE International Joint Conference on Neural Networks (2005)
8. Jiang, S., Hong, Y., Fu, C., Qian, Y., Han, L.: Function-level obfuscation detection method based on graph convolutional networks. J. Inf. Secur. Appl. **61**, 102953 (2021)
9. Kingma, D.P., Welling, M.: Auto-encoding variational bayes (2014)
10. Kipf, T.N., Welling, M.: Variational graph auto-encoders. arXiv:1611.07308 [cs, stat] (2016)
11. Kipf, T.N., Welling, M.: Semi-supervised classification with graph convolutional networks (2017)
12. Li, Y., Gu, C., Dullien, T., Vinyals, O., Kohli, P.: Graph matching networks for learning the similarity of graph structured objects (2019)
13. Li, Y., Tarlow, D., Brockschmidt, M., Zemel, R.: Gated graph sequence neural networks (2015)
14. Liu, C., Lin, Z., Lou, J.G., Wen, L., Zhang, D.: Can neural clone detection generalize to unseen functionalities*f*. In: 2021 36th IEEE/ACM International Conference on Automated Software Engineering (ASE), pp. 617–629 (2021)
15. Liu, S.: A unified framework to learn program semantics with graph neural networks. In: 2020 35th IEEE/ACM International Conference on Automated Software Engineering (ASE) (2020)
16. Ma, G., Ahmed, N.K., Willke, T.L., Yu, P.S.: Deep graph similarity learning: a survey. Data Min. Knowl. Disc. **35**(3), 688–725 (2021)
17. McInnes, L., Healy, J., Melville, J.: UMAP: uniform manifold approximation and projection for dimension reduction. arXiv:1802.03426 (2020)
18. Mehrotra, N., Agarwal, N., Gupta, P., Anand, S., Lo, D., Purandare, R.: Modeling functional similarity in source code with graph-based siamese networks. arXiv:2011.11228 [cs] (2020)

19. Merlo, E., Antoniol, G., Di Penta, M., Rollo, V.: Linear complexity object-oriented similarity for clone detection and software evolution analyses. In: Proceedings of the 20th IEEE International Conference on Software Maintenance, pp. 412–416 (2004)

20. Nair, A., Roy, A., Meinke, K.: funcGNN: a graph neural network approach to program similarity. In: Proceedings of the 14th ACM/IEEE International Symposium on Empirical Software Engineering and Measurement (ESEM), pp. 1–11 (2020). arXiv: 2007.13239

21. Nguyen, V.A., Nguyen, D.Q., Nguyen, V., Le, T., Tran, Q.H., Phung, D.: ReGVD: revisiting graph neural networks for vulnerability detection. In: 2022 IEEE/ACM 44th International Conference on Software Engineering: Companion Proceedings (2022)

22. Pan, S., Hu, R., Long, G., Jiang, J., Yao, L., Zhang, C.: Adversarially regularized graph autoencoder for graph embedding. In: Proceedings of the 27th International Joint Conference on Artificial Intelligence, IJCAI 2018. AAAI Press (2018)

23. Park, J., Lee, M., Chang, H., Lee, K., Choi, J.: Symmetric graph convolutional autoencoder for unsupervised graph representation learning. In: 2019 IEEE/CVF International Conference on Computer Vision (ICCV) (2019)

24. Paszke, A., et al.: Pytorch: an imperative style, high-performance deep learning library. In: Advances in Neural Information Processing Systems, vol. 32. Curran Associates, Inc. (2019)

25. Roy, C.K., Cordy, J.R., Koschke, R.: Comparison and evaluation of code clone detection techniques and tools: a qualitative approach. Sci. Comput. Program. **74**(7), 470–495 (2009)

26. Rozi, M.F., Ban, T., Ozawa, S., Kim, S., Takahashi, T., Inoue, D.: JStrack: enriching malicious JavaScript detection based on AST graph analysis and attention mechanism. In: Neural Information Processing: ICONIP (2021)

27. Scarselli, F., Gori, M., Tsoi, A.C., Hagenbuchner, M., Monfardini, G.: The graph neural network model. IEEE Trans. Neural Netw. **20**(1), 61–80 (2009)

28. Siow, J.K., Liu, S., Xie, X., Meng, G., Liu, Y.: Learning program semantics with code representations: an empirical study. In: 2022 IEEE International Conference on Software Analysis, Evolution and Reengineering (SANER), pp. 554–565 (2022)

29. Tai, K.S., Socher, R., Manning, C.D.: Improved semantic representations from tree-structured long short-term memory networks. In: Proceedings of the 53rd Annual Meeting of the Association for Computational Linguistics, Beijing, China, pp. 1556–1566. Association for Computational Linguistics (2015)

30. Wang, L., et al.: Inductive and unsupervised representation learning on graph structured objects. In: International Conference on Learning Representations (2020)

31. Wang, W., Li, G., Ma, B., Xia, X., Jin, Z.: Detecting code clones with graph neural network and flow-augmented abstract syntax tree. In: 2020 IEEE 27th International Conference on Software Analysis, Evolution and Reengineering (SANER), pp. 261–271 (2020)

32. Wei, H., Li, M.: Supervised deep features for software functional clone detection by exploiting lexical and syntactical information in source code. In: Proceedings of the 26th International Joint Conference on Artificial Intelligence, IJCAI 2017 (2017)

33. White, M., Tufano, M., Vendome, C., Poshyvanyk, D.: Deep learning code fragments for code clone detection. In: 2016 31st IEEE/ACM International Conference on Automated Software Engineering (ASE), pp. 87–98 (2016)

34. Wu, Z., Pan, S., Chen, F., Long, G., Zhang, C., Yu, P.S.: A comprehensive survey on graph neural networks. IEEE Trans. Neural Netw. Learn. Syst. **32**, 4–24 (2020)
35. Yahya, M.A., Kim, D.K.: CLCD-I: cross-language clone detection by using deep learning with infercode. Computers **12**(1) (2023)
36. Yu, H., Lam, W., Chen, L., Li, G., Xie, T., Wang, Q.: Neural detection of semantic code clones via tree-based convolution. In: 2019 IEEE/ACM 27th International Conference on Program Comprehension (ICPC), pp. 70–80 (2019)
37. Zeng, J., Ben, K., Li, X., Zhang, X.: Fast code clone detection based on weighted recursive autoencoders. IEEE Access **7**, 125062–125078 (2019)
38. Zhang, J., Wang, X., Zhang, H., Sun, H., Wang, K., Liu, X.: A novel neural source code representation based on abstract syntax tree. In: 2019 IEEE/ACM 41st International Conference on Software Engineering (ICSE), pp. 783–794 (2019)
39. Zhou, J., Cui, G., Zhang, Z., Yang, C., Liu, Z., Sun, M.: Graph neural networks: a review of methods and applications. AI Open **1**, 57–81 (2020)

Time Series and Forecasting

A Universal Approach for Post-correcting Time Series Forecasts: Reducing Long-Term Errors in Multistep Scenarios

Dennis Slepov⬤, Arunas Kalinauskas⬤, and Hadi Fanaee-T$^{(\boxtimes)}$⬤

Halmstad University, Halmstad, Sweden
{densle18,arukal18}@student.hh.se, hadi.fanaee@hh.se

Abstract. Time series forecasting is an important problem with various applications in different domains. Improving forecast performance has been the center of investigation in the last decades. Several research studies have shown that old statistical method, such as ARIMA, are still state-of-the-art in many domains and applications. However, one of the main limitations of these methods is their low performance in longer horizons in multistep scenarios. We attack this problem from an entirely new perspective. We propose a new universal post-correction approach that can be applied to fix the problematic forecasts of any forecasting model, including ARIMA. The idea is intuitive: We query the last window of observations plus the given forecast, searching for similar "shapes" in the history, and using the future shape of the nearest neighbor, we post-correct the estimates. To ensure that post-correction is adequate, we train a meta-model on the successfulness of post-corrections on the training set. Our experiments on three diverse time series datasets show that the proposed method effectively improves forecasts for 30 steps ahead and beyond.

Keywords: Multi-steap Time series Forecasting · Post-correction

1 Introduction

Time series forecasting is the process of predicting future values of a series of data points that are indexed or ordered in time. This is typically done by analyzing historical data and developing a model to capture patterns from that data. The model is then used to project future values of the series based on the assumptions and parameters of the model.

Many methods exist for time series forecasting, including statistical methods such as Auto-regressive Integrated Moving Average *(ARIMA)*, Exponential Smoothing, and machine learning methods such as neural networks. Depending on the goal of the analysis, these models can be employed to perform either

Supported by Swedish Knowledge Foundation through Project #280033.

short or long-term forecasts. Statistical models usually provide a linear relationship equation between future and past observations. They are known to be very robust and accurate for short-term forecasting. However, their performance degrades progressively by enlarging the forecast horizon. This problem of statistical methods is our focus in this research. In particular, we are interested in a universal post-processing approach that can correct the forecast of any model in long-term forecasting. We aim to investigate the effectiveness of shape search ideas in time series mining. Our central hypothesis is that when the forecast horizon gets longer, the shape of the recent observation plus the forecasted part can be an essential piece of information to correct the forecast. To the best of our knowledge, we are the first group that investigates this idea.

1.1 Research Questions

Our fundamental hypothesis is that if the forecast is valid, we should be able to find a similar shape of a query (last window of observation + prediction window) in the historical set. Otherwise, we assume the forecast is inaccurate, and we post-correct it according to the nearest shape match. According to this assumption, we set our two main research questions as follows.

- Is shape post-correction an effective idea for improving time series forecasting?
- How we ensure that shape post-correction would result in a forecasting improvement?

1.2 Contribution

Our contributions are the following:

- **Novel Methodology.** This work introduces a new universal approach to improving the accuracy of a time series long-term forecast without modifying the forecasting model that performed the estimates. To the best of our knowledge, forecasting post-correction has not been studied before.
- **Performance Analysis across Different Datasets.** Extensive performance analysis across different types of time series containing various characteristics. This analysis provides valuable insights into the proposed method's robustness and adaptability to different data characteristics.
- **Evaluation of Impact of Hyperparameters.** This research offers an in-depth understanding of how different important hyperparameters affect the performance of the proposed method.

The expected outcome of this research is to improve the accuracy of time series long-term forecasts by providing a way to identify, validate and post-correct the estimates.

2   Proposed Method: Time Series Post-correction (TSPC)

2.1   Overview

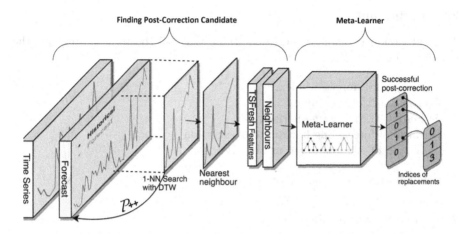

Fig. 1. High-Level system architecture of the TSPC

Figure 1 illustrates the complete data-flow process of TSPC. We start by producing a forecast for time series data with any arbitrary model. Next, we create
our query, which is composed of the last window of observation plus the forecast
window. For instance, if the sliding window size is 20 and the forecast horizon
is 30, the query will have a length of 50. To allow a query to become independent of scale, we also perform a z-normalization on the query and all sliding
windows of the same size. The purpose is to find a similar shape in the past,
not a similar sequence of values. Once we found the most similar shape in the
historical data, we denormalize the shape using the normalization parameters
of the last observation window. With this trick, we project back the shape into
the more up-to-date scale of the recent data. We perform our shape search using
nearest shape search (1NN) with Dynamic Time Warping (DTW) as a similarity
measure.

The next step is to create a meta-learning decision model to tell us whether
we should perform post-correction or not on an unseen example. Once we find
the nearest shape, we check if the future part of the denormalized nearest match
provides a better accuracy than the original forecast. So, we will have two outcomes, either we have improvement (successful correction) or no improvement
(unsuccessful correction). This is how we generate auto-labels. Then from the
query, we extract time series features using *tsfresh* [2], which will later serve as
the input. *tsfresh* is a feature extraction tool for time series that extracts tens
of features, including statistical measures, time-domain characteristics, Fourier
transformations, etc. Once we created features using *tsfresh* we train a meta-
learner on the training set using *tsfresh* features as the input and auto-labels as

the output. The trained model later will be served as a decision model to tell us whether we do post-correction or not based on a given new query on the test set.

2.2   Definitions

This subsection explains several definitions that will be used throughout the remainder of Sect. 2.

- $\mathcal{P}(start)$: First forecast index.
- $\mathcal{P}(end)$: Last forecast index.
- $\mathcal{F}$: Length of forecasting horizon.
- $\mathcal{D}$: Ratio of forecasting horizon to observation
- $\mathcal{K}$: Length of window of recent observations

$\mathcal{K} + \mathcal{F}$ determines the query size.

2.3   Nearest Shape Search

As it is depicted in Fig. 2, after we are given of a forecast of length, $\mathcal{F}$ we concatenate it with last $\mathcal{K}$ recent values. The new time series forms our target query. The search is always performed from the earliest data point in the time series up to $\mathcal{P}$ - $(\mathcal{K}+\mathcal{F})$. The subtraction of $\mathcal{K}$ and $\mathcal{F}$ is made to avoid matching the searched shape with itself. The lowest DTW distance determines the NN match. The last $\mathcal{F}$ data point of the best match is the potential correction for the original forecast. Therefore, they are evaluated against the ground truth. In the case of the best match, having a lower error than the original forecast, a label "1" is assigned, representing the case when post-correction improved the forecast, and "0": when post-correction failed to improve the forecast. The method is repeated from $\mathcal{P}(start)$ to $\mathcal{P}(end)$.

2.4   Z-Normalization of Sliding Windows

We use Z-normalization to normalize the sliding window by subtracting the mean and dividing it by the standard deviation. Specifically, z-normalization involves the following formula: $z = (x - \mu)/\sigma$,where z is the standardized value, x is the original value, μ is the mean of the window, and σ is the standard deviation of the window.

The rationale behind Z-normalization is to make the shape search independent of the magnitude of the values. It is a quite typical technique used in time series mining and methods such as SAX.

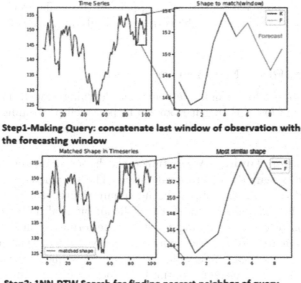

Step1-Making Query: concatenate last window of observation with the forecasting window

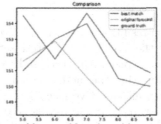

Step2: 1NN-DTW Search for finding nearest neighbor of query

Step 3: Comparison of future of found neighbour with ground truth

Fig. 2. 1NN-DTW search process

2.5   Meta-learning for Predicting the Success of Post-correction

Sometimes the effect of post-correction might have an inverse effect. To reduce this effect, we train a meta-model on the performance of post-correction on the train set. The following phase of the method involves generating meta-features that will be utilized for building a decision model for predicting the success of post-correction. These meta-features are composed of information such as the nearest shape, its measured DTW distance, $\mathcal{P}$ value, and $\mathcal{K}$ value. Additional relevant information collected from the *tsfresh* is included in the dataset to help collect more features from the candidate window. Note that not all features extracted by tsfresh are always relevant. So, we have to filter out some redundant and irrelevant features by using hypothesis tests, which dramatically reduces the number of features (e.g., in our case, from 783 to 105). We also standardize meta-features since their scale is different. Based on the outcome of MAE evaluations on the train set, binary labels are then appended to the dataset. These binary labels are indicative of whether the correction would improve the results (1) or not (0).

2.6 Random Forest Meta-learner

We use the random forest model as our meta-learning model. The reason for using the random forest is that it is known to be a well-performing classifier for tabular data, particularly for small or medium size data sets.

Hyperparameter Tuning. We optimize the hyperparameters of a random forest using grid search with cross-validation. The parameter grid consists of the number of trees in the forest, the maximum depth of each tree, and the minimum number of samples required to split an internal node. We use 5-fold cross-validation to evaluate the meta-learner's performance on the training set with each combination of hyperparameters. The best set of hyperparameters is selected based on the mean cross-validation score. Subsequently, the random forest meta-learner is fitted using the optimized hyperparameters on the normalized training set.

Threshold Tuning. The resulting classification model is then used to predict post-correction status of the validation and test sets. To increase the chance of a successful forecast post-correction, threshold tuning focuses on improving the precision score, meaning that the model primarily focuses on predicting the true positives and avoiding false positives accurately. For each set, the predicted labels are adjusted using a threshold selected based on the precision score on the validation set. The precision score is computed for various thresholds ranging from 0.5 to 0.7 with a step size 0.05. The threshold that yielded the highest precision score on the validation set was selected as the optimal threshold and was subsequently used to adjust the predicted labels on the test set. The resulting labels are then used to evaluate the meta-learner's performance on the test set.

3 Experimental Evaluation

After the meta-learning model generated predictions on where the post-corrections should be made (See Fig. 3), we are ready to evaluate the effectiveness of the post-correction method. This is accomplished by calculating the total error of all original forecasts in the test data. Next, the errors of the original forecasts are replaced with the post-corrected errors in the predicted indices Fig. 3. Lastly, the difference is measured in percentage.

3.1 Datasets

We use the following time series datasets to evaluate our proposed method.

- M4 Competition [5].
- Stock data from yfinance [1].
- Kaggle Climate dataset [8].

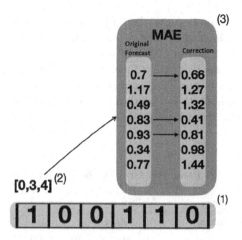

Fig. 3. Only those forecasts that are predicted to be effective by meta-model will be replaced by the correction obtained from 1NN-DTW search

3.2 Experimental Configuration

The following section explains various paradigms used to experiment with this study. This study will test and evaluate three main parts: 1NN search performance, the meta-learner's ability to accurately identify when the corrections should occur, and the effect of using *tsfresh* to select features.

Hyperparameters of Nearest Shape Search. The parameters that are used in the 1NN search are: $\mathcal{P}$, $\mathcal{F}$, and $\mathcal{D}$. The range of $\mathcal{P}(start)$ and $\mathcal{P}(end)$ determines the forecast horizon. This mainly contributes to the meta-learner in the form of more training data. The $\mathcal{P}(start)$ also needs to have some historical data for the search, and $\mathcal{K}(end)$ is chosen with consideration regarding time, since with greater forecasting horizons, the process of finding the best match becomes more computationally heavy. Hence, for the majority of the test, the $\mathcal{P}(start)$ is chosen to be 400 and $\mathcal{P}(end)$ to 700. The dataset size could affect these parameters.

The range of parameters $\mathcal{F}$ is chosen to be between 2 and 50 with intervals of 4. This will provide insight into when TSPC is having a positive effect. The increment of 4 is applied to reduce the whole process's time.

The last parameter $\mathcal{D}$ is chosen to be 0.5, meaning 50% of the searched window size will be the historical data, and 50% will be the forecast. A lower percentage would be beneficial for the lower forecasting ranges. However, working with high forecasting lengths would be too computationally heavy.

Meta-learner's Hyperparameters. As previously mentioned, there will be tests of the 1NN search with varying dataset sizes, directly impacting the meta-learner in the form of training data size. This section will compare the meta-learner's performance to its absence, meaning all forecasting errors in the test data are replaced with the best matches. The meta-learner is hyper-tuned with a temporal-wise split, which maintains the temporal order during cross-validation with five splits. The parameters for grid search are shown in Table 1. The main objective of meta-learner testing is to evaluate its performance in different scenarios and attempt to improve its correction score, done by threshold tuning. After the model is trained, the validation dataset is used to find a more optimal threshold, and the threshold is searched between 0.5 (default) and 0.7 with 0.05 step increments.

Table 1. Parameter Grid for the Random Forest

Parameter	Values							
n_estimators	50	100	200	-	-	-	-	-
max_depth	10	20	30	40	50	60	70	80
min_samples_split	2	3	4	5	6	7	8	9

Tsfresh's Hyperparameters. The *tsfresh* needs to be tuned too. This is done by adjusting the False Discovery Rate (FDR) level to see how significant an impact this method has on TSPC. FDR is the expected proportion of falsely rejected null hypotheses among all rejected null hypotheses. The tested are tested are: 0.01, 0.05, and 0.1.

3.3 Evaluation Metrics

The model's performance is evaluated with Mean Absolute Error (MAE), and Mean Absolute Percentage Error (MAPE). These metrics have been known and widely accepted in the community and are used as metrics in the popular M4 time series competition challenge [6].

4 Result

The obtained results are presented in Fig. 4. As can be seen, the TSPC method is effective in successfully improving forecasting in horizons, at least greater than 30 steps ahead and beyond. This observation holds for all three datasets tested and kernel size distributions.

Figure 5 also compares the successful post-correction ratio in test sets. As we can see, we can confidently say that after 30 steps, we have an increasing trend in the number of successful post-corrections.

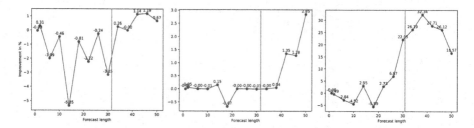

Fig. 4. Forecasting Improvement using TSPC method (Left: Financial Dataset; Middle: Climate Dataset; Right: M4 Dataset)

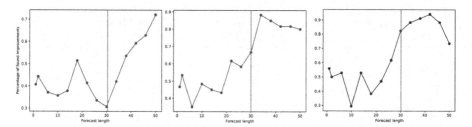

Fig. 5. Successful Ratio of Post-correction in Test set (Left: Financial Dataset; Middle: Climate Dataset; Right: M4 Dataset)

4.1 Sensitivity Analysis

Tsfresh's FDR Level. To see how tsfresh's parameter on FDR-levels tests influences the post-correction, we perform an experiment on the climate dataset.

The first appearance of new features in the meta-learner is seen after forecast length 18, seen in the Fig. 6, and with few found features, the meta-learner performed worse. In cases with extended forecasting range and with more selected features, the *tsfresh* feature extraction was beneficial for the meta-learner (See Fig. 7).

Forecast Ratio. To see how the forecast ratio influences the post-correction, we perform an experiment on the climate dataset. The result is presented in Fig. 8. As can be seen, we observe a more or less similar pattern across different forecast ratios. At least on 30 steps and beyond, TSPC still provides a positive improvement irrespective of the forecast ratio.

4.2 Analysis of the Results

The initial plan was to comprehensively search for the optimal kernel sizes for each forecast length. However, due to the high computational costs associated with this approach, a percent-wise distribution of the kernel and forecast was used instead. It was anticipated that measurement of the DTW distance would

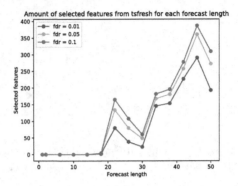

Fig. 6. Amount of selected features with different fdr-levels

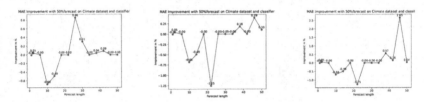

Fig. 7. Effect of FDR-level: left: 0.01, middle: 0.05, right: 0.10

indicate the likelihood of the best match being an improvement, but this assumption did not hold. Therefore, an attempt was made to identify new parameters that would improve the meta-learner's predictive accuracy, leading to the implementation of *tsfresh*. *Tsfresh* effectively extracted relevant features, mainly when working with the M4 dataset. However, when applied to other datasets, the number of extracted features decreased significantly or, in some cases, was nonexistent.

When analyzing the results, the increment of the forecast length decreases the percent-wise distribution of positive corrections, leading to an uneven data distribution during meta-learner training. This holds for all three datasets. However, when analyzing the results from the M4 dataset, its clearly seen that the training/validation and test vary drastically compared to the other datasets. This could indicate that the TSPC method is able to find more corrections when having more data to search through.

4.3   Post-correction Meta-learner

The evaluation of the Random Tree Forest model reveals that the model is not overfitting or underfitting the training data. However, there is a low correlation between the Dynamic Time Warping (DTW) distance of the best match and its ability to outperform the ARIMA. To solve this problem, we try a precision threshold tuning with validation data, resulting in improved results. However,

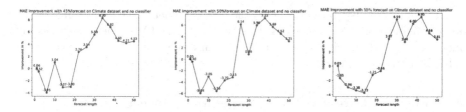

Fig. 8. Effect of forecast ratio: left: 0.45, middle: 0.50, right: 0.55

in some scenarios, the number of corrections must be higher to yield significant results. Another issue observed during the meta-learner phase is that the results are negative when the number of true and false positives are equal. The meta-learner can detect cases where the correction should not occur, resulting in improved results with low forecasting length compared to results where the meta-learner was not used. The results regarding high forecast length show that the meta-learner's predictions have low confidence values, resulting in many missed opportunities for beneficial corrections.

4.4 Persistent Pattern Among Different Datasets

The TSPC method exhibited similar patterns in terms of forecast improvement based on forecast length. The TSPC model was effective in improving longer forecast ranges on all tree datasets but was not so effective on the shorter forecast lengths. This could be because ARIMA is effective in short-term forecasting but falls off as we go on and increase the forecast length. This is where the TSPC method is able to come in and post-correct the inaccurate long-term forecasts generated by the ARIMA, which was the selected forecast model in the experimental setup.

4.5 Improving Original Forecast

Fitting data to the ARIMA model with arbitrary parameters might not always produce an accurate forecast. That is why some experiments were also carried out by an Auto-ARIMA model, which automatically selects the optimal parameters based on the lowest AIC and BIC criteria. Using a model more closely fitted to the data can produce a more accurate forecast, making the TCPS method even more effective. Since Auto-ARIMA's forecast is more accurate when compared to the arbitrary ARIMA, the TSPC model can find a better replacement candidate when performing post-correction, resulting in an overall model improvement.

4.6 Importance of the Distribution Parameter

The results from testing the parameter $\mathcal{D}$ showed expected results, where it is preferred to use more of the historical data when working with longer forecasts and the opposite distribution benefits the TSPC model in the lower range of the forecasting length. The reason might be that the ARIMA is a better forecasting model when working with the near future. Meanwhile, the TCPC can outperform ARIMA on the other end.

4.7   Effect of Tsfresh Features on the Meta-learner

The results of using *tsfresh* to select features indicates that there is an impact from the selected features on the performance of the meta-learner. Since the number of selected features relies on the window size, there are no changes in the results for low forecasting lengths. However, when the number of selected features increases to approximately 150, the meta-learner can find more true positives and improve the results.

5   Related Work

To our knowledge, forecast post-correction is a new problem in time series fore- casting. A similar idea has been proposed for higher-order tensors, particularly for tensor completion problems [3]. However, in this work, naturally, the concept of shape which is only relevant in time series was not considered, and a naive nearest neighbor was proposed to solve the problem. However, this work was the main inspiration for our research. In the time series domain, most related works use nearest neighbor and dynamic time warping in forecasting, and not particularly post-correction of forecasts. However, 1NN-DTW is quite popular in time series classification, and the nearest neighbor idea has already been used for forecasting. The studies presented in this section will discuss similar problems and solutions associated with these lines of research.

In [7], the authors developed a methodology for applying KNN in time series. Several preprocessing techniques are analyzed, which deal with problems such as trends, seasonality, and outliers. Data differencing with one lag is a common technique to remove trends from the time series. Normalization is also applied when dealing with outliers. Also, the authors proposed how to choose the optimal K-value and apply the method of using average results from multiple models with different K-values.

A different kNN-based methodology using weights is applied in [4], which pro- posed a model that uses a weighted Euclidean distance to calculate the similarity between the current input and historical data and then uses a weighted averag- ing method to obtain the final forecast. The authors also propose a method for finding the optimal kernel window m. The method compared the distance with a x datapoint and its candidate neighbor n between $x+1$ and neighbor $n+1$. If the second distance is larger, then x and i are considered false neighbors. Then a parameter m is chosen, which minimizes the number of false neighbors. Finally, authors in [9] developed several algorithms that tackle the problems with the time series prediction task using modified cross-validation to Weighted Near- est Neighbor, mainly computational power when using a sliding kernel that is too large and the number of neighbors. The authors also discuss several data-preprocessing techniques which tackle outliers and time series trends.

6 Conclusion

We propose a post-correction method, called TSPC to improve the long-term forecasting performance of statistical methods. Our findings show that our method is effective in the improvement of the forecast, at least 30 steps ahead and beyond. Interestingly, the pattern of forecast improvement has been observed to be consistent across the three diverse datasets tested in this research. The performance of our proposed method even can become more competitive if the original single-step forecast is more accurate. For instance, we noticed that TSPC is more effective with hyperparameter-tuned ARIMA than its non-tuned counterpart.

7 Future Work

For instance, several techniques could be explored to improve the model's performance in finding the best match. One option is to apply a weighted search, increasing the chances of finding a match that outperforms the original forecast. Although 1NN has been proven to be more accurate in time series, contrary to other problems, it would still be relevant to examine the effectiveness of KNN, which allows the meta-learner to train on multiple best matches. Another direction can be testing different heuristics for determining the kernel size for each forecasting horizon length. Another option for the meta-learner would be to use part of the proposed correction as validation. If the correction fails to improve the forecast in the early stages, then the correction should not be used.

References

1. Bordino, I., Kourtellis, N., Laptev, N., Billawala, Y.: Stock trade volume prediction with yahoo finance user browsing behavior. In: 2014 IEEE 30th International Conference on Data Engineering, pp. 1168–1173. IEEE (2014)
2. Christ, M., Braun, N., Neuffer, J., Kempa-Liehr, A.W.: Time series feature extraction on basis of scalable hypothesis tests (tsfresh-a python package). Neurocomputing **307**, 72–77 (2018)
3. Fanaee-T, H.: Tensor completion post-correction. In: Bouadi, T., Fromont, E., Hüllermeier, E. (eds.) IDA 2022. LNCS, vol. 13205, pp. 89–101. Springer, Cham (2022). https://doi.org/10.1007/978-3-031-01333-1_8
4. Lora, A.T., Santos, J.M.R., Expósito, A.G., Ramos, J.L.M., Santos, J.C.R.: Electricity market price forecasting based on weighted nearest neighbors techniques. IEEE Trans. Power Syst. **22**(3), 1294–1301 (2007)
5. Makridakis, S., Spiliotis, E., Assimakopoulos, V.: The m4 competition: results, findings, conclusion and way forward. Int. J. Forecast. **34**(4), 802–808 (2018)
6. Makridakis, S., Spiliotis, E., Assimakopoulos, V.: The M4 competition: 100,000 time series and 61 forecasting methods. Int. J. Forecast. **36**(1), 54–74 (2020)

7. Martínez, F., Frías, M.P., Pérez, M.D., Rivera, A.J.: A methodology for applying k-nearest neighbor to time series forecasting. Artif. Intell. Rev. **52**(3) (2019)
8. Sumanthvrao: Daily climate time series data. Kaggle (2019)
9. Tajmouati, S., Wahbi, B.E., Bedoui, A., Abarda, A., Dakkoun, M.: Applying k-nearest neighbors to time series forecasting: two new approaches. arXiv preprint arXiv:2103.14200 (2021)

Explainable Deep Learning-Based Solar Flare Prediction with Post Hoc Attention for Operational Forecasting

Chetraj Pandey[1]([⊠])(iD), Rafal A. Angryk[1](iD), Manolis K. Georgoulis[2](iD), and Berkay Aydin[1](iD)

[1] Georgia State University, Atlanta, GA, USA
{cpandey1,rangryk,baydin2}@gsu.edu
[2] Research Center for Astronomy and Applied Mathematics, Academy of Athens, Athens, Greece
manolis.georgoulis@academyofathens.gr

Abstract. This paper presents a post hoc analysis of a deep learning-based full-disk solar flare prediction model. We used hourly full-disk line-of-sight magnetogram images and selected binary prediction mode to predict the occurrence of $\geq$M1.0-class flares within 24 h. We leveraged custom data augmentation and sample weighting to counter the inherent class-imbalance problem and used true skill statistic and Heidke skill score as evaluation metrics. Recent advancements in gradient-based attention methods allow us to interpret models by sending gradient signals to assign the burden of the decision on the input features. We interpret our model using three post hoc attention methods: (i) Guided Gradient-weighted Class Activation Mapping, (ii) Deep Shapley Additive Explanations, and (iii) Integrated Gradients. Our analysis shows that full-disk predictions of solar flares align with characteristics related to the active regions. The key findings of this study are: (1) We demonstrate that our full disk model can tangibly locate and predict near-limb solar flares, which is a critical feature for operational flare forecasting, (2) Our candidate model achieves an average TSS=0.51±0.05 and HSS=0.38±0.08, and (3) Our evaluation suggests that these models can learn conspicuous features corresponding to active regions from full-disk magnetograms.

Keywords: Solar flares · Deep learning · xAI · Interpretability

1 Introduction

Solar flares are transient solar events of central importance to space weather forecasting, manifested as the sudden large eruption of electromagnetic radiation on the outermost atmosphere of the Sun. They are classified according to their peak X-ray flux level into the following five categories by National Oceanic and Atmospheric Administration (NOAA): X ($\geq 10^{-4} Wm^{-2}$), M

© The Author(s), under exclusive license to Springer Nature Switzerland AG 2023
A. Bifet et al. (Eds.): DS 2023, LNAI 14276, pp. 567–581, 2023.
https://doi.org/10.1007/978-3-031-45275-8_38

($\geq 10^{-5} Wm^{-2}$), C ($\geq 10^{-6} Wm^{-2}$), B ($\geq 10^{-7} Wm^{-2}$), and A ($\geq 10^{-8} Wm^{-2}$), where, X>M>C>B>A [6]. These flare classes are on a logarithmic scale, meaning that each class represents a tenfold increase in X-ray flux compared to the previous class. Large flares (M- and X-class) are scarce events that are more likely to incur a terrestrial impact and, therefore, the classes of interest that gather the attention of researchers. These flares may potentially disrupt the electricity supply chain, airline industry, and satellite communications, and pose radiation hazards to astronauts in space. To mitigate these risks, the necessity of a precise and reliable flare prediction model becomes imperative.

Active regions (ARs) on the Sun are places characterized by the largest accumulations of dipolar magnetic flux in the solar atmosphere. Most operational flare forecasts target these regions of interest and issue predictions for individual ARs, which are the main initiators of space weather events. To issue a full-disk forecast with an AR-based model, the output flare probabilities for each active region are usually aggregated using a heuristic function as mentioned in [20]. The heuristic function used to aggregate the final forecast operates under the assumption of conditional independence among ARs and that all ARs contribute equally to the aggregate forecast. This uniform weighting scheme may not accurately reflect the true influence of each AR on full-disk flare prediction probability. It is important to highlight that the weights of these ARs are generally unknown; there are no established methods to accurately determine them, nor are there any prior assumptions that guide the assignment of these weights.

Furthermore, the magnetic field measurements, employed by the AR-based forecasting techniques, are susceptible to severe projection effects as ARs get closer to limbs (to the degree that after $\pm 60°$ the magnetic field readings are distorted [5]); therefore, the aggregated full-disk flare probability is in fact, restrictive (i.e., from ARs in central locations) as the data in itself is limited to ARs located within $\pm 45°$ [11] to $\pm 70°$ [9] and in some cases, even $\pm 30°$ [8] due to severe projection effects [7]. As AR-based models include data up to $\pm 70°$, in the context of this paper, this upper limit ($\pm 70°$) is used as a boundary for central (within $\pm 70°$) and near-limb regions (beyond $\pm 70°$).

In contrast to AR-based models, which use individual AR data from central locations, full-disk models use complete magnetogram images corresponding to the entire disk. These images are typically compressed JP2 (JPEG 2000) 8-bit representations (i.e., pixel values ranging from 0 to 255) derived from original magnetogram rasters which contain magnetic field strength values ranging from $\sim \pm 4500G$. The compressed magnetogram images are used for shape-based parameters, e.g., size, directionality, borders, and inversion lines. Although projection effects still prevail in these images, full-disk models can learn from the near-limb areas. Thus, incorporating a full-disk model is essential to supplement AR-based models, enabling the prediction of flares in the Sun's near-limb areas and enhancing operational flare forecasting systems.

With recent advancements in machine learning and deep learning methods, their application in predicting solar flares has demonstrated great experimental success and accelerated the efforts of many interdisciplinary researchers

[8,11,15–19,32]. Although deep learning methods have significantly enhanced solutions to image classification and computer vision problems, these models learn highly complex data representations, rendering them as black-box models. Consequently, the decision-making process within these models remains obscured, presenting a critical challenge for operational forecasting communities that rely on transparency to make informed decisions. Recently several empirical methods have been developed to explain and interpret the decisions made by deep neural networks. These are post hoc analysis methods (attribution methods) [12], meaning they focus on the analysis of trained models and do not contribute to the models' parameters while training. In this work, we primarily focus on developing a CNN-based full-disk model for solar flare prediction of ≥M1.0-class flares and evaluate and explain our model's performance by using three of the attribution methods: (i) Guided Gradient-weighted Class Activation Mapping (Guided Grad-CAM) [25], (ii) Deep Shapley Additive Explanations (Deep SHAP) [13], and (iii) Integrated Gradients (IG) [31]. More specifically, we show that our model's decisions are based on the characteristics corresponding to ARs, and our models can tackle the flares appearing on near-limb regions.

The rest of this paper is organized as follows. In Sect. 2, we present the related work on flare forecasting. In Sect. 3, we present our methodology with data preparation and model architecture. In Sect. 4 we provide the description of all three post hoc explanation methods utilized in this work. In Sect. 5, we present our experimental settings, and model evaluation, and discuss the interpretation of our models, and in Sect. 6, we present our conclusions and future work.

2   Related Work

There have been several attempts to predict solar flares using machine learning and deep learning models. A multi-layer perceptron-based model was applied for solar flare prediction of ≥C1.0- and ≥M1.0-class flares in [15] by utilizing 79 manually selected physical precursors extracted from multi-modal solar observations. A CNN-based flare forecasting model trained with solar AR patches extracted from line-of-sight (LoS) magnetograms within ±30° of the central meridian to predict ≥C1.0-, ≥M1.0-, and ≥X1.0-class flares was presented in [8]. Similarly, [11] also used a CNN-based model to issue binary class predictions for both ≥C1.0- and ≥M1.0-class flares within 24 h using AR patches located within ±45° of the central meridian. It is important to note that both of these models [8,11] are limited to a small portion of the observable disk in central locations (within ±30° to ±45°) and thus possess the limited operational capability.

More recently, we presented a deep learning-based binary full-disk flare prediction model to predict ≥M1.0-class flares in [17] and to predict ≥C4.0- and ≥M1.0-class flares in [18] using bi-daily observations (i.e., two magnetograms per day) of full-disk LoS magnetograms. It is important to note that in [18] all the instances that fall between the ≥C4.0- and ≥M1.0-class flares were excluded in both training and validation sets. These particular sets of instances lie on the border of two binary outcomes and can be considered the harder-to-predict instances. These models are still black-box and do not provide explanations on any global and local variable importance. These explanations are important to understand the capabilities of full-disk models in near-limb regions and improve their trustworthiness in operational settings. In solar flare prediction, [2] used an occlusion-based method to interpret a CNN-based solar flare prediction model trained with AR patches. Similarly, [33] presented a deep learning-based flare prediction model for predicting C-, M-, and X-class flares and provided visual explanations using Grad-CAM [25], and Guided Backpropagation [28]. They used daily observations of solar full-disk LoS magnetograms at 00:00 UT, and their models show limitations for the near-limb flares. Moreover, in [30], DeepLIFT [27] and IG [31] were evaluated for explaining CNN-based flare prediction model trained using tracked AR patches within ±70°.

This paper presents a CNN-based model to predict ≥M1.0-class flares, trained with full-disk LoS magnetograms images. The novel contributions of this paper are as follows: (i) We show an improved overall performance of a full-disk solar flare prediction model, (ii) We utilized contemporary attribution methods to explain and interpret the decisions of our deep learning model, and (iii) More importantly, we show that our models can predict solar flares appearing on difficult-to-predict near-limb regions of the Sun.

3   Data and Model

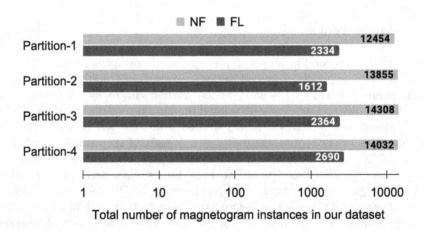

Fig. 1. Data distribution used in this study with four tri-monthly partitions for training ≥M1.0-class flare prediction models. Note: The length of the bars is in logarithmic scale.

We used full-disk LoS solar magnetograms obtained from the Helioseismic and Magnetic Imager (HMI) [24] instrument onboard Solar Dynamics Observatory (SDO) [21] available as compressed JP2 images in near real-time publicly via

Helioviewer[1]. To enhance computational efficiency for training the deep learning model, these compressed images are resized to a smaller resolution of 512×512 pixels. We sampled hourly instances of magnetogram images at [00:00, 01:00, ..., 23:00] each day from Dec 2010 to Dec 2018. We labeled our data with a prediction window of 24 h. The images are labeled based on the maximum peak X-ray flux (converted to NOAA flare classes) within the next 24 h. We collect a total of 63,649 images and label them such that if the maximum X-ray intensity of flare is weaker than M1.0, the observations are labeled as "No Flare" (NF: <M1.0) and ≥M1.0 ones are labeled as "Flare" (FL: ≥M1.0). This results in 54,649 instances for the NF-class and 9,000 instances (8,120 instances of M-class and 880 instances of X-class flares) for the FL-class.

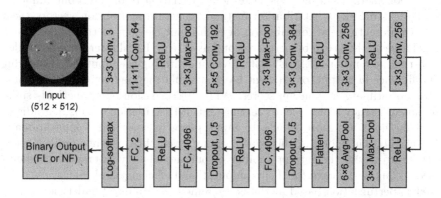

Fig. 2. The architecture of our full-disk flare prediction model.

We finally split our data into four temporally non-overlapping tri-monthly partitions for the cross-validation experiments. This partitioning of the dataset is created by dividing the data timeline from Dec 2010 to Dec 2018 into four partitions, where Partition-1 contains data from Jan to Mar, Partition-2 contains data from Apr to Jun, Partition-3 contains data from Jul to Sep, and finally, Partition-4 contains data from Oct to Dec as shown in Fig. 1. As a result of the infrequent occurrence of ≥M1.0-class flares, the dataset exhibits a significant imbalance, with the ratio of FL to NF class being approximately 1:6.

In this work, we extend the AlexNet [10] model by concatenating a convolutional layer at the beginning of the network to make use of the pre-trained weights for our 1-channel input magnetogram images as the pre-trained model requires a 3-channel image as input to the network. Our added convolutional layer uses a 3×3 kernel, size-1 stride, and outputs a 3-channel feature map which is then integrated into the standard AlexNet architecture as shown in Fig. 2. Furthermore, to efficiently utilize the pre-trained weights regardless of the architecture of the AlexNet model, which expects 224×224, 3-channel image

---
[1] Helioviewer: https://api.helioviewer.org

as input, we use the adaptive average pooling after feature extraction before the fully-connected layer to match the dimension on our 1-channel, 512×512 magnetogram image. Overall, our model has six convolutional layers, three max-pool layers, one average-pool layer, and two fully-connected layers.

4   Interpretation Methods

Deep learning models are often deemed black-box due to their complex representations, resulting in interpretability, transparency, and consistency challenges concerning the patterns they learn [12]. To address this, various methods [34] have been proposed to interpret CNNs. One common approach is using attribution methods, which visualize how specific parts of the input influence the model's decisions. Attribution methods generate attribution vectors (heat maps) representing the contribution of each input element to the model's decision. These methods can be perturbation-based (e.g., Local Interpretable Model-Agnostic Explanations (LIME) [23]), involving altering the input and measuring the difference in output, or gradient-based, calculating gradients via backpropagation to estimate attribution scores. While perturbation-based methods suffer from inconsistency issues due to creating Out-of-Distribution data [22], gradient-based methods are more robust to input perturbations and computationally efficient [14]. Therefore, in this work, we employed three recent gradient-based methods to assess the interpretability of our models. By leveraging gradient-based techniques, known for their computational efficiency and robustness compared to perturbation-based methods, we aimed to visualize the decisions made by our model and gain insights into the specific characteristics in a magnetogram image that trigger the models' decisions. These methods allowed us to cross-validate and ensure the consistency of the explanations provided by our models, contributing to a more reliable and robust interpretation.

Guided Grad-CAM: The Guided Gradient-weighted Class Activation Mapping (Guided Grad-CAM) method [25] leverages the benefits of the Grad-CAM and guided backpropagation [28]. Grad-CAM is a model-agnostic method that uses the class-specific gradient information flowing into the final convolutional layer of a CNN to produce a coarse localization map of the important regions in the image. Guided Backpropagation is based on the premise that the neurons act as detectors of certain image features, so it computes the gradient of the output with respect to the input, except that when propagating through ReLU functions, it only backpropagates the non-negative gradients and highlights the pixels that are important in the image. Attributions from Grad-CAM are class-discriminative and localize relevant image regions; however, do not highlight the fine-grained pixel importance as guided backpropagation [3]. Guided Grad-CAM combines the fine-grained details of guided backpropagation with the course localization advantages of Grad-CAM and is computed as the element-wise product of guided backpropagation with the upsampled Grad-CAM attributions.

Deep SHAP: SHAP values (SHapley Additive exPlanations) [13] is a method based on cooperative game theory [26] and used to increase the transparency and interpretability of machine learning models. SHAP shows the contribution of each feature to the prediction of the model, it does not evaluate the quality of the prediction itself. The contribution of each feature is calculated using cooperative game theory and Shapley values to assess how much each feature adds to the difference between the actual prediction and the average prediction. For deep-learning models, Deep SHAP [13] is considered an enhanced version of the DeepLIFT algorithm [27], where we approximate the conditional expectations of SHAP values using a selection of baseline samples from the dataset. The baselines typically contain a set of representative samples from the same distribution as the input data. For each input sample, it computes DeepLIFT attribution with respect to each baseline and averages resulting attributions. This method assumes that input features are independent of one another, and the explanations are modeled through the additive composition of feature effects.

Integrated Gradients: The last method we will analyze in this study is Integrated Gradients (IG) [31], which quantifies feature attributions by integrating the gradients of the model's output along a straight-line path from a baseline reference to the input feature under consideration. This method requires an extra input as the baseline, representing the non-appearance of the feature in the original image which is typically an all-zero vector. IG is favored for its completeness property, where the sum of integrated gradients for all features precisely equals the difference between the model's output for the given input and the baseline input values. This property ensures that the feature attributions accurately represent each feature's individual contribution to the model output, allowing us to reliably recover the model's output value by summing these contributions [29].

5   Experimental Evaluation

5.1   Experimental Settings

We trained a full-disk flare prediction model with stochastic gradient descent (SGD) as an optimizer and negative log-likelihood (NLL) as the objective function. Our model is initialized with pre-trained weights of AlexNet Model [10], and then we make use of a dynamic learning rate (initialized at 0.0099 and reduced 5%) to further train the model to 40 epochs with a batch size of 64. We address the class-imbalance issue using data augmentation and class weights to the loss function. We use three augmentation techniques: vertical flipping, horizontal flipping, and $+5°$ to $-5°$ rotations. We augment the data for both classes (where the entire FL-class data are augmented three times with three augmentation techniques and NF-class is augmented once randomly). We then adjust class weights inversely proportional to the class frequencies after augmentations. The use of class weights penalizes the misclassification made in the minority class. Our models are trained as 4-fold cross-validation experiments with each fold representing a different partition serving as the test set. Specifically, Fold-1 corresponds to Partition-1, Fold-2 corresponds to Partition-2, and so on.

We evaluate the performance of our models using two widely-used forecast skills scores: True Skill Statistics (TSS, in Eq. 1) and Heidke Skill Score (HSS, in Eq. 2), derived from the elements of confusion matrix: True Positives (TP), True Negatives (TN), False Positives (FP), and False Negatives (FN). In the context of our paper, the FL class is the positive outcome and NF is the negative.

$$TSS = \frac{TP}{TP + FN} - \frac{FP}{FP + TN} \tag{1}$$

$$HSS = 2 \times \frac{TP \times TN - FN \times FP}{((P \times (FN + TN) + (TP + FP) \times N))}, \tag{2}$$

where $N = TN + FP$ and $P = TP + FN$.

$$Recall = \frac{TP}{TP + FN} \tag{3}$$

TSS and HSS values range from -1 to 1, where 1 indicates all correct predictions, -1 represents all incorrect predictions, and 0 represents no skill. In contrast to TSS, HSS is an imbalance-aware metric, and it is common practice to use HSS in combination with TSS for the solar flare prediction models due to the high class-imbalance ratio present in the datasets. For a balanced dataset, these metrics are equivalent [1]. In solar flare prediction, TSS and HSS are the preferred choices of evaluation metrics compared to commonly used metrics in image classification (e.g., accuracy) as they ensure a comprehensive and reliable evaluation of predictive capabilities, especially in scenarios with imbalanced class distributions. Lastly, we report the subclass and overall recall (shown in Eq. 3) for flaring instances (M- and X-class) to assess the prediction sensitivity of our models in central and near-limb regions. To reproduce this work, the source code and experimental results can be accessed from our open-source repository [4].

5.2   Model Evaluation

Our models have on average TSS~0.51±0.05 and HSS~0.38±0.08, which improves over the performance of [17] by ~4% in terms of TSS (reported 0.47±0.06) and by ~3% in terms of HSS (reported 0.35±0.05) [2]. The detailed experimental results for each fold are shown in Table 1.

In addition, we evaluate our results for correctly predicted and missed flare counts for class-specific flares (X-class and M-class) in central locations (within ±70°) and near-limb locations (beyond ±70°) of the Sun as shown in Table 2. We observe that our models made correct predictions for ~95% of the X-class flares

---

[2] While there are several other works (mentioned in Sect. 2) in solar flare prediction, the results of these models are not directly comparable since they employ different datasets, data timelines, and data partitioning strategies.

Table 1. A comprehensive overview of 4-fold cross-validation experiments, showing all the four outcomes of confusion matrices (TP, FP, TN, FN) evaluated on the test sets, and performance of our models in terms of two skill scores (TSS and HSS).

Folds	TP	FP	TN	FN	TSS	HSS
Fold-1	1,720	1,943	10,511	614	0.58	0.47
Fold-2	1,155	3,083	10,772	457	0.49	0.29
Fold-3	1,585	2,668	11,640	779	0.48	0.36
Fold-4	1,706	2,241	11,791	984	0.47	0.40
Aggregated	6,166	9,935	44,714	2,834	**0.51±0.05**	**0.38±0.08**

Table 2. Counts of correctly (TP) and incorrectly (FN) classified X- and M-class flares in central ($|longitude| \leq \pm 70°$) and near-limb locations. The recall across different location groups is also presented. Counts are aggregated across folds.

	Within ±70°			Beyond ±70°		
Flare-Class	TP	FN	Recall	TP	FN	Recall
X-Class	637	31	0.95	157	55	0.74
M-Class	4,229	1,601	0.73	1,143	1,147	0.50
Total (X&M)	4,866	1,632	0.75	1,300	1,202	0.52

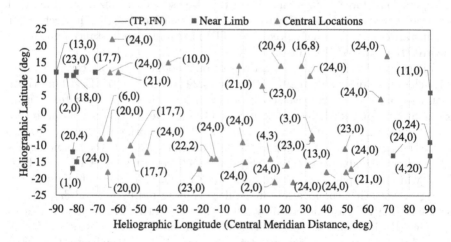

Fig. 3. A scatterplot to quantify the performance of our models in terms of True Positives (TP) and False Negatives (FN) for X-class flares grouped by flare locations. The flare events beyond ±70° longitude are represented as near-limb events. Note: (i) Red marker is for locations with zero TP. (ii) For some locations, TP+FN<24, given that we used hourly instances, is due to the unavailable instances from the source.

and ∼73% of the M-class flares in central locations. Similarly, our models show a compelling performance for flares appearing on near-limb locations of the Sun, where ∼74% of the X-class and ∼50% of the M-class flares are predicted correctly.

This is important because, to our knowledge, the prediction of near-limb flares is often overlooked. More false positives in M-class are expected because of the model's inability to distinguish bordering class flares (C4.0 to C9.9) from ≥M1.0-class flares, which we have observed empirically in our prior work [18] as well. Overall, we observed that ∼90% and ∼66% of the X-class and M-class flares, respectively, are predicted correctly by our models.

Furthermore, given that we sample our data with a 1-hour cadence resulting in 24 instances per day unless there are gaps due to unavailable data instances, any given flare instance is expected to be in the prediction window of 24 instances. X-class flares are relatively large flares that often dominate the prediction window. Therefore, we analyzed the predictions on X-class flares and observed that from a total of 45 X-class flare locations, our models correctly predict the occurrence of a flare at least once for 44 of them, as shown in Fig. 3. In particular, we show that the full-disk model presented in this paper can predict flares appearing on near-limb locations of the Sun at great accuracy, which provides a crucial addition to operational flare forecasting systems.

5.3 Model Interpretation

In this section, we present a case study, interpreting the visual explanations generated by our model, and also discuss the implications of these explanations in the operational forecasting scenario. For this, we use the visualizations generated using all three post hoc explanation methods mentioned earlier in Sect. 4 for two instances: (i) a correctly predicted (TP) near-limb flare instance and (ii) an incorrectly predicted (FP) instance.

Firstly, we interpret the predictions of our model for a correctly predicted X1.4-class flare observed on 2011-09-22 at 10:29:00 UTC on the East limb (note that East and West are reversed in solar coordinates). We generate a visual explanation using all three attribution methods. We utilized an input image from 2011-09-22 05:00:00 UTC (approximately 5.5 h prior to the flare event) where the sunspot corresponding to the flare becomes visible in the magnetogram image. Interestingly, we observed that the pixels covering the AR on the East limb, which is responsible for the eventual X1.4 flare, are activated, as shown in Fig. 4. Note that the location of the flare is indicated by a green flag and all visible NOAA ARs are indicated by red flags in Fig. 4 (b). The model focuses on specific ARs, including the relatively smaller AR on the East limb, even though other ARs are present in the magnetogram image. The visualization of attribution maps suggests that, for this particular prediction, the region responsible for the flare event is attributed as important, contributing to the consequent decision. This finding is consistent across all three methods, corroborating the explanation's reliability. However, Guided Grad-CAM and Deep SHAP provide finer details by suppressing noise compared to IG.

Similarly, to analyze a false positive case, we present an example of a C7.1 flare observed on 2014-01-06 at 00:08:00 UTC. To explain the result, we used an input magnetogram instance from 2014-01-05 06:00:00 UTC (∼18 h prior to the event). The model's prediction probability for this instance being an

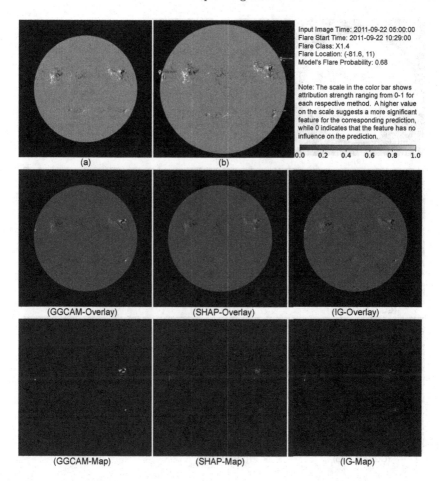

Fig. 4. A visual explanation for a correctly predicted near-limb FL-class instance. (a) Actual magnetogram from the dataset used as the input image. (b) Annotated full-disk magnetogram at flare start time, showing flare location (green flag) and NOAA ARs (red flags). Overlays (GGCAM, SHAP, IG) depict the input image overlayed with attributions, and Maps (GGCAM, SHAP, IG) showcase the attribution maps obtained from Guided Grad-CAM, Deep SHAP, and Integrated Gradients respectively. (Color figure online)

FL-class is ~0.97. Therefore, we seek a visual explanation of this prediction using all three interpretation methods. Upon analysis, we observed that the prediction mainly relies on only one AR, which indeed corresponds to the location of the eventual C7.1 flare (indicated by the green flag) when visualized with all three attribution methods, as shown in Fig. 5. This incorrect prediction can be attributed to the interference of the bordering class flares mentioned in [18]. Such interference poses a problem for binary flare prediction models. We noticed that out of 25,150 C-class flares, 9,240 flares led to incorrect predictions, accounting for approximately 37% of the total C-class flares in our dataset.

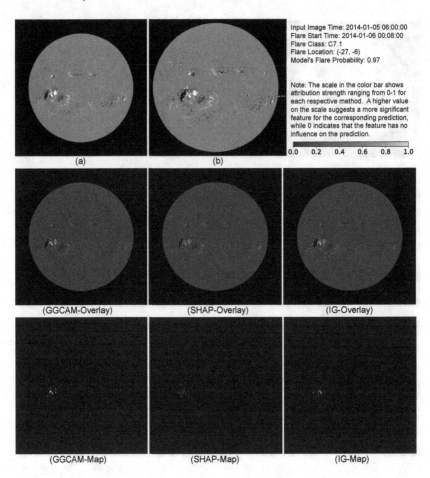

Fig. 5. A visual explanation for an incorrectly predicted NF-class instance. (a) Actual magnetogram from the dataset used as the input image. (b) Annotated full-disk magnetogram at flare start time, showing flare location (green flag) and NOAA ARs (red flags). Overlays (GGCAM, SHAP, IG) depict the input image overlayed with attributions, and Maps (GGCAM, SHAP, IG) showcase the attribution maps obtained from Guided Grad-CAM, Deep SHAP, and Integrated Gradients respectively. (Color figure online)

These two examples, although not exhaustive, carry significant implications for operational forecasting systems. By incorporating visual explanations into the forecasting process, in addition to providing a full-disk flare prediction probability, we have the capability to identify potential flare event locations among all visible ARs precisely. This is invaluable for improving the accuracy and reliability of solar flare forecasts, aiding in effective risk assessment and mitigation strategies. Furthermore, it provides a deeper understanding of the underlying factors contributing to flare occurrences, empowering researchers and space weather experts to make more informed decisions and take timely actions to safeguard critical infrastructure and space assets.

6 Conclusion and Future Work

In this work, we used three recent gradient-based methods to interpret the predictions of our AlexNet-based binary flare prediction model trained for the prediction of ≥M1.0-class flares. We addressed the highly overlooked problem of flares appearing in near-limb locations of the Sun, and our model shows a compelling performance for such events. Furthermore, we evaluated our model's predictions with visual explanations, showing that the decisions are primarily capturing characteristics corresponding to the active regions in the magnetogram instance. Although our model shows improved capability, still suffers from high false positives attributed to high C-class flares. As an extension, we plan to study the individual class characteristics to obtain a better way of segregating these flare classes considering the background flux and generate a new set of labels that can better address the issue with border class flares. Furthermore, at this point, the models are only looking at the spatial patterns in our data, and we intend to widen this work toward spatiotemporal models to improve the performance.

Acknowledgement. This project is supported in part under two NSF awards #2104004 and #1931555, jointly by the Office of Advanced Cyberinfrastructure within the Directorate for Computer and Information Science and Engineering, the Division of Astronomical Sciences within the Directorate for Mathematical and Physical Sciences, and the Solar Terrestrial Physics Program and the Division of Integrative and Collaborative Education and Research within the Directorate for Geosciences. This work is also partially supported by the National Aeronautics and Space Administration (NASA) grant award #80NSSC22K0272. Data used in this study is a courtesy of NASA/SDO and the AIA, EVE, and HMI science teams and NOAA National Geophysical Data Center (NGDC).

References

1. Ahmadzadeh, A., Aydin, B., Georgoulis, M., Kempton, D., Mahajan, S., Angryk, R.: How to train your flare prediction model: revisiting robust sampling of rare events. APJ Suppl. Ser. **254**(2), 23 (2021)
2. Bhattacharjee, S., Alshehhi, R., Dhuri, D.B., Hanasoge, S.M.: Supervised convolutional neural networks for classification of flaring and nonflaring active regions using line-of-sight magnetograms. APJ **898**(2), 98 (2020)
3. Chattopadhay, A., Sarkar, A., Howlader, P., Balasubramanian, V.N.: Gradcam++: Generalized gradient-based visual explanations for deep convolutional networks. In: 2018 IEEE Winter Conference on Applications of Computer Vision (WACV). IEEE (2018). https://doi.org/10.1109/wacv.2018.00097
4. DMLab: Source Code. https://bitbucket.org/gsudmlab/explainingfulldisk/src/main/
5. Falconer, D.A., Tiwari, S.K., Moore, R.L., Khazanov, I.: A new method to quantify and reduce the net projection error in whole-solar-active-region parameters measured from vector magnetograms. APJ **833**(2), L31 (2016)
6. Fletcher, L., et al.: An observational overview of solar flares. Space Sci. Rev. **159**(1–4), 19–106 (2011)

7. Hoeksema, J.T., et al.: The helioseismic and magnetic imager (HMI) vector magnetic field pipeline: overview and performance. Sol. Phys. **289**(9), 3483–3530 (2014)

8. Huang, X., Wang, H., Xu, L., Liu, J., Li, R., Dai, X.: Deep learning based solar flare forecasting model. I. results for line-of-sight magnetograms. APJ **856**(1), 7 (2018). https://doi.org/10.3847/1538-4357/aaae00

9. Ji, A., Aydin, B., Georgoulis, M.K., Angryk, R.: All-clear flare prediction using interval-based time series classifiers. In: 2020 IEEE International Conference on Big Data (Big Data), pp. 4218–4225. IEEE (2020)

10. Krizhevsky, A.: One weird trick for parallelizing convolutional neural networks (2014)

11. Li, X., Zheng, Y., Wang, X., Wang, L.: Predicting solar flares using a novel deep convolutional neural network. APJ **891**(1), 10 (2020)

12. Linardatos, P., Papastefanopoulos, V., Kotsiantis, S.: Explainable AI: a review of machine learning interpretability methods. Entropy **23**(1), 18 (2020)

13. Lundberg, S.M., Lee, S.I.: A unified approach to interpreting model predictions. In: Proceedings of the 31st International Conference on Neural Information Processing Systems. pp. 4768–4777. NIPS 2017, Curran Associates Inc., Red Hook, NY, USA (2017)

14. Nielsen, I.E., Dera, D., Rasool, G., Ramachandran, R.P., Bouaynaya, N.C.: Robust explainability: a tutorial on gradient-based attribution methods for deep neural networks. IEEE Signal Process. Mag. **39**(4), 73–84 (2022)

15. Nishizuka, N., Sugiura, K., Kubo, Y., Den, M., Ishii, M.: Deep flare net (DeFN) model for solar flare prediction. APJ **858**(2), 113 (2018)

16. Nishizuka, N., Sugiura, K., Kubo, Y., Den, M., Watari, S., Ishii, M.: Solar flare prediction model with three machine-learning algorithms using ultraviolet brightening and vector magnetograms. APJ **835**(2), 156 (2017)

17. Pandey, C., Angryk, R.A., Aydin, B.: Solar flare forecasting with deep neural networks using compressed full-disk HMI magnetograms. In: 2021 IEEE International Conference on Big Data (Big Data), pp. 1725–1730. IEEE (2021). https://doi.org/10.1109/bigdata52589.2021.9671322

18. Pandey, C., Angryk, R.A., Aydin, B.: Deep neural networks based solar flare prediction using compressed full-disk line-of-sight magnetograms. In: Lossio-Ventura, J.A., et al. (eds.) Information Management and Big Data, SIMBig 2021. Communications in Computer and Information Science, vol. 1577, pp. 380–396. Springer, Cham (2022). https://doi.org/10.1007/978-3-031-04447-2_26

19. Pandey, C., Angryk, R.A., Aydin, B.: Explaining full-disk deep learning model for solar flare prediction using attribution methods (2023). https://arxiv.org/abs/2307.15878

20. Pandey, C., Ji, A., Angryk, R.A., Georgoulis, M.K., Aydin, B.: Towards coupling full-disk and active region-based flare prediction for operational space weather forecasting. Front. Astron. Space Sci. **9**, 897301 (2022). https://doi.org/10.3389/fspas.2022.897301

21. Pesnell, W., Thompson, B.J., Chamberlin, P.C.: The solar dynamics observatory (SDO). Sol. Phys. **275**(1–2), 3–15 (2011)

22. Qiu, L., et al.: Generating perturbation-based explanations with robustness to out-of-distribution data. In: Proceedings of the ACM Web Conference 2022. ACM (2022). https://doi.org/10.1145/3485447.3512254

23. Ribeiro, M.T., Singh, S., Guestrin, C.: "why should i trust you?". In: Proceedings of the 22nd ACM SIGKDD International Conference on Knowledge Discovery and Data Mining. ACM (2016). https://doi.org/10.1145/2939672.2939778

24. Schou, J., et al.: Design and ground calibration of the helioseismic and magnetic imager (HMI) instrument on the solar dynamics observatory (SDO). Sol. Phys. **275**(1–2), 229–259 (2011)
25. Selvaraju, R.R., Cogswell, M., Das, A., Vedantam, R., Parikh, D., Batra, D.: Grad-CAM: visual explanations from deep networks via gradient-based localization. In: 2017 IEEE International Conference on Computer Vision (ICCV). IEEE (2017). https://doi.org/10.1109/iccv.2017.74
26. Shapley, L.: A Value for N-Person Games. RAND Corporation, Santa Monica (1952)
27. Shrikumar, A., Greenside, P., Kundaje, A.: Learning important features through propagating activation differences (2019)
28. Springenberg, J.T., Dosovitskiy, A., Brox, T., Riedmiller, M.: Striving for simplicity: the all convolutional net (2014). https://arxiv.org/abs/1412.6806
29. Sturmfels, P., Lundberg, S., Lee, S.I.: Visualizing the impact of feature attribution baselines. Distill **5**(1), e22 (2020). https://doi.org/10.23915/distill.00022 https://doi.org/10.23915/distill.00022
30. Sun, Z., et al.: Predicting solar flares using CNN and LSTM on two solar cycles of active region data. APJ **931**(2), 163 (2022). https://doi.org/10.3847/1538-4357/ac64a6
31. Sundararajan, M., Taly, A., Yan, Q.: Axiomatic attribution for deep networks (2017). https://arxiv.org/abs/1703.01365
32. Whitman, K., et al.: Review of solar energetic particle models. Adv. Space Res. (2022). https://doi.org/10.1016/j.asr.2022.08.006
33. Yi, K., Moon, Y.J., Lim, D., Park, E., Lee, H.: Visual explanation of a deep learning solar flare forecast model and its relationship to physical parameters. APJ **910**(1), 8 (2021). https://doi.org/10.3847/1538-4357/abdebe
34. Zhou, J., Gandomi, A.H., Chen, F., Holzinger, A.: Evaluating the quality of machine learning explanations: a survey on methods and metrics. Electronics **10**(5), 593 (2021). https://doi.org/10.3390/electronics10050593

Pseudo Session-Based Recommendation with Hierarchical Embedding and Session Attributes

Yuta Sumiya[1]($\boxtimes$)(ID), Ryusei Numata[2], and Satoshi Takahashi[1](ID)

[1] The University of Electro-Communications, Tokyo, Japan
{`sumiya,stakahashi`}`@uec.ac.jp`
[2] The Japan Research Institute Limited, Tokyo, Japan
`numata.ryusei@jri.co.jp`

Abstract. Recently, electronic commerce (EC) websites have been unable to provide an identification number (user ID) for each transaction data entry because of privacy issues. Because most recommendation methods assume that all data are assigned a user ID, they cannot be applied to the data without user IDs. Recently, session-based recommendation (SBR) based on session information, which is short-term behavioral information of users, has been studied. A general SBR uses only information about the item of interest to make a recommendation (e.g., item ID for an EC site). Particularly in the case of EC sites, the data recorded include the name of the item being purchased, the price of the item, the category hierarchy, and the gender and region of the user. In this study, we define a pseudo-session for the purchase history data of an EC site without user IDs and session IDs. Finally, we propose an SBR with a co-guided heterogeneous hypergraph and globalgraph network plus, called CoHHGN+. The results show that our CoHHGN+ can recommend items with higher performance than other methods.

Keywords: Session-Based Recommendation · Pseudo Session ID · Session information · Auxiliary information · Heterogeneous hypergraph network · Global Graph · Co-guided Learning

1 Introduction

In electronic commerce (EC) markets, the effective recommendation of items and services based on individual user preferences and interests is an important factor in improving customer satisfaction and sales, and several previous studies have focused on recommendation systems. A recommendation system is a technology that suggests items based on a user's past actions and online behavior. However, in recent years, user IDs are not assigned to users to protect their privacy. Under such circumstances, it is difficult to identify users; therefore, conventional effective recommendation systems that need user IDs cannot be used.

Session-based recommendation (SBR), which makes recommendations without focusing on user IDs, is currently attracting attention. SBR is a method

© The Author(s), under exclusive license to Springer Nature Switzerland AG 2023
A. Bifet et al. (Eds.): DS 2023, LNAI 14276, pp. 582–596, 2023.
https://doi.org/10.1007/978-3-031-45275-8_39

of providing recommendations based on session IDs assigned to short-term user actions. They are assigned when a user logs into an EC site, and are advantageous in that users cannot be uniquely identified as they are assigned different IDs depending on the time of day. However, even if session ID management is inadequate, there is a risk that the session ID of a logged-in user may be illegally obtained to gain access. To prevent this, we propose a new method for recommending items without using either user or session IDs. Specifically, for purchase history data to which user and session IDs are not assigned, records with consecutive user attributes, such as gender and place of residence, are defined as pseudo-sessions, and the next item to be purchased in the pseudo-session is predicted. In this manner, items that anonymous users place in their carts in chronological order can be recommended for their next purchase without using session IDs.

Generally, existing SBRs are often graph neural network (GNN)-based [5] methods that consider only item transactions within a session. However, in the case of purchase history, other features such as item price and category tend to be observed as well. The existing method CoHHN [10] shows that price information and categories are effective in recommending items. In this study, we propose a new GNN model called the co-guided heterogeneous hypergraph and globalgraph network plus (CoHHGN+), which consider not only the purchase transition and price of items, but also the category hierarchy of items and auxiliary information of sessions; our model also learns the co-occurrence relationships with other sessions within the same features, and takes into account the importance of embeddings between different features and same features. In summary, our key contributions are as follows:

1. A pseudo session-based high-accuracy recommendation system is proposed.
2. We exploit session information about users and time series sales.
3. Item hierarchies and co-occurrence relationships of the same features are considered.

2 Related Work

Rendle et al. proposed a Markov chain-based SBR model, called factorized personalized Markov chains (FPMC) [6]. FPMC is a hybrid method that combines Markov chains and matrix factorization to capture sequential patterns and long-term user preferences. The method is based on a Markov chain that focuses on two adjacent states between items and is adaptable to anonymous SBRs. However, a major problem with Markov chain-based models is that they combine past components independently, which restricts their predictive accuracy.

Hidasi et al. proposed a recurrent neural network (RNN)-based SBR model called GRU4Rec [4]. GRU4Rec models transition between items using gated recurrent units (GRUs) for inputs represented as graphs.

The purchase transitions of an EC site can be represented by a graph structure, which is a homogeneous or heterogeneous graph depending on whether the attributes of the nodes are singular or plural. Homogeneous graphs are graphs that represent relationships by only one type of node and edge and are used to represent relationships in social networks. In contrast, heterogeneous graphs are graphs that contain multiple and diverse nodes and edges and are used to represent relationships between stores and customers.

Wu et al. proposed SR-GNN [9], which uses a GNN to predict the next item to be purchased in a session based on a homogeneous graph of items constructed across sessions. Using GNNs, we obtain item embeddings that are useful for predicting by introducing attention mechanism to the continuously observed item information. Currently, SBRs based on this GNN have shown more effective results than other methods, and several extended methods based on SR-GNN have been proposed. Wang et al. proposed GCE-GNN [8], which embeds not only the current session but also item transitions of other sessions in the graph.

Existing methods, such as SR-GNN and GCE-GNN are models that learn item-only transitions; however, sessions may also include item prices and categorical features. To construct a model that takes these into account, it is necessary to use heterogeneous graphs. However, when using graphs to represent the relationship between auxiliary information such as price and items, the graph becomes more complex as the number of items in a particular price range increases. Therefore, we apply an extended heterogeneous hypergraph to allow the edges to be connected to multiple nodes. This makes it possible to understand complex higher-order dependencies between nodes, especially in recommendation tasks [10]. Zhang et al. proposed CoHHN [10], which embeds not only item transitions, but also item prices and categories. While CoHHN can consider price and item dependencies, it does not consider the hierarchical features of categories or sales information and user attributes observed during the sessions. It also does not embed the global information that represents item purchase transitions in other sessions. Therefore, we propose a new GNN model that embeds global information as in GCE-GNN, and considers item category hierarchy, user attributes, and sale information.

3   Preliminaries

Let τ be a feature type that changes within a given session. Let $\mathcal{V}^\tau = \{v_1^\tau, v_2^\tau, \cdots, v_{n^\tau}^\tau\}$ be a unique set of feature τ and n^τ be their size. We consider four items: item ID, price, and hierarchical category of item (large and middle); we subsequently denote its item set as $\mathcal{V}^{\mathrm{id}}$, $\mathcal{V}^{\mathrm{pri}}$, $\mathcal{V}^{\mathrm{lrg}}$, and $\mathcal{V}^{\mathrm{mid}}$, respectively. Note that the prices are discretized into several price ranges according to a logistic distribution [2, 10], taking into account the market price of each item.

Let $S_a^\tau = [v_1^{a,\tau}, v_2^{a,\tau}, \cdots, v_s^{a,\tau}]$ be a sequence of the feature τ for a pseudo-session and s be its length. Note that each element $v_i^{a,\tau}$ of S_a^τ is belongs to $\mathcal{V}^\tau$. The objective of SBR is to recommend the top k items from $\mathcal{V}^{\mathrm{id}}$ that are most likely to be purchased or clicked next by the user in the current session a.

3.1 Heterogeneous Hypergraph and Global Graph

To learn the transitions of items in a pseudo-session, two different graphs are constructed from all available sessions.

We construct heterogeneous hypergraphs $\mathcal{G}^{\tau_1,\tau_2} = (\mathcal{V}^{\tau_1}, \mathcal{E}_h^{\tau_2})$ to consider the relationships between different features. Let $\mathcal{E}_h^{\tau_2}$ be a set of hyperedges for feature τ_2. Each hyperedge $e_h^{\tau_2} \in \mathcal{E}_h^{\tau_2}$ can be connected to multiple nodes $v_i^{\tau_1} \in \mathcal{V}^{\tau_1}$ in the graph. This means that a node $v_i^{\tau_1}$ is connected to a hyperedge e^{τ_2} when the features τ_1 and τ_2 are observed in the same record. If several nodes are contained in the same hyperedge, they are considered to be adjacent.

Heterogeneous hypergraphs are a method of constructing graphs with reference to different features; however, transition regarding information about features of the same type is not considered. Additionally, item purchase transitions may include items that are not relevant to prediction. Thus, we construct the global graph shown below.

The global graph captures the relationship between items of the same type that co-occur with an item for all sessions. According to [8], the global graph is constructed based on ε-neighborhood set of an item for all sessions. Assuming that a and b are different arbitrary session, we define the ε-neighborhood set as follows.

$$\mathcal{N}_\varepsilon(v_i^{a,\tau}) = \left\{ v_j^{b,\tau} | v_i^{a,\tau} = v_{i'}^{b,\tau} \in S_a^\tau \cap S_b^\tau; v_j^{a,\tau} \in S_b^\tau; j \in [i' - \varepsilon, i' + \varepsilon]; a \neq b \right\}, \tag{1}$$

where i' is an index of $v_i^{a,\tau}$ in S_b^τ and ε is a parameter that controls how close items are considered from the position of i' in session B. Consider that $\mathcal{G}_g = (\mathcal{V}^\tau, \mathcal{E}_g^\tau)$ is a global graph where $\mathcal{E}_g^\tau$ is an edge set and $e_g^\tau \in \mathcal{E}_g^\tau$ connects two vertices $v_i^\tau \in \mathcal{V}^\tau$ and $v_j^\tau \in \mathcal{N}_\varepsilon(v_i^\tau)$. Notably, the global graph only shows the relationship between identical features, and the adjacency conditions between nodes are not affected by other features.

4 Proposed Method

From the perspective of privacy protection, we propose a pseudo session-based recommendation method using a heterogeneous hypergraph constructed from a set of features including a categorical hierarchy, a global graph for item and price features, and additional session attribute information. Figure 1 shows an overview of our proposed method. To consider the interactions and importance between features, our model learns feature embeddings in two steps. In the first step of aggregation, the intermediate embedding of each feature is learned from a heterogeneous hypergraph which consider the interrelationships among different features. In the second step, the final feature embedding vector obtained by aggregating the intermediate embedding in accordance with their respective importance. To address the problem of the heterogeneous hypergraph not being able to learn purchase transitions within the same feature, a global graph is used to incorporate co-occurrence relationships within the same feature into learning. Finally, we propose learning of purchase transitions within a session by considering the features of the session itself, in addition to existing methods.

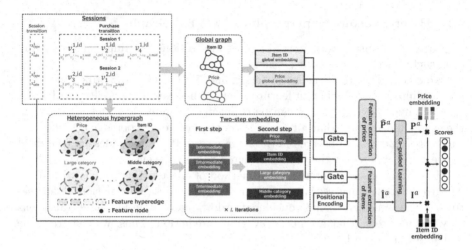

Fig. 1. Overview of the proposed system. First, heterogeneous hypergraphs and global graphs are constructed for all training sessions. In two-step embedding training, embeddings within and between graphs are iteratively trained to obtain multiple feature embeddings, including categorical hierarchies. Then, using the item and price embeddings, we apply co-guided Learning [10] to predict the next item to be purchased by extracting features that account for transitions within the session and the interaction between the two.

4.1 Two-Step Embedding with Category Hierarchy

Based on intra-type and inter-type aggregating method in CoHHN [10], we extend it to multiple categorical hierarchies. We obtain the item ID, price, large category, and middle category embedding vectors from the two-step learning method. In the first step of embedding, the embedding of a feature is learned from a heterogeneous hypergraph in which the feature is a node and others are hyperedges. For example, if the item ID is a node, price, large category, and middle category correspond to the hyperedges. In this case, multiple intermediate embeddings are obtained depending on the type of feature, i.e., the hyperedge. In the second step, these embeddings are used to learn the final node embeddings by aggregating them based on their importance. Each learning step is repeated for all L iterations.

First Step. We learn a first-step embedding for a feature t from a heterogeneous hypergraph, where the target feature t is a node and another feature τ is a hyperedge. First, we define the embedding of a node $v_i^t \in \mathcal{V}^t$ as $\mathbf{h}_{l,i}^{\text{hyper},t} \in \mathbb{R}^d$. Here, l denotes the location of the training iteration. In the initial state $l = 0$, the parameters are initialized using He's method [3]. Let $\mathcal{N}_\tau^t(v_i^t)$ be the adjacent node set of v_i^t. Then, the intermediate embedding of v_i^t in the l-iteration is given by

$$\mathbf{m}_{\tau,i}^t = \sum_{v_j^t \in \mathcal{N}_\tau^t(v_i^t)} \alpha_j \mathbf{h}_{l-1,j}^{\text{hyper},t}, \tag{2}$$

$$\alpha_j = \text{Softmax}_j \left(\left[\mathbf{u}_t^\top \mathbf{h}_{l-1,k}^{\text{hyper},t} \mid v_k^t \in \mathcal{N}_\tau^t(v_i^t) \right] \right), \tag{3}$$

where $\boldsymbol{u}_t^\top$ is an attention vector that determines the importance of $\mathbf{h}_{l-1,j}^{\text{hyper},t}$. The function Softmax_i is defined as

$$\text{Softmax}_i\left([\boldsymbol{a}_1, \cdots, \boldsymbol{a}_s]\right) = \frac{\exp\left(\boldsymbol{a}_i\right)}{\sum_{j=1}^{s}\exp\left(\boldsymbol{a}_j\right)}. \tag{4}$$

Here, $\mathbf{m}_{\tau,i}^t \in \mathbb{R}^d$ represents an intermediate embedding of the feature t when τ is a type of hyperedge. In the first step of embedding, we learn the features to focus on when embedding t.

Second Step. Let us assume that $\mathbf{m}_{\tau_1,i}^t$, $\mathbf{m}_{\tau_2,i}^t$, and $\mathbf{m}_{\tau_3,i}^t$ are intermediate embeddings for a feature t when τ_1, τ_2, τ_3 are types of hyperedge, respectively. By aggregating the embeddings of the first step, we obtain the embedding of v_i^t shown in the following equation.

$$\mathbf{h}_{l,i}^{\text{hyper},t} = \beta_1 * \mathbf{h}_{l-1,i}^{\text{hyper},t} + \sum_{j=2}^{4}\beta_j * \mathbf{m}_{\tau_{j-1},i}^t, \tag{5}$$

$$\beta_j = \text{Softmax}_j\left(\left[\boldsymbol{W}^t\mathbf{h}_{l-1,i}^{\text{hyper},t}, \boldsymbol{W}_{\tau_1}^t\mathbf{m}_{\tau_1,i}^t, \boldsymbol{W}_{\tau_2}^t\mathbf{m}_{\tau_2,i}^t, \boldsymbol{W}_{\tau_3}^t\mathbf{m}_{\tau_3,i}^t\right]\right), \tag{6}$$

where $\boldsymbol{W}^t, \boldsymbol{W}_{\tau_1}^t, \boldsymbol{W}_{\tau_2}^t, \boldsymbol{W}_{\tau_3}^t \in \mathbb{R}^{d\times d}$ are learnable parameters, and $*$ denotes the element-wise items of the vectors. Further, β_j is a parameter that computes the importance between the embedding vectors and aggregates the previous and intermediate iteration embeddings.

4.2 Embedding of Global Graph

Since heterogeneous hypergraph does not consider the co-occurrence relationships or counts between sessions related to the same feature, we use the learning of embedding global graphs in a GCE-GNN [8] with two configurations: propagation and aggregation of information.

Information Propagation. The ε-neighborhood of each feature from the global graph for feature t are embedded. Because the number of features of interest within a neighborhood is considered to be different for each user, based on the attention score shown in the following equation, the neighborhood embedding $\mathbf{h}_{\mathcal{N}_\varepsilon(v_i^t)}$ is first learned.

$$\mathbf{h}_{\mathcal{N}_\varepsilon(v_i^t)} = \sum_{v_j^t \in \mathcal{N}_\varepsilon(v_i^t)} \pi(v_i^t, v_j^t)\mathbf{h}_{l-1,j}^{\text{global},t}, \tag{7}$$

$$\pi(v_i^t, v_j^t) = \text{Softmax}_j\left(\left[a(v_i^t, v_k^t) \mid v_k^t \in \mathcal{N}_\varepsilon(v_i^t)\right]\right), \tag{8}$$

$$a(v_i^t, v_j^t) = \boldsymbol{q}^\top\text{LeakyRelu}\left(\boldsymbol{W}_1\left[\boldsymbol{s} * \mathbf{h}_{l-1,j}^{\text{global},t}\right]; w_{ij}\right), \tag{9}$$

where $\mathbf{h}_{l-1}^{\text{global},t}$ is an embedding of the global graph for the feature j on the $l-1$-th learning iteration, and $\pi(v_i^t, v_j^t)$ is an attention weight that considers the importance of neighborhood node embedding. The attention score $a(v_i^t, v_j^t)$ employs LeakyRelu. In LeakyRelu, $w_{ij} \in \mathbb{R}$ is the weight of an edge (v_i^t, v_j^t) in the global graph that represents the number of co-occurrences with features v_j^t, and ; is a concatenation operator. Further, $\boldsymbol{W}_1 \in \mathbb{R}^{(d+1)\times(d+1)}$ and $\dot{\boldsymbol{q}} \in \mathbb{R}^{d+1}$ are learnable parameters, and $\boldsymbol{s}$ is the average embedding of the session to which v_i^t belongs, defined as

$$s = \frac{1}{s} \sum_{v_i^t \in S_a^t} \mathbf{h}_{l-1,i}^{\text{global},t}. \tag{10}$$

Information Aggregation. For a feature v^t to be learned, the l-iteration embedding $\mathbf{h}_l^{\text{global},t}$ is obtained by aggregating the $(l-1)$-iteration embedding and the neighborhood embeddings using the following formula:

$$\mathbf{h}_{l,i}^{\text{global},t} = \text{ReLU}\left(\boldsymbol{W}_2 \left[\mathbf{h}_{l-1,i}^{\text{global},t}; \mathbf{h}_{\mathcal{N}_\varepsilon(v_i^t)}\right]\right), \tag{11}$$

where $\boldsymbol{W}_2 \in \mathbb{R}^{d\times 2d}$ denotes a learnable parameter. In global graph embedding, highly relevant item information can be incorporated throughout the session by aggregating the reference features and their ε-neighborhoods.

4.3 Embedding Feature Nodes

For the feature node v_i^t, the final embedding is obtained from the embedding of heterogeneous hypergraphs considering the category hierarchy and the embedding of global graphs by the following gate mechanism:

$$g_i^t = \sigma(\boldsymbol{W}_3 \mathbf{h}_{L,i}^{\text{hyper},t} + \boldsymbol{W}_4 \mathbf{h}_{L,i}^{\text{global},t}), \tag{12}$$

$$\mathbf{h}_i^t = g_i^t * \mathbf{h}_{L,i}^{\text{hyper},t} + (1 - g_i^t) * \mathbf{h}_{L,i}^{\text{global},t}, \tag{13}$$

where σ is a sigmoid function, $\boldsymbol{W}_3 \in \mathbb{R}^{d\times d}$ and $\boldsymbol{W}_4 \in \mathbb{R}^{d\times d}$ are learnable parameters, and L is the final iteration of graph embedding. g_i^t is learned to consider the importance of embedding heterogeneous hypergraphs and embedding global graphs. The final feature node embedding is required only for the item ID and price based on the training of the next item.

4.4 Feature Extraction Considering Session Attributes

To enhance the recommendation accuracy in pseudo-sessions based on the learned node embeddings, we propose an extraction method of features related to the user's items and prices in each session.

Feature Extraction of Items. The embedding of an item node in session a is given by the sequence $[\mathbf{h}_1^{a,\mathrm{id}}, \cdots, \mathbf{h}_s^{a,\mathrm{id}}]$. In addition to items, user attribute information, time-series information, and EC site sale information, among others, may be observed in each session. Therefore, we considered this information and learned to capture the session-by-session characteristics associated with the items. Let d_{sale} be the number of types of sale information and $\boldsymbol{x}_{\mathrm{sale}}^a \in \{0,1\}^{d_{\mathrm{sale}}}$ items be given per session. Each dimension of this vector represents the type of sale, with a value of 1 if it is during a particular sale period and a value of 0 if it is outside that period. Similarly, if the number of types of attribute information is d_{type}, then $\boldsymbol{x}_{\mathrm{type}}^a \in \{0,1\}^{d_{\mathrm{type}}}$ is a vector representing user attributes.

For items and sales, we also consider time-series location information. The item location information defines a location encoding $\boldsymbol{pos\_item}_i \in \mathbb{R}^d$ as in [7]. Furthermore, for the location information of the sale, the week information to which the current session belongs is encoded by the following formula:

$$pos\_time_{2k-1}^a = \sin\left(\frac{2m\pi}{52k}\right), \tag{14}$$

$$pos\_time_{2k}^a = \cos\left(\frac{2m\pi}{52k}\right), \tag{15}$$

where $\boldsymbol{pos\_time}^a \in \mathbb{R}^c$ is the location encoding associated with the week information of the session a, $m \in \mathbb{Z}$ represents the week, and k is the embedding dimension. Because a year comprise 52 weeks, the trigonometric function argument is divided by 52. Based on the above, item embedding in a session is defined as follows:

$$\mathbf{v}_i^{a,\mathrm{id}} = \tanh\left(\boldsymbol{W}_5\left[\mathbf{h}_i^{a,\mathrm{id}}; \boldsymbol{pos\_item}_i\right] + \boldsymbol{W}_6\left[\boldsymbol{x}_{\mathrm{sale}}^a; \boldsymbol{pos\_time}^a\right] + \boldsymbol{W}_7 \boldsymbol{x}_{\mathrm{type}}^a + \boldsymbol{b}_1\right), \tag{16}$$

where $\boldsymbol{W}_5 \in \mathbb{R}^{d \times 2d}$, $\boldsymbol{W}_6 \in \mathbb{R}^{d \times (d_{\mathrm{sale}}+c)}$, $\boldsymbol{W}_7 \in \mathbb{R}^{d \times d_{\mathrm{type}}}$, $\boldsymbol{b}_1 \in \mathbb{R}^d$ are trainable parameters, $\mathbf{v}_i^{a,\mathrm{id}}$ is the i-th item embedding in session a. The item preferences $\widehat{\mathbf{I}}^a$ of a user in a session are determined according to [10] as follows:

$$\widehat{\mathbf{I}}^a = \sum_{i=1}^s \beta_i \mathbf{h}_i^{a,\mathrm{id}}, \tag{17}$$

$$\beta_i = \boldsymbol{u}^\top \sigma(\boldsymbol{W}_8 \mathbf{v}_i^{a,\mathrm{id}} + \boldsymbol{W}_9 \bar{\mathbf{v}}^{a,\mathrm{id}} + \boldsymbol{b}_2), \tag{18}$$

where $\boldsymbol{W}_8, \boldsymbol{W}_9 \in \mathbb{R}^{d \times d}$, $\boldsymbol{b}_2 \in \mathbb{R}^d$ are learnable parameters, $\boldsymbol{u}^\top \in \mathbb{R}^d$ is the attention vector. Additionally, $\bar{\mathbf{v}}^{a,\mathrm{id}} = \frac{1}{s}\sum_{i=1}^s \mathbf{v}_i^{a,\mathrm{id}}$.

Feature Extraction of Prices. The price hyperedge in session a is given by $[\mathbf{h}_1^{a,\mathrm{p}}, \cdots, \mathbf{h}_s^{a,\mathrm{p}}]$. To estimate price preferences with respect to users, we follow [10] and learn the features of the price series using multi-head attention as shown in the following equation:

$$\mathbf{E}^{a,\mathrm{p}} = [\mathbf{h}_1^{a,\mathrm{p}}; \cdots ; \mathbf{h}_s^{a,\mathrm{p}}], \tag{19}$$

$$\mathbf{M}_i^{a,\mathrm{p}} = [\mathit{head}_1^a; \cdots ; \mathit{head}_h^a], \tag{20}$$

$$\mathit{head}_i^a = \mathit{Attention}(\boldsymbol{W}_i^Q \mathbf{E}^{a,\mathrm{p}}, \boldsymbol{W}_i^K \mathbf{E}^{a,\mathrm{p}}, \boldsymbol{W}_i^V \mathbf{E}^{a,\mathrm{p}}), \tag{21}$$

where h is the number of blocks of self-attention, $\boldsymbol{W}_i^Q, \boldsymbol{W}_i^K, \boldsymbol{W}_i^V \in \mathbb{R}^{\frac{d}{h} \times d}$ are parameters that map item i in session a to query and key, value, and $\mathit{head}_i^a \in \mathbb{R}^{\frac{d}{h}}$ is the embedding vector of each block of multi-head-attention for item i. Further, $\mathbf{E}^{a,\mathrm{p}} \in \mathbb{R}^{dm}, \mathbf{M}_i^{a,\mathrm{p}} \in \mathbb{R}^d$ and the embedded price series is $[\mathbf{M}_1^{a,\mathrm{p}}, \cdots , \mathbf{M}_s^{a,\mathrm{p}}]$.

Because the last price embedding is considered to be the most relevant to the next item price in the price series, we determine the user's price preference $\widehat{\mathbf{P}}^a = \mathbf{M}_s^{a,\mathrm{p}}$ in the session.

4.5   Predicting and Learning About the Next Item

The user's item preferences $\widehat{\mathbf{I}}^a$ and price preferences $\widehat{\mathbf{P}}^a$ are transformed into $\mathbf{I}^a$ and $\mathbf{P}^a$ respectively by co-guided learning [10], considering mutual dependency relations. When an item $v_i^{a,\mathrm{id}} \in \mathcal{V}^{\mathrm{id}}$ and a price range $v_i^{a,\mathrm{p}} \in \mathcal{V}^{\mathrm{p}}$ are observed in session a, the next item to view and purchase is given by the score of the following Softmax function:

$$\widehat{y}_i = \mathrm{Softmax}_i \left([q_1, \cdots , q_{n^{\mathrm{id}}}]\right), \tag{22}$$

$$q_i = \mathbf{P}^{a\top} \mathbf{h}_i^{a,\mathrm{p}} + \mathbf{I}^{a\top} \mathbf{h}_i^{a,\mathrm{id}}. \tag{23}$$

At the training time, this score is used to compute the cross-entropy loss.

$$\mathcal{L}(\boldsymbol{y}, \widehat{\boldsymbol{y}}) = - \sum_{j=1}^{n^{\mathrm{id}}} \left(y_j \log \left(\widehat{y}_j\right) + (1 - y_j) \log \left(1 - \widehat{y}_j\right)\right), \tag{24}$$

where $\boldsymbol{y} \in \{0, 1\}^{n^{\mathrm{id}}}$ is the objective variable that indicates whether the user has viewed and purchased item v_i^{id}. $\widehat{\boldsymbol{y}} \in \mathbb{R}^{n^{\mathrm{id}}}$ is the score for all items.

5   Experiments

We evaluate our proposed method using purchasing history data of an EC market. The dataset comprises the purchasing history of 100,000 people randomly selected by age group which are obtained from the users registered in 2019–20 in the Rakuten [1] market, which is a portal site for multiple EC sites. We consider four age groups: 21–35, 36–50, 51–65, and 66–80. Each purchasing history comprises the category name of the purchased item (large, middle, small), week (week 1–105), gender (male or female), residence (nine provinces in Japan), and price segment (separated by thousands of JPY). The user ID and session information are not recorded. Note that this dataset is provided at the 2022 Data Analysis Competition organized by Joint Association Study Group of Management Science and is not open to the public.

Table 1. Statistical information of data set.

Age group	21–35	36–50	51–65	66–80
# of price range	10	10	10	10
# of large categories	36	36	36	35
# of middle categories	342	354	340	322
# of small categories	2,800	2,975	2,763	2,327
# of interaction	727,655	1,033,405	712,894	452,496
# of sessions	326,110	462,290	323,184	203,725
Avg. session length	2.24	2.24	2.21	2.24

5.1 Preprocessing

Our method recommends a small category name as the item ID. Additionally, the proposed model also considers session attributes, such as purchaser gender, region of residence, and EC site sales. As specific sale information, we include two types of sales that are regularly held at the Rakuten market. Sale 1 is held once every three months for one week, during which many item prices are reduced by up to half or less. Sale 2 is held for a period of one week each month, and more points are awarded for shopping for items on the EC site. Each session attribute is represented by a discrete label. When learning, we treat each gender, region, and sale as a vector with the observed value as 1 and all other values as 0. The price intervals are converted to price range labels by applying a logistic distribution [2].

In each transformed dataset, consecutive purchase intervals with the same gender and residential area are labeled as pseudo-sessions. Based on the assigned pseudo session ID, records with a session length of less than 2 or frequency of occurrence of less than 10 are deleted, according to [10]. Within each session, the last observed item ID is used as the prediction target, and the other series are used for training. In dividing the data, weeks 1 through 101 are used as training data, and the remaining weeks 102 through 105 are used as test data. Additionally, 10% of the training data re used as validation data for hyperparameter tuning of the model. The statistical details of the four datasets are listed in Table 1.

5.2 Evaluation Criteria

We employ the following criteria to evaluate the recommendation accuracy:

- **P@k (Precision)** : The percentage of the top k recommended items that are actually purchased.
- **M@k (Mean Reciprocal Rank)** : The mean value for the inverse of the rank of the items actually recommended for purchase. If the rank exceeds k, it is 0.

The precision does not consider the ranking of recommended items; however, the mean reciprocal rank is a criterion that considers ranking, implying that the higher the value, the higher the item actually purchased in the ranking. In our experiment, we set $k = 10, 20$.

5.3 Comparative Model

To verify the effectiveness of the proposed method, we compare it with the following five models.

- **FPMC** [6]: By combining matrix factorization and Markov chains, this method can capture both time-series effects and user preferences. As the dataset is not assigned an ID to identify the user, the observations for each session are estimated as if they were separate users.
- **GRU4Rec** [4]: An SBR based on RNN with GRU when recommending items for each session.
- **SR-GNN** [9]: An SBR that constructs a session graph and captures transitions between items using a GNN.
- **GCE-GNN** [8]: An SBR that builds a session graph and global graph, and captures transitions between items by a GNN while considering their importance.
- **CoHHN** [10]: An SBR that constructs a heterogeneous hypergraph regarding sessions that considers information other than items and captures transitions between items with a GNN.

5.4 Parameter Setting

To fairly evaluate the performance of the model, we use many of the same parameters for each model. For all models, the size of the embedding vector is set to 128, the number of epochs to 10, and the batch size to 100. For the optimization method, GRU4Rec uses Adagrad (learning rate 0.01) based on the results of previous studies, while the GNN method uses Adam (learning rate 0.001) with a weight decay of 0.1 applied every three epochs. The coefficients of the L2-norm regularity are set to 10^{-5}. Additionally, in GCE-GNN and our model CoHHGN+, the size of the neighborhood item-set ε in the global graph is set to 12. Furthermore, in CoHHN and our model, the number of self-attention heads is set to 4 ($h = 4$), and the number of price ranges to 10. Finally, the number of GNN iterations and percentage of dropouts used in the architecture are determined by grid search for each model using the validation data. We have released the source code of our model online[1].

6 Results and Discussion

6.1 Performance Comparison

Tables 2 and 3 show the results of evaluating the five existing methods and the proposed method CoHHGN+ on the four selected datasets. CoHHGN+ obtains

---

[1] https://github.com/sumugit/CoHHGN_plus.

Table 2. Precision of CoHHGN+ and comparative methods. The most accurate value for each dataset is shown in bold, and the second most accurate value is underlined. Each value is the average of three experiments conducted to account for variations due to random numbers. For CoHHGN+ and the other most accurate models, a t-test is performed to confirm statistical significance, and a p-value of less than 0.01 is marked with an asterisk (*).

Dataset	age 21–35		age 36–50		age 51–65		age 66–80	
Method	P@10	P@20	P@10	P@20	P@10	P@20	P@10	P@20
FPMC	3.84	6.22	4.00	6.56	0.66	2.83	1.13	3.46
GRU4Rec	1.72	2.57	1.73	2.60	1.81	2.82	1.63	2.73
SR-GNN	15.06	20.92	13.71	20.11	13.78	20.34	14.30	22.55
GCE-GNN	15.16	20.88	13.72	20.11	13.87	20.46	<u>14.42</u>	22.46
CoHHN	<u>15.19</u>	<u>21.06</u>	<u>13.96</u>	<u>20.22</u>	<u>13.93</u>	<u>20.73</u>	14.36	<u>22.49</u>
CoHHGN+	**15.92***	**22.28***	**14.75***	**22.01***	**15.16***	**22.57***	**15.55***	**23.84***

Table 3. Mean reciprocal rank of CoHHGN+ and comparative methods. The symbols attached to the values are the same as those in the table 2.

Dataset	age 21–35		age 36–50		age 51–65		age 66–80	
Method	M@10	M@20	M@10	M@20	M@10	M@20	M@10	M@20
FPMC	0.88	1.04	1.14	1.31	0.15	0.28	0.38	0.54
GRU4Rec	0.78	0.84	0.75	0.81	0.71	0.78	0.59	0.66
SR-GNN	6.56	6.95	5.95	6.38	5.51	5.97	5.24	5.80
GCE-GNN	6.65	7.04	**6.02**	**6.45**	5.58	6.04	5.21	5.75
CoHHN	<u>6.67</u>	<u>7.06</u>	<u>6.01</u>	<u>6.44</u>	<u>5.62</u>	<u>6.08</u>	<u>5.27</u>	<u>5.81</u>
CoHHGN+	**6.89***	**7.32***	5.93	6.42	**5.83***	**6.34***	**5.81***	**6.37***

the most accurate results for all datasets with precision for $k = 10, 20$. The mean reciprocal rank is also the most accurate, except for the data for the 36–50 age group. For the 36–50 year age group dataset, the precision is higher than that for the other models, while the mean reciprocal rank shows the highest accuracy for GCE-GNN. However, there is no statistically significant difference in the prediction accuracy between CoHHGN+ and GCE-GNN in this dataset. Thus, it can be inferred that there is no clear difference in prediction accuracy. This confirms the effectiveness of the proposed method for all the data.

In the comparison method, a large discrepancy in accuracy between the GNN-based method, which introduces an attention mechanism in the purchase series, and the other methods is noted. Overall, the GRU4Rec without attention mechanism results in the lowest accuracy, suggesting that the results were not sufficiently accurate for data with a small number of sessions. This is because the model focuses only on purchase transitions between adjacent items. Similarly, for FPMC, although the accuracy is improved compared to GRU4Rec, modeling with Markov chains and matrix factorization is not effective for purchase data with pseudo-sessions. Moreover, SR-GNN, GCE-GNN, CoHHN, and CoHHGN+ using graphs of purchase transitions between sessions show a significant improvement in accuracy and are able to learn the purchase trends of non-adjacent items as well.

Among the compared methods, CoHHN, which considered price and large category information in addition to item ID information, tends to have a higher prediction accuracy overall. The number of series per session is generally small for purchase history data, and it can be said that higher accuracy can be obtained by learning data involving multiple features, including items. GCE-GNN, which also considers the features of other sessions, shows the second highest prediction accuracy after CoHHN. When using purchase history data with short session lengths, it is more accurate to learn embedding vectors by considering items that have co-occurrence relationships with other sessions, in addition to series within sessions. The SR-GNN that has learned only from item ID transitions is inferior to the GCE-GNN in terms of overall accuracy among GNN-based systems, although it is more accurate than the GCE-GNN for some datasets. Therefore, it can be considered that adopting features other than the item ID and other session information will lead to improved recommendation accuracy.

We confirm that the proposed method improves accuracy not only by considering auxiliary information in the purchase transition of items, but also by learning methods for its embedding vectors and including additional features that change from session to session. Furthermore, the embedding vector obtained from the global graph of the item of interest works well for a series with short session lengths.

6.2 Impact of Each Model Extension

Next, we conduct additional experiments on four datasets to evaluate the effectiveness of embedding item category hierarchies and accounting for session attributes, as well as global-level features. Particularly, we design the following two comparative models:

- CoHHGN (H): A model that incorporates hierarchical embedding of three or more features that vary within a session.
- CoHHGN (HS): A model that considers the hierarchical embedding of three or more features and session attributes in the proposed method.

To compare the performance with existing methods, we use the most accurate values of the existing methods shown in Tables 2 and 3 as the baselines. Tables 4 and 5 show the prediction results of the comparison model. For both precision and Mean Reciprocal Rank, CoHHGN+, which incorporates all the proposed methods, performs better overall than the other two models. For Precision, the accuracy of CoHHGN (HS) is higher for P@10 in the 21–35 year age group dataset. However, because the accuracy of CoHHGN+ is higher than that of other methods in P@20, we believe that considering the embedding of global

Table 4. Comparison of the precision accuracy for each model extension. The most accurate values for each dataset are shown in bold. Each value is the average of three experiments conducted to account for random number variation. A t-test was conducted to confirm the statistical significance of the accuracy between the baseline and the proposed method, and an astarisk (*) is added if the p-value is less than 0.01.

Dataset	age 21–35		age 36–50		age 51–65		age 66–80	
Method	P@10	P@20	P@10	P@20	P@10	P@20	P@10	P@20
Baseline	15.19	21.06	13.96	20.22	13.93	20.73	14.42	22.49
CoHHGN (H)	15.24	21.13	14.10	21.13	13.98	20.69	14.24	22.56
CoHHGN (HS)	**15.95***	22.11*	14.66*	21.97*	13.97	20.57	15.13*	23.51*
CoHHGN+	15.92*	**22.28***	**14.75***	**22.01***	**15.16***	**22.57***	**15.55***	**23.84***

Table 5. Comparison of mean reciprocal rank accuracy for each model extension. The symbols attached to the values are the same as those in the table 4.

Dataset	age 21–35		age 36–50		age 51–65		age 66–80	
Method	M@10	M@20	M@10	M@20	M@10	M@20	M@10	M@20
Baseline	6.67	7.06	6.02	6.45	5.62	6.08	5.27	5.81
CoHHGN (H)	6.63	7.02	**6.02**	6.45	5.66	6.12	5.32	5.88
CoHHGN (HS)	6.77	7.19	5.94	**6.45**	5.65	6.11	5.69*	6.24*
CoHHGN+	**6.89***	**7.32***	5.93	6.42	**5.83***	**6.34***	**5.81***	**6.37***

graph features will improve the accuracy in a stable manner. For CoHHGN (H), although the accuracy is improved over the baseline in several datasets, no statistically significant differences are identified. However, extending the model to CoHHGN (HS), which also considers session attributes, results in a significant difference in precision in all datasets, except for the age group 51–65.

Further, considering the mean reciprocal rank, although the recommendation accuracy tends to improve as the model is extended to CoHHGN (H) and CoHHGN (HS), the only dataset in which statistically significant differences can be confirmed is that for the 66–80 age group. However, when extended to CoHHGN+, which incorporates all the proposed methods, the overall prediction accuracy is higher and significant differences are confirmed. This confirms that the recommendation accuracy of the item ID can be improved by simultaneously considering features that vary between sessions and attributes of other sessions, in addition to features that vary within sessions.

7   Conclusion

In this study, we developed CoHHGN+ based on CoHHN, which is an SBR considering various features, and GCE-GNN considering global graphs, for purchase history data of EC sites. Moreover, we considered global time-series information, sale information, and user information. The application of the proposed model to pseudo-session data with no user IDs shows that the GNN-based method exhibits significantly higher accuracy than those for the other methods, and that our proposed CoHHGN+ is the most accurate method on the dataset.

Although incorporating several types of data improves the prediction accuracy, there are still issues from the viewpoint of feature selection for data with more types of information recorded. If there are n types of heterogeneous information, the number of heterogeneous hypergraphs used to embed heterogeneous information is 2^n. Therefore, selecting and integrating heterogeneous information remains an issue.

Future work on issues related to more efficient feature selection and methods for integrating heterogeneous information will lead to the development of models with even higher accuracy. We would also like to expand the scope of application of CoHHGN+ proposed in this study and attempt to provide useful recommendations in other domains as well.

Acknowledgments. We would like to thank the sponsor of the Data Analysis Competition, Joint Association Study Group of Management sCience (JASMAC), and Rakuten Group, Inc. for providing us with the data. This work was also supported by JSPS KAKENHI Grant Number JP20H04146.

References

1. Rakuten ichiba homepage. https://www.rakuten.co.jp/. Accessed 1 May 2023
2. Greenstein-Messica, A., Rokach, L.: Personal price aware multi-seller recommender system: evidence from eBay **150**, 14–26 (2018)
3. He, K., Zhang, X., Ren, S., Sun, J.: Delving deep into rectifiers: surpassing human-level performance on ImageNet classification. In: Proceedings of the IEEE International Conference on Computer Vision, pp. 1026–1034 (2015)
4. Hidasi, B., Karatzoglou, A., Baltrunas, L., Tikk, D.: Session-based recommendations with recurrent neural networks. In: 4th International Conference on Learning Representations (2016)
5. Li, Y., Tarlow, D., Brockschmidt, M., Zemel, R.S.: Gated graph sequence neural networks. In: 4th International Conference on Learning Representations (2016)
6. Rendle, S., Freudenthaler, C., Schmidt-Thieme, L.: Factorizing personalized Markov chains for next-basket recommendation. In: Proceedings of the 19th International Conference on World Wide Web, pp. 811–820 (2010)
7. Vaswani, A., et al.: Attention is all you need. In: Advances in Neural Information Processing Systems, vol. 30 (2017)
8. Wang, Z., Wei, W., Cong, G., Li, X.L., Mao, X.L., Qiu, M.: Global context enhanced graph neural networks for session-based recommendation. In: Proceedings of the 43rd International ACM SIGIR Conference on Research and Development in Information Retrieval, pp. 169–178 (2020)
9. Wu, S., Tang, Y., Zhu, Y., Wang, L., Xie, X., Tan, T.: Session-based recommendation with graph neural networks. AAAI **33**(1), 346–353 (2019)
10. Zhang, X., et al.: Price does matter! modeling price and interest preferences in session-based recommendation. In: Proceedings of the 45th International ACM SIGIR Conference on Research and Development in Information Retrieval, pp. 1684–1693 (2022)

Healthcare and Biological Data Analysis

Chance and the Predictive Limit in Basketball (Both College and Professional)

Albrecht Zimmermann$^{(\boxtimes)}$ (iD)

UNICAEN, ENSICAEN, CNRS – UMR GREYC, 14000 Caen, France
`albrecht.zimmermann@unicaen.fr`

Abstract. There seems to be an upper limit to predicting the outcome of matches in (semi-)professional sports. A number of works have proposed that this is due to chance and attempts have been made to simulate the distribution of win percentages to identify the most likely proportion of matches decided by chance. We argue that the approach that has been chosen so far makes some simplifying assumptions that cause its result to be of limited practical value, especially for settings where teams do not play all possible opponents. Instead, we propose to use clustering of statistical team profiles and observed scheduling information to derive limits on the predictive accuracy for particular seasons, which can be used to assess the performance of predictive models on those seasons. Using NCAA basketball data, we show that the resulting simulated distributions are much closer to the observed distributions and give higher assessments of chance and tighter limits on predictive accuracy. We also show similar results for the NBA.

1 Introduction

In prior work on the topic of NCAA basketball [14], we speculated about the existence of a "glass ceiling" in (semi-)professional sports match outcome prediction, noting that season-long accuracies in the mid-seventies seemed to be the best that could be achieved for college basketball, with similar results for other sports. One possible explanation for this phenomenon is that we are lacking the attributes to properly describe sports teams, having difficulties to capture player experience or synergies, for instance. While this is the focus of on-going work in the community, especially for "under-described" sports such as European soccer or NFL football, we consider a different question in this paper: *the influence of chance on match outcomes.*

Even if we were able to accurately describe sports teams in terms of their performance statistics, the fact remains that athletes are humans, who might make mistakes and/or have a particularly good/bad day, that matches are refereed by humans, see before, that injuries might happen during the match, that the interaction of balls with obstacles off which they ricochet quickly becomes too complex to even model etc. Each of these can affect the match outcome to

A. Bifet et al. (Eds.): DS 2023, LNAI 14276, pp. 599–613, 2023.
https://doi.org/10.1007/978-3-031-45275-8_40

varying degrees and especially if we have only static information from before the match available, it will be impossible to take them into account during prediction.

While this may be annoying from the perspective of a researcher in sports analytics, from the perspective of sports leagues and betting operators, this is a feature, not a bug. Matches of which the outcome is effectively known beforehand do not create a lot of excitement among fans, nor will they motivate bettors to take risks.

Intuitively, we would expect that chance has a stronger effect on the outcome of a match if the two opponents are roughly of the same quality, and if scoring is relatively rare: since a single goal can decide a soccer match, one (un)lucky bounce is all it needs for a weaker team to beat a stronger one. In a fast-paced basketball game, in which the total number of points can number in the two or even three hundreds, a single basket might be the deciding event between two evenly matched teams but probably not if the skill difference is large.

For match outcome predictions, a potential question is then: *"How strong is the impact of chance for a particular league?"*, in particular since quantifying the impact of chance also allows to identify the "glass ceiling" for predictions. The topic has been explored for the NFL in [2–4], which reports

> The actual observed distribution of win-loss records in the NFL is indistinguishable from a league in which 52.5% of the games are decided at random and not by the comparative strength of each opponent.

Using the same methodology, Weissbock *et al.* [13] derive that 76% of matches in the NHL are decided by chance. As we will argue in the following section, however, the approach used in those works is not applicable to NCAA basketball.

Before we continue, a short remark on terminology: Burke uses the term "luck" but we prefer the term "chance" since "luck" implies a positive outcome, whereas "chance" is meant to indicate randomness.

2 Identifying the Impact of Chance by Monte Carlo Simulations

The general idea used by Burke and Weissbock[1] is the following:

1. A chance value $c \in [0,1]$ is chosen.
2. Each out of a set of virtual teams is randomly assigned a strength rating.
3. For each match-up, a value $v \in [0,1]$ is randomly drawn from a uniform distribution.
   - If $v \geq c$, the stronger team wins.
   - Otherwise, the winner is decided by flipping an unweighted coin.
4. The simulation is re-iterated a large number of times (e.g. $10,000$) to smooth results.

---

[1] For details for Weissbock's work, we direct the reader to [12].

Figure 1 shows the distribution of win percentages for 340 teams, 40 matches per team (roughly the settings of an NCAA basketball season including playoffs), and 10,000 iterations for $c = 0.0$ (pure skill), $c = 1.0$ (pure chance), and $c = 0.5$.

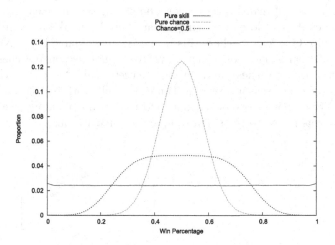

Fig. 1. MC simulated win percentage distributions for different amounts of chance

By using a goodness of fit test – χ^2 in the case of Burke's work, F-$Test$ in the case of Weissbock's – the c-value is identified for which the simulated distribution fits the empirically observed one best, leading to the values reproduced in the introduction. The identified c-value can then be used to calculate the upper limit on predictive accuracy in the sport: since in $1 - c$ cases the stronger team wins, and a predictor that predicts the stronger team to win can be expected to be correct in half the remaining cases in the long run, the upper limit lies at:

$$(1 - c) + c/2,$$

leading in the case of

– the NFL to: $0.475 + 0.2625 = 0.7375$, and
– the NHL to: $0.24 + 0.36 = 0.62$

Any predictive accuracy that lies above those limits is due to the statistical quirks of the observed season: theoretically it is possible that chance always favors the stronger team, in which case predictive accuracy would actually be 1.0. As we will argue in the following section, however, NCAA seasons (and not only they) are likely to be quirky indeed.

3   Limitations of the MC Simulation for NCAA Basketball

A remarkable feature of Fig. 1 is the symmetry and smoothness of the resulting curves. This is an artifact of the distribution assumed to model the theoretical distribution of win percentages – the Binomial distribution – together with the large number of iterations. This can be best illustrated in the "pure skill" setting: even if the stronger team were *always guaranteed* to win a match, real-world sports schedules do not always guarantee that any team actually plays against a representative mix of teams both weaker and stronger than itself. A reasonably strong team could still lose every single match, and a weak one could win at a reasonable clip. One league where this is almost unavoidable is the NFL, which consists of 32 teams, each of which plays 16 regular season matches (plus at most 4 post-season matches), and ranking "easiest" and "hardest" schedules in the NFL is an every-season exercise. Burke himself worked with an empirical distribution that showed two peaks, one before 0.5 win percentage, one after. He argued that this is due to the small sample size (five seasons).

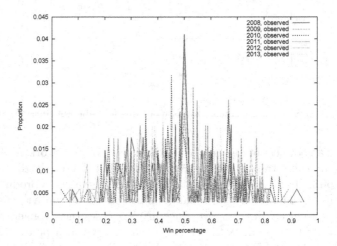

Fig. 2. Observed distribution of win percentages in the NCAA, 2008–2013

The situation is even more pronounced in NCAA basketball, where 340+ Division I teams play at most 40 matches each. Figure 2 shows the empirical distribution for win percentages in NCAA basketball for six season (2008–2013).[2] While there is a pronounced peak for a win percentage of 0.5 for 2008 and 2012, the situation is different for 2009, 2010, 2011, and 2013. Even for the former two seasons, the rest of the distribution does not have the shape of a Binomial distribution. Instead it seems to be that of a *mix* of distributions – e.g. "pure skill" for match-ups with large strength disparities overlaid over "pure chance" for approximately evenly matched teams.

NCAA scheduling is subject to conference memberships and teams will try to pad out their schedules with relatively easy wins, violating the implicit assumptions made for the sake of MC simulations. This also means that the "statistical

---

[2] The choice of seasons is due to presentation concerns, especially in the case of visualizations. Other/additional seasons exhibit similar phenomena.

quirks" mentioned above are often the norm for any given season, not the exception. Thought to its logical conclusion, the results that can be derived from the Monte Carlo simulation described above are purely theoretical: if one could observe an **effectively unlimited** number of seasons, during which schedules are **not systematically imbalanced**, the overall attainable predictive accuracy were bound by the limit than can be derived by the simulation. For a given season, however, and the question how well a learned model performed w.r.t. the specificities of that season, this limit might be too high (or too low).

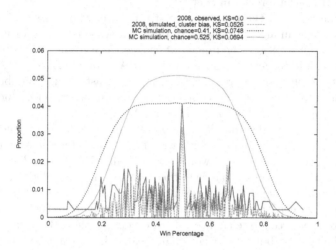

Fig. 3. Distribution of win percentages 2008

As an illustration, consider Fig. 3, showing several season simulations as well as observed winning percentages for the 2008 season.[3] Differing from Burke and Weissbock, we use neither the χ^2 test (which requires frequency binning), to compare distributions, nor the *F-Test*, which only compares variances. Instead, we chose the *Kolmogorov-Smirnov Test*, which returns the maximum distances between two distributions' cumulative distribution functions (CDF).

The MC simulation that matches the observed proportion of teams having a win percentage of 0.5 is derived by setting $c = 0.42$, implying that a predictive accuracy of 0.79 should be possible. The MC simulation that fits the observed distribution best, according to the Kolmogorov-Smirnov (KS) test (which overestimates the proportion of teams having a win percentage of 0.5 along the way), is derived from $c = 0.525$ (same as Burke's NFL analysis), setting the predictive limit to 0.7375. Both curves have visually nothing in common with the observed distribution, yet the null hypothesis – that both samples derive from the same distribution – is not rejected at the 0.001 level by the KS test for sample comparison. This hints at the weakness of using such tests to establish similarity:

---

[3] Other seasons show similar behavior, so we treat 2008 as a representative example.

CDFs and standard deviations might simply not provide enough information to decide whether a distribution is appropriate.

4 Related Work

As mentioned above, Weissbock [12,13] and Burke [2–4] use a Monte Carlo simulation to identify winning percentage distributions most similar to an observed one using a comparison measure and then derive the chance value. We have given the details of their approach in Sect. 2.

Academic work on the subject is relatively rare. Mauboussin [8] compares the variance of winning percentages of observed seasons to the variance of an all-chance season to derive the impact of chance. As a result of this analysis, he claims that the influence of chance on the NBA is only 12% because the skill difference between teams were so large. This is an astounding number since it would imply that one can achieve a predictive accuracy of **94%**! Incidentally, this would make lucrative betting on the NBA impossible.

Aoki *et al.* [1] consider the distribution of points won at home and away for teams (for soccer, for instance, 3 for a win, 1 for a tie, 0 for a loss), which in an equal-skill (pure-chance) league would be normally distributed. By comparing the observed distribution's variance to the expected one, they derive a coefficient placing different leagues on a skill-chance spectrum. They report basketball as being least influenced by chance, followed by volleyball, soccer, and handball. They do not report concrete chance values, except for the NBA where they estimate it as being 35%, but explore how many teams that are much better (or much worse) than the rest have to be removed to end up with a pure-chance league. Basketball (50%) and volleyball (40%) require many such removals, whereas in soccer (19%) and handball (14%) there are only few outlier teams.

Gilbert and Wells [7] place "luck" in the context of ludology, the study of complex games. Defining two measures, , they consider individual MLB (baseball), NFL, NHL, and NBA matches. They find NBA matches to be less affected by chance than NHL and MLB ones (due to larger skill differences in the NBA). They also report NFL matches to be least affected by discuss that the shorter season seems to cancel this out.

Sarkar and Kamath [11] consider soccer, and aim to establish whether there is a difference in "X-factors" between the first and last six teams in a season's ranking, and to what degree chance affects the final rankings. They compare expected (predicted) points, goals etc. to observed ones to derive what they term "X-factors". By comparing the mean of X-factors between consecutive positions in the ranking, they find that the X-factor has explanatory power for the first two positions, and that chance has no effect on the top-six.

Csurilla *et al.* [5] consider the results of 3×3 and 5×5 basketball world cups, using the final ranking as ground truth. Using four chance measures, they find that women's basketball competitions are less influenced by chance than men's, and that 3×3 is more influenced by chance than 5×5. They do not report concrete chance values.

5   Deriving Limits for Specific Seasons

The ideal case derived from the MC simulation does not help us very much in assessing how close a predictive model comes to the best possible prediction. Instead of trying to answer the theoretical question: *What is the expected limit to predictive accuracy for a given league?*,
we therefore want to answer the practical question: *Given a specific season, what was the highest possible predictive accuracy?*.

To this end, we still need to find a way of estimating the impact of chance on match outcomes, while *taking the specificities of scheduling into account*. The problem with estimating the impact of chance stays the same, however: for any given match, we need to know the relative strength of the two teams but if we knew that, we would have no need to learn a predictive model in the first place. If one team has a lower adjusted <u>offensive</u> efficiency than the other (i.e. scoring less), for example, but also a lower adjusted <u>defensive</u> efficiency (i.e. giving up fewer points), should it be considered weaker, stronger, or of the same strength?

Learning a model for relative strength and using it to assess chance would therefore feed the models potential errors back into that estimate. What we *can* attempt to identify, however, is which teams are *similar*.

5.1   Clustering Team Profiles and Deriving Match-Up Settings

Table 1. Statistics used to described teams for clustering, definitions can be found at www.basketball-reference.com/glossary

Offensive stats		Defensive stats	
AdjOEff	Points per 100 possessions scored, adjusted for opponent's strength	AdjDEff	Points per 100 possessions allowed, adjusted for opponent's strength
OeFG%	Effective field goal percentage	DeFG%	eFG% allowed
OTOR	Turnover rate	DTOR	TOR forced
OORR	Offensive rebound rate	DORR	ORR allowed
OFTR	Free throw rate	DFTR	FTR allowed

We describe each team in terms of their adjusted efficiencies, and their Four Factors, adopting Ken Pomeroy's representation [10]. Each statistic is present both in its offensive form – how well the team performed, and in its defensive form – how well it allowed its opponents to perform (Table 1). We use the averaged end-of-season statistics, leaving us with approximately 340 data points per season. Clustering daily team profiles, to identify finer-grained relationships, and teams' development over the course of the season, is left as future work. As a clustering algorithm, we used the WEKA [6] implementation of the EM algorithm with

Table 2. Number of clusters per season and clusters represented in the NCAA tournament (second line)

Season	2008	2009	2010	2011	2012	2013
Number of Clusters (EM)	5	4	6	7	4	3
Cluster IDs in Tournament	1,5	4	2,6	1,2,5	3,4	2
# Clusters (Optimized EM, Sect. 7)	20	4	19	20	14	13

default parameters. This involves EM selecting the appropriate number of clusters by internal cross validation, with the second row of Table 2 showing how many clusters have been found per season.

As can be seen, depending on the season, the EM algorithm does not separate the 340 teams into many different statistical profiles. Additionally, as the third row shows, only certain clusters, representing relatively strong teams, make it into the NCAA tournament, with the chance to eventually play for the national championship (and one cluster dominates, like Cluster 5 in 2008). These are strong indications that the clustering algorithm does indeed discover similarities among teams that allow us to abstract "relative strength". Using the clustering results, we can re-encode a season's matches in terms of the clusters to which the playing teams belong, capturing the specificities of the season's schedule.

Table 3. Wins / total matches for pairings of teams of the clusters indicated in the row and column of each cell, 2008

	Cluster 1	Cluster 2	Cluster 3	Cluster 4	Cluster 5	Weaker opponent
Cluster 1	76/114	161/203	52/53	168/176	65/141	381/687 (0.5545)
Cluster 2	100/176	298/458	176/205	429/491	91/216	705/1546 (0.4560)
Cluster 3	7/32	55/170	47/77	119/194	4/40	119/513 (0.2320)
Cluster 4	22/79	161/379	117/185	463/769	28/145	117/1557 (0.0751)
Cluster 5	117/154	232/280	78/83	232/247	121/198	659/962 (0.6850)

Table 3 summarizes the re-encoded schedule for 2008.[4] The re-encoding allows us to flesh out the intuition mentioned in the introduction some more: teams from the same cluster can be expected to have approximately the same strength, increasing the impact of chance on the outcome. Since we want to take all non-chance effects into account, we encode pairings in terms of which team has home court. The left margin indicates which team has home court in the pairing: this means, for instance, that while teams from Cluster 1 beat teams from Cluster 2 almost 80% of the time when they have home court advantage, teams from Cluster 2 prevail in almost 57% of the time when home court advantage is

---

[4] The results for other seasons can be found in Appendix A.

theirs. The effect of home court advantage is particularly pronounced on the diagonal, where unconditional winning percentages by definition should be at approximately 50%. Instead, home court advantage pushes them always above 60%. One can also see that many teams — all those in clusters 3 and 4 — mainly play against stronger teams, and the teams in cluster 5 mainly against weaker ones. In those cases, chance would need to intervene rather strongly to alter match outcomes.

Table 3 is the empirical instantiation of our remark in Sect. 3: instead of a single distribution, 2008 seems to have been a weighted mixture of 25 distributions.[5] None of these specificities can be captured by the unbiased MC simulation.

5.2 Estimating Chance

The re-encoded schedule includes all the information we need to assess the effects of chance. The win percentage for a particular cluster pairing indicates which of the two clusters should be considered the stronger one in those circumstances, and from those matches that are lost by the stronger team, we can calculate the chance involved.

Consider, for instance, the pairing *Cluster 5 – Cluster 2*. When playing at home, teams from Cluster 5 win this match-up in 82.85% of the cases! This is the practical limit to predictive accuracy in this setting for a model that always predicts the stronger team to win, and in the same way we used c to calculate that limit above, we can now invert the process: $c = 2 * (1 - 0.8285) = 0.343$. When teams from Cluster 5 welcomed teams from Cluster 2 on their home court in 2008, the overall outcome is indistinguishable from 34.3% of matches having been decided by chance.

The impact of chance for each cluster pairing, and the number of matches that have been played in particular settings, finally, allows us to calculate the effect of chance on the entire season, and using this result, the upper limit for predictive accuracy that could have been reached for a particular season.

Table 4. Effects of chance on different seasons' matches and limit on predictive accuracy (for team encoding shown in Table 1)

Season	2008	2009	2010	2011	2012	2013
Unconstrained EM						
KS	0.0526	0.0307	0.0506	0.0327	0.0539	0.0429
Chance	0.5736	0.5341	0.5066	0.5343	0.5486	0.5322
Limit for predictive accuracy	0.7132	0.7329	0.7467	0.7329	0.7257	0.7339
Optimized EM (Sect. 7)						
KS	0.0236	0.0307	0.0396	0.0327	0.0315	0.0410
Chance	0.4779	0.5341	0.4704	0.5343	0.4853	0.5311
Limit	0.7610	0.7329	0.7648	0.7329	0.7573	0.7345
KenPom prediction accuracy	0.7105	0.7112	0.7244	0.7148	0.7307	0.7035
ANN prediction accuracy from [14]	0.7136	0.7357	0.7115	0.7248	0.7120	0.7187

---

[5] Although some might be similar enough to be merged.

The upper part of Table 4 shows the resulting effects of chance and the limits regarding predictive accuracy for the six seasons under consideration. Notably, the last two rows show the predictive accuracy when using the method described on [10] and results we have reported in [14], using an artificial neural network model. Ken Pomeroy's method uses the Log5-method, with Pythagorean expectation to derive each team's win probability, and the adjusted efficiencies of the home (away) team improved (degraded) by 1.4%. This method effectively always predicts the stronger team to win and should therefore show similar behavior as the observed outcomes. Its accuracy is always close to the limit and in one case (2012) actually exceeds it. One could explain this by the use of daily instead of end-of-season statistics but there is also another aspect in play. To describe that aspect, we need to discuss simulating seasons.

6   Simulating Seasons

With the scheduling information and the impact of chance for different pairings, we can simulate seasons in a similar manner to the Monte Carlo simulations we have discussed above, but with results that are much closer to the distribution of observed seasons:

1. A chance value $c \in [0, 1]$ is chosen.
2. A set of virtual teams is assigned cluster labels as derived by the EM clustering in such a manner that the distribution of instances to clusters matches the observed distribution.
3. For each match-up, a value $v \in [0, 1]$ is randomly drawn from a uniform distribution.
   - If $v \geq c$, the stronger team wins.
   - Otherwise, the winner is decided by flipping a *weighted* coin, with the weight derived from the observed win probability for the cluster pairing.
4. The simulation is re-iterated a large number of times (e.g. 10, 000) to smooth results.

Figure 3 shows that while the simulated distribution is not equivalent to the observed one, it shows very similar trends. In addition, while the KS test does not reject any of the three simulated distributions, the distance of the one resulting from our approach to the observed one is lower than for the two Monte Carlo simulated ones.

The figure shows the result of simulating the season 10, 000 times, leading to the stabilization of the distribution. For fewer iterations, e.g. 100 or fewer, distributions that diverge more from the observed season can be created. In particular, this allows the exploration of counterfactuals: if certain outcomes were due to chance, how would the model change if they came out differently? Finally, the information encoded in the different clusters – means of statistics and covariance matrices – allows the generation of synthetic team instances that fit the cluster (similar to value imputation), which in combination with scheduling information could be used to generate wholly synthetic seasons to augment the training data used for learning predictive models. We plan to explore this direction in future work.

7   Finding a Good Clustering

Coming back to predictive limits, there is no guarantee that the number of clusters found by the unconstrained EM will actually result in a distribution of win percentages that is necessarily close to the observed one. Instead, we can use the approach outlined in the preceding section to find a good clustering to base our chance and predictive accuracy limits on:

1. We let EM cluster teams for a fixed number of clusters (we evaluated 4–20)
2. For a derived clustering, we simulate 10,000 seasons
3. The resulting distribution is compared to the observed one using the Kolmogorov-Smirnov score

Table 5. KS similarity between observed and simulated distributions for different numbers of clusters, lower is better, best values are indicated in **bold**

k	Season					
	2008	2009	2010	2011	2012	2013
4	0.0838	**0.0307**	0.0560	0.0702	0.0539	0.0644
5	0.0710	0.0356	0.0543	0.0624	0.0609	0.0468
6	0.0488	0.0377	0.0506	0.0528	0.0404	0.0448
7	0.0494	0.0382	0.0433	0.0327	0.0390	0.0432
8	0.0466	0.0381	0.0417	0.0355	0.0404	0.0435
9	0.0534	0.0338	0.0424	0.0396	0.0398	0.0414
10	0.0564	0.0332	0.0421	0.0349	0.0370	0.0503
11	0.0478	0.0342	0.0447	0.0336	0.0325	0.0433
12	0.0326	0.0436	0.0545	0.0390	0.0378	0.0410
13	0.0357	0.0432	0.0559	0.0367	0.0351	**0.0402**
14	0.0374	0.0384	0.0511	0.0342	**0.0315**	0.0439
15	0.0385	0.0449	0.0578	0.0380	0.0350	0.0412
16	0.0388	0.0456	0.0570	0.0364	0.0361	0.0527
17	0.0269	0.0437	0.0480	0.0433	0.0441	0.0464
18	0.0293	0.0413	0.0441	0.0409	0.0327	0.0449
19	0.0276	0.0420	**0.0396**	0.0392	0.0462	0.0462
20	**0.0236**	0.0387	0.0460	**0.0289**	0.0371	0.0493

The results of this optimization are shown in Table 5. What is interesting to see is that a) increasing the number of clusters does not automatically lead to a better fit with the observed distribution, and b) clusterings with different numbers of clusters occasionally lead to the same KS, validating our comment in Footnote 5.

Based on the clustering with the lowest KS, we calculate chance and predictive limit and show them in the second set of rows of Table 4. EM already found the optimal assignment of teams to clusters for 2009 but for other seasons, there are quite a few more clusters. Generally speaking, optimizing the fit allows to lower the KS quite a bit and leads to lower estimated chance and higher predictive limits. For both categories, however, the fact remains that different seasons were influenced by chance to differing degrees and therefore different limits exist. Furthermore, the limits we have found stay significantly below 80% and are different from the limits than can be derived from MC simulation.

Those results obviously come with some caveats:

1. Teams were described in terms of adjusted efficiencies and Four Factors – adding or removing statistics could lead to different numbers of clusters and different cluster memberships.
2. Predictive models that use additional information, e.g. experience of players, or networks models for drawing comparisons between teams that did not play each other, can exceed the limits reported in Table 4.

The table also indicates that it might be less than ideal to learn from preceding seasons to predict the current one (the approach we have chosen in our previous work): having a larger element of chance (e.g. 2009) could bias the learner against relatively stronger teams and lead it to underestimate a team's chances in a more regular season (e.g. 2010).

8   NBA Results

Finally, we apply the same approach to NBA seasons. The NBA contains much fewer teams (30), and plays many more matches (82), with the result that every team plays every other one at least twice.

Table 6. Effects of chance on different NBA seasons' matches and limits on predictive accuracy

Season	2008	2009	2010	2011	2012	2013	2014	2015
Optimized EM (Sect. 7)								
Number of clusters	5	6	5	9	6	10	9	7
KS	0.0749	0.0693	0.0848	0.0802	0.0618	0.0822	0.0512	0.0840
Chance	0.5821	0.5779	0.5930	0.5802	0.6091	0.5841	0.6005	0.6066
Limit	0.7090	0.7110	0.7035	0.7099	0.6954	0.7080	0.6998	0.6967
KenPom prediction	0.6725	0.6920	0.6700	0.6697	0.6601	0.6530	0.6475	0.6590
NB prediction accuracy from [14]	0.6608	0.6494	0.6506	0.6331	0.6187	0.6240	0.6438	0.6545

The results can be seen in Table 6. Compared to Table 4, chance values are higher, and predictive limits lower. This is expected since talent differences between NBA teams are less than for NCAAB teams (only the best NCAAB players end up in the NBA after all), which allows chance to have a bigger effect, as we discussed in the introduction. We also see that the prediction accuracies of the predictive models are lower, most notably for Ken Pomeroy's model, which, as mentioned before, effectively always predicts the best team to win.

9   Summary and Conclusions

In this paper, we have considered the question of the impact of chance on the outcome of (semi-)professional sports matches in more detail. In particular, we have shown that the unbiased MC simulations used to assess chance in the NFL and NHL are not applicable to the college basketball setting. We have argued that the resulting limits on predictive accuracy rest on simplifying and idealized assumptions and therefore do not help in assessing the performance of a predictive model on a particular season.

As an alternative, we propose clustering teams' statistical profiles and re-encoding a season's schedule in terms of which clusters play against each other. Using this approach, we have shown that college basketball seasons violate the assumptions of the unbiased MC simulation, given higher estimates for chance, as well as tighter limits for predictive accuracy.

There are several directions that we intend to pursue in the future. First, as we have argued above, NCAA basketball is not the only setting in which imbalanced schedules occur. We would expect similar effects in the NFL, and soccer, which, as mentioned, is lower-scoring and where teams typically only play each other twice. What is needed to explore this question is a good statistical representation of teams, something that is easier to achieve for basketball than football/soccer teams.

In addition, as we have mentioned in Sect. 6, the exploration of counterfactuals and generation of synthetic data should help in analyzing sports better. We find a recent paper [9] particularly inspirational, in that the authors used a detailed simulation of substitution and activity patterns to explore alternative outcomes for an NBA playoff series.

Finally, since we can identify different cluster pairings and the differing of chance therein, separating those cases and training classifiers independently for each could improve classification accuracy. To achieve this, however, we will need solve the problem of clustering statistical profiles over the entire season – which should also allow to identify certain trends over the course of seasons.

A   Clustered schedules for different seasons, unconstrained EM

See Tables 7, 8, 9, 10 and 11.

Table 7. Wins and total matches for different cluster pairings, 2009

	Cluster 1	Cluster 2	Cluster 3	Cluster 4	Weaker opponent
Cluster 1	133/197	46/182	105/272	1/45	0/696 (0.0000)
Cluster 2	210/227	231/352	262/374	76/247	472/1200 (0.3933)
Cluster 3	261/308	192/357	409/663	56/261	453/1589 (0.2851)
Cluster 4	210/211	341/374	424/448	515/818	975/1851 (0.5267)

Table 8. Wins and total matches for different cluster pairings, 2010

	Cluster 1	Cluster 2	Cluster 3	Cluster 4	Cluster 5	Cluster 6	Weaker opponent
Cluster 1	129/204	18/104	6/14	47/126	33/145	0/18	0/611 (0.0000)
Cluster 2	163/167	269/437	76/105	255/292	134/195	73/249	628/1445 (0.4346)
Cluster 3	29/34	64/95	12/18	49/58	21/41	30/87	163/333 (0.4895)
Cluster 4	109/136	71/240	19/46	159/232	55/119	6/87	109/860 (0.1267)
Cluster 5	147/163	87/166	14/23	101/123	71/118	17/57	349/650 (0.5369)
Cluster 6	120/120	336/361	100/117	169/172	133/141	360/579	858/1490 (0.5758)

Table 9. Wins and total matches for different cluster pairings, 2011

	Cluster 1	Cluster 2	Cluster 3	Cluster 4	Cluster 5	Cluster 6	Cluster 7	Weaker opponent
Cluster 1	89/138	140/174	40/40	69/73	93/185	97/103	86/99	525/812 (0.6466)
Cluster 2	66/148	235/369	70/71	141/167	29/121	166/206	118/176	495/1258 (0.3935)
Cluster 3	2/14	15/55	29/39	16/42	0/8	20/85	4/31	0/274 (0.0000)
Cluster 4	10/48	48/151	36/40	42/85	2/28	55/100	28/68	91/520 (0.1750)
Cluster 5	166/217	187/206	43/43	79/80	205/339	80/83	148/160	703/1128 (0.6232)
Cluster 6	11/49	76/178	77/88	72/97	7/47	94/151	34/65	183/675 (0.2711)
Cluster 7	29/82	97/160	57/58	59/72	30/125	74/92	79/127	287/716 (0.4008)

Table 10. Wins and total matches for different cluster pairings, 2012

	Cluster 1	Cluster 2	Cluster 3	Cluster 4	Weaker opponent
Cluster 1	108/201	110/320	20/119	19/121	0/761 (0.0000)
Cluster 2	362/416	610/960	105/354	175/394	362/2124 (0.1704)
Cluster 3	197/197	458/500	264/418	191/251	846/1366 (0.6193)
Cluster 4	179/191	373/454	111/245	163/258	552/1148 (0.4808)

Table 11. Wins and total matches for different cluster pairings, 2013

	Cluster 1	Cluster 2	Cluster 3	Weaker opponent
Cluster 1	507/807	89/374	272/567	0/1748 (0.0000)
Cluster 2	569/607	622/967	518/578	1087/2152 (0.5051)
Cluster 3	435/611	119/381	358/572	435/1564 (0.2781)

References

1. Aoki, R.Y., Assuncao, R.M., Vaz de Melo, P.O.: Luck is hard to beat: the difficulty of sports prediction. In: Proceedings of the 23rd ACM SIGKDD International Conference on Knowledge Discovery and Data Mining, pp. 1367–1376 (2017)
2. Burke, B.: Luck and NFL outcomes 1. http://www.advancedfootballanalytics.com/2007/08/luck-and-nfl-outcomes.html. Accessed 15 May 2023
3. Burke, B.: Luck and NFL outcomes 1. http://www.advancedfootballanalytics.com/2007/08/luck-and-nfl-outcomes-2.html. Accessed 15 May 2023
4. Burke, B.: Luck and NFL outcomes 1. http://www.advancedfootballanalytics.com/2007/08/luck-and-nfl-outcomes-3.html. Accessed 15 May 2023
5. Csurilla, G., Boros, Z., Fűrész, D.I., Gyimesi, A., Raab, M., Sterbenz, T.: How much is winning a matter of luck? a comparison of 3×3 and 5v5 basketball. Int. J. Environ. Res. Public Health **20**(4), 2911 (2023)
6. Frank, E., Witten, I.H.: Data Mining: Practical Machine Learning Tools and Techniques with Java Implementations. Morgan Kaufmann, Burlington (1999)
7. Gilbert, D.E., Wells, M.T.: Ludometrics: luck, and how to measure it. J. Quant. Anal. Sports **15**(3), 225–237 (2019)
8. Mauboussin, M.J.: The Success Equation: Untangling Skill and Luck in Business, Sports, and Investing. Harvard Business Review Press, Brighton (2012)
9. Oh, M.H., Keshri, S., Iyengar, G.: Graphical model for basketball match simulation. In: MIT Sloan Sports Analytics Conference (2015)
10. Pomeroy, K.: Advanced analysis of college basketball. http://kenpom.com
11. Sarkar, S., Kamath, S.: Does luck play a role in the determination of the rank positions in football leagues? A study of Europe's 'big five'. Ann. Oper. Res. **325**, 1–16 (2021)
12. Weissbock, J.: Theoretical predictions in machine learning for the NHL: Part ii. http://nhlnumbers.com/2013/8/6/theoretical-predictions-in-machine-learning-for-the-nhl-part-ii. Accessed 22 June 2015
13. Weissbock, J., Inkpen, D.: Combining textual pre-game reports and statistical data for predicting success in the national hockey league. In: Sokolova, M., van Beek, P. (eds.) AI 2014. LNCS (LNAI), vol. 8436, pp. 251–262. Springer, Cham (2014). https://doi.org/10.1007/978-3-319-06483-3_22
14. Zimmermann, A.: Basketball predictions in the NCAAB and NBA: similarities and differences. Stat. Anal. Data Min. **9**(5), 350–364 (2016)

Exploring Label Correlations
for Quantification of ICD Codes

Isabel Coutinho[1,2]([⊠]) [iD] and Bruno Martins[1,2,3] [iD]

[1] INESC-ID, Lisbon, Portugal
[2] Instituto Superior Técnico, University of Lisbon, Lisbon, Portugal
isabel.coutinho@tecnico.ulisboa.pt
[3] LUMLIS (Lisbon ELLIS Unit), Lisbon, Portugal

Abstract. The International Classification of Diseases (ICD) has been
adopted worldwide in the healthcare domain, e.g. to summarize the key
information in clinical documents. Since manual ICD coding is very
expensive, time-consuming, and error-prone, deep learning algorithms
have been proposed to automate this task. However, the final goal of
ICD coding often lays not in determining the codes associated with individual documents, but instead in quantifying the prevalence of each code
within sets of documents. In this work, we experimentally assess different
quantification methods in connection to ICD coding, including a simple
learning-based approach that leverages associations between the codes, in
order to predict their relative frequencies more accurately. Experiments
show that the proposed approach can effectively explore existing associations between ICD codes, improving the quantification performance
over baseline methods that deal with each code independently.

Keywords: Clinical Text Processing · ICD Coding · Quantification

1 Introduction

The International Classification of Diseases (ICD)[1] coding system corresponds
to a standardised way of indicating diagnoses and procedures, supporting a
variety of analyses over clinical data (e.g., in the context of administrative
processes or public health studies). Despite being adopted worldwide in the
healthcare domain, manually assigning ICD codes to clinical documents is both
time-consuming and error-prone, and it represents a huge monetary burden for
health facilities [3,16]. Noting the aforementioned problems, many efforts have
been placed on the design of deep learning methods to automatically assign ICD
codes to clinical text, usually formulating the task as a multi-label classification problem. Recent work has proposed the use of Transformer-based language
models [28]. Still, although these approaches have become the state-of-the-art
in many Natural Language Processing (NLP) problems, several challenges have
not yet been overcome for the ICD coding task, which requires the processing of
long narratives that involve a large and domain-specific vocabulary [12].

---

[1] https://www.who.int/standards/classifications/classification-of-diseases.

© The Author(s), under exclusive license to Springer Nature Switzerland AG 2023
A. Bifet et al. (Eds.): DS 2023, LNAI 14276, pp. 614–627, 2023.
https://doi.org/10.1007/978-3-031-45275-8_41

On the other hand, a line of work that has not been explored concerns using the results of a classifier to estimate the prevalence (i.e., the relative frequency) of ICD codes in a given dataset. In other domains, this task is usually referred to as the problem of text quantification [10,17,25]. In fact, in most practical applications for ICD coding, we want to estimate the prevalence of an ICD code (or group of codes) in a dataset, rather than simply estimating codes for individual documents. For instance, epidemiologists are often interested in monitoring the prevalence of specific diseases, and this may be done through clinical NLP. Still, more than classifying documents associated to specific individuals (e.g., death certificates, hospital discharge summaries, etc.), this requires the analysis of sets of documents representing a population under a period of analysis.

Previous work has already proposed several methods for text quantification that outperform the standard "classify and count" procedure [17,22]. However, scenarios that include multi-label classifiers and/or large label spaces have received less attention. A straightforward solution to the multi-label quantification problem could simply consist of recasting the problem as a set of independent binary quantification problems [21]. Although simple, this solution is not entirely satisfactory when the independence assumption between the target labels is not verified, which happens naturally in the case of ICD coding due to frequent comorbidities or due to the association between diagnoses and procedures.

This work presents experiments using different quantification approaches, including a simple learning-based method that leverages relations between the ICD codes in order to predict their relative frequencies more accurately. Our tests were conducted using the MIMIC-III dataset [15] for multi-label text classification, which is a large and extensively used publicly available Electronic Health Record (EHR) dataset. We specifically considered a subset of hospital discharge summaries with the 50 most frequent codes, named MIMIC-III-50 [23], and the recently released dataset splits named MIMIC-III-clean [6].

We show that the proposed approach can effectively explore the existing associations between ICD codes, refining probabilistic "classify and count" estimates from the multi-label classifier, so that the quantification errors are in some cases reduced by a considerable margin.

2   ICD Coding of Clinical Text

This section describes implementation details regarding the ICD coding task. Section 2.1 presents the characterization of the dataset used in our experiments, which have also been used in previous work focusing on automated ICD coding. In turn, Sect. 2.2 explores the neural network architecture used in our approach for multi-label ICD classification.

2.1   Clinical Text Dataset Splits

The MIMIC-III dataset [15] is a large critical care database comprising information of over 40,000 patients admitted to intensive care units. It includes different types of clinical notes, such as hospital discharge summaries. This last type of documents constitute the clinical text considered in most previous studies for the ICD coding task, as these condense all the information during a patient visit into one document. Each document is tagged with a set of ICD-9-CM codes, describing both diagnoses and procedures made during the patient visit.

Our experiments used the exact same data splits from the work of Mullenbach et al. [23], which were also used in many of the subsequent studies in the area. Similarly to recent studies exploring the use of Transformer-based models for automated ICD coding [4], we considered a version of the data with the subset of the top-50 most frequent codes, referred to as MIMIC-III-50. This subset consists of 11,368 hospital discharge summaries, in which 8,066, 1,573, and 1,729 documents are used as training, validation, and test sets, respectively.

Additionally, we performed experiments using recently released new dataset splits, named MIMIC-III-clean [6]. This subset considers 3,681 unique ICD-9-CM codes, thus representing a more challenging classification problem. It contains 52,712 hospital discharge summaries, in which 38,401, 5,577, and 8,734 documents are used as training, validation, and test sets, respectively. Table 1 presents a comparison of the two MIMIC-III dataset splits considered in the experiments.

The different data splits from MIMIC-III were used in the ICD coding task (see Sect. 2.2), and the documents in the validation/test sets were used to assess quantification performance, according to the setup in Sect. 4.1. Notice that, in the multi-label setting associated with MIMIC-III, one can expect label associations to play a particular importance in the performance of classification and quantification methods (e.g., codes corresponding to common comorbidities, or common disease/procedure pairs, are expected to have a similar prevalence).

2.2   Neural Network Architecture

In this work, ICD coding of MIMIC-III clinical texts mostly relied on a language model based on the Transformer architecture [28], namely a Longformer [2] for clinical text. Contrarily to standard Transformer-based models like BERT [5], which are unable to process long sequences due to their quadratic self-attention operation, the Longformer overcomes this limitation by introducing a sparse

Table 1. Comparison of the MIMIC-III dataset splits.

	MIMMIC-III-50	MIMIC-III-clean
Number of unique codes	50	3,681
Median number of codes per document	5	14
Number of training documents	8,066	38,401
Number of validation documents	1,573	5,577
Number of test documents	1,729	8,734

attention mechanism that scales linearly with sequence length, making it possible to process longer documents (e.g., up to 4,096 tokens).

Specifically, we used a previously proposed Clinical-Longformer model from Li et al. [18], which corresponds to a clinical knowledge-enriched version of the Longformer that was pre-trained using clinical text. The Clinical-Longformer was initialised from a Longformer-base model and was further pre-trained on all clinical notes from the MIMIC-III dataset. This model is publicly available in association to the HuggingFace[2] Transformers library [30].

The approach used to train the model for multi-label classification involves minimising the Binary Cross-Entropy (BCE) loss between the predicted and true ICD code assigments, as follows:

$$\mathcal{L}_{\mathrm{BCE}}(\mathbf{p}, \mathbf{y}) = -\frac{1}{L} \sum_{i=1}^{L} \left[y_i \log(p_i) + (1 - y_i) \log(1 - p_i) \right], \tag{1}$$

where $\mathbf{p}$ and $\mathbf{y} \in \mathbb{R}^L$. The variable $y_i \in \{0, 1\}$ is the ground truth for label i, p_i is the probability of the label i being true as given by the classifier, and L is the number of different ICD labels.

We fine-tuned two different Clinical-Longformer models using the data splits mentioned in the previous section. Experiments on MIMIC-III-50 showed that this classifier achieves a macro-F1 of 63.2%, a micro-F1 of 68.4%, and a precision@5 of 64.8%. These results are on par with those from other previous publications on the topic of ICD coding [4]. In the case of the MIMIC-III-clean dataset splits, the classifier achieves a macro-F1 of 6.5%, a micro-F1 of 46.2%, and a precision@15 of 47.8%, slightly lower than previously reported scores [6].

Additionally, we also considered the LAAT classifier [29] for our experiments, which is one of the state-of-the-art models on MIMIC-III. This model consists of a Bi-LSTM network producing latent feature representations of all the input tokens in a clinical note. These representations are then fed to a label attention layer, in order to learn label-specific vectors that represent the important clinical text fragments relating to certain labels, this way producing label-specific weight vectors. Each one of these vectors is finally used to build a binary classifier for each of the labels. Edin et al. [6] reported on a revised model comparison over the MIMIC-III-clean dataset, and the corresponding source code and model parameters were made publicly available, allowing for result reproducibility. Table 2 summarizes the classification results for the different methods.

3   Quantification

Classify and Count (CC) is perhaps the simplest quantification method, which we use as a baseline in our study. Each class i is handled independently, and the method consists of simply counting the number of documents classified as i and dividing by the total number of documents in a sample. Given a classifier c and a sample of documents ϵ, CC for each ICD code i is defined as follows:

---

[2] https://github.com/huggingface/transformers.

$$\hat{p}_\epsilon^{CC}(i) = \frac{|\{\mathbf{x} \in \epsilon | c_i(\mathbf{x}) = 1\}|}{|\epsilon|}. \tag{2}$$

A variant that corresponds to a stronger baseline is Probabilistic Classify and Count (PCC). Instead of counting the number documents classified as i, we can use the posterior probabilities returned by the classifier, as follows:

$$\hat{p}_\epsilon^{PCC}(i) = \frac{1}{|\epsilon|} \sum_{\mathbf{x} \in \epsilon} p_i(\mathbf{x}). \tag{3}$$

Although conceptually simple, the CC and PCC methods correspond to very strong baselines, often outperforming more sophisticated quantification methods [22]. We initially also considered the Adjusted Classify and Count (ACC) and Probabilistic Adjusted Classify and Count (PACC) approaches [1,8,9], which use estimates for the True Positive and False Positive Rates (TPR/FPR) of the underlying classifier to make adjustments over the CC/PCC methods. However, initial experimental results were discouraging. Previous studies have noted that the performance of these methods is highly sensitive to any errors in the TPR/FPR estimates (whose computation, in the case of multi-label classification, also poses additional challenges [11]), and that ACC and PACC can degrade severely when the training class distribution is highly imbalanced [9].

The main approach explored in this study involves training a Multi-Layer Perceptron (MLP) in order to obtain an adjusted estimate of the prevalence of ICD codes in the input dataset, accounting with associations between the labels. Our MLP, which is illustrated in Fig. 1, takes its inspiration on undercomplete denoising auto-encoders, and also on previous work focused on improving extreme multi-label classification through label correlations [31]. We use a single bottleneck hidden layer that forces a compressed knowledge representation of a noisy input, in an attempt to capture the underlying structure of the data (i.e., the correlations between the different classes, which correspond to the input features). This structure is learned and consequently used when forcing the inputs through the bottleneck, in order to predict the adjusted class prevalence values (i.e., the denoised versions of the inputs).

More formally, the input corresponds to $\hat{\mathbf{p}}_\epsilon^{PCC}$ estimates, while the output is an adjusted quantification, $\hat{\mathbf{p}}_\epsilon^{MLP}$. We thus obtain:

Table 2. Results for the different classification methods, over the considered dataset splits. Results for methods with † were taken directly from a previous publication.

Dataset splits	Models	AUC		F1		Precision@k		
		Macro	Micro	Macro	Micro	5	8	15
MIMIC-III-50	Longformer	89.8	92.2	63.2	68.4	64.8	52.6	34.5
MIMIC-III-clean	Longformer	91.5	97.7	6.5	46.2	72.7	63.6	47.8
MIMIC-III-clean	LAAT [6]†	94.0	98.6	22.6	57.8	—	70.1	54.8

$$\hat{\mathbf{p}}_\epsilon^{\text{MLP}} = \sigma(\mathbf{W}_2\sigma(\mathbf{W}_1\hat{\mathbf{p}}_\epsilon^{\text{PCC}} + \mathbf{b}_1) + \mathbf{b}_2), \tag{4}$$

where $\mathbf{W}_1 \in \mathbb{R}^{D \times L}$ and $\mathbf{W}_2 \in \mathbb{R}^{L \times D}$ are the learned weights, $\mathbf{b}_1 \in \mathbb{R}^D$ and $\mathbf{b}_2 \in \mathbb{R}^L$ are the biases, D is the dimensionality of the hidden layer, and σ is a non-linear activation function applied after the linear transformations (in our case, the logistic sigmoid function).

Model training is done by minimising the Mean Squared Error (MSE) over a quantification training set, as follows:

$$\mathcal{L}_{\text{MSE}}(\hat{\mathbf{p}}_\epsilon^{\text{MLP}}, \mathbf{p}_\epsilon) = \sum_{i=1}^{L} |\hat{p}_\epsilon^{\text{MLP}}(i) - p_\epsilon(i)|^2, \tag{5}$$

where $\hat{\mathbf{p}}_\epsilon^{\text{MLP}}$ and $\mathbf{p}_\epsilon \in \mathbb{R}^L$. The function $p_\epsilon(i)$ corresponds to the ground-truth quantification for the ICD label i, i.e., the actual class prevalence values defined as the number of documents where $y_i = 1$ divided by the total number of documents in sample ϵ.

4   Experimental Evaluation

This section presents the experimental evaluation of the proposed methods. Section 4.1 details the model implementation and the experimental setup, while Sect. 4.2 focuses on the evaluation metrics that were adopted to assess result quality. Finally, Sect. 4.3 presents and discusses the obtained results.

4.1   Experimental Setup

Our tests involved clinical text classification for the assignment of ICD codes to individual documents, followed by the training and/or assessment of quantification methods over sets of documents.

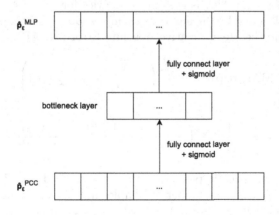

Fig. 1. The proposed MLP architecture for quantification.

On what regards multi-label classification of the MIMIC-III documents, we fine-tuned two different Clinical-Longformer models, as described in Sect. 2.2, with an effective batch size of 16 instances, a learning rate of 2e-5, and a maximum of 30 training epochs together with an early stopping patience of 5, monitoring the micro-F1 score over the validation split.

Regarding multi-label quantification, for each experiment, our setup involved the creation of training and testing collections, respectively by sampling documents from the original MIMIC-III-50/MIMIC-III-clean validation and testing splits, and considering samples with different sizes [19]. Each sample ϵ consists of a set of documents randomly chosen from the corresponding split, with $|\epsilon|$ taking a random value for each sample between one and the number of documents in the split. A total of 5,000 samples were generated for training our MLP quantifier, and 1,000 samples were generated for testing all the quantification approaches (including the unsupervised baselines). We aimed at the creation of reasonably sized datasets, with sufficient samples for model training without under-fitting, and for stable model evaluation. Notice that the MIMIC-III-clean data splits feature a much larger number of documents (i.e., more than four times the number of training documents in MIMIC-III-50). Given the much larger label space and dataset size, we also considered a setting for the quantification experiments featuring a set of training samples that is four times larger (i.e., 20,000 training samples), built with the same general methodology. We then trained another MLP quantifier with these samples named MLPX4.

For training the MLPs for quantification, we used a batch size of 32 samples, a learning rate of 1e-3, and a maximum of 1,000 training epochs, stopping if the training loss does not decrease for five consecutive epochs. We considered a hidden layer dimensionality of $D = 32$ for the case of MIMIC-III-50, and $D = 3072$ for MIMIC-III-clean, i.e., slightly less than the number of ICD classes.

4.2 Evaluation Metrics

Following previous work [21,26], we use the Absolute Error (AE) and the Relative Absolute Error (RAE) as evaluation metrics for quantification. We additionally divide the error by the number of samples, obtaining the Mean Absolute Error (MAE) and the Mean Relative Absolute Error (MRAE):

$$\mathrm{MAE}(\mathbf{p}, \hat{\mathbf{p}}) = \frac{1}{N} \sum_{j=1}^{N} \left(\frac{1}{L} \sum_{i=1}^{L} |p_{\epsilon_j}(i) - \hat{p}_{\epsilon_j}(i)| \right), \tag{6}$$

$$\mathrm{MRAE}(\mathbf{p}, \hat{\mathbf{p}}) = \frac{1}{N} \sum_{j=1}^{N} \left[\frac{1}{2L} \sum_{i=1}^{L} \left(\frac{|p_{\epsilon_j}(i) - \hat{p}_{\epsilon_j}(i)|}{p_{\epsilon_j}(i)} + \frac{|(1 - p_{\epsilon_j}(i)) - (1 - \hat{p}_{\epsilon_j}(i))|}{(1 - p_{\epsilon_j}(i))} \right) \right], \tag{7}$$

where $\mathbf{p}$ is the ground-truth, $\hat{\mathbf{p}}$ is a quantification method (CC, PCC, or MLP), $\mathbf{p}, \hat{\mathbf{p}} \in \mathbb{R}^{N \times L}$, and N is the number of samples in the evaluation dataset. Since the MRAE is undefined when $p_{\epsilon_j}(i) = 0$ or $p_{\epsilon_j}(i) = 1$, we smooth the probability distributions $\mathbf{p}_\epsilon$ and $\hat{\mathbf{p}}_\epsilon$ via additive smoothing, as follows:

$$s(\mathbf{p}_{\epsilon_j}) = \frac{\gamma + \mathbf{p}_{\epsilon_j}}{2\gamma + 1}, \tag{8}$$

with $\gamma = (2|\epsilon_j|)^{-1}$ as the smoothing factor.

4.3   Results and Discussion

Table 3 presents the results obtained for the different quantification experiments. As a simple baseline, we show the performance of a "lazy" quantification method, which returns a constant result corresponding to the expected values for the prevalence of each ICD class within the MLP training samples (i.e., we measure class prevalence on each training sample, and make a constant prediction for all the testing samples with the average of these results). We also use CC and PCC as additional unsupervised baselines, as these models do not leverage relations between ICD codes (i.e., they treat each code independently), comparing their results against the use of the proposed MLP approach to adjust the values of PCC. In the case of MIMIC-III-clean, we present an additional variant of the MLP, which consists of using a four times larger set of training samples, named MLPX4. We repeated the experiments with the MLP method five times with different random seeds for parameter initialization, and we report the mean and standard deviation of these five runs.

Table 3. Results for the different quantification methods, over the different dataset splits and classification models that were considered. The values in bold represent the best-in-class performance in terms of the MAE and MRAE metrics.

Dataset splits	Models	Methods	MAE	MRAE
MIMIC-III-50	Longformer	Lazy	0.01413	0.07935
		CC	0.01952	0.10609
		PCC	0.01375	0.08227
		MLP	**0.01094** ± 0.00032	**0.06510** ± 0.00177
MIMIC-III-clean	Longformer	Lazy	0.00085	0.22805
		CC	0.00222	0.32510
		PCC	**0.00078**	**0.20785**
		MLP	0.00079 ± 0.00001	0.21565 ± 0.00091
		MLPX4	**0.00078** ± 0.00000	0.21951 ± 0.00102
MIMIC-III-clean	LAAT	Lazy	0.00085	0.22805
		CC	0.00154	0.26203
		PCC	0.00088	0.24575
		MLP	**0.00077** ± 0.00001	0.21251 ± 0.00082
		MLPX4	**0.00077** ± 0.00000	**0.21140** ± 0.00107

On what regards the experiments with MIMIC-III-50, the "lazy" baseline actually achieves results that are better than those of the CC method. Using the probabilities returned by the classifier, rather than counting the number of documents, results in a significant improvement in quantification accuracy (an MAE drop of 0.00577). With the MLP approach, the quantification performance also increases (an MAE drop of 0.00281 when comparing PCC against MLP). These results show that, while relatively simple, our learning-based approach can effectively improve over standard quantification methods when label associations are an important factor to consider, as is the case for ICD coding.

Regarding the tests with MIMIC-III-clean, we also can notice that the "lazy" baseline is stronger than CC, and that there is a significant performance improvement when using PCC instead of CC. However, the results are also slightly different when using different classification models. In the case of the Clinical-Longformer, the MLP approach actually failed to improve the quantification performance over PCC. In contrast, for the the LAAT model, the MLP gives an improved performance when compared to PCC (an MAE decrease of 0.00011 in both MLP variants). Regardless of the classification approach, using a four times larger collection of examples for training the MLP does not appear to increase performance (with the exception of LAAT, where the MRAE decreases, but the change is not very significant). Interestingly, although the classification performance of the Clinical-Longformer is significantly worse than that of LAAT (see Sect. 2.2), the quantification performance on the PCC baseline is superior, which suggests that the model is better calibrated and, consequently, the impact of the MLP approach for adjusting the PCC estimates is negligible.

Figure 2 presents a comparison, in terms of the AE for each of the MIMIC-III-50 classes, between the PCC and MLP methods. We also present the relative frequencies of the ICD codes over the quantification test set, and the F1-score of the underlying classifier over the MIMIC-III-50 classification test split. The MLP method improves results for the majority of the classes. For instance, for the ICD code 401.9, which is the most frequent on the test set, the difference between the AE in PCC and MLP is one of the most significant ones (i.e., the error decreases to less than half). There are however no noticeable patterns in the association between the quantification performance, the class prevalence over the testing data, and the overall accuracy of the underlying classifier.

Figure 3 presents scatter-plots with quantification results for the two ICD codes that occur more frequently in the data (i.e., 401.9 and 272.4, which respectively correspond to *unspecified essential hypertension* and *other and unspecified hyperlipidemia*), and for the two rarest codes in MIMIC-III-50 (i.e., 37.23 and 99.04, respectively corresponding to *combined right and left heart cardiac catheterization* and *transfusion of packed cells*). The x axis corresponds to the real prevalence values, while the y axis corresponds to the estimated prevalence values (i.e., an ideal quantification method would produce results in the diagonal of the scatter-plots). In all cases, the figures show that the PCC method frequently overestimates the prevalence values of these ICD codes, with the MLP method adjusting the results towards better estimates.

5   Related Work

Previous literature has separately addressed the problems of automatic ICD coding [14] and text quantification [25].

Regarding ICD coding, several approaches based on Convolutional Neural Networks (CNNs) and Recurrent Neural Networks (RNNs) have been successfully explored. Mullenbach et al. [23] proposed Convolutional Attention for Multi-Label classification (CAML), i.e., a CNN-based model that employs a label-wise attention mechanism, allowing for the selection of the most relevant segments of a given document for each of the possible codes. The authors made their dataset splits publicly available, so the method became a milestone for reproducibility in terms of methods for the ICD coding task. Ji et al. [13] proposed Gated CNNs and a novel Note-Code Interaction (GatedCNN-NCI) method. This approach simultaneously captures the lengthy and rich semantic information of clinical notes, and simultaneously exploits the interaction between notes and codes. Vu et al. [29] introduced LAAT, which we considered as a classification baseline in this work. This model combines a RNN-based encoder and a new label attention mechanism for ICD coding, aiming to handle the various lengths and the interdependence between different text segments related to ICD codes. Additionally, the authors proposed a hierarchical joint learning mechanism that enables the label attention model to handle the class-imbalance issue.

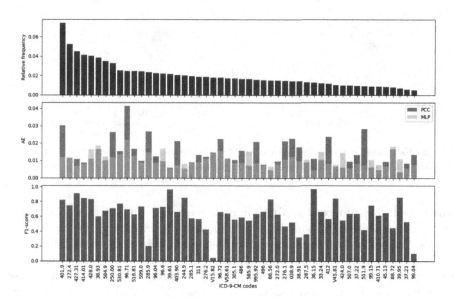

Fig. 2. Relative frequency, AE, and F1-score for each ICD code over MIMIC-III-50. Notice that, in the middle part of the figure, the bars for the PCC and MLP methods are overlapped, meaning that the color in the bottom part of the bars corresponds to the overlap of the two methods. If the color in the top part of the bar corresponds to PCC, the error of this method is higher than that of the MLP (and vice versa).

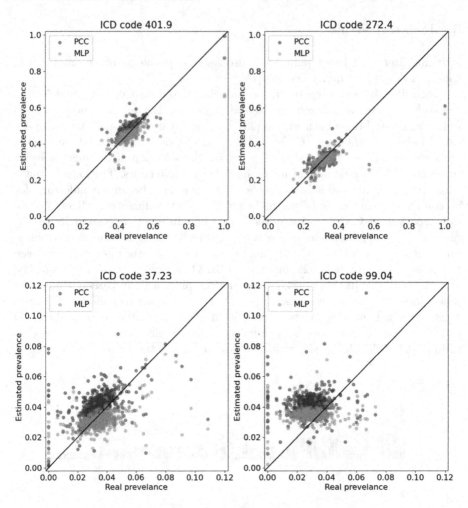

Fig. 3. Estimated versus real prevalence for the two most frequent (top) or rarest (bottom) ICD codes in the MIMIC-III-50 dataset.

Yuan et al. [32] proposed a Multiple Synonyms Matching Network (MSMN). Instead of exploiting the code hierarchy, the authors leverage synonyms for better code representation learning and, consequently, help in the coding task.

More recently, research has shifted towards Transformer-based language models, focusing on two main approaches for processing long documents: sparse attention Transformers, such as the Longformer [2] or BigBird [33], and hierarchical Transformers [24]. Dai et al. [4] compared different Transformer-based models on long document classification, which aim to mitigate the computational overhead of standard Transformers when encoding longer text. Michalopoulos et al. [20] introduced ICDBigBird, i.e., a BigBird-based model that can integrate a Graph Convolutional Network (GCN), taking advantage of the relations between ICD codes in order to create enriched representations of their embeddings. Zhang and Jankowski [34] proposed a hierarchical BERT model for ICD code assignment, named MDBERT, which is the first BERT-based approach with performance comparable to state-of-the-art methods that rely on CNNs or RNNs.

On what regards text quantification, previous work has explored the application to different domains, including sentiment quantification over tweets or in the analysis of product reviews [7,21]. Although many methods have been proposed for binary or multi-class problems, the multi-label setting remains a challenge. One possible solution is simply to recast the problem as a set of independent binary quantification problems. While simple, this solution is not, is most cases, satisfactory, since the independence assumption between the labels can not be verified. Moreo et al. [21] proposed the first truly multi-label quantification methods, leveraging the dependencies among the classes when inferring class prevalence. The proposed methods take inspiration on approaches for adapting multi-label classification into simpler multi-class problems, e.g. considering sets of ICD codes as labels.

In the standard binary or multi-class quantification settings, other studies have already proposed methods that go beyond the simplest baselines, such as the CC approach, showing that, while strong, the baselines can be outperformed [22]. Levin et al. [17] presented two enhanced PCC methods that focus on improving the quantification accuracy, by employing another supervised learning phase. While Levin et al. [17] evaluated their approach on a multi-label dataset of short text comments, the proposed method treats each label independently. Specifically, these authors extended the PCC approach by calibrating a given classifier's posterior probabilities for the specific task of text quantification, rather than merely using the posterior probabilities of the classifier which was adjusted for the classification task. The idea has some similarity to what is done in this work, but our method leverages associations between the labels, as Moreo et al. [21] proposed.

6   Conclusions and Future Work

This work assessed different text quantification approaches in connection to ICD coding of clinical documents. No previous work had addressed quantification of ICD codes and, apart from a few exceptions [17,21], quantification in association to multi-label classification had also not been previously explored. While simple and computational inexpensive, our learning-based approach can effectively explore the relations between the ICD codes, in some cases outperforming classical methods by a significant margin on texts from the MIMIC-III dataset.

Despite the interesting results, there are still many opportunities for improvement. One can, for instance, consider the use of different classification models, including approaches based on recent very large language models [27], aiming to improve the classification performance. Another possibility is to consider stronger baselines for the quantification task, including methods based on adjusting the classification thresholds according to a selection policy that maximizes quantification performance [9]. Additionally, we aim to explore more sophisticated architectures for learning-based quantification [7], e.g. within end-to-end models capable of simultaneously doing classification and quantification.

Acknowledgements. This research was supported by Fundação para a Ciência e Tecnologia (FCT), through the project with reference DSAIPA/DS/0133/2020 (DETECT) and the PhD scholarship with reference 2022.09649.BD, as well as the INESC-ID multi-annual funding from the PIDDAC programme with reference UIDB/50021/2020. We also gratefully acknowledge the support of NVIDIA Corporation, with the donation of the two Titan Xp GPUs used in our experiments.

References

1. Bella, A., Ferri, C., Hernández-Orallo, J., Ramirez-Quintana, M.J.: Quantification via probability estimators. In: Proceedings of the IEEE International Conference on Data Mining (2010)
2. Beltagy, I., Peters, M.E., Cohan, A.: Longformer: the long-document transformer. arXiv preprint arXiv:2004.05150 (2020)
3. Coutinho, I., Martins, B.: Transformer-based models for ICD-10 coding of death certificates with Portuguese text. J. Biomed. Inform. **136**, 104232 (2022)
4. Dai, X., Chalkidis, I., Darkner, S., Elliott, D.: Revisiting transformer-based models for long document classification. arXiv preprint arXiv:2204.06683 (2022)
5. Devlin, J., Chang, M.W., Lee, K., Toutanova, K.: BERT: pre-training of deep bidirectional transformers for language understanding. In: Proceedings of the Conference of the North American Chapter of the Association for Computational Linguistics (2019)
6. Edin, J., et al.: Automated medical coding on MIMIC-III and MIMIC-IV: a critical review and replicability study. arXiv preprint arXiv:2304.10909 (2023)
7. Esuli, A., Moreo Fernández, A., Sebastiani, F.: A recurrent neural network for sentiment quantification. In: Proceedings of the ACM International Conference on Information and Knowledge Management (2018)
8. Forman, G.: Counting positives accurately despite inaccurate classification. In: Proceedings of the European Conference on Machine Learning (2005)
9. Forman, G.: Quantifying counts and costs via classification. Data Min. Knowl. Disc. **17**, 164–206 (2008)
10. González, P., Castaño, A., Chawla, N.V., Coz, J.J.D.: A review on quantification learning. ACM Comput. Surv. **50**(5), 1–40 (2017)
11. Heydarian, M., Doyle, T.E., Samavi, R.: MLCM: multi-label confusion matrix. IEEE Access **10**, 19083–19095 (2022)
12. Ji, S., Hölttä, M., Marttinen, P.: Does the magic of BERT apply to medical code assignment? A quantitative study. Comput. Biol. Med. **139**, 104998 (2021)
13. Ji, S., Pan, S., Marttinen, P.: Medical code assignment with gated convolution and note-code interaction. In: Findings of the Association for Computational Linguistics: ACL-IJCNLP (2021)
14. Ji, S., Sun, W., Dong, H., Wu, H., Marttinen, P.: A unified review of deep learning for automated medical coding. arXiv preprint arXiv:2201.02797 (2022)
15. Johnson, A.E., et al.: MIMIC-III, a freely accessible critical care database. Sci. Data **3**(1), 1–9 (2016)

16. Kaur, R., Ginige, J.A., Obst, O.: A systematic literature review of automated ICD coding and classification systems using discharge summaries. arXiv preprint arXiv:2107.10652 (2021)
17. Levin, R., Roitman, H.: Enhanced probabilistic classify and count methods for multi-label text quantification. In: Proceedings of the ACM SIGIR International Conference on the Theory of Information Retrieval (2017)
18. Li, Y., Wehbe, R.M., Ahmad, F.S., Wang, H., Luo, Y.: Clinical-longformer and clinical-BigBird: transformers for long clinical sequences. arXiv preprint arXiv:2201.11838 (2022)
19. Maletzke, A.G., Hassan, W., dos Reis, D.M., Batista, G.E.: The importance of the test set size in quantification assessment. In: Proceedings of the International Joint Conferences on Artificial Intelligence Organization (2020)
20. Michalopoulos, G., Malyska, M., Sahar, N., Wong, A., Chen, H.: ICDBigBird: a contextual embedding model for ICD code classification. In: Proceedings of the ACL Workshop on Biomedical Language Processing (2022)
21. Moreo, A., Francisco, M., Sebastiani, F.: Multi-label quantification. arXiv preprint arXiv:2211.08063 (2022)
22. Moreo, A., Sebastiani, F.: Re-assessing the "classify and count" quantification method. In: Proceedings of the European Conference on Information Retrieval (2021)
23. Mullenbach, J., Wiegreffe, S., Duke, J., Sun, J., Eisenstein, J.: Explainable prediction of medical codes from clinical text. In: Proceedings of the Conference of the North American Chapter of the Association for Computational Linguistics (2018)
24. Nawrot, P., et al.: Hierarchical transformers are more efficient language models. arXiv preprint arXiv:2110.13711 (2021)
25. Sebastiani, F.: Text quantification. In: Proceedings of the Conference on Empirical Methods in Natural Language Processing: Tutorial Abstracts (2014)
26. Sebastiani, F.: Evaluation measures for quantification: an axiomatic approach. Inf. Retr. J. **23**(3), 255–288 (2020)
27. Touvron, H., et al.: Llama 2: open foundation and fine-tuned chat models. arXiv preprint arXiv:2307.09288 (2023)
28. Vaswani, A., et al.: Attention is all you need. In: Proceedings of the Annual Conference on Advances in Neural Information Processing Systems (2017)
29. Vu, T., Nguyen, D.Q., Nguyen, A.: A label attention model for ICD coding from clinical text. In: Proceedings of the International Joint Conference on Artificial Intelligence (2021)
30. Wolf, T., et al.: Transformers: state-of-the-art natural language processing. In: Proceedings of the Conference on Empirical Methods in Natural Language Processing: System Demonstrations (2020)
31. Xun, G., Jha, K., Sun, J., Zhang, A.: Correlation networks for extreme multi-label text classification. In: Proceedings of the ACM SIGKDD International Conference on Knowledge Discovery and Data Mining (2020)
32. Yuan, Z., Tan, C., Huang, S.: Code synonyms do matter: multiple synonyms matching network for automatic ICD coding. In: Proceedings of the Annual Meeting of the Association for Computational Linguistics (Volume 2: Short Papers) (2022)
33. Zaheer, M., et al.: Big bird: transformers for longer sequences. In: Proceedings of the Annual Conference on Advances in Neural Information Processing Systems (2020)
34. Zhang, N., Jankowski, M.: Hierarchical BERT for medical document understanding. arXiv preprint arXiv:2204.09600 (2022)

LGEM$^+$: A First-Order Logic Framework for Automated Improvement of Metabolic Network Models Through Abduction

Alexander H. Gower[1]([✉])[iD], Konstantin Korovin[2][iD], Daniel Brunnsåker[1][iD], Ievgeniia A. Tiukova[1,3][iD], and Ross D. King[1,4,5][iD]

[1] Chalmers University of Technology, Gothenburg, Sweden
{gower,danbru,tiukova,rossk}@chalmers.se
[2] The University of Manchester, Manchester, UK
Konstantin.Korovin@manchester.ac.uk
[3] KTH Royal Institute of Technology, Stockholm, Sweden
[4] Cambridge University, Cambridge, UK
[5] Alan Turing Institute, London, UK

Abstract. Scientific discovery in biology is difficult due to the complexity of the systems involved and the expense of obtaining high quality experimental data. Automated techniques are a promising way to make scientific discoveries at the scale and pace required to model large biological systems. A key problem for 21st century biology is to build a computational model of the eukaryotic cell. The yeast *Saccharomyces cerevisiae* is the best understood eukaryote, and genome-scale metabolic models (GEMs) are rich sources of background knowledge that we can use as a basis for automated inference and investigation.

We present LGEM$^+$, a system for automated abductive improvement of GEMs consisting of: a compartmentalised first-order logic framework for describing biochemical pathways (using curated GEMs as the expert knowledge source); and a two-stage hypothesis abduction procedure.

We demonstrate that deductive inference on logical theories created using LGEM$^+$, using the automated theorem prover iProver, can predict growth/no-growth of *S. cerevisiae* strains in minimal media. LGEM$^+$ proposed 2094 unique candidate hypotheses for model improvement. We assess the value of the generated hypotheses using two criteria: (a) genome-wide single-gene essentiality prediction, and (b) constraint of flux-balance analysis (FBA) simulations. For (b) we developed an algorithm to integrate FBA with the logic model. We rank and filter the hypotheses using these assessments. We intend to test these hypotheses using the robot scientist Genesis, which is based around chemostat cultivation and high-throughput metabolomics.

Keywords: Scientific discovery · artificial intelligence · systems biology · metabolic modelling · first-order logic · automated theorem proving

© The Author(s) 2023
A. Bifet et al. (Eds.): DS 2023, LNAI 14276, pp. 628–643, 2023.
https://doi.org/10.1007/978-3-031-45275-8_42

1 Introduction

An important aspect of modern biology is improving our understanding of cellular processes, and the complex interactions between genes, proteins and chemical species. Systems biology is the research discipline that tackles this complexity. *Saccharomyces cerevisiae*, commonly known as "baker's yeast", is an excellent model organism used for the study of eukaryote biology. This is due to the availability of tools for easy genetic manipulation, and low cultivation cost, enabling targeted experiments to characterise the system. *S. cerevisiae*'s was the first eukaryotic genome to be fully sequenced [10] and there is a wealth of knowledge about the gene functions, many of which are conserved or expected to have equivalents in other eukaryotes, including humans [5]. Metabolic network models (MNMs) represent the cellular biochemistry of an organism and the related action of enzymatic genes; such models which seek to integrate knowledge from the entire organism are known as genome-scale metabolic models (GEMs).

The scientific discovery problem we address is to add knowledge to or reduce *S. cerevisiae* GEMs such that quality is increased. Model quality in GEMs is multi-faceted—desirable properties of a model include: predictive power; metabolic network coverage; and parsimony. There are trade-offs between different desirable properties [11]. Foremost, however, is the predictive power of the GEM. Ultimately the aim is to understand the entities, mechanisms and adaptations that govern yeast growth in different environments.

Given a draft model, improvement consists broadly of three stages: hypothesise refinements to the model; conversion of refined model to a format suitable for simulation; and evaluation based on experimental evidence and internal consistency [24]. Repetition of these stages consists a scientific discovery process. Evaluation is dependent on executing simulations using a mathematical formalism, however optimising a model for a specific formalism is not the objective—any improvements that are made to a GEM within a certain framework should translate to improvements in the underlying knowledge.

Challenges for the future of genome-scale modelling of *S. cerevisiae* include: improving annotation; removing noise from low-confidence components; and adding reactions to eliminate so-called "dead-end" compounds [1]. To multiply the efforts of human researchers, previous work has investigated automating parts of the scientific method. GrowMatch was a technique developed to resolve inconsistencies between predictions and experimental observations of single-gene mutant strains of *Escherichia coli* [15]. Other approaches to metabolic network gap-filling have exploited answer-set programming, the most complete of which is MENECO which is designed to efficiently identify candidate additions to draft network models [19].

Logical inference can be applied to generate and improve metabolic models: induction allows us to generalise models from data; given a theory we can draw conclusions using deduction; and abduction enables us to form hypotheses to improve consistency with empirical data. In this work we use first-order logic (FOL) to simulate the metabolic network, an approach first proposed in 2001 [20]. A FOL model was used to generate functional genomics hypotheses then tested by a robot scientist [13]; logical induction and abduction was applied to identify inhibition in metabolic pathways after introduction of toxins [23]; and an FOL model constructed in Prolog using the GEM iFF708 [7] as the background knowledge source was used to predict single-gene essentiality [25]. Huginn is a tool that uses abductive logic programming (ALP), and demonstrates the ability to improve metabolic models and suggest *in vivo* experiments [21].

A core advantage of our model—both over these previous FOL approaches that used Prolog, and over bespoke algorithmic methods such as MENECO—is that we use first order theorem provers (FOTPs) to perform deductive and abductive inference. This removes a large part of the burden of abductive algorithm design and simulation. For the reasoning tasks we use the FOTP iProver [14]. We extended iProver to include abduction inference. iProver is a saturation-based theorem prover that saturates via consequence finding algorithms which are well-suited to abduction [22]. Other declarative programming techniques that we tried, for example Prolog, and SAT solvers based on backtrack search algorithms (e.g. CDCL), lacked certain features that enable abduction. Using FOTPs will also allow us to combine different deduction and abduction strategies.

Furthermore, our model is capable of deductive and abductive reasoning at scales far greater than previous FOL approaches. The ability to reason at scale is particularly important for the automation of scientific discovery in eukaryotic biology where the domain is complex and data are expensive to generate.

One current limitation of our FOL framework is that we do not include information on reaction stoichiometry. To integrate quantitative modelling, we propose in this paper a method to combine flux balance analysis (FBA) and logical inference to validate metabolic pathway configurations found by LGEM$^+$.

The main contributions of LGEM$^+$ as presented in this paper are: (1) a compartmentalised FOL model of yeast metabolism; (2) a two-stage method for the abduction of novel hypotheses on improved models; (3) scalable methods for evaluating these models and hypotheses; and (4) an algorithm to integrate FBA with abductive reasoning.

2   Methods

2.1   The First-Order Logic Framework

We chose FOL as the language to express the mechanics of the biochemical pathways. FOL allows for a rich expression of knowledge about biological processes, such as reactions and enzyme catalysis. We use FOL to express our knowledge about how entities are known to interact, for example that a reaction has substrates and products, and possibly some required enzyme. By contrast, a propositional logic framework would be unable to express these higher level concepts and as such would be less suitable for abduction. The method and model we design is independent of the specific network, meaning that although here we apply LGEM$^+$ to *S. cerevisiae*, this modelling framework could equally well be applied to other organisms.

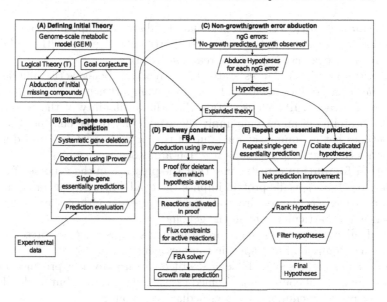

Fig. 1. Processes in LGEM$^+$. **(A)** defining the logical theory, including abduction of missing compounds to enable viability of base strain; **(B)** single-gene essentiality prediction; **(C)** abduction of hypotheses from ngG errors; **(D)** using FBA to assess viability of each hypothesis; and **(E)** repeating single-gene deletion to assess viability of each hypothesis.

We define five predicates in the first-order language: met\2, gn\1, pro\1, enz\1, and rxn\1. The semantic interpretation of these predicates is outlined in Table 1. Here a cellular "compartment" refers to a component of the cellular anatomy, e.g. mitochondrion, nucleus or cytosol.

Table 1. Predicates used in the logical theory of yeast metabolism. Forward and reverse reactions are represented separately in the model, thus a "positive flux" through a reversed reaction indicates the reaction flux is negative.

Predicate	Arguments	Natural language interpretation
met\2	metabolite, compartment	"Metabolite X is present in cellular compartment Y"
gn\1	gene identifier	"Gene X is expressed"
pro\1	protein complex identifier	"Protein complex X is available (in every cellular compartment)"
enz\1	enzyme category identifier	"Enzyme category X is available"
rxn\1	reaction	"There is positive flux through reaction X"

Clauses in our model are one of seven types, each expressing relationships between entities in terms of the predicates given above. These types of clauses are listed below, and we provide a graphical overview and example statements in Fig. 2.

- **Reaction activation** clauses state that all substrate compounds for a specific reaction being present in the correct compartments, together with availability of a relevant enzyme, implies the reaction is active.
- **Reaction product** clauses state that a reaction being active implies the presence of a product compound in a given compartment.
- **Enzyme availability** clauses state that the availability of the constituent parts (proteins) of an enzyme imply the availability of the enzyme. Enzymes sometimes act in complexes made up of two or more proteins, and different enzymes that catalyse the same reaction are called isoenzymes.
- **Protein formation** clauses state that the presence in the genome of a gene that codes for a specific protein implies the availability of that protein.
- **Gene presence** clauses are statements expressing either the presence or absence of a particular gene in the genome.
- **Metabolite presence** clauses are statements expressing the presence of a particular compound in a specific compartment.
- **Goal** clauses represent a biological objective, usually the presence in the cytosol of a set of compounds deemed essential for growth, but could also be another pathway endpoint or intermediary compound.

2.2   Assessing Growth and Production of Compounds

Yeast growth is dependent on the production of essential chemical products—intermediary points or endpoints of biochemical pathways within the organism. The core of these biochemical pathways is the enzymatic reactions, and they are facilitated by diffusion of chemicals within cellular compartments, including the cytosol, and passive or active transport across compartment boundaries or the cell membrane. Certain products are deemed essential for growth, so if production of these compounds is inhibited then the organism is inviable.

Logical inference was performed using the automated theorem proving software iProver (v3.7) which was chosen due to its performance and scalability as well as completeness for first-order theorem finding. The general formulation of the problem provided to iProver is to identify whether a theory, T, "entails" a goal, G. In other words that the goal is a logical consequence of the theory ($T \vDash G$). Here T is a set of logical axioms that encode, using the formalism defined in Sect. 2.1: knowledge from the GEM; the medium in which the yeast is growing, represented by axioms in the theory for the presence of compounds in the extracellular space; the availability of ubiquitous compounds in each cellular compartment and the extracellular space; and the presence and expression of genes. Deduction can be used to analyse pathways and reachable metabolites. In the case of growth/no-growth simulations, G represents the availability of all the essential compounds in the cytoplasm. So if $T \vDash G$ we say that there is growth, otherwise not. Other goals used here are the availability of other endpoints of biochemical pathways. T and G are provided to iProver in plain text files and plaintext proofs are output. The logical proofs (that the goal is reachable) found by iProver correspond to detected biochemical pathways.

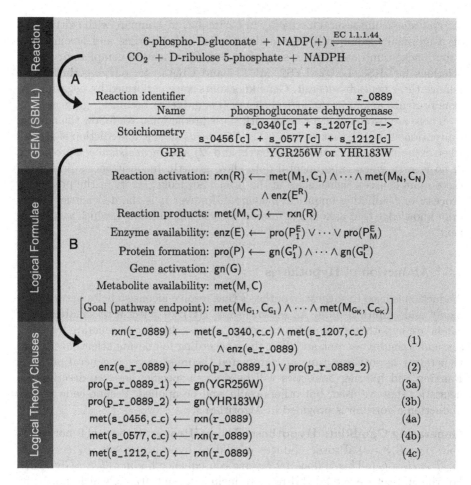

Fig. 2. Conversion of genome-scale metabolic model provided in SBML to logical theory. **(A)** A reaction is encoded in SBML using identifiers to represent the substrates and products, and a logical rule for enzyme availability (GPR = "gene-protein-reaction rule"). **(B)** The information contained on each reaction is encoded using logical formulae into a set of clauses; predicate definitions are provided in Table 1. Here equation (1) is the reaction activation clause. "∧" is a conjunction symbol ("AND"), meaning all of the literals in the expression must be true for the RHS of the clause to be true; "∨" is a disjunction symbol ("OR"). So we can read (1) as: "reaction r_0889 is active if all of the metabolites in the set {s_0340, s_1207} are present in the cytoplasm and at least one of the isoenzymes is present". Similarly equation (2) describes the condition for a relevant enzyme to be present; equations (3a,b) describe the conditions for each of these isoenzymes to be formed; and equations (4a-c) are the reaction product clauses and state that "if reaction r_0889 is active then each of its products is present".

Single-Gene Essentiality Prediction. Here we seek to predict genes without which *S. cerevisiae* cannot grow. We compare predictions against lists of viable and inviable strains from a genome-wide deletion mutant cultivation for

S. cerevisiae using several media [9]. In particular, we compare with cultivations on a minimal medium with the addition of uracil, histidine and leucine. The strain background used in this study was S288C, which has complete or partial deletions for HIS3, LEU2, LYS2, MET17 and URA3—for our experiments we remove these genes by default. Gene knockouts were performed by negating the gene presence axiom in the logical theory (i.e. gn(gene) becomes ¬gn(gene)).

There are two basic error types with these predictions. We follow the naming convention as in [15], that we have: (1) *gNG inconsistency:* a prediction of growth when experimental data show no growth; and (2) *ngG inconsistency:* a prediction of no growth when experimental data show growth. Inconsistencies arise from three main sources: deficiencies in the prior knowledge; errors in the prediction process; or conflicting empirical evidence. However it is the deficiencies in the prior knowledge that are of most interest for scientific discovery, which we explore next.

2.3 Abduction of Hypotheses

Abduction is used to suggest hypotheses that resolve inconsistencies between our model and empirical data. As shown in Fig. 1(C) we select a reasonable set of candidate hypotheses through a two-stage process: firstly, we generate hypotheses; and secondly, we rank and filter these according to relevant scientific criteria. Generating hypotheses using an automated theorem prover is general purpose. Ranking and filtering heuristics will be domain-specific; here we describe the heuristics that we used, but others could well be applied. Pseudo-code for the abduction algorithm is provided in Algorithm 1.

Generating Candidate Hypotheses Using iProver. If the goal is not reachable (i.e. $T \nvDash G$) iProver abduces candidate hypotheses: sets H_i such that $\forall i \ (T \wedge H_i \vDash G)$. This is done by reverse consequence finding ($T \wedge \neg G \vDash \neg H_i$). For this project we extended iProver to include these features, which, not being specific to biochemical reaction networks, could be used for automated discovery in other scientific domains by constructing an appropriate FOL model. The form of the hypotheses, H_i, is a set of clauses expressed in terms of the predicates described above in Sect. 2.1. It is possible to restrict or guide the reverse consequence finding algorithm in iProver to seek certain types of hypotheses. For example a hypothesis could be: met(compound, compartment), that compound is available in compartment. Such hypotheses are challenging to discover because of the complexity of interaction in these networks.

None of the logical theories resultant from the conversion from Yeast8, iMM904 and iFF708 was viable given the minimal medium and ubiquitous compounds, even without any gene deletions, meaning one or more of the essential compounds was not produced. iProver abduced hypotheses consisting of combinations of compounds whose presence would enable viability of the base strain (deletions for HIS3, LEU2, LYS2, MET17 and URA3), as shown in Fig. 1(A). We chose the hypothesis with the fewest additional compounds.

For ngG inconsistencies there exists a set of essential metabolites not being produced that empirical data indicate will be produced given the specified genotype and conditions—in some sense the pathways in the model are incomplete. Hypotheses in this scenario are those that repair an incomplete pathway: additional reactions; annotation of an isoenzyme for knocked out genes; or removal of reaction annotations. For gNG inconsistencies there is a pathway in the model that empirical data suggest should be interrupted but is not. Thus hypotheses in this scenario will be those that interrupt a complete pathway: annotation of a pathway-critical reaction with a gene that is in the set of knocked out genes; removal of an isoenzyme annotation; or removal of reactions.

Heuristics for Ranking and Filtering Hypotheses. We filter hypotheses to only include either: (a) addition of one or more compounds (i.e. containing only atoms using the met predicate); or (b) the presence of one or more particular enzyme groups for a reaction (i.e. containing only atoms using the enz predicate). The motivation is that the subsequent model improvement step (to repair the pathway) for case (a) would be to add reactions to the model that produce the hypothesised metabolites, and for case (b) to either identify an isoenzyme for hypothesised groups or remove the annotation for the deleted gene for one of these reactions. We also remove hypotheses that introduced availability of one or more of the target compounds in the cytosol, as this would directly ensure the goal was reached but is of no scientific value.

We applied two criteria to assess the merit of each hypothesis. Firstly, by using our FBA constraint method, as shown in Fig. 1(D) and described in Sect. 2.4. Around half of the hypotheses resulted in infeasible solutions or very small growth—this means perhaps there might be something else that is missing from the model, and so we have not got a reasonable hypothesis. The second criteria was evaluating the impact each hypothesis had on the overall error in single-gene essentiality prediction, as shown in Fig. 1(E). If the total number of ngG errors fixed is greater than the number of gNG errors introduced then this is a good hypothesis. Another, more conservative, approach would be to only add hypotheses to the model that do not introduce any gNG errors.

A final heuristic was whether hypotheses contained compounds that were not produced by any reaction in the GEM, meaning adding a suitable reaction that produces this compound would repair the error. These hypotheses could be tested experimentally by constructing a deletion mutant, cultivating with minimal medium and after observing growth, using metabolomic analysis (e.g. with mass spectrometry) to identify if the hypothesised intermediary metabolite set is present. If there were a reaction already in the GEM that produced the compound there could be other deficiencies in the model that need addressing first, for example gene annotation for those reactions. In this case iProver abduces hypotheses of case (b) above. Currently LGEM$^+$ can hypothesise to remove gene annotation, but this could be extended to include a search for an isoenzyme based on similarity (e.g. sequence similarity) to the knocked out gene.

Algorithm 1. Abduction using LGEM$^+$

1: **procedure** ABDUCTIONSINGLEGENE
2:     $\mathcal{H} \leftarrow \emptyset$
3:     **for** gene in all genes in theory **do**
4:         $\widetilde{T} \leftarrow T$ ▷ Make a copy of the base theory
5:         $\widetilde{T} \leftarrow \widetilde{T} \setminus \{\mathsf{gn(gene)}\} \cup \{\neg\mathsf{gn(gene)}\}$ ▷ Construct deletant
6:         Use iProver to deduce if goal is reachable by identifying if $\widetilde{T} \vDash G$
7:         **if** $\widetilde{T} \vDash G$ **then**   ▷ Growth prediction
8:             **continue**
9:         **else if** $\widetilde{T} \nvDash G$ **then**   ▷ Non-growth prediction
10:             **if** gene is essential **then**   ▷ No growth observed; no error
11:                 **continue**
12:             **else if** gene is not essential **then**   ▷ Growth observed; ngG error
13:                 Abduction of potential hypotheses set $\mathcal{H}_{\mathsf{gene}}$ using iProver
14:                 $\mathcal{H} \leftarrow \mathcal{H} \cup \mathcal{H}_{\mathsf{gene}}$
15:             **end if**
16:         **end if**
17:     **end for**
18:     Filter and rank $\mathcal{H} = \bigcup\limits_{\mathsf{gene} \in \mathsf{theory}} \mathcal{H}_{\mathsf{gene}}$, according to heuristics, e.g. Section 2.3
19: **end procedure**

2.4 Constraining Flux Balance Analysis Simulations Using Proofs

Flux balance analysis (FBA) finds a reaction flux distribution, $\boldsymbol{\nu}$, given stoichiometric constraints from the GEM and a biologically relevant optimisation objective, $f(\boldsymbol{\nu})$, for example maximisation of biomass production [8,18]. FBA assumes the metabolism is in steady state, resulting in the constraint $S\boldsymbol{\nu} = \mathbf{0}$, where S is the stoichiometric matrix for the metabolic network and $\boldsymbol{\nu}$ is the reaction flux vector ($S \in \mathbb{Z}^{m \times n}$, where m is the number of compounds and n is the number of reactions in the metabolic network).

$$\begin{aligned} \underset{\boldsymbol{\nu} \in \mathbb{R}^n}{\text{maximize}} \quad & f(\nu_1, \ldots, \nu_n) \\ \text{subject to} \quad & S\boldsymbol{\nu} = \mathbf{0} \\ & \nu_i^{\mathrm{LB}} \le \nu_i \le \nu_i^{\mathrm{UB}}, \quad i = 1, \ldots, n. \end{aligned}$$

Whilst the stoichiometric matrix is fixed, the upper and lower bounds for each reaction can be set to achieve relevant results. Existing methods to set these bounds include integrating experimental measurements of fluxes, or using enzyme turnover rates and availability [4]. We use FBA to assess the feasibility of proofs found using iProver by: setting reaction bounds based on pathways activated in the proof; and then solving the resultant optimisation problem. We are able to do this neatly as both use the same GEM as the knowledge source. The procedure is outlined in Algorithm 2.

Flux values are measured in mmol $\mathrm{g_{DW}^{-1}h^{-1}}$ and metabolite concentrations vary substantially between compounds, so finding a forcing threshold which is appropriate for all reactions is not straightforward. For our FBA simulations we used the Python package `cobrapy` (version 0.26.3) [6]; in the absence of relevant documentation on a suitable threshold, we found in a discussion for a MATLAB implementation of COBRA that a suitable threshold should be set at 1×10^{-9} [2].

Algorithm 2. Constraining FBA solution given a logical theory T and a goal G

1: **function** FBACONSTRAIN(GEM, T, G, ν_0)   ▷ ν_0 is minimum flux threshold for activation
2:     Use iProver to find proof of $T \vDash G$ ▷ The goal is reachable
3:     $i \leftarrow 1$
4:     **while** $i \leq N$ **do**   ▷ N is the number of reactions in the GEM
5:         **if** r_i active in the proof in the forward direction **then**
6:             $\nu_i^{LB} \leftarrow \nu_0$ ▷ Force reactions to have positive flux
7:         **else if** r_i active in the proof in the reverse direction **then**
8:             $\nu_i^{UB} \leftarrow -\nu_0$
9:         **end if**
10:         $i \leftarrow i + 1$
11:     **end while**
12:     Solve FBA problem ($S\nu = \mathbf{0}$) with resultant flux bounds
13:     **return** $(\nu, \text{growthValue}, \text{solutionStatus}) \in \mathbb{R}^N \times \mathbb{R} \times \{\text{optimal}, \text{infeasible}\}$
14: **end function**

2.5 Sources of Knowledge

The primary source of the knowledge about reactions and associated genes is the GEM Yeast8 (v8.46.4.46.2) [16]. This was chosen due to its broad coverage of the reactions and gene associations as well as its specificity to the organism *S. cerevisiae*. The other two GEMs used were: iMM904 [17] and iFF708 [7]. (We include iFF708 as a background knowledge source partly to enable comparison with previous logical modelling approach [25].) The models are stored using Systems Biology Markup Language (SBML). The software written to convert a GEM SBML file to a logical knowledge base is available in the supporting material, and follows the process described below and shown in Fig. 2.

We use three reference lists of compounds from [25]; these are shown in the first column of the files on the LGEM$^+$ GitHub repository[1] corresponding to: (1) all compounds deemed essential for growth in *S. cerevisiae*[2]; (2) compounds assumed ubiquitous during growth assumed to be present throughout the cell regardless of initial conditions, such as H_2O and O_2[3]; and (3) the growth media for the experiments, in this case yeast nitrogen base (YNB) with addition of ammonium, glucose and three amino acids (uracil, histidine and leucine)[4].

---

[1] https://github.com/AlecGower/LGEMPlus.
[2] `src/model-files/essential-compounds-{model}.tsv`.
[3] `src/model-files/ubiquitous-compounds-{model}.tsv`.
[4] `src/model-files/ynb-compounds-{model}.tsv`.

Each compound in these lists has an associated Kyoto Encyclopedia of Genes and Genomes (KEGG) [12] identifier. We matched compounds in the curated GEMs based firstly on KEGG ID, otherwise using the species name or synonyms. Some of the compounds we wish to include do not have corresponding entities in the GEMs used as background knowledge. Therefore there are discrepancies between the reference lists and the compiled lists.

3   Results

Automated Theorem Proving Software can be Used to Estimate Single-Gene Essentiality given a Prior Network Model. Using three GEMs—Yeast8, iMM904 and iFF708—as background knowledge sources we conducted single-gene deletant simulations to assess essentiality of each gene and compared against a genome-wide deletion mutant cultivation [9]. Detailed descriptions of these methods are provided in Sect. 2, and context in the overall method in Fig. 1(B). A summary of the single-gene essentiality prediction results is provided in Table 2.

When compared to previous qualitative methods our method showed state of the art results [25,26]. Yet quantitative prediction using FBA achieves a higher precision and recall. These error rates indicate how much is still to be learnt about yeast metabolism. We also found that gene essentiality predictions vary somewhat depending on the prior.

Simulation times for gene knockouts also appear to scale linearly with the size of the network. Comparing network size to average gene knockout simulation times for the three GEMs tested, we see that the mean (± 1 s.d.) times for one knockout simulation were: 0.52 s $\pm$ 0.09 s for iFF708 (1379 reactions); 0.67 s $\pm$ 0.12 s for iMM904 (1577 reactions); and 1.46 s $\pm$ 0.32 s for Yeast8 (4058 reactions).

Abductive Reasoning Allows for Identification of Possible Missing Reactions. We apply the LGEM$^+$ abduction procedure to model improvement, here demonstrated on the Yeast8 model. For each of the 41 ngG errors in the single-gene deletion task, we generated candidate hypotheses according to methods described in Sect. 2.3. In total we generated 2094 unique hypotheses; some hypotheses would result in an error correction for several genes. We ranked and filtered these hypotheses according to domain-specific heuristics, finding 681 of these were valid, i.e. only containing met (633) or enz (48) predicates. The FBA evaluation outlined in Sect. 2.4 indicated 534 hypotheses that could be balanced by the reactions forced in the model, 118 of which were valid. There were 14 hypotheses that were valid and also resulted in a net improvement on the single-gene prediction task.

Strict Essentiality Criteria and Incomplete Annotation may Explain ngG and gNG Inconsistencies. If just one essential compound is not produced we have no growth. One result of this setup is a relatively low precision in the single-gene essentiality prediction. Of the 72 deletions predicted inviable

Table 2. Comparative prediction results for single-gene essentiality using LGEM$^+$ across three background knowledge sources: Yeast8 (v8.46.4.46.2); iMM904; and iFF708, with comparison to: (a) an FBA-simulation with a viability threshold on growth rate set at $1 \times 10^{-6} h^{-1}$ (according to [16]); and (b) another qualitative prediction method, the "synthetic accessibility" approach taken by Wunderlich et. al. [26]. The empirical data used as truth data for these statistics were taken from a genome-wide screening study using a minimal medium [9]. The FOL model performance represents an improvement on previous qualitative method.

Base GEM	Yeast8	iMM904	iFF708	Yeast8 (FBA)	*Syn. Acc.* [26]
# predictions (#genes in GEM)	1056 (1150)	827 (905)	566 (619)	1068 (1150)	*682*
NG Recall (*ngNG/*NG*)	0.193 (31/161)	0.266 (33/124)	0.140 (14/100)	0.447 (72/161)	*0.119 (14/118)*
NG Precision (*ngNG/ng**)	0.431 (31/72)	0.478 (33/69)	0.778 (14/18)	0.459 (72/157)	*0.292 (14/48)*
gNG Rate (*gNG/*NG*)	0.807 (130/161)	0.734 (91/124)	0.860 (86/100)	0.553 (89/161)	*0.881 (104/118)*
ngG Rate (*ngG/*G*)	0.046 (41/895)	0.051 (36/703)	0.009 (4/466)	0.094 (85/907)	*0.060 (34/564)*
F1 score	0.266	0.342	0.237	0.453	*0.169*

Shorthand: *NG-observed no growth; *G-observed growth; ng*-predicted no growth. (Note that the performance statistics for the synthetic accessibility method are taken directly from the authors' report so there may be a difference in truth data to those used to evaluate our model.)

by our model, 41 of these are shown to result in experimentally viable mutant strains (*ngG* errors).

For several genes in the L-arginine biosynthesis pathway the only essential metabolite not reachable in the model was L-arginine. These resulted in *ngG* errors despite the pathway structure and previous empirical evidence showing that null mutants for genes in this pathway (e.g. for *ARG1* [3]) are auxotrophic for L-arginine (i.e. L-arginine was not produced). These results demonstrate that the model can successfully identify behaviour of the metabolic network consistent with other experimental evidence and not the genome-wide screen results [9]. These cases are candidates for experimental testing, and highlight the potential of such models to inform laboratory experimental design and research direction.

In the Yeast8 model there are 4058 reactions, 1425 (35%) of which have no enzyme annotation and 540 (13%) are annotated with a set of isoenzymes that do not have a specific gene in common. Thus nearly half of all reactions will not be affected by single-gene deletions, which is likely to account for a portion of the 130 *gNG* inconsistencies in LGEM$^+$ single-gene essentiality predictions.

Pathways Output from LGEM$^+$ Overlap with FBA Simulations. In the case of predicting growth, LGEM$^+$ outputs reaction pathways. FBA simulations output a reaction flux distribution, and from this we can use a flux threshold for reaction activate to obtain reaction pathways. When comparing reaction pathways obtained from both methods, for each deletant simulation just over 50% of reactions in the LGEM$^+$ derived pathways are also active in the FBA pathways. However, only around 30% of reactions in FBA derived pathways are also active in the LGEM$^+$ derived pathways.

Using pathways derived from the FBA constraint method described in Sect. 2.4, we investigated the *gNG* errors. Of the 130 errors, 50 of them resulted in pathways that the FBA method indicated were unfeasible (i.e., they resulted in low or zero growth). This would mean that by including this constraint method in the LGEM$^+$ framework we could eliminate these errors. However doing so would also falsely predict 56 viable deletant strains as inviable (new ngG errors).

4   Discussion and Conclusion

Scientific discovery in biology is difficult due to the complexity of the systems involved and the expense of obtaining high quality experimental data. Automated techniques that make good use of background knowledge, of which GEMs are prime examples, will have a strong starting point. LGEM$^+$ seeks to do just that by using FOL combined with a powerful theorem prover, iProver.

We efficiently predicted single-gene essentiality in *S. cerevisiae* using a first-order logic (FOL) model. Our method showed state of the art results compared to previous qualitative methods, yet quantitative prediction using FBA achieves a higher precision and recall.

We designed and implemented an algorithm for the abduction of hypotheses for improvement of a GEM. We found 633 hypotheses proposing availability of compounds in specific compartments, and therefore indicate possible missing reactions, 118 of which were validated through FBA constraint and 14 of which resulted in improvements in the single-gene essentiality prediction task. These heuristics help to select more promising hypotheses for experimentation; further selection will be informed by viability or cost of experiment design. We intend to test these hypotheses using the robot scientist Genesis, which is based around chemostat cultivation and high-throughput metabolomics. As we scale the system we can adjust parameters in the heuristics, or introduce new heuristics, to return only the most promising hypotheses.

Measuring performance statistics relative to the number of genes in a model, rather than the number of genes in the organism, presents some challenges when designing a learning process to improve this performance (e.g. GrowMatch [15]). This highlights the need for better model assessment criteria to drive abduction. We have attempted here to provide an example with the constraint of FBA solutions. Future work could certainly be directed to defining such criteria and integrating them into LGEM$^+$.

The logical theory developed here was focused on efficient inference on biochemical pathways. A challenge for future development is to extend the first-order vocabulary to improve the power and performance of LGEM$^+$. Extending the vocabulary could mean: including more predicates, increasing the arity (number of arguments) of predicates, and introducing other logical clause forms. All to better encode biological processes, for example more detail regarding enzyme availability, integration of gene regulation and signalling or introducing time-dependent processes. Aligning the logic more closely with existing ontologies, for example the Systems Biology Ontology (SBO), would ensure the theory remains useful and semantically precise as it is extended. This is a common challenge across the scientific discovery community as we move further toward joint teams of human and robot scientists—ontologies provide a common language. Using FOL allows us to work toward connecting LGEM$^+$ with external knowledge bases.

The best way to test hypotheses is through *in vivo* experimentation. Integrating LGEM$^+$ into an automated experimental design process would enable the next generation of robot scientists.

Acknowledgements. The authors thank the King Group at Chalmers University of Technology for valuable discussion and feedback. This work was partially supported by the Wallenberg AI, Autonomous Systems and Software Program (WASP) funded by the Alice Wallenberg Foundation. Funding was also provided by the Chalmers AI Research Centre and the UK Engineering and Physical Sciences Research Council (EPSRC) grant nos: EP/R022925/2 and EP/W004801/1, as well as the Swedish Research Council Formas (2020-01690).

Code and Data Availability. Code and data used in this study, including the tables for essential compounds, ubiquitous compounds and minimal media, are available at https://github.com/AlecGower/LGEMPlus.

References

1. Chen, Y., Li, F., Nielsen, J.: Genome-scale modeling of yeast metabolism: Retrospectives and perspectives. FEMS Yeast Res. **22**(1), foac003 (2022). https://doi.org/10.1093/femsyr/foac003
2. Cobra-Toolbox: what is the minimum flux computed by flux balance analysis or the accuracy of FBA? https://groups.google.com/g/cobra-toolbox/c/9xmP1VcrWL0
3. Crabeel, M., Seneca, S., Devos, K., Glansdorff, N.: Arginine repression of the Saccharomyces cerevisiae ARG1 gene. Comparison of the ARG1 and ARG3 control regions. Curr. Genet. **13**(2), 113–124 (1988). https://doi.org/10.1007/BF00365645
4. Domenzain, I., Sánchez, B., Anton, M., et al.: Reconstruction of a catalogue of genome-scale metabolic models with enzymatic constraints using GECKO 2.0. Nat. Commun. **13**(1), 3766 (2022). https://doi.org/10.1038/s41467-022-31421-1
5. Dujon, B.: Yeast evolutionary genomics. Nat. Rev. Genet. **11**(7), 512–524 (2010). https://doi.org/10.1038/nrg2811
6. Ebrahim, A., Lerman, J.A., Palsson, B.O., Hyduke, D.R.: COBRApy: constraints-based reconstruction and analysis for python. BMC Syst. Biol. **7**(1), 74 (2013). https://doi.org/10.1186/1752-0509-7-74
7. Förster, J., Famili, I., Fu, P., Palsson, B.Ø., Nielsen, J.: Genome-scale reconstruction of the Saccharomyces cerevisiae metabolic network. Genome Res. **13**(2), 244–253 (2003). https://doi.org/10.1101/gr.234503
8. García Sánchez, C.E., Torres Sáez, R.G.: Comparison and analysis of objective functions in flux balance analysis. Biotechnol. Prog. **30**(5), 985–991 (2014). https://doi.org/10.1002/btpr.1949
9. Giaever, G., Chu, A.M., Ni, L., et al.: Functional profiling of the Saccharomyces cerevisiae genome. Nature **418**(6896), 387–391 (2002). https://doi.org/10.1038/nature00935
10. Goffeau, A., Barrell, B.G., Bussey, H., et al.: Life with 6000 genes. Science **274**(5287), 546–567 (1996). https://doi.org/10.1126/science.274.5287.546

11. Heavner, B.D., Price, N.D.: Comparative analysis of yeast metabolic network models highlights progress, opportunities for metabolic reconstruction. PLoS Comput. Biol. **11**(11), e1004530 (2015). https://doi.org/10.1371/journal.pcbi.1004530

12. Kanehisa, M.: KEGG: Kyoto encyclopedia of genes and genomes. Nucleic Acids Res. **28**(1), 27–30 (2000). https://doi.org/10.1093/nar/28.1.27

13. King, R.D., Whelan, K.E., Jones, F.M., et al.: Functional genomic hypothesis generation and experimentation by a robot scientist. Nature **427**(6971), 247–252 (2004). https://doi.org/10.1038/nature02236

14. Korovin, K.: iProver – an instantiation-based theorem prover for first-order logic (system description). In: Armando, A., Baumgartner, P., Dowek, G. (eds.) IJCAR 2008. LNCS (LNAI), vol. 5195, pp. 292–298. Springer, Heidelberg (2008). https://doi.org/10.1007/978-3-540-71070-7_24

15. Kumar, V.S., Maranas, C.D.: GrowMatch: an automated method for reconciling in silico/in vivo growth predictions. PLoS Comput. Biol. **5**(3), e1000308 (2009). https://doi.org/10.1371/journal.pcbi.1000308

16. Lu, H., Li, F., Sánchez, B.J., et al.: A consensus S. cerevisiae metabolic model Yeast8 and its ecosystem for comprehensively probing cellular metabolism. Nat. Commun. **10**(1), 3586 (2019). https://doi.org/10.1038/s41467-019-11581-3

17. Mo, M.L., Palsson, B., Herrgård, M.J.: Connecting extracellular metabolomic measurements to intracellular flux states in yeast. BMC Syst. Biol. **3**, 1–17 (2009). https://doi.org/10.1186/1752-0509-3-37

18. Orth, J.D., Thiele, I., Palsson, B.Ø.: What is flux balance analysis? Nat. Biotechnol. **28**(3), 245–248 (2010). https://doi.org/10.1038/nbt.1614

19. Prigent, S., Frioux, C., Dittami, S.M., et al.: Meneco, a topology-based gap-filling tool applicable to degraded genome-wide metabolic networks. PLoS Comput. Biol. **13**(1), e1005276 (2017). https://doi.org/10.1371/journal.pcbi.1005276

20. Reiser, P.G.K., King, R.D., Muggleton, S.H.: Developing a logical model of yeast metabolism. Electron. Trans. Artif. Intell. **5**(B), 223–244 (2001)

21. Rozanski, R., Bragaglia, S., Ray, O., King, R.: Automating the development of metabolic network models. In: Roux, O., Bourdon, J. (eds.) CMSB 2015. LNCS, vol. 9308, pp. 145–156. Springer, Cham (2015). https://doi.org/10.1007/978-3-319-23401-4_13

22. Simon, L., del Val, A.: Efficient consequence finding. In: Nebel, B. (ed.) Proceedings of the Seventeenth International Joint Conference on Artificial Intelligence, IJCAI 2001, Seattle, Washington, USA, 4–10 August 2001, pp. 359–370. Morgan Kaufmann (2001)

23. Tamaddoni-Nezhad, A., Chaleil, R., Kakas, A., Muggleton, S.: Abduction and induction for learning models of inhibition in metabolic networks. In: Fourth International Conference on Machine Learning and Applications (ICMLA 2005), p. 6 (2005). https://doi.org/10.1109/ICMLA.2005.6

24. Thiele, I., Palsson, B.Ø.: A protocol for generating a high-quality genome-scale metabolic reconstruction. Nat. Protoc. **5**(1), 93–121 (2010). https://doi.org/10.1038/nprot.2009.203

25. Whelan, K.E., King, R.D.: Using a logical model to predict the growth of yeast. BMC Bioinf **9**, 97 (2008). https://doi.org/10.1186/1471-2105-9-97

26. Wunderlich, Z., Mirny, L.A.: Using the topology of metabolic networks to predict viability of mutant strains. Biophys. J . **91**(6), 2304–2311 (2006). https://doi.org/10.1529/biophysj.105.080572

Predicting Age from Human Lung Tissue Through Multi-modal Data Integration

Athos Moraes[1,2] , Marta Moreno[1,2] , Rogério Ribeiro[1,2] ,
and Pedro G. Ferreira[1,2(✉)]

[1] Department of Computer Science, Faculty of Sciences, University of Porto,
Rua do Campo Alegre 1055, 4169-007 Porto, Portugal
pgferreira@fc.up.pt
[2] Laboratory of Artificial Intelligence and Decision Support, Institute for Systems
and Computer Engineering, Technology and Science, Rua Dr. Roberto Frias,
4200-465 Porto, Portugal

Abstract. The accurate prediction of biological age can bring important benefits in promoting therapeutic and behavioural strategies for healthy aging. We propose the development of age prediction models using multi-modal datasets, including transcriptomics, methylation and histological images from lung tissue samples of 793 human donors. From a technical point of view this is a challenging problem since not all donors are covered by the same data modalities and the datasets have a very high feature dimensionality with a relatively smaller number of samples. To fairly compare performance across different data types, we've created a test set including donors represented in each modality. Given the unique characteristics of the data distribution, we developed gradient boosting tree and convolutional neural network models for each dataset. The performance of the models can be affected by several covariates, including smoking history, and, most importantly, by a skewed distribution of age. Data-centric approaches, including feature engineering, feature selection, data stratification and resampling, proved fundamental in building models that were optimally adapted for each data modality, resulting in significant improvements in model performance for imbalanced regression. The models were then applied to the test set independently, and later combined into a multi-modal ensemble through a voting strategy, predicting age with a median absolute error of 4 years. Even if prediction accuracy remains a challenge, in this work we provide insights to address the difficulties of multi-modal data integration and imbalanced data prediction.

Keywords: Regression · Multi-modal data · Bioinformatics · Computational Biology · Health applications

1 Introduction

Aging is a time-dependent process that leads to a decline in body functions and an increased incidence of several diseases and conditions. The identification of

---

A. Moraes, M. Moreno and R. Ribeiro—Equal contribution.

molecular biomarkers can help to predict the longevity of an individual, promoting therapeutic and behavioral strategies for healthy aging. However, to assess an individual's health status, it is necessary to determine both chronological age and biological age. The former refers to the time passed since birth, and occurs at the same rate for everyone. The latter varies with the former, but it is a more complex concept to define and therefore harder to determine. Biological age is related to molecular changes that occur in cells and influence the physiological and functional state of organisms. The rate of biological aging is highly variable between individuals and depends on genetic factors, environmental exposure, and lifestyle. The identification of reliable biomarkers for biological age is a long-standing research question.

Technological advances have allowed high-throughput profiling of multiple biological layers, also called omics. These include: the amount of RNA molecules in the cell (transcriptome); DNA methylation patterns (methylome); protein abundance (proteome); or genetic variation (genome). Immunology and oncology are at the forefront of multi-omics profiling. Histopathology provides microscopic images of stained tissue samples. These images can provide important visual information on the structure, composition and cellular characteristics of tissues.

Aging has been shown to affect methylation patterns and gene expression [7,18], and both data types have been previously leveraged for age prediction tasks [16,38]. However, to our knowledge, studies reporting associations between histological images and aging or incorporating multiple data modalities in age prediction are lacking in the literature.

Multi-modal datasets represent substantial analysis challenges, including a sample size (typically dozens) much lower than the feature space (typically tens of thousands); partial coverage among samples of the different data modalities; differences in data distributions with different signal to noise ratios; or shared and specific latent variables and factors between data modalities [1,39].

The Genotype-Tissue Expression (GTEx) project [12–14,23] has built a comprehensive resource to study tissue-specific gene expression and regulation. Samples were collected from 54 non-diseased *post-mortem* tissue sites across nearly 1000 individuals, summing more than 17,000 RNA-seq samples profiling 58,219 genes per sample. Histology images are available for each sample. The DNA methylation status of selected tissues from GTEx has been profiled by other initiatives [32]. The sampled population is highly heterogeneous, meaning that individuals might be affected by different technical and biological covariates.

We set out to investigate the development of computational models to predict biological age in lung samples from different data modalities, namely gene expression, methylation status and histological images. We focused on lung tissue, since it is one of the few GTEx tissue sites with the three data types available with a relatively large sample size. In this work we aim to contribute to three research questions: a) which data modality provides better age predictions? b) do the data types provide complementary information and improve their predictions when integrated? c) can we technically overcome the incomplete and imbalanced distribution of the response variable, age?

2   Results

Because not all samples are covered by the same types of data (only roughly 20% are covered by the three types, see Fig. 1a, we trained and optimized independent models for each data modality. For evaluation purposes we held out a test set with n = 45 samples from individuals covered by the three data types, spanning as uniformly as possible across age strata, see Fig. 1b. The goal of this common test set was to compare the performance of the independent models, which were later combined into a multi-modal ensemble, see Fig. 1c.

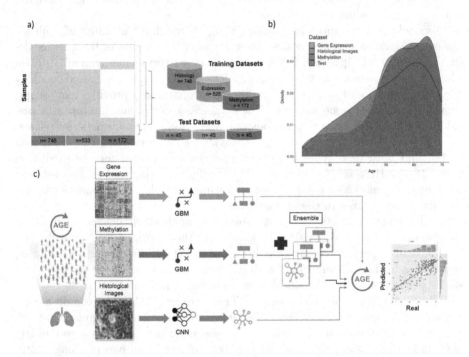

Fig. 1. Development of age prediction models. **(a)** Sample distribution of data modalities. Not all samples are covered by the same modalities. Test sets for final evaluation are derived from samples containing all the data modes, while the training sets are modality specific. **(b)** Age distribution in the training and test datasets. **(c)** Individual models are developed for each data modality and their final predictions combined into a multi-modal ensemble model.

Table 1 summarizes the results for the top-performing models of each data modality. The performance was evaluated based on four key metrics: the coefficient of determination (r^2), which indicates how well regression predictions fit the actual data; Root Mean Square Error (RMSE) and Mean Absolute Error

(MAE), which measure the average magnitude of prediction errors, with RMSE being more sensitive to outliers due to squaring's error amplification; and the Median Absolute Error (MED), a robust measure that mitigates outlier effects. Higher r^2 values signify better predictions, while lower RMSE, MAE, and MED indicate fewer errors. Gene expression and methylation models were trained with cross-validation of, respectively, 10-fold and 5-fold.

Table 1. Performance metrics for the top-performing models. Gene expression features are the expression levels for 150 genes, while the methylation model uses 30 CpG sites. Histological images are segmented into tiles of $256 \times 256 \times 3$ pixels, with tile number varying per sample. All models were optimized for RMSE and r^2. Sample sizes for training and test data are indicated in parentheses. For gene expression and methylation, training metrics represent the average $\pm$ standard deviation derived from 10- and 5-fold cross-validation, respectively.

Data Type	Dataset	r^2	RMSE	MAE	MED
Gene Expression (n = 570)	Training (n = 525)	0.65 ± 0.045	6.88 ± 1.64	5.56 ± 0.24	5.11 ± 0.41
	Test (n = 45)	0.64	7.57	6.09	5.38
Methylation (n = 217)	Training (n = 172)	0.76 ± 0.05	5.07 ± 0.75	4.18 ± 0.46	3.65 ± 0.85
	Test (n = 45)	0.82	5.30	3.94	2.97
Histological Images (n = 793)	Training (n = 748)	0.62	7.34	6.03	5.50
	Test (n = 45)	0.35	10.14	8.14	8.16
Ensemble	Test (n = 45)	0.84	5.02	4.00	3.39

2.1 Data Retrieval and Code Availability

All data used in this work, with the exception of donor smoking status, is publicly available. Donor information (including gender, age and other variables), processed gene expression tables (version 8), and histological images are available via the GTEx data portal (http://gtexportal.org/). Methylation data generated by Oliva et al. [32] was downloaded from GEO (GSE213478).

Code developed for this project and detailed description of the pipelines used to train the prediction models can be found in the following repository: https://github.com/PedroGFerreira/MultiModalHumanLungAgePrediction

2.2 Predicting Age from Gene Expression

RNA sequencing (RNA-seq) is a high-throughput method used to quantify gene expression by counting reads obtained from sequencing biological samples [34].

Read counts are normalized into TPM (transcripts per million) [37], which accounts for variations in sequencing depth (*i.e.* the total number of reads per sample) and gene length across samples. The gene expression data table consists of 570 lung samples and 58,219 genes.

We filtered for protein-coding genes (19,291) and applied a $\log_2$-transformation to reduce the range of values and stabilize the variance. The data was then split into a training set (n = 525) and a common test set (n = 45). A quantile transformation was applied to minimize distribution right-skewness in the expression values. The transformation was fitted on the training data and applied to both training and test sets to ensure consistent scaling. This effectively spreads out the values concentrated around 0 and reduces the impact of outliers.

To further reduce the number of features we tested several approaches using 10-fold cross-validation (CV) [11] and fit models based on gradient boosting trees (LightGBM [19]) with default hyperparameters. The best performing approach was an embedded method [21], set to yield a maximum of 150 features based on the coefficients of a Bayesian Automatic Relevance Determination (ARD) model [26]. To address the skew towards older individuals, see Fig. 1b, we assessed the impact of a pre-processing strategy for imbalanced regression [3]. Unfortunately, since there were no improvements, this approach was not considered. A hyperparameter random search [2] was performed on the training set without data rebalancing (number of iterations = 1000) and evaluated through 5×2 nested CV [8]. The inner 2 folds select a set of hyperparameters, and the outer 5 folds evaluate the model produced in the inner folds. The selected final model was optimized using the previously selected top 150 features.

The model presents stable results between training and test sets as shown in Table 1. Figure 2a depicts the fit on the test data, indicating an overestimation of younger individuals and an underestimation of older ones (MED: 5.11 ± 0.41).

Feature importance, based on information gain, identifies the most influential genes for the model, see Fig. 3a. Information gain balances the frequency of each feature with the magnitude of their contribution. Among the top features we have *EDA2R* (r = 0.40, p-value = 1.4e−23, cf. Figure 4a), an essential gene for ectodermal tissue development, with a possible impact on lung aging and progression of Chronic Obstructive Pulmonary Disease [36]. Furthermore, *EDA2R* has been previously identified by us as a gene with increased expression with age in multiple tissues [28]. Other influential features include *CDKN2A* (r = 0.42, p-value = 8.5e−26), a biomarker for cellular senescence, associated with cognitive decline and aging [25,29], and *TCHD4* (r = 0.38, p-value = 4.4e−21), also identified as a biomarker for cellular senescence [4]. *NEFH* has a decreasing expression with age (r = −0.38, p-value = 2.7e−21). The significant correlation of these features with age underscores their potential importance in the aging process, as captured by our machine learning model.

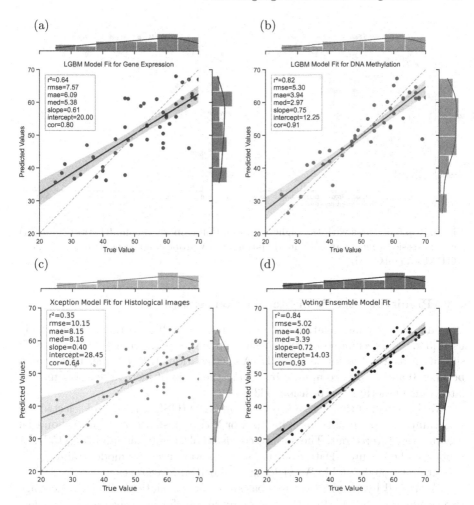

Fig. 2. Scatter plots of true vs predicted age. The dotted gray line represents a perfect x = y linear model fit. Predicted age values were obtained on the test set (n = 45) using the best performing model for **(a)** (Color figure online) gene expression, **(b)** methylation, **(c)** histology, and **(d)** a voting ensemble of the results for the three datasets. **LGBM** - LightGBM.

Finally, to investigate possible cohort bias, we calculated the Spearman correlation between the absolute difference of the predicted and true age, and several donor characteristics, including lung-related clinical conditions. No significant associations were found.

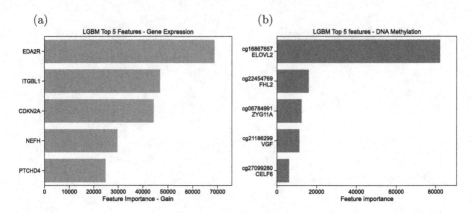

Fig. 3. Top 5 features with the highest information gain for the models trained on **(a)** gene expression (genes) and **(b)** methylation (DNA probes and associated genes) data. **LGBM** - LightGBM.

2.3 Predicting Age from Methylation Arrays

Profiling of DNA methylation patterns using the EPIC array has been widely used [33]. This technique targets specific CpG sites across the genome. At each site, it measures the intensity of DNA methylation, the β value, that ranges between 0 and 1. β is computed by taking the ratio of methylated over methylated plus unmethylated positions [33].

We downloaded the methylation data in GEO (GSE213478) and selected the lung samples. CpG probes mapping to X or Y chromosomes or the mitochondrial genome were filtered out. The dataset consisted of 217 samples and 754,115 DNA probes as the features. Data from 172 samples were used for model training and validation through 5-fold CV, yielding r^2 and RMSE averages.

We started by filtering features based on the correlation with biological age ($|$spearman rho$| > 0.4$). The data was quantile transformed, several feature selection methods were tested, and the resulting datasets were used to fit a gradient boosted trees (LightGBM) model [19] with fine tuning of several parameters. This procedure was repeated for each of the five CV folds. The best performing approach was achieved with the top 30 features based on the coefficients of an ARD regression [26] (r^2: 0.73 ± 0.04 and RMSE: 5.40 ± 0.59).

In the case of methylation, the application of an imbalanced data preprocessing strategy [3] on the unscaled data after feature selection improved the performance and was applied during the training of the final model. Table 1 and Fig. 2b indicate an average error of 3.9 years for the prediction of the 45 individuals of the test set. Feature importance analysis (Fig. 3b) highlights CpG probe cg16867657, located in the promoter on ELOVL2, a known aging biomarker [17] and previously used to predict biological age [31], see Fig. 4b. Interestingly, all the top 5 CpG probes have previously been implicated in aging [10,35]. Congestion, a condition that increases with age given the decreased ability of the lung to clear mucus [24] was the only significantly correlated covariate with the error of the predictions (rho = 0.45, p-value = 0.0018).

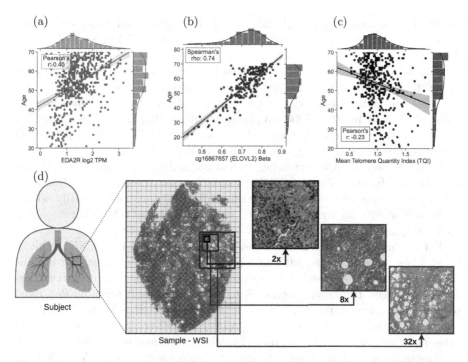

Fig. 4. Correlation between age and the top predictive features for **(a)** gene expression (*EDA2R*) and **(b)** DNA methylation (CpG probe cg16867657). **(c)** Correlation between relative telomere length (measured as the mean Telomere Quality Index for each probe) and age. **(d)** Image resolution at different downsampling values has an important impact on model performance when making predictions with histopathological images.

2.4   Predicting Age from Histological Images

Histological images are stained microscope slides that provide high resolution representations of tissue samples, and allow trained pathologists to detect changes in cellular morphology and tissue structure. Histopathology has been used for decades for disease diagnosis and characterization of complex phenotypes. Given the recent advances in image pattern analysis with machine learning, we postulated if the lung histological images of nearly 800 donors could be used to predict their biological age.

We started by evaluating a feature-based regression analysis approach. Haralick Features (HR) [22], a set of statistical measures that quantify texture properties of an image, were extracted. Dimensionality reduction based on UMAP [27] suggests a lack of distinct age patterns that may indicate low discriminative power for HR features for the analyzed data (data not shown).

To automatically derive the features and perform minimal image processing, we applied convolutional neural networks (CNN). Four models were tested: a custom CNN-based model built and trained from scratch for this dataset, and three pre-trained models (VGG16, VGG19 and Xception [5]) fine-tuned for our data.

The histological images were downloaded from the GTEx data portal as whole slide images (WSI), and then segmented into tiles using *PyHist* [30]. This segmentation process is necessary for analyzing WSI, which can have dimensions upwards of $40000 \times 40000 \times 3$ pixels, to ensure that they fit in the GPU's memory. Thus, each donor has a set of associated tiles (256×256 x 3 pixels), all inheriting the same label from their respective WSI.

Examples were split into training, validation, and test sets. We built a custom stratification function for the training-validation and training-test splits: if an age value has enough samples for stratification (three or more), stratify with a proportion of 80% for training and 20% for validation/test sets; otherwise, keep those samples in the training set only.

For the image-based regression model, we trained the four CNN architectures. These were enhanced with data augmentation, a batch generator module for efficient feeding of tiles, and a sample weights module that prioritized less represented ages. Hyperparameters (*Batch size*, *Epochs* and *Learning Rate*) and parameters (*Augmentation*, *Sample Weights* and *Resampling Method*), were systematically monitored and logged using *Mlflow* [41] for comparative analysis and subsequent model selection. Overall, 173 different models were tested.

Through our analysis of selected metrics, particularly r^2 and RMSE, we discovered that oversampling by repeating images did not yield any benefits and, in fact, led to worse results. On the other hand, undersampling consistently demonstrated positive effects, optimizing our chosen metrics. Although data augmentation techniques like rotations, reflections, and translations generally provided some improvement, their impact was not significant. Our hypothesis regarding the poor performance of oversampling is based on the presence of numerous non-informative tiles in the dataset. By duplicating tiles randomly and increasing the number of tiles per subject, the non-informative tiles introduce more noise into the learning process. As a result, we decided to focus on undersampling and data augmentation, as they proved more beneficial in optimizing our selected metrics and improve model performance. After optimization of hyperparameters, the best performing model was applied to the common test set ($r^2 = 0.35$ and RMSE = 10.1, see Table 1 and Fig. 2c).

Our analysis revealed that the primary factor hindering the performance of image-based models is the inadequate representation of the neighborhood, particularly isolated donors from age groups with few or no neighbors. The lack of representativity in the neighborhood has a more significant impact on model performance than the number of samples per individual. Figure 5 illustrates that the prediction errors are considerably larger for lower ages where there are larger gaps in the number of donors per age, even if the number of tiles per donor remains approximately the same across all ages, as shown in Fig. 5b.

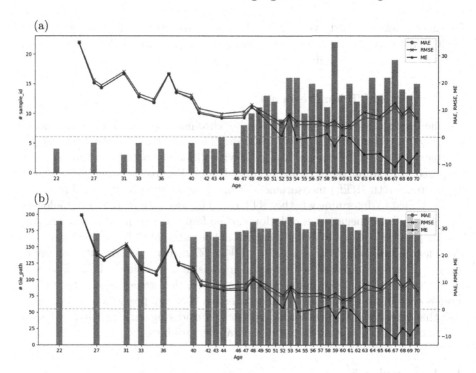

Fig. 5. Distribution of **(a)** the number of samples and **(b)** the number of tiles per age in the test set (blue barplots) and the corresponding metrics (lines) in the test set. The shift between the barplots and the points in the lines results from the application in the train and test sets. Both figures correspond to the same experiment. Metrics are the Mean Error (ME), Mean Absolute Error (MAE) and Root Mean Square Error (RMSE).

In conclusion, relying solely on data augmentation and oversampling strategies in the less represented age groups is not sufficient to achieve better performance in our histological dataset. Ensuring adequate representation of the neighborhood, with a balanced distribution of donors across age groups, is essential for obtaining reliable and high-performing image-based models.

2.5 Ensemble Prediction Model

In an attempt to enhance prediction accuracy and robustness, we combined the three individual pre-trained models using an ensemble approach. Briefly, we employed a weighted voting strategy using different combinations of data modalities. The weights were optimized by minimizing a custom loss function, $(\frac{1}{2} \times \frac{RMSE}{10}) - (\frac{1}{2} \times r^2)$, to achieve a balanced control over RMSE and r^2 on age predictions from the common test set (n = 45). This approach yielded a model with marginally better performance than the methylation model for the optimized metrics ($r^2 = 0.84$, RMSE = 5.01, see Fig. 1). The main contribution for this model stemmed from the methylation predictor, while gene expression made a notable but secondary contribution (76.9% and 23.1%, respectively). Histology data did not influence the final prediction.

2.6   Aging and Telomere Length

Telomeres are protective caps at the end of chromosomes that maintain genomic stability and integrity during cell division. They tend to shorten with age [6]. Telomere length (TL) can be measured with different techniques and has been proposed as a biomarker of biological age. The GTEx project provides a Relative Telomere Length (RTL) measurement using a high-throughput technique.

We aimed to determine whether RTL could be used as a predictor of biological age. A significant negative correlation can be found between RTL and age (r = -0.23, p-value = 6.7e$-$07, cf. Figure 4c), where age could be approximated by the linear association: $Age \sim 64.46 - 10.94 \times RTL$. However, despite employing linear and gradient boosting regression models and combining RTL data with other relevant clinical and demographic variables, the predictive accuracy of the models was found to be poor. This emphasizes that age prediction is a complex task that requires a multifactorial approach, considering various other factors in addition to RTL and clinical and demographic variables.

3   Discussion

As technology continues to advance, the generation of extensive biomedical datasets containing multiple molecular and phenotypic measurements from the same sample is expected to become increasingly common. This poses significant difficulties in terms of data analysis. This work highlights the challenges of building prediction models based on such datasets, which are characterized by their heterogeneous nature, high dimensionality, and non-uniform distributions.

In our study, we conducted an analysis to determine which data modality that yields the most accurate age predictions. DNA methylation exhibited the highest predictive performance, achieving a median error of 2.97 years and a mean absolute error of 3.9 years when evaluated on a test set of 45 samples. Gene expression data also showed promising results, albeit with slightly lower accuracy. Histological images, while providing valuable insights at the tissue and cellular level, had reduced accuracy when compared to the other modalities.

Next, we investigated whether integrating multiple predictive models could enhance age estimation and reveal potential complementarity among different data modalities. Our final ensemble model leveraged DNA methylation and gene expression, but not histological images, to slightly improve the methylation predictor in terms of RMSE and r^2 (Table 1).

We hypothesize that the limited accuracy of the histological model can be attributed to the highly imbalanced distribution of the data, particularly the missing data for lower age ranges, as shown in Fig. 1b. In an attempt to address this issue, we applied data augmentation techniques such as rotations, scaling, and flipping transformations to increase variability and balance the representation of different age groups. However, these efforts had minimal impact on the model's performance. As part of our future work, we plan to explore the application of recent strategies, such as the approach described in [40], which tackles imbalanced regression based on the generation of continuous and smooth distribution of the age. Moreover, we would like to combine HR features, which were not used in this work, with other more relevant image properties [15,20].

Employing a data rebalancing technique that undersampled overrepresented age groups and oversampled underrepresented ones [3], led to improved predictions for methylation, but the impact for gene expression was not significant. We posit that this may in part be explained by the considerable differences in training set size for these two modalities, with gene expression having access to three times more samples, see Table 1.

The disparity between biological age and chronological age can be substantial, and stems from a combination of genetic, lifestyle, and environmental factors. Furthermore, the divergence in these two aging categories can be biased by biological or technical covariates. In our study, we focused on optimizing our models to predict chronological age, although it is likely that the data reflects biological age instead. The largest prediction errors were observed in the lower age range, where less training data is available. This highlights the impact of imbalanced age distribution, a predictive modelling problem which also makes it difficult to understand if prediction deviations result from model errors or differences between biological and chronological age.

To assess whether demographic and clinical covariates, such as smoking habits, do not introduce bias, we have tested the association between the regression residuals and the covariates. Overall, no significant associations were found. As future work, we aim to address potential bias by employing more general techniques. One such approach involves incorporating the available covariates into the linear models of expression and methylation. The corresponding residuals can then be used as phenotypes of expression or methylation.

In addition to age prediction, our analysis revealed several genes of interest for age biology. Notably, some of these genes, including *EDA2R*, *CDKN2A*, and *PTCHD4*, have previously been associated with age-related conditions.

The GTEx samples used in this study were obtained *post-mortem*, where the ischemic time may have an impact in the molecular fingerprint [9]. How these models will perform on data sampled from living tissue is a question that requires further evaluation and investigation.

4   Conclusion

This study applied state-of-the-art machine learning approaches for predicting age from multi-modal biological data, including gene expression, DNA methylation, and histological images. The datasets posed significant analysis challenges given their highly imbalanced age distribution, including missing data at the lower end ages, and the very high dimensionality of the molecular data when compared with the number of available samples. Despite that, the DNA methylation model and the voting ensemble model are promising strategies for predicting age, with the former having a median absolute error of approximately 4 years. These findings provide important insights for future explorations into biological age prediction and age-related conditions, underscoring the potential of integrating diverse biological data types for complex phenotype predictions.

Acknowledgments and Funding. This work is supported in part by funds from the Portuguese National Science Foundation - FCT: RNCA:2022.15770.CPCA.A1 to PGF and also through Ph.D. Fellowships SFRH/BD/145707/2019 to MM and SFRH/BD/07092/2021 to RR.

Author contributions. PGF designed the study with contributions from MM, AM and RR. The prediction models were developed by MM, AM and RR. All authors have drafted, edited, revised and approved the final version of the manuscript.

References

1. Argelaguet, R., et al.: Multi-omics factor analysis-a framework for unsupervised integration of multi-omics data sets. Mol. Syst. Biol. **14**(6), e8124 (2018)
2. Bergstra, J., Bengio, Y.: Random search for hyper-parameter optimization. J. Mach. Learn. Res. **13**(2), 281–305 (2012)
3. Branco, P., Torgo, L., Ribeiro, R.P.: SMOGN: a pre-processing approach for imbalanced regression. In: First International Workshop on Learning With Imbalanced Domains: Theory and Applications, pp. 36–50. PMLR (2017)
4. Casella, G., et al.: Transcriptome signature of cellular senescence. Nucleic Acids Res. **47**(14), 7294–7305 (2019)
5. Chollet, F.: Deep Learning with Python. Simon and Schuster, New York (2021)
6. Codd, V., et al.: Measurement and initial characterization of leukocyte telomere length in 474,074 participants in UK biobank. Nat. Aging **2**(2), 170–179 (2022)
7. De Magalhães, J.P., Curado, J., Church, G.M.: Meta-analysis of age-related gene expression profiles identifies common signatures of aging. Bioinformatics **25**(7), 875–881 (2009)
8. Dietterich, T.G.: Approximate statistical tests for comparing supervised classification learning algorithms. Neural Comput. **10**(7), 1895–1923 (1998)
9. Ferreira, P.G.: The effects of death and post-mortem cold ischemia on human tissue transcriptomes. Nat. Commun. **9**(1), 490 (2018)
10. Florath, I., Butterbach, K., Müller, H., Bewerunge-Hudler, M., Brenner, H.: Cross-sectional and longitudinal changes in DNA methylation with age: an epigenome-wide analysis revealing over 60 novel age-associated CPG sites. Hum. Mol. Genet. **23**(5), 1186–1201 (2014)
11. Fushiki, T.: Estimation of prediction error by using K-fold cross-validation. Stat. Comput. **21**, 137–146 (2011)
12. GTEx Consortium: Genetic effects on gene expression across human tissues. Nature **550**(7675), 204–213 (2017)
13. GTEx Consortium: The GTEx consortium atlas of genetic regulatory effects across human tissues. Science **369**(6509), 1318–1330 (2020)

14. GTEx Consortium, Ardlie, K.G., Deluca, D.S., Segrè, A.V., Sullivan, T.J., Young, T.R., Gelfand, E.T., Trowbridge, C.A., Maller, J.B., Tukiainen, T., et al.: The genotype-tissue expression (GTEx) pilot analysis: multitissue gene regulation in humans. Science **348**(6235), 648–660 (2015)

15. Hoffman, R.A., Kothari, S., Phan, J.H., Wang, M.D.: A high-resolution tile-based approach for classifying biological regions in whole-slide histopathological images. In: Zhang, Y.-T. (ed.) The International Conference on Health Informatics. IP, vol. 42, pp. 280–283. Springer, Cham (2014). https://doi.org/10.1007/978-3-319-03005-0_71

16. Horvath, S.: DNA methylation age of human tissues and cell types. Genome Biol. **14**(10), 1–20 (2013)

17. Johansson, Å., Enroth, S., Gyllensten, U.: Continuous aging of the human DNA methylome throughout the human lifespan. PLoS ONE **8**(6), e67378 (2013)

18. Jung, M., Pfeifer, G.P.: Aging and DNA methylation. BMC Biol. **13**(1), 1–8 (2015)

19. Ke, G., et al.: LightGBM: a highly efficient gradient boosting decision tree. In: Advances in Neural Information Processing Systems, vol. 30 (2017)

20. Kothari, S., Phan, J.H., Osunkoya, A.O., Wang, M.D.: Biological interpretation of morphological patterns in histopathological whole-slide images. In: Proceedings of the ACM Conference on Bioinformatics, Computational Biology and Biomedicine, pp. 218–225 (2012)

21. Lal, T.N., Chapelle, O., Weston, J., Elisseeff, A.: Embedded methods. In: Guyon, I., Nikravesh, M., Gunn, S., Zadeh, L.A. (eds.) Feature Extraction. Studies in Fuzziness and Soft Computing, vol. 207, pp. 137–165. Springer, Heidelberg (2006). https://doi.org/10.1007/978-3-540-35488-8_6

22. Löfstedt, T., Brynolfsson, P., Asklund, T., Nyholm, T., Garpebring, A.: Gray-level invariant Haralick texture features. PLoS ONE **14**(2), e0212110 (2019)

23. Lonsdale, J., et al.: The genotype-tissue expression (GTEx) project. Nat. Genet. **45**(6), 580–585 (2013)

24. Lowery, E.M., Brubaker, A.L., Kuhlmann, E., Kovacs, E.J.: The aging lung. Clin. Interv. Aging **8**, 1489–1496 (2013)

25. Lye, J.J., et al.: Astrocyte senescence may drive alterations in GFAPα, CDKN2A p14 ARF, and TAU3 transcript expression and contribute to cognitive decline. Geroscience **41**, 561–573 (2019)

26. MacKay, D.J., et al.: Astrocyte senescence may drive alterations in GFAPα, CDKN2A p14 ARF, and TAU3 transcript expression and contribute to cognitive decline. ASHRAE Trans. **100**(2), 1053–1062 (1994)

27. McInnes, L., Healy, J., Melville, J.: UMAP: uniform manifold approximation and projection for dimension reduction. arXiv preprint arXiv:1802.03426 (2018)

28. Melé, M., Ferreira, P.G., Reverter, F., DeLuca, D.S., Monlong, J., Sammeth, M., Young, T.R., Goldmann, J.M., Pervouchine, D.D., Sullivan, T.J., et al.: The human transcriptome across tissues and individuals. Science **348**(6235), 660–665 (2015)

29. Melzer, D., Pilling, L.C., Ferrucci, L.: The genetics of human ageing. Nat. Rev. Genet. **21**(2), 88–101 (2020)

30. Muñoz-Aguirre, M., Ntasis, V.F., Rojas, S., Guigó, R.: Pyhist: a histological image segmentation tool. PLoS Comput. Biol. **16**(10), e1008349 (2020)

31. Naue, J., et al.: Chronological age prediction based on DNA methylation: massive parallel sequencing and random forest regression. Forensic Sci. Int. Genet. **31**, 19–28 (2017)

32. Oliva, M.: DNA methylation QTL mapping across diverse human tissues provides molecular links between genetic variation and complex traits. Nat. Genet. **55**(1), 112–122 (2023)

33. Pidsley, R.: Critical evaluation of the illumina methylationepic beadchip microarray for whole-genome DNA methylation profiling. Genome Biol. **17**(1), 1–17 (2016)
34. Stark, R., Grzelak, M., Hadfield, J.: RNA sequencing: the teenage years. Nat. Rev. Genet. **20**(11), 631–656 (2019)
35. Tajuddin, S.M., et al.: Novel age-associated DNA methylation changes and epigenetic age acceleration in middle-aged African Americans and whites. Clin. Epigenetics **11**(1), 1–16 (2019)
36. de Vries, M., et al.: Lung tissue gene-expression signature for the ageing lung in COPD. Thorax **73**(7), 609–617 (2018)
37. Wagner, G.P., Kin, K., Lynch, V.J.: Measurement of mRNA abundance using RNA-SEQ data: RPKM measure is inconsistent among samples. Theory Biosci. **131**, 281–285 (2012)
38. Wang, F., et al.: Improved human age prediction by using gene expression profiles from multiple tissues. Front. Genet. **11**, 1025 (2020)
39. Welch, J.D., Kozareva, V., Ferreira, A., Vanderburg, C., Martin, C., Macosko, E.Z.: Single-cell multi-omic integration compares and contrasts features of brain cell identity. Cell **177**(7), 1873–1887 (2019)
40. Yang, Y., Zha, K., Chen, Y., Wang, H., Katabi, D.: Delving into deep imbalanced regression. In: International Conference on Machine Learning, pp. 11842–11851. PMLR (2021)
41. Zaharia, M., et al.: Accelerating the machine learning lifecycle with MLflow. IEEE Data Eng. Bull. **41**(4), 39–45 (2018)

Anomaly, Outlier and Novelty Detection

Error Analysis on Industry Data: Using Weak Segment Detection for Local Model Agnostic Prediction Intervals

Rafael Mamede[1,4]([✉]) [ID], Nuno Paiva[2,4] [ID], and João Gama[2,3] [ID]

[1] Faculdade de Ciências, Universidade do Porto, Porto, Portugal
up201706701@fc.up.pt
[2] INESC TEC, 4200-465 Porto, Portugal
{nuno.paiva,joao.gama}@inesctec.pt
[3] Faculdade de Economia, Universidade do Porto, Porto, Portugal
jgama@fep.up.pt
[4] NOS, Comunicações S.A., Lisbon, Portugal
{rafael.mamede,Nuno.Paiva}@nos.pt

Abstract. Machine Learning has been overtaken by a growing necessity to explain and understand decisions made by trained models as regulation and consumer awareness have increased. Alongside understanding the inner workings of a model comes the task of verifying how adequately we can model a problem with the learned functions. Traditional global assessment functions lack the granularity required to understand local differences in performance in different regions of the feature space, where the model can have problems adapting. Residual Analysis adds a layer of model understanding by interpreting prediction residuals in an exploratory manner. However, this task can be unfeasible for high-dimensionality datasets through hypotheses and visualizations alone.

In this work, we use weak interpretable learners to identify regions of high prediction error in the feature space. We achieve this by examining the absolute residuals of predictions made by trained regressors. This methodology retains the interpretability of the identified regions. It allows practitioners to have tools to formulate hypotheses surrounding model failure on particular regions for future model tunning, data collection, or data augmentation on critical cohorts of data. We present a way of including information on different levels of model uncertainty in the feature space through the use of locally fitted Model Agnostic Prediction Intervals (MAPIE) in the identified regions, comparing this approach with other common forms of conformal predictions which do not take into account findings from weak segment identification, by assessing local and global coverage of the prediction intervals.

To demonstrate the practical application of our approach, we present a real-world industry use case in the context of inbound retention call-centre operations for a Telecom Provider to determine optimal pairing between a customer and an available assistant through the prediction of contracted revenue.

Keywords: xAI · Uncertainty Quantification · Error Analysis

A. Bifet et al. (Eds.): DS 2023, LNAI 14276, pp. 661–672, 2023.
https://doi.org/10.1007/978-3-031-45275-8_44

1   Introduction

As the world evolves to be increasingly more data-driven, a need for a widespread understanding of machine learning models and their decision-making processes arises for practitioners and everyone affected by them. In recent years, the increasing regulation of data products reflects this necessity. For example, the European Union General Data Protection Regulation, in Recital 71, states that "[the data subject should have] the right ... to obtain an explanation of the decision reached", effectively providing a legal right to an explanation.

Even without considering the regulation, one can easily see the gap that exists if no efforts are made to make decisions made by machine learning algorithms explainable to the end user. When used in sensitive domains such as medical applications, banking, security, law, and many others, traditional performance metrics alone don't allow for an accurate understanding of the decisions made and whether they are being affected by undesired factors such as biases in the training dataset.

Here is where the growing domain of Explainable AI (xAI) enters to bridge the model and the people (ML practitioners and non-practitioners) affected by the model. There have been many efforts to formalize the objectives of this emerging field, with one popular definition focusing on the following 4 pillars for the necessity of xAI [1]:

- explain to justify - the decisions made by utilising an underlying model should be explained to increase their justifiability;
- explain to control - explanations should enhance the transparency of a model and its functioning, allowing its debugging and the identification of potential flaws;
- explain to improve - explanations should help scholars improve the accuracy and efficiency of their models;
- explains to discover - explanations should support the extraction of novel knowledge and learning relationships and patterns.

Even with well-defined goals and motivations, the use of xAI in industry use cases is rarely formulaic, depending highly on the context of the data and the necessities of the owners and users of the data project. Fellous et al. [2] propose a division of explainable efforts regarding the data product's development stage: Pre-modeling, In-modeling and Post-modeling. A true implementation of explainability efforts in a data product should reflect efforts in each stage.

In industry applications, data products are rarely one-shot projects. Instead, they need to be maintained and updated by adjusting to possible concept drifts over their lifetime, reflecting the reality seen in production, or by adding or removing features for newer iterations of a certain data product. As the process of producing and delivering analytics models is dynamic, so should the steps to guarantee transparency in model decisions. As such, the results of the Post-modeling stages can motivate further Pre-modeling analysis and choices of algorithms in the In-modeling phase.

This paper focused on a Post-modeling approach with a goal centred on controlling and improving machine learning models. We will explore the benefits of Error Analysis by segmenting the feature space in regions of different model performances and how this can contribute to a more transparent decision process through prediction intervals. As such, the **contributions of this paper** can be summarized as follows:

- Expanding current methodology to analyse the local error in machine learning models for high-impact industry applications;
- Integration between weak performance segments (data cohorts) and estimation of prediction error for regression problems;
- Application of described methodology in a high-impact real-world industry use case in the context of inbound retention call-centre operations for a Telecom Provider.

2 Error Analysis

Uncertainty in machine learning models refers to the lack of complete certainty or precision in their predictions. ML models are often trained on limited or noisy data, making it challenging to make definitive predictions. Uncertainty can arise due to factors such as inherent variability in the data, model assumptions, parameter estimation, and inherent stochasticity in the underlying processes [9]. Quantifying and understanding uncertainty in ML models is important as it provides insights into the reliability and robustness of predictions. Understanding and quantifying each source of uncertainty in real-world applications can be a nearly impossible. As such, a way of obtaining predictions that reflect uncertainty can be achieved alternatively by using Error Analysis for segmenting our feature space and employing prediction interval estimates. We explain this process in detail in this and the following section.

Error Analysis (EA) is a method that falls into the category of a Post-modeling approach, with a goal centred on controlling and improving a certain machine learning model. A thorough EA seeks to go beyond aggregate performance metrics since single-score evaluation may hide important conditions of inaccuracies. This occurs because, by only considering global performance metrics, we neglect that the error is often not uniformly distributed across the feature space. In general, models will perform better in certain regions, either by the amount of data present to train or for ease of fit of the decision function according to model assumptions. By contrast, for hard-to-model or unexpected inputs, the model performance might be significantly worse than the aggregate performance.

Although EA methodologies are loosely defined, different sources converge into the idea of the segmentation of a given feature space (or a feature space with meta-information) [4,6,8], S, into n sub-regions, $R_i \in S$ ($R_i \in \{R_1, ..., R_n\}$ with $\bigcup_{i=1}^{n} R_i = S$, and $R_i \bigcap R_j = \emptyset$ for $i \neq j$), and extract local behaviour of the model in the regions. The segmentation can be either using only the features available for inference or meta-information about the observation that is not accessible to the model. The latter can be either information that is deemed sensitive, such as gender or ethnicity, not included in the training for fairness concerns or information regarding characteristics of the data that are not seen by the model (such as the presence of a cat in a binary image classifier that identifies images of dogs). These two complementary ways of approaching EA can lead to different steps for model improvement.

Identifying regions using boundaries or rules in the feature space is the more direct way to find improvements for an existing model. One way to do so is to train a different model for data in the feature space's high error region, or regions, to complement the existing model. Another way to improve performance in underperforming regions is to increase the training observations in the area, either by obtaining fresh data or performing data augmentation. Our approach is not focused on improving the model performance but rather reflects model uncertainty in the high error regions in the predictions.

We can use Error Analysis to evaluate and construct our prediction intervals that estimate uncertainty in an observation. We can assess how well a prediction interval estimate behaves locally by identifying the regions of high error in a test set. Ideally, we should construct prediction intervals that retain the same level of adequacy across all identified regions. For constructing prediction intervals with information extracted from EA, we can perform local fits of agnostic predictors in a calibration set's identified regions to improve local coverage.

3   Model Agnostic Prediction Intervals (MAPIE)

MAPIE is a framework, proposed in [10], for estimating uncertainties associated with the predictions of Machine Learning models to better understand the robustness and predictive power of predictions via conformal prediction methods. These methods don't require any considerations about the model, utilizing only the exchangeability of the data as a requirement, which in the case of tabular regression, is not a harsh assumption.

As this work is inserted in a larger study of interpretable models in industry scenarios, there is a strong necessity for an agnostic approach for obtaining prediction intervals to maintain compatibility with the trained modelling approaches. The choice of this framework is mainly due to the versatility of the conformal methods employed by it.

Mathematically, the problem can be defined as follows: given a training dataset $(X, y) = (x_1, y_1), ..., (x_n, y_n)$, we seek to create an estimator for the prediction interval, $\hat{C}_\alpha$, such that the probability that the real value of the target of a new observation falls on the estimated interval is $P\left(y_{n+1} \in \hat{C}_\alpha(x_{n+1})\right) = 1 - \alpha$. The value of $1 - \alpha$ is defined as the target coverage and is a user-determined parameter based on desired risk α, which depends on how lenient the constructed interval is on error.

Current implementations achieve this by either split-conformal or cross-conformal methods. First, we define a conformity score function that assesses how incorrect a prediction is. Typically, for regression problems, we can use absolute residuals of a prediction. For split approaches, we train the desired ML model on a train set and use the conformal scores on a calibration set. For the chosen risk level α, we determine the estimated quantile of the conformity score distribution associated and use this value to construct boundaries around the prediction. Cross-conformal approaches base themselves on a similar principle, fitting a desired model on each of the folds and assessing out-of-fold conformity scores using the selected conformity score function. The prediction interval is then constructed from estimating the quantile associated with risk α from the conformity scores.

One of the limitations of these methods is the lack of ability to correctly undertake heteroscedastic noise. Expanding on this weakness, Conformalized Quantile Regression [11] offers an alternative approach to regular conformal methods, better adapting the width of prediction intervals to regions of higher noise. This implementation requires compatible models, that is, models that take a quantile regression as the machine learning task.

Our approach tries to bridge the knowledge extracted through the error analysis to produce locally fitted split-conformal predictions sensitive to different error regions in the model without requiring a quantile regression learning task.

4 Related Work

The necessity to identify and diagnose machine learning model failures has led to the development of tools for practitioners' use, such as Uber's MANIFOLD [5,7] or Microsoft's Error Analysis [4], that allow performing this task in an exploratory way. The approach followed in this work for detecting high error cohorts is similar to the latter, using shallow regression trees trained on the absolute residuals of model predictions.

Alternatively to finding weak segments using features of the model, using meta-information of the observations can lead to less direct ways of improving existing models since new observations might lack the feature at the time of inference, or even the use of this information might collide with fairness concerns. As such, ways to improve the developed model with this type of EA tend to follow an exploratory path, auditing the model by framing hypotheses on why it under-performs in the region [3] and complementing the work with other xAI approaches to understand the difference between local and global behaviour. There have been approaches suggested for automating the process of finding patterns that lead to higher error in applications related to image and text applications [6], such as using a pipeline based on meta-feature extraction from examples and a rules-based classifier for the prediction of errors.

Also of note, in uncertainty quantification, IBM's Uncertainty Quantification 360 [12] provides alternatives to obtaining prediction intervals by employing models that intrinsically produce uncertainty estimates and offering alternative post-hoc methodologies for obtaining prediction intervals on already trained models.

5   Methodology

In this work, we used an interpretable regressor - shallow regression trees - trained on the absolute prediction residuals for identifying high error regions. Given a model, $f : S \to \mathbb{R}$, and data points, (X, y) with $X \in S$, $y \in \mathbb{R}$, the absolute residuals are calculated as $|f(X) - y|$.

We split the available data into 3 distinct sets: train, test, and calibration. The train set is used to train each of the models used and, through cross-validation, tune the hyperparameters of each model using grid-search. The test set assesses local performance by fitting our residual trees and verifying Mean Absolute Error (MAE) and coverage of prediction intervals in each cohort identified. The calibration set is used for fitting our prediction interval estimators in a split-conformal fashion. We experiment using 2 distinct approaches, Fig. 1, for obtaining these intervals:

- Global Fitting - by using the entire calibration set for the estimator.
- Local Fitting - by training a residual tree on the calibration set, to segment it into areas of different performance, and for each area create an associated interval estimator.

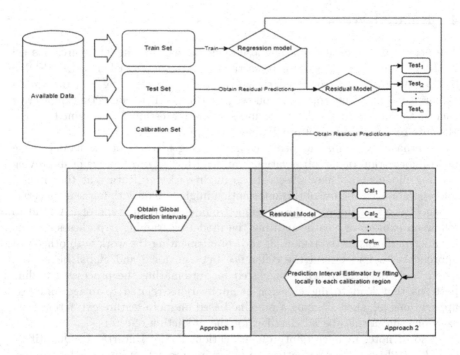

Fig. 1. Experimental approach for obtaining prediction intervals

The residual trees on the test set were trained with a maximum depth of 3, and a minimum number of samples required to be at a leaf node of 500 to avoid creating outlier cohorts, instead focusing on obtaining larger groups of higher error. In the calibration set we used a slightly deeper estimator with a maximum depth of 5 and all other parameters equal.

For each of the cohorts in the test set, we assess local coverage of both of our prediction interval estimates.

For the choice of modeling approaches, since this work is framed in a larger study on the impact of explainable models in industry use cases, we used a range of interpretable additive models and a state-of-the-art black box model. For our interpretable models, we have a Ridge Regression, a Generalized Linear Model (GLM) with a logarithmic link function [13], a Generalized Additive Model (GAM) with an identity link function and non-interacting terms [14], and an Explainable Boosting Machine (EBM) [15] that expands on the GAM by adding the most relevant pair-wise interactions between features. We used Extreme Gradient Boosting [16] for our black-box model.

6 Smart Pairing Use-Case

6.1 Use-Case Description

Smart Pairing is a project developed by a TelCo and media provider that seeks to optimize the pairing of operators and customers, calling a retention line regarding the possibility of changes to their current telecommunications plan. The clients calling these retention lines can fall into several categories, such as customers that are seeking to change their current plan for a more favourable one, customers scanning the market to assess how competitive the offers provided by the company are, or customers unhappy with their current service. In either case, there is a risk of churn (customer abandonment) associated with these requests.

To decrease the risk of client churn, as well as increase overall customer satisfaction with the retention service provided, Smart Pairing proposes a pairing between operators and clients based on analytics. For a given customer, we seek to recommend the best operator, where an ideal pairing system minimizes churn and maximizes contracted revenue.

The implementation strategy of Smart Paring consists of two parts:

- A **regression problem**, where given a pair of an operator and a customer we predict contracted revenue (our target variable), given by the number of contracted months multiplied by the monthly value contracted;
- A **ranking problem**, where given a model to predict contracted revenue, we predict the target variable for each pairing possible for a certain customer and return a custom list of the best operators to address the customer.

In this specific use case, expanding a traditional regression problem into a conformal prediction, yielding prediction intervals rather than point predictions, can be viewed as a measure to mitigate fairness issues that arise in the ranking stage. If, for a specific customer, the model would yield overlapping prediction intervals between two or more operators, enforcing a hierarchy between them might favour certain operators with performance similar to their peers. By giving them the same priority, we can better reflect the actual prediction of the model and not fall into the pit-trap of giving an advantage to individuals based on marginal non-significant differences.

For model variables for the task, we possess historical operator information regarding the number of calls for different technologies and other service segments and the number of calls by outcome (for example, calls that ended in customer churn) over different aggregation periods. We also have available customer information regarding the types of service included in their TelCo package and the volume of calls made to the company operations over different aggregation periods.

6.2   Experimental Results

After training and tuning each of our models, we assess local performance and segment the test cohorts. We seek to guarantee that the coverage of our prediction intervals is stable over these regions.

By fitting our shallow regression trees on the residuals of the predictions for each model on the test data, we noticed that our models seem to underperform for customers with high-paying packages and high number of voice and mobile cards. This is a scenario that occurred regardless of the model used which indicates that those customers are hard to model with the available data, and suggests that improvements could be made by acquiring more examples in these critical regions.

In Fig. 2 we expose a visual representation of the regression tree of the residuals for the EBM model to illustrate the insights we can extract with this analysis. Here we see that the most important factor for different errors is a higher value of the customer spending, *arpu_in_amt*, with variables regarding operator background and performance (*ohp*), such as the direction of their teams and the volume of calls that resulted in churn, also contributing to the segmentation. In this example, the highest error cohort is defined as observations that reflect calls by higher paying customers and that were picked by teams of B2C (Business to Consumer), with a local error that is 3.05 times larger than the lowest error cohort.

Even though these regions are hard to model, if instead of point predictions, we had an interval that reflected this local inaccuracy of the modelling approaches, we could still make decisions in these areas for the ranking stage. This could be achieved by observing overlaps, or lack thereof, between prediction intervals and constructing a ranking system that only favours a certain operator over another if and only if its predicted value is greater and no overlap of their prediction intervals occurs.

The local error in each test cohort, alongside cumulative error, for each model can be found in Fig. 3.

An ideal prediction interval model would yield not only a global coverage near the user-defined target coverage parameter, but also maintain local coverage in each cohort near said target. For this work we selected a target coverage of 90%,

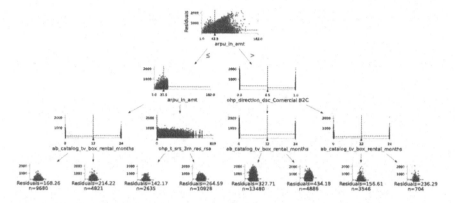

Fig. 2. Residual tree in the test set for the EBM model. The leaves represent each data cohort used for assessing local error and coverage of prediction intervals.

meaning it is expected that globally, 90% of the targets of new observations will fall on the predicted intervals. In Fig. 4, we can observe the limitations of using the traditional split-conformal approach of MAPIE, which leads to prediction intervals that fail to maintain the local coverage near the desired target since the prediction interval width does not vary. As such, these prediction intervals grossly overestimate uncertainty in regions where model error is low, leading to coverages above the target (valid but overly conservative), and under-estimate in regions of high model error.

In contrast, our approach to locally fit an estimator in each error cohort in the calibration set leads to more granular interval estimates that locally maintain the coverage near the desired target in each test cohort.

In Table 1 we present the average absolute deviation between measured coverage and desired coverage over each cohort, for each of the trained models in both approaches. We conclude that our approach leads to a significant average improvement over a global fitting approach.

Table 1. Mean absolute deviation between target coverage and measured coverage over each cohort (lower is better), for each prediction interval approach.

Approach	Models				
	Ridge	GLM	GAM	EBM	XGBoost
Global MAPIE	7.4%	8.2%	7.5%	7.4%	8.0%
Local MAPIE	**2.2%**	**2.8%**	**2.2%**	**1.9%**	**4.8%**

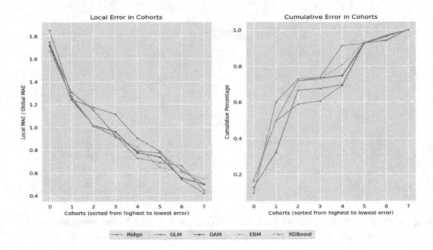

Fig. 3. Local Error (cohort MAE/Global MAE) and cumulative error for each identified cohort

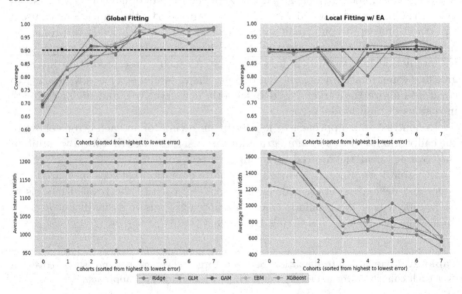

Fig. 4. Local coverage and mean interval width assessment using both described strategies. The dashed line represents the desired coverage selected, 90%. We can visually verify a higher stability of coverages using our locally fitted approach, as well as a more granular interval width over test error cohorts.

7   Conclusion

In this work we explored the creation of prediction intervals through local fitting of split-conformal methods for each region of different error detected in a calibration set, and validated their quality by assessing the coverage over each region of different error in a test set. We concluded, through our use-case, that this approach yields more stable coverages over the different regions than a global split-conformal estimator, being a better solution for use cases where error has very different local behaviours. Our approach, by being based on an interpretable regressor for segmentation of error regions on the calibration set, also maintains high transparency as the path to each leaf can be used to comprehend the different width of the intervals for a prediction.

Our approach for verifying coverage in error regions can also be extended for other prediction interval methodologies. We believe this assessment can be a useful benchmark to guarantee prediction intervals are being generated in a way that takes into consideration the inherent variability of prediction error in the feature space.

8   Future Work and Limitations

In our approach, we did not include a comparison between our methodology and quantile MAPIE models, which include a more resilient way to deal with error changes in different regions. The requirement to have a quantile regression estimator for this more adaptable method can lead to additional considerations on the modelling step that our approach circumvents. Nevertheless, for future work, we recommend a comparison between our approach and other more adaptable ways of estimating prediction intervals.

Cross-conformal methods implemented by MAPIE were not explored in depth during this work, as preliminary results reflected poor results for local coverage while maintaining a high computational cost. Either way, a more in-depth comparison of our local methodology with these approaches can also be fruitful in comprehending how they are fair in comparison.

Other ways to deal with heteroscedastic noise in the context of prediction intervals, outside the MAPIE ecosystem also could be explored in the future, expanding on the work of Johansson *et al.* [17].

Lastly, the methodology used in this paper will continue to be explored for creating a fairer decision process in the ranking segment of this use case, continuing to push for the positive impact of the xAI methodologies on industry applications.

References

1. Vilone, G., Longo, L.: Notions of explainability and evaluation approaches for explainable artificial intelligence. Inf. Fusion **76**, 89–106 (2021). https://doi.org/10.1016/j.inffus.2021.05.009
2. Fellous, J.M., Sapiro, G., Rossi, A., Mayberg, H., Ferrante, M.: Explainable artificial intelligence for neuroscience: behavioral neurostimulation. Front. Neurosci. **13**, 1346 (2019). https://doi.org/10.3389/fnins.2019.01346
3. MLOps Notes 3.2: error analysis for machine learning models. https://pub.towardsai.net/mlops-3-2-error-analysis-750ef03c96e9. Accessed 13 Apr 2023

4. Error Analysis (Responsible AI toolkits). https://erroranalysis.ai. Accessed 13 Apr 2023

5. Manifold (A model-agnostic visual debugging tool for machine learning). https://github.com/uber/manifold

6. Gao, T., Singh, S., Mooney, R.J.: Towards automated error analysis: learning to characterize errors (2022). http://arxiv.org/abs/2201.05017

7. Zhang, J., Wang, Y., Molino, P., Li, L., Ebert, D.S.: Manifold: a model-agnostic framework for interpretation and diagnosis of machine learning models. IEEE Trans. Vis. Comput. Graph. **25**, 364–373 (2018). https://doi.org/10.1109/TVCG.2018.2864499

8. Wu, W.: Identify error-sensitive patterns by decision tree. In: Perner, P. (ed.) ICDM 2015. LNCS, vol. 9165, pp. 81–93. Springer, Cham (2015). https://doi.org/10.1007/978-3-319-20910-4_7

9. Gillmann, C., Saur, D., Scheuermann, G.: How to deal with uncertainty in machine learning for medical imaging? (2021)

10. Taquet, V., Blot, V., Morzadec, T., Lacombe, L., Brunel, N.: MAPIE: an open-source library for distribution-free uncertainty quantification (2022). http://arxiv.org/abs/2207.12274

11. Romano, Y., Patterson, E., Candès, E.J.: Conformalized quantile regression (2019). http://arxiv.org/abs/1905.03222

12. Ghosh, S., et al.: Uncertainty quantification 360: a holistic toolkit for quantifying and communicating the uncertainty of AI (2021). http://arxiv.org/abs/2106.01410

13. McCullagh, P., Nelder, J.A.: Generalized Linear Models. Chapman & Hall, Boca Raton (1992)

14. Servén, D., Brummitt, C.: pyGAM: generalized additive models in python. Zenodo (2018). https://doi.org/10.5281/zenodo.1208723

15. Lou, Y., Caruana, R., Gehrke, J., Hooker, G.: Accurate intelligible models with pairwise interactions. In Proceedings of the 19th ACM SIGKDD International Conference on Knowledge Discovery and Data Mining, pp. 623–631 (2013)

16. Chen T., Guestrin C.: XGBoost: a scalable tree boosting system (2016). https://doi.org/10.1145/2939672.2939785

17. Johansson, U., Linusson, H., Löfström, T., Boström, H.: Interpretable regression trees using conformal prediction. Expert Syst. Appl. **97**, 394–404 (2018). https://doi.org/10.1016/j.eswa.2017.12.041

HEART: Heterogeneous Log Anomaly Detection Using Robust Transformers

Paul K. Mvula$^{(\boxtimes)}$, Paula Branco , Guy-Vincent Jourdan ,
and Herna L. Viktor

School of Electrical Engineering and Computer Science (EECS),
University of Ottawa, 800 King Edward Avenue, Ottawa K1N 6N5, ON, Canada
{pmvul089,pbranco,gjourdan,hviktor}@uottawa.ca

Abstract. Log sequences generated by heterogeneous systems are critical for understanding computer system behaviour and ensuring operational and security integrity. However, the diverse formats, structures, and content of logs pose challenges for traditional log anomaly detection approaches that rely on log parsing, which can be imperfect and incomplete in information extraction. To address these challenges, we propose HEART (**HE**terogeneous Log **A**nomaly Detection using **R**obust **T**ransformers), an end-to-end framework for log-based anomaly detection. HEART eliminates the need for log parsing and leverages Transfer Learning (TL) and Transformer models to operate directly on raw log events from multiple systems. We enhance existing tokenizers with domain-specific tokens, applied to `BERT` and `RoBERTa`, and introduce two novel Transformer models, `LogAnBERT` and `LogBERTa`, trained from scratch on log events. We comprehensively evaluate HEART in intra-system and cross-system scenarios, demonstrating its competitive performance with enhanced anomaly detection using fewer training parameters. Our findings highlight the importance of adapting Transformers and tokenizers for log anomaly detection, enabling improved system monitoring and security across domains. HEART is a significant contribution, being the first end-to-end TL framework for log-based anomaly detection in heterogeneous systems.

Keywords: log anomaly detection · transformers · transfer learning

1 Introduction

System logs, which are commonly employed for troubleshooting in large computer systems, typically exhibit diverse formats and structures, making it challenging to use a universal log parser. Traditional approaches to log anomaly detection often rely on techniques such as keyword matching or regular expressions to identify abnormal log entries. However, these methods may fall short in detecting attacks or anomalies that involve logs from different systems or logs exhibiting different formats and structures [10].

© The Author(s), under exclusive license to Springer Nature Switzerland AG 2023
A. Bifet et al. (Eds.): DS 2023, LNAI 14276, pp. 673–687, 2023.
https://doi.org/10.1007/978-3-031-45275-8_45

Log parsing is a common approach in log anomaly detection, where the log sequences are initially transformed into a standardized format called "*templates*" [9]. Log parsing involves extracting structured information from unstructured or semi-structured log data, typically using pattern matching or rule-based techniques, to create these log templates. Each sequence can then be mapped to a specific template, enabling further analysis and detection of anomalies using ML/DL approaches. Despite being a common approach in log anomaly detection, log parsing has some limitations [10]. First, log parsing relies on predefined log templates, which may not cover all possible log patterns, leading to potential parsing errors or missing relevant log information. Second, log parsing may not effectively handle variations in log formats or entries that deviate from the predefined log templates, resulting in incomplete or inaccurate parsing results. Third, log parsing may not be suitable for handling dynamic or evolving log data, where the formats or patterns change over time, requiring frequent updates to the log templates. Lastly, log parsing may not be efficient in processing large volumes of log data in real-time or near real-time, potentially leading to processing delays or resource limitations in high-velocity logging environments [10].

In this work, we propose a novel methodology called HEART (**HE**terogeneous Log **A**nomaly Detection using **R**obust **T**ransformers) to address the limitations of traditional log anomaly detection approaches. HEART eliminates the need for log parsing and harnesses the power of Transfer Learning (TL) [17] in our cross-system setup. Cross-system and intra-system scenarios encompass different contexts for log data analysis. The intra-system scenarios focus on detecting anomalies within a single system or application to identify system-specific anomalies. In contrast, cross-system scenarios involve analyzing log data from multiple systems, building a model based on this diverse data, and applying it to detect anomalies in a target system [17]. HEART addresses the challenges of both cross-system and intra-system log analysis. By directly operating on raw log events and employing Transformer models, HEART captures intricate patterns and dependencies within log data, enabling effective anomaly detection within a single system. This eliminates the need for predefined log templates and enhances the detection of complex anomalies within the system itself. Moreover, HEART's TL ability extends to the cross-system setup, where log data from multiple systems with different formats and structures are analyzed together. HEART leverages the knowledge learned from multiple systems and applies it to a target, novel, system, enhancing anomaly detection performance in cross-system scenarios and enabling a comprehensive analysis of log data in a heterogeneous environment.

By emphasizing both cross-system and intra-system notions, our work highlights the versatility and effectiveness of HEART in addressing anomaly detection challenges in various contexts. The utilization of TL, Transformer models, and elimination of log parsing capability make HEART a robust, end-to-end methodology for log-based anomaly detection not only within a single system but also across different systems and applications. To the best of our knowledge, our work represents the first comprehensive analysis of cross-system scenarios in log data, that harnesses the power of TL and Transformer models trained from scratch for log-based anomaly detection while making use of multiple datasets with different log formats and structures.

Our main contributions can be summarized as follows:

1. We propose the HEART framework, a novel, end-to-end methodology for log-based anomaly detection. HEART eliminates the need for log parsing while benefiting from the power of TL and Transformer models, allowing us to operate directly on raw log events from multiple systems.
2. We introduce a method to dynamically extend the tokenizers of pre-trained Language Models (LMs), customizing them for targeting log data processing. This approach is evaluated on widely-used architectures such as Bidirectional Encoder Representations from Transformers (`BERT`) [5] and Robustly Optimized `BERT` approach (`RoBERTa`) [12] that were originally designed for Natural Language Processing (NLP) tasks.
3. We present `LogAnBERT` and `LogBERTa`, two domain-specific models tailored for log-based anomaly detection, trained from scratch for Masked Language Modelling (MLM) on a large-scale log dataset for 3.7 million steps. These architectures are built upon the foundation of `BERT` and `RoBERTa` but are customized and fine-tuned to effectively handle log data.
4. We conduct an extensive experimental study to assess the performance of both the state-of-the-art publicly available approaches and our proposed HEART framework. The evaluation encompasses various log datasets with diverse formats and structures, covering both intra-system and cross-system scenarios.

The remainder of the manuscript is organized as follows: Sect. 2 reviews related work on log-based anomaly detection using parsers and TL. Section 3 introduces our proposed HEART framework and its components. Section 4 provides details on the experimental datasets, scenarios, and our evaluation framework. Next, in Sect. 5, we analyze and compare the results of our proposed models. Finally, Sect. 6 offers conclusions on the topic, along with suggestions for future work.

2   Related Work

2.1   Log Anomaly Detection with Templates and Transformers

Several recent studies have utilized contextual representations from Transformers for log anomaly detection. For instance, LogBERT [8] uses self-supervised learning objectives to detect anomalies, while NeuralLog [10] directly processes log messages with Transformers without log parsing. BERT-Log [4] integrates an event template extractor for structured log processing, and LogFiT [1] uses Longformer to handle longer log sequences. Other anomaly detection approaches, such as Prog-BERT-LSTM [16], PoSBERT [19], and CAT [21], combine Transformers with other techniques, including parsing and word-embeddings.

It is worth pointing out that NeuralLog stands out in its direct utilization of Transformers, bypassing the need for explicit log parsing [10]. However, NeuralLog still necessitates the extraction of the log messages from the log events, which can be considered a form of parsing. Our proposed HEART framework, in contrast, operates directly on raw log events without the requirement for parsing or log message extraction. By harnessing the power of Transformers with customized domain-specific models, HEART offers a novel approach to log-based anomaly detection that overcomes the limitations associated with traditional parsing techniques.

2.2 Transfer Learning for Log Anomaly Detection

Transfer learning can be defined as improving a model's performance in a target dataset by utilizing the knowledge from "related" source domains [17]. Specifically, in the context of log anomaly detection, TL involves harnessing the knowledge gained from one or multiple source systems to facilitate the adaptation of models to another target system's log data. This adaptation process allows tackling the challenges posed by variations in log formats, structures, and content, ultimately enhancing the performance and applicability of anomaly detection models in real-world scenarios [3].

LogTransfer [3] is an example of a TL framework for log anomaly detection. It focuses on detecting anomalies in logs generated by different software systems. The framework utilizes pre-training and fine-tuning, where a Long-Short Term Memory (LSTM) model is first pre-trained on templates, extracted with FT-tree [20], from a single source system and then fine-tuned on a target system's events. This transfer of knowledge has been shown to improve the model's ability to detect anomalies in the target system when faced with diverse log formats. While LogTransfer makes use of the power of TL, its adaptability to diverse datasets from different systems is still limited due to its dependence on FT-tree for template extraction, i.e., log parsing. In contrast, our HEART framework provides a significant advantage by being parser-independent, enabling greater flexibility in employing multiple datasets. By decoupling from the constraints of log template extraction, HEART seamlessly accommodates diverse log data while still leveraging TL with diverse datasets, resulting in improved adaptability for log anomaly detection.

3 HEART Framework, LogAnBERT and LogBERTa

This section provides an overview of the proposed specialized tokenizers designed specifically for log data and the novel, domain-specific models we built from scratch, namely LogAnBERT and LogBERTa, tailored for log anomaly detection. Finally, we present the integration of these models and tokenizers within our novel HEART[1] framework for evaluating solutions for log anomaly detection within a TL context.

---

[1] The code, datasets, and Transformer models developed and evaluated in this work are available upon reasonable request to the corresponding author.

3.1   Special-Purpose Tokenizers

To deal with the inherent diversity present in logs originating from multiple sources, we propose a novel approach that dynamically expands the vocabulary of the tokenizers used in the classification process. This method enables us to effectively handle the unique characteristics and variations within log data. We have tested this approach on both BERT's WordPiece [5] and RoBERTa's ByteLevelBPE (BBPE) [12] tokenizers for adaptability to the log anomaly detection task.

The proposed technique first tokenizes the training data using the tokenizer and computes the Term Frequency-Inverse Document Frequency (TF-IDF) scores for the tokens present in the corpus using the TfidfVectorizer from scikit-learn[2]. Then, the tokens are ranked according to their TF-IDF scores and those not present in the current tokenizer's vocabulary are identified as most representative. Finally, the top t new tokens are added to the tokenizer, allowing for the incorporation of domain-specific terminology and leveraging the power of TL. This approach offers flexibility in adapting the tokenizer to different tasks by dynamically updating its vocabulary with relevant tokens. In this work, we experimented with a value of $t = 1000$.

3.2   LogAnBERT

To enhance the training process for log anomaly detection, we developed a custom Transformer model and tokenizer from scratch for MLM, namely LogAnBERT (Log Anomaly Detection using the BERT architecture). LogAnBERT is a variant of BERT, specifically designed for logs. It is a 6-layer Transformer model trained on log events from publicly available sources for 3.7 million steps, using a batch size of 16 samples. To ensure effective preprocessing, we employed the WordPiece tokenizer for LogAnBERT. LogAnBERT's tokenizer was trained from scratch on the same log events, resulting in a vocabulary size of 30,522. During MLM, we randomly masked 15% of the tokens in the events to create a context prediction task. The model then learned to predict the masked tokens based on the surrounding context, effectively capturing the contextual information within the events. The motivation behind developing LogAnBERT was to tailor it specifically to log data, enabling it to accurately understand and identify anomalies in log events.

3.3   LogBERTa

In addition to LogAnBERT, we also developed another custom Transformer model and tokenizer called LogBERTa. LogBERTa is a variant of RoBERTa, designed specifically for log anomaly detection and also trained for MLM. Similar to LogAnBERT, LogBERTa is a 6-layer model trained on the same log events. For preprocessing, we utilized the BBPE tokenizer for LogBERTa. LogBERTa's tokenizer was trained from scratch on the same log events, also resulting in a vocabulary size of 30,522. By building LogBERTa from scratch and employing a dedicated

---

[2] https://scikit-learn.org/stable/.

tokenizer, we aimed to enhance its ability to effectively analyze log events and identify anomalies with improved accuracy.

Our motivation behind building two Transformer models, LogAnBERT and LogBERTa, with different tokenizers, stems from the unique characteristics of log data and the specific requirements of log anomaly detection. Log events originating from different systems exhibit diverse patterns and structures necessitating an effective means to capture and represent these patterns. By employing distinct tokenization strategies, we aim to better handle the heterogeneity of log data, optimize the models' performance in understanding and identifying anomalies and harness the power of TL. LogAnBERT utilizes WordPiece, well-suited for capturing subword-level information [5]. On the other hand, LogBERTa employs BBPE, which operates at the byte level and can effectively handle non-standard encodings [12]. This choice of tokenizers allows us to leverage their strengths when dealing with diverse log datasets, enhancing HEART's adaptability.

3.4   HEART Framework

Our proposed classification framework, HEART, is shown in Fig. 1 (the solid arrow represents the core workflow). HEART combines the power of pre-trained models, BERT and RoBERTa, from the HuggingFace transformers library with our novel domain-specific models, LogAnBERT and LogBERTa, for log classification.

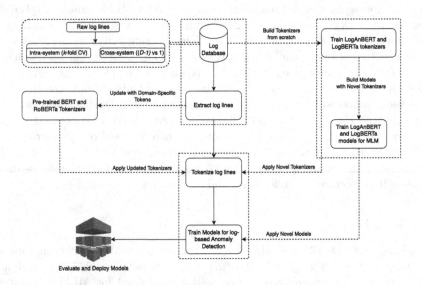

Fig. 1. Workflow of the HEART Framework

In the HEART framework, we augment the existing pre-trained models' (BERT and RoBERTa) tokenizers with domain-specific tokens, to enable them to handle log data. These updated models are then utilized to tokenize the log lines and trained for sequence classification. On the other hand, the right-hand side of the framework focuses on making use of all log lines from the database to train the novel tokenizers and models (`LogAnBERT` and `LogBERTa`) for MLM. In the classification stage, the tokenized events are passed through a fully connected linear layer, following a dropout layer to prevent overfitting. The output of this linear layer is then passed through a sigmoid activation function to produce the binary anomaly scores, indicating the probability of log events being anomalous or normal. For evaluation, HEART encompasses both intra- (with k-fold cross-validation) as well as cross-system (($\mathcal{D} - 1$) vs 1 for a total of $\mathcal{D}$ datasets) performance evaluations. By incorporating these aspects, HEART provides a thorough assessment of performance across different configurations, ensuring its readiness for both intra- and cross-system performance evaluation. By combining the power of TL with custom, domain-specific models and tokenizers, HEART offers a comprehensive and end-to-end anomaly detection framework.

4   Experimental Setup

This section presents the details of the datasets, scenarios, evaluation metrics, and hardware employed in our experiments to assess HEART's performance.

4.1   Datasets

In this research, we evaluate the effectiveness of our approach using four widely used publicly available labelled datasets downloaded from the loghub repository[3]. Table 1 presents the statistics of the datasets.

Table 1. Datasets Statistics

System	#Events	#Anomalies	#Non-anomalies	Data Size
BGL	4,747,963	348,460	4,399,503	708.76 MB
Hadoop	393,433	25,285	368,148	48.61 MB
HDFS	11,175,629	288,250	10,887,379	1.47 GB
OpenStack	207,820	189,386	18,434	58.61 MB

BlueGene/L (BGL) [14]: The BGL dataset consists of logs collected from a BlueGene/L supercomputer system at Lawrence Livermore National Labs (LLNL) in California. The system has 131,072 processors and 32.7TB of memory. The logs contain labelled alert and non-alert events.

Hadoop [11]: In this dataset, logs were generated from a Hadoop cluster consisting of five machines with a total of 46 cores, each machine equipped with an Intel(R) Core(TM) i7-3770 CPU and 16GB RAM. The cluster was used to execute two testing applications, WordCount and PageRank. To simulate real-world production environment failures, various deployment failures were manually injected during the applications' run-time.

---

[3] https://github.com/logpai/loghub.

Hadoop Distributed File System (HDFS) [18]: The HDFS dataset was generated in a Hadoop-based map-reduce cloud environment using benchmark workloads, and annotated manually with handcrafted rules to identify anomalies. The logs are segmented into traces based on block IDs, each with a ground truth label (normal/anomaly).

OpenStack [6]: OpenStack[4] is a cloud operating system used for managing compute, storage, and networking resources in data centers. The dataset was generated on a flexible scientific infrastructure for cloud computing research. It includes both normal logs and abnormal log events.

4.2   Scenarios and Motivation

Our experiments comprised two scenarios: a cross-system scenario and an intra-system scenario. In the traditional experimental setting of intra-system evaluation, it has been common practice to employ fixed train-test-validation splits without incorporating cross-validation [10]. This approach involves dividing each system's log data into fixed portions, typically allocating 70% for training, 15% for validation, and 15% for testing. However, this method may not fully capture the variability within the dataset and can lead to overfitting or limited generalization to unseen data. To address these limitations and enhance the validity, reliability, and robustness of our evaluation, we have chosen to incorporate stratified k-fold cross-validation into our experimental design. This allows us to obtain a more comprehensive understanding of our models' performance across various data configurations and minimize potential biases introduced by fixed splits [2].

Additionally, to assess the generalizability of our models across different systems, we have introduced a cross-system scenario. In this setup, our models were trained on labelled log events from three out of the four different datasets, each representing a distinct system. The fourth dataset was reserved exclusively as the test set, allowing us to evaluate the models' performance in a more diverse and challenging environment. This cross-system evaluation is a form of TL, as it capitalizes on knowledge from log events from multiple systems to improve the models' performance on a target system [17]. It should be emphasized that our exploration of the cross-system scenario in log data is novel, as it involves multiple datasets from different systems. Similar approaches utilizing TL have been investigated in the context of recommender systems to address the data sparsity challenge in cross-system scenarios [22], demonstrating promising results.

To provide a fair comparison in the intra-system scenario, we compare our approach to NeuralLog [10], which utilizes Transformers (BERT, GPT2 [15], and RoBERTa) for log-based anomaly detection without log parsing. NeuralLog combines positional encoding and the Transformer encoder to learn attention patterns and make predictions from log "*messages*" extracted from the log sequences. Our approach differs in that it directly utilizes the raw log "*events*" (or "*lines*") from the sequences without parsing steps to extract the log messages. However, NeuralLog was only developed and evaluated on two of the four datasets we used in our work (HDFS and BGL). To ensure a fair comparison, we reran the

---

[4] https://www.openstack.org.

experiments for NeuralLog using the same k-fold cross-validation setup, which was previously lacking in the original implementation[5].

Both cross-system and intra-system scenarios are part of the HEART framework and offer valuable insights. The cross-system scenario evaluates the model's performance in detecting log anomalies across diverse systems, simulating an analyst's assessment of a novel system using a model trained on multiple systems' log data. This evaluation offers an initial assessment of the framework's performance in a new system for which there is limited or no prior knowledge about the nature of anomalies. On the other hand, the intra-system scenario assesses the model's performance in detecting log anomalies within each individual system, providing insights into its effectiveness in specific system contexts. While our binary approach of distinguishing "anomalous" and "non-anomalous" events is valuable, we recognize that teams may have different classification tasks or domain-specific requirements. For instance, they might aim to detect specific types of anomalies or address unique security concerns such as malware infections or brute force attacks. To accommodate such scenarios, we include the intra-system performance on the "anomalous" versus "non-anomalous" classification task, showcasing our approach's flexibility for system-specific customization.

4.3   Hardware Setup and Evaluation Metrics

The experiments were conducted on the Digital Research Alliance of Canada (the Alliance) clusters utilizing 4 nodes, each equipped with 4 NVIDIA V100 GPUs (with 16GB memory each). The models were trained for 5 epochs with a batch size of 16 samples. In evaluating the performance, we considered the precision, recall and F_1-score as evaluation metrics. In the cross-system scenarios, we examined the performance at the class-wise level, which allowed us to assess the performance of each individual class and understand the model's ability to detect anomalies in different system contexts.

In the intra-system setup, we primarily focus on the macro F_1-score as it provides a balanced measure of precision and recall, accounting for the accurate detection of anomalies and minimizing false positives. This is particularly important in the context of imbalanced datasets, where accuracy alone can be misleading [13]. We also constructed a robust stratified 10-fold cross-validation framework to report reliable metrics and account for potential sampling bias [2]. Finally, to assess significant differences among our experimental results (in both intra-system and cross-system setups), considering the macro F_1-score, we used the Friedman test [7]. NeuralLog was excluded from our intra-system analysis as it was not developed for the Hadoop and OpenStack datasets.

5   Results and Discussion

This section presents the results of the experiments we conducted in both intra- and cross-system setups, including statistical tests. We further discuss the obtained results, their implications, and potential avenues for further exploration.

---

[5] https://github.com/vanhoanglepsa/NeuralLog.

5.1 Intra-system Results

We conducted a 10-fold stratified cross-validation on data from each of the four systems in the intra-system setup. Table 2 presents the average results obtained from the 10 runs conducted across the systems, with the standard deviation indicated in subscript next to the average values. NeuralLog's results are presented as NL-B, NL-R, and NL-G, indicating the LMs used: B for BERT, G for GPT2, and R for RoBERTa. It is noteworthy that our implementation offers the advantage of system independence, making it adaptable to various systems and datasets, underscoring its wide applicability.

As seen in Table 2, our novel, yet smaller LMs (LogAnBERT and LogBERTa) achieve comparable results to the existing models (NeuralLog, BERT and RoBERTa) which contain more layers and parameters while taking less training time. It is evident that despite having fewer layers and parameters and taking the entire log events, our domain-specific models achieve performance on par with the larger counterparts. This highlights the efficiency and effectiveness of our approach, making it a compelling choice for cybersecurity applications where reduced training time is of paramount importance.

The Friedman test yielded a Q-statistic of 6.8399, accompanied by a corresponding p-value of 0.0771 and a Critical Distance (CD) of 2.3452. These results indicate that there is no statistically significant difference among the measures at a confidence level, α, of 0.05. Notably, our approach utilizing BERT with an updated tokenizer achieved the highest rank, although it did not exhibit a statistically significant difference compared to the other models. Conversely, both LogBERTa and RoBERTa obtained the same rank, while LogAnBERT ranked last.

Table 2. Intra-system results (average and standard deviation).

System	M[1]	LBa[2]	RoBERTa	LBT[3]	BERT	NL-B	NL-G	NL-R
BGL	F	$1.0_{\pm0.0}$	$1.0_{\pm0.01}$	$0.99_{\pm0.0}$	$1.0_{\pm0.0}$	$0.99_{\pm0.0}$	$0.99_{\pm0.0}$	$0.98_{\pm0.0}$
	P	$1.0_{\pm0.0}$	$1.0_{\pm0.0}$	$0.99_{\pm0.0}$	$1.0_{\pm0.0}$	$0.99_{\pm0.0}$	$0.99_{\pm0.0}$	$0.98_{\pm0.0}$
	R	$1.0_{\pm0.0}$	$1.0_{\pm0.0}$	$0.99_{\pm0.0}$	$1.0_{\pm0.0}$	$0.99_{\pm0.0}$	$0.99_{\pm0.0}$	$0.98_{\pm0.0}$
Hadoop	F	$0.87_{\pm0.01}$	$0.89_{\pm0.04}$	$0.86_{\pm0.01}$	$0.92_{\pm0.01}$	$-^4$	$-^4$	$-^4$
	P	$0.84_{\pm0.02}$	$0.86_{\pm0.04}$	$0.85_{\pm0.01}$	$0.90_{\pm0.01}$	$-^4$	$-^4$	$-^4$
	R	$0.91_{\pm0.01}$	$0.91_{\pm0.05}$	$0.88_{\pm0.0}$	$0.95_{\pm0.01}$	$-^4$	$-^4$	$-^4$
HDFS	F	$0.97_{\pm0.0}$	$0.96_{\pm0.0}$	$0.96_{\pm0.01}$	$0.98_{\pm0.01}$	$0.99_{\pm0.0}$	$0.99_{\pm0.0}$	$0.98_{\pm0.0}$
	P	$0.95_{\pm0.0}$	$0.94_{\pm0.0}$	$0.95_{\pm0.0}$	$0.97_{\pm0.01}$	$0.99_{\pm0.0}$	$0.98_{\pm0.0}$	$0.98_{\pm0.0}$
	R	$0.98_{\pm0.0}$	$0.98_{\pm0.0}$	$0.98_{\pm0.02}$	$0.98_{\pm0.0}$	$0.99_{\pm0.0}$	$0.98_{\pm0.0}$	$0.96_{\pm0.0}$
OpenStack	F	$1.0_{\pm0.0}$	$1.0_{\pm0.0}$	$1.0_{\pm0.0}$	$1.0_{\pm0.0}$	$-^4$	$-^4$	$-^4$
	P	$1.0_{\pm0.0}$	$1.0_{\pm0.0}$	$1.0_{\pm0.0}$	$1.0_{\pm0.0}$	$-^4$	$-^4$	$-^4$
	R	$1.0_{\pm0.0}$	$1.0_{\pm0.0}$	$1.0_{\pm0.0}$	$1.0_{\pm0.0}$	$-^4$	$-^4$	$-^4$

[1] M: Metric – F: Macro F_1-score, P: Macro Precision, R: Macro Recall; [2] LBa: LogBERTa; [3] LBT: LogAnBERT; [4]NeuralLog method not developed for this dataset.

5.2   Cross-System Results

In the cross-system setup, we utilized $\mathcal{D}-1$ datasets from the available $\mathcal{D}$ datasets for training the models, while reserving one dataset for testing. Specifically, we trained each model on three datasets and evaluated its performance on the remaining dataset. The results of this evaluation are shown in Table 3. To the best of our knowledge, this study represents the first exploration of such a heterogeneous experiment, employing TL and involving multiple datasets. In contrast, LogTransfer solely utilizes a single system as the source data and assesses its performance on another system as the target data. Therefore, we utilized our implementations of the BERT and RoBERTa models, along with updated tokenizers, as baselines for comparison against the novel LogAnBERT and LogBERTa models.

Table 3 provides a comprehensive view of the performance of different models, including our proposed domain-specific models (LogAnBERT and LogBERTa), as well as the widely used BERT and RoBERTa models, across various systems. The class-wise cross-system results in the table highlight the models' ability to distinguish between non-anomalies (class 0) and anomalies (class 1). We observe varying degrees of success on the different datasets with the different models. For the BGL dataset, only RoBERTa demonstrated exceptional performance while the others' performances were not effective for anomaly detection. However, for the Hadoop dataset, the performance of all models was relatively lower, with only the BERT yielding an F_1-score for class 1 (anomalies) of 72%. Conversely, on the OpenStack dataset, only our novel models LogAnBERT and LogBERTa yielded F_1-scores of 95% on class 0 (non-anomalies), while the other two yielded poor results for both classes. Interestingly, our proposed domain-specific models

Table 3. Cross-system results

System	M[1]	BERT		RoBERTa		LBT[2]		LBa[3]	
		0	1	0	1	0	1	0	1
BGL	F	0.63	0.19	1.00	1.00	0.02	0.14	0.02	0.14
	P	0.97	0.11	1.00	1.00	0.99	0.07	0.88	0.07
	R	0.47	0.79	1.00	1.00	0.01	1.00	0.01	0.98
Hadoop	F	0.23	0.72	0.12	0.07	0.12	0.00	0.12	0.23
	P	0.13	1.00	0.07	0.98	0.06	0.00	0.07	0.95
	R	0.96	0.56	0.99	0.04	1.00	0.00	0.89	0.13
HDFS	F	0.00	0.06	0.00	0.06	0.99	0.30	0.99	0.54
	P	0.00	0.03	0.00	0.03	0.98	1.00	0.98	1.00
	R	0.00	1.00	0.00	1.00	1.00	0.18	1.00	0.37
OpenStack	F	0.00	0.16	0.00	0.16	0.95	0.00	0.95	0.00
	P	0.00	0.09	0.00	0.09	0.91	0.00	0.91	0.00
	R	0.00	1.00	0.00	1.00	1.00	0.00	1.00	0.00

[1]M: Class-wise Metric, F: F_1-score, P: Precision, R: Recall, [2]LBT: LogAnBERT, [3]LBa: LogBERTa

showed promising results on the HDFS dataset, with `LogBERTa` achieving an F_1-score of 54% for class 1 (anomalies). This performance surpasses both `BERT` and `RoBERTa`, which yielded similar, relatively poor results and achieved an F_1-score of 0 for class 0 (non-anomalies), indicating a complete failure to correctly predict instances of the majority class. Overall, the cross-system results highlight the potential of our proposed models in adapting to diverse log formats on the HDFS dataset, while also underscoring the challenges in handling the specific patterns present in those datasets. These findings emphasize the necessity for additional optimization to achieve improved performance in scenarios involving diverse formats.

Based on the obtained macro F_1-scores, we calculated a Q-statistic of 0.5675, accompanied by a corresponding p-value of 0.9038 and a CD of 2.3451. Similar to the intra-system scenarios, no statistically significant difference was observed among the measures at $\alpha = 0.05$. However, in this case, both `LogBERTa` and `BERT` with the updated tokenizer achieved the highest rank, followed by `RoBERTa` in second place, while `LogAnBERT` retained its position as the lowest performer.

5.3   Discussion

The results obtained from the intra- and cross-system scenarios (Tables 2 & 3) show interesting performance trends. In intra-system scenarios, our models demonstrate relatively strong performance, achieving the lowest F_1-score of 86% on the Hadoop dataset using `LogAnBERT`. However, in cross-system scenarios, the performance becomes more challenging. Despite our proposed `LogBERTa` and `LogAnBERT` models achieving relatively high macro F_1-scores of 76% and 64% on the HDFS dataset, the statistical analysis we conducted showed that there is no significant difference when considering results from all datasets in the cross-system scenarios. These findings highlight the inherent difficulties of effectively handling log events from multiple systems that exhibit diverse formats and structures, while also indicating that our novel, more compact models can achieve comparable performance to the larger `BERT` and `RoBERTa` models.

Examining the log events from the four experimental datasets offers valuable insights into the challenges encountered and HEART's effectiveness in both intra- and cross-system scenarios. Notably, Table 4 illustrates distinct patterns in the log events from the datasets. In the BGL dataset, there is a clear differentiation between normal and anomalous events. Anomalies in this dataset consistently begin with keywords belonging to the 41 identified classes, as established by Oliner and Stearley [14], such as KERNDTLB, KERNSTOR, and APPSEV. These keywords are associated with various aspects of the system, including the kernel's handling of the Translation Lookaside Buffer (TLB), storage and critical application-level events. In contrast, normal events consistently begin with numerical values. This provides a plausible explanation for the high performance achieved by all models in intra-system scenarios on the BGL dataset.

The Hadoop dataset lacks the distinct characteristics found in the BGL dataset, resulting in an F_1-score range of 87% to 92% in intra-system scenarios. The dataset's relatively small size and high class imbalance contribute to the

Table 4. Example log events from the experimental datasets

System	Label	Log Event
BGL	Normal	1118766826 2005.06.14 R25-M0-N3-C:J09-U11 2005-06-14-09 ...
	Anomaly	KERNSTOR 1118766822 2005.06.14 R24-M0-NC-C:J02-U11 ...
HDFS	Normal	081110 003838 13 INFO dfs.DataBlockScanner: ...
	Anomaly	081110 003838 13 INFO dfs.DataBlockScanner: Verification ...
Hadoop	Normal	2015-10-18 21:36:41,236 INFO [main] org.apache.hadoop...
	Anomaly	2015-10-18 18:25:44,584 WARN [LeaseRenewer:msrabi@...
Openstack	Normal	nova-api.log.1.2017-05-16_13:53:08 2017-05-16 06:25:02.870 ...
	Anomaly	nova-api.log.2017-05-14_21:27:04 2017-05-14 19:39:01.445 ...

performance variation. Similar trends are observed in the OpenStack and HDFS datasets, but the distinctive length of log events in OpenStack and the larger size of HDFS assist our models in achieving F_1-scores ranging from 97% to 100% in intra-system scenarios. During testing in the cross-system setup, our models face challenges when processing log events from the target dataset. We posit this difficulty arises due to discrepancies in log events across the datasets, resulting in sub-optimal performance. To address this issue, the concept of Heterogeneous TL (HTL) can be further explored. HTL involves extracting features from both the source domain and target domain and it has been proven effective in similar scenarios [17]. Updating the tokenizers with tokens specific to the target system may potentially improve the models' performance by aligning them more closely with the characteristics of the target domain.

6   Conclusion

In this work, we demonstrate the competitiveness and efficiency of our novel HEART framework that utilizes our domain-specific LMs, `LogAnBERT` and `LogBERTa`, for log-based anomaly detection without log parsing in cybersecurity applications when used in intra-system scenarios. Through comprehensive evaluations and comparisons with existing models, including NeuralLog, `BERT`, and `RoBERTa` with updated tokenizers, we have shown that our smaller models achieve comparable results while requiring less training parameters, thus less training time. This highlights the efficiency and effectiveness of our domain-specific models in processing entire, raw, log events. By leveraging the domain-specific knowledge embedded in our models, we can effectively capture the semantics of log events from single systems and achieve performance on par with larger counterparts. However, in cross-system settings, although leveraging TL showed improvements on specific datasets, with `LogBERTa` and `LogAnBERT` models yielding macro F_1-scores of 76% and 64% on the HDFS dataset; our novel models generally fell short compared to our implementations of the larger counterparts with updated tokenizers. Nonetheless, our research contributes valuable insights into log-based anomaly detection and enhances streamlined and time-efficient log analysis, benefiting the field of cybersecurity.

In order to further enhance our HEART framework, we plan to explore data augmentation to enhance training data diversity, incorporate additional datasets to increase variability and investigate HTL and unsupervised approaches for anomaly detection. By addressing these areas, we aim to improve HEART's robustness, generalization, and scalability and advance the log-based anomaly detection techniques, especially in cross-system scenarios.

Acknowledgements. This work was supported by the Natural Sciences and Engineering Research Council of Canada (NSERC), the Vector Institute, and The IBM Center for Advanced Studies (CAS) Canada within Project 1059. We are also grateful to the Digital Research Alliance of Canada (the Alliance) for their continuous support and access to their High-Performance Computing clusters.

References

1. Almodovar, C., Sabrina, F., Karimi, S., Azad, S.: Can language models help in system security? Investigating log anomaly detection using BERT. In: Proceedings of the The 20th Annual Workshop of the Australasian Language Technology Association, pp. 139–147. Australasian Language Technology Association, Adelaide, Australia (2022). https://aclanthology.org/2022.alta-1.19
2. Arp, D., et al.: Dos and don'ts of machine learning in computer security. In: Proceedings of the USENIX Security Symposium (2022). https://doi.org/10.48550/arXiv.2010.09470
3. Chen, R., et al.: LogTransfer: Cross-system log anomaly detection for software systems with transfer learning. In: 2020 IEEE 31st International Symposium on Software Reliability Engineering (ISSRE) pp. 37–47 (2020). https://doi.org/10.1109/ISSRE5003.2020.00013. ISSN: 2332-6549
4. Chen, S., Liao, H.: Bert-log: anomaly detection for system logs based on pretrained language model. Appl. Artif. Intell. **36**(1), 2145642 (2022). https://doi.org/10.1080/08839514.2022.2145642
5. Devlin, J., Chang, M.W., Lee, K., Toutanova, K.: BERT: pre-training of deep bidirectional transformers for language understanding (2019). https://doi.org/10.48550/arXiv.1810.04805
6. Du, M., Li, F., Zheng, G., Srikumar, V.: DeepLog: anomaly detection and diagnosis from system logs through deep learning. In: Proceedings of the 2017 ACM SIGSAC Conference on Computer and Communications Security, pp. 1285–1298. ACM (2017). https://doi.org/10.1145/3133956.3134015
7. Friedman, M.: The use of ranks to avoid the assumption of normality implicit in the analysis of variance. J. Am. Stat. Assoc. **32**(200), 675–701 (1937). https://doi.org/10.1080/01621459.1937.10503522
8. Guo, H., Yuan, S., Wu, X.: LogBERT: log anomaly detection via BERT. In: 2021 International Joint Conference on Neural Networks (IJCNN), pp. 1–8 (2021). https://doi.org/10.1109/IJCNN52387.2021.9534113. ISSN: 2161-4407
9. He, P., Zhu, J., Zheng, Z., Lyu, M.R.: Drain: an online log parsing approach with fixed depth tree. In: 2017 IEEE International Conference on Web Services (ICWS), pp. 33–40. IEEE (2017). https://doi.org/10.1109/ICWS.2017.13
10. Le, V.H., Zhang, H.: Log-based anomaly detection without log parsing. In: 2021 36th IEEE/ACM International Conference on Automated Software Engineering (ASE), pp. 492–504 (2021). https://doi.org/10.1109/ASE51524.2021.9678773. ISSN: 2643-1572

11. Lin, Q., Zhang, H., Lou, J.G., Zhang, Y., Chen, X.: Log clustering based problem identification for online service systems. In: 2016 IEEE/ACM 38th International Conference on Software Engineering Companion (ICSE-C), pp. 102–111 (2016). https://doi.org/10.1145/2889160.2889232

12. Liu, Y., et al.: Roberta: a robustly optimized BERT pretraining approach. ArXiv abs/1907.11692 (2019). https://doi.org/10.48550/arXiv.1907.11692

13. Mvula, P.K., Branco, P., Jourdan, G.V., Viktor, H.L.: A systematic literature review of cyber-security data repositories and performance assessment metrics for semi-supervised learning. Discov. Data 1(1), 4 (2023). https://doi.org/10.1007/s44248-023-00003-x

14. Oliner, A., Stearley, J.: What supercomputers say: a study of five system logs. In: 37th Annual IEEE/IFIP International Conference on Dependable Systems and Networks (DSN 2007), pp. 575–584. IEEE (2007). https://doi.org/10.1109/DSN.2007.103

15. Radford, A., Wu, J., Child, R., Luan, D., Amodei, D., Sutskever, I.: Language models are unsupervised multitask learners (2019)

16. Shao, Y., et al.: Log anomaly detection method based on BERT model optimization. In: 2022 7th International Conference on Cloud Computing and Big Data Analytics (ICCCBDA), pp. 161–166 (2022). https://doi.org/10.1109/ICCCBDA55098.2022.9778900

17. Weiss, K., Khoshgoftaar, T.M., Wang, D.D.: A survey of transfer learning. J. Big Data 3(1), 1–40 (2016). https://doi.org/10.1186/s40537-016-0043-6

18. Xu, W., Huang, L., Fox, A., Patterson, D., Jordan, M.I.: Detecting large-scale system problems by mining console logs. In: Proceedings of the ACM SIGOPS 22nd Symposium on Operating Systems Principles - SOSP 2009, p. 117. ACM Press (2009). https://doi.org/10.1145/1629575.1629587

19. Zhang, J., Li, Z., Zhang, X., Lin, F., Wang, C., Cai, X.: PoSBert: log classification via modified BERT based on part-of-speech weight. In: 2022 5th International Conference on Pattern Recognition and Artificial Intelligence (PRAI), pp. 979–983 (2022). https://doi.org/10.1109/PRAI55851.2022.9904207

20. Zhang, S., et al.: Syslog processing for switch failure diagnosis and prediction in datacenter networks. In: 2017 IEEE/ACM 25th International Symposium on Quality of Service (IWQoS), pp. 1–10 (2017). https://doi.org/10.1109/IWQoS.2017.7969130

21. Zhang, S., Liu, Y., Zhang, X., Cheng, W., Chen, H., Xiong, H.: CAT: beyond efficient transformer for content-aware anomaly detection in event sequences. In: Proceedings of the 28th ACM SIGKDD Conference on Knowledge Discovery and Data Mining, pp. 4541–4550 (2022). https://doi.org/10.1145/3534678.3539155

22. Zhao, L., Pan, S., Xiang, E., Zhong, E., Lu, Z., Yang, Q.: Active transfer learning for cross-system recommendation. Proc. AAAI Conf. Artif. Intell. 27(1), 1205–1211 (2013). https://doi.org/10.1609/aaai.v27i1.8458

Multi-kernel Times Series Outlier Detection

Florian Kalinke[✉] ⓘ, Edouard Fouché ⓘ, Haiko Thiessen, and Klemens Böhm

Karlsruhe Institute of Technology (KIT), Karlsruhe, Germany
{florian.kalinke,edouard.fouche,klemens.boehm}@kit.edu

Abstract. Time series are sequences of observations ordered by time. Detecting outliers in a set of time series is very important for many use cases, including fraud detection and predictive maintenance. However, this task continues to be difficult: First, time series may be of different lengths and conventional distance measures like the Euclidean distance can not capture their similarity well. Workarounds like feature engineering require domain knowledge and render solutions domain-specific. Second, many existing techniques are supervised, but training labels are expensive if not impossible to obtain. In this paper, we propose Multi-Kernel Times Series Outlier Detection (MK-TSOD), a method that combines the Fourier Transform, Global Alignment Kernels, and Multiple Kernel Learning with Support Vector Data Description. We describe its specifics, and show that MK-TSOD outperforms existing methods on standard benchmark data.

Keywords: Time Series · Outlier Detection · Global Alignment Kernel · Fourier Transform · Support Vector Data Description

1 Introduction

Outlier detection is of fundamental importance for many real-world applications, such as fraud detection or predictive maintenance. In such settings, data is often collected over time; the data has the form of time series. In the literature on time series, "outlier" either refers to anomalous subsequences [24] or to anomalous full sequences [14]. This article addresses the latter, that is, detecting few outlying time series from a set of time series.

Outlier detection in time series continues to be challenging for two reasons: (1) First, most outlier detection algorithms rely on a notion of **distance** to quantify data dissimilarity. Yet, time series may have different lengths and be shifted in time, which makes classic distance measures (e.g., the Euclidean distance) inadequate. As a workaround, many existing techniques rely on extracted features instead of directly comparing the series by a distance measure. However, extracting features limits the applicability of respective algorithms and generally leads to a loss of information. (2) Second, the outlier detection problem is **unsupervised** in nature and typically imbalanced, that is, outliers are rare, so that optimizing the parameters of outlier detectors is hardly feasible in practice.

A. Bifet et al. (Eds.): DS 2023, LNAI 14276, pp. 688–702, 2023.
https://doi.org/10.1007/978-3-031-45275-8_46

One common way to tackle both problems is using dynamic time warping (DTW; [20]) together with Support Vector Data Description (SVDD; [23]). SVDD is a kernel-based approach that encloses a predefined share of the data within a hypersphere of minimal volume; points outside the sphere are outliers. The kernel function quantifies the dissimilarity of the data in an implicit feature space. However, DTW does not yield a valid kernel function; the theory [21,22] supporting support vector-based approaches does not hold when using DTW with SVDD [8].

In this paper, we propose a kernel-based method for time series outlier detection that addresses all challenges identified above:

We propose Multi-Kernel Time Series Outlier Detection (MK-TSOD). Our idea is to combine SVDD with multiple kernels, which can capture frequency information of time series with fast Fourier transform, and time information with Global Alignment Kernels (GAK; [8]). Unlike DTW, GAK is guaranteed to work with SVDD. We combine the time and frequency information in an optimal way with Multiple Kernel Learning (MKL; [18]). MK-TSOD has one parameter, the expected outlier ratio, which is intuitive to set.

We run extensive experiments on standard benchmark data. They reveal that the proposed method outperforms the existing approaches on 9 out of 15 data sets with the balanced accuracy metric. We release the implementation together with our experiments on GitHub.[1]

Paper outline: Sect. 2 presents related work. Section 3 presents the definitions and the existing elements our method uses. Section 4 introduces the proposed approach. The experiments are in Sect. 5 and Sect. 6 concludes.

2 Related Work

While outlier detection has been well addressed for numerous types of data, e.g., numerical, categorical, mixed, or text data, detecting outliers from time series remains particularly challenging.

Due to the lack of proper distance measures for time series, most outlier detectors use extracted features instead. Examples are Highest Density Regions (HDR; [15]), and α-hull [15]: HDR extracts features of the time series and then applies Principal Component Analysis (PCA) to project the features to the first two principal components. It then estimates the local density of each observation. Observations whose density is below a threshold are the outliers. α-hull is similar to HDR as it also uses PCA. However, instead of using a density-based approach, it relies on α-convex hulls. Both algorithms use the same set of extracted features.

Finding a good set of features tends to be difficult. Established approaches to find such sets are either expert- or algorithm-based. The expert-based ones are costly and require domain knowledge. The algorithm-based ones, e.g., [6,16], only target classification and regression, i.e., supervised settings.

DOTS (Detection of Outlier Time Series; [3]) does not rely on extracted features, but clusters the data based on DTW and then uses the entropy to find an

---

[1] https://github.com/flopska/mk-tsod/.

optimal positioning of cluster centers. It ranks the "outlierness" of observations based on their distances to the clusters. DOTS has more free parameters than our approach, and several of them are difficult to optimize in an unsupervised setting, e.g., the regularization parameter λ and the number of clusters k.

ADSL (Anomaly Detection algorithm with shapelet-based Feature Learning; [2]) is so far the approach most related to ours, as it also bases on SVDD. The main difference is that ADSL applies SVDD to a learned intermediate representation, namely shapelets, but not explicitly to time series data, as we do. It then classifies outliers based on their distance to the shapelets. However, this approach is only successful in cases where shapelets are indeed a meaningful representation.

The DeepSVDD [19] approach combines ideas from neural networks with the outlier detection paradigm of SVDD, that is, it learns a hypersphere encompassing the networks representation of most observations with minimal volume. Similarly to SVDD, DeepSVDD labels points outside this sphere as outliers. However, the algorithm is limited to data with fixed length and thus not applicable to time series.

While theoretically unsound, support vector-based approaches with DTW-based kernels can work in practice [13]. DTW-SVDD, ADSL, DOTS, HDR, and α-hull form a set of strong baselines against which we compare in Sect. 5.

3   Background

This section summarizes the definitions (Sect. 3.1). To render the article fully self-contained, we recall SVDD (Sect. 3.2) and GAK (Sect. 3.3).

3.1   Definitions

A function $k : \mathcal{X} \times \mathcal{X} \to \mathbb{R}$ is a kernel on an input space $\mathcal{X}$ if there exists a real Hilbert space $\mathcal{H}$ and a map $\phi : \mathcal{X} \to \mathcal{H}$ such that $k(x, x') = \langle \phi(x), \phi(x') \rangle$ for all $x, x' \in \mathcal{X}$. We call ϕ the feature map and $\mathcal{H}$ the feature space of k [22]. The corresponding Gram matrix for a subset $\{x_1, \ldots, x_l\} \in \mathcal{X}^l$ is the symmetric $l \times l$ matrix $\mathbf{K}_k = [k(x_i, x_j)]_{i,j=1}^{l} \in \mathbb{R}^{l \times l}$.

A time series x with length n is a sequence $x := (x_m)_{m=1}^{n}$ with $x_m \in \mathbb{R}$ for $m = 1, \ldots, n$. In what follows, we consider the input space $\mathcal{X} = \{x_1, \ldots, x_l\}$, i.e., a set of l time series with potentially different lengths. The proposed method is straightforward to extend to $\mathbb{R}^D$ ($D \in \mathbb{N}$), but for clarity, we consider observations taking values in $\mathbb{R}$ in what follows.

With those definitions, detecting outlying time series can be seen as finding the $l \cdot \theta$ time series that are the most dissimilar within a set of time series $\mathcal{X}$, where the parameter $\theta \in (0, 1)$ specifies the expected ratio of outliers in that set. Since outliers are rare, θ is typically small.

3.2 Support Vector Data Description (SVDD)

The construction of SVDD is similar to the well-known SVM (Support-Vector Machine; [7]). In short, the goal is to solve the constrained optimization problem

$$\min R^2 + C \sum_{i=1}^{l} \xi_i, \quad \text{s.t.} \quad \|\phi(x_i) - a\|^2 \le R^2 + \xi_i, \ \xi_i \ge 0,$$

for $i = 1, \ldots, l$, that is, to find a sphere with center a and radius R^2 so that most observations are enclosed. The slack variables ξ_i allow points to lie outside the sphere with a penalty controlled by parameter C. [23] recommends setting

$$C = 1/(l \cdot \theta), \tag{1}$$

where θ is the expected ratio of outliers in the data, and l the size of the data set. Hence, θ can be chosen intuitively. The corresponding dual problem is

$$\max_{\alpha} \sum_{i=1}^{l} \alpha_i k(x_i, x_i) - \sum_{i,j=1}^{l} k(x_i, x_j),$$

$$\text{s.t.} \sum_{i=1}^{l} \alpha_i = 1, \ 0 \le \alpha_i \le C, \tag{2}$$

for all $i = 1, \ldots, l$ and with Lagrange multipliers $\boldsymbol{\alpha} = (\alpha_1, \ldots, \alpha_l)^\mathsf{T}$.

Having obtained a solution to (2), a time series $z \in \mathcal{X}$ is an outlier if and only if

$$\|\phi(z) - a\|^2 = k(z, z) - 2 \sum_{i=1}^{l} \alpha_i k(z, x_i) + \sum_{i,j=1}^{l} \alpha_i \alpha_j k(x_i, x_j) > R^2, \tag{3}$$

with the radius R^2 computed as

$$R^2 = k(x_k, x_k) - 2 \sum_{i=1}^{l} \alpha_i k(x_i, x_k) + \sum_{i,j=1}^{l} \alpha_i \alpha_j k(x_i, x_k), \tag{4}$$

with any $x_k \in \mathcal{X}$ for which the corresponding Lagrange multiplier α_k fulfills $0 < \alpha_k < C$.

3.3 Global Alignment Kernels (GAK)

GAKs [8] extend DTW to the kernel setting. The definition of GAK bases on the notion of alignment: An alignment π of length p between $x, y \in \mathcal{X}$ of lengths n, n' is a pair (π_1, π_2) that fulfills the following conditions:

Boundary & Monotonicity. The first observation in x must map to the first observation in y and analogously for the last observations. Also, the alignment must be increasing. Formally, one has

$$1 = \pi_1(1) \leq \cdots \leq \pi_1(p) = n,$$
$$1 = \pi_2(1) \leq \cdots \leq \pi_2(p) = n'. \tag{5}$$

Continuity. There must not be any gap in the alignment path, i.e., each observation must map to at least one other observation. Further, there must not be any repetition. Formally, for all $1 \leq i, j \leq p - 1$,

$$\pi_1(i+1) \leq \pi_1(i) + 1, \pi_2(j+1) \leq \pi_2(j) + 1,$$
$$(\pi_1(i+1) - \pi_1(i)) + (\pi_2(i+1) - \pi_2(i)) \geq 1. \tag{6}$$

Adjustment Window. Given an observation $x_i \in \mathbb{R}$, $i = 1, \ldots, n$ of time series $x \in \mathcal{X}$ and parameter T, x_i must map to an observation $y_i \in \mathbb{R}$, $i = 1, \ldots, n'$ of $y \in \mathcal{X}$ that is "sufficiently close", i.e., strictly less than T steps away and vice versa. Formally, for all $1 \leq i \leq p - 1$

$$|\pi_1(i) - \pi_2(i)| < T. \tag{7}$$

While not strictly necessary, the adjustment window condition speeds up the computation by reducing the number of alignments considered without impacting result quality by much [8]; we confirm this in our experiments.

The kernel k_{GAK} sums all distances computed over alignments that satisfy (5), (6), and (7):

$$k_{GAK}(x, y) = \sum_{(x', y') \in \mathcal{M}(n, n')} k(x', y'), \tag{8}$$

with $\mathcal{M}(n, n') = \{(x'_{\pi_1}, y'_{\pi_2}) \mid \pi = (\pi_1, \pi_2) \in \mathcal{A}(n, n')\}$, where $\mathcal{A}(n, n')$ is the set of all valid alignments, and where $k(x'_{\pi_1}, y'_{\pi_2}) = \prod_{i=1}^{|\pi|} \kappa(x'_{\pi_1(i)}, y'_{\pi_2(i)})$ for a so-called local kernel κ.

[8] shows that k_{GAK} is not positive definite for all such κ. This is problematic, as (2) is then non-convex, and the global optimum might be not be found. Additionally, the theory that supports kernel functions does not hold in such cases. However, [8] proves that $\kappa/(1 + \kappa)$ being positive definite is a sufficient condition to guarantee that k_{GAK} is positive definite and show that this holds for the local kernel,

$$\kappa(x, y) = \exp\left\{ -\frac{\|x - y\|^2}{2\sigma^2} - \log\left(2 - e^{-\frac{\|x-y\|^2}{2\sigma^2}}\right) \right\},$$

where, by abuse of notation, $x, y \in \mathbb{R}$ in our case, and $\|\cdot\|$ the Euclidean distance.

In turn, DTW is defined as the minimum distance over all valid alignments

$$\mathrm{DTW}(x, y) = \min_{\pi \in \mathcal{A}(n, n')} \sum_{i=1}^{|\pi|} \left\|x_{\pi_1(i)} - y_{\pi_2(i)}\right\|^2, \quad x, y \in \mathcal{X},$$

with the corresponding DTW kernel

$$k_{DTW}(x,y) = \exp\{-\gamma \cdot \mathrm{DTW}(x,y)\}. \tag{9}$$

As DTW does not fulfill the triangle inequality, the kernel k_{DTW} is not guaranteed to be positive definite, a problem one avoids with GAK.

GAK and DTW have a recursive formulation that one can compute with dynamic programming. So their complexity when comparing two time series of lengths n and n' and dimensionality d is $\mathcal{O}(dnn')$. As GAK only considers alignments within a band of width T, its runtime reduces to $\mathcal{O}(dT \min(n, n'))$.

GAK and DTW only consider the time information of the respective time series. But it is known that considering the frequency information can prove beneficial when working with time series. The proposed method that we present next builds upon this observation.

4   Multi-kernel Time Series Outlier Detection

Depending on the characteristics of a time signal that one wishes to highlight, it is common to represent the signal in the time or in the frequency domain. Accordingly, we propose a kernel k_{FFT} (Sect. 4.1) that considers similarities in the frequency domain, which we then combine with k_{GAK} in an optimal fashion with Multiple Kernel Learning (MKL; [18]). This guarantees that the proposed method (Sect. 4.2) detects outliers by taking both time and frequency information into account. We analyze the runtime complexity of MK-TSOD in Sect. 4.3.

4.1   Fast Fourier Transform Kernels

The Fourier transformation of a time series $x = (x_m)_{m=1}^{n}$ is the sequence $X = (X_k)_{k=1}^{n}$ of the Fourier coefficients

$$X_k = \sum_{m=1}^{n} x_m \exp\left\{-2\pi i \frac{(k-1)(m-1)}{n}\right\}, \quad k = 1, \ldots, n,$$

with $i^2 = -1$ the imaginary number. Let $x, y \in \mathcal{X}$ be time series of lengths n, n', having Fourier coefficients $X = (X_k)_{k=1}^{n}$, $Y = (Y_k)_{k=1}^{n'}$, respectively. To compare x and y, we propose k_{FFT} as a modified Gaussian kernel that truncates the sequence of Fourier coefficients, that is,

$$k_{FFT}(x,y) := \exp\left\{-\gamma \sum_{j=1}^{t} (X_j - Y_j)^2\right\}, \tag{10}$$

with smoothing parameter γ, and $1 \le t \le \min(n, n')$. Hence, parameter t controls the quality of the approximation by restricting the number of coefficients.

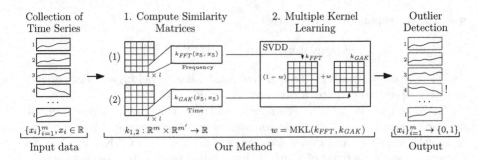

Fig. 1. Schematic representation of the proposed outlier detection method.

To select the bandwidth parameter γ in an unsupervised fashion, we use an argument from [12], which states that dissimilarities in the input space and dissimilarities in the feature space behave similarly:

$$\frac{\delta_1}{\delta_2} \approx \frac{\exp\left(-\gamma\delta_1^2\right)}{\exp\left(-\gamma\delta_2^2\right)},$$

where we denote by $\delta_i = \|\cdot\|$ ($i \in \{1,2\}$) the Euclidean distance between the truncated Fourier transformations of two arbitrary observations. One solves for γ and sets

$$\gamma = \frac{-\ln\left(\frac{\delta_{\min}}{\delta_{\mathrm{avg}}}\right)}{\delta_{\mathrm{avg}}^2 - \delta_{\min}^2}, \tag{11}$$

with the quantities $\delta_{\min} := \|x_q - x_{1-\mathrm{NN}(q)}\|$, $\delta_{\mathrm{avg}} := \frac{1}{n-1}\sum_{i\neq q}\|x_q - x_i\|$, and $q := \arg\min_{1\leq i\leq n}\|x_i - x_{1-\mathrm{NN}(i)}\|$. Here, $1 - \mathrm{NN}(k)$ denotes the index of the nearest neighbor of x_k, i.e., the transformation with the smallest distance in the frequency domain to x_k. Hence, x_q is the time series with the smallest distance to its nearest neighbor. $\delta_{\min}$ is the smallest distance between the Fourier coefficients of any two time series, and δ_{avg} is the average distance of all time series to x_q w.r.t. their Fourier coefficients. (11) allows accounting for the characteristics of the frequencies observed.

4.2   MK-TSOD Algorithm

Fig. 1 provides an intuitive schematic representation of the proposed algorithm, which we elaborate in what follows.

To merge kernel k_{GAK} and the proposed kernel k_{FFT}, we first recall a property of kernels [22, Lemma 4.5] that allows their combination. We then detail how we adapt MKL to SVDD in order to optimize over the free parameter that results from the kernel combination, and conclude the section with the presentation and runtime analysis of the full algorithm.

Lemma 1 (Additivity). *Let $\mathcal{X}$ be a set, $\beta \geq 0$, and k, k_1, and k_2 be kernels on $\mathcal{X}$. Then βk and $k_1 + k_2$ are kernels on $\mathcal{X}$ as well.*

With Lemma 1, a convex combination with weight $w \in [0, 1]$ of GAK kernel k_{GAK} and the proposed kernel k_{FFT} is a valid kernel that takes the form

$$k(x, y) = w \cdot k_{GAK}(x, y) + (1 - w) \cdot k_{FFT}(x, y),$$

and incorporates information of both the time and the frequency domain of $x, y \in \mathcal{X}$.

More generally, the MKL problem [18] is to find the Lagrange multipliers α_i of a kernel machine and the weights $\mathbf{w} = (w_1, \ldots, w_M)^\mathsf{T}$ for a convex combination k of kernels k_m given by

$$k(x, y) = \sum_{m=1}^{M} w_m k_m(x, y), \text{ s.t. } w_m \geq 0 \quad \wedge \quad \sum_{m=1}^{M} w_m = 1. \tag{12}$$

It follows from Lemma 1 and an induction argument that (12) defines a valid kernel. To find the solution, we proceed as follows:

The Lagrangian of (2) is

$$L = \sum_{i,j} \alpha_i \alpha_j k(x_i, x_j) - \sum_i \alpha_i k(x_i, x_i).$$

Combining this with $k(x, y)$ from (12), we obtain the MKL problem for SVDD

$$L = \sum_{i,j=1}^{l} \alpha_i \alpha_j \sum_{m=1}^{M} w_m k_m(x_i, x_j) - \sum_{i=1}^{l} \alpha_i \sum_{m=1}^{M} w_m k_m(x_i, x_j).$$

To optimize w.r.t. $\mathbf{w}$, [18] propose SimpleMKL, a gradient descent-based approach. Hence, we compute the partial derivative w.r.t. w_m, which for SVDD takes the form

$$\frac{\partial L}{\partial w_m} = \boldsymbol{\alpha}^\mathsf{T} \mathbf{K}_m \boldsymbol{\alpha} - \boldsymbol{\alpha}^\mathsf{T} \mathrm{diag}(\mathbf{K}_m)$$

with Gram matrix $\mathbf{K}_m$ associated with kernel k_m, and then apply their framework: In the present case, $M = 2$, $k_1 = k_{GAK}$, and $k_2 = k_{FFT}$. Performing the gradient descent optimization yields a weight w so that the volume of the hypersphere is again minimized.

Algorithm 1 presents MK-TSOD in full. The method has a total of five parameters. We recommend values for T, σ^2, and t in Sect. 5.1. Parameter γ is set according to (11); C is set by (1).

4.3   Complexity Analysis

The runtime of MK-TSOD depends on that of computing the Gram matrices for kernels k_{FFT}, k_{GAK}, and on that of solving the MKL problem. For a worst-case scenario, we assume that the longest time series is of length n, and that

---

Algorithm 1. MK-TSOD

Require: Time series $\mathcal{X} = \{x_1, \ldots, x_l\}$, outlier ratio θ

1: $C \leftarrow 1/(l \cdot \theta)$ ▷ Equation (1)
2: $\mathbf{K}_{k_{FFT}} = [k_{FFT}(x_i, x_j)]_{ij}$ for $i, j = 1, \ldots, l$ ▷ Equation (10)
3: $\mathbf{K}_{k_{GAK}} = [k_{GAK}(x_i, x_j)]_{ij}$ for $i, j = 1, \ldots, l$ ▷ Equation (8)
4: $\mathbf{w}, \boldsymbol{\alpha} \leftarrow \text{MKL}(\mathbf{K}_{k_{FFT}}, \mathbf{K}_{k_{GAK}}, C)$ ▷ Equation (12)
5: $\mathbf{K} \leftarrow w_1 \cdot \mathbf{K}_{k_{FFT}} + (1 - w_1) \cdot \mathbf{K}_{k_{GAK}}$
6: $R^2 \leftarrow (\mathbf{K})_{kk} - 2\sum_{i=1}^{l} \alpha_i(\mathbf{K})_{ik} + \boldsymbol{\alpha}^{\mathsf{T}}\mathbf{K}\boldsymbol{\alpha}$ ▷ Equation (4)
7: outliers $\leftarrow \emptyset$
8: **for** $x_i \in \mathcal{X}$ **do**
9:      **if** $(\mathbf{K})_{ii} - 2\sum_{j=1}^{l} \alpha_j(\mathbf{K})_{ij} + \boldsymbol{\alpha}^{\mathsf{T}}\mathbf{K}\boldsymbol{\alpha} > R^2$ **then**   ▷ Equation (3)
10:         outliers $\leftarrow$ outliers $\cup\, x_i$
11: **return** outliers

---

one observes l time series. Then the runtime of k_{GAK} per pair of observations is in $\mathcal{O}(n^2)$ [8]. As the Gram matrix computes all pairwise combinations, its computational cost is $\mathcal{O}(n^2 l^2)$. Computing the Fourier coefficients of a time series of length n has a complexity of $\mathcal{O}(n \log(n))$, and, by the same reasoning as before, obtaining the corresponding Gram matrix costs $\mathcal{O}(n \log(n) l^2)$. The worst-case bound for an optimal solution of SVDD is $\mathcal{O}(l^3)$ [4]. As the number of SimpleMKL iterations is bounded and does not depend on l [18], running SimpleMKL does not affect the worst-case estimate. Putting the previous estimates together, we obtain a total runtime complexity of $\mathcal{O}(n^2 l^2 + n \log(n) l^2 + l^3)$.

While the worst-case complexity is relatively high, the actual runtime is reasonable for practical applications and often lower than that of competitors, as our experiments show. In practice, one typically uses an approximate solver, such as sequential minimal optimization (SMO; [17]), which yields a solution to the SVDD problem (2) in $\mathcal{O}(l^2)$; this reduces the runtime cost.

5 Experiments

In our experiments, we compare the proposed technique to the state of the art, both in terms of outlier detection quality and runtime; we also conduct a parameter sensitivity and ablation analysis. We start by describing the experiment setup (Sect. 5.1), collect the results w.r.t. balanced accuracy, runtime, and parameter sensitivity in Sect. 5.2, and compare to ablations in Sect. 5.3.

5.1 Setup

Metrics and Evaluation. Our experiments evaluate the balanced accuracy (BA), which is commonly used for outlier detection tasks. We repeat each experiment 10 times, keeping the normal data but sampling a different set of outliers, and report the mean score and standard deviation. We run all algorithms on a server running Ubuntu 20.04 with 124 GB RAM, and 32 cores with 2 GHz each.

Table 1. Summary of the 15 data sets. *Length* is the length of the time series in the respective data set, *#N / #O* is the absolute count of normal and outlying observations, and *#C (N)* is the number of classes in the original data set together with the class set as normal.

Data set	Length	#N / #O	#C (N)
ArrowHead	251	65 / 3	3 (2)
CBF	128	310 / 16	3 (2)
Ch.Concent	166	1000 / 52	3 (1)
ECG200	96	133 / 7	2 (1)
ECGFiveDays	136	442 / 23	2 (1)
GunPoint	150	100 / 5	2 (1)
Ham	431	103 / 5	2 (1)
Herring	512	77 / 4	2 (1)
Lightning2	637	73 / 3	2 (1)
MoteStrain	84	685 / 36	2 (1)
Strawberry	235	351 / 18	2 (1)
ToeSeg1	277	140 / 7	2 (0)
ToeSeg2	343	124 / 6	2 (0)
Wafer	152	6402 / 336	2 (1)
Wine	234	57 / 3	2 (1)

Data Sets and Data Preparation. We follow the approach by [10,11] and adapt time series classification data sets from the UCR repository [1,9] to our setting. To improve comparability, our process mirrors the selection and pre-processing from [2] but we exclude data sets with fewer than 50 time series [2, Table 1], due to their small size. Specifically, for binary classification problems, we choose the majority class as "normal" class, and for multiclass classification problems, we set the class that is visually the most distinct as "normal". We then sample 5% of the observations from the respective other class(es), which constitute the "outliers". This yields 15 diverse data sets. Training sets include the outliers, as this is closer to real-world settings, but the outliers are regenerated between runs. Table 1 summarizes the respective characteristics of the data sets.[2]

Configurations. In the following, we detail the parameter settings for each algorithm. We start with a recommendation for the parameters of our algorithm.

MK-TSOD. We set the regularization parameter C as in (1) with an expected outlier ratio $\theta = 0.05$. For k_{GAK}, we follow the recommendation of [8] and set $\sigma^2 = a^2 \cdot \text{median}(\|\mathbf{x} - \mathbf{y}\|) \cdot \sqrt{\text{median}(|\mathbf{x}|)}$, with $a = 1.5$, and

---

[2] We abbreviate the data sets "ChlorineConcentration", "ToeSegmentation1", and "ToeSegmentation2" as "Ch.Concent.", "ToeSeg1", and "ToeSeg2", respectively.

$T = b \cdot \text{median}(|\mathbf{x}|)$, with $b = 0.5$. We vary factors a, b in the parameter sensitivity analysis. To solve the optimization problem (2), we use libsvm[3], which we adapt to use precomputed Gram matrices with SVDD. We set the smoothing parameter γ of the Fourier transform-based kernel using (11), and set $t = 20$, based on the parameter analysis we present at the end of the section. The code for reproducing our experiments is available on GitHub.[4]

SVDD with DTW. As a baseline, we use k_{DTW} (9) with SVDD; recall that while the approach is theoretically unsound, it has shown good results in practice. Because of the absence of a heuristic, we set $\gamma = 1$. We set the regularization parameter C as in MK-TSOD.

HDR and α-hull. We use the reference implementations provided by the authors together with the recommended parameters.[5]

DOTS. We set the regularization parameter λ to 0.045, and the number of medoids k to the number of classes per data set, as recommended by the authors, and use their reference implementation.[6] To compute the BA, we cut off the ranking based on the expected ratio of outliers, which is 0.05.

ADSL. We set the maximum number of iterations to 1000, $k = 0.02$, and $l = 0.2$, as in [2]. We obtained the code from the authors.

LOF with DTW. As an additional baseline, we combine the well-known Local Outlier Factor (LOF; [5]) with DTW in place of the Euclidean distance. As LOF is sensitive to the amount of neighbors n to consider, we set $n \in \{5, 10, 20\}$ and report the best results.

5.2   Results

Performance. Table 2 shows the average Balanced Accuracy (BA).[7] N/A indicates that the respective algorithm did not complete a single run in 24 h.

One sees that MK-TSOD achieves the best score on 9 out of 15 data sets with the BA metric, and that LOF-DTW performs second best, that is, it performs better than the competitors.

Runtime. We measure the absolute runtime of each algorithm w.r.t. the number and length of times series. We use the data set featuring the longest time series, Lightning2, and simulate different input configurations. To vary their length, we sub- or oversample the measured points. To vary the size of the set, we sub- or oversample the time series themselves. When oversampling, we add Gaussian noise with a standard deviation of 10^{-3}. This mimics that real-world data does not consist of duplicates only.

Figure 2 shows our results. MK-TSOD is slower than HDR and α-hull but faster than ADSL and DOTS w.r.t. the number of time series. However, the

---

[3] https://www.csie.ntu.edu.tw/~cjlin/libsvm/.

[4] https://github.com/flopska/mk-tsod/.

[5] https://github.com/robjhyndman/anomalous-acm.

[6] https://github.com/B-Seif/anomaly-detection-time-series.

[7] Here, MK denotes MK-TSOD; DTW denotes DTW-SVDD.

Table 2. Mean BA over 10 runs. Bold print highlights the best results.

Data set	MK	DTW	HDR	DOTS	α-hull	ADSL	LOF-DTW
ArrowHead	**0.70 ± 0.2**	0.58 ± 0.2	0.67 ± 0.1	0.51 ± 0.1	0.67 ± 0.2	0.49 ± 0.0	0.52 ± 0.1
CBF	**0.66 ± 0.0**	0.49 ± 0.1	0.50 ± 0.0	0.49 ± 0.0	0.50 ± 0.0	0.50 ± 0.0	0.65 ± 0.1
Ch.Concent	0.49 ± 0.0	0.48 ± 0.0	0.50 ± 0.0	0.50 ± 0.0	0.50 ± 0.0	0.50 ± 0.0	**0.63 ± 0.0**
cre ECG200	**0.67 ± 0.1**	0.55 ± 0.1	0.50 ± 0.0	0.55 ± 0.1	0.50 ± 0.1	0.52 ± 0.0	0.65 ± 0.1
ECGFiveDays	0.64 ± 0.0	0.58 ± 0.0	0.52 ± 0.0	0.54 ± 0.0	0.52 ± 0.0	0.50 ± 0.0	**0.77 ± 0.0**
GunPoint	**0.72 ± 0.1**	0.61 ± 0.1	0.49 ± 0.0	0.64 ± 0.1	0.50 ± 0.0	0.62 ± 0.1	0.70 ± 0.1
Ham	**0.51 ± 0.1**	0.48 ± 0.1	0.49 ± 0.0	0.48 ± 0.0	0.49 ± 0.0	0.49 ± 0.0	0.49 ± 0.0
Herring	**0.52 ± 0.1**	0.51 ± 0.1	0.50 ± 0.1	0.50 ± 0.1	0.47 ± 0.0	0.50 ± 0.0	0.50 ± 0.1
Lightning2	0.57 ± 0.2	0.49 ± 0.2	0.48 ± 0.0	0.50 ± 0.1	0.51 ± 0.1	0.64 ± 0.1	**0.72 ± 0.2**
MoteStrain	**0.70 ± 0.0**	0.62 ± 0.1	0.52 ± 0.0	0.61 ± 0.0	0.52 ± 0.0	0.51 ± 0.0	0.55 ± 0.0
Strawberry	0.69 ± 0.1	0.70 ± 0.0	0.47 ± 0.0	0.68 ± 0.0	0.48 ± 0.0	0.56 ± 0.0	**0.76 ± 0.0**
ToeSeg1	0.65 ± 0.1	0.50 ± 0.1	0.49 ± 0.0	0.47 ± 0.0	0.48 ± 0.0	0.61 ± 0.0	**0.73 ± 0.1**
ToeSeg2	**0.67 ± 0.1**	0.48 ± 0.1	0.51 ± 0.0	0.48 ± 0.0	0.52 ± 0.0	0.60 ± 0.0	0.61 ± 0.1
Wafer	**0.65 ± 0.0**	0.64 ± 0.0	0.49 ± 0.0	N/A	0.49 ± 0.0	0.50 ± 0.0	0.56 ± 0.0
Wine	0.48 ± 0.1	0.50 ± 0.1	0.60 ± 0.1	0.56 ± 0.1	**0.65 ± 0.2**	0.54 ± 0.1	0.58 ± 0.1

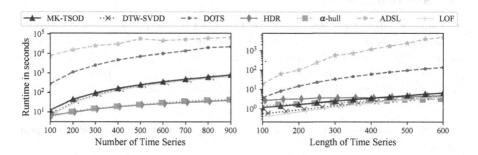

Fig. 2. Runtime analysis. We report the median runtime of five independent runs.

slope of MK-TSOD and DOTS is similar, so differences in runtime might be due to implementation details. The difference to DTW-SVDD and LOF-DTW is negligible. Regarding the length of time series, the figure shows that MK-TSOD and DTW-SVDD scale better than ADSL and DOTS, but worse than HDR and α-hull. Again, LOF-DTW scales similar to MK-TSOD, as expected.

Parameter Sensitivity Analysis. We study the sensitivity of MK-TSOD w.r.t. parameters σ, T (the smoothness and the width of the window of k_{GAK}), and t (the number of Fourier coefficients for k_{FFT}). Figure 3 shows the average results obtained over all data sets from Table 1. When varying one parameter, we keep the others fixed at their recommended values.

We see that for changes in σ, BA stays nearly constant from $x = 1.5$. The figure also shows that the width T of the band considered for alignments does not influence the result by much. However, we see a slight increase for the BA metric at $T = 0.2$. This indicates that focusing on local similarities proves beneficial for the data sets considered. For the number of Fourier coefficients t, we see

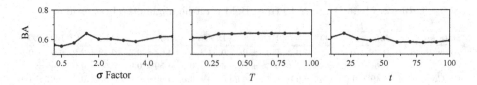

Fig. 3. Influence of the factors a, b for the median heuristics for σ, T, and influence of parameter t. We report the median BA of five independent runs.

that the best performance is obtained for $t = 20$, with a slight decline for larger values. We hypothesize that using more than 20 coefficients approximates the time series too closely, and the algorithm cannot generalize, that is, it overfits. Altogether, we see that MK-TSOD is robust w.r.t. its parameters.

5.3   Ablation Analysis

We consider alternative designs of the proposed method. Instead of combining multiple kernels, we run SVDD with the individual kernels k_{FFT} (FFT-SVDD) and k_{GAK} (GAK-SVDD) and compare their results to the ones obtained with MK-TSOD in terms of the average balanced accuracy over five draws of outliers. The settings of the individual kernels are the same as in Sect. 5.1.

Table 3. Ablation analysis. Mean BA over five runs. Bold print highlights the best results.

Data set	MK-TSOD	FFT-SVDD	GAK-SVDD
ArrowHead	**0.70 ± 0.2**	0.65 ± 0.1	0.61 ± 0.2
CBF	**0.66 ± 0.0**	0.60 ± 0.1	**0.66 ± 0.0**
Ch.Concent	0.49 ± 0.0	**0.52 ± 0.0**	0.48 ± 0.0
ECG200	**0.67 ± 0.1**	0.63 ± 0.1	0.62 ± 0.1
ECGFiveDays	0.64 ± 0.0	**0.65 ± 0.0**	0.62 ± 0.1
GunPoint	**0.72 ± 0.1**	0.65 ± 0.1	0.64 ± 0.1
Ham	**0.51 ± 0.1**	0.47 ± 0.1	0.50 ± 0.1
Herring	**0.52 ± 0.1**	0.49 ± 0.1	0.51 ± 0.1
Lightning2	0.57 ± 0.2	**0.67 ± 0.1**	0.47 ± 0.1
MoteStrain	**0.70 ± 0.0**	0.62 ± 0.0	0.67 ± 0.1
Strawberry	0.69 ± 0.1	0.71 ± 0.1	**0.73 ± 0.0**
ToeSeg1	**0.65 ± 0.1**	**0.65 ± 0.1**	0.62 ± 0.1
ToeSeg2	**0.67 ± 0.1**	0.55 ± 0.1	0.57 ± 0.1
Wafer	**0.65 ± 0.0**	0.62 ± 0.0	**0.65 ± 0.0**
Wine	0.48 ± 0.1	**0.54 ± 0.1**	0.42 ± 0.0

Table 3 shows the results (with the standard deviation) of our ablation study. MK-TSOD achieves the largest BA on 10 out of 15 datasets; SVDD with the k_{FFT} kernel performs best on 5 datasets, and SVDD with the k_{GAK} kernel has the best score on 3 data sets (including ties). The proposed algorithm can leverage the respective strengths of the kernels. Considering the unsupervised setting, where parameter optimization — including kernel selection — is typically infeasible in practice, this result indicates that MK-TSOD provides a good default choice, often improving performance over employing a single kernel.

6   Conclusions

This paper tackles the long-standing problem of detecting outliers in a set of time series, for which we propose a new method, MK-TSOD. It builds on SVDD and combines global alignment and Fourier transform kernels, taking the time and frequency information of time series into account. The parameters of MK-TSOD are either intuitive to set or we recommend heuristics. Our evaluation shows that MK-TSOD achieves state-of-the-art performance and outperforms existing approaches w.r.t. the balanced accuracy metric on 9 out of 15 standard benchmark data sets.

Acknowledgements. This work was supported by the DFG Research Training Group 2153: "Energy Status Data — Informatics Methods for its Collection, Analysis and Exploitation".

References

1. Bagnall, A., Lines, J., Bostrom, A., Large, J., Keogh, E.: The great time series classification bake off: a review and experimental evaluation of recent algorithmic advances. Data Min. Knowl. Disc. **31**, 606–660 (2017)
2. Beggel, L., Kausler, B.X., Schiegg, M., Pfeiffer, M., Bischl, B.: Time series anomaly detection based on Shapelet learning. Comput. Statistics **34**, 945–976 (2019)
3. Benkabou, S., Benabdeslem, K., Canitia, B.: Unsupervised outlier detection for time series by entropy and dynamic time warping. Knowl. Inf. Syst. **54**(2), 463–486 (2018)
4. Bordes, A., Ertekin, S., Weston, J., Bottou, L.: Fast kernel classifiers with online and active learning. J. Mach. Learn. Res. (JMLR) **6**, 1579–1619 (2005)
5. Breunig, M.M., Kriegel, H., Ng, R.T., Sander, J.: LOF: identifying density-based local outliers. In: SIGMOD Conference, pp. 93–104 (2000)
6. Christ, M., Braun, N., Neuffer, J., Kempa-Liehr, A.W.: Time series feature extraction on basis of scalable hypothesis tests (tsfresh - a python package). Neurocomputing **307**, 72–77 (2018)
7. Cortes, C., Vapnik, V.: Support-vector networks. Mach. Learn. **20**(3), 273–297 (1995)
8. Cuturi, M.: Fast global alignment kernels. In: International Conference on Machine Learning (ICML). pp. 929–936 (2011)
9. Dau, H.A., Keogh, E., et al.: The UCR time series classification archive (2018)

10. Emmott, A., Das, S., Dietterich, T.G., Fern, A., Wong, W.: Systematic construction of anomaly detection benchmarks from real data. Technical report (2015). http://arxiv.org/abs/1503.01158
11. Emmott, A.F., Das, S., Dietterich, T., Fern, A., Wong, W.K.: Systematic construction of anomaly detection benchmarks from real data. In: ACM SIGKDD Workshop on Outlier Detection and Description, pp. 16–21 (2013)
12. Ghafoori, Z., Erfani, S.M., Rajasegarar, S., Bezdek, J.C., Karunasekera, S., Leckie, C.: Efficient unsupervised parameter estimation for one-class support vector machines. Neural Netw. Learn. Syst. **29**(10), 5057–5070 (2018)
13. Gudmundsson, S., Runarsson, T.P., Sigurdsson, S.: Support vector machines and dynamic time warping for time series. In: International Joint Conference on Neural Networks (IJCNN), pp. 2772–2776 (2008)
14. Gupta, M., Gao, J., Aggarwal, C.C., Han, J.: Outlier detection for temporal data: a survey. Knowl. Data Eng. **26**(9), 2250–2267 (2013)
15. Hyndman, R.J., Wang, E., Laptev, N.: Large-scale unusual time series detection. In: International Conference on Data Mining Workshop (ICDMW), pp. 1616–1619 (2015)
16. Patel, D., et al.: FLOps: on learning important time series features for real-valued prediction. In: IEEE BigData 2020, pp. 1624–1633 (2020)
17. Platt, J.: Sequential minimal optimization: a fast algorithm for training support vector machines (1998). https://www.microsoft.com/en-us/research/publication/sequential-minimal-optimization-a-fast-algorithm-for-training-support-vector-machines/
18. Rakotomamonjy, A., Bach, F., et al.: SimpleMKL. J. Mach. Learn. Res. (JMLR) **9**, 2491–2521 (2008)
19. Ruff, L., et al.: Deep one-class classification. In: International Conference on Machine Learning (ICML), vol. 80, pp. 4390–4399 (2018)
20. Sakoe, H., Chiba, S.: Dynamic programming algorithm optimization for spoken word recognition. Acoust. Speech Signal Process. **26**(1), 43–49 (1978)
21. Schölkopf, B., Smola, A.: Learning with Kernels: Support Vector Machines. Regularization, Optimization, and Beyond. MIT Press, Cambridge (2002)
22. Steinwart, I., Christmann, A.: Support Vector Machines. Springer, New York (2008). https://doi.org/10.1007/978-0-387-77242-4
23. Tax, D.M.J., Duin, R.P.W.: Support vector data description. Mach. Learn. **54**(1), 45–66 (2004)
24. Vercruyssen, V., Meert, W., Davis, J.: "Now you see it, now you don't!" detecting suspicious pattern absences in continuous time series. In: SIAM International Conference on Data Mining (SDM), pp. 127–135 (2020)

Toward Streamlining the Evaluation of Novelty Detection in Data Streams

Jean-Gabriel Gaudreault[(⊠)] and Paula Branco

University of Ottawa, Ottawa, ON K1N6N5, Canada
{j.gaudreault,pbranco}@uottawa.ca

Abstract. While batch machine learning algorithms typically assume that all the concepts are available at training, the reality is often different when dealing with continuous streams of data, where new concepts can emerge and existing ones change over time. The task of novelty detection is an increasingly popular field that tackles this problem by trying to recognize these formerly unidentified concepts that fall outside the decision boundary of the models. Although there have been numerous works discussing the implementation of such algorithms, studies covering their adequate performance evaluation are still scarce. In this paper, we present an evaluation framework that aims to streamline the evaluation of novelty detection algorithms. This framework irons out the shortcomings we identified in the domain, allowing us to obtain a more robust assessment of the performance. Specifically, we propose novel metrics to complement the existing ones, and we incorporate the temporal aspect of data streams within the evaluation. We empirically test the impact of intrinsic data streams' characteristics when using our proposed framework. We show the added value of this novel framework with experiments carried out on both artificial and real-world data sets.

Keywords: Novelty detection · Data streams · Evaluation framework

1 Introduction

Contrary to typical uses of machine learning, where we generally assume that the concepts learned at training time will be constant throughout the use of the model, many real-world scenarios such as sensor networks, medical diagnosis, or network intrusions, present the data as data streams (DSs) rather than datasets. DSs present many challenges, for example, the constant change in data distribution, as well as the emergence of new concepts which were previously unknown. Novelty detection (ND) is a research area that aims to detect, classify, and potentially learn these unknown concepts. In this regard, many algorithms have been created to accomplish this task effectively. However, the evaluation of these algorithms has historically been inconsistent, as some works have adapted the use of batch learning metrics to ND (e.g. [10,11]) while others have used novel, domain-specific metrics (e.g. [7,9]). Moreover, many works that assume the emergence of multiple different classes have used binary metrics, which do not represent the misclassification between novel classes in the evaluation. While there has been some research in the direction of properly evaluating multi-class ND scenarios [6], there are still many open challenges, including not considering the temporal aspect and other characteristics of DSs, as well as the complexity of comparing multiple models between themselves due to the need to track several metrics. It is thus necessary to develop a standardized framework to evaluate ND algorithms in a simple and clear way, that takes into account the classification of multiple novel classes and also considers other DSs' characteristics.

In this study, we tackle these gaps. Our main contributions are summarized as follows: (i) we introduce a novel metric for binary classification scenarios that allows incorporating multiple metrics previously defined in the field; (ii) we propose the use of a metric for multi-classification scenarios that allows for better comparability between models and adds a baseline for random classifiers; (iii) we present a novel measure that incorporates the time necessary to detect novel classes, an important aspect for many ND real-world deployment scenarios; and (iv) we study the impact of different data characteristics of DSs and whether or not they should be reported when evaluating such algorithms.

The rest of this paper is structured as follows: Sect. 2 reviews concepts and existing metrics for ND; Sect. 3 presents some of the drawbacks of the existing literature as well as our proposed solutions and framework; Sect. 4 compares and discusses the empirical evidence; and Sect. 5 concludes this paper and mentions potential future works.

2   Background

In this section, we characterize some of the common concepts within the field of ND in DSs that are going to be used throughout this paper. We also formulate the ND task and describe existing evaluation methods that have been proposed in the literature.

2.1   Concept Definitions

Data Stream. Formally, a DS is characterized by a sequential, continuous flow of samples that arrives in an online fashion and is potentially unbounded [8]. The samples within these streams usually represent different concepts (classes) and are not necessarily dependent.

Concept Drift. DSs are non-stationary, as new data becomes available, the probability distribution of the concepts in the stream might change [5], this is a phenomenon referred to as concept drift, and can either happen gradually or suddenly [11].

Concept Evolution. While known concepts can shift over time (concept drift), new ones can also appear in DSs as novel classes, named concept evolutions [5]. These new concepts are typically outside the decision boundary of the classes learned by the models and need to be detected properly in order to perform the ND task.

Novelty Pattern. A Novelty Pattern (NP) is a set of related samples identified within the stream which fall outside the known classes of the ND algorithm [6]. A novel class can be detected as multiple NPs by the model, although a minimal number of NPs per novel class is preferable.

2.2 Novelty Detection Task

The ND task consists in detecting samples corresponding to novel classes which were not seen by the algorithm during its training phase, in other words, detecting concept evolutions. Contrary to anomaly detection, where one sample differing from the known distribution might be enough to classify an anomaly, ND focuses on the detection of a cohesive aggregation of these samples which fall outside the model's decision boundary and therefore form a NP [5]. There are commonly two ways to approach this task, whether as a binary classification problem, where all the NPs are detected as a single novel class, or as a multiclass problem, where the NPs are separated and detected as multiple different classes.

To perform this task, the algorithms implementing ND typically consist of two phases, namely an offline and an online phase. In the former, similar to batch learning, the model is trained on a set of examples representing the known classes, while in the latter, the model is used in an online fashion on a DS and tries to classify the known classes, while also detecting NPs. To detect these NPs in the stream, numerous algorithms implement the concept of a buffer, where samples that fall outside their decision boundary are temporarily stored and classified as unknown to eventually cluster them in NPs if there is a sufficient number of neighboring samples. The resulting confusion matrix is rectangular, as the number of detected NPs does not necessarily equal the number of actual classes, nor do the NPs labels match with the actual class labels [6].

2.3 Analysis of Metrics

In this section, we present and analyze the metrics for ND evaluation that have been proposed in the literature. We start with the introduction of some notation. We highlight that for the binary case, we consider the novelty class to be the positive label (1), and the known class to be the negative label (0). The following terms will be used in the definition of the metrics:

- **True Positives** (TP): Number of detected novelties correctly classified
- **False Positives** (FP): Number of known class samples incorrectly classified as novelties
- **False Negatives** (FN): Number of novelties incorrectly classified as known
- **True Negatives** (TN): Number of known class samples correctly classified
- N: Number of samples
- N_c: Number of novel samples in the stream
- FE: Known class instances misclassified (other than FP)

Using this notation, an example of a confusion matrix for a ND multi-class problem, as first proposed by Faria et al. [6], can be seen in Table 1, where C_i represents the class i, NP_i is the ith NP discovered by the algorithm, and the number of known classes k is lower than the number of actual classes m.

Table 1. Confusion matrix of a multi-class novelty detection problem.

		Predicted Values								
		Known Classes				Novelty Patterns				
		C_0	C_1	...	C_k	NP_1	NP_2	...	NP_j	Unk
	C_0	TN	FE	FE	FE					
Known C_1	C_1	FE	TN	FE	FE			FP		
Actual Classes ...	...	FE	FE	TN	FE					
Values	C_k	FE	FE	FE	TN					
	C_{k+1}									
Novel ...	...			FN				TP		
Classes	C_m									

As mentioned in Sect. 2.2, the ND task can be either considered as a binary problem, where the goal is to distinguish between known classes and novel concepts, or as a multi-class scenario where each known and each novel class represents different target classes. Numerous algorithms (e.g. [1,9,12]) can detect multiple classes of novelties but still use binary evaluation metrics for their evaluation which, as we will demonstrate in the next section, present several issues for the proper evaluation of the algorithms.

Binary Metrics

$\mathbf{M_{new}}$ *and* $\mathbf{F_{new}}$. M_{new} (c.f. Eq. 1) and F_{new} (c.f. Eq. 2) represent, respectively, the percentage of novel class samples misclassified as known, and the percentage of known class samples misclassified as novel. While these are commonly used metrics, they present some downsides, as they do not consider classification errors within the novelty and novel classes, and need the tracking of two separate measures for the evaluation. Hence, these two metrics are useful in the specific scenario where one would only care about correctly distinguishing novel classes from known ones.

$$M_{new} = \frac{FN}{N_c} \cdot 100 \quad (1) \qquad\qquad F_{new} = \frac{FP}{N - N_c} \cdot 100 \quad (2)$$

Accuracy and Error Rate. Accuracy (c.f. Eq. 3) and its complement the Error Rate (Err) (c.f. Eq. 4) are also sometimes used in the context of ND. While the error rate metric considers the misclassification within the known classes (FE), similarly to M_{new} and F_{new}, both metrics do not consider misclassification within the novel classes, i.e., NPs that are classified as the wrong novel class. For this reason, they are considered binary metrics.

It is also worth noting that in imbalanced domains, where one or multiple classes of interest are less predominant than others, accuracy and error rate have been demonstrated to be unsuitable, as the less prevalent classes will have a lesser impact on the metrics [3]. Considering that ND often deals with such imbalanced domains, where some of the classes such as the novel classes might be underrepresented compared to others, these metrics might not be adequate to evaluate the models' performance.

$$Accuracy = \frac{TP + TN}{N} \cdot 100 \quad (3) \qquad Err = \frac{(FP + FN + FE)}{N} \cdot 100 \quad (4)$$

Others. A number of other metrics that are already defined for batch learning, such as the F_β-Measure, and ROC Area Under the Curve (ROC AUC) have also been used in ND, but also present similar issues as the aforementioned metrics, as they consider the problem as binary, and regard the samples labeled as unknown as correctly classified novelties.

Multi-class Metrics. Considering the limitations presented in the previous section, Faria et al. [6] proposed one of the few frameworks available to evaluate ND algorithms in multi-classification scenarios. As mentioned in Sect. 2.2, the NPs identified by ND algorithms do not directly correspond to the true labels of the dataset, which makes it impossible to compute typical multi-class evaluation metrics made for batch scenarios. To resolve this issue, the authors propose to associate each NP to a true class, by taking the most occurring true label in each NP and associating that NP to that specific class. Once this association is performed, the authors suggest plotting the three following metrics over the course of the DS to follow the performance of the model.

Combined Error (CER). Defined as the weighted average of False Positive Rate (FPR) and False Negative Rate (FNR) (c.f. Eq. 5), the Combined Error (CER) (c.f. Eq. 6) is defined for M classes and does not include the samples labeled as unknown within the metric's computation.

$$FPR_i = \frac{FP_i}{FP_i + TN_i} \quad FNR_i = \frac{FN_i}{FN_i + TP_i} \qquad (5)$$

$$CER = \frac{1}{2} \sum_{i=1}^{M} \frac{\#ExC_i}{\#Ex}(FPR_i + FNR_i) \qquad (6)$$

where $\#ExC_i$ represents the number of samples of class C_i and $\#Ex$ the total number of samples.

Akaike Information Criterion (AIC). The complexity of the model needs to also be taken into account in its evaluation, as a model clustering novelties into a bigger amount of clusters could have a better classification performance than a model with fewer clusters but would display an undesirable behavior (multiple independent clusters). To resolve this issue, the authors propose the use of an adaptation of the Akaike Information Criterion [14] (AIC) (c.f. Eq. 7) which punishes a high number of NPs.

$$AIC = -2\ln(1 - CER) + \frac{2p}{\ln(N)} \tag{7}$$

where p is the number of classes detected by the ND algorithm (NPs and known classes) and N is the total number of samples without those labeled as unknown.

Unknown Rate (UnkRate). Finally, to resolve the issue of the samples labeled as unknown being considered as a correctly labeled novel class, as discussed in Sect. 2.3, the authors propose to compute those samples independently of the other metrics, by using the Unknown Rate (UnkRate) (c.f. Eq. 8) which computes the average amount of samples classified as unknown over the total number of samples of each class i.

$$UnkR = \frac{1}{M} \sum_{i=1}^{M} \frac{\#Unk_i}{\#ExC_i} \tag{8}$$

where $\#Unk_i$ represents the number of samples of class i classified as unknown and $\#ExC_i$ the total number of samples of class i.

3   Proposed Framework

The evaluation framework by Faria et al. [6] presented in Sect. 2.3 is a good initial base to properly evaluate ND scenarios. However, there are several ways in which the performance assessment can be enhanced and a number of gaps that can be addressed. This motivates the proposed evaluation framework that we present next.

3.1   Novel Metric for the Binary Scenario

End-users may be interested in correctly classifying if a sample is novel or not, in which case they would run into some of the issues mentioned in Sect. 2.3 such as the need to track multiple metrics (M_{new} and F_{new}) to properly evaluate the model. For the cases where the user is only concerned with correctly differentiating novelties from known samples (such as in an active learning system), we propose using a new metric: the M_{β}-Measure (c.f. Eq. 9), which is also the harmonic mean of the inverse of M_{new} and F_{new} when $\beta = 1$. The complement of both metrics is used as we want M_{β} to display a poor performance if any of the two metrics is as well, which requires them to have their optimal value at 1 and worst at 0. The idea behind this metric is akin to the combination of precision and recall with F_{β}, as it incorporates the correct classification of both classes in a single, easy-to-track metric. Moreover, similarly to F_{β}, it allows increasing the weight given to correctly classified novel class samples (Inverted M_{new}) using $\beta < 1$, or correctly classified known class samples (Inverted F_{new}) using $\beta > 1$.

$$M_\beta = (1 + \beta^2) \cdot \frac{(1 - M_{new}) \cdot (1 - F_{new})}{(\beta^2 \cdot (1 - M_{new})) + (1 - F_{new})} \qquad (9)$$

3.2 Facing the Undesirable Properties of AIC

The use of AIC for the ranking of models' complexity presents some undesirable properties. In particular, AIC measures the quality of a model relative to others [14]. The value of AIC will increase with the number of classes of a dataset, considering its factor p. Therefore, there is no baseline value representing a random model for any dataset, which means it can't be used to compare models between different datasets. Moreover, the metric is not defined for the worst-case scenario where $CER = 1$, and since the metric scales a lot more with the model's complexity than its accuracy, it requires the user to track two different metrics: one for the model's performance (CER) and one for its complexity (AIC). While DSs necessitate tracking multiple metrics due to their evolving nature, we believe that, since these two aspects are directly related, combining them reduces the analysis complexity and allows tracking other aspects not included in current frameworks.

Given these issues, we suggest replacing the use of both CER and AIC for the evaluation of models' performance and complexity with the use of the Adjusted Mutual Information (AMI) (c.f. Eq. 10). While this metric is commonly used in the context of unsupervised learning when ground truth labels are available, to the best of our knowledge it hasn't been applied to the field of ND. We chose this metric as it is adjusted for chance, which means that it will give a value close to 0 for random clustering or multiple very small, independent clusters. Moreover, it has been previously demonstrated [13] to be preferable to other similar metrics such as the adjusted random index when dealing with an imbalance within the clusters, a phenomenon which is likely in ND scenarios since novelty concepts are often more scarce than known ones.

We believe that the use of the AMI has several advantages, as it offers a baseline score for random labeling and an upper bound of 1 for perfect agreement, it is not affected by permutation of the samples in different clusters, it uses a widely establish metric, and it combines both the classification performance and model's complexity in a single metric which is easier to track and compare between algorithms.

$$AMI(U, V) = \frac{MI - E[MI]}{mean(H(U), H(V)) - E[MI]} \qquad (10)$$

where U and V are two clusters (partitions), MI is the mutual information of those partitions, $E[MI]$ is the expected mutual information, and $H(x)$ is the entropy of x.

3.3 Inclusion of the Temporal Aspect of DSs

While the metrics discussed are reported over the course of the DS, which shows the evolution of the model over time, there are currently no metrics reporting directly the time needed for the model to detect novel classes as novel when they first appear in the stream. This is a very important aspect, considering that in the fields typically pertaining to ND, such as intrusion detection, fraud detection, medical examination, and many others, the difference between detecting a novelty very quickly or not could be critical. As such, we propose the inclusion of a novel metric named Time To Detection (TTD) (c.f. Eq. 11). This metric computes, for each novel class, the number of samples of that specific class necessary for the class to be detected as a NP (other than known or unknown).

$$TTD_i = \sum_{j=TFA_i}^{TFD_i} [c_j = i] \tag{11}$$

where TFA_i represents the time of first appearance of the novel class i, TFD_i the time when it was first detected as a NP, and c_j the true class label of the jth sample of the stream.

3.4 Inclusion of Intrinsic Data Characteristics of DSs

Lastly, we believe that there exist multiple intrinsic data characteristics of DSs such as the time of arrival, data sampling, the density of novel samples, and others, which could affect the performance of a ND model greatly. Therefore, this implies that for the proper evaluation of ND algorithms, these data characteristics should be reported by researchers. However, most propositions within the field do not display these properties, as is common to do so in batch learning with train and test splits, for example. Hence, we want to extend our proposed evaluation framework with the inclusion of these characteristics. We established three main characteristics to further test and report on, which will be described in more detail in the next section, and are as follows: (i) time between the appearance of novel classes; (ii) ratio of offline samples; and (iii) ratio of known classes.

4 Experiments

This section presents the results of the experiments carried out. We tested our proposed improvements and new metrics in multiple scenarios, including by varying each of the data characteristics of the stream, and compared the results to previously mentioned metrics both in the binary and multi-class setting. This allowed us to confirm which of the extracted DS characteristics have an impact on the model's performance, as well as demonstrate how the proposed improvements of the evaluation framework actually perform. We first start by describing the evaluation methodology used in the next section and analyze the results in the subsequent ones.

4.1 Methodology

All experiments were performed using both synthetic and real datasets, the former, which we named SynthRBF, was generated using the RandomRBF generator of the MOA framework [2], a known framework for machine learning in DSs. For the real-world dataset, we selected the Forest Cover Type dataset (FCT) [4] as it possesses numerous classes and is widely used in the field of ND. Some experiments, including on other datasets, are not included due to space constraints but are freely available with all the code and hyperparameters to reproduce the experiments in an online repository[1].

Since we needed a ND algorithm that supported multi-classification of known classes and novelty concepts, can run without external feedback (availability of the true labels), and can predict unknown samples, our choice was somewhat limited. Hence, we selected the MINAS algorithm [7] to perform the experiments since it corresponded to our requirements and implemented the ND process using clustering, a common way of implementing ND algorithms.

While it is possible to evaluate each metric at every new sample treated by the algorithm, it is highly inefficient to do so as it requires considerable computing power and offers very marginal benefits in our tests. Consequently, we decided to log each metric at a period of every 100 samples for all experiments, which provided a high definition of tracking for the metrics, while limiting the computing requirements.

4.2 Baseline

As a baseline for our experiments, we selected the following parameters for each of the data characteristics tested, based on other studies that have been performed in the field of ND: no wait time between the arrival of different novel classes (order of samples is randomized), 10% of the dataset is used as the offline phase and 30% of classes used as known classes.

A simple example of the results of this baseline on the FCT dataset is demonstrated in Fig. 1 as well as part of Table 2. We must highlight that with M_{new}, F_{new}, CER, AIC, and TTD, a better score is represented by a lower value, while a higher value (up to 1) represents a better score for the other two metrics proposed in our framework (M_β and AMI). We can also observe some advantages of our proposed methodology, in the binary scenario, the usage of M_β allows us to easily follow the model's ability to distinguish between novel and known classes without the use of two metrics and efficiently compare the performance of multiple models/datasets. For the multi-class case, we must draw attention to the second y-axis, where the value of AIC is displayed. As previously mentioned, the value of AIC is unbounded, hence a second y-axis is displayed on the right-hand side. Since it does not have a baseline for a random model, it can not be used to compare models' performance between different datasets.

---

[1] https://github.com/jgaud/StreamliningEvaluationNDDataStreams.

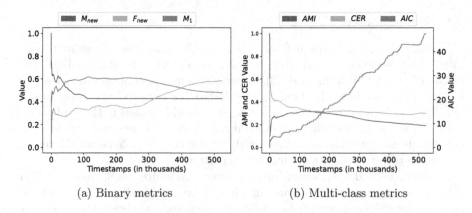

(a) Binary metrics   (b) Multi-class metrics

Fig. 1. Baseline metrics for the FCT dataset

4.3   Time Between the Appearance of Novel Classes

The first data characteristic we tested is the time between the arrival of a new class within the stream. Indeed, while in real-world scenarios, novel concepts could arrive at any time, randomly $(t = 0)$, some works in the literature (e.g. [1, 12]) consider that novel classes arrive one after the other. Therefore, we wanted to not only test whether or not classes that arrived in an orderly fashion versus randomly affected the performance, but also if the number of samples between the arrival of novel classes did. We tested three different parameters including our baseline where novel classes appeared randomly $(t = 0)$, a scenario where they appeared at an intermediate interval $(t = 5000)$, and one with a long interval $(t = 20000)$.

The results for the FCT and SynthRBF datasets can be seen in Figs. 2 and 3, respectively. Looking at the binary results, it is clear that a longer period between the arrival of novel classes allows the ND model to perform a better classification between the novel and known classes over the long term. Indeed, on the FCT dataset, we can see the model's ability to classify the novel classes as NPs on the longest interval largely improving once a few novel classes have arrived, represented by a lower M_{new} and consequently, a higher M_1 score. The results on SynthRBF also demonstrated that a longer period between the arrival of novel classes allowed the model to distinguish novel concepts from known ones better, but in this case mainly due to a better detection of the known classes (F_{new}). This difference could be explained by the FCT dataset's novel classes being harder to distinguish from the known ones compared to the synthetic data. The results are also similar to the multi-class metrics for both the synthetic and FCT dataset, as both the AMI and CER show a slight improvement in classification performance using a longer period between novel classes, as well as ending up with a model slightly less complex for the largest time.

Looking at the time to detect each novel class in Table 2, we can see that the addition of a time between the arrival of novel classes allowed to lower the TTD

Table 2. TTD experimental results on different data streams' characteristics.

Characteristic	Data Stream		Class					
			2	3	4	5	6	7
Baseline	FCT		–	–	270	1083	258	451
	SynthRBF		234	251	262	–	–	–
Time between the appearance of novel classes	FCT	$t = 5000$	–	–	0	931	19	169
		$t = 20000$	–	–	4	9	5	211
	SynthRBF	$t = 5000$	14	1	6	–	–	–
		$t = 20000$	0	1	0	–	–	–
Ratio of offline samples	FCT	$r = 0.4$	–	–	337	1020	363	365
		$r = 0.7$	–	–	525	1076	555	478
	SynthRBF	$r = 0.4$	235	254	265	–	–	–
		$r = 0.7$	240	257	265	–	–	–
Number of known classes	FCT	$n = 0.5$	–	–	–	867	278	280
		$n = 0.7$	–	–	–	–	400	882
	SynthRBF	$n = 0.5$	–	348	356	–	–	–
		$n = 0.7$	–	–	585	–	–	–

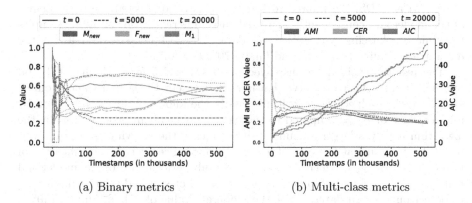

(a) Binary metrics   (b) Multi-class metrics

Fig. 2. Effect of the time between the appearance of novel classes for the FCT dataset

quite drastically with all classes, while the addition of a longer period ($t = 5000$ vs $t = 20000$) allowed to also accelerate the detection of some novel classes, albeit less excessively. This is following our expectations, as adding a period between the novel classes allows the model to better understand the decision boundary of its known classes and therefore quickly identify samples that are located outside and report them as novel. These results show the importance of reporting on the DS's characteristics, as they can affect the performance of the models significantly. Moreover, it also highlights the usefulness of our proposed metrics.

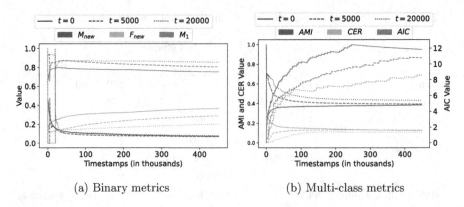

(a) Binary metrics   (b) Multi-class metrics

Fig. 3. Effect of the time between the appearance of novel classes for the SynthRBF dataset

4.4   Ratio of Offline Samples

In this scenario, we tested the impact of the ratio of offline to online samples. Similar to train/test sets in batch learning, ND algorithms typically use a portion of the dataset to learn the known classes in the offline phase and then use the other part as a DS. We wanted to test how this ratio affected the performance of the model and consequently analyzed three different ratios of offline samples to online: our baseline with 10% of the dataset used offline ($r = 0.1$), 40% of the dataset used offline ($r = 0.4$), and finally 70% ($r = 0.7$).

The results obtained are displayed in Figs. 4 and 5 for both the FCT and SynthRBF datasets. We observe that in both datasets, the largest value of offline samples (70%) helped the algorithm classify the known classes, represented by a lower value of F_{new} compared to the baseline, which can be explained by the algorithm having a larger amount of data to learn the known classes. However, on FCT, this came at the cost of a lower performance over the classification of novel classes (M_{new}), which lowered the overall performance of the model and could be explained by the higher number of novel classes and them being harder to detect on the real dataset. For the moderate value of offline samples (40%), the algorithm performed worse in all cases, which is demonstrated by an overall lower M_1 score and AMI. It is also important to note that while the number of samples in the offline phase increases with a higher ratio, the number of known classes stays the same. This means that there are fewer known class samples in the online phase and that the density of novel class samples is increased. The obtained results suggest that this higher density of novel samples, combined with a shorter time to learn them in the online phase, hinders the model from properly learning the novel concepts.

The TTD included in Table 2 did not show any specific trends for the synthetic data, but did show an increase in time to detect classes 4 and 6 in FCT when increasing the ratio of offline samples. This is in accordance with the other results displaying some of the novel classes, especially on FCT, being harder

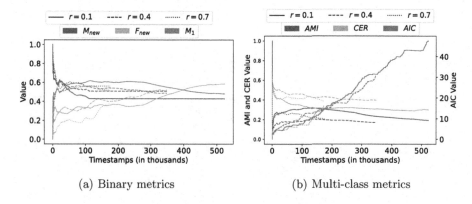

(a) Binary metrics

(b) Multi-class metrics

Fig. 4. Effect of the ratio of offline samples for the FCT dataset.

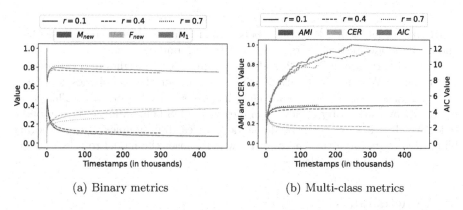

(a) Binary metrics

(b) Multi-class metrics

Fig. 5. Effect of the ratio of offline samples for the SynthRBF dataset.

to detect. It also highlights another useful usage of our proposed metric, as it allows us to see which specific classes might be problematic to detect with a single scalar metric.

4.5 Number of Known Classes

Lastly, we tested the effect of the ratio of the number of known classes, i.e., the percentage of all classes that will be shown to the model during the offline phase. We selected 30% as our baseline ($n = 0.3$), 50% ($n = 0.5$), and 70% ($n = 0.7$).

The results can be seen in Fig. 6 for the FCT dataset and are included in the aforementioned repository for the SynthRBF dataset since it presented a similar behavior, and due to space constraints. Both show that a higher ratio of known classes tends to lead to better results in terms of the distinction between novel and known classes (M_1), as well as similar or better performance in terms of separation within NPs and known concepts (AMI). These results are per our

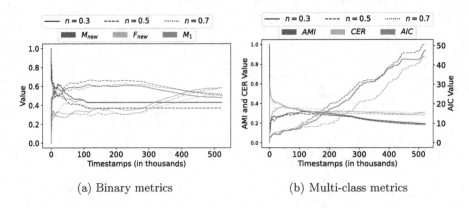

(a) Binary metrics   (b) Multi-class metrics

Fig. 6. Effect of the ratio of known classes for the FCT dataset.

expectations, as having the true labels of more classes during the offline phase allows the model to better create the clusters around each class.

The TTD metric in Table 2 is also providing useful insights into the models' performance. We can observe that, when the number of novel classes gets fairly low (< 2), the time to detect each class increases quite importantly. This could be caused by the density of novel samples in the DS being a lot lower compared to known classes. Consequently, the model is discarding samples in its temporary buffer due to its forgetting mechanism, which causes the model to take a longer time to detect the novel classes.

5   Conclusion

This paper proposes an evaluation framework for ND in DSs based on existing and novel concepts with the goal of streamlining and facilitating the evaluation of algorithms in the field and allowing for better comparability between works. Most works in the literature evaluate the ND problem as a binary classification. However, this does not allow us to evaluate if the model properly classifies novel concepts between themselves. Moreover, existing frameworks require tracking of numerous metrics, which can make the comparison of models difficult, or display undesirable properties such as the lack of upper bound and baseline. In our framework, we suggest the use of a novel proposed metric, M_β, for the binary scenario, which combines existing metrics, and the use of AMI for the multi-class scenario, a metric well-known in the field of unsupervised learning which can combine two other metrics. We also proposed a novel metric, TTD, to include the temporal aspect of DSs within the evaluation, an important facet for multiple ND applications. Lastly, we tested and demonstrated the use of our proposed metrics as well as the impact of numerous data characteristics of DSs, which we suggest should be reported when testing ND algorithms since, as we demonstrated, they can influence greatly the results and can show biased performance results.

In our future work, we plan to extend this framework by exploring other data characteristics of DSs such as the density of novel samples or concept drift, increase significantly the number of datasets and algorithms tested, and perform an analytical study of our proposed framework compared to existing solutions.

References

1. Al-Khateeb, T., Masud, M.M., Khan, L., Aggarwal, C., Han, J., Thuraisingham, B.: Stream classification with recurring and novel class detection using class-based ensemble. In: 2012 IEEE 12th International Conference on Data Mining, pp. 31–40 (2012). https://doi.org/10.1109/ICDM.2012.125
2. Bifet, A., Holmes, G., Kirkby, R., Pfahringer, B.: MOA: massive online analysis. J. Mach. Learn. Res. **11**, 1601–1604 (2010). https://doi.org/10.5555/1756006.1859903
3. Branco, P., Torgo, L., Ribeiro, R.P.: A survey of predictive modeling on imbalanced domains. ACM Comput. Surv. **49**(2), 1–50 (2016). https://doi.org/10.1145/2907070
4. Dua, D., Graff, C.: UCI machine learning repository (2017). http://archive.ics.uci.edu/ml
5. Faria, E.R., Gonçalves, I.J.C.R., de Carvalho, A.C.P.L.F., Gama, J.: Novelty detection in data streams. Artif. Intell. Rev. **45**(2), 235–269 (2015). https://doi.org/10.1007/s10462-015-9444-8
6. de Faria, E.R., Gonçalves, I.R., Gama, J., de Leon Ferreira Carvalho, A.C.P.: Evaluation of multiclass novelty detection algorithms for data streams. IEEE TKDE **27**(11), 2961–2973 (2015). https://doi.org/10.1109/TKDE.2015.2441713
7. de Faria, E.R., Ponce de Leon Ferreira Carvalho, A.C., Gama, J.: MINAS: multiclass learning algorithm for novelty detection in data streams. Data Min. Knowl. Disc. **30**(3), 640–680 (2015). https://doi.org/10.1007/s10618-015-0433-y
8. Gama, J.: Knowledge Discovery from Data Streams, 1st edn. Chapman & Hall/CRC (2010)
9. Haque, A., Khan, L., Baron, M.: SAND: semi-supervised adaptive novel class detection and classification over data stream. In: Proceedings of the AAAI Conference on Artificial Intelligence, vol. 30, no. 1 (2016). https://doi.org/10.1609/aaai.v30i1.10283
10. Krawczyk, B., Woźniak, M.: Incremental learning and forgetting in one-class classifiers for data streams. In: Burduk, R., Jackowski, K., Kurzynski, M., Wozniak, M., Zolnierek, A. (eds.) CORES 2013, vol. 226, pp. 319–328. Springer, Heidelberg (2013). https://doi.org/10.1007/978-3-319-00969-8_31
11. Krawczyk, B., Woźniak, M.: One-class classifiers with incremental learning and forgetting for data streams with concept drift. Soft. Comput. **19**(12), 3387–3400 (2014). https://doi.org/10.1007/s00500-014-1492-5
12. Masud, M., Gao, J., Khan, L., Han, J., Thuraisingham, B.M.: Classification and novel class detection in concept-drifting data streams under time constraints. IEEE TKDE **23**(6), 859–874 (2011). https://doi.org/10.1109/TKDE.2010.61
13. Romano, S., Vinh, N.X., Bailey, J., Verspoor, K.: Adjusting for chance clustering comparison measures. JMLR **17**(1), 4635–4666 (2016)
14. Stoica, P., Selen, Y.: Model-order selection: a review of information criterion rules. IEEE Sig. Process. Mag. **21**(4), 36–47 (2004). https://doi.org/10.1109/MSP.2004.1311138

Author Index

Printed in the United States
by Baker & Taylor Publisher Services